Social Sciences

Life Sciences

General Interest

FREE with this text: an acclaimed student Web site and access to the InfoTrac® College Edition *online library*

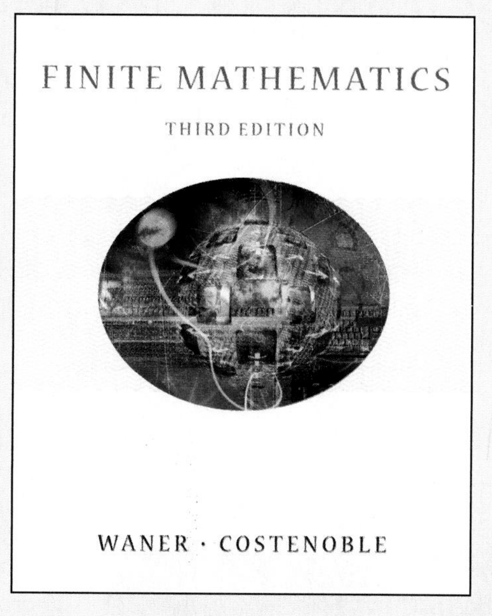

Stefan Waner and Steven R. Costenoble, both of Hofstra University

P R E V I E W

MOTIVATING YOUR STUDENTS BEGINS RIGHT HERE—

WITH REAL WORLD EXAMPLES, TODAY'S TECHNOLOGY, PEDAGOGICAL TOOLS, AND ROCK SOLID CONCEPTUAL DEVELOPMENT.

Stefan Waner and Steven R. Costenoble have been motivating students with ***Finite Mathematics*** for more than a decade. They've been successful because the text is readable, relevant to the interests of business, life science, and social science students, and stocked with interesting examples and exercises. At the same time, because motivation alone doesn't guarantee mastery in the course, the authors have also provided the sound conceptual development that students need.

Finite Mathematics has been honed and enhanced over the years, from the careful integration of such technologies as graphing calculators and Microsoft® Excel to the development of a highly regarded Web site. These features support the interactive techniques that have worked well with Waner and Costenoble's own students, even those who are less than enthusiastic about the course. The refinement has continued in this Third Edition, with interesting new applications and improved learning aids for your students, all showcased more effectively in a new four-color design.

Here's what you'll find in this Visual Preview:

- Examples of new and updated applications (pages VP-2 and VP-3)
- Examples of integrated discussions of spreadsheets and graphing calculators, expanded in this edition (pages VP-4 and VP-5)
- Samples of highly effective pedagogy, including more **Quick Examples**, more **Question and Answer Dialog** boxes, exercises now presented in increasing level of difficulty, and new **Guidelines** to assist in problem solving (pages VP-6 and VP-7)
- An introduction to the popular and widely acclaimed Web site that features interactive tutorials, exercises, and online utilities to support each chapter of the text (pages VP-8 and VP-9)
- A review of the ancillary package that accompanies the text—including additional teaching support for you (pages VP-10 and VP-11)

"I thought **Finite Mathematics** was innovative in many respects and I liked some of the innovations enough that I will permanently adopt them as part of my approach to teaching this course. I think the greatest strength of this book is the realistic nature of the examples and exercises. ...I also think the students appreciate books that make an earnest effort to communicate with them in a way they will find useful."

— **E. Arthur Robinson, Jr.**
George Washington University

APPLICATIONS USE REAL DATA,

MODIFIED JUST ENOUGH TO CAPTURE STUDENTS' INTEREST AND UNDERSTANDING.

Users consistently praise the text's diversity, breadth, and abundance of examples and exercises, many of which are based on material referenced from business, economics, life sciences, and social sciences. Examples and exercises in the Third Edition feature new data-based scenarios, maintaining the currency and relevance of the material.

A Case Study framework for chapter topics puts students into a realistic problem situation at the start of each chapter, then revisits the problem in an extended application at the chapter's end. Designed to challenge students and cement their understanding of critical concepts, these varied case studies are both engaging and practical.

CASE STUDY

Spotting Tax Fraud with Benford's Law

You are a tax-fraud specialist working for the Internal Revenue Service (IRS), and you have just been handed a portion of the tax return from Colossal Conglomerate. The IRS suspects that the portion you were handed may be fraudulent and would like your opinion. Is there any mathematical test, you wonder, that can point to a suspicious tax return based on nothing more than the numbers entered?

Case Study Tax Fraud wi

Notice also the effect of multiplying by the sample size n: The l
more likely that the discrepancy between the $P(x_i)$ and the $P(y_i)$
Substituting the numbers gives[12]

$$\text{SSE} \approx 625\left[\frac{(.29 - .30)^2}{.30} + \frac{(.1 - .18)^2}{.18} + \cdots + \frac{(.01 -}{.0}\right.$$

Question The value of SSE does seem quite large. But how can I us
port? I would like to say something impressive, such as "Based
Colossal Conglomerate tax return analyzed, one can be 95% certa
anomalous."

Answer Statisticians use the error SSE to answer exactly such a quest
do is compare this figure to the largest SSE we would have expect
95 out of 100 selections of data that *do* satisfy Benford's law. This "
puted using a *chi-squared* distribution and can be found in Excel

```
=CHIINV(0.05,8)
```

Here, the 0.05 is 1 − 0.95, encoding the 95% certainty, and the 8 i
degrees of freedom = number of outcomes (9) minus 1.

You now find, using Excel, that the chi-squared figure is 15
largest SSE that you could have expected purely by chance is 15.5
glomerate's error is much larger at 552, you can now justifiably sa
there is a 95% certainty that the figures are anomalous.[13]

EXERCISES

State whether you would expect each of the following lists of dat
law. If the answer is no, give a reason.

1. Distances between cities in France, measured in kilometers
2. Distances between cities in France, measured in miles
3. The grades (0–100) in your math instructor's grade book
4. The Dow Jones averages for the past 100 years
5. Verbal SAT scores of college-bound high school seniors
6. Life spans of companies

Use a spreadsheet to determine whether the given distribution o
95% certainty, to follow Benford's law.

7. Good Neighbor Inc.'s tax return (n = 1000)

y	1	2	3	4	5	6	7
$P(Y = y)$	.31	.16	.13	.11	.07	.07	.05

CASE STUDY **Spotting Tax Fraud with Benford's Law[10]**

You are a tax-fraud specialist working for the IRS, and you have just been handed a portion of the tax return from Colossal Conglomerate (CC). The IRS suspects that the portion you were handed may be fraudulent and would like your opinion. Is there any mathematical test, you wonder, that can point to a suspicious tax return based on nothing more than the numbers entered?

You decide, on an impulse, to make a list of the first digits of all the numbers entered in the portion of the CC tax return (there are 625 of them). You reason that, if the tax return is an honest one, the first digits of the numbers should be uniformly distributed. More precisely, if the experiment consists of selecting a number at random from the tax return and the random variable X is defined to be the first digit of the selected number, then X should have the following probability distribution:

x	1	2	3	4	5	6	7	8	9
$P(X = x)$	$\frac{1}{9}$	$\frac{1}{9}$	$\frac{1}{9}$	$\frac{1}{9}$	$\frac{1}{9}$	$\frac{1}{9}$	$\frac{1}{9}$	$\frac{1}{9}$	$\frac{1}{9}$

You then do a quick calculation based on this probability distribution and find an expected value of $E(X) = 5$. Next, you turn to the CC tax return data and calculate the relative frequency (estimated probability) of the actual numbers in the tax return. You find the following results:

Colossal Conglomerate Return

y	1	2	3	4	5	6	7	8	9
$P(Y = y)$	.29	.10	.04	.15	.31	.08	.01	.01	.01

It certainly does look suspicious! For one thing, the digits 1 and 5 seem to occur a lot more often than any of the other digits and roughly three times what you predicted. Moreover, when you compute the expected value, you obtain $E(Y) = 3.48$, considerably lower than the value of 5 you predicted. "Gotcha!" you exclaim.

You are about to file a report recommending a detailed audit of Colossal Conglomerate when you recall an article you once read about first digits in lists of numbers. The article dealt with a remarkable discovery in 1938 by Frank Benford, a physicist at General Electric. What Benford noticed was that the pages of logarithm tables that listed numbers starting with the digits 1 and 2 tended to be more soiled and dog-eared than the pages that listed numbers starting with higher digits—say, 8. For some reason, numbers that start with low digits seemed more prevalent than numbers that start with high digits. He subsequently analyzed more than 20,000 sets of numbers, such as tables of baseball statistics, listings of widths of rivers, half-lives of radioactive elements, street addresses, and numbers in magazine articles. The result was always the same: Inexplicably, numbers that start with low digits tended to appear more frequently than those that start with high ones, with numbers beginning with the digit 1 most prevalent of all.[11] Moreover, the expected value of the first digit was not the expected 5, but 3.44.

[10] The discussion is based on the article "Following Benford's Law, or Looking Out for No. 1," Malcolm W. Browne, *New York Times*, August 4, 1998, p. F4. The use of Benford's law in detecting tax evasion is discussed in a Ph.D. dissertation by Dr. Mark J. Nigrini (Southern Methodist University, Dallas).

[11] The does not apply to all lists of numbers. For instance, a list of randomly chosen numbers between 100 and 999 will have first digits uniformly distributed between 1 and 9.

NEW AND UPDATED APPLICATIONS

INCLUDE SUV SALES, MEDICARE SPENDING, "DOT COM" EMPLOYMENT, AND INTERNET USAGE.

Here are a few of the hundreds of real world examples and exercises that motivate students' understanding.

Example 1 • Web Searches

In May 2002 a search using the Web search engine AltaVista for the phrase "NASA space station" yielded 800 Web sites containing that phrase. A search for the phrase "NASA mars mission" yielded 201 sites. A search for sites containing both phrases yielded only one site. How many Web sites contained either "NASA space station" or "NASA mars mission" or both?

Solution Let A be the set of sites containing "NASA space station" and let B be the set of sites containing "NASA mars mission". We are told that

$n(A) = 800$

$n(B) = 201$

$n(A \cap B) = 1$ "NASA space station" AND "NASA mars mission"

The formula for the cardinality of the union tells us that

$$n(A \cup B) = n(A) + n(B) - n(A \cap B) = 800 + 201 - 1 = 1000$$

So, 1000 sites in the AltaVista database contained one or both of the phrases "NASA space station" and "NASA mars mission". (This was subsequently confirmed by doing a search for "NASA space station" or "NASA mars mission".)

Before we go on . . . Each search engine has a different way of specifying a search for a union or an intersection. At AltaVista or Google, a search for "NASA space station" OR "NASA mars mission" finds sites containing either phrase (or both), so finds $A \cup B$. Many search engines also support the *and* operator, so you can find
for "NASA space station" AND "NASA mars mission".
Although the formula $n(A \cup B) = n(A) + n(B) - n(A \cap B)$
may sometimes find that, in an actual search, the numbers don't ad
plains that the number reported may be only estimates of the actual
their database containing the search words.[2]

Question What about $n(A \cap B)$?

Answer The formula for the cardinality of a union can also be thou
for the cardinality of an intersection. We can solve for $n(A \cap B)$ to

$$n(A \cap B) = n(A) + n(B) - n(A \cup B)$$

In fact, we can think of this formula as an equation relating four qu
any three of them, we can use the equation to find the fourth (see

Question What about complements?

Answer We can get a formula for the cardinality of a complement as
universal set and $A \subseteq S$, then S is the disjoint union of A and its co

$$S = A \cup A' \quad \text{and} \quad A \cap A' = \varnothing$$

Applying the cardinality formula for a disjoint union, we get

$$n(S) = n(A) + n(A')$$

[2] Email correspondence January 2002.

24. A: Neither die is 1 or 6. B: The sum is even.

25. A: Neither die is 1. B: Exactly one die is 2.

26. A: Both dice are 1. B: Neither die is 2.

27. If a coin is tossed 11 times, find the probability of the sequence H, T, T, H, H, H, T, H, H, T, T.

28. If a die is rolled four times, find the probability of the sequence 4, 3, 2, 1.

APPLICATIONS

29. Movies In 1998 the probability that a randomly selected movie ticket in France was not for a French film was approximately .7, and the probability that it was for a U.S. film was .6. What is the probability that a randomly selected movie ticket was for a U.S. film, given that it was not for a French film?
Source: Center National de la Cinématographie/*New York Times*, December 14, 1999, p. E1.

30. Movies In 1997 the probability that a randomly selected movie ticket in France was not for a U.S. film was approximately .5. Furthermore, the probability that the ticket was for a French film, given that it was not for a U.S. film, was .6. What is the probability that a randomly selected movie ticket was for a French film?
Source: Center National de la Cinématographie/*New York Times*, December 14, 1999, p. E1.

31. Road Safety In 1999 the probability that a randomly selected vehicle would be involved in a deadly tire-related accident was approximately 3×10^{-6}, whereas the probability that a tire-related accident would prove deadly was .02. What was the probability that a vehicle would be involved in a tire-related accident?
Source: *New York Times* analysis of National Traffic Safety Administration crash data/Polk Company vehicle registration data/*New York Times*, November 22, 2000, p. C5. The original data reported three tire-related deaths per million vehicles.

32. Road Safety In 1998 the probability that a randomly selected vehicle would be involved in a deadly tire-related accident was approximately 2.8×10^{-6}, while the probability that a tire-related accident would prove deadly was .016. What was the probability that a vehicle would be involved in a tire-related accident?
Source: *New York Times* analysis of National Traffic Safety Administration crash data/Polk Company vehicle registration data/*New York Times*, November 22, 2000, p. C5. The original data reported 2.8 tire-related deaths per million vehicles.

ood Safety According to a University of Maryland study of
amples of ground meats, the probability that a sample
ntaminated by salmonella was .20. The probability
nonella-contaminated sample was contaminated by
stant to at least three antibiotics was .53. What
bility that a ground meat sample was contami-
in of salmonella resistant to at least three an-

s, October 16, 2001, p. A12.

34. Food Safety According the study mentioned in Exercise 33, the probability that a ground meat sample was contaminated by salmonella was .20. The probability that a salmonella-contaminated sample was contaminated by a strain resistant to at least one antibiotic was .84. What was the probability that a ground meat sample was contaminated by a strain of salmonella resistant to at least one antibiotic?
Source: *New York Times*, October 16, 2001, p. A12.

Publishing Exercises 35–42 are based on the following table, which shows the results of a survey of 100 authors by a publishing company.

	New Authors	Established Authors	Total
Successful	5	25	30
Unsuccessful	15	55	70
Total	20	80	100

Compute the following conditional probabilities:

35. An author is established, given that she is successful.

36. An author is successful, given that he is established.

37. An author is unsuccessful, given that he is a new author.

38. An author is a new author, given that she is unsuccessful.

39. An author is unsuccessful, given that she is established.

40. An author is established, given that he is unsuccessful.

41. An unsuccessful author is established.

42. An established author is successful.

43. Internet Use The following pie chart shows the percentage of the population that uses the Internet, broken down further by family income, based on a survey taken in August, 2000:

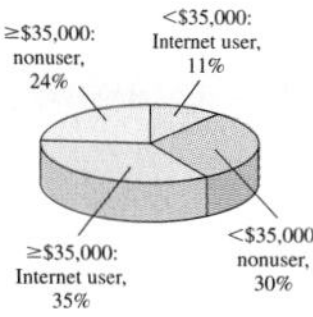

Source: "Falling through the Net: Toward Digital Inclusion, A Report on Americans' Access to Technology Tools," U.S. Department of Commerce, http://www.ntia.doc.gov/ntiahome/fttn00/contents00.html, October 2000.

a. Determine the probability that a randomly chosen person was an Internet user, given that his or her family income was at least \$35,000.

b. Based on the data, was a person more likely to be an Internet user if his or her family income was less than \$35,000 or \$35,000 or more? (Support your answer by citing the relevant conditional probabilities.)

"I have been using Waner/Costenoble for several years now and have enjoyed their use of humor and useful application exercises. ...Their integration of real world problems helps answer students' questions of 'When am I ever going to use this?' Rare is such a question when using this book. The addition of Excel is a great tool for those using computers in their courses."

— **James LaGrone**
South Plains College

A P P L I C A T I O N S

THE TEXT'S INTEGRATED DISCUSSIONS

OF SPREADSHEETS AND GRAPHING CALCULATORS AS OPTIONAL TOOLS HAVE BEEN EXPANDED.

TECHNOLOGY TOOLS

Throughout the text, the authors consistently present opportunities for using technology to solve problems, without losing sight of the accurate development of mathematical concepts. Unique integration of graphing calculators, spreadsheets (using Microsoft Excel), and the Web is color coded for flexibility, and supports a wide range of instructional settings.

NEW! The Third Edition includes additional graphing calculator and Microsoft Excel instructions. The examples and exercises shown here are among the many that guide students in the use of these technology tools.

Example 4 • Graphing More Complicated Piecewise-Defined Functions

Graph the function f specified by

$$f(x) = \begin{cases} -1 & \text{if } -4 \le x < -1 \\ x & \text{if } -1 \le x \le 1 \\ x^2 - 1 & \text{if } 1 < x \le 2 \end{cases}$$

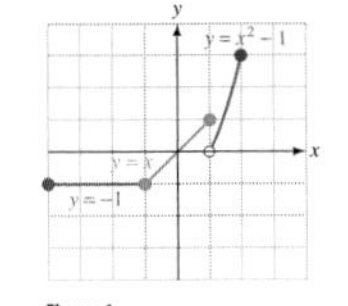

Figure 6

Solution The domain of f is $[-4, 2]$, since $f(x)$ is only specified when $-4 \le x \le 2$. Further, the function changes formulas when $x = -1$ and $x = 1$.

To sketch the graph by hand, we first sketch the three graphs $y = -1$, $y = x$, and $y = x^2 - 1$ and then use the appropriate portion of each (Figure 6).

Question What are solid dots and open dots doing there?

Answer Solid dots indicate points on the graph, whereas open dots indicate points *not* on the graph. For example, when $x = 1$, the inequalities in the formula tell us that we are to use the middle formula (x) rather than the bottom one ($x^2 - 1$). Thus, $f(1) = 1$, not 0, so we place a solid dot at (1, 1) and an open dot at (1, 0).

Graphing Calculator On the TI-83 you can enter this function as

```
(-1)*(X<-1)+X*(-1≤X and X≤1)+(X²-1)*(1<X)
```

First formula — Second formula — Third formula

The logical operator and is found in the `TEST LOGIC` menu. The following alternative formula will also work:

```
(-1)*(X<-1)+X*(-1≤X)*(X≤1)+(X²-1)*(1<X)
```

In the graph the TI-83 will not handle the transition at $x = 1$ correctly; it will connect the two parts of the graph with a spurious line segment.

Excel

For Excel you can use the following formula (the same as the alternative TI-83 formula):[19]

```
=(-1)*(x<-1)+x*(-1<=x)*(x<=1)+(x^2-1)*(1<x)
```

First formula — Second formula — Third formula

Since one of the formulas (the third) specifying the function f is not linear, we need to plot many points in Excel to get a smooth graph. Here is one possible setup (for a smoother curve, plot more points):

	A	B	C	D	E	F
1	x	y				
2	-4	=(-1)*(A2<-1)+A2*(-1<=A2)*(A2<=1)+(A2^2-1)*(1<A2)				
3	-3.9					
4						
61	1.9					
62	2					

[19] We could also use a nested IF statement to accomplish the same result: `=IF(x<-1,-1,IF(x<=1,x,x^2-1))`.

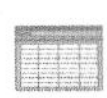

Using Logical Expressions with Technology

The following technology formula defines the function U … Excel, and the Web site (follow Web Site → Online Utilities → F… pher) as well as several other technologies:

```
(x<=10)*(76-3.3*x)+(x>10)*(18+2.5*x)
(X≤10)*(76-3.3*X)+(X>10)*(18+2.5*X)
```

In the TI-83 the logical operators (≤ and >, for example) c… [2nd] [TEST]. When x is less than or equal to 10, the logical ex… ates to 1 because it is true, and the expression (x<10) evaluate… The value of the function is given by the expression (76-3.3*… 10, the expression (x<=10) evaluates to 0, and the expressio… so the value of the function is given by the expression (18+2.5…

As in Example 3, you can use the Table feature to compute s… tion at once.

In Excel we can set up a worksheet as shown:

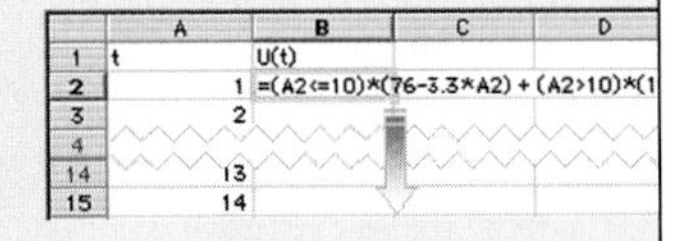

	A	B	C	D
1	t	U(t)		
2	1	=(A2<=10)*(76-3.3*A2) + (A2>10)*(1		
3	2			
4				
14	13			
15	14			

Using the IF Function in Excel

The following worksheet shows how we can get the same res… in Excel:

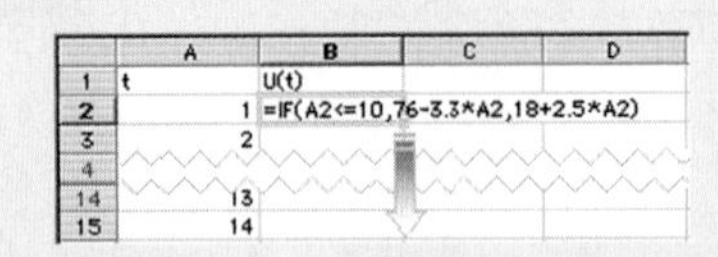

	A	B	C	D
1	t	U(t)		
2	1	=IF(A2<=10,76-3.3*A2,18+2.5*A2)		
3	2			
4				
14	13			
15	14			

The IF function evaluates its first argument, which tests to see if the value of t is in the range $t \le 10$. If the first argument is true, IF returns the result of evaluating its second argument; if not, it returns the result of evaluating its third argume

NEW! A new four-color design enhances the pedagogical effect by clarifying graphics and highlighting key points. Also note the **Before We Go On** and **Question-and-Answer Dialog** boxes (described on page VP-6), which further support students' learning.

136 **Chapter 3** Matrix Algebra and Applications

Graphing Calculator: Entering a Matrix

On the TI-83, matrices are referred to as [A], [B], and so on. To enter a matrix, press MATRX to bring up the matrix menu, select EDIT, select a matrix, and press ENTER. Then enter the dimensions of the matrix followed by its entries. When you want to use a matrix, press MATRX, select the matrix, and press ENTER.

Excel: Entering a Matrix

To enter a matrix in a spreadsheet, we put its entries in any convenient block of cells. For example, the matrix A in the Quick Example above might look like this:

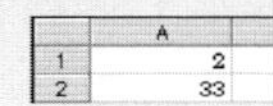

	A
1	2
2	33

Web Site

The path

Web site → Online Utilities → Ma

will take you to an online computational to

trices.

The path

Web site → Online Tutorials → Ex

Scalar Multiplication

will give you a downloadable Excel tutorial

examples in this section.

Referring to the Entries of a Matrix

There is a systematic way of referring to p

numbers, then the entry in the ith row an

ijth **entry** of A. We usually write this entr

we would write its ijth entry as b_{ij} or B_{ij}.)

tion: The row number is specified first an

Quick Example

With $A = \begin{bmatrix} 2 & 0 & 1 \\ 33 & -22 & 0 \end{bmatrix}$,

$a_{13} = 1$ Firs

$a_{21} = 33$ Sec

According to the labeling convention, t

$$A = \begin{bmatrix} a_{11} & a_{12} & a_{13} \\ a_{21} & a_{22} & a_{23} \end{bmatrix}$$

64 **Chapter 1** Functions and Linear Models

In column D we compute the squares of the residues using the Excel formula

```
=(B2-C2)^2
```

Here is the completed worksheet, with SSE in cell F4:

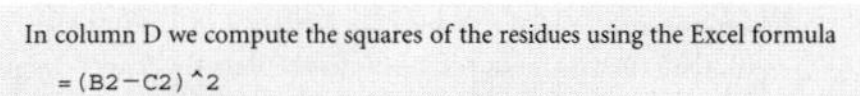

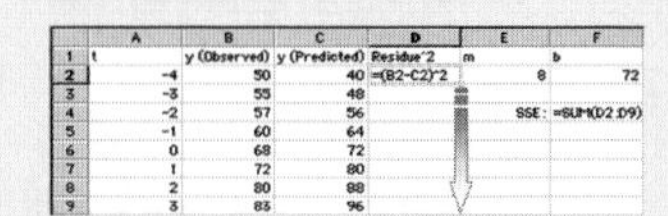

As we have seen, SSE = 479. Changing m to 5 and b to 68 gives SSE for the second model, SSE = 23.

Before we go on . . . You can use your preferred mode of technology to plot the original data points together with the two lines.

To plot the points using the TI-83, you need to turn PLOT1 on in the STAT PLOT window, obtained by pressing 2nd STAT PLOT. To show the lines, enter them in the Y= screen as usual. To obtain a convenient window showing all the points and the lines, press ZOOM and choose option 9: ZoomStat.

In Excel use a scatter plot to graph the data in columns A through C in the worksheet discussed above. Figure 23 shows the two models with the original data points, as shown in Excel.

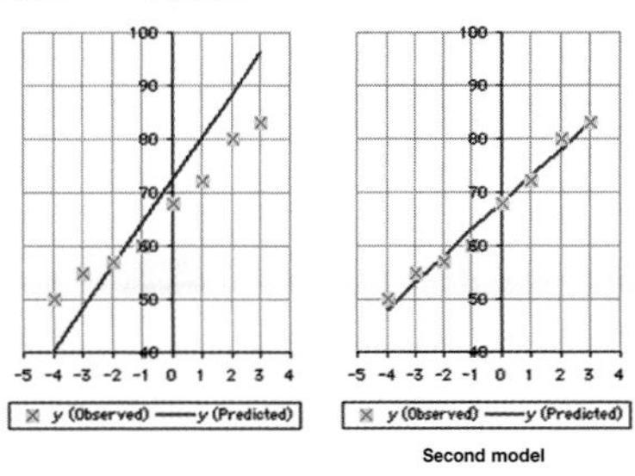

Figure 23

Question It seems clear from the figure that the second model gives a better fit. Why bother to compute SSE to tell me this?

Answer The difference between the two models we chose is so dramatic that it is clear from the graph which is the better fit. However, if we used a third model with $m = 5$ and $b = 68.1$, then its graph would be almost indistinguishable from that of the second but a better fit as measured by SSE = 22.88.

"In general, I believe that technology is the present and future of mathematics instruction and methods. **Finite Mathematics** is ahead of the others in this respect."

—**Joseph Erdeky**
Prince George Community College

TECHNOLOGY TOOLS

IN-TEXT LEARNING AIDS—SHOWCASED
IN A NEW FOUR-COLOR DESIGN—WORK BETTER THAN EVER TO AID STUDENT COMPREHENSION.

PEDAGOGY

Guideline: Recognizing a Linear Programming Problem, Setting Up Inequalities, and Dealing with Unbounded Regions

Question How do I recognize when an application leads to an LP problem as opposed to a system of linear equations?

Answer Here are some cues that suggest an LP problem:

- Key phrases suggesting inequalities rather than equalities, like *at most, up to, no more than, at least,* and *or more.*
- A quantity that is being maximized or minimized (this will be the objective). Key phrases are *maximum, minimum, most, least, largest, greatest, smallest, as large as possible,* and *as small as possible.*

Question How do I deal with tricky phrases like "there should be no more than twice as many nuts as bolts" or "at least 50% of the total should be bolts"?

Answer The easiest way to deal with phrases like this is to use the technique we discussed in Chapter 2: Reword the phrases using "the number of," as in

The number of nuts (x) is no more than twice the number of bolts (y). $x \le 2y$

The number of bolts is at least 50% of the total. $y \ge 0.50(x + y)$

Question Do I always have to add a rectangle to deal with unbounded regions?

Answer Under some circumstances you can tell right away whether optimal solutions exist, even when the feasible region is unbounded:

Note that the following apply only when we have the constraints $x \ge 0$ and $y \ge 0$.

1. If you are minimizing $c = ax + by$ with a and b nonnegative, then optimal solutions always exist. [Examples 4(a) and 5 are of this type.]
2. If you are maximizing $p = ax + by$ with a and b nonnegative (and not both zero), then there is no optimal solution unless the feasible region is bounded.

Do you see why statements 1 and 2 are true?

NEW! A new feature called **Guidelines** appears at the ends of sections to assist students in problem solving. **Guidelines** employ the effective question-and-answer technique used elsewhere in the text.

"I love the Question and Answer paragraphs. They pinpoint key ideas and offer clear explanations of why things are done in a particular way."

— **Professor Terry Nyman**
University of Wisconsin, Fox Valley

NEW! More **Question-and-Answer Dialog** boxes anticipate student questions, and guide them through development of new concepts. More **Quick Examples** have been added, too. Most definition boxes include one or more of these straightforward examples, which allow students to solidify their understanding of each new concept as soon as it is encountered.

The Product Row × Column

The **product** AB of a row matrix A and a column matrix B is a 1×1 matrix. The length of the row in A must match the length of the column in B for the product to be defined. To find the product, multiply each entry in A (going from left to right) by the corresponding entry in B (going from top to bottom) and then add the results.

Quick Examples

1. $\begin{bmatrix} 2 & 1 \end{bmatrix}\begin{bmatrix} -3 \\ 1 \end{bmatrix} = [2 \times (-3) + 1 \times 1] = [-6 + 1] = [-5]$
2. $\begin{bmatrix} 2 & 4 & 1 \end{bmatrix}\begin{bmatrix} 2 \\ 10 \\ -1 \end{bmatrix} = [2 \times 2 + 4 \times 10 + 1 \times (-1)] = [4 + 40 + (-1)] = [43]$

Minimization Problems

We convert a minimization problem into a maximization problem by taking the negative of the objective function. All the constraints remain unchanged.

Quick Example

Minimization Problem

Minimize $c = 10x - 30y$
subject to $2x + y \le 160$
$x + 3y \ge 120$
$x \ge 0, y \ge 0$

Maximization Problem

Maximize $p = -10x + 30y$
subject to $2x + y \le 160$
$x + 3y \ge 120$
$x \ge 0, y \ge 0$

Before We Go On supplementary discussions follow most examples, and include a check on the answer, discussion on the feasibility of the solution, or an in-depth look at what the solution means.

NEW! The number of exercises has been increased by 20%, with additional easier computational exercises and a greater number of **Communication and Reasoning** exercises. Exercises are now arranged by increasing level of difficulty.

7.5 EXERCISES

In Exercises 1–16, use the given information to find the indicated probability.

1. $P(A) = .1$; $P(B) = .6$; $P(A \cap B) = .05$. Find $P(A \cup B)$.
2. $P(A) = .3$; $P(B) = .4$; $P(A \cap B) = .02$. Find $P(A \cup B)$.
3. $A \cap B = \varnothing$; $P(A) = .3$; $P(A \cup B) = .4$. Find $P(B)$.
4. $A \cap B = \varnothing$; $P(B) = .8$; $P(A \cup B) = .8$. Find $P(A)$.
5. $A \cap B = \varnothing$; $P(A) = .3$; $P(B) = .4$. Find $P(A \cup B)$.
6. $A \cap B = \varnothing$; $P(A) = .2$; $P(B) = .3$. Find $P(A \cup B)$.
7. $P(A \cup B) = .9$; $P(B) = .6$; $P(A \cap B) = .1$. Find $P(A)$.
8. $P(A \cup B) = 1.0$; $P(A) = .6$; $P(A \cap B) = .1$. Find $P(B)$.
9. $P(A) = .75$. Find $P(A')$.
10. $P(A) = .22$. Find $P(A')$.
11. A, B, and C are mutually exclusive. $P(A) = .3$; $P(B) = .4$; $P(C) = .3$. Find $P(A \cup B \cup C)$.
12. A, B, and C are mutually exclusive. $P(A) = .2$; $P(B) = .6$; $P(C) = .1$. Find $P(A \cup B \cup C)$.
13. A and B are mutually exclusive. $P(A) = .3$ and $P(B) = .4$. Find $P[(A \cup B)']$.
14. A and B are mutually exclusive. $P(A) = .4$ and $P(B) = .4$. Find $P[(A \cup B)']$.
15. $A \cup B = S$ and $A \cap B = \varnothing$. Find $P(A) + P(B)$.
16. $P(A \cup B) = .3$ and $P(A \cap B) = .1$. Find $P(A) + P(B)$.

In Exercises 17–24, determine whether the information shown is consistent with a probability distribution. If not, say why.

17.

Outcome	a	b	c	d	e
Probability	0	0	.65	.3	.05

18.

Outcome	a	b	c	d	e
Probability	.1	−.1	.65	.3	.05

19. $P(A) = .2$; $P(B) = .1$; $P(A \cup B) = .4$
20. $P(A) = .2$; $P(B) = .4$; $P(A \cup B) = .2$
21. $P(A) = .2$; $P(B) = .4$; $P(A \cap B) = .2$
22. $P(A) = .2$; $P(B) = .4$; $P(A \cap B) = .3$
23. $P(A) = .1$; $P(B) = 0$; $P(A \cup B) = 0$
24. $P(A) = .1$; $P(B) = 0$; $P(A \cap B) = 0$

APPLICATIONS

25. **Internet Use** The following pie chart shows the percentage of the population that uses the Internet, broken down by family income, based on a survey taken in August 2000:

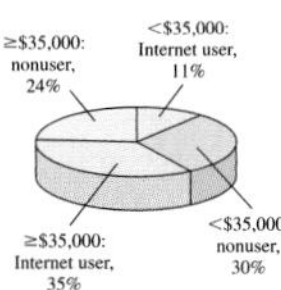

Source: "Falling through the Net: Toward Digital Inclusion, A Report on Americans' Access to Technology Tools," U.S. Department of Commerce, http://www.ntia.doc.gov/ntiahome/fttn00/contents00.html, October 2000.

What is the probability that a randomly chosen person was an Internet user?

26. **Internet Use** Repeat Exercise 25, using the following pie chart that shows the results of a similar survey taken in September 2001:

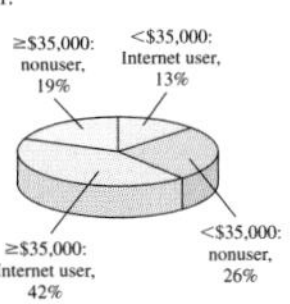

Source: "A Nation Online: How Americans Are Expanding Their Use of the Internet," U.S. Department of Commerce, http://www.ntia.doc.gov/ntiahome/dn/index.html, February 2002.

27. **Astrology** The astrology software package, Turbo Kismet,[18] works by first generating random number sequences and then interpreting them numerologically. When I ran it yesterday, it informed me that there was a 1/3 probability that I would meet a tall dark stranger this month, a 2/3 probability that I would travel this month, and a 1/6 probability that I would meet a tall dark stranger and also travel this month. What is the probability that I will either meet a tall dark stranger or that I will travel this month?

[18]The name and concept were borrowed from a hilarious (as yet unpublished) novel by the science fiction writer William Orr, who also happens to be a faculty member at Hofstra University.

astrology software package, Java Kismet, day traders choose stocks based on the ets and constellations. When I ran it yes- me that there was a .5 probability that go up this afternoon, a .2 probability ll go up this afternoon, and a .2 chance p this afternoon. What is the probability .com or Yahoo.com will go up this after-

In 1999 the probability that a consumer iday gifts at a discount department store bability that a consumer would shop for catalogs was .42. Assuming that 90% of l from one or the other, what percentage

tment; Deloitte & Touche Survey/*New York Times*, November

In 1997 the probability that a consumer iday gifts at a discount department store bability that a consumer would shop for catalogs was .32. Assuming that 75% of l from one or the other, what percentage

tment; Deloitte & Touche Survey/*New York Times*, November

s In 2001 the probability that a randomly household (a household connected to the access was .11, and the probability that an ad DSL access was .05. What percentage ls had neither cable nor DSL access? (As- bility that an online household had both egligible.)

mber 24, 2001, p. C1. The 5% figure is an estimate based on a

s In 2001 6.1% of all U.S. households the Internet via cable, and 2.7% of them the internet through DSL. What percent- olds did not have high-speed (cable or the Internet? (Assume that the percent- vith both cable and DSL access is negligi- ble.)

33. **Fast-Food Stores** In 2000 the top 100 chain restaurants in the United States owned a total of approximately 130,000 outlets. Of these, the three largest (in numbers of outlets) were McDonald's, Subway, and Burger King, owning among them 26% of all of the outlets. The two hamburger companies, McDonald's and Burger King, together owned approximately 16% of all outlets, and the two largest, McDonald's and Subway, together owned 19% of the outlets. What was the probability that a randomly chosen restaurant was a McDonald's?

Source: "Technomic 2001 Top 100 Report," Technomic, Inc. Information obtained from their Web site, www.technomic.com.

34. **Fast-Food Revenues** In 2000 the top 100 chain restaurants in the United States earned a total of approximately \$123 billion in sales. The three largest (in numbers of outlets) were McDonald's, Subway, and Burger King, earning between them 26% of all sales. The two hamburger companies, McDonald's and Burger King, together earned approximately 23% of all sales, and the two largest (in number of outlets), McDonald's and Subway, together earned 19% of sales. What was the probability that a randomly selected dollar in sales was earned by McDonald's?

Source: "Technomic 2001 Top 100 Report," Technomic, Inc. Information obtained from their Web site, www.technomic.com.

35. **Opinion Polls** A *New York Times*/CBS News poll of 1368 people interviewed in March 1993 showed that 85% of all respondents favored a national law requiring a 7-day waiting period for handgun purchases, 13% opposed it, and 2% were undecided. Considering the poll as an experiment, find the probability that a randomly chosen resident was not opposed to such a law.

Source: *New York Times*, August 15, 1993, sect. 4, p. 4

36. **Opinion Polls** The opinion poll referred to in Exercise 35 also showed that 41% of the respondents would favor an outright ban on the sale of handguns (law enforcement officers excepted), 55% would oppose such a ban, and 4% were undecided. Considering the poll as an experiment, find the probability that a randomly chosen resident was not opposed to such a law.

37. **Greek Life** The ΤΦΦ Sorority has a tough pledging program—it requires its pledges to master the Greek alphabet forward, backward, and "sideways." During the last pledge period, two-thirds of the pledges failed to learn it backward, and three-quarters of them failed to learn it sideways; 5 of the 12 pledges failed to master it either backward or sideways. Since admission into the sisterhood requires both backward and sideways mastery, what fraction of the pledges were disqualified on this basis?

38. **Swords and Sorcery** Lance the Wizard has been informed that tomorrow there will be a 50% chance of encountering the evil Myrmidons and a 20% chance of meeting up with the dreadful Balrog. Moreover, Hugo the elf has predicted that there is a 10% chance of encountering both tomorrow. What is the probability that Lance will be lucky tomorrow and encounter neither the Myrmidons nor the Balrog?

d not

tics and Holden- tics (Or-

wing cho- quit- prob- omly

ation of the total n ei-

ation

hat it was not a Mitsubishi

MW 3-series and a to explain, it is len this year. ll get stolen

will get ffect the

lity of a .17, and osen de- n a judge endant was

en defendant. sen defendant. uit a randomly cho-

has been vaccinated against the Martian ague, but 4% of this group gets this disease anyway. If 10% of the total population gets this disease, what is the probability that a randomly selected person has been neither vaccinated nor has contracted Martian ague?

COMMUNICATION AND REASONING EXERCISES

47. Complete the following sentence. The probability of the union of two events is the sum of the probabilities of the two events if ___.
48. If you know $P(E)$ and $P(F)$, what additional information would you need to calculate $P(E \cap F)$, and how would you calculate it?
49. Give an example of a sample space S, a probability distribution on S, and two events A and B with the property that A and B are not mutually exclusive and yet $P(A) + P(B) = P(A \cup B)$.

"The exercises are outstanding. There are enough mechanical problems to let the student perfect technique, yet the real strength is the application problems which, by their nature, provide understandable insight into how mathematics permeates modern life. ...Being able to see why a technique is useful is often just the motivation some students need to really try to understand it.

The communication exercises are well thought out. They can make students think more deeply about the ideas and processes and learn to express themselves in the terminology of mathematics."

— **Professor Terry Nyman**
University of Wisconsin, Fox Valley

PEDAGOGY

A RICH WEB SITE

CONNECTS WITH TODAY'S INTERNET-SAVVY STUDENTS.

www.FiniteMath.com

Throughout the text, students are directed to the acclaimed Waner and Costenoble Web site. This extensive, yet easy-to-use site supports the text with interactive tutorials, utilities, review materials, and more, all working to enhance students' interest, learning, and comprehension. If you use the Web, or if you've been considering the idea of integrating this technology into your course, the authors offer all the support your students will need.

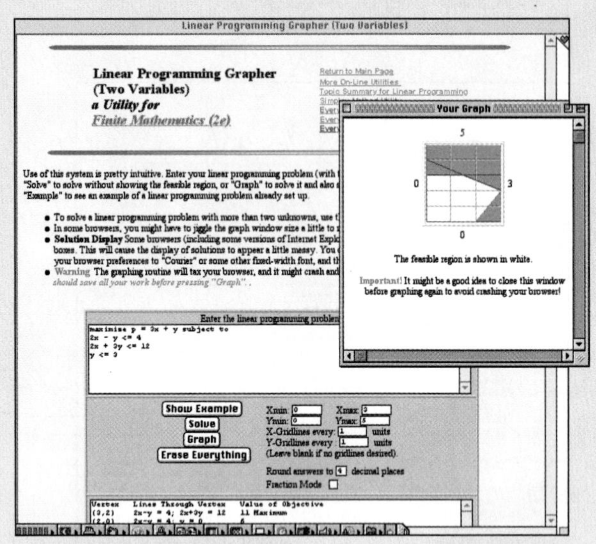

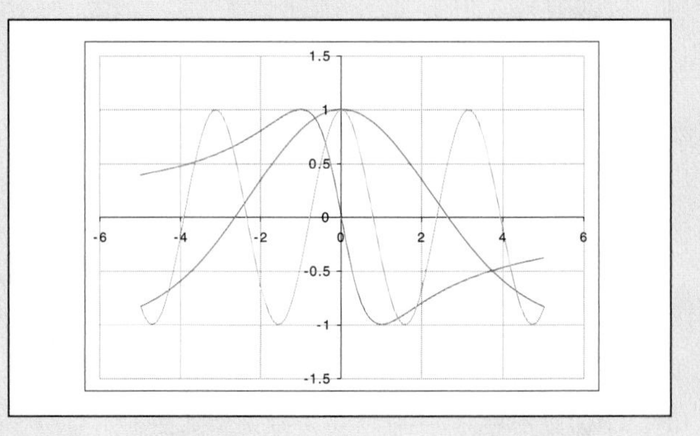

WWW.FINITEMATH.COM

Web Site Features

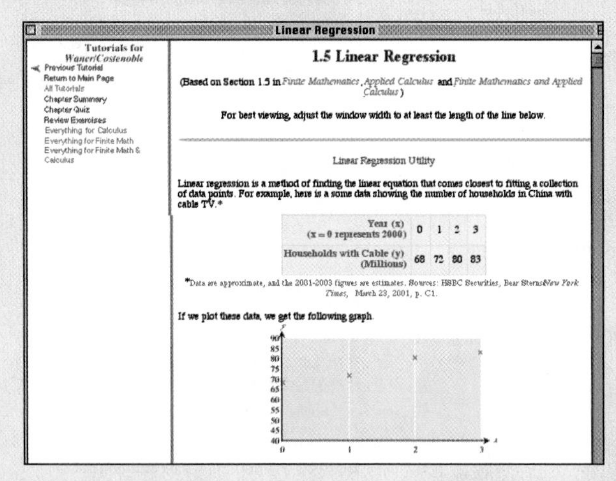

- **Online Tutorials** present the material from individual chapter sections in greater depth. Highly interactive, they're designed to help familiarize students with each new topic rather than simply to summarize the main ideas in a chapter.

- **Topic Summaries** provide detailed summaries of all the chapters in the book. These complete outlines of the subject material, accompanied by examples (many interactive), are ideal for review purposes or as a basic introduction to each subject. Summary pages also include links to online utilities, quizzes, and review exercises for particular topics.

- **Online Utilities** include a large collection of Java-, JavaScript-, and Microsoft Excel-based online utilities for everything from regression, time value of money, and graphing to matrix manipulation, and for solving linear programming problems both graphically and algebraically. Links to the appropriate utilities are placed throughout the resources on the site.

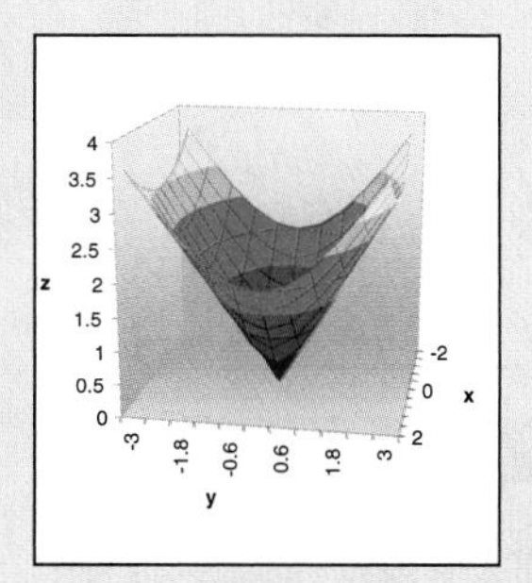

- **Downloadable Microsoft Excel Tutorials** walk students through textual examples, exploring concepts from the perspective of spreadsheet technology. Your students will find these spreadsheet tutorials helpful in building their command of mathematics, while also preparing them to use the tools they will see in business.

- **Online Text** presents complete text material covering many optional topics not in the book, along with complete exercise sets and answers to odd-numbered exercises. As with the topic summaries and tutorials, much of the online text material is interactive.

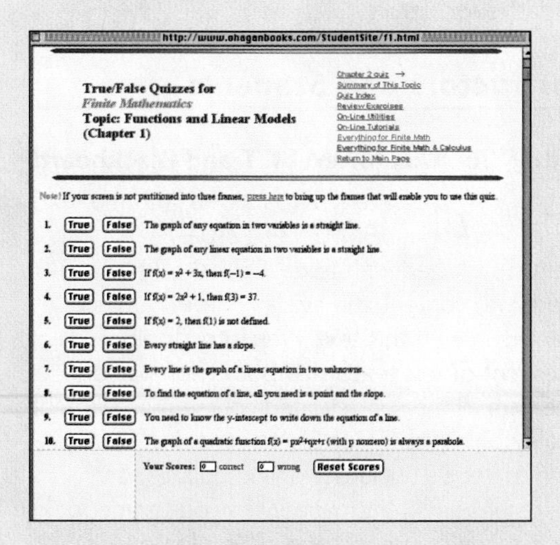

- **True-False Quizzes** test basic comprehension of concepts presented in the book chapters. Lengthier **Review Exercises** test students' grasp of chapter material more thoroughly than the **True-False Quizzes**.

"Thank you for the online math tutorials. They are great! They make learning a lot easier."

—**Jeff Mason**, Student
Brigham Young University

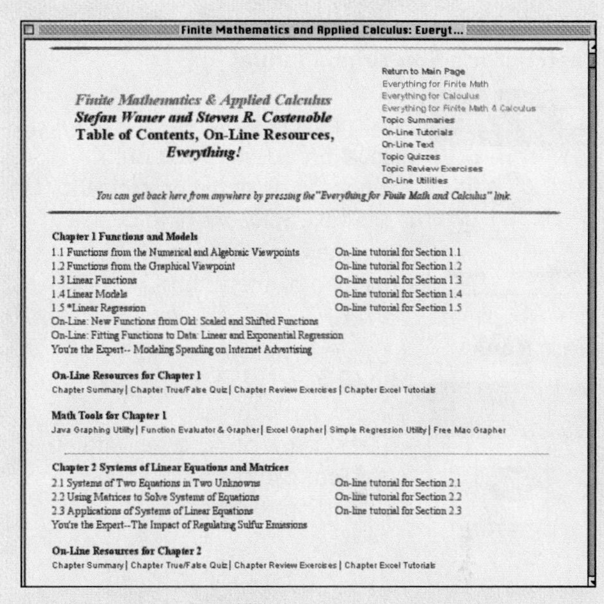

- **Live Tables of Contents** provide central locations from which everything pertinent to each chapter can be accessed, making navigation of this powerful Web site simple and logical.

W W W . F I N I T E M A T H . C O M

IN ADDITION TO THE WEB SITE,

BROOKS/COLE OFFERS A COMPLETE PACKAGE OF ANCILLARY PRODUCTS.

For Instructors

Instructor's Suite CD-ROM

ISBN: 0-534-41952-6

This CD-ROM contains complete solutions to every exercise in Microsoft® Word format, Microsoft PowerPoint® slides, and test items in Word Format.

Instructor's Solutions Manual

ISBN: 0-534-41950-X

The complete solutions manual provides worked out solutions to all of the problems in the text. Solutions have been completely rewritten and expanded by the authors.

Test Bank

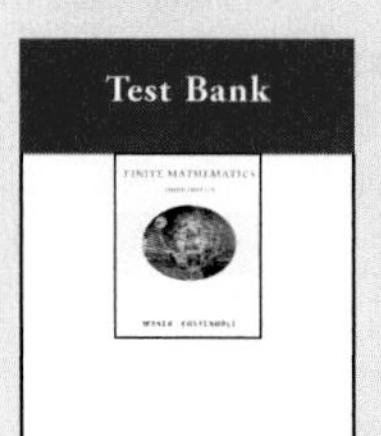

ISBN: 0-534-41953-4

The test bank contains 110 test items per chapter with a problem grid breaking down the questions by section.

BCA Testing

ISBN: 0-534-41954-2

BCA Testing is a revolutionary text-specific testing suite that allows you to customize exams and track your students' progress—all in an accessible, browser-based format. A key feature of this fully integrated testing and course management tool is automatic homework grading that flows directly to your gradebook.

MyCourse 2.1

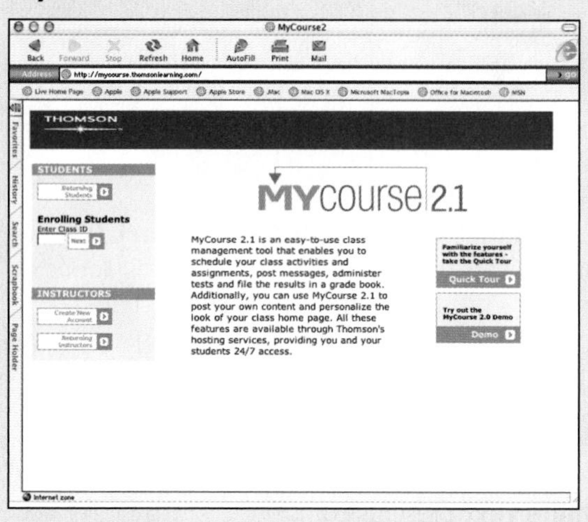

Ask us about our new FREE online course builder! Brooks/Cole offers you a simple solution for a custom course Web site that allows you to assign, track, and report on student progress; load your syllabus; and more. Contact your Thomson•Brooks/Cole representative for details.

For Instructors and Students

WebTutor™ ToolBox for WebCT and Blackboard

WebCT (packaged with the text) ISBN: 0-534-12104-7

Blackboard (packaged with text) ISBN: 0-534-12113-6

Preloaded with content and available free via pincode when packaged with this text, WebTutor ToolBox pairs all the content of this text's rich Book Companion Web Site with all the sophisticated course management functionality of a WebCT or Blackboard product. You can assign materials (including online quizzes) and have the results flow automatically to your gradebook. ToolBox is ready to use as soon as you log on—or, you can customize its content by uploading images and other resources, adding Web links, or creating your own practice materials. Students only have access to student resources on the Web site. Instructors can enter a pincode to access password-protected Instructor Resources.

Automatically packaged free with the text!

InfoTrac® College Edition

http://infotrac.thomsonlearning.com

You and your students receive four months of anytime, anywhere access to InfoTrac College Edition. This vast online library offers the full text of articles from almost 4,000 scholarly and popular publications, updated daily and going back as far as 22 years. It's a great resource for exploring, completing assignments, or simply catching up on the news.

Visit Us on the Web!

http://mathematics.brookscole.com

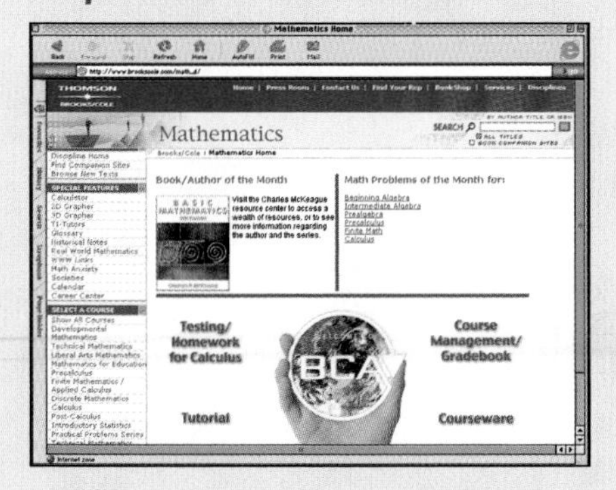

Here, you'll find information on the full range of mathematics texts available from Brooks/Cole, as well as discipline-related resources for you and your students. For instance, NewsEdge offers daily feeds of the latest news in your field. You can also link to the text-specific Web site for all of Waner and Costenoble's books.

For Students

Student Solutions Manual

ISBN: 0-534-41951-8
To package with the text, use
ISBN: 0-534-06857-X
The student solutions manual provides worked out solutions to the odd-numbered problems in the text as well as complete solutions to all the chapter review tests. Solutions have been completely rewritten and expanded by the authors.

Graphing Calculator Manual

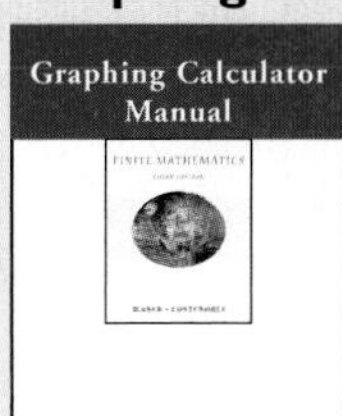

ISBN: 0-534-41956-9
To package with the text, use
ISBN: 0-534-06866-9
This text-specific graphing calculator manual guides students in the use of the TI-83 Plus and TI-86 calculators.

Microsoft® Excel Manual

ISBN: 0-534-41955-0
To package with the text, use
ISBN: 0-534-06875-8
This distinctive, text-specific manual uses Microsoft Excel instructions and formulas to reinforce vital concepts.

Chapter 1: Functions and Linear Models

Print Resources	Media Resources for Instructors	Media Resources for Students
For Instructors **Test Bank Questions** • Chapter 1—Featuring 110 test questions **Instructor's Solutions Manual** • Chapter 1 For Students **Student Solutions Manual** • Chapter 1 **Graphing Calculator Manual** Provides instruction for the TI-83 Plus & TI-86 • Chapter 1 **Microsoft® Excel Manual** • Chapter 1 **Finite Math on the Web by Pilant, Bollinger, Epstein, Hall, Hester, and Strader** • Chapter 1: Pages 1–19 • Chapter 2: Pages 20–36	**BCA Testing** http://bca.brookscole.com **Web Site** http://mathematics.brookscole.com **Microsoft PowerPoint® Presentation on Instructor's Suite CD-ROM** **Electronic Test Bank files on Instructor's Suite CD-ROM** **Electronic files for the Instructor's Solutions Manual on Instructor's Suite CD-ROM**	**Web Sites** http://mathematics.brookscole.com http://www.FiniteMath.com

Chapter 2: Systems of Linear Equations and Matrices

Print Resources	Media Resources for Instructors	Media Resources for Students
For Instructors **Test Bank Questions** • Chapter 2—Featuring 110 test questions **Instructor's Solutions Manual** • Chapter 2 For Students **Student Solutions Manual** • Chapter 2 **Graphing Calculator Manual** Provides instruction for the TI-83 Plus & TI-86 • Chapter 2 **Microsoft Excel Manual** • Chapter 2 **Finite Math on the Web by Pilant, Bollinger, Epstein, Hall, Hester, and Strader** • Chapter 3: Pages 37–68	**BCA Testing** http://bca.brookscole.com **Web Site** http://mathematics.brookscole.com **Microsoft PowerPoint Presentation on Instructor's Suite CD-ROM** **Electronic Test Bank files on Instructor's Suite CD-ROM** **Electronic files for the Instructor's Solutions Manual on Instructor's Suite CD-ROM**	**Web Sites** http://mathematics.brookscole.com http://www.FiniteMath.com

Chapter 3: Matrix Algebra and Applications

Print Resources	Media Resources for Instructors	Media Resources for Students
For Instructors Test Bank Questions • Chapter 3—Featuring 110 test questions Instructor's Solutions Manual • Chapter 3 For Students Student Solutions Manual • Chapter 3 Graphing Calculator Manual Provides instruction for the TI-83 Plus & TI-86 • Chapter 3 Microsoft Excel Manual • Chapter 3 Finite Math on the Web by Pilant, Bollinger, Epstein, Hall, Hester, and Strader • Chapter 3: Pages 37–68	BCA Testing http://bca.brookscole.com Web Site http://mathematics.brookscole.com Microsoft PowerPoint Presentation on Instructor's Suite CD-ROM Electronic Test Bank files on Instructor's Suite CD-ROM Electronic files for the Instructor's Solutions Manual on Instructor's Suite CD-ROM	Web Sites http://mathematics.brookscole.com http://www.FiniteMath.com

Chapter 4: Linear Programming

Print Resources	Media Resources for Instructors	Media Resources for Students
For Instructors Test Bank Questions • Chapter 4—Featuring 110 test questions Instructor's Solutions Manual • Chapter 4 For Students Student Solutions Manual • Chapter 4 Graphing Calculator Manual Provides instruction for the TI-83 Plus & TI-86 • Chapter 4 Microsoft Excel Manual • Chapter 4 Finite Math on the Web by Pilant, Bollinger, Epstein, Hall, Hester, and Strader • Chapter 4: Pages 69–84	BCA Testing http://bca.brookscole.com Web Site http://mathematics.brookscole.com Microsoft PowerPoint Presentation on Instructor's Suite CD-ROM Electronic Test Bank files on Instructor's Suite CD-ROM Electronic files for the Instructor's Solutions Manual on Instructor's Suite CD-ROM	Web Sites http://mathematics.brookscole.com http://www.FiniteMath.com

Chapter 5: The Mathematics of Finance

Print Resources	Media Resources for Instructors	Media Resources for Students
For Instructors **Test Bank Questions** • Chapter 5—Featuring 110 test questions **Instructor's Solutions Manual** • Chapter 5 For Students **Student Solutions Manual** • Chapter 5 **Graphing Calculator Manual** Provides instruction for the TI-83 Plus & TI-86 • Chapter 5 **Microsoft Excel Manual** • Chapter 5 **Finite Math on the Web by Pilant, Bollinger, Epstein, Hall, Hester, and Strader** • Chapter 10: Pages 167–186	**BCA Testing** http://bca.brookscole.com **Web Site** http://mathematics.brookscole.com **Microsoft PowerPoint Presentation on Instructor's Suite CD-ROM** **Electronic Test Bank files on Instructor's Suite CD-ROM** **Electronic files for the Instructor's Solutions Manual on Instructor's Suite CD-ROM**	**Web Sites** http://mathematics.brookscole.com http://www.FiniteMath.com

Chapter 6: Sets and Counting

Print Resources	Media Resources for Instructors	Media Resources for Students
For Instructors **Test Bank Questions** • Chapter 6—Featuring 110 test questions **Instructor's Solutions Manual** • Chapter 6 For Students **Student Solutions Manual** • Chapter 6 **Graphing Calculator Manual** Provides instruction for the TI-83 Plus & TI-86 • Chapter 6 **Microsoft Excel Manual** • Chapter 6 **Finite Math on the Web by Pilant, Bollinger, Epstein, Hall, Hester, and Strader** • Chapter 5: Pages 85–99 • Chapter 6: Pages 101–115	**BCA Testing** http://bca.brookscole.com **Web Site** http://mathematics.brookscole.com **Microsoft PowerPoint Presentation on Instructor's Suite CD-ROM** **Electronic Test Bank files on Instructor's Suite CD-ROM** **Electronic files for the Instructor's Solutions Manual on Instructor's Suite CD-ROM**	**Web Sites** http://mathematics.brookscole.com http://www.FiniteMath.com

Chapter 7: Probability

Print Resources	Media Resources for Instructors	Media Resources for Students
For Instructors Test Bank Questions • Chapter 7—Featuring 110 test questions Instructor's Solutions Manual • Chapter 7 For Students Student Solutions Manual • Chapter 7 Graphing Calculator Manual Provides instruction for the TI-83 Plus & TI-86 • Chapter 7 Microsoft Excel Manual • Chapter 7 Finite Math on the Web by Pilant, Bollinger, Epstein, Hall, Hester, and Strader • Chapter 6: Pages 101–115	BCA Testing http://bca.brookscole.com Web Site http://mathematics.brookscole.com Microsoft PowerPoint Presentation on Instructor's Suite CD-ROM Electronic Test Bank files on Instructor's Suite CD-ROM Electronic files for the Instructor's Solutions Manual on Instructor's Suite CD-ROM	Web Sites http://mathematics.brookscole.com http://www.FiniteMath.com

Chapter 8: Random Variables and Statistics

Print Resources	Media Resources for Instructors	Media Resources for Students
For Instructors Test Bank Questions • Chapter 8—Featuring 110 test questions Instructor's Solutions Manual • Chapter 8 For Students Student Solutions Manual • Chapter 8 Graphing Calculator Manual Provides instruction for the TI-83 Plus & TI-86 • Chapter 8 Microsoft Excel Manual • Chapter 8 Finite Math on the Web by Pilant, Bollinger, Epstein, Hall, Hester, and Strader • Chapter 8: Pages 133–146	BCA Testing http://bca.brookscole.com Web Site http://mathematics.brookscole.com Microsoft PowerPoint Presentation on Instructor's Suite CD-ROM Electronic Test Bank files on Instructor's Suite CD-ROM Electronic files for the Instructor's Solutions Manual on Instructor's Suite CD-ROM	Web Sites http://mathematics.brookscole.com http://www.FiniteMath.com

Chapter 9: Markov Systems		
Print Resources	**Media Resources for Instructors**	**Media Resources for Students**
For Instructors **Test Bank Questions** • Chapter 9—Featuring 110 test questions **Instructor's Solutions Manual** • Chapter 9 For Students **Student Solutions Manual** • Chapter 9 **Graphing Calculator Manual** Provides instruction for the TI-83 Plus & TI-86 • Chapter 9 **Microsoft Excel Manual** • Chapter 9 **Finite Math on the Web by Pilant, Bollinger, Epstein, Hall, Hester, and Strader** • Chapter 6: Page 101–115	**BCA Testing** http://bca.brookscole.com **Web Site** http://mathematics.brookscole.com **Microsoft PowerPoint Presentation on Instructor's Suite CD-ROM** **Electronic Test Bank files on Instructor's Suite CD-ROM** **Electronic files for the Instructor's Solutions Manual on Instructor's Suite CD-ROM**	**Web Sites** http://mathematics.brookscole.com http://www.FiniteMath.com

Finite Mathematics

THIRD EDITION

Titles of Related Interest

Albright, Winston, & Zappe *Data Analysis and Decision Making*
Albright, Winston, & Zappe *Data Analysis for Managers*
Berk & Carey *Data Analysis with Microsoft Excel*
Clemen & Reilly *Making Hard Decisions with Decision Tools*
Giordano et al. *A First Course in Mathematical Modeling*
Goodman *The Mathematics of Finance: Modeling and Hedging*
Gordon *Succeeding in Applied Calculus: Algebra Essentials*
Graybill *Matrices with Applications in Statistics*
Hoerl & Snee *Statistical Thinking: Improving Business Performance*
Howell *Fundamental Statistics for the Behavioral Sciences*
Kao *Introduction to Stochastic Processes*
Keller & Warrack *Statistics for Management and Economics*
Lapin & Whisler *Quantitative Decision Making with Spreadsheet Applications*
Middleton *Data Analysis Using Microsoft Excel*
Minh *Applied Probability Models*
Neuwirth & Arganbright *The Active Modeler: Mathematical Modeling with Excel*
Pagano & Gauvreau *Principles of Biostatistics*
Pilant et al. *Finite Math on the Web*
Rosner *Fundamentals of Biostatistics*
Sandefur *Elementary Mathematical Modeling—A Dynamic Approach*
Savage *Decision Making with Insight*
Scheaffer *Introduction to Probability and Its Applications*
Schrage *Optimization Modeling Using LINDO*
Taylor *Excel Essentials: Using Microsoft Excel for Data Analysis and Decision Making*
Waner & Costenoble *Applied Calculus*
Waner & Costenoble *Finite Mathematics and Applied Calculus*
Weiers *Introduction to Business Statistics*
Winston *Introduction to Probability Models*
Winston & Albright *Practical Management Science*
Winston & Venkataramanan *Introduction to Mathematical Programming*

Finite Mathematics

THIRD EDITION

STEFAN WANER
HOFSTRA UNIVERSITY

STEVEN R. COSTENOBLE
HOFSTRA UNIVERSITY

Australia • Canada • Mexico • Singapore • Spain
United Kingdom • United States

THOMSON
BROOKS/COLE

Publisher: Curt Hinrichs
Development Editor: Cheryll Linthicum
Assistant Editor: Ann Day
Editorial Assistant: Katherine Brayton
Technology Project Manager: Earl Perry
Marketing Manager: Karin Sandberg
Marketing Assistant: Jennifer Gee
Advertising Project Manager: Bryan Vann
Project Manager, Editorial Production: Sandra Craig
Print/Media Buyer: Jessica Reed
Permissions Editor: Kiely Sexton
Production: Cecile Joyner, The Cooper Company
Text Designer: Kathleen Cunningham
Photo Researcher: Terri Wright
Copy Editor: Betty Duncan
Indexer: Julie Shawvan
Proofreader: Amy Mayfield
Cover Designer: Rob Hugel
Cover Image: Getty Images
Cover Printer: The Lehigh Press, Inc.
Composition, Illustrations: Graphic World
Printer: R.R. Donnelley/Willard

Printed in the United States of America
1 2 3 4 5 6 7 07 06 05 04 03

For more information about our products, contact us at:
Thomson Learning Academic Resource Center
1-800-423-0563
For permission to use material from this text, contact us by:
Phone: 1-800-730-2214
Fax: 1-800-730-2215
Web: http://www.thomsonrights.com

Library of Congress Control Number: 2002116967

ISBN 0-534-41949-6

Instructor's Edition: ISBN 0-534-41966-6

Brooks/Cole–Thomson Learning
10 Davis Drive
Belmont, CA 94002
USA

Asia
Thomson Learning
5 Shenton Way #01-01
UIC Building
Singapore 068808

Australia/New Zealand
Thomson Learning
102 Dodds Street
Southbank, Victoria 3006
Australia

Canada
Nelson
1120 Birchmount Road
Toronto, Ontario M1K 5G4
Canada

Europe/Middle East/Africa
Thomson Learning
High Holborn House
50/51 Bedford Row
London WC1R 4LR
United Kingdom

Latin America
Thomson Learning
Seneca, 53
Colonia Polanco
11560 Mexico D.F.
Mexico

Contents

Preface

Finite Mathematics, Third Edition, is a careful revision of the second edition. This book is intended for a one-term finite mathematics course for students majoring in business, the social sciences, or the liberal arts. Like the earlier editions, the third edition is designed to address the considerable challenge of generating enthusiasm and developing mathematical sophistication in an audience that is often underprepared for and disaffected by traditional mathematics courses. Like the second edition, this book is supported by and linked with an extensive and highly developed Web site containing interactive tutorials, online technology, optional supplementary material, and much more.

Rationale for the Third Edition

While preparing the third edition, we have updated and increased the number of examples and exercises based on real data. We have expanded the collections of communication and reasoning exercises designed to help students articulate mathematical concepts and ideas. We have also responded to the growing use of technology in the classroom by expanding the discussions of Excel and graphing calculators as optional tools. These discussions are *seamlessly integrated into the text for those instructors who desire them but carefully delineated so that instructors can easily skip all or parts of them without interrupting the flow of the material.* Thus, this book supports a wide range of instructional paradigms: from settings incorporating little or no technology to courses taught in computerized classrooms and from classes in which a single form of technology is used exclusively to those incorporating several technologies. Moreover—and unlike any other text on the market to date—this book fully supports an instructor who wants to exploit the familiarity with Excel of students who have taken or will take courses in a business school, where its importance and use are well established.

Organization

The following chart shows the logical dependence of the chapters. Notice that Chapter 5 (Mathematics of Finance) can be covered at any time after Chapter 1 and that Chapter 4 does not depend on Chapter 3. Notice also the complete independence of Chapter L and the sequence Chapter 6, Chapter 7, and Chapter 8. Finally, note that Chapter 9 does not assume the detailed study of probability given in Chapter 7.

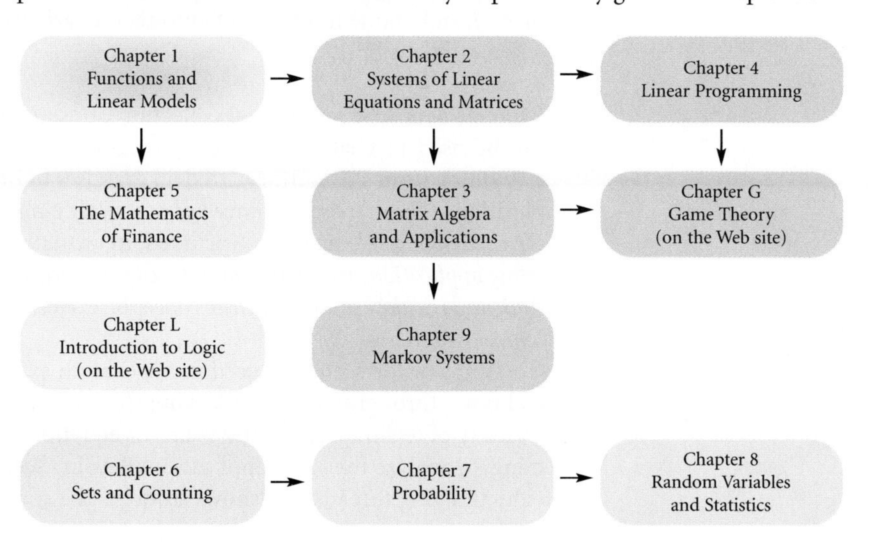

Our Approach and Hallmark Features

Real-World Orientation We are particularly proud of the diversity, breadth, and abundance of examples and exercises we have been able to include in this edition. *A large number of these are based on real, referenced data from business, economics, the life sciences, and the social sciences.* Examples and exercises based on dated information have generally been replaced by more current versions; applications based on unique or historically interesting data have been kept.

Adapting real data for pedagogical use can be tricky; available data can be numerically complex, intimidating for students, or incomplete. We have modified and streamlined many of the real-world applications, rendering them as tractable as any "made-up" application. At the same time, we have been careful to strike a pedagogically sound balance between applications based on real data and more traditional "generic" applications. Thus, the density and selection of real data-based applications has been tailored to the pedagogical goals and appropriate difficulty level for each section.

Readability We would like students to read this book. We would like students to *enjoy* reading this book. Thus, we have written the book in a conversational and student-oriented style and have made frequent use of question-and-answer dialogs to encourage the development of the student's mathematical curiosity and intuition. We hope that this text will give the student insight into how a mathematician develops and thinks about mathematical ideas and their applications.

Five Elements of Mathematical Pedagogy The "Rule of Three" is a common theme in reform-oriented texts. Adapting this approach, we discuss many of the central concepts numerically, graphically, and algebraically and go to some lengths to clearly delineate these distinctions. As a fourth element, *we incorporate verbal communication of mathematical concepts through our emphasis on verbalizing mathematical concepts,* our discussions on translating English sentences into mathematical statements, and our communication and reasoning exercises at the end of each section. The fifth element, interactivity, is implemented within the printed text through expanded use of question-and-answer dialogs but is seen most dramatically in the Web site. At the Web site, students can interact with the material in several ways: through true/false quizzes on every topic, interactive review exercises and questions and answers in the tutorials (including multiple-choice items with detailed feedback and Javascript-based interactive elements), and online utilities that automate a variety of tasks, from graphing to regression and matrix algebra.

Exercise Sets The substantial collection of exercises provides a wealth of material that can be used to challenge students at almost every level of preparation and includes everything from straightforward drill exercises to interesting and rather challenging applications. *The exercise sets have been carefully graded to move from the straightforward to the challenging.* We have also included, in virtually every section of every chapter, *interesting applications based on real data, communication and reasoning exercises that help the student articulate mathematical concepts, exercises ideal for the use of technology, and amusing exercises.*

Many of the scenarios used in application examples and exercises are revisited several times throughout the book. Thus, for instance, students will find themselves using a variety of techniques, from systems of equations to matrix algebra to linear programming, to analyze the same application. Reusing scenarios and important functions provides unifying threads and shows students the complex texture of real-life problems.

New to This Edition

- Chapter 1 has been revised to include expanded coverage of linear regression.
- The integrated discussions of Excel and graphing calculators as optional tools have been expanded, boxed, and color coded for easy reference or to easily skip them altogether.
- The successful pedagogy from the second edition has been expanded with more Quick Examples, more Question and Answer dialog, additional Excel and graphing calculator instructions, and additional Web-site information.
- A new boxed feature called Guidelines appears at the ends of sections to reinforce concepts.
- The number of exercises has been increased by over 15% with additional easier computational exercises at the start of many exercise sets and significantly more communication and reasoning exercises. In addition, exercises and examples based on dated information have generally been replaced by more current versions.
- A new arrangement of exercise sets reflects increasing level of difficulty.
- A new four-color design clearly delineates the pedagogical tools.

Additional Features

- **Case Studies** Each chapter begins with the statement of an interesting problem that is returned to at the end of that chapter in a section titled Case Study. This extended application uses and illustrates the central ideas of the chapter. The themes of these applications are varied, and they are designed to be as nonintimidating as possible. We avoid pulling complicated formulas out of thin air but focus instead on the development of mathematical models appropriate to the topics. These applications are ideal for assignment as projects, and to this end we have included groups of exercises at the end of each.
- **Question-and-Answer Dialog** We frequently use informal question-and-answer dialogs that anticipate the kind of questions that may occur to the student and also guide the student through the development of new concepts.
- **Before We Go On** Most examples are followed by supplementary interpretive discussions under the heading "Before We Go On." These discussions may include a check on the answer, a discussion of the feasibility and significance of a solution, or an in-depth look at what the solution means.
- **Quick Examples** Most definition boxes include one or more straightforward examples that a student can use to solidify each new concept as soon as it is encountered.
- **Communication and Reasoning Exercises for Writing and Discussion** These are exercises designed to broaden the student's grasp of the mathematical concepts. They include exercises in which the student is asked to provide his or her own examples to illustrate a point or design an application with a given solution. They also include fill-in-the-blank exercises and exercises that invite discussion and debate. These exercises often have no single correct answer.
- **Footnotes** We use footnotes throughout the text to provide interesting background, extended discussion, and various asides.
- **Thorough Integration of Spreadsheet and Graphing Technology** Guidance on the use of spreadsheets (we use Microsoft® Excel) and graphing calculators (we use the TI-83) is thoroughly integrated throughout the discussion, examples, and exercise sets. At the same time, this material is clearly delineated so that it can be skipped if desired. In many examples we include a discussion of the use of spreadsheets or graphing technology to aid in the solution. Groups of exercises for which the use of technology is suggested or required appear throughout the exercise sets.

The Web Site: www.FiniteMath.com

Our site at www.FiniteMath.com has been evolving for several years with growing recognition. Students, raised in an environment in which computers suffuse both work and play, can use their familiar Web browsers to engage the material in an active way. At the Web site students and faculty can find:

- **Interactive Tutorials** Highly interactive tutorials are included on many major topics, with guided exercises that parallel the text.
- **Interactive True/False Chapter Quizzes** Interactive true/false quizzes based on the material in each chapter help the student review all the pertinent concepts and avoid common pitfalls.
- **Chapter Review Exercises** The site includes a constantly growing collection of review questions, taken from tests and exams, that do not appear in the printed text.
- **Detailed Chapter Summaries** Comprehensive summaries with interactive elements review all the basic definitions and problem solving techniques discussed in each chapter. The summaries include additional examples for review and can be easily printed.
- **Downloadable Excel Tutorials** Detailed Excel tutorials are available for almost every section of the book. These interactive tutorials expand on the examples given in the text.
- **Online Utilities** Our collection of easy-to-use online utilities, written in Java™ and Javascript, allow students to solve many of the technology-based application exercises directly on the Web page. The utilities available include a function grapher, a function evaluator, and matrix manipulation utilities, as well as utilities to help with regression, Markov chains, game theory, and probability. These utilities require nothing more than a standard, Java-capable Web browser such as the current versions of Netscape Navigator and Microsoft Explorer.
- **Downloadable Software** In addition to the Web-based utilities, the site offers a suite of free and intuitive stand-alone Macintosh® programs for the Gauss–Jordan algorithm, the simplex method, matrix algebra, and function graphing.
- **Supplemental Topics** We include complete interactive text and exercise sets for a selection of topics not ordinarily included in printed texts but often requested by instructors. The text refers to these topics at appropriate points, letting instructors decide whether to include this material in their courses.
- **Optional Internet Chapters** The Web site includes three complete additional chapters on Game Theory, Logic, and Calculus Applied to Probability. These chapters incorporate the same pedagogy as the printed material and can be readily printed and distributed.

Supplemental Material

For Students

Student Solutions Manual ***by Waner and Costenoble***
ISBN: 0-534-41951-8
The student solutions manual provides worked-out solutions to the odd-numbered problems in the text as well as complete solutions to all the chapter review tests. Solutions have been completely rewritten and expanded by the authors.

Microsoft Excel Manual ***by Edwin C. Hackleman and Larry J. Stephans***
ISBN: 0-534-41955-0
This distinctive, text-specific manual uses Excel instructions and formulas to reinforce vital concepts.

Graphing Calculator Manual ***by Mark Stevenson and Florence Chambers***
ISBN: 0-534-41956-9
This text-specific graphing calculator manual guides students in using the TI-83 Plus and TI-86.

For Instructors

Instructor's Suite CD-ROM
ISBN: 0-534-41952-6
This CD-ROM contains Complete Solutions to every exercise in Microsoft Word and PDF format, Microsoft® PowerPoint® slides, and test items in Word format.

Instructor's Solutions Manual ***by Waner and Costenoble***
ISBN: 0-534-41950-X
The complete solutions manual provides worked-out solutions to all of the problems in the text. Solutions have been completely rewritten and expanded by the authors.

Test Bank ***by James Ball***
ISBN: 0-534-41953-4
The test bank contains 110 test items per chapter with a problem grid breaking down the questions by section.

BCA Testing
ISBN: 0-534-41954-2
BCA Testing is a revolutionary text-specific testing suite that allows you to customize exams and track your students' progress—all in an accessible, browser-based format. The Internet-ready tool offers full algorithmic generation of problems and free response mathematics, with automatic homework grading that flows directly to your gradebook.

MyCourse 2.1
Ask us about our new FREE online course builder!
Brooks/Cole offers you a simple solution for a custom course Web site that allows you to assign, track, and report on student progress; load your syllabus; and more. Contact your Thomson•Brooks/Cole representative for details. In addition to the Web site, Brooks/Cole offers a complete ancillary package.

For Instructors and Students

InfoTrac® College Edition http://infotrac.thomsonlearning.com
Automatically packaged free with the text!
You and your students receive four months of anytime, anywhere access to InfoTrac College Edition. This vast online library offers the full text of articles from almost 4000 scholarly and popular publications, updated daily and going back as far as 22 years. It's a great resource for exploring, completing assignments, or catching up on the news.

Visit Us on the Web!
http://mathematics.brookscole.com
Here, you'll find information on the full range of mathematics texts available from Brooks/Cole, as well as discipline-related resources for you and your students. For instance, *NewsEdge* offers daily feeds of the latest news in your field. You can also link to the text-specific Web site for all of Waner and Costenoble's books.

Acknowledgments

This project would not have been possible without the contributions and suggestions of numerous colleagues, students, and friends. We are particularly grateful to our colleagues at Hofstra and elsewhere who used and gave us useful feedback on the second edition. We are also grateful to everyone at Brooks/Cole for their encouragement and guidance throughout the project. Specifically, we would like to thank Curt Hinrichs for his unflagging enthusiasm, Cheryll Linthicum for whipping the book into shape, and Karin Sandberg for letting the world know about it.

We would also like to thank the authors who contributed to the ancillaries associated with this text: James Ball, Florence Chambers, Edwin C. Hackleman, Larry J. Stephens, Mark Stevenson, and Patrick C. Ward. Finally, we would like to thank accuracy checker Jerrold W. Grossman for his meticulous reading of the manuscript and the numerous reviewers who provided many helpful suggestions that have shaped the development of this book.

Mark Clark
Palomar University

Jon Cole
St. John's University

Matt Coleman
Fairfield University

Casey Cremins
University of Maryland

James Czachor
Fordham University

Julie Daberkow
University of Maryland

Deborah Denvir
Marshall University

Michael Ecker
Pennsylvania State University

Janice Epstein
Texas A & M University

Joseph Erdeky
Prince George Community College

Candy Giovanni
Michigan State University

Joe Guthrie
University of Texas–El Paso

Bennette Harris
University of Wisconsin–Whitewater

Loek Helminck
North Carolina State University

Victor Kaftal
University of Cincinnati

Thomas Keller
Southwest Texas State University

Thomas Kelley
Metropolitan State College of Denver

Keith Kendig
Cleveland State University

Greg Klein
Texas A & M University

Michael Moses
George Washington University

Terry A. Nyman
University of Wisconsin–Fox Valley

Ralph Oberste-Vorth
University of South Florida

James Osterburg
University of Cincinnati

James Parks
State University of New York–Potsdam

Joanne Peeples
El Paso Community College

Mihaela Poplicher
University of Cincinnati

Timothy Ray
Southeast Missouri State University

E. Arthur Robinson, Jr.
George Washington University

Gordon Savin
University of Utah

Daniel Scanlon
Orange Coast College

Brad Shelton
University of Oregon

Cynthia Siegel
University of Missouri–St. Louis

Stephen Stuckwisch
Auburn University

Stephen Suen
University of South Florida

Doug Ulmer
University of Arizona

Jackie Vogel
Pellissippi State Community College

Will Watkins
University of Texas–Pan American

Denise Widup
University of Wisconsin–Parkside

Janet E. Yi
Ball State University

Stefan Waner
Steven R. Costenoble

Finite Mathematics

THIRD EDITION

1

FUNCTIONS AND LINEAR MODELS

CASE STUDY

Modeling Spending on Internet Advertising

You are the new director of Impact Advertising's Internet division, which has enjoyed a steady 0.25% of the Internet advertising market. You have drawn up an ambitious proposal to expand your division in light of your anticipation that Internet advertising will continue to skyrocket. The vice president in charge of Financial Affairs feels that current projections (based on a linear model) do not warrant the level of expansion you propose. How can you persuade the vice president that those projections do not fit the data convincingly?

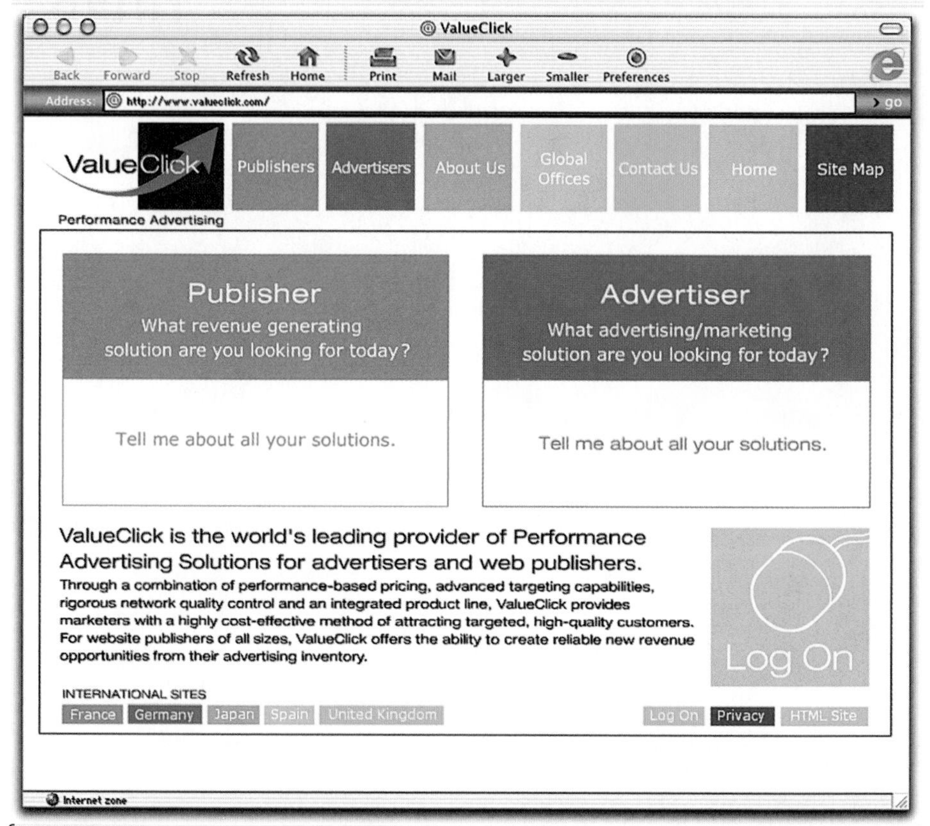

COURTESY VALUECLICK.COM

INTERNET RESOURCES FOR THIS CHAPTER

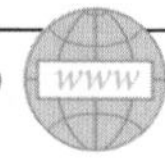

At the Web site, follow the path

Web Site → Everything for Finite Math → Chapter 1

where you will find a detailed chapter summary you can print out, a true/false quiz, and a collection of review exercises. You will also find downloadable Excel tutorials for each section, an online grapher, an online regression utility, and other resources. In addition, complete interactive text and exercises have been placed on the Web site, covering the following optional topic:

New Functions from Old: Scaled and Shifted Functions

Introduction

To analyze recent trends in spending on Internet advertising and to make reasonable projections, we need a mathematical model of this spending. Where do we start? To apply mathematics to real-world situations like this, we need a good understanding of basic mathematical concepts. Perhaps the most fundamental of these concepts is that of a function: a relationship that shows how one quantity depends on another. Functions may be described numerically and, often, algebraically. They can also be described graphically—a viewpoint that is extremely useful.

The simplest functions—the ones with the simplest formulas and the simplest graphs—are linear functions. Because of their simplicity, they are also among the most useful functions and can often be used to model real-world situations, at least over short periods of time. In discussing linear functions, we will meet the concepts of slope and rate of change, which are the starting point of the mathematics of change.

In the last section of this chapter, we discuss *simple linear regression:* construction of linear functions that best fit given collections of data. Regression is used extensively in applied mathematics, statistics, and quantitative methods in business. The inclusion of regression utilities in computer spreadsheets like Excel® makes this powerful mathematical tool readily available for anyone to use.

Algebra Review

For this chapter you should be familiar with real numbers and intervals. To review this material, see Appendix A.

1.1 *Functions from the Numerical and Algebraic Viewpoints*

The following table gives the weights of a particular child at various ages in her first year:

Age (mo)	0	2	3	4	5	6	9	12
Weight (lb)	8	9	13	14	16	17	18	19

Let's write $W(0)$ for the child's weight at birth (in pounds), $W(2)$ for her weight at 2 months, and so on [we read $W(0)$ as "W of 0"]. Thus, $W(0) = 8$, $W(2) = 9$, $W(3) = 13, \ldots, W(12) = 19$. More generally, if we write t for the age of the child (in months) at any time during her first year, then we write $W(t)$ for the weight of the child at age t. We call W a **function** of the variable t, meaning that for each value of t between 0 and 12, W gives us a single corresponding number $W(t)$ (the weight of the child at that age).

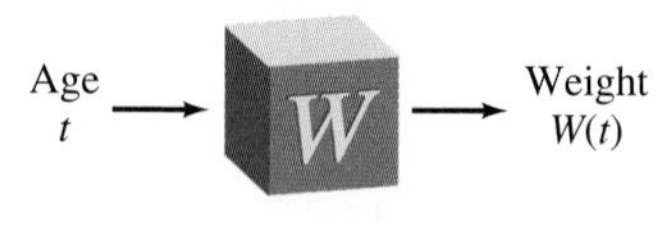

Figure 1

In general, we think of a function as a way of producing new objects from old ones. The functions we deal with in this text produce new numbers from old numbers. The numbers we have in mind are the *real* numbers, including not only positive and negative integers and fractions but also numbers like $\sqrt{2}$ or π (see Appendix A for more on real numbers). For this reason, the functions we use are called **real-valued functions of a real variable.** For example, the function W takes the child's age in months and returns her weight in pounds at that age (Figure 1).

If we introduce another variable—y, say—that stands for the weight of the child, then we can write $y = W(t)$. The function W then tells us exactly how y depends on t. We call y the **dependent variable,** since its value depends on the value of t, the **independent variable.**

A function may be specified in several different ways. It may be specified **numerically,** by giving the values of the function for a number of values of the independent variable, as in the preceding table. It may be specified **verbally,** as in "Let $W(t)$ be the weight of the child at age t months in her first year."[1] In some cases we may be able to use an algebraic formula to calculate the function, and we say that the function is specified **algebraically.** In Section 1.2 we will see that a function may also be specified **graphically.**

Question For which values of t does it make sense to ask for $W(t)$? In other words, for which ages t is the function W defined?

Answer Since $W(t)$ refers to the weight of the child at age t months *in her first year,* $W(t)$ is defined when t is any number between 0 and 12—that is, when $0 \le t \le 12$. Using interval notation (see Appendix A), we can say that $W(t)$ is defined when t is in the interval $[0, 12]$. The set of values of the independent variable for which a function is defined is called its **domain** and is a necessary part of the definition of the function. Notice that the preceding table gives the value of $W(t)$ at only some of the infinitely many possible values in the domain $[0, 12]$.

Here is a summary of the terms we've just introduced.

Functions

A **real-valued function f of a real-valued variable x** assigns to each real number x in a specified set of numbers, called the **domain** of f, a unique real number $f(x)$, read "f of x."

The variable x is called the **independent variable.** If $y = f(x)$, we call y the **dependent variable.**

Note on Domains

The domain of a function is not always specified explicitly; if no domain is specified for the function f, we take the domain to be the largest set of numbers x for which $f(x)$ makes sense. This "largest possible domain" is sometimes called the **natural domain.**

Quick Examples

1. Let $W(t)$ be the weight (in pounds) at age t months of a particular child during her first year. The independent variable is t. If we write $y = W(t)$, then the dependent variable is y, the child's weight. The domain of W is $[0, 12]$ because it was specified that W gives the child's weight during her first year.

[1]Specifying a function verbally in this way is useful for understanding what the function is doing, but it gives no numerical information.

2. Let $f(x) = 1/x$. The function f is specified algebraically. Some specific values of f are

$$f(2) = \frac{1}{2} \qquad f(3) = \frac{1}{3} \qquad f(-1) = \frac{1}{-1} = -1$$

Here, $f(0)$ is not defined because there is no such number as 1/0. The natural domain of f consists of all real numbers except zero because $f(x)$ makes sense for all values of x other than $x = 0$.

Example 1 • A Numerically Specified Function: Airline Profits

The following table shows the combined third-quarter operating profits of U.S. airlines each year from 1996 to 2001, with $t = 6$ representing 1996:

t **(years since 1990)**	6	7	8	9	10	11
P **($ billions)**	5.3	6.0	7.8	7.2	6.0	−3.5

Profits are for domestic travel only, excluding commuter airlines and small regional carriers.
SOURCE: Bureau of Transportation Statistics, http://www.bts.gov/oai/indicators/yrlyopfinan.html, 2002.

Viewing P as a function of t, give its domain and the values $P(6)$, $P(10)$, and $P(11)$. Estimate and interpret the value $P(8.5)$.

Solution The domain of P is the set of numbers t, with $6 \le t \le 11$—that is, [6, 11]. From the table, we have

$P(6) = 5.3$ — $5.3 billion profits in the third quarter of 1996

$P(10) = 6.0$ — $6.0 billion profits in the third quarter of 2000

$P(11) = -3.5$ — $3.5 billion loss in the third quarter of 2001

What about $P(8.5)$? Since $P(8) = 7.8$ and $P(9) = 7.2$, we estimate that

$P(8.5) \approx 7.5$ — 7.5 is midway between 7.8 and 7.2.

We call the process of estimating values for a function between points where it is already known **interpolation.**

Question How should we interpret $P(8.5)$?

Answer $P(8)$ represents the third-quarter profits for the year beginning January 1998, and $P(9)$ represents the third-quarter profits for the year beginning January 1999. Thus, the most logical interpretation of $P(8.5)$ is that it represents the third-quarter profits for the year beginning July 1998—that is, the first-quarter profits for 1999.[2]

[2]Airline profits were actually $7.9 billion in the first quarter of 1999. The fact that this is higher than the interpolated estimate probably results from the peak travel season in January.

Before we go on . . .

Question Can we use the table to estimate $P(t)$ for values of t *outside* the domain—say, $t = 15$?

Answer Strictly speaking, $P(t)$ is defined only when $6 \leq t \leq 11$. However, we could consider a function with a larger domain, like the function that gives the third-quarter profits each year from 1996 to 2005. Estimating values for a function outside a range where it is already known is called **extrapolation.** As a general rule, extrapolation is far less reliable than interpolation: Predicting the future from current data is difficult.

The two functions we have looked at so far were both specified **numerically,** meaning that we were given numerical values of the function evaluated at *certain* values of the independent variable. It would be more useful if we had a formula that would allow us to calculate the value of the function for *any* value of the independent variable we wished. If a function is specified by a formula, we say that it is specified **algebraically.**

Example 2 • An Algebraically Defined Function

Let f be the function specified by

$$f(x) = -0.4x^2 + 7x - 23$$

with domain $(-2, 10]$. This formula gives an approximation of the airline profit function P in Example 1. Use the formula to calculate $f(0)$, $f(10)$, $f(-1)$, $f(a)$, and $f(x + h)$. Is $f(-2)$ defined?

Solution Let's check first that the values we are asked to calculate are all defined. Since the domain is stated to be $(-2, 10]$, the quantities $f(0)$, $f(10)$, and $f(-1)$ are all defined. The quantities $f(a)$ and $f(x + h)$ will also be defined if a and $x + h$ are understood to be in $(-2, 10]$. However, $f(-2)$ is not defined, since -2 is not in the domain $(-2, 10]$.

If we take the formula for $f(x)$ and substitute 0 for x (replace x everywhere it occurs by 0), we get

$$f(0) = -0.4(0)^2 + 7(0) - 23 = -23$$

so $f(0) = -23$. Similarly,

$$f(10) = -0.4(10)^2 + 7(10) - 23 = -40 + 70 - 23 = 7$$

$$f(-1) = -0.4(-1)^2 + 7(-1) - 23 = -0.4 - 7 - 23 = -30.4$$

$$f(a) = -0.4a^2 + 7a - 23 \qquad \text{Substitute } a \text{ for } x.$$

$$f(x + h) = -0.4(x + h)^2 + 7(x + h) - 23 \qquad \text{Substitute } (x + h) \text{ for } x.$$

$$= -0.4x^2 - 0.8xh - 0.4h^2 + 7x + 7h - 23$$

Before we go on . . . Had we not specified anything about the domain of f, we would have used the natural domain of f. In this case the natural domain is the set of all real numbers, since $-0.4x^2 + 7x - 23$ is defined for every real number x.

We said that the function f given in this example is an approximation of the profit function P of Example 1. The following table compares some of their values:

x	8	9	10
$P(x)$	7.8	7.2	6.0
$f(x)$	7.4	7.6	7.0

The function f is a "best-fit," or regression quadratic, curve based on the data in Example 1 (coefficients are rounded and the third-quarter 2001 figure is excluded). We will learn more about regression later in this chapter.

We call the algebraic function f an **algebraic model** of U.S. airline third-quarter profits since it models, or represents (approximately), the third-quarter profits, using an algebraic formula. The particular kind of algebraic model we used is called a **quadratic model** (see the end of this section for the names of some commonly used models).

Notes

1. Instead of using *function notation*

   $f(x) = -0.4x^2 + 7x - 23$ Function notation

   we could use *equation notation*

   $y = -0.4x^2 + 7x - 23$ Equation notation

   (the choice of the letter y is conventional), and we say that "y is a function of x." Note also that there is nothing magical about the letter x. We might just as well say

   $f(t) = -0.4t^2 + 7t - 23$

   which defines *exactly the same function* as $f(x) = -0.4x^2 + 7x - 23$. For example, to calculate $f(10)$ from the formula for $f(t)$, we would substitute 10 for t, getting $f(10) = 7$, just as we did using the formula for $f(x)$.

2. It is important to place parentheses around the number at which you are evaluating a function. For instance, if $g(x) = x^2 + 1$, then

   $g(-2) = (-2)^2 + 1 = 4 + 1 = 5$ ✓ Not $-2^2 + 1 = -4 + 1 = -3$ ✗

   $g(x + h) = (x + h)^2 + 1$ ✓ Not $x^2 + h + 1$ or $x + h^2 + 1$ ✗

T *Example 3 • Evaluating a Function with Technology*

Evaluate the function $f(x) = -0.4x^2 + 7x - 23$ of Example 2 for $x = 0, 1, 2, \ldots, 10$.

Solution This task is tedious to do by hand, but various technologies can make evaluating a function easier.

Graphing Calculator

There are several ways to evaluate an algebraically defined function on a graphing calculator such as the TI-83. First, enter the function in the `Y=` screen, as

`Y1 = -0.4*X^2+7*X−23` — Negative (-) and minus (−) are different keys on the TI-83.

or `Y1 = -0.4X²+7X−23`

(See Appendix A for a discussion of technology formulas.) Then, to evaluate $f(0)$, for example, enter the following in the home screen:

`Y1(0)` — This evaluates the function Y_1 at 0.

Alternatively, you can use the Table feature: After entering the function under Y_1, press [2nd] [TABLE] and set `Indpnt` to `Ask`. (You do this once and for all; it will permit you to specify values for x in the table screen.) Then, press [2nd] [TABLE], and you will be able to evaluate the function at several values of x. Whichever method you use, you should obtain the following set of values:

x	0	1	2	3	4	5	6	7	8	9	10
$f(x)$	−23	−16.4	−10.6	−5.6	−1.4	2	4.6	6.4	7.4	7.6	7

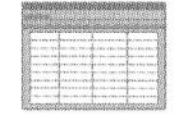

Excel

To create a table of values of f using Excel, first set up two columns—one for the values of x and one for the values of $f(x)$. To enter the sequence of values 0, 1, . . . , 10 in the x column, start by entering the first two values, `0` and `1`, highlight both of them, and drag the *fill handle* (the little dot at the lower right-hand corner of the selection) down until you reach row 12. (Why 12?)

	A	B
1	x	f(x)
2	0	
3	1	
4		
5		
6		
7		

Now enter the formula for f in cell B2. The technology formula (see Appendix A) for f is

`-0.4*x^2+7*x-23` — Technology formula

To get the formula to use for Excel, replace each occurrence of x by the name of the cell holding the value of x (cell A2 in this case) to obtain

`=−0.4*A2^2+7*A2−23` — A2 refers to the cell containing x.

(The formula

`=−0.4*x^2+7*x−23`

will also work in many versions of Excel, provided you have entered the heading `x` in cell A1 as shown. Try it!)

Enter this formula in cell B2 and drag it down (using the fill handle) to cell B12, as shown below (on the left), to obtain the result shown on the right.

	A	B	C
1	x	f(x)	
2	0	=-0.4*A2^2+7*A2-23	
3	1		
4	2		
5	3		
6	4		
7	5		
8	6		
9	7		
10	8		
11	9		
12	10		

	A	B
1	x	f(x)
2	0	-23
3	1	-16.4
4	2	-10.6
5	3	-5.6
6	4	-1.4
7	5	2
8	6	4.6
9	7	6.4
10	8	7.4
11	9	7.6
12	10	7

Web Site At the Web site, follow the path

Web site → Online Utilities → Function Evaluator & Grapher

Enter

`-0.4*x^2+7*x-23`

or `-0.4x^2+7x-23`

in the `f(x)` = box, enter the values of x in the `Values of x` boxes (you can use the tab key to move from one box to the next), and press Evaluate.

Question In Example 3 the values of $f(x)$ differ slightly from those of $P(x)$. Is this the best we can do with an algebraic model? Can't we get the airline profit data exactly?

Answer It is possible to find algebraic formulas that give the same exact values as Example 1, but such formulas would be far more complicated than the one given and quite possibly less useful.[3]

Question How do we measure how closely the algebraic model approximates the actual data? For instance, I notice that the approximation is close when $x = 8$ but not so close when $x = 10$.

Answer In the discussion on regression at the end of this chapter, we describe precise ways of measuring this "goodness of fit." These measurements are based on a simple procedure: Take the differences between the actual values and the values given by the model, square them, and then add the results.

---

[3]Methods for obtaining such formulas are described in most numerical analysis textbooks. One reason that more complex formulas are often less realistic than simple ones is that it is often random phenomena in the real world, rather than algebraic relationships, that cause data to fluctuate. Attempting to model these random fluctuations using algebraic formulas amounts to imposing mathematical structure where structure does not exist.

Excel Warning

In interpreting a negative sign at the start of an expression, Excel uses a different convention from the usual mathematical one used by almost all other technology and programming languages:

Excel Formula	Usual Interpretation	Excel Interpretation		
`-x^2`	$-x^2$	$(-x)^2$	Same as x^2	Different
`2-x^2`	$2 - x^2$	$2 - x^2$	The same	
`-1*x^2`	$-x^2$	$-x^2$	The same	
`-(x^2)`	$-x^2$	$-x^2$	The same	

Thus, if a formula begins with $-x^2$, you should enter it in Excel as

`=-1*x^2` or `-(x^2)` With x replaced by the cell holding x

Quick Examples

1. To enter $-x^2 + 4x - 3$ in Excel, type `=-1*x^2+4*x-3`.
2. To enter $-3x^2 + 4x - 3$ in Excel, type `=-3*x^2+4*x-3`.
3. To enter $4x - x^2$ in Excel, type `=4*x-x^2`.

In short, you need to be careful only when the expression you want to use begins with a negative sign in front of an x.

Sometimes, as in Example 4, we need to use several formulas to specify a single function.

Example 4 • A Piecewise-Defined Function: Semiconductors

The percentage $U(t)$ of semiconductor equipment manufactured in the United States during the period 1981–1994 can be approximated by the following function of time t in years ($t = 0$ represents 1980):[4]

$$U(t) = \begin{cases} 76 - 3.3t & \text{if } 1 \le t \le 10 \\ 18 + 2.5t & \text{if } 10 < t \le 14 \end{cases}$$

What percentage of semiconductor equipment was manufactured in the United States in the years 1982, 1990, and 1992?

Solution The years 1982, 1990, and 1992 correspond, respectively, to $t = 2$, 10, and 12.

$t = 2$: $U(2) = 76 - 3.3(2) = 69.4$ We use the first formula since $1 \le t \le 10$.

$t = 10$: $U(10) = 76 - 3.3(10) = 43$ We use the first formula since $1 \le t \le 10$.

$t = 12$: $U(12) = 18 + 2.5(12) = 48$ We use the second formula since $10 < t \le 14$.

Thus, the percentage of semiconductor equipment that was manufactured in the United States was 69.4% in 1982, 43% in 1990, and 48% in 1992.

[4]Source for data: VLSI Research/*New York Times,* October 9, 1994, sec. 3, p. 2.

Using Logical Expressions with Technology

The following technology formula defines the function U in graphing calculators, Excel, and the Web site (follow Web Site → Online Utilities → Function Evaluator & Grapher) as well as several other technologies:

```
(x<=10)*(76-3.3*x)+(x>10)*(18+2.5*x)        Excel, Web site

(X≤10)*(76-3.3*X)+(X>10)*(18+2.5*X)         TI-83
```

In the TI-83 the logical operators ($\leq$ and $>$, for example) can be found by pressing [2nd] [TEST]. When x is less than or equal to 10, the logical expression `(x<=10)` evaluates to 1 because it is true, and the expression `(x>10)` evaluates to 0 because it is false. The value of the function is given by the expression `(76-3.3*x)`. When x is greater than 10, the expression `(x<=10)` evaluates to 0, and the expression `(x>10)` evaluates to 1, so the value of the function is given by the expression `(18+2.5*x)`.

As in Example 3, you can use the Table feature to compute several values of the function at once.

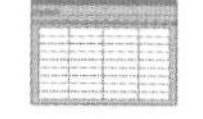

In Excel we can set up a worksheet as shown:

	A	B	C	D	E
1	t	U(t)			
2	1	=(A2<=10)*(76-3.3*A2) + (A2>10)*(18+2.5*A2)			
3	2				
4					
14	13				
15	14				

Using the IF Function in Excel

The following worksheet shows how we can get the same result using the IF function in Excel:

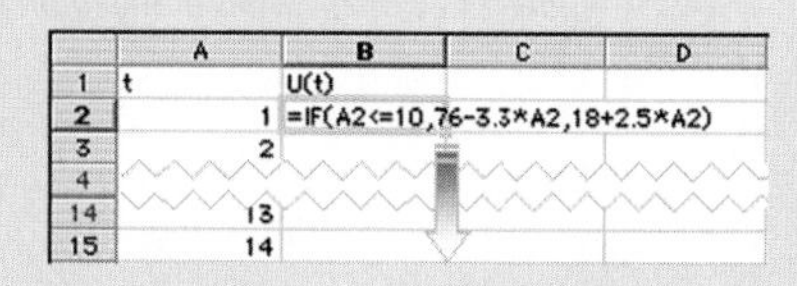

	A	B	C	D
1	t	U(t)		
2	1	=IF(A2<=10,76-3.3*A2,18+2.5*A2)		
3	2			
4				
14	13			
15	14			

The IF function evaluates its first argument, which tests to see if the value of t is in the range $t \leq 10$. If the first argument is true, IF returns the result of evaluating its second argument; if not, it returns the result of evaluating its third argument.

The functions we used in Examples 1–4 are **mathematical models** of real-life situations, since they model, or represent, situations in mathematical terms.

Mathematical Modeling

To *mathematically model* a situation means to represent it in mathematical terms. The particular representation used is called a **mathematical model** of the situation. Mathematical models do not always represent a situation perfectly or completely. Some (like that in Example 2) represent a situation only approximately, whereas others represent only some aspects of the situation.

Quick Examples

Situation	**Model**
1. Albano's bank balance is twice Bravo's.	$a = 2b$ (a = Albano's balance; b = Bravo's)
2. The temperature is now 10° F and increasing by 20° per hour.	$T(t) = 10 + 20t$ (t = time in hours; T = temperature)
3. The volume of a rectangular solid with square base is obtained by multiplying the area of its base by its height.	$V = x^2h$ (h = height; x = length of a side of the base)
4. U.S. airline profits	The table in Example 1 is a **numerical model** of U.S. airline profits. The function in Example 2 is an **algebraic model** of U.S. airline profits.
5. Semiconductor manufacture in the United States	Example 4 gives a **piecewise algebraic model** of the percentage of semiconductor equipment manufactured in the United States.

Table 1 lists some common types of functions that are often used to model real-world situations.

Table 1 • Common Types of Algebraic Functions

	Type of Function	**Example**
Linear	$f(x) = mx + b$ m, b constant	$f(x) = 3x - 2$ `3*x-2`
Quadratic	$f(x) = ax^2 + bx + c$ a, b, c constant $(a \neq 0)$	$f(x) = -3x^2 + x - 1$ `-3*x^2+x-1`
Cubic	$f(x) = ax^3 + bx^2 + cx + d$ a, b, c, d constant $(a \neq 0)$	$f(x) = 2x^3 - 3x^2 + x - 1$ `2*x^3-3*x^2+x-1`
Polynomial	$f(x) = ax^n + bx^{n-1} + \cdots + rx + s$ $a, b, \ldots, r, s$ constant (Includes all of the above functions)	All of the above and $f(x) = x^6 - x^4 + x - 3$ `x^6-x^4+x-3`
Exponential	$f(x) = Ab^x$ A, b constant (b positive)	$f(x) = 3(2^x)$ `3*2^x`
Rational	$f(x) = \dfrac{P(x)}{Q(x)}$ $P(x)$ and $Q(x)$ polynomials	$f(x) = \dfrac{x^2 - 1}{2x + 5}$ `(x^2-1)/(2*x+5)`

Functions and models other than linear ones are called **nonlinear.**

Guideline: How Many Decimal Places?

Question When I use technology to evaluate a function, I often get numbers with many decimals, like 2.034 239 847 2. How many decimal places should I round to?

Answer General rules of thumb are these:

1. When dealing with a mathematical model in which the coefficients are already rounded to a certain number of digits, round your answer to the same number of digits, since the additional digits are meaningless. (See Appendix A for a discussion on significant digits.) For instance, if

$$f(x) = -0.62x^2 + 4.3x - 2.1$$

has rounded coefficients, then $f(2.1) \approx 4.1958$, which we should round to two digits: 4.2.

2. If the coefficients in the model are exact, then retain as many digits as practically convenient or as called for in the problem.
3. Never round the answers to intermediate steps in your calculation. *Only round at the end of the calculation.*

1.1 EXERCISES

In Exercises 1–4, evaluate or estimate each expression based on the following table:

x	-3	-2	-1	0	1	2	3
$f(x)$	1	2	4	2	1	0.5	0.25

1. a. $f(0)$ **b.** $f(2)$ **2. a.** $f(-1)$ **b.** $f(1)$

3. a. $f(2) - f(-2)$ **b.** $f(-1)f(-2)$ **c.** $-2f(-1)$

4. a. $f(1) - f(-1)$ **b.** $f(1)f(-2)$ **c.** $3f(-2)$

5. Given $f(x) = 4x - 3$, find **a.** $f(-1)$ **b.** $f(0)$ **c.** $f(1)$ **d.** $f(y)$ **e.** $f(a + b)$

6. Given $f(x) = -3x + 4$, find **a.** $f(-1)$ **b.** $f(0)$ **c.** $f(1)$ **d.** $f(y)$ **e.** $f(a + b)$

7. Given $f(x) = x^2 + 2x + 3$, find **a.** $f(0)$ **b.** $f(1)$ **c.** $f(-1)$ **d.** $f(-3)$ **e.** $f(a)$ **f.** $f(x + h)$

8. Given $g(x) = 2x^2 - x + 1$, find **a.** $g(0)$ **b.** $g(-1)$ **c.** $g(r)$ **d.** $g(x+h)$

9. Given $g(s) = s^2 + \frac{1}{s}$, find **a.** $g(1)$ **b.** $g(-1)$ **c.** $g(4)$ **d.** $g(x)$ **e.** $g(s + h)$ **f.** $g(s + h) - g(s)$

10. Given $h(r) = \frac{1}{r + 4}$, find **a.** $h(0)$ **b.** $h(-3)$ **c.** $h(-5)$ **d.** $h(x^2)$ **e.** $h(x^2 + 1)$ **f.** $h(x^2) + 1$

11. Given

$$f(t) = \begin{cases} -t & \text{if } t < 0 \\ t^2 & \text{if } 0 \le t < 4 \\ t & \text{if } t \ge 4 \end{cases}$$

find **a.** $f(-1)$ **b.** $f(1)$ **c.** $f(4) - f(2)$ **d.** $f(3)f(-3)$

12. Given

$$f(t) = \begin{cases} t - 1 & \text{if } t \le 1 \\ 2t & \text{if } 1 < t < 5 \\ t^3 & \text{if } t \ge 5 \end{cases}$$

find **a.** $f(0)$ **b.** $f(1)$ **c.** $f(4) - f(2)$ **d.** $f(5) + f(-5)$

In Exercises 13–16, say whether $f(x)$ is defined for the given values of x. If it is defined, give its value.

13. $f(x) = x - \frac{1}{x^2}$, with domain $(0, +\infty)$ **a.** $x = 4$ **b.** $x = 0$ **c.** $x = -1$

14. $f(x) = \frac{2}{x} - x^2$, with domain $[2, +\infty)$ **a.** $x = 4$ **b.** $x = 0$ **c.** $x = 1$

15. $f(x) = \sqrt{x + 10}$, with domain $[-10, 0)$ **a.** $x = 0$ **b.** $x = 9$ **c.** $x = -10$

16. $f(x) = \sqrt{9 - x^2}$, with domain $(-3, 3)$ **a.** $x = 0$ **b.** $x = 3$ **c.** $x = -3$

In Exercises 17–20, find and simplify **a.** $f(x + h) - f(x)$ **b.** $\frac{f(x + h) - f(x)}{h}$

17. $f(x) = x^2$ **18.** $f(x) = 3x - 1$

19. $f(x) = 2 - x^2$ **20.** $f(x) = x^2 + x$

T In Exercises 21–24, first give the technology formula for the given function and then use technology to evaluate the function for the given values of x (when defined there).

21. $f(x) = 0.1x^2 - 4x + 5$; $x = 0, 1, \ldots, 10$

22. $g(x) = 0.4x^2 - 6x - 0.1$; $x = -5, -4, \ldots, 4, 5$

23. $h(x) = \dfrac{x^2 - 1}{x^2 + 1}$; $x = 0.5, 1.5, 2.5, \ldots, 10.5$ (Round all answers to four decimal places.)

24. $r(x) = \dfrac{2x^2 + 1}{2x^2 - 1}$; $x = -1, 0, 1, \ldots, 9$ (Round all answers to four decimal places.)

APPLICATIONS

25. Employment The following table lists the approximate number of people employed in the United States during the period 1995–2001, on July 1 of each year ($t = 5$ represents 1995):

Year, t	5	6	7	8	9	10	11
Employment, $P(t)$ (millions)	117	120	123	125	130	132	132

The given values represent nonfarm employment and are approximate.
SOURCE: Bureau of Labor Statistics/*New York Times*, December 17, 2001, p. C3.

a. Find or estimate $P(5)$, $P(10)$, and $P(9.5)$. Interpret your answers.
b. What is the domain of P?

26. Cell Phone Sales The following table lists the net sales (after-tax revenue) at the Finnish cell phone company Nokia for each year in the period 1995–2001 ($t = 5$ represents 1995):

Year, t	5	6	7	8	9	10	11
Nokia Net Sales, $P(t)$ ($ billions)	8	8	10	16	20	27	28

SOURCE: Nokia/*New York Times*, February 6, 2002, p. A3.

a. Find or estimate $P(5)$, $P(10)$, and $P(7.5)$. Interpret your answers.
b. What is the domain of P?

27. Coffee Shops The number $C(t)$ of coffee shops and related enterprises in the United States can be approximated by the following function of time t in years since 1990:[5]

$$C(t) = \begin{cases} 500t + 800 & \text{if } 0 \le t \le 4 \\ 1300t - 2400 & \text{if } 4 < t \le 10 \end{cases}$$

a. Evaluate $C(0)$, $C(4)$, and $C(5)$ and interpret the results.
b. Use the model to estimate when there were 5400 coffee shops in the United States.

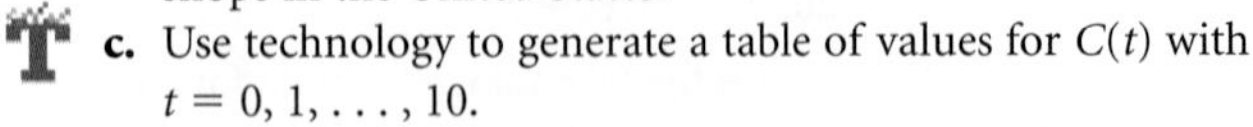

c. Use technology to generate a table of values for $C(t)$ with $t = 0, 1, \ldots, 10$.

[5]SOURCE: Specialty Coffee Association of America/*New York Times*, August 13, 1995, p. F10.

28. Semiconductors The percentage $J(t)$ of semiconductor equipment manufactured in Japan during the period 1981–1994 can be approximated by the following function of time t in years since 1980:[6]

$$J(t) = \begin{cases} 17 + 3.1t & \text{if } 1 \le t \le 10 \\ 68 - 2t & \text{if } 10 < t \le 14 \end{cases}$$

a. Evaluate $J(1)$, $J(10)$, and $J(12)$ and interpret the results.
b. Use the model to estimate, to the closest year, when Japan manufactured 40% of all semiconductor equipment.
c. Use technology to generate a table of values for $J(t)$ with $t = 1, 2, \ldots, 14$.

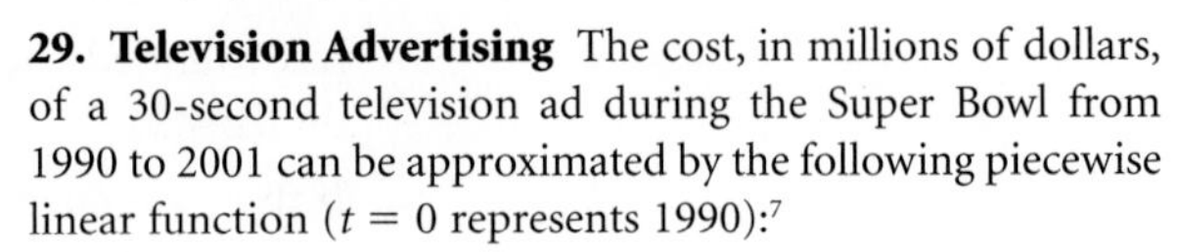

29. Television Advertising The cost, in millions of dollars, of a 30-second television ad during the Super Bowl from 1990 to 2001 can be approximated by the following piecewise linear function ($t = 0$ represents 1990):[7]

$$C(t) = \begin{cases} 0.08t + 0.6 & \text{if } 0 \le t < 8 \\ 0.355t - 1.6 & \text{if } 8 \le t \le 11 \end{cases}$$

a. Give the technology formula for C and complete the following table of values of the function C:

t	0	1	2	3	4	5	6	7	8	9	10	11
$C(t)$												

b. Between 1998 and 2000, the cost of a Super Bowl ad was increasing at a rate of $____ million per year.

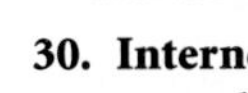

30. Internet Purchases The percentage $p(t)$ of buyers of new cars who used the Internet for research or purchase since 1997 is given by the following function ($t = 0$ represents 1997):[8]

$$p(t) = \begin{cases} 10t + 15 & \text{if } 0 \le t < 1 \\ 15t + 10 & \text{if } 1 \le t \le 4 \end{cases}$$

a. Give the technology formula for p and complete the following table of values of the function p:

t	0	0.5	1	1.5	2	2.5	3	3.5	4
$p(t)$									

b. Between 1998 and 2000, the percentage of buyers of new cars who used the Internet for research or purchase was increasing at a rate of ____% per year.

[6]SOURCE for data: VLSI Research/*New York Times*, October 9, 1994, sec. 3, p. 2.

[7]SOURCE: *New York Times*, January 26, 2001, p. C1.

[8]Model is based on data through 2000 (the 2000 value was estimated). SOURCE: J. D. Power Associates/*New York Times*, January 25, 2000, p. C1.

31. Income Taxes The U.S. federal income tax is a function of taxable income. Write T for the tax owed on a taxable income of I dollars. In 2001 the function T for a single taxpayer was specified as follows:

If your taxable income was		Your tax is	Of the amount
Over—	But not over—		over—
$0	$6,000	----------10%	**$0**
6,000	27,050	**$600.00 + 15%**	**6,000**
27,050	65,550	**3,757.50 + 27%**	**27,050**
65,550	136,750	**14,152.50 + 30%**	**65,550**
136,750	297,350	**35,512.50 + 35%**	**136,750**
297,350	----------	**91,722.50 + 38.6%**	**297,350**

These were the tax rates that went into effect as of July 1, 2001. The lowest bracket, 10%, was implemented not as a tax bracket but as a midyear rebate to the taxpayer of (in most cases) 5% of $6000, or $300, representing the difference between the old tax of 15% in this range and the new tax of 10%.

What was the tax owed by a single taxpayer on a taxable income of $26,000? On a taxable income of $65,000?

32. Income Taxes The income tax function T in Exercise 31 can also be written in the following form:

$$T(I) = \begin{cases} 0.10I & \text{if } 0 < I \le 6000 \\ 600 + 0.15(I - 6000) & \text{if } 6000 < I \le 27{,}050 \\ 3757.50 + 0.27(I - 27{,}050) & \text{if } 27{,}050 < I \le 65{,}550 \\ 14{,}152.50 + 0.30(I - 65{,}550) & \text{if } 65{,}550 < I \le 136{,}750 \\ 35{,}512.50 + 0.35(I - 136{,}750) & \text{if } 136{,}750 < I \le 297{,}350 \\ 91{,}722.50 + 0.386(I - 297{,}350) & \text{if } I > 297{,}350 \end{cases}$$

What was the tax owed by a single taxpayer on a taxable income of $25,000? On a taxable income of $125,000?

33. Demand The demand for Sigma Mu Fraternity plastic brownie dishes is

$$q(p) = 361{,}201 - (p + 1)^2$$

where q represents the number of brownie dishes Sigma Mu can sell each month at a price of p¢. Use this function to determine

a. the number of brownie dishes Sigma Mu can sell each month if the price is set at 50¢.

b. the number of brownie dishes they can unload each month if they give them away.

c. the lowest price at which Sigma Mu will be unable to sell any dishes.

34. Revenue The total weekly revenue earned at Royal Ruby Retailers is given by

$$R(p) = -\frac{4}{3}p^2 + 80p$$

where p is the price (in dollars) RRR charges per ruby. Use this function to determine

a. the weekly revenue, to the nearest dollar, when the price is set at $20/ruby.

b. the weekly revenue, to the nearest dollar, when the price is set at $200/ruby (interpret your result).

c. the price RRR should charge in order to obtain a weekly revenue of $1200.

35. Investments in South Africa The number of U.S. companies that invested in South Africa from 1986 through 1994 closely followed the function

$$n(t) = 5t^2 - 49t + 232$$

Here, t is the number of years since 1986, and $n(t)$ is the number of U.S. companies that own at least 50% of their South African subsidiaries and employ 1000 or more people.[9]

a. Find the appropriate domain of n.

b. Is $t \ge 0$ an appropriate domain? Give reasons for your answer.

36. Sony Net Income The annual net income for Sony Corporation from 1989 through 1994 can be approximated by the function

$$I(t) = -77t^2 + 301t + 524$$

Here, t is the number of years since 1989, and $I(t)$ is Sony's net income in millions of dollars for the corresponding fiscal year.[10]

a. Find the appropriate domain of I.

b. Is $[0, +\infty)$ an appropriate domain? Give reasons for your answer.

37. Spending on Corrections The following table shows the annual spending by all states in the United States on corrections ($t = 0$ represents the year 1990):

Year, t	0	1	2	3	4	5	6	7
Spending, S ($ billions)	16	18	18	20	22	26	28	30

SOURCE: National Association of State Budget Officers/*New York Times*, February 28, 1999, p. A1. Data are rounded.

a. Which of the following functions best fits the given data? (*Warning:* None of them fits exactly, but one fits more closely than the others.)

(1) $S(t) = -0.2t^2 + t + 16$

(2) $S(t) = 0.2t^2 + t + 16$

(3) $S(t) = t + 16$

b. Use your answer to part (a) to "predict" spending on corrections in 1998, assuming that the trend continued.

---

[9] The model is the authors' (least-squares quadratic regression with coefficients rounded to nearest integer). SOURCE: Investor Responsibility Research Center Inc., Fleming Martin/*New York Times*, June 7, 1994, p. D1.

[10] The model is the authors' (least-squares quadratic regression with coefficients rounded to nearest integer). SOURCE: Sony Corporation/*New York Times*, May 20, 1994, p. D1.

38. Spending on Corrections Repeat Exercise 37, this time choosing from the following functions:

$$(1)\ S(t) = 16 + 2t$$

$$(2)\ S(t) = 16 + t + 0.5t^2$$

$$(3)\ S(t) = 16 + t - 0.5t^2$$

39. Toxic Waste Treatment The cost of treating waste by removing PCPs rises rapidly as the quantity of PCPs removed increases. Here is a possible model:

$$C(q) = 2000 + 100q^2$$

where q is the reduction in toxicity (in pounds of PCPs removed per day) and $C(q)$ is the daily cost (in dollars) of this reduction.

a. Find the cost of removing 10 pounds/day of PCPs.

b. Government subsidies for toxic waste cleanup amount to

$$S(q) = 500q$$

where q is as above and $S(q)$ is the daily dollar subsidy. Calculate the net cost function $N(q)$ (the cost of removing q pounds of PCPs per day after the subsidy is taken into account) given the cost function and subsidy above, and find the net cost of removing 20 pounds/day of PCPs.

40. Dental Plans A company pays for its employees' dental coverage at an annual cost C given by

$$C(q) = 1000 + 100\sqrt{q}$$

where q is the number of employees covered and $C(q)$ is the annual cost in dollars.

a. If the company has 100 employees, find its annual outlay for dental coverage.

b. Assuming that the government subsidizes coverage by an annual dollar amount of

$$S(q) = 200q$$

calculate the net cost function $N(q)$ to the company and calculate the net cost of subsidizing its 100 employees. Comment on your answer.

41. Acquisition of Language The percentage $p(t)$ of children who can speak in at least single words by the age of t months can be approximated by the equation[11]

$$p(t) = 100\left(1 - \frac{12{,}200}{t^{4.48}}\right) \qquad (t \geq 8.5)$$

a. Give a technology formula for p.

b. Create a table of values of p for $t = 9, 10, \ldots, 20$ (rounding answers to one decimal place).

c. What percentage of children can speak in at least single words by the age of 12 months?

d. By what age are 90% or more children speaking in at least single words?

42. Acquisition of Language The percentage $p(t)$ of children who can speak in sentences of five or more words by the age of t months can be approximated by the equation[12]

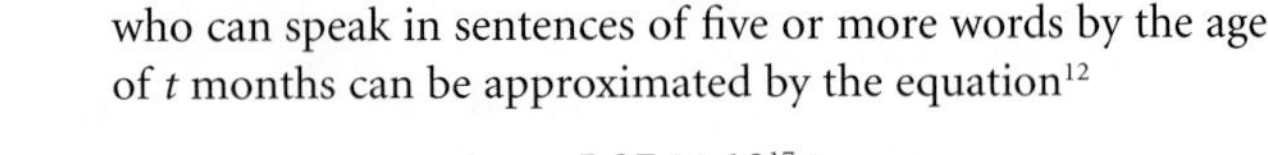

$$p(t) = 100\left(1 - \frac{5.27 \times 10^{17}}{t^{12}}\right) \qquad (t \geq 30)$$

a. Give a technology formula for p.

b. Create a table of values of p for $t = 30, 31, \ldots, 40$ (rounding answers to one decimal place).

c. What percentage of children can speak in sentences of five or more words by the age of 36 months?

d. By what age are 75% or more children speaking in sentences of five or more words?

COMMUNICATION AND REASONING EXERCISES

43. Complete the following: If the market price m of gold varies with time t, then the independent variable is ____, and the dependent variable is ____.

44. Complete the following: If weekly profit P is specified as a function of selling price s, then the independent variable is ____, and the dependent variable is ____.

45. Complete the following: The function notation for the equation $y = 4x^2 - 2$ is ____.

46. Complete the following: The equation notation for $C(t) = -0.34t^2 + 0.1t$ is ____.

47. You now have 200 sound files on your hard drive, and this number is increasing by 10 sound files each day. Find a mathematical model for this situation.

48. The amount of free space left on your hard drive is now 50 gigabytes (GB) and is decreasing by 5 GB/month. Find a mathematical model for this situation.

49. Why is the following assertion false? "If $f(x) = x^2 - 1$, then $f(x + h) = x^2 + h - 1$."

50. Why is the following assertion false? "If $f(2) = 2$ and $f(4) = 4$, then $f(3) = 3$."

51. True or false: Every function can be specified numerically.

52. Which supplies more information about a situation: a numerical model or an algebraic model?

---

[11] The model is the authors' and is based on data presented in the article "The Emergence of Intelligence" by William H. Calvin, *Scientific American* (October 1994): 101–107.

[12] Ibid.

1.2 Functions from the Graphical Viewpoint

Consider again the function W discussed in Section 1.1, giving a child's weight during her first year. If we represent the data given in Section 1.1 graphically by plotting the given pairs of numbers $(t, W(t))$, we get Figure 2. (We have connected successive points by line segments.)

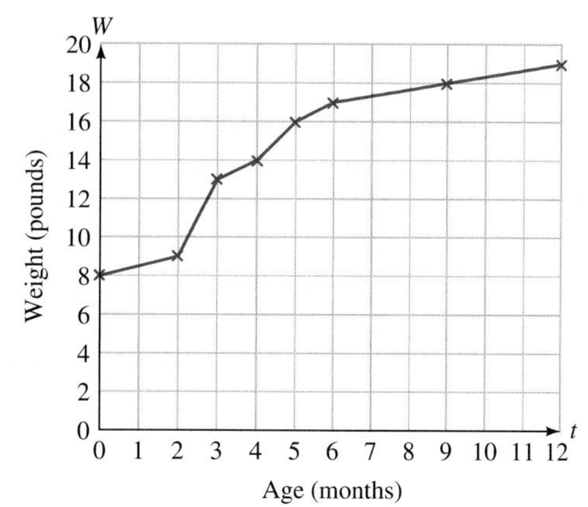

Figure 2

Suppose now that we had only the graph without the table of data given in Section 1.1. We could use the graph to find values of W. For instance, to find $W(9)$ from the graph, we do the following:

1. Find the desired value of t at the bottom of the graph ($t = 9$ in this case).

2. Estimate the height (W coordinate) of the corresponding point on the graph (18 in this case).

Thus, $W(9) \approx 18$ pounds.[13]

We say that Figure 1 specifies the function W **graphically.** The graph is not a very accurate specification of W; the actual weight of the child would follow a smooth curve rather than a jagged line. However, the jagged line is useful in that it permits us to interpolate: For instance, we can estimate that $W(1) \approx 8.5$ pounds.

Example 1 • A Function Specified Graphically: Magazine Circulation

Figure 3 shows the approximate annual circulation of *Outside Magazine* for the period 1993–2000 ($t = 0$ represents December 1993; $C(t)$ represents the circulation in the year ending at time t). Estimate and interpret $C(0)$, $C(2)$, and $C(4.5)$. What is the domain of C?

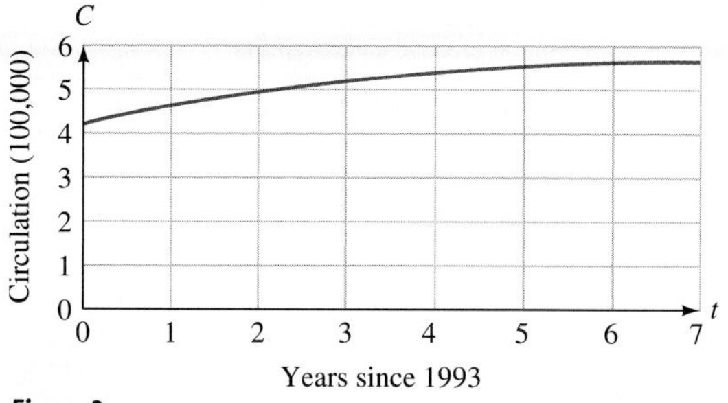

Figure 3

SOURCES: 1993–1997 data: Audit Bureau of Circulations/ Publishers Information Bureau/*New York Times*, October 27, 1997, p. D13. 2000 data: http://writenews.com/2000/030600_gearunlimited_outside.htm.

---

[13]In a graphically defined function, we can never know the y coordinates of points exactly; no matter how accurately a graph is drawn, we can only obtain *approximate* values of the coordinates of points. That is why we have been using the word *estimate* rather than *calculate* and why we say "$W(9) \approx 18$" rather than "$W(9) = 18$."

Solution We carefully estimate the C coordinates of the points with t coordinates 0, 2, and 4.5:

$$C(0) \approx 4.2$$

meaning that the circulation in the year ending December 1993 ($t = 0$) was approximately 420,000 copies;

$$C(2) \approx 5.0$$

meaning that the circulation in the year ending December 1995 ($t = 2$) was approximately 500,000 copies;

$$C(4.5) \approx 5.5$$

meaning that the circulation in the year ending June 1998 ($t = 4.5$) was approximately 550,000 copies.

The domain of C is the set of all values of t for which $C(t)$ is defined: $0 \le t \le 7$, or $[0, 7]$.

Sometimes we are interested in drawing the graph of a function that has been specified in some other way—perhaps numerically or algebraically. We do this by plotting points with coordinates $(x, f(x))$.[14] Here is the formal definition of a graph.

Graph of a Function

The **graph of the function** f is the set of all points $(x, f(x))$ in the xy plane, where *we restrict the values of* x *to lie in the domain of* f.

To obtain the graph of a function, plot points of the form $(x, f(x))$ for several values of x in the domain of f. The shape of the entire graph can usually be inferred from sufficiently many points. (Calculus can give us information that allows us to draw a graph with relatively few points plotted.)

Quick Example

To sketch the graph of the function

$f(x) = x^2$ Function notation

or $y = x^2$ Equation notation

with domain the set of all real numbers, first choose some convenient values of x in the domain and compute the corresponding y-coordinates.

x	-3	-2	-1	0	1	2	3
$y = x^2$	9	4	1	0	1	4	9

Plotting these points gives the picture on the left, suggesting the graph on the right.*

*If you plot more points, you will find that they lie on a smooth curve as shown. That is why we did not use line segments to connect the points.

[14] Graphing utilities typically draw graphs by plotting and joining a large number of points.

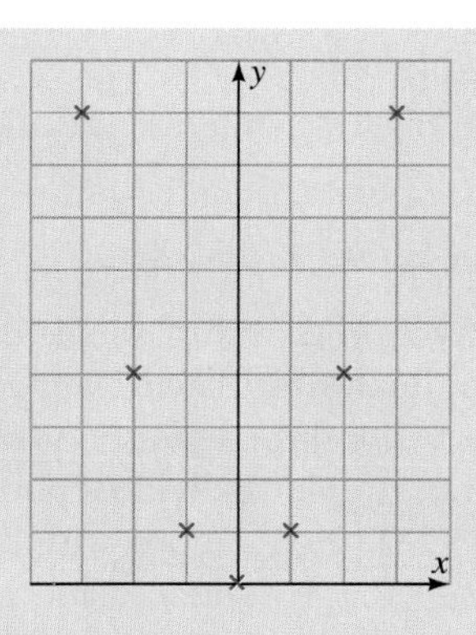

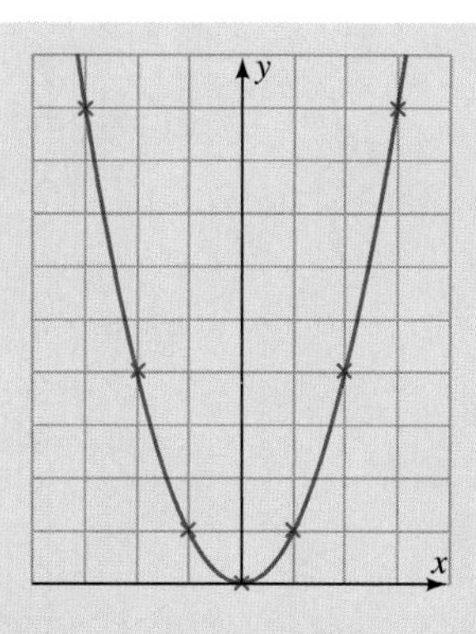

(This particular curve happens to be called a **parabola,** and its lowest point, at the origin, is called its **vertex.**)

Example 2 • Drawing the Graph of a Function: Web-Site Revenue

The monthly revenue[15] R from users logging on to your gaming site depends on the monthly access fee p you charge according to the formula

$$R(p) = -5600p^2 + 14{,}000p \qquad (0 \le p \le 2.5)$$

(R and p are in dollars.) Sketch the graph of R. Find the access fee that will result in the largest monthly revenue.

Solution To sketch the graph of R by hand, we plot points of the form $(x, R(x))$ for several values of x in the domain $[0, 2.5]$ of R. First, we calculate several points:

p	0	0.5	1	1.5	2	2.5
$R(p) = -5600p^2 + 14{,}000p$	0	5600	8400	8400	5600	0

Graphing these points gives the graph shown in Figure 4(a), suggesting the parabola shown in Figure 4(b).

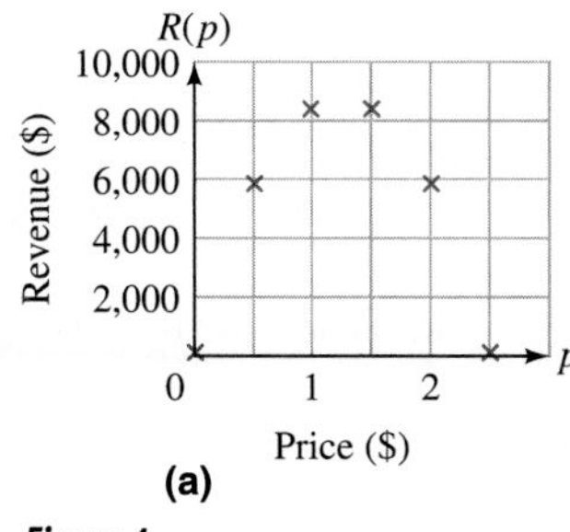

(a)

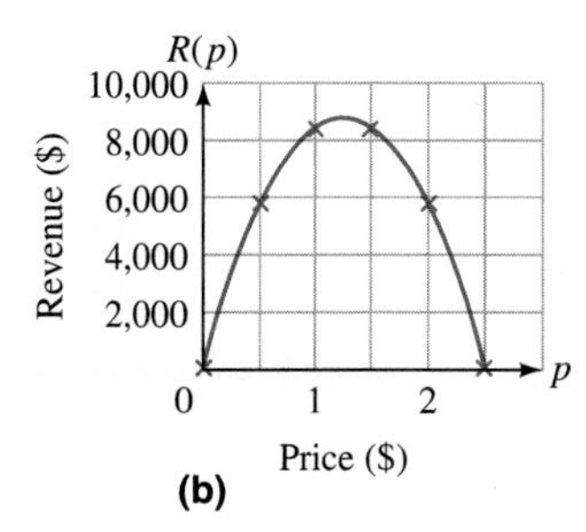

(b)

Figure 4

The revenue graph appears to reach its highest point when $p = 1.25$, so setting the access fee at \$1.25 appears to result in the largest monthly revenue.[16]

[15] The **revenue** resulting from one or more business transactions is the total payment received, sometimes called the gross proceeds.

[16] We are hedging our language with words like *suggesting* and *appears* because the few points we have plotted don't, by themselves, allow us to draw these conclusions with certainty.

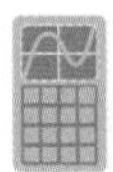

Graphing Calculator

As the name suggests, graphing calculators are designed for graphing functions. If you enter

`Y₁ = -5600*X^2+14000*X`

in the `Y=` screen, you can reproduce the graph shown in Figure 4(b) by setting the window coordinates to match Figure 4(a): `Xmin = 0`, `Xmax = 2.5`, `Ymin = 0`, `Ymax = 10000`, and then pressing GRAPH.

If you want to plot individual points [as in Figure 4(a)] on the TI-83, enter the data in the STAT `EDIT` mode with the values of p in L_1 and the value of $R(p)$ in L_2. Then go to the `Y=` window and turn `Plot1` on by selecting it and pressing ENTER. Now select `ZoomStat` ZOOM 9 to obtain the plot.

Excel

Here's how to plot this graph using Excel. As in Example 3 of Section 1.1, we create a table of values for the function by entering the values of the independent variable p in column A and the formula for the function in cell B2, then copying the formula into the cells beneath it as shown:

	A	B	C
1	p	R	
2	0	=-5600*A2^2+14000*A2	
3	0.5		
4	1		
5	1.5		
6	2		
7	2.5		

To draw the graph shown in Figure 4(a), select (highlight) both columns of data and then ask Excel to insert a chart. When it asks you to specify what type of chart, select the `XY (Scatter)` option. This tells the program that your data specify the x and y coordinates of a sequence of points. In the same dialog box, select the option that shows points connected by lines. Select `Next` to bring up a new dialog called `Data Type`, where you should make sure that the `Series in Columns` option is selected, telling the program that the x and y coordinates are arranged vertically, down columns. You can then set various other options, add labels, and otherwise fiddle with it until it looks nice.

Before we go on . . . To get a smoother curve in Excel, you need to plot many more points. Here is a method of plotting 100 points (in addition to the starting point), similar to the Excel worksheet posted on the Web site at
Web Site → Everything for Finite Math → Chapter 1 → Excel Tutorials → Section 1.2: Functions from the Graphical Viewpoint
Moreover, if you decide to follow this method, you can save the resulting spreadsheet and use it as an "Excel graphing calculator" to graph any other function (see next page).

Excel Graphing Worksheet

Set up your worksheet as follows:

	A	B	C	D	E	F
1	p	R				
2	=$F$2	= -5600*A2^2 + 14000*A2			Xmin	0
3	=A2+$F$5				Xmax	2.5
4					Points	100
5					Delta X	=(F3-F2)/F4
6						
7						
100						
101						
102						

(Columns C and D are empty in case you want to add additional functions to graph.) The 101 values of the x coordinate (the price p) will appear in column A. The corresponding values of the y coordinate (the revenue R) will appear in column B. In column F you see some settings: Xmin = 0 and Xmax = 2.5 (see Figure 4). Delta X (in cell F5) is the amount by which the x coordinate is increased as you go from one x value in column A to the next, starting with Xmin in A2. Enter the formula for $R(p)$ in cell B2 and copy it into the other cells in column B, as shown.

When done, graph the data in columns A and B, choosing the scatter-plot option with subtype `Points connected by lines with no markers`. You should see a curve similar to that in Figure 4(b).

Save this worksheet with its graph, and you can use it to graph new functions as follows:

1. Enter the new values for Xmin and Xmax in column F.
2. Enter the new function in cell B2 (using A2 in place of x).
3. Copy the contents of cell B2 to cells B3–B102.

The graph will be updated automatically.

Web Site

Follow

Web site → Online Utilities

and then click on any of the following:

Function Evaluator & Grapher	Gives a small graph and also values.
Java Graphing Utility	A high-quality Java grapher
Excel Graphing Utility	An Excel sheet that graphs[17]

Instructions for these graphers are available on the Web pages.

---

[17]Since the grapher requires macros, make sure that macros are *enabled* when Excel prompts you. If macros are disabled, the grapher will not work.

Note: Equation and Function Form

As we discussed after Example 2 in Section 1.1, we could write the function R in the example above in equation form as

$$R = -5600p^2 + 14{,}000p \qquad \text{Equation form}$$

The independent variable is p, and the dependent variable is R. Function notation and equation notation, using the same letter for the function name and the dependent variable, are used interchangeably throughout the literature. It is important to be able to switch back and forth from function notation to equation notation easily.

Vertical-Line Test

Every point in the graph of a function has the form $(x, f(x))$ for some x in the domain of f. Since f assigns a *single* value $f(x)$ to each value of x in the domain, it follows that, in the graph of f, there should be only one y corresponding to any such value of x—namely, $y = f(x)$. In other words, *the graph of a function cannot contain two or more points with the same x coordinate—that is, two or more points on the same vertical line.* On the other hand, a vertical line at a value of x not in the domain will not contain any points in the graph. This gives us the following rule.

Vertical-Line Test

For a graph to be the graph of a function, every vertical line must intersect the graph in *at most* one point.

Quick Examples

As illustrated below, only graph B passes the vertical-line test, so only graph B is the graph of a function.

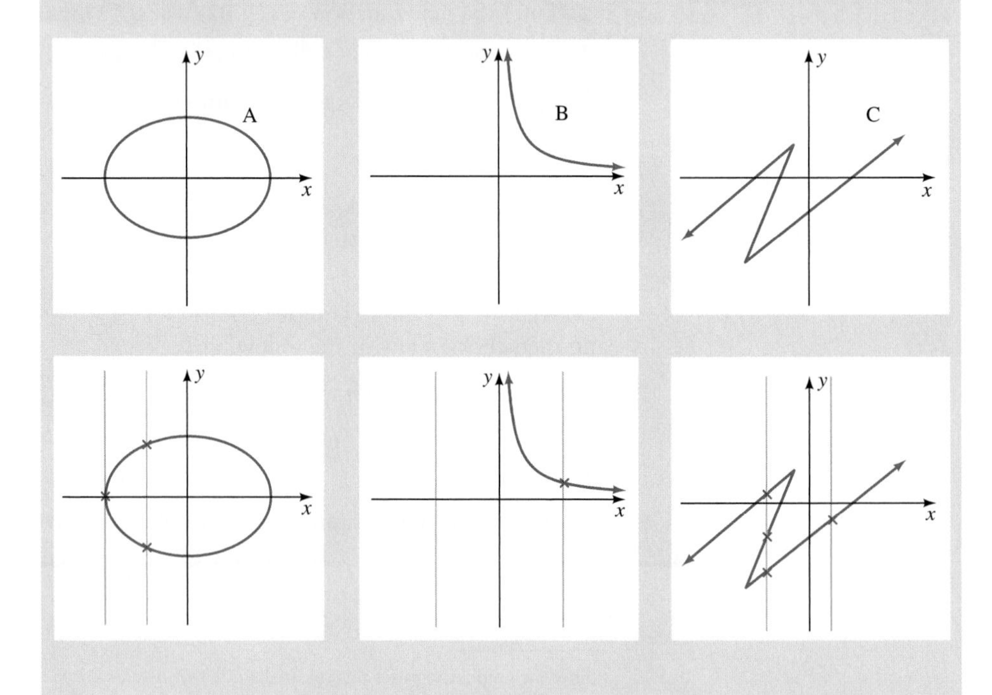

Graphing Piecewise-Defined Functions

Let's revisit the semiconductor example from Section 1.1.

Example 3 • Graphing a Piecewise-Defined Function: Semiconductors

The percentage $U(t)$ of semiconductor equipment manufactured in the United States during the period 1981–1994 can be approximated by the following function of time t in years ($t = 0$ represents 1980):[18]

$$U(t) = \begin{cases} 76 - 3.3t & \text{if } 1 \le t \le 10 \\ 18 + 2.5t & \text{if } 10 < t \le 14 \end{cases}$$

Graph the function U.

Solution As in Example 2, we can sketch the graph of U by hand by computing $U(t)$ for a number of values of t, plotting these points on the graph, and then connecting them.

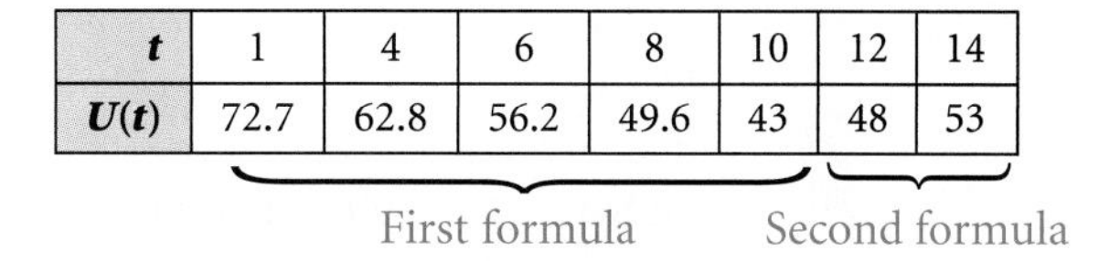

t	1	4	6	8	10	12	14
$U(t)$	72.7	62.8	56.2	49.6	43	48	53

First formula (t = 1 to 10); Second formula (t = 12, 14)

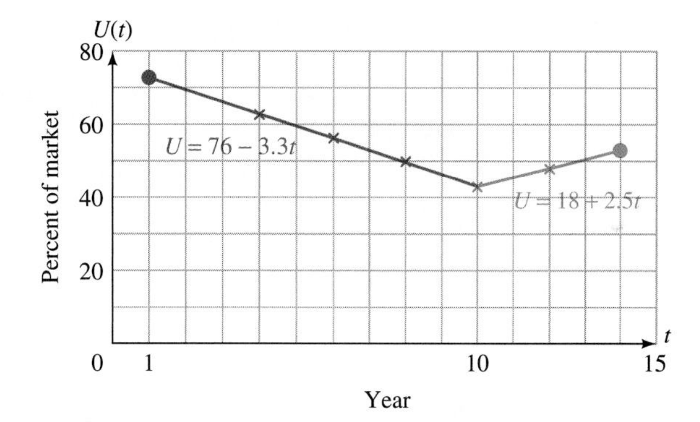

Figure 5

The graph (Figure 5) has the following features:

1. The first formula (the descending line) is used for $1 \le t \le 10$.
2. The second formula (ascending line) is used for $10 < t \le 14$.
3. The domain is $[1, 14]$, so the graph is cut off at $t = 1$ and $t = 14$.
4. The heavy dots at the ends indicate the endpoints of the domain.

T To graph the function U using technology, consult Example 4 of Section 1.1, which shows how to enter this piecewise-defined function and obtain a table of values. Example 2 shows how to then draw the graph (highlight the table of values and use an `XY (Scatter)` plot).

[18]Source for data: VLSI Research/*New York Times*, October 9, 1994, sec. 3, p. 2.

Example 4 • Graphing More Complicated Piecewise-Defined Functions

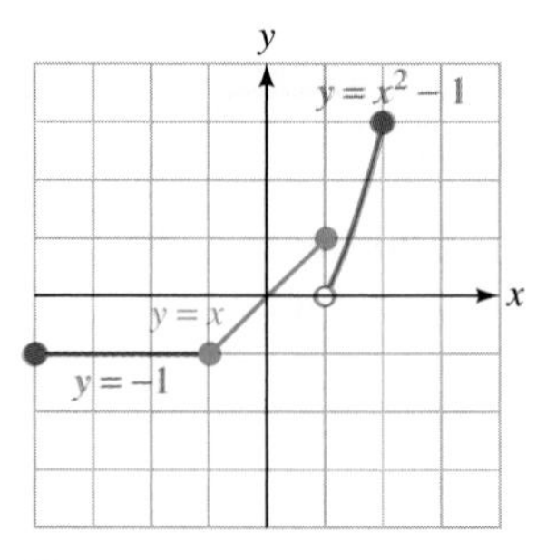

Figure 6

Graph the function f specified by

$$f(x) = \begin{cases} -1 & \text{if } -4 \le x < -1 \\ x & \text{if } -1 \le x \le 1 \\ x^2 - 1 & \text{if } 1 < x \le 2 \end{cases}$$

Solution The domain of f is $[-4, 2]$, since $f(x)$ is only specified when $-4 \le x \le 2$. Further, the function changes formulas when $x = -1$ and $x = 1$.

To sketch the graph by hand, we first sketch the three graphs $y = -1$, $y = x$, and $y = x^2 - 1$ and then use the appropriate portion of each (Figure 6).

Question What are solid dots and open dots doing there?

Answer Solid dots indicate points on the graph, whereas open dots indicate points *not* on the graph. For example, when $x = 1$, the inequalities in the formula tell us that we are to use the middle formula (x) rather than the bottom one ($x^2 - 1$). Thus, $f(1) = 1$, not 0, so we place a solid dot at $(1, 1)$ and an open dot at $(1, 0)$.

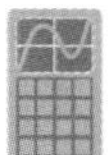

Graphing Calculator

On the TI-83 you can enter this function as

```
(-1)*(X<-1)+X*(-1≤X and X≤1)+(X²-1)*(1<X)
```

First formula — Second formula — Third formula

The logical operator `and` is found in the `TEST LOGIC` menu. The following alternative formula will also work:

```
(-1)*(X<-1)+X*(-1≤X)*(X≤1)+(X²-1)*(1<X)
```

In the graph the TI-83 will not handle the transition at $x = 1$ correctly; it will connect the two parts of the graph with a spurious line segment.

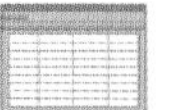

Excel

For Excel you can use the following formula (the same as the alternative TI-83 formula):[19]

```
=(-1)*(x<-1)+x*(-1<=x)*(x<=1)+(x^2-1)*(1<x)
```

First formula — Second formula — Third formula

Since one of the formulas (the third) specifying the function f is not linear, we need to plot many points in Excel to get a smooth graph. Here is one possible setup (for a smoother curve, plot more points):

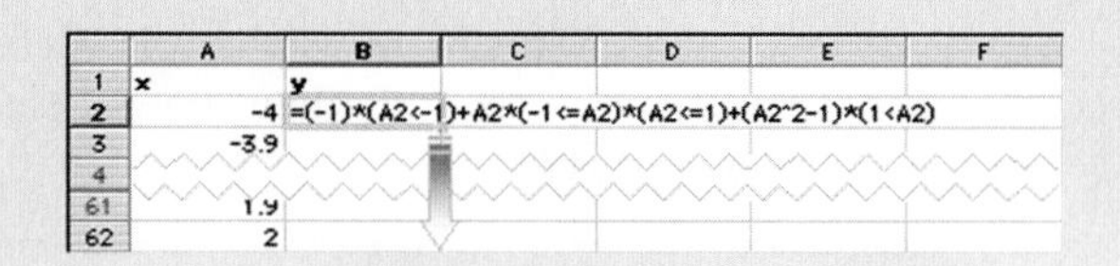

	A	B	C	D	E	F
1	x	y				
2	-4	=(-1)*(A2<-1)+A2*(-1<=A2)*(A2<=1)+(A2^2-1)*(1<A2)				
3	-3.9					
4						
61	1.9					
62	2					

[19]We could also use a nested IF statement to accomplish the same result: `=IF(x<-1,-1,IF(x<=1,x,x^2-1))`.

This results in the graph shown in Figure 7. (Notice that, like the graphing calculator, Excel does not handle the transition at $x = 1$ correctly and connects the two parts of the graph with a spurious line segment.)

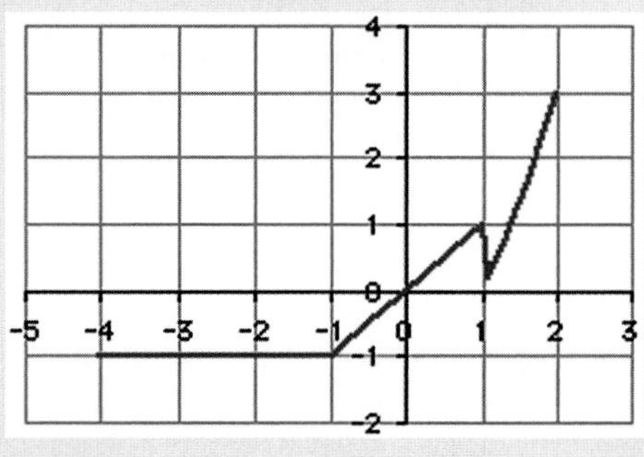

Figure 7

We end this section with a list of some useful types of functions and their graphs (Table 2).

Table 2 • Functions and Their Graphs

Type of Function	Examples	
Linear $f(x) = mx + b$ m, b constant Graphs of linear functions are straight lines.	$y = x$	$y = -2x + 2$
Quadratic $f(x) = ax^2 + bx + c$ a, b, c constant $(a \neq 0)$ Graphs of quadratic functions are called **parabolas.**	$y = x^2$	$y = -2x^2 + 2x + 4$
Technology formulas:	`x^2`	`-2*x^2+2*x+4`
Cubic $f(x) = ax^3 + bx^2 + cx + d$ a, b, c, d constant $(a \neq 0)$	$y = x^3$	$y = -x^3 + 3x^2 + 1$
Technology formulas:	`x^3`	`-(x^3)+3*x^2+1`

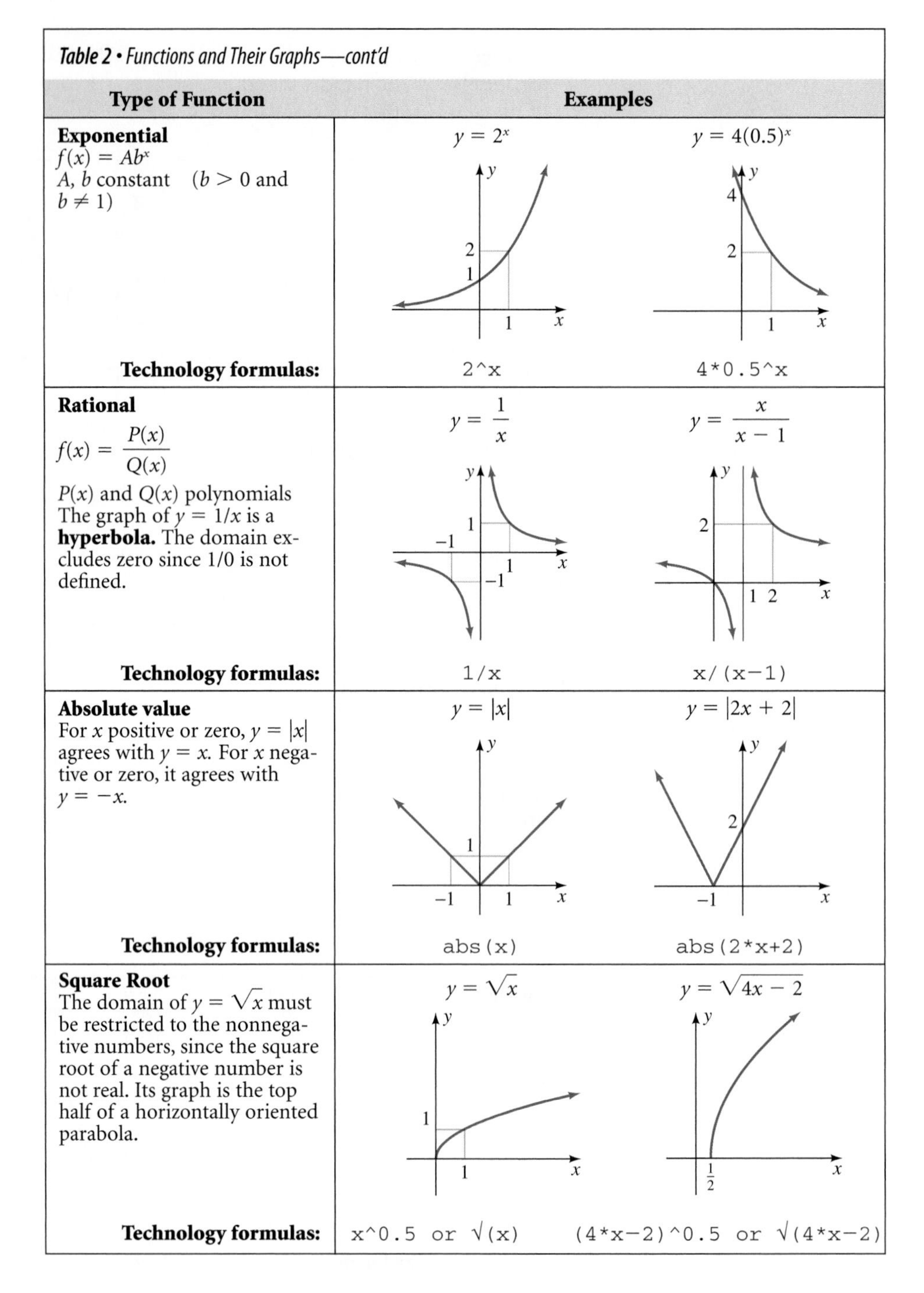

Table 2 • Functions and Their Graphs—cont'd

Type of Function	Examples	
Exponential $f(x) = Ab^x$ A, b constant $(b > 0$ and $b \neq 1)$	$y = 2^x$	$y = 4(0.5)^x$
Technology formulas:	`2^x`	`4*0.5^x`
Rational $f(x) = \dfrac{P(x)}{Q(x)}$ $P(x)$ and $Q(x)$ polynomials The graph of $y = 1/x$ is a **hyperbola.** The domain excludes zero since $1/0$ is not defined.	$y = \dfrac{1}{x}$	$y = \dfrac{x}{x-1}$
Technology formulas:	`1/x`	`x/(x-1)`
Absolute value For x positive or zero, $y = \|x\|$ agrees with $y = x$. For x negative or zero, it agrees with $y = -x$.	$y = \|x\|$	$y = \|2x + 2\|$
Technology formulas:	`abs(x)`	`abs(2*x+2)`
Square Root The domain of $y = \sqrt{x}$ must be restricted to the nonnegative numbers, since the square root of a negative number is not real. Its graph is the top half of a horizontally oriented parabola.	$y = \sqrt{x}$	$y = \sqrt{4x - 2}$
Technology formulas:	`x^0.5 or √(x)`	`(4*x-2)^0.5 or √(4*x-2)`

New Functions from Old: Scaled and Shifted Functions (Online Optional Section)

If you follow the path

Web site → Everything for Calculus → Chapter 1

you will find a link to this section, with complete online interactive text, examples, and exercises on scaling and translating the graph of a function by changing the formula.

1.2 EXERCISES

In Exercises 1–4, use the graph of the function f to find approximations of the given values.

1.

y
40
30
20
10
0 1 2 x

a. $f(1)$ **b.** $f(2)$
c. $f(3)$ **d.** $f(5)$
e. $f(3) - f(2)$

2.

y
40
30
20
10
0 1 2 x

a. $f(1)$ **b.** $f(2)$
c. $f(3)$ **d.** $f(5)$
e. $f(3) - f(2)$

3.

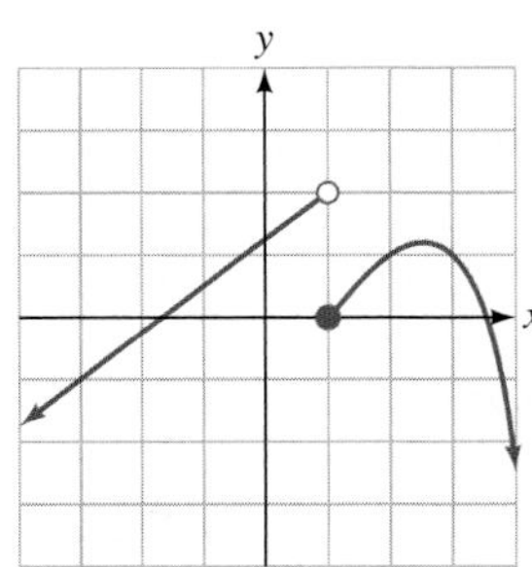

a. $f(-3)$ **b.** $f(0)$
c. $f(1)$ **d.** $f(2)$
e. $\dfrac{f(3) - f(2)}{3 - 2}$

4.

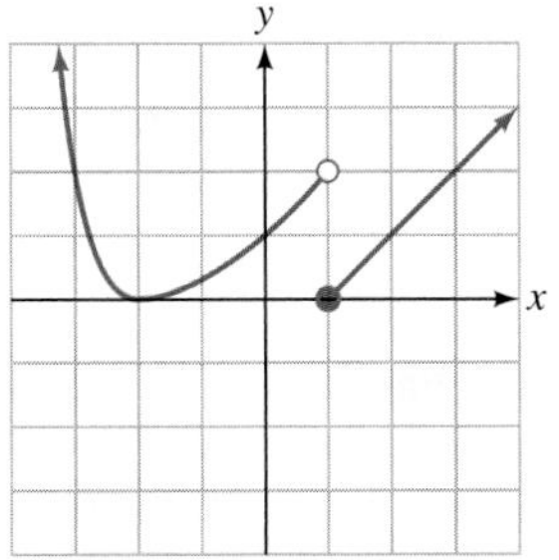

a. $f(-2)$ **b.** $f(0)$
c. $f(1)$ **d.** $f(3)$
e. $\dfrac{f(3) - f(1)}{3 - 1}$

In Exercises 5 and 6, match the functions to the graphs. Using technology to draw the graphs is suggested but not required.

5. a. $f(x) = x \quad (-1 \le x \le 1)$
b. $f(x) = -x \quad (-1 \le x \le 1)$
c. $f(x) = \sqrt{x} \quad (0 < x < 4)$
d. $f(x) = x + \dfrac{1}{x} - 2 \quad (0 < x < 4)$
e. $f(x) = |x| \quad (-1 \le x \le 1)$
f. $f(x) = x - 1 \quad (-1 \le x \le 1)$

(I)

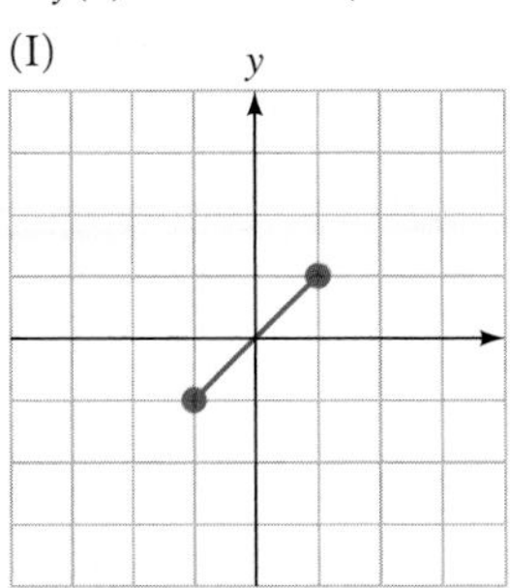

(II)

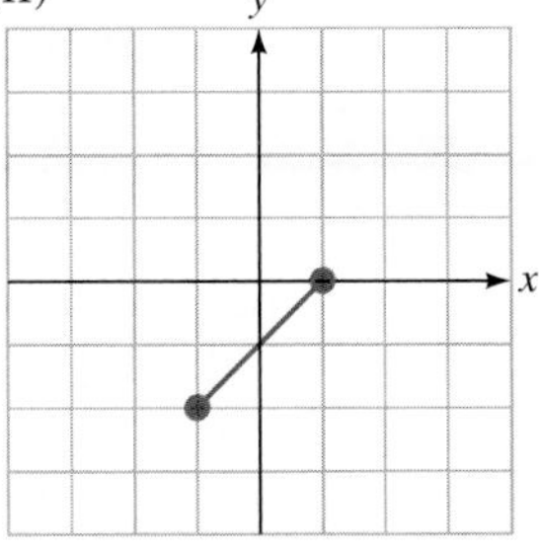

(III)

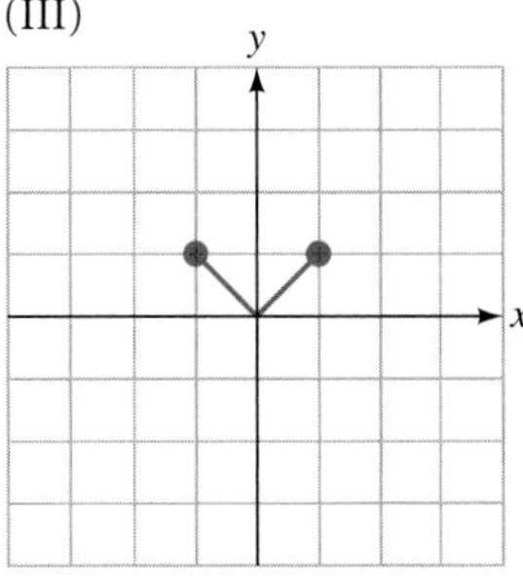

(IV)

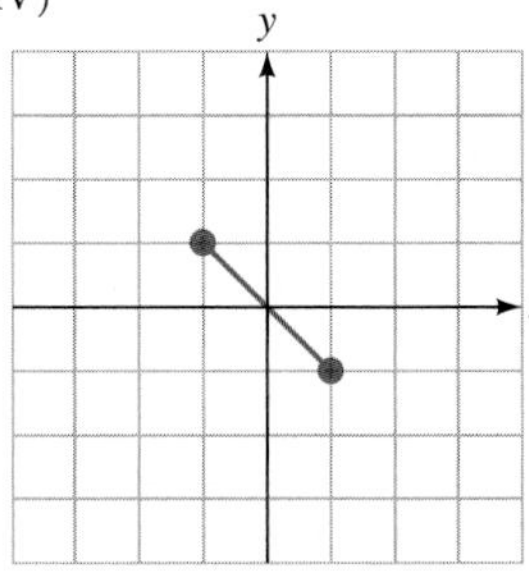

(V)

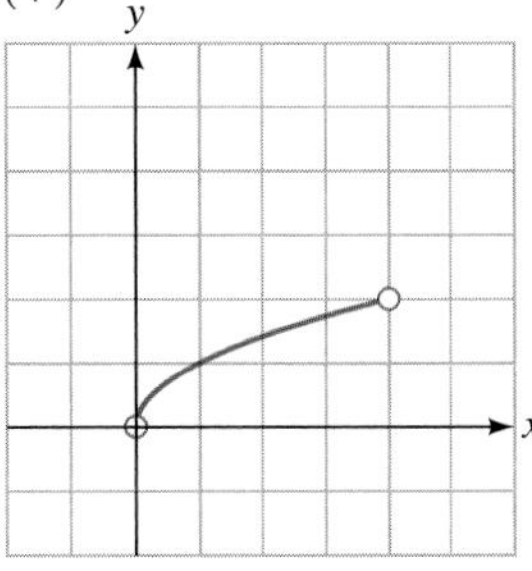

(VI)

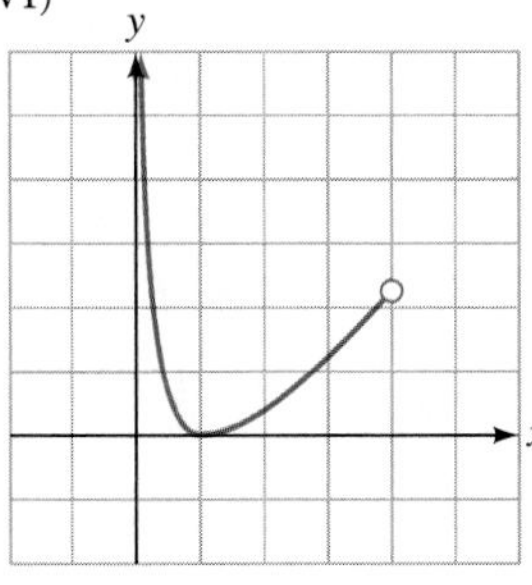

6. a. $f(x) = -x + 4 \quad (0 < x \le 4)$
b. $f(x) = 2 - |x| \quad (-2 < x \le 2)$
c. $f(x) = \sqrt{x + 2} \quad (-2 < x \le 2)$
d. $f(x) = -x^2 + 2 \quad (-2 < x \le 2)$
e. $f(x) = \dfrac{1}{x} - 1 \quad (0 < x \le 4)$
f. $f(x) = x^2 - 1 \quad (-2 < x \le 2)$

(I)

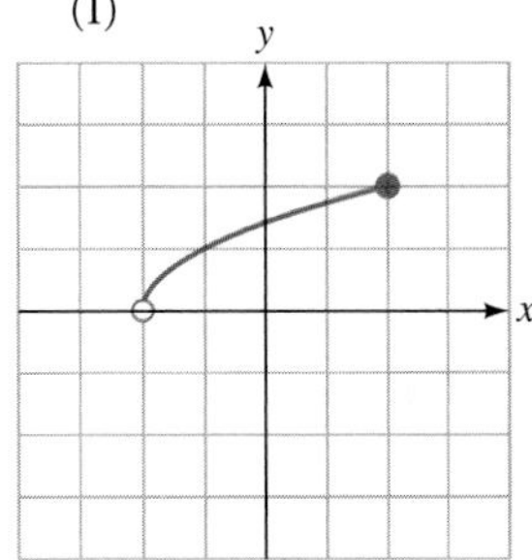

(II)

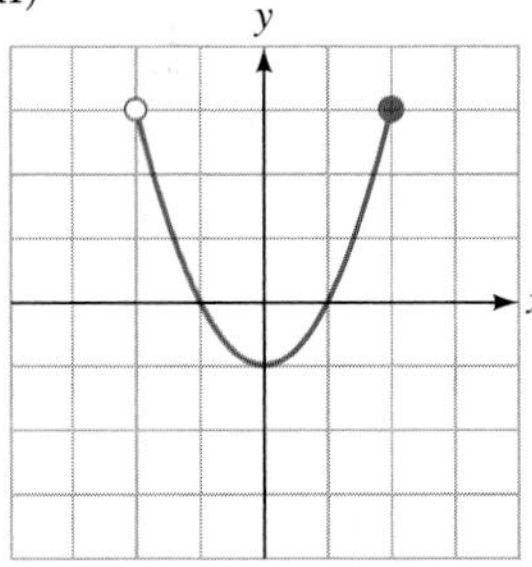

(III)

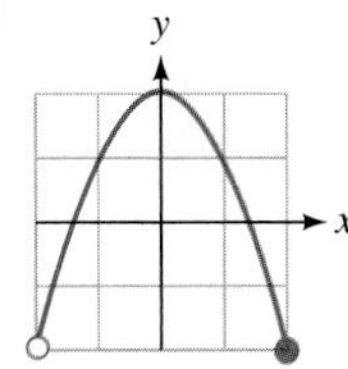

(IV)

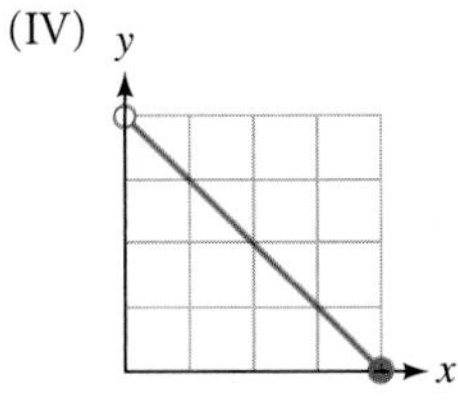

(V)

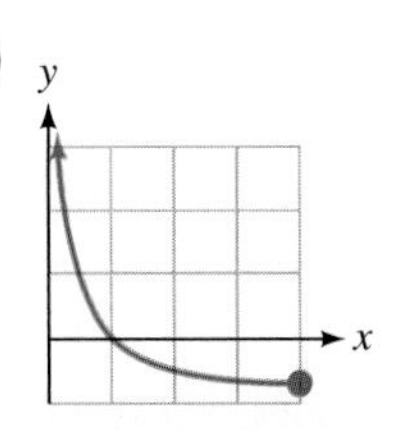

(VI)

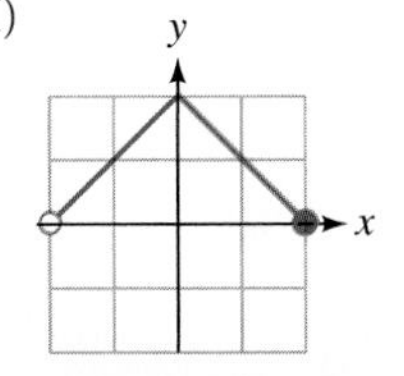

In Exercises 7–12, graph the given functions. Give the technology formula and use technology to check your graph. We suggest that you become familiar with these graphs in addition to those in Table 2.

7. $f(x) = -x^3$ (domain $(-\infty, +\infty)$)

8. $f(x) = x^3$ (domain $[0, +\infty)$)

9. $f(x) = x^4$ (domain $(-\infty, +\infty)$)

10. $f(x) = \sqrt[3]{x}$ (domain $(-\infty, +\infty)$)

11. $f(x) = \dfrac{1}{x^2}$ $(x \neq 0)$

12. $f(x) = x + \dfrac{1}{x}$ $(x \neq 0)$

In Exercises 13–18, sketch the graph of the given function, evaluate the given expressions, and then use technology to duplicate the graphs. Give the technology formula.

13. $f(x) = \begin{cases} x & \text{if } -4 \le x < 0 \\ 2 & \text{if } 0 \le x \le 4 \end{cases}$

a. $f(-1)$ **b.** $f(0)$ **c.** $f(1)$

14. $f(x) = \begin{cases} -1 & \text{if } -4 \le x \le 0 \\ x & \text{if } 0 < x \le 4 \end{cases}$

a. $f(-1)$ **b.** $f(0)$ **c.** $f(1)$

15. $f(x) = \begin{cases} x^2 & \text{if } -2 < x \le 0 \\ \dfrac{1}{x} & \text{if } 0 < x \le 4 \end{cases}$

a. $f(-1)$ **b.** $f(0)$ **c.** $f(1)$

16. $f(x) = \begin{cases} -x^2 & \text{if } -2 < x \le 0 \\ \sqrt{x} & \text{if } 0 < x < 4 \end{cases}$

a. $f(-1)$ **b.** $f(0)$ **c.** $f(1)$

17. $f(x) = \begin{cases} x & \text{if } -1 < x \le 0 \\ x + 1 & \text{if } 0 < x \le 2 \\ x & \text{if } 2 < x \le 4 \end{cases}$

a. $f(0)$ **b.** $f(1)$ **c.** $f(2)$ **d.** $f(3)$

18. $f(x) = \begin{cases} -x & \text{if } -1 < x < 0 \\ x - 2 & \text{if } 0 \le x \le 2 \\ -x & \text{if } 2 < x \le 4 \end{cases}$

a. $f(0)$ **b.** $f(1)$ **c.** $f(2)$ **d.** $f(3)$

APPLICATIONS

Sales of Sport Utility Vehicles Exercises 19–22 refer to the following graph, which shows the number $f(t)$ of sports utility vehicles (SUVs) sold in the United States each year from 1990 through 2003 ($t = 0$ represents 1990, and $f(t)$ represents sales in year t in thousands of vehicles).

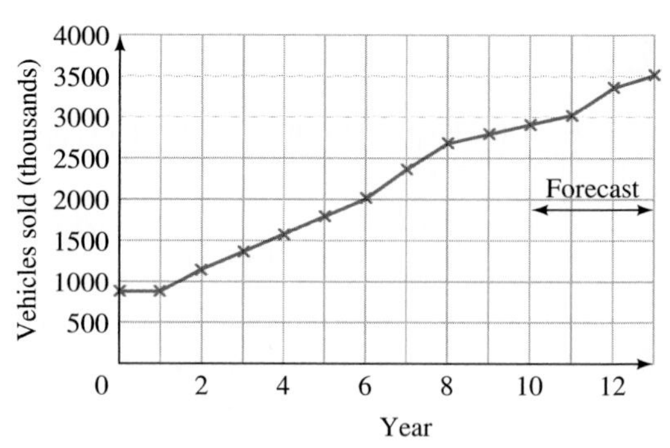

SOURCES: Oak Ridge National Laboratory, Light Vehicle MPG and Market Shares System, AutoPacific, *The US Car and Light Truck Market*, 1999, pp. 24, 120, 121.

19. Estimate $f(6)$, $f(9)$, and $f(7.5)$ and interpret your answers.

20. Estimate $f(5)$, $f(11)$, and $f(1.5)$ and interpret your answers.

21. Which is larger: $f(6) - f(5)$ or $f(10) - f(9)$? Interpret the answer.

22. Which is larger: $f(10) - f(8)$ or $f(13) - f(11)$? Interpret the answer.

23. Employment The following graph shows the number $N(t)$ of people, in millions, employed in the United States (t is time in years, and $t = 0$ represents January 2000).

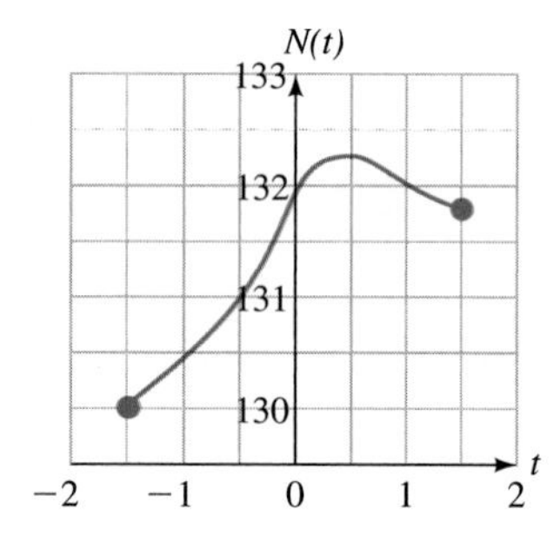

SOURCE: Haver Analytics: The Conference Board/*New York Times*, November 24, 2001.

a. What is the domain of N?

b. Estimate $N(-0.5)$, $N(0)$, and $N(1)$ and interpret your answers.

c. On which interval is $N(t)$ falling? Interpret the result.

24. Productivity The following graph shows an index $P(t)$ of productivity in the United States (t is time in years, and $t = 0$ represents January 2000).

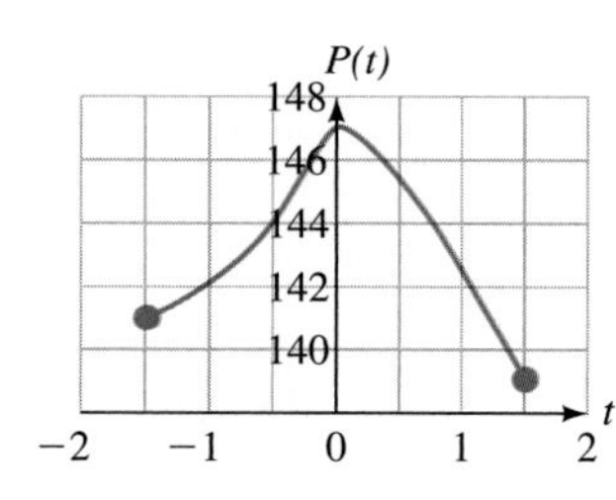

SOURCE: Haver Analytics: The Conference Board/*New York Times*, November 24, 2001.

a. What is the domain of P?

b. Estimate $P(-0.5)$, $P(0)$, and $P(1.5)$ and interpret your answers.

c. On which interval is $P(t) \ge 144$? Interpret the result.

25. Soccer Gear The East Coast College soccer team is planning to buy new gear for its road trip to California. The cost per shirt depends on the number of shirts the team orders, as shown in the following table:

x (shirts ordered)	5	25	40	100	125
$A(x)$ (cost/shirt, \$)	22.91	21.81	21.25	21.25	22.31

a. Which of the following functions best models the data?
(A) $A(x) = 0.005x + 20.75$
(B) $A(x) = 0.01x + 20 + \dfrac{25}{x}$
(C) $A(x) = 0.0005x^2 - 0.07x + 23.25$
(D) $A(x) = 25.5(1.08)^{(x-5)}$

b. Graph the model you chose in part (a) for $10 \le x \le 100$. Use your graph to estimate the lowest cost per shirt and the number of shirts the team should order to obtain the lowest price per shirt.

26. Hockey Gear The South Coast College hockey team wants to purchase wool hats for its road trip to Alaska. The cost per hat depends on the number of hats the team orders, as shown in the following table:

x (hats ordered)	5	25	40	100	125
$A(x)$ (cost/hat, \$)	25.50	23.50	24.63	30.25	32.70

a. Which of the following functions best models the data?
(A) $A(x) = 0.05x + 20.75$
(B) $A(x) = 0.1x + 20 + \dfrac{25}{x}$
(C) $A(x) = 0.0008x^2 - 0.07x + 23.25$
(D) $A(x) = 25.5(1.08)^{(x-5)}$

b. Graph the model you chose in part (a) with $5 \le x \le 30$. Use your graph to estimate the lowest cost per hat and the number of hats the team should order to obtain the lowest price per hat.

27. Hawaiian Tourism The following table shows the number of visitors to Hawaii during three different years:

Year	1985	1991	1994
Visitors (millions)	4.9	7.0	6.5

SOURCE: Hawaii Visitors Bureau/*New York Times,* September 5, 1995, p. A12.

Which of the following kind of model would best fit the given data? Explain your choice of model. (A, a, b, c, and m are constants.)
(A) Linear: $n(t) = mt + b$
(B) Quadratic: $n(t) = at^2 + bt + c$
(C) Exponential: $n(t) = Ab^t$

28. Magazine Advertising The following table shows the number of advertising pages in *Outside Magazine* for the period 1993–1997 ($t = 0$ represents December 1993):

Year	1993	1994	1995
Advertising Pages	1000	1040	1080

SOURCE: Audit Bureau of Circulations/Publishers Information Bureau/*New York Times,* October 27, 1997, p. D13.

Which of the following kind of model would best fit the given data? Explain your choice of model. (A, a, b, c, and m are constants.)
(A) Linear: $n(t) = mt + b$
(B) Quadratic: $n(t) = at^2 + bt + c$
(C) Exponential: $n(t) = Ab^t$

29. Acquisition of Language The percentage $p(t)$ of children who can speak in at least single words by the age of t months can be approximated by the equation[20]

$$p(t) = 100\left(1 - \frac{12{,}200}{t^{4.48}}\right) \qquad (t \ge 8.5)$$

a. Give a technology formula for p.
b. Graph p for $8.5 \le t \le 20$ and $0 \le p \le 100$. Use your graph to answer parts (c) and (d).
c. What percentage of children can speak in at least single words by the age of 12 months? (Round your answer to the nearest percentage point.)
d. By what age are 90% of children speaking in at least single words? (Round your answer to the nearest month.)

30. Acquisition of Language The percentage $p(t)$ of children who can speak in sentences of five or more words by the age of t months can be approximated by the equation[21]

$$p(t) = 100\left(1 - \frac{5.27 \times 10^{17}}{t^{12}}\right) \qquad (t \ge 30)$$

a. Give a technology formula for p.
b. Graph p for $30 \le t \le 45$ and $0 \le p \le 100$. Use your graph to answer parts (c) and (d).
c. What percentage of children can speak in sentences of five or more words by the age of 36 months? (Round your answer to the nearest percentage point.)
d. By what age are 75% of children speaking in sentences of five or more words? (Round your answer to the nearest month.)

[20]The model is the authors' and is based on data presented in the article "The Emergence of Intelligence" by William H. Calvin, *Scientific American* (October 1994): 101–107.

[21]Ibid.

31. Coffee Shops (Compare Exercise 27 in Section 1.1.) The number $C(t)$ of coffee shops and related enterprises in the United States can be approximated by the following function of time t in years since 1990:[22]

$$C(t) = \begin{cases} 500t + 800 & \text{if } 0 \le t \le 4 \\ 1300t - 2400 & \text{if } 4 < t \le 10 \end{cases}$$

Sketch the graph of C and use your graph to estimate, to the nearest year, when there were 9000 coffee shops in the United States.

32. Semiconductors (Compare Exercise 28 in Section 1.1.) The percentage $J(t)$ of semiconductor equipment manufactured in Japan during the period 1981–1994 can be approximated by the following function of time t in years since 1980:[23]

$$J(t) = \begin{cases} 17 + 3.1t & \text{if } 1 \le t \le 10 \\ 68 - 2t & \text{if } 10 < t \le 14 \end{cases}$$

Sketch the graph of J and use your graph to estimate, to the nearest year, when Japan manufactured 40% of all semiconductor equipment.

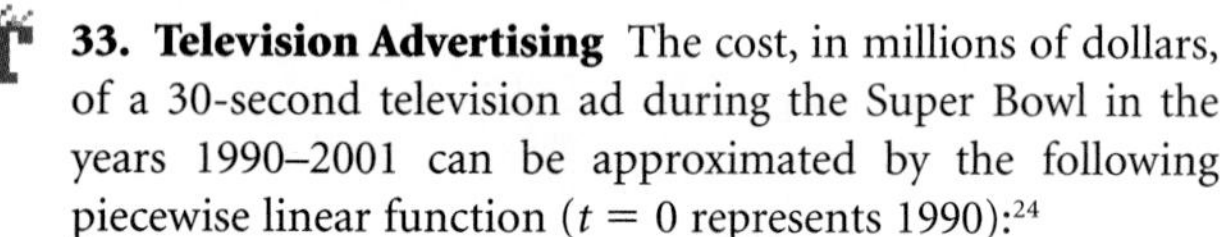

33. Television Advertising The cost, in millions of dollars, of a 30-second television ad during the Super Bowl in the years 1990–2001 can be approximated by the following piecewise linear function ($t = 0$ represents 1990):[24]

$$C(t) = \begin{cases} 0.08t + 0.6 & \text{if } 0 \le t < 8 \\ 0.355t - 1.6 & \text{if } 8 \le t \le 11 \end{cases}$$

a. Give a technology formula for C and use technology to graph the function C.

b. Based on the graph, a Super Bowl ad first exceeded \$2 million in what year?

[22]SOURCE: Specialty Coffee Association of America/*New York Times*, August 13, 1995, p. F10.

[23]SOURCE: VLSI Research/*New York Times*, October 9, 1994, sec. 3, p. 2.

[24]SOURCE: *New York Times*, January 26, 2001, p. C1.

34. Internet Purchases The percentage $p(t)$ of buyers of new cars who used the Internet for research or purchase each year since 1997 is given by the following function ($t = 0$ represents 1997):[25]

$$p(t) = \begin{cases} 10t + 15 & \text{if } 0 \le t < 1 \\ 15t + 10 & \text{if } 1 \le t \le 4 \end{cases}$$

a. Give a technology formula for p and use technology to graph the function p.

b. Based on the graph, 50% or more of all new car buyers used the Internet for research or purchase in what years?

COMMUNICATION AND REASONING EXERCISES

35. True or false: Every graphically specified function can also be specified numerically. Explain.

36. True or false: Every algebraically specified function can also be specified graphically. Explain.

37. True or false: Every numerically specified function with domain [0, 10] can also be specified graphically. Explain.

38. True or false: Every graphically specified function can also be specified algebraically. Explain.

39. How do the graphs of two functions differ if they are specified by the same formula but have different domains?

40. How do the graphs of two functions $f(x)$ and $g(x)$ differ if $g(x) = f(x) + 10$? (Try an example.)

41. How do the graphs of two functions $f(x)$ and $g(x)$ differ if $g(x) = f(x - 5)$? (Try an example.)

42. How do the graphs of two functions $f(x)$ and $g(x)$ differ if $g(x) = f(-x)$? (Try an example.)

[25]Model is based on data through 2000 (the 2000 value was estimated). SOURCE: J. D. Power Associates/*New York Times*, January 25, 2000, p. C1.

1.3 Linear Functions

Linear functions are among the simplest functions and are perhaps the most useful of all mathematical functions.

Linear Function

A **linear function** is one that can be written in the form

$$f(x) = mx + b \qquad \text{Function form}$$

or

$$y = mx + b \qquad \text{Equation form}$$

where m and b are fixed numbers (the names m and b are traditional).*

Quick Examples

$f(x) = 3x - 1$

$y = 3x - 1$

*Actually, c is sometimes used instead of b. As for m, there has even been some research lately into the question of its origin, but no one knows exactly why the letter m is used.

Linear Functions from the Numerical and Graphical Point of View

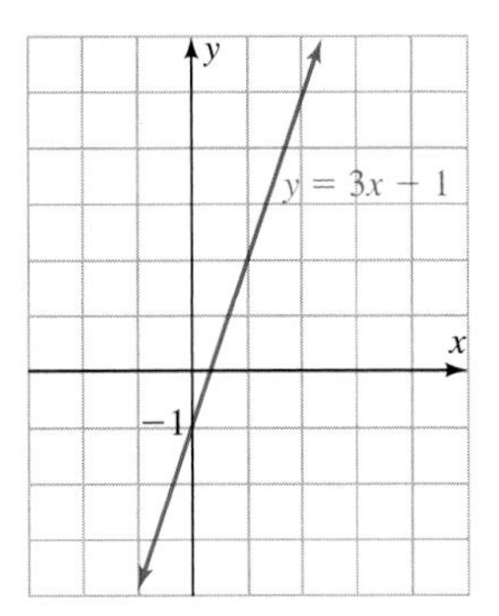

Figure 8

The following table shows values of $y = 3x - 1$ ($m = 3$, $b = -1$) for some values of x:

x	-4	-3	-2	-1	0	1	2	3	4
y	-13	-10	-7	-4	-1	2	5	8	11

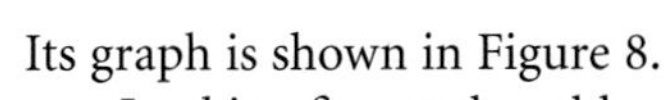

Its graph is shown in Figure 8.

Looking first at the table, notice that setting $x = 0$ gives $y = -1$, the value of b.

Numerically, b is the value of y when x = 0.

On the graph, the corresponding point $(0, -1)$ is the point where the graph crosses the y axis, and we say that $b = -1$ is the ***y* intercept** of the graph (Figure 9).

Graphically, b is the y intercept of the graph.

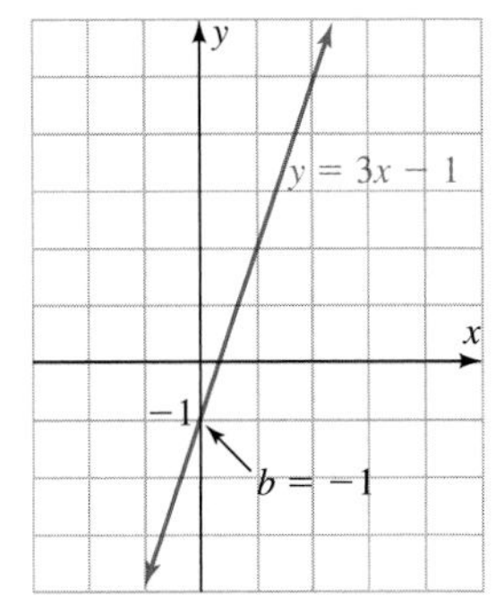

Figure 9
y intercept $= b = -1$

What about m? Looking once again at the table, notice that y increases by $m = 3$ units for every increase of 1 unit in x. This is caused by the term $3x$ in the formula: For every increase of 1 in x, we get an increase of $3 \times 1 = 3$ in y.

Numerically, y increases by m units for every 1-unit increase of x.

Likewise, for every increase of 2 in x, we get an increase of $3 \times 2 = 6$ in y. In general, if x increases by some amount, y will increase by three times that amount. We write

$$\text{change in } y = 3 \times \text{change in } x$$

The Change in a Quantity: Delta Notation

If a quantity q changes from q_1 to q_2, the **change in *q*** is just the difference:

$$\begin{aligned}\text{change in } q &= \text{second value} - \text{first value}\\ &= q_2 - q_1\end{aligned}$$

Mathematicians traditionally use Δ (delta, the Greek equivalent of the Roman letter D) to stand for change and write the change in q as Δq:

$$\Delta q = \text{change in } q = q_2 - q_1$$

Quick Examples

1. If x is changed from 1 to 3, we write

$$\Delta x = \text{second value} - \text{first value} = 3 - 1 = 2$$

2. Looking at our linear function, we see that, when x changes from 1 to 3, y changes from 2 to 8. So,

$$\Delta y = \text{second value} - \text{first value} = 8 - 2 = 6$$

Using delta notation, we can now write, for our linear function,

$$\Delta y = 3\Delta x \qquad \text{change in } y = 3 \times \text{change in } x$$

$$\frac{\Delta y}{\Delta x} = 3$$

Question How does the number $m = 3$ show up in the graph?

Answer Because the value of y increases by exactly 3 units for every increase of 1 unit in x, the graph is a straight line rising by 3 units for every 1 unit we go to the right. We say that we have a **rise** of 3 units for each **run** of 1 unit. Because the value of y changes by $\Delta y = 3\Delta x$ units for every change of Δx units in x, in general we have a rise of $\Delta y = 3\Delta x$ units for each run of Δx units (Figure 10). Thus, we have a rise of 6 for a run of 2, a rise of 9 for a run of 3, and so on. So, $m = 3$ is a measure of the steepness of the line; we call m the **slope of the line:**

$$\text{slope} = m = \frac{\Delta y}{\Delta x} = \frac{\text{rise}}{\text{run}}$$

Graphically, m is the slope of the graph.

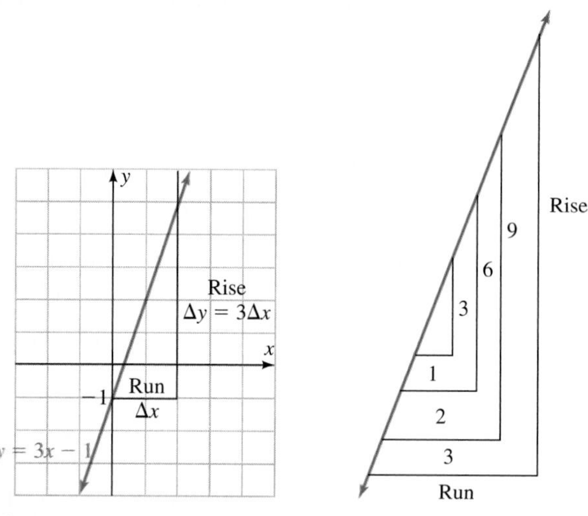

***Figure 10* • *Slope* = *m* = 3**

In general (replace the number 3 by a general number m), we can say the following.

The Roles of m and b in the Linear Function f(x) = mx + b

Role of m

Numerically If $y = mx + b$, then y changes by m units for every 1-unit change in x. A change of Δx units in x results in a change of $\Delta y = m\Delta x$ units in y. Thus,

$$m = \frac{\Delta y}{\Delta x} = \frac{\text{change in } y}{\text{change in } x} \qquad m = \frac{\Delta f}{\Delta x} = \frac{\text{change in } f}{\text{change in } x}$$

Equation form ($y = mx + b$) Function form ($f(x) = mx + b$)

Graphically m is the slope of the line $y = mx + b$:

$$m = \frac{\Delta y}{\Delta x} = \frac{\text{rise}}{\text{run}} = \text{slope}$$

For positive m the graph rises m units for every 1-unit move to the right and rises $\Delta y = m\Delta x$ units for every Δx units moved to the right. For negative m the graph drops $|m|$ units for every 1-unit move to the right and drops $|m|\Delta x$ units for every Δx units moved to the right.

Graph of $y = mx + b$

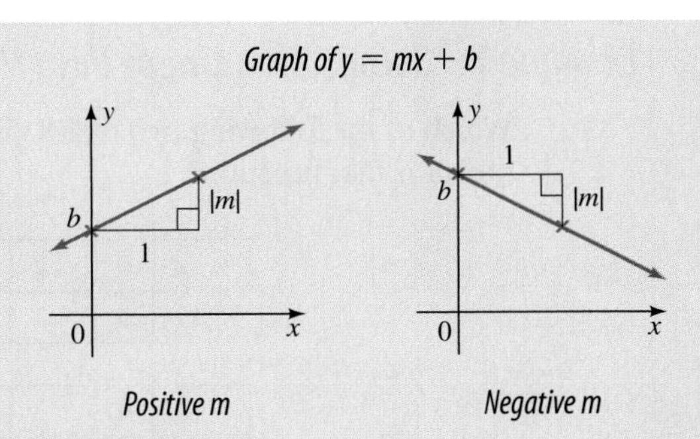

Positive m *Negative m*

Role of b

Numerically When $x = 0, y = b$ Equation form

$f(0) = b$ Function form

Graphically b is the y intercept of the line $y = mx + b$.

Quick Examples

1. $f(x) = 2x + 1$ has slope $m = 2$ and y intercept $b = 1$. To sketch the graph, we start at the y intercept $b = 1$ on the y axis and then move 1 unit to the right and up $m = 2$ units to arrive at a second point on the graph. Now we connect the two points to obtain the graph on the left.

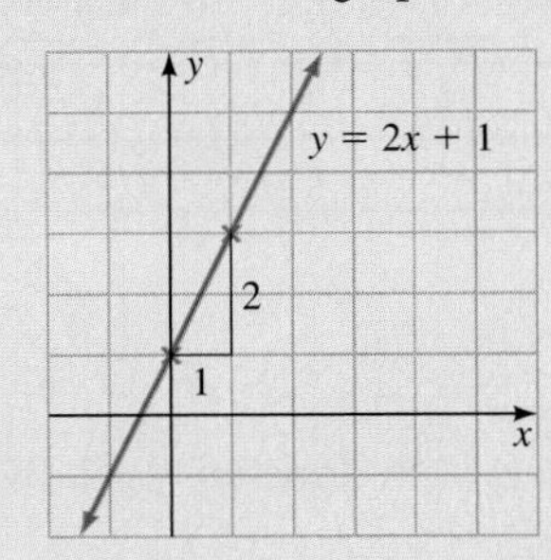

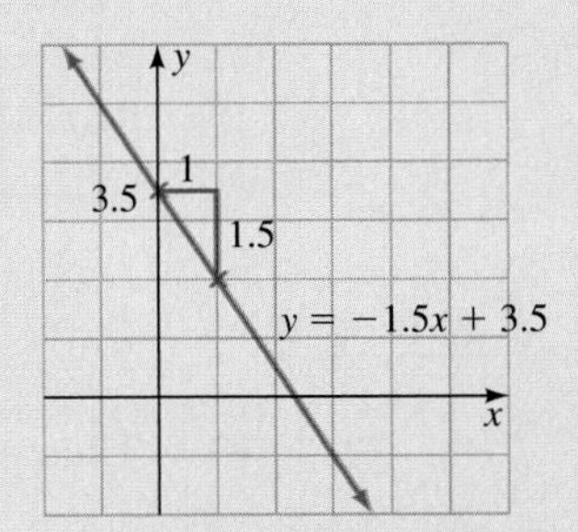

2. The line $y = -1.5x + 3.5$ has slope $m = -1.5$ and y intercept $b = 3.5$. Since the slope is negative, the graph (above right) goes *down* 1.5 units for every 1 unit it moves to the right.
3. Some slopes:

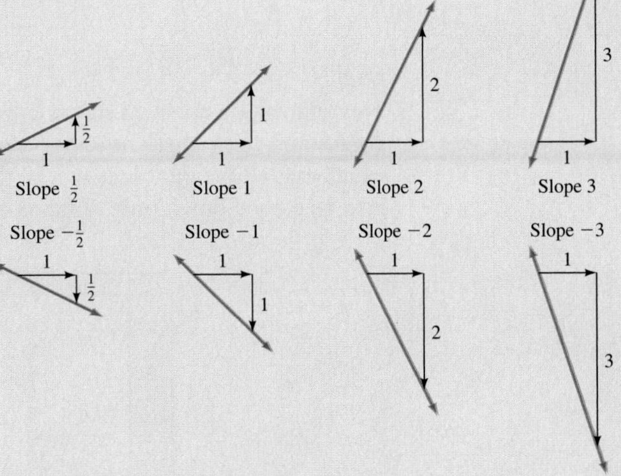

(Notice that the larger the absolute value of the slope, the steeper is the line.)

Example 1 • *Recognizing Linear Data Numerically and Graphically*

Which of the following two tables gives the values of a linear function? What is the formula for that function?

x	0	2	4	6	8	10	12
$f(x)$	3	−1	−3	−6	−8	−13	−15

x	0	2	4	6	8	10	12
$g(x)$	3	−1	−5	−9	−13	−17	−21

Solution The function f cannot be linear: If it were, we would have $\Delta f = m\Delta x$ for some fixed number m. However, although the change in x between successive entries in the table is $\Delta x = 2$ each time, the change in f is not the same each time. Thus, the ratio $\Delta f/\Delta x$ is not the same for every successive pair of points.

On the other hand, the ratio $\Delta g/\Delta x$ is the same each time, namely,

$$\frac{\Delta g}{\Delta x} = \frac{-4}{2} = -2$$

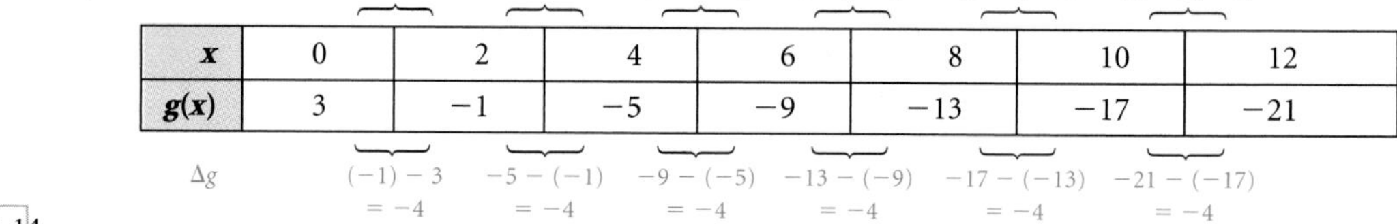

Δx		$2-0=2$	$4-2=2$	$6-4=2$	$8-6=2$	$10-8=2$	$12-10=2$
x	0	2	4	6	8	10	12
$g(x)$	3	−1	−5	−9	−13	−17	−21
Δg		$(-1)-3 = -4$	$-5-(-1) = -4$	$-9-(-5) = -4$	$-13-(-9) = -4$	$-17-(-13) = -4$	$-21-(-17) = -4$

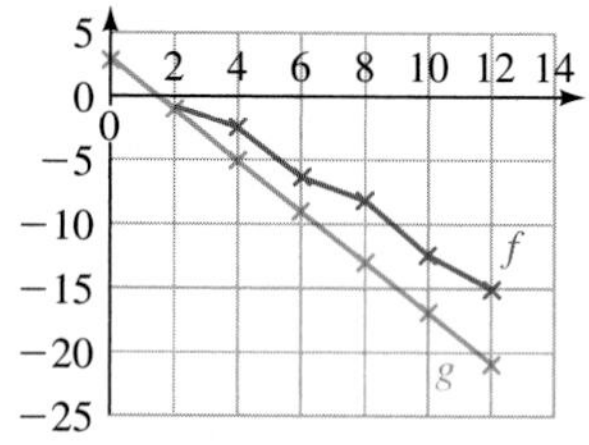

Figure 11

Thus, g is linear with slope $m = -2$. By the table, $g(0) = 3$; hence, $b = 3$. Therefore,

$$g(x) = -2x + 3 \qquad \text{Check that this formula gives the values in the table.}$$

If you graph the points in the tables defining f and g, it becomes easy to see that g is linear and f is not; the points of g lie on a straight line (with slope −2), whereas the points of f do not lie on a straight line (Figure 11).

Excel

The following worksheet shows how you can compute the successive quotients $m = \Delta y/\Delta x$ and then check whether a given set of data shows a linear relationship, in which case all the quotients will be the same. (The shading indicates that the formula is to be copied down only as far as cell C7. [Why not cell C8?])

	A	B	C	D
1	x	f	m	
2	0	3	=(B3-B2)/(A3-A2)	
3	2	-1		
4	4	-3		
5	6	-6		
6	8	-8		
7	10	-13		
8	12	-15		

Example 2 • Graphing a Linear Equation by Hand: Intercepts

Graph the equation $x + 2y = 4$. Where does the line cross the x and y axes?

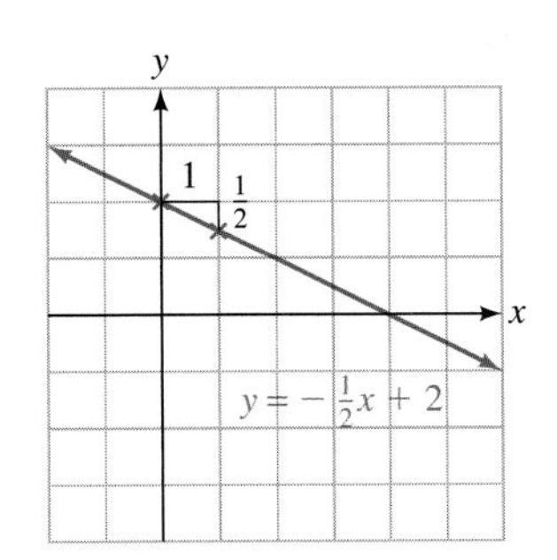

Figure 12

Solution We first write y as a linear function of x by solving the equation for y:

$$2y = -x + 4$$

$$y = -\frac{1}{2}x + 2$$

Now we can see that the graph is a straight line with a slope of $-\frac{1}{2}$ and a y intercept of 2. We start at 2 on the y axis and go down $\frac{1}{2}$ unit for every 1 unit we go to the right. The graph is shown in Figure 12.

We already know that the line crosses the y axis at 2. Where does it cross the x axis? Wherever that is, we know that the y coordinate will be 0 at that point. So, we set $y = 0$ and solve for x. It's most convenient to use the equation we were originally given:

$$x + 2(0) = 4$$

$$x = 4$$

The line crosses the x axis at 4.

Before we go on . . . We could have graphed the equation in another way by first finding the intercepts. Once we know that the line crosses the y axis at 2 and the x-axis at 4, we can draw those two points and then draw the line connecting them.

It's worth noting what we did to find the intercepts in Example 2.

Finding the Intercepts

The **x intercept** of a line is where it crosses the x axis. To find it, set $y = 0$ and solve for x. The **y intercept** is where it crosses the y axis. If the equation of the line is written as $y = mx + b$, then b is the y intercept. Otherwise, set $x = 0$ and solve for y.

Quick Example

Consider the equation $3x - 2y = 6$. To find its x intercept, set $y = 0$ to find $x = \frac{6}{3} = 2$. To find its y intercept, set $x = 0$ to find $y = \frac{6}{-2} = -3$. The line crosses the x axis at 2 and the y axis at -3.

Computing the Slope of a Line

We know that the slope of a line is given by

$$\text{slope} = m = \frac{\text{rise}}{\text{run}} = \frac{\Delta y}{\Delta x}$$

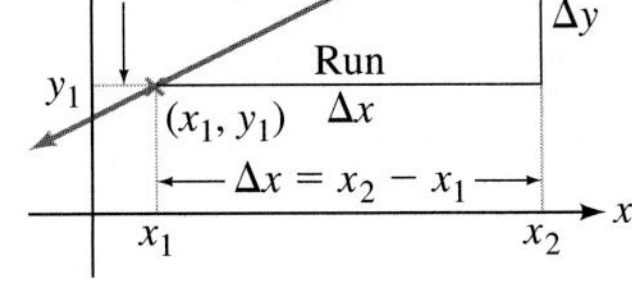

Figure 13

Question Two points—say, (x_1, y_1) and (x_2, y_2)—determine a line in the xy plane. How do we find its slope?

Answer As you can see in Figure 13, the rise is $\Delta y = y_2 - y_1$, the change in the y coordinate from the first point to the second; the run is $\Delta x = x_2 - x_1$, the change in the x coordinate.

Computing the Slope of a Line

We can compute the slope m of the line through the points (x_1, y_1) and (x_2, y_2) using

$$m = \frac{\Delta y}{\Delta x} = \frac{y_2 - y_1}{x_2 - x_1}$$

Quick Examples

1. The slope of the line through $(x_1, y_1) = (1, 3)$ and $(x_2, y_2) = (5, 11)$ is

$$m = \frac{\Delta y}{\Delta x} = \frac{y_2 - y_1}{x_2 - x_1} = \frac{11 - 3}{5 - 1} = \frac{8}{4} = 2$$

2. The slope of the line through $(x_1, y_1) = (1, 2)$ and $(x_2, y_2) = (2, 1)$ is

$$m = \frac{\Delta y}{\Delta x} = \frac{y_2 - y_1}{x_2 - x_1} = \frac{1 - 2}{2 - 1} = \frac{-1}{1} = -1$$

3. Here is an Excel worksheet to compute the slope of the line through two given points (such as those in Quick Example 1):

	A	B	C	D
1	x	y	m	
2	1	3	=(B3-B2)/(A3-A2)	
3	5	11		

4. The line through (2, 3) and (−1, 3) has slope

$$m = \frac{\Delta y}{\Delta x} = \frac{3 - 3}{-1 - 2} = \frac{0}{-3} = 0$$

A line of slope 0 has a 0 rise, so it is a *horizontal* line, as shown below by the graph on the left. On the other hand, the line through (3, 2) and (3, −1) has

$$m = \frac{\Delta y}{\Delta x} = \frac{-1 - 2}{3 - 3} = \frac{-3}{0}$$

which is undefined. If we plot the two points, we see that the line passing through them is *vertical,* as shown by the graph on the right.

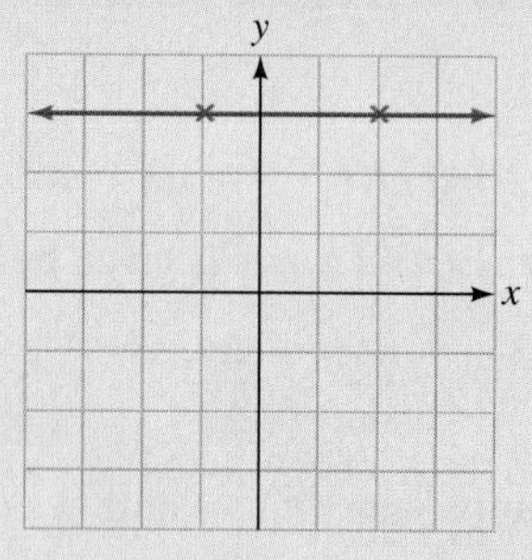

The line through (2, 3) and (−1, 3); m = 0.

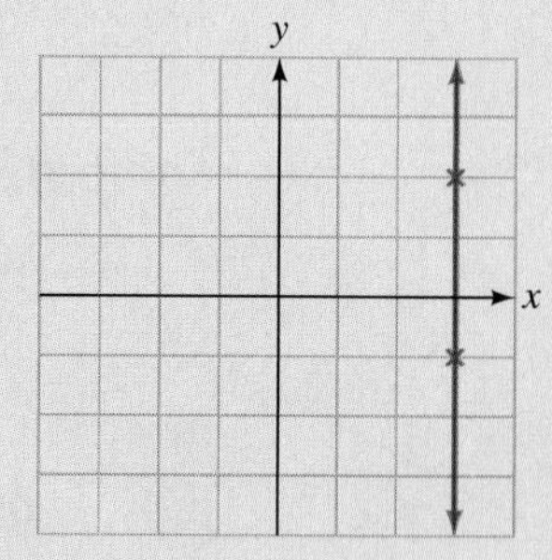

The line through (3, 2) and (3, −1); m is undefined.

Question What if we had chosen to list the two points in Quick Example 1 in reverse order? That is, suppose we had taken $(x_1, y_1) = (5, 11)$ and $(x_2, y_2) = (1, 3)$. What would have happened?

Answer We would have found

$$m = \frac{\Delta y}{\Delta x} = \frac{y_2 - y_1}{x_2 - x_1} = \frac{3 - 11}{1 - 5} = \frac{-8}{-4} = 2$$

the same answer. The order in which we take the points is not important, *as long as we use the same order in the numerator and the denominator.*

Finding a Linear Equation from Data: How to Make a Linear Model

If we happen to know the slope and y intercept of a line, writing down its equation is straightforward. For example, if we know that the slope is 3 and the y intercept is -1, then the equation is $y = 3x - 1$. Sadly, the information we are given is seldom so convenient. For instance, we may know the slope and a point other than the y intercept, two points on the line, or other information.

We describe a straightforward method for finding the equation of a line: the **point-slope** method. As the name suggests, we need two pieces of information:

- The *slope* m (which specifies the direction of the line)
- A *point* (x_1, y_1) on the line (which pins down its location in the plane)

Question What is the equation of the line through the point (x_1, y_1) with slope m?

Answer Since its slope is m, it must have the form

$$y = mx + b$$

for some (unknown) number b. To determine b we use the fact that the line must pass through the point (x_1, y_1), so that

$$y_1 = mx_1 + b \qquad (x_1, y_1) \text{ satisfies the equation } y = mx + b.$$

Solving for b gives

$$b = y_1 - mx_1$$

The following provides a summary.

The Point-Slope Formula

An equation of the line through the point (x_1, y_1) with slope m is given by

$$y = mx + b \qquad \text{Equation form}$$

where

$$b = y_1 - mx_1$$

When to Apply the Point-Slope Formula

Apply the point-slope formula to find the equation of a line whenever you are given information about a point and the slope of the line. The formula does not apply if the slope is undefined [as in a vertical line; see Example 3(d)].

Quick Examples

1. The line through (2, 3) with slope 4 has equation

$$y = 4x + b$$

where $b = 3 - (4)(2) = -5$, so

$$y = 4x - 5$$

2. Here is an Excel worksheet to compute the equation of the line through any point with a given slope:

	A	B	C	D
1	x	y	m	b
2	2	3	4	=B2-C2*A2

Example 3 • Using the Point-Slope Formula

Find equations for the following straight lines:

a. Through the points (1, 2) and (3, −1)

b. Through (2, −2) and parallel to the line $3x + 4y = 5$

c. Horizontal and through (−9, 5)

d. Vertical and through (−9, 5)

Solution In each case other than (d), we apply the point-slope formula.

a. To apply the point-slope formula, we need

- **Point** We have two to choose from, so we take the first, $(x_1, y_1) = (1, 2)$.
- **Slope** Not given *directly*, but we do have enough information to calculate it. Since we are given two points on the line, we can use the slope formula:

$$m = \frac{y_2 - y_1}{x_2 - x_1} = \frac{-1 - 2}{3 - 1} = -\frac{3}{2}$$

An equation of the line is therefore

$$y = -\frac{3}{2}x + b$$

where $b = y_1 - mx_1 = 2 - \left(-\frac{3}{2}\right)(1) = \frac{7}{2}$, so

$$y = -\frac{3}{2}x + \frac{7}{2}$$

b. Proceeding as before,

- **Point** Given here as (2, −2).

- **Slope** We use the fact that *parallel lines have the same slope.* (Why?) We can find the slope of $3x + 4y = 5$ by solving for y and then looking at the coefficient of x:

$$y = -\frac{3}{4}x + \frac{5}{4}$$ To find the slope, solve for y.

so the slope is $-\frac{3}{4}$.

An equation for the desired line is

$$y = -\frac{3}{4}\,x + b$$

where $b = y_1 - mx_1 = -2 - \left(-\frac{3}{4}\right)(2) = -\frac{1}{2}$, so

$$y = -\frac{3}{4}\,x - \frac{1}{2}$$

c. We are given a point: $(-9, 5)$. Furthermore, we are told that the line is horizontal, which tells us that the slope is 0. Therefore, we get

$$y = 0x + b = b$$

where $b = y_1 - mx_1 = 5 - (0)(-9) = 5$, so

$$y = 5$$

d. We are given a point: $(-9, 5)$. This time we are told that the line is vertical, which means that the slope is undefined. Thus, we can't use the point-slope formula. (That formula only makes sense when the slope of the line is defined.) What can we do? Well, here are some points on the desired line:

$$(-9, 1), (-9, 2), (-9, 3), \ldots,$$

so $x = -9$ and $y = $ *anything.* If we simply say that $x = -9$, then these points are all solutions, so the equation is $x = -9$.

Excel

Here is an Excel worksheet that computes m and b for the line passing through the two points given in part (a).

	A	B	C	D
1	x	y	m	b
2	1	2	=(B3-B2)/(A3-A2)	=B2-C2*A2
3	3	-1	Slope	Intercept

1.3 EXERCISES

In Exercises 1–6, a table of values for a linear function is given. Fill in the missing value and calculate m in each case.

1.

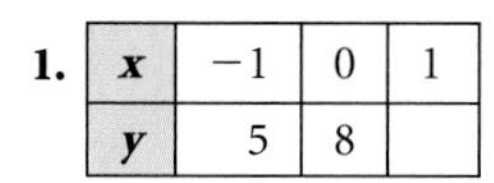

x	−1	0	1
y	5	8	

2.

x	−1	0	1
y	−1	−3	

3.

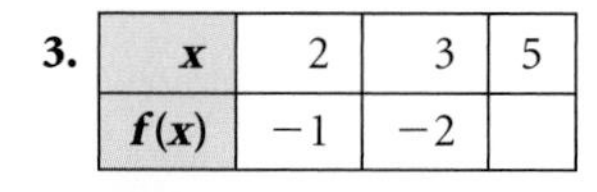

x	2	3	5
$f(x)$	−1	−2	

4.

x	2	4	5
$f(x)$	−1	−2	

5.

x	−2	0	2
$f(x)$	4		10

6.

x	0	3	6
$f(x)$	−1		−5

In Exercises 7–10, first find $f(0)$, if not supplied, and then find an equation of the given linear function.

7.

x	−2	0	2	4
$f(x)$	−1	−2	−3	−4

8.

x	−6	−3	0	3
$f(x)$	1	2	3	4

9.

x	−4	−3	−2	−1
$f(x)$	−1	−2	−3	−4

10.

x	1	2	3	4
$f(x)$	4	6	8	10

In each of Exercises 11–14, decide which of the two given functions is linear and find its equation.

11.

x	0	1	2	3	4
$f(x)$	6	10	14	18	22
$g(x)$	8	10	12	16	22

12.

x	−10	0	10	20	30
$f(x)$	−1.5	0	1.5	2.5	3.5
$g(x)$	−9	−4	1	6	11

13.

x	0	3	6	10	15
$f(x)$	0	3	5	7	9
$g(x)$	−1	5	11	19	29

14.

x	0	3	5	6	9
$f(x)$	2	6	9	12	15
$g(x)$	−1	8	14	17	26

In Exercises 15–24, find the slope of the given line, if it is defined.

15. $y = -\dfrac{3}{2}x - 4$ **16.** $y = \dfrac{2x}{3} + 4$

17. $y = \dfrac{x+1}{6}$ **18.** $y = -\dfrac{2x-1}{3}$

19. $3x + 1 = 0$ **20.** $8x - 2y = 1$

21. $3y + 1 = 0$ **22.** $2x + 3 = 0$

23. $4x + 3y = 7$ **24.** $2y + 3 = 0$

In Exercises 25–38, graph the given equations.

25. $y = 2x - 1$ **26.** $y = x - 3$

27. $y = -\frac{2}{3}x + 2$ **28.** $y = -\frac{1}{2}x + 3$

29. $y + \frac{1}{4}x = -4$ **30.** $y - \frac{1}{4}x = -2$

31. $7x - 2y = 7$ **32.** $2x - 3y = 1$

33. $3x = 8$ **34.** $2x = -7$

35. $6y = 9$ **36.** $3y = 4$

37. $2x = 3y$ **38.** $3x = -2y$

In Exercises 39–54, calculate the slope, if defined, of the straight line through the given pair of points. Try to do as many as you can without writing anything down except the answer.

39. $(0, 0)$ and $(1, 2)$ **40.** $(0, 0)$ and $(-1, 2)$

41. $(-1, -2)$ and $(0, 0)$ **42.** $(2, 1)$ and $(0, 0)$

43. $(4, 3)$ and $(5, 1)$ **44.** $(4, 3)$ and $(4, 1)$

45. $(1, -1)$ and $(1, -2)$ **46.** $(-2, 2)$ and $(-1, -1)$

47. $(2, 3.5)$ and $(4, 6.5)$ **48.** $(10, -3.5)$ and $(0, -1.5)$

49. $(300, 20.2)$ and $(400, 11.2)$ **50.** $(1, -20.2)$ and $(2, 3.2)$

51. $(0, 1)$ and $(-\frac{1}{2}, \frac{3}{4})$ **52.** $(\frac{1}{2}, 1)$ and $(-\frac{1}{2}, \frac{3}{4})$

53. (a, b) and (c, d) $\quad (a \neq c)$ **54.** (a, b) and (c, b) $\quad (a \neq c)$

55. In the following figure, estimate the slopes of all line segments.

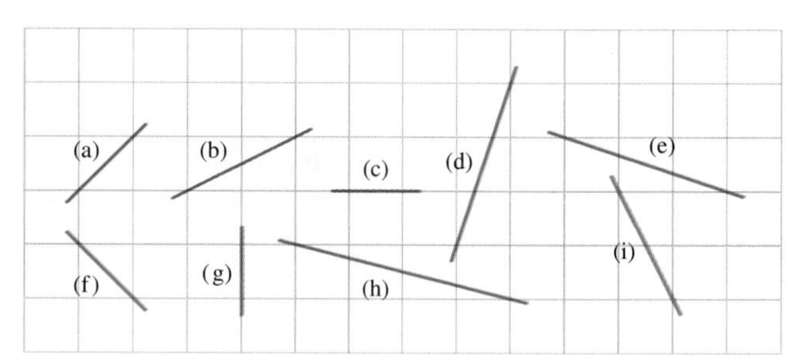

56. In the following figure, estimate the slopes of all line segments.

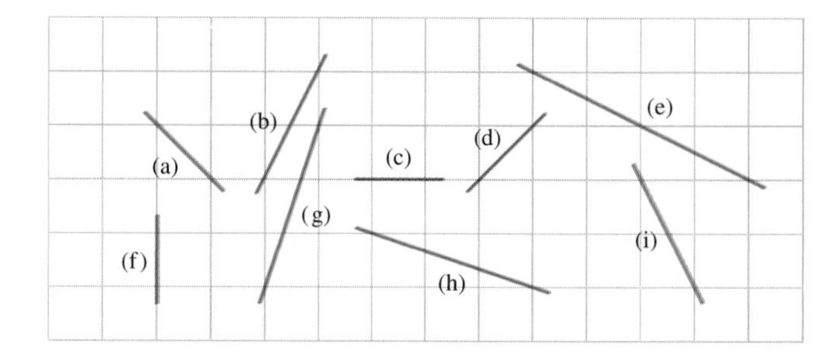

In Exercises 57–70, find linear equations whose graphs are the straight lines.

57. Through $(1, 3)$ with slope 3

58. Through $(2, 1)$ with slope 2

59. Through $(1, -\frac{3}{4})$ with slope $\frac{1}{4}$

60. Through $(0, -\frac{1}{3})$ with slope $\frac{1}{3}$

61. Through $(20, -3.5)$ and increasing at a rate of 10 units of y per unit of x

62. Through $(3.5, -10)$ and increasing at a rate of 1 unit of y per 2 units of x

63. Through $(2, -4)$ and $(1, 1)$

64. Through $(1, -4)$ and $(-1, -1)$

65. Through $(1, -0.75)$ and $(0.5, 0.75)$

66. Through $(0.5, -0.75)$ and $(1, -3.75)$

67. Through $(6, 6)$ and parallel to the line $x + y = 4$

68. Through $(\frac{1}{3}, -1)$ and parallel to the line $3x - 4y = 8$

69. Through $(0.5, 5)$ and parallel to the line $4x - 2y = 11$

70. Through $(\frac{1}{3}, 0)$ and parallel to the line $6x - 2y = 11$

COMMUNICATION AND REASONING EXERCISES

71. How would you test a table of values of x and y to see if it comes from a linear function?

72. You have ascertained that a table of values of x and y corresponds to a linear function. How do you find an equation for that linear function?

73. To what linear function of x does the linear equation $ax + by = c$ $(b \neq 0)$ correspond? Why did we specify $b \neq 0$?

74. Complete the following. The slope of the line with equation $y = mx + b$ is the number of units that ______ increases per unit increase in ______.

75. Complete the following. If, in a straight line, y is increasing three times as fast as x, then its ______ is ______.

76. Suppose that y is decreasing at a rate of 4 units per 3-unit increase of x. What can we say about the slope of the linear relationship between x and y? What can we say about the intercept?

77. If y and x are related by the linear expression $y = mx + b$, how will y change as x changes if m is positive? Negative? Zero?

78. Your friend April tells you that $y = f(x)$ has the property that, whenever x is changed by Δx, the corresponding change in y is $\Delta y = -\Delta x$. What can you tell her about f?

79. Consider the following worksheet [as in Example 3(a)]:

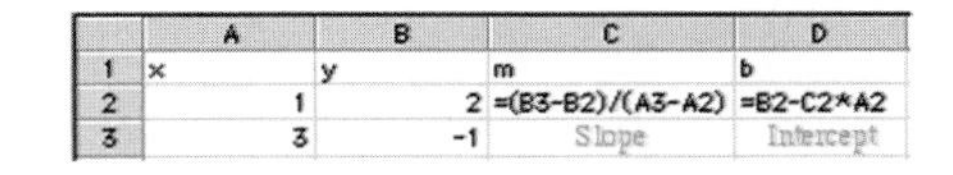

	A	B	C	D
1	x	y	m	b
2	1	2	=(B3-B2)/(A3-A2)	=B2-C2*A2
3	3	-1	Slope	Intercept

What is the effect on the slope of increasing the y coordinate of the second point (the point whose coordinates are in row 3)? Explain.

80. Referring to the worksheet in Exercise 79, what is the effect on the slope of increasing the x coordinate of the second point (the point whose coordinates are in row 3)? Explain.

1.4 Linear Models

Using linear functions to describe or approximate relationships in the real world is called **linear modeling.** We start with some examples involving cost, revenue, and profit.

Cost, Revenue, and Profit Functions

Example 1 • Linear Cost Function

As of April 2002, Yellow Cab Chicago's rates were \$1.90 on entering the cab plus \$1.60 for each mile.[26]

a. Find the cost C of an x-mile trip.

b. Use your answer to calculate the cost of a 40-mile trip.

c. What is the cost of the second mile? The tenth mile?

d. Graph C as a function of x.

---

[26]According to their Web site at http://www.yellowcabchicago.com/.

Solution

a. We are being asked to find how the cost C depends on the length x of the trip, or to find C as a function of x. Here is the cost in a few cases:

Cost of a 1-mile trip: $C = 1.60(1) + 1.90 = \$3.50$ 1 mile @ \$1.60 per mile plus \$1.90

Cost of a 2-mile trip: $C = 1.60(2) + 1.90 = \$5.10$ 2 miles @ \$1.60 per mile plus \$1.90

Cost of a 3-mile trip: $C = 1.60(3) + 1.90 = \$6.70$ 3 miles @ \$1.60 per mile plus \$1.90

Do you see the pattern? The cost of an x-mile trip is given by the linear function

$$C = 1.60x + 1.90 \qquad \text{Equation form}$$

$$C(x) = 1.60x + 1.90 \qquad \text{Function form}$$

Notice that the slope 1.60 is the incremental cost per mile. In this context we call 1.60 the **marginal cost;** the varying quantity $1.60x$ is called the **variable cost.** The C intercept 1.90 is the cost to enter the cab, which we call the **fixed cost.** In general, a linear cost function has the following form.

$$C(x) = \overbrace{mx}^{\text{Variable cost}} + b$$

$\uparrow$ Marginal cost $\uparrow$ Fixed cost

b. We can use the formula for the cost function to calculate the cost of a 40-mile trip as

$$C(40) = 1.60(40) + 1.90 = \$65.90$$

c. To calculate the cost of the second mile, we *could* proceed as follows:

Find the cost of a 1-mile trip: $C(1) = 1.60(1) + 1.90 = \$3.50$.

Find the cost of a 2-mile trip: $C(2) = 1.60(2) + 1.90 = \$5.10$.

Therefore, the cost of the second mile is $\$5.10 - \$3.50 = \$1.60$.

But notice that this is just the marginal cost. In fact, the marginal cost is the cost of each additional mile, so we could have done this more simply:

$$\text{cost of second mile} = \text{cost of tenth mile} = \text{marginal cost} = \$1.60$$

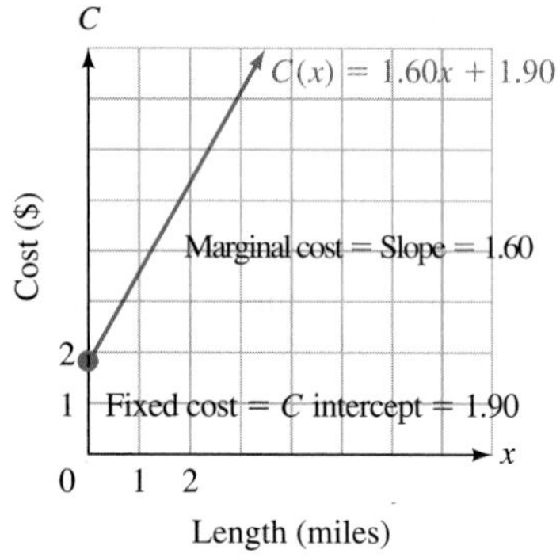

Figure 14

d. Figure 14 shows the graph of the cost function, which we can interpret as a *cost-versus-miles* graph. The fixed cost is the starting height on the left, and the marginal cost is the slope of the line.

Before we go on . . . In general, the slope m measures the number of units of change in y per 1-unit change in x, so *we measure m in units of y per unit of x:*

$$\text{units of slope} = \text{units of } y \text{ per unit of } x$$

In this example, y is the cost C, measured in dollars, and x is the length of a trip, measured in miles. Hence,

$$\text{units of slope} = \text{units of } y \text{ per unit of } x = \text{dollars per mile}$$

The y intercept b, being a value of y, is measured in the same units as y. In this example, b is measured in dollars.

Here is a summary of the terms used in this example, along with an introduction to some new terms.

Cost, Revenue, and Profit Functions

A **cost function** specifies the cost C as a function of the number of items x. Thus, $C(x)$ is the cost of x items. A cost function of the form

$$C(x) = mx + b$$

is called a **linear cost function.** The quantity mx is called the **variable cost,** and the intercept b is called the **fixed cost.** The slope m, the **marginal cost,** measures the incremental cost per item.

The **revenue** resulting from one or more business transactions is the total payment received, sometimes called the gross proceeds. If $R(x)$ is the revenue from selling x items at a price of m each, then R is the linear function $R(x) = mx$, and the selling price m can also be called the **marginal revenue.**

The **profit,** on the other hand, is the *net* proceeds, or what remains of the revenue when costs are subtracted. If the profit depends linearly on the number of items, the slope m is called the **marginal profit.** Profit, revenue, and cost are related by the following formula:

$$\text{profit} = \text{revenue} - \text{cost}$$

$$P = R - C$$

If the profit is negative, say $-\$500$, we refer to a **loss** (of \$500 in this case). To **break even** means to make neither a profit nor a loss. Thus, break even occurs when $P = 0$, or

$$R = C \qquad \text{Break even}$$

The **break-even point** is the number of items x at which break even occurs.

Quick Example

If the daily cost (including operating costs) of manufacturing x T-shirts is $C(x) = 8x + 100$ and the revenue obtained by selling x T-shirts is $R(x) = 10x$, then the daily profit resulting from the manufacture and sale of x T-shirts is

$$P(x) = R(x) - C(x) = 10x - (8x + 100) = 2x - 100$$

Break even occurs when $P(x) = 0$, or $x = 50$.

Example 2 • Cost, Revenue, and Profit

The manager of the FrozenAir Refrigerator factory notices that on Monday it cost the company a total of \$25,000 to build 30 refrigerators and on Tuesday it cost \$30,000 to build 40 refrigerators.

a. Find a linear cost function based on this information. What is the daily fixed cost, and what is the marginal cost?

b. FrozenAir sells its refrigerators for \$1500 each. What is the revenue function?

c. What is the profit function? How many refrigerators must FrozenAir sell in a day in order to break even for that day? What will happen if it sells fewer refrigerators? If it sells more?

Solution

a. We are seeking C as a linear function of x, the number of refrigerators sold:

$$C = mx + b$$

We are told that $C = 25{,}000$ when $x = 30$, and this amounts to being told that $(30, 25{,}000)$ is a point on the graph of the cost function. Similarly, $(40, 30{,}000)$ is another point on the line (Figure 15).

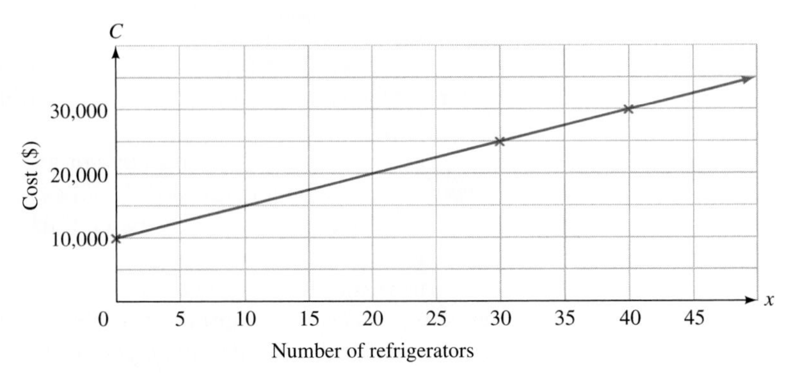

Figure 15

We can now use the point-slope formula to construct a linear cost equation. Recall that we need two items of information: a point on the line and the slope.

- **Point** Let's use the first point: $(x_1, C_1) = (30, 25{,}000)$ — C plays the role of y.
- **Slope** $m = \dfrac{C_2 - C_1}{x_2 - x_1} = \dfrac{30{,}000 - 25{,}000}{40 - 30} = 500$ — Marginal cost = \$500

The cost function is therefore

$$C = 500x + b$$

where $b = C_1 - mx_1 = 25{,}000 - (500)(30) = 10{,}000$ — Fixed cost = \$10,000

so $C = 500x + 10{,}000$ — Cost function: equation notation

or $C(x) = 500x + 10{,}000$ — Cost function: function notation

Since $m = 500$ and $b = 10{,}000$, the factory's fixed cost is \$10,000 each day, and its marginal cost is \$500 per refrigerator.

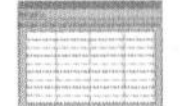

Excel

We can also use Excel to compute the cost equation from the given data:

	A	B	C	D
1	x	C	m	b
2	30	25000	=(B3-B2)/(A3-A2)	=B2-C2*A2
3	40	30000	Slope	Intercept

b. The revenue FrozenAir obtains from the sale of a single refrigerator is \$1500. So, if it sells x refrigerators, it earns a revenue of

$$R(x) = 1500x$$

c. For the profit, we use the formula

$$\text{profit} = \text{revenue} - \text{cost}$$

For the cost and revenue, we can substitute the answers from parts (a) and (b) and obtain

$$\begin{aligned} P(x) &= R(x) - C(x) && \text{Formula for profit} \\ &= 1500x - (500x + 10{,}000) && \text{Substitute } R(x) \text{ and } C(x). \\ &= 1000x - 10{,}000 \end{aligned}$$

Here, $P(x)$ is the daily profit FrozenAir makes by making and selling x refrigerators.

Finally, to break even means to make zero profit. So, we need to find the x such that $P(x) = 0$. All we have to do is set $P(x) = 0$ and solve for x:

$$1000x - 10{,}000 = 0$$

giving

$$x = \frac{10{,}000}{1000} = 10$$

To break even, FrozenAir needs to manufacture and sell 10 refrigerators in a day.

For values of x less than the break-even point, 10, $P(x)$ is negative, so the company will have a loss. For values of x greater than the break-even point, $P(x)$ is positive, so the company will make a profit. This is the reason we are interested in the point where $P(x) = 0$. Since $P(x) = R(x) - C(x)$, we can also look at the break-even point as the point where revenue = cost: $R(x) = C(x)$ (see Figure 16).

Before we go on . . . We can graph the cost and revenue function and find the break-even point graphically (Figure 16):

$$\text{Cost: } C(x) = 500x + 10{,}000$$

$$\text{Revenue: } R(x) = 1500x$$

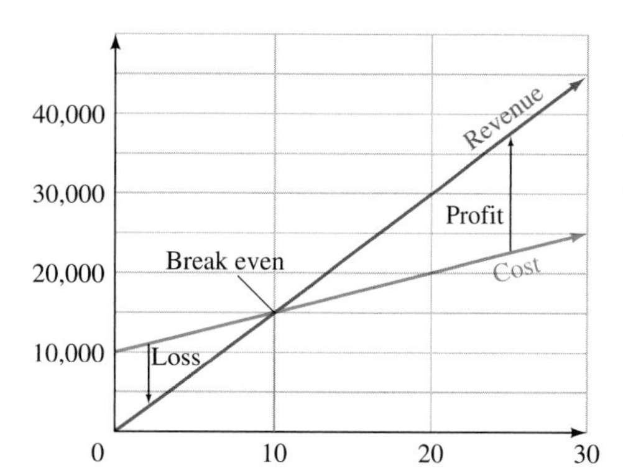

Figure 16 • Break even occurs at the point of intersection of the graphs of revenue and cost.

The break-even point is the point where the revenue and cost are equal—that is, where the graphs of cost and revenue cross. Figure 16 confirms that break even occurs when $x = 10$ refrigerators. Or we can use the graph to find the break-even point in the first place. If we use technology, we can zoom in for an accurate estimate of the point of intersection.

Excel

Excel also has an interesting feature called Goal Seek that can be used to find the point of intersection of two lines numerically (rather than graphically). The downloadable Excel tutorial for this section contains detailed instructions on using Goal Seek to find break-even points.

Demand and Supply Functions

The demand for a commodity usually goes down as its price goes up. It is traditional to use the letter q for the (quantity of) demand, as measured, for example, in weekly sales. Consider the following example.

Example 3 • Linear Demand Function

You run a small supermarket and must determine how much to charge for Hot'n'Spicy brand baked beans. The following chart shows weekly sales figures for Hot'n'Spicy at two different prices:

Price/Can, *p* ($)	0.50	0.75
Demand, *q* (cans sold/week)	400	350

a. Model the data by expressing the demand q as a linear function of the price p.

b. How do we interpret the slope? The q intercept?

c. How much should you charge for a can of Hot'n'Spicy beans if you want the demand to increase to 410 cans per week?

Solution

a. A **demand equation** or **demand function** expresses demand q (in this case, the number of cans of beans sold per week) as a function of the unit price p (in this case, price per can). We model the demand using the two points we are given: (0.50, 400) and (0.75, 350):

- **Point** (0.50, 400)
- **Slope** $m = \dfrac{q_2 - q_1}{p_2 - p_1} = \dfrac{350 - 400}{0.75 - 0.50} = \dfrac{-50}{0.25} = -200$

Thus, the demand equation is

$$\begin{aligned} q &= mp + b \\ &= -200p + [400 - (-200)(0.50)] \\ &= -200p + 500 \end{aligned}$$

Figure 17 shows the data points and the linear model.

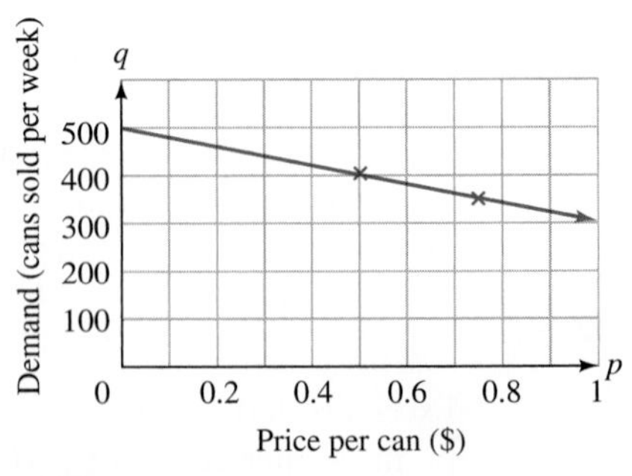

Figure 17

b. The key to interpreting the slope, $m = -200$, is to recall (see Example 1) that we measure the slope in units of y per unit of x. In this example, we mean units of q per unit of p, or the number of cans sold per dollar change in the price. Since m is negative, we see that the number of cans sold decreases as the price increases. We conclude that the weekly sales will drop by 200 cans per \$1-increase in the price.

To interpret the q intercept, recall that it gives the q coordinate when $p = 0$. Hence, it is the number of cans you can "sell" every week if you were to give them away.[27]

c. If we want the demand to increase to 410 cans per week, we set $q = 410$ and solve for p:

$$410 = -200p + 500$$

$$200p = 90$$

$$p = \frac{90}{200} = \$0.45/\text{can}$$

Before we go on . . .

Question Just how reliable *is* the linear model?

Answer Look at it this way: The *actual* demand graph could in principle be obtained by tabulating new sales figures for a large number of different prices. If the resulting points were plotted on the pq plane, they would probably suggest a curve and not a straight line. However, if you looked at a small enough portion of this curve, you could closely *approximate* it by a straight line. In other words, *over a small range of values of p, the linear model is accurate.* Linear models of real-world situations are generally reliable only for small ranges of the variables. (This point will come up again in some of the exercises.)

Graphing Calculator

In part (c) we could also find p numerically using the Table feature in the TI-83. First make sure that the table settings permit you to enter values of x: Press [2nd] [TBLSET] and set `Indpnt` to `Ask`. Now enter

`Y1=-200*X+500` Demand equation

and press [2nd] [TABLE]. You will now be able to adjust the price (x) until you find the value at which the demand equals 410.

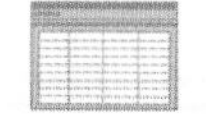

Excel

In Excel we can use the worksheet from Example 2 to compute m and b for part (a):

	A	B	C	D
1	x	y	m	b
2	0.5	400	=(B3-B2)/(A3-A2)	=B2-C2*A2
3	0.75	350	Formula for Slope	Formula for Interept

[27]Does this seem realistic? Demand is not always unlimited if items were given away. For instance, campus newspapers are sometimes given away, and yet piles of them are often left untaken. Also see the "Before we go on" discussion at the end of this example.

In part (c) we could find p numerically by using a worksheet like the following to compute the demand for many prices starting at $p = \$0.00$, until we find the price at which the demand equals 410.

	A	B	C
1	p	q	
2	0	=-200*A2+500	
3	0.01	498	
4	0.02	496	
5			
101	[illegible]	[illegible]	
102	0.44	412	
103	0.45	410	
104	0.46	408	

Demand Function

A **demand equation** or **demand function** expresses demand q (the number of items demanded) as a function of the unit price p (the price per item). A **linear demand function** has the form

$$q(p) = mp + b$$

Interpretation of m

The (usually negative) slope m measures the change in demand per unit change in price. For instance, if p is measured in dollars and q in monthly sales and $m = -400$, then each \$1 increase in the price per item will result in a drop in sales of 400 items per month.

Interpretation of b

The y intercept b gives the demand if the items were given away.

Quick Example

If the demand for T-shirts, measured in daily sales, is given by $q(p) = -4p + 90$, where p is the sale price in dollars, then daily sales drop by four T-shirts for every \$1 increase in price. If the T-shirts were given away, the demand would be 90 T-shirts per day.

We have seen that a demand function gives the number of items consumers are willing to buy at a given price, and a higher price generally results in a lower demand. However, as the price rises, suppliers will be more inclined to produce these items (as opposed to spending their time and money on other products), so supply will generally rise. A **supply function** gives q, the number of items suppliers are willing to make available for sale, as a function of p, the price per item.[28]

[28]Although a bit confusing at first, it is traditional to use the same letter q for the quantity of supply and the quantity of demand, particularly when we want to compare them, as in Example 4.

Example 4 • Demand, Supply, and Equilibrium Price

Continuing with Example 3, consider the following chart, which shows weekly sales figures for Hot'n'Spicy at two different prices (the demand), as well as the number of cans per week that you are prepared to place on sale at these prices (the supply).

Price/Can ($)	0.50	0.75
Demand (cans sold/week)	400	350
Supply (cans placed on sale/week)	300	500

a. Model these data with linear demand and supply functions.

b. How much should you charge per can of Hot'n'Spicy beans if you want the demand to equal the supply? How many cans will you sell at that price, known as the **equilibrium price**? What happens if you charge more than the equilibrium price? If you charge less?

Solution

a. We have already modeled the demand function in Example 3:

$$q = -200p + 500$$

To model the supply, we use the first and third rows of the table. We are again given two points: (0.50, 300) and (0.75, 500).

- **Point** (0.50, 300)
- **Slope** $m = \dfrac{q_2 - q_1}{p_2 - p_1} = \dfrac{500 - 300}{0.75 - 0.50} = \dfrac{200}{0.25} = 800$

So, the supply equation is

$$\begin{aligned} q &= mp + b \\ &= 800p + [300 - (800)(0.50)] \\ &= 800p - 100 \end{aligned}$$

b. To find where the demand equals the supply, we equate the two functions:

$$\begin{aligned} \text{demand} &= \text{supply} \\ -200p + 500 &= 800p - 100 \\ -1000p &= -600 \end{aligned}$$

so $$p = \frac{-600}{-1000} = \$0.60$$

This is the equilibrium price. We can find the corresponding demand by substituting 0.60 for p in the demand (or supply) equation.

$$\text{equilibrium demand} = -200(0.60) + 500 = 380 \text{ cans/week}$$

So, to balance supply and demand, you should charge \$0.60 per can of Hot'n'Spicy beans, and you should place 380 cans on sale each week.

If we graph supply and demand on the same set of axes, we obtain the graphs shown in Figure 18.

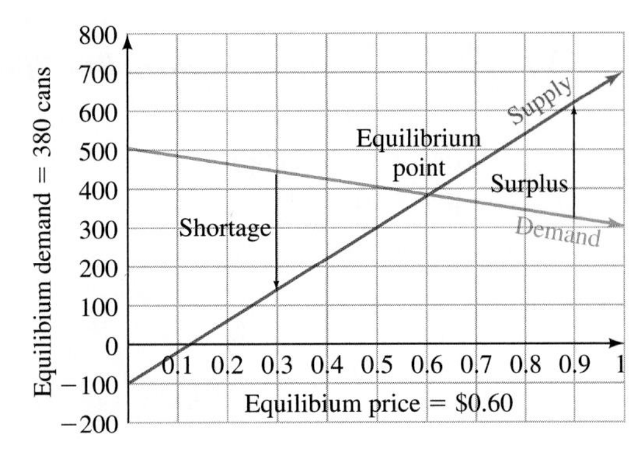

Figure 18 • *Equilibrium occurs at the point of intersection of the graphs of supply and demand.*

Figure 18 shows what happens if you charge prices other than the equilibrium price. If you charge more—say, \$0.90 per can ($p = 0.90$)—then the supply will be larger than demand and there will be a weekly surplus. Similarly, if you charge less—say \$0.30 per can—then the supply will be less than the demand and there will be a shortage of Hot'n'Spicy beans.

Before we go on . . . We just saw that if you charge less than the equilibrium price there will be a shortage. If you were to raise your price toward the equilibrium, you would sell more items and increase revenue, since it is the supply equation—not the demand equation—that determines what you can sell below the equilibrium price. On the other hand, if you charge more than the equilibrium price, you will be left with a possibly costly surplus of unsold items (and will want to lower prices to reduce the surplus). Prices tend to move toward the equilibrium, so supply tends to equal demand. When supply equals demand, we say that the market **clears.**

Supply Function and Equilibrium Price

A **supply equation** or **supply function** expresses supply q (the number of items a supplier is willing to make available) as a function of the unit price p (the price per item). A **linear supply function** has the form

$$q(p) = mp + b$$

It is usually the case that supply increases as the unit price increases, so m is usually positive.

Demand and supply are said to be in **equilibrium** when demand equals supply. The corresponding values of p and q are called the **equilibrium price** and **equilibrium demand.** To find the equilibrium price, set demand equal to supply and solve for the unit price p. To find the equilibrium demand, evaluate the demand (or supply) function at the equilibrium price.

Change Over Time

Things around us change with time. There are many quantities, such as your income or the temperature in Honolulu, that it is natural to think of as functions of time.

Example 5 • Growth of Sales

The U.S. Air Force's satellite-based Global Positioning System (GPS) allows people with radio receivers to determine their exact location anywhere on Earth. The following table shows the estimated total sales of U.S.-made products that use the GPS:

Year	1994	2000
Sales ($ billions)	0.8	8.3

Data estimated from published graph.
SOURCE: U.S. Global Positioning System Industry Council/*New York Times*, March 5, 1996, p. D1.

a. Use these data to model total sales of GPS-based products as a linear function of time t measured in years since 1994. What is the significance of the slope?

b. Use the model to predict when sales of GPS-based products will reach $13.3 billion, assuming they continue to grow at the same rate.

Solution

a. First, notice that 1994 corresponds to $t = 0$ and 2000 to $t = 6$. Thus, we are given the coordinates of two points on the graph of sales s as a function of time t: $(0, 0.8)$ and $(6, 8.3)$. The slope is

$$m = \frac{s_2 - s_1}{t_2 - t_1} = \frac{8.3 - 0.8}{6 - 0} = \frac{7.5}{6} = 1.25$$

Using the point $(0, 0.8)$, we get

$$\begin{aligned} s &= mt + b \\ &= 1.25t + 0.8 - (1.25)(0) \\ &= 1.25t + 0.8 \end{aligned}$$

Notice that we calculated the slope as the ratio (change in sales)/(change in time). Thus, m is the *rate of change* of sales and is measured in units of sales per unit of time, or billions of dollars per year. In other words, to say that $m = 1.25$ is to say that sales are increasing by $1.25 billion per year.

b. Our model of sales as a function of time is

$$s = 1.25t + 0.8$$

Sales of GPS-based products will reach $13.3 billion when $s = 13.3$, or

$$13.3 = 1.25t + 0.8$$

Solving for t,

$$\begin{aligned} 1.25t &= 13.3 - 0.8 = 12.5 \\ t &= \frac{12.5}{1.25} = 10 \end{aligned}$$

In 2004 ($t = 10$), sales are predicted to reach $13.3 billion.

Example 6 • Velocity

You are driving down the Ohio Turnpike, watching the mileage markers to stay awake. Measuring time in hours after you see the 20-mile marker, you see the following markers each half hour:

Time (h)	0	0.5	1	1.5	2
Marker (mi)	20	47	74	101	128

Find your location s as a function of t, the number of hours you have been driving. (The number s is also called your **position** or **displacement.**)

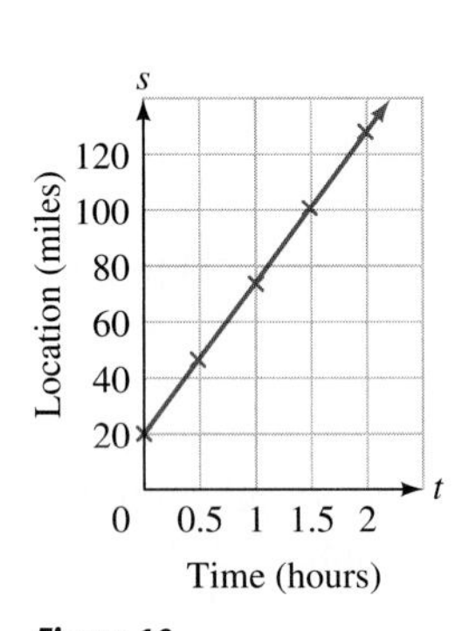

Figure 19

Solution If we plot the location s versus the time t, the five markers listed give us the graph in Figure 19. These points appear to lie along a straight line. We can verify this by calculating how far you traveled in each half hour. In the first half hour, you traveled $47 - 20 = 27$ miles. In the second half hour, you traveled $74 - 47 = 27$ miles also. In fact, you traveled exactly 27 miles each half hour. The points we plotted lie on a straight line that rises 27 units for every 0.5 unit we go to the right, for a slope of $27/0.5 = 54$.

To get the equation of that line, notice that we have the s intercept, which is the starting marker of 20. From the slope-intercept form (using s in place of y and t in place of x), we get

$$s(t) = 54t + 20$$

Before we go on . . . Notice the significance of the slope: For every hour you travel, you drive a distance of 54 miles. In other words, you are traveling at a constant velocity of 54 miles per hour. We have uncovered a very important principle:

In the graph of displacement versus time, velocity is given by the slope.

Linear Change Over Time

If a quantity q is a linear function of time t, so that

$$q(t) = mt + b$$

then the slope m measures the **rate of change** of q, and b is the quantity at time $t = 0$, the **initial quantity.** If q represents the position of a moving object, then the rate of change is also called the **velocity.**

Units of m and b

The units of measurement of m are units of q per unit of time; for instance, if q is income in dollars and t is time in years, then the rate of change m is measured in dollars per year.

The units of b are units of q; for instance, if q is income in dollars and t is time in years, then b is measured in dollars.

Quick Example

If the accumulated revenue from sales of your video-game software is given by $R(t) = 2000t + 500$ dollars, where t is time in years from now, then you have earned \$500 in revenue so far, and the accumulated revenue is increasing at a rate of \$2000 per year.

Examples 1–6 share the following common theme.

General Linear Models

If $y = mx + b$ is a linear model of changing quantities x and y, then the slope m is the rate at which y is increasing per unit increase in x, and the y intercept b is the value of y that corresponds to $x = 0$.

Units of m and b

The slope m is measured in units of y per unit of x, and the intercept b is measured in units of y.

Quick Example

If the number n of spectators at a soccer game is related to the number g of goals your team has scored so far by the equation $n = 20g + 4$, then you can expect 4 spectators if no goals have been scored and 20 additional spectators per additional goal scored.

Guideline: What to Use as x and y and How to Interpret a Linear Model

Question In a problem where I must find a linear relationship between two quantities, which quantity do I use as x and which do I use as y?

Answer The key is to decide which of the two quantities is the independent variable and which is the dependent variable. Then use the independent variable as x and the dependent variable as y. In other words, *y depends on x.*

Here are examples of phrases that convey this information, usually of the form *"Find y (dependent variable) in terms of x (independent variable)."*

Find the cost in terms of the number of items.	y = cost; x = number of items
How does color depend on wavelength?	y = color; x = wavelength

If no information is conveyed about which variable is intended to be independent, then you can use whichever is convenient.

Question How do I interpret a general linear model $y = mx + b$?

Answer The key to interpreting a linear model is to remember the units we use to measure m and b:

The slope m is measured in units of y per unit of x; the intercept b is measured in units of y.

For instance, if $y = 4.3x + 8.1$ and you know that x is measured in feet and y in kilograms, then you can already say, "y is 8.1 kilograms when $x = 0$ feet and increases at a rate of 4.3 kilograms per foot" without even knowing anything more about the situation!

1.4 EXERCISES

APPLICATIONS

1. **Cost** A piano manufacturer has a daily fixed cost of $1200 and a marginal cost of $1500 per piano. Find the cost $C(x)$ of manufacturing x pianos in one day. Use your function to answer the following questions:
   a. On a given day, what is the cost of manufacturing 3 pianos?
   b. What is the cost of manufacturing the third piano that day?
   c. What is the cost of manufacturing the eleventh piano that day?

2. **Cost** The cost of renting tuxes for the Choral Society's formal is $20 down, plus $88 per tux. Express the cost C as a function of x, the number of tuxedos rented. Use your function to answer the following questions.
   a. What is the cost of renting 2 tuxes?
   b. What is the cost of the second tux?
   c. What is the cost of the 4098th tux?
   d. What is the marginal cost per tux?

3. **Cost** The RideEm Bicycles factory can produce 100 bicycles in a day at a total cost of $10,500, and it can produce 120 bicycles in a day at a total cost of $11,000. What are the company's daily fixed costs, and what is the marginal cost per bicycle?

4. **Cost** A soft-drink manufacturer can produce 1000 cases of soda in a week at a total cost of $6000 and 1500 cases of soda at a total cost of $8500. Find the manufacturer's weekly fixed costs and marginal cost per case of soda.

5. **Break-Even Analysis** Your college newspaper, *The Collegiate Investigator,* has fixed production costs of $70 per edition and marginal printing and distribution costs of 40¢/copy. *The Collegiate Investigator* sells for 50¢/copy.
   a. Write down the associated cost, revenue, and profit functions.
   b. What profit (or loss) results from the sale of 500 copies of *The Collegiate Investigator*?
   c. How many copies should be sold in order to break even?

6. **Break-Even Analysis** The Audubon Society at Enormous State University (ESU) is planning its annual fund-raising "Eatathon." The society will charge students 50¢/serving of pasta. The only expenses the society will incur are the cost of the pasta, estimated at 15¢/serving, and the $350 cost of renting the facility for the evening.
   a. Write down the associated cost, revenue, and profit functions.
   b. How many servings of pasta must the Audubon Society sell in order to break even?
   c. What profit (or loss) results from the sale of 1500 servings of pasta?

7. **Demand** Sales figures show that your company sold 1960 pen sets each week when they were priced at $1/pen set and 1800 pen sets each week when they were priced at $5/pen set. What is the linear demand function for your pen sets?

8. **Demand** A large department store is prepared to buy 3950 of your neon-colored shower curtains per month for $5 each, but only 3700 shower curtains per month for $10 each. What is the linear demand function for your neon-colored shower curtains?

9. **Demand for Microprocessors** The following table shows the worldwide quarterly sales of microprocessors and average wholesale price:

Year	1997 (2nd quarter)	1997 (3rd quarter)	1998 (1st quarter)
Wholesale Price ($)	235	215	210
Sales (millions)	21	24	23

Source: International Data Corporation/*New York Times,* "Computer Economics 101," June 9, 1998, p. D6. Data are rounded.

Use the 1997 second- and third-quarter data to obtain a linear demand function for microprocessors. Is the 1998 first-quarter data consistent with the demand equation? Explain.

10. **Demand for Microprocessors** Referring to the data in Exercise 9, use the 1997 second-quarter data and the 1998 first-quarter data to obtain a linear demand function for microprocessors. Use your model to give an estimate of worldwide sales if the price changes to $215. Can all three data points be represented by the same linear demand function?

11. **Equilibrium Price** You can sell 90 Chia pets each week if they are marked at $1/Chia but only 30 each week if they are marked at $2/Chia. Your Chia supplier is prepared to sell you 20 Chias each week if they are marked at $1/Chia and 100 each week if they are marked at $2/Chia.
    a. Write down the associated linear demand and supply functions.
    b. At what price should the Chias be marked so that there is neither a surplus nor a shortage of Chias?

12. **Equilibrium Price** The demand for your college newspaper is 2000 copies each week if the paper is given away free of charge and drops to 1000 each week if the charge is 10¢/copy. However, the university is only prepared to supply 600 copies each week free of charge but will supply 1400 each week at 20¢/copy.
    a. Write down the associated linear demand and supply functions.
    b. At what price should the college newspapers be sold so that there is neither a surplus nor a shortage of papers?

13. **Swimming Pool Sales** The following graph shows approximate annual sales of new in-ground swimming pools in the United States:

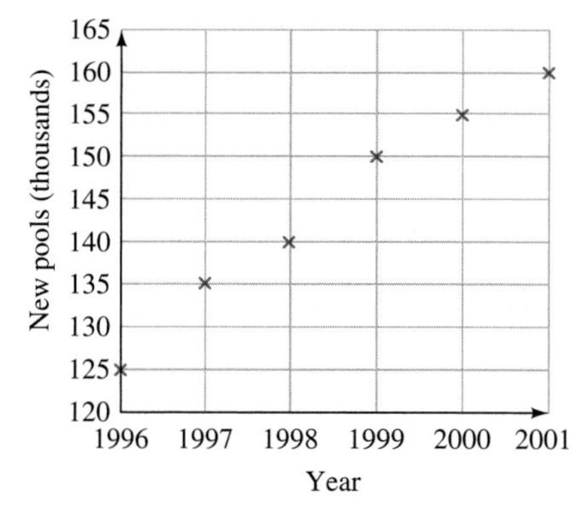

SOURCE: PK Data/*New York Times,* July 5, 2001, p. C1. 2001 figure was estimated.

a. Find two points on the graph such that the slope of the line segment joining them is the largest possible.

b. What does your answer tell you about swimming pool sales?

14. Swimming Pool Sales Repeat Exercise 13, using the following graph for sales of new above-ground swimming pools in the United States.

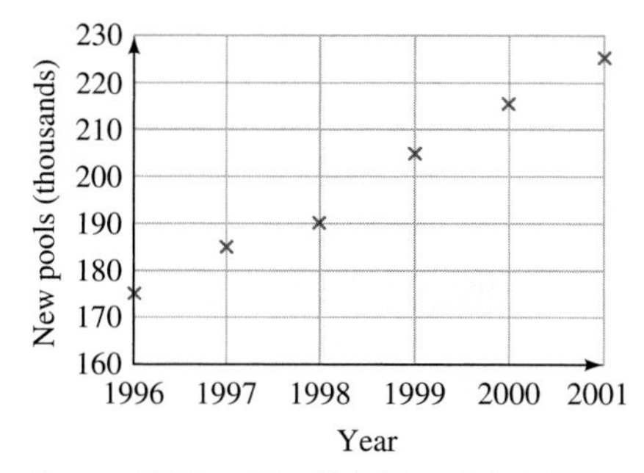

SOURCE: PK Data/*New York Times,* July 5, 2001, p. C1. 2001 figure was estimated.

15. Online Shopping The number of online shopping transactions in the United States increased from 350 million transactions in 1999 to 450 million transactions in 2001.[29] Find a linear model for the number N of online shopping transactions in year t, with $t = 0$ corresponding to 2000. What is the significance of the slope? What are its units?

16. Online Shopping The percentage of people in the United States who have ever purchased anything online increased from 20% in January 2000 to 35% in January 2002.[30] Find a linear model for the percentage P of people in the United States who have ever purchased anything online in year t, with $t = 0$ corresponding to January 2000. What is the significance of the slope? What are its units?

17. Medicare Spending Annual federal spending on Medicare (in constant 2000 dollars) was projected to increase from \$240 billion in 2000 to \$600 billion in 2025.[31]

a. Use this information to express s, the annual spending on Medicare (in billions of dollars), as a linear function of t, the number of years since 2000. How fast is Medicare predicted to rise in the coming years?

b. Use your model to predict Medicare spending in 2040, assuming the spending trend continues.

18. Pasta Imports During the period 1990–2001, U.S. imports of pasta increased from 290 million pounds in 1990 ($t = 0$) by an average of 40 million pounds/year.[32]

a. Use these data to express q, the annual U.S. imports of pasta (in millions of pounds), as a linear function of t, the number of years since 1990.

b. Use your model to estimate U.S. pasta imports in 2005, assuming the import trend continues.

19. Velocity The position of a model train, in feet along a railroad track, is given by

$$s(t) = 2.5t + 10$$

after t seconds.

a. How fast is the train moving?

b. Where is the train after 4 seconds?

c. When will it be 25 feet along the track?

20. Velocity The height of a falling sheet of paper, in feet from the ground, is given by

$$s(t) = -1.8t + 9$$

after t seconds.

a. What is the velocity of the sheet of paper?

b. How high is it after 4 seconds?

c. When will it reach the ground?

21. Fast Cars A police car was traveling down Ocean Parkway in a high-speed chase from Jones Beach. It was at Jones Beach at exactly 10 P.M. ($t = 10$) and was at Oak Beach, 13 miles from Jones Beach, at exactly 10:06 P.M.

a. How fast was the police car traveling?

b. How far was the police car from Jones Beach at time t?

22. Fast Cars The car that was being pursued by the police in Exercise 21 was at Jones Beach at exactly 9:54 P.M. ($t = 9.9$) and passed Oak Beach (13 miles from Jones Beach) at exactly 10:06 P.M., where it was overtaken by the police.

a. How fast was the car traveling?

b. How far was the car from Jones Beach at time t?

[29]Second half of 2001 data was estimated. SOURCE: Odyssey Research/*New York Times,* November 5, 2001, p. C1.

[30]Ibid. January 2002 data was estimated.

[31]Data are rounded. SOURCE: The Urban Institute's Analysis of the 1999 Trustee's Report, http://www.urban.org.

[32]Data are rounded. SOURCES: Department of Commerce/*New York Times,* September 5, 1995, p. D4; International Trade Administration, http://www.ita.doc.gov/, March 31, 2002.

23. **Fahrenheit and Celsius** In the Fahrenheit temperature scale, water freezes at 32°F and boils at 212°F. In the Celsius scale, water freezes at 0°C and boils at 100°C. Assuming that the Fahrenheit temperature F and the Celsius temperature C are related by a linear equation, find F in terms of C. Use your equation to find the Fahrenheit temperatures corresponding to 30°C, 22°C, −10°C, and −14°C, to the nearest degree.

24. **Fahrenheit and Celsius** Use the information about Celsius and Fahrenheit given in Exercise 23 to obtain a linear equation for C in terms of F and use your equation to find the Celsius temperatures corresponding to 104°F, 77°F, 14°F, and −40°F, to the nearest degree.

25. **Income** The well-known romance novelist Celestine A. Lafleur (a.k.a. Bertha Snodgrass) has decided to sell the screen rights to her latest book, *Henrietta's Heaving Heart,* to Boxoffice Success Productions for $50,000. In addition, the contract ensures Ms. Lafleur royalties of 5% of the net profits.[33] Express her income I as a function of the net profit N and determine the net profit necessary to bring her an income of $100,000. What is her marginal income (share of each dollar of net profit)?

26. **Income** Due to the enormous success of the movie *Henrietta's Heaving Heart* based on a novel by Celestine A. Lafleur (see Exercise 25), Boxoffice Success Productions decides to film the sequel, *Henrietta, Oh Henrietta.* At this point, Bertha Snodgrass (whose novels now top the best-seller lists) feels she is in a position to demand $100,000 for the screen rights and royalties of 8% of the net profits. Express her income I as a function of the net profit N and determine the net profit necessary to bring her an income of $1,000,000. What is her marginal income (share of each dollar of net profit)?

27. **Newspaper Circulation** The following table gives the average daily circulation of all the newspapers of the Gannett Company and Newhouse Newspapers, Inc., two major newspaper companies.

	Gannett	Newhouse
Number of Newspapers	82	26
Daily Circulation (millions)	5.8	3.0

Circulation figures are rounded to the nearest 0.1 million and reflect average daily circulation as of September 30, 1994. SOURCE: Newspaper Association of America/*New York Times,* July 30, 1995, p. E6.

a. Use these data to express a company's daily circulation in millions c as a linear function of the number n of newspapers it publishes.
b. What information does the slope give about newspaper publishers?
c. Gannett Company was planning to add 11 new newspapers in 1995. According to your model, how would this affect daily circulation?

28. **Newspaper Circulation** The following table gives the average daily circulation of all the newspapers of Knight Ridder and Dow Jones, two major publishers of newspapers.

	Knight Ridder	Dow Jones
Number of Newspapers	27	22
Daily Circulation (millions)	3.6	2.4

Circulation figures are rounded to the nearest 0.1 million and reflect average daily circulation as of September 30, 1994. SOURCE: Newspaper Association of America/*New York Times,* July 30, 1995, p. E6.

a. Use these data to express a company's daily circulation c as a linear function of the number n of newspapers it publishes.
b. The Times Mirror Company publishes 11 newspapers and has a daily circulation of 2.6 million. To what extent is this consistent with the model?
c. Find a domain for your model that is consistent with a circulation of between 0 and 4.8 million.

Exercises 29–32 are based on the following data, showing the total amount of milk and cheese, in billions of pounds, produced in 13 western and 12 north-central states in 1999 and 2000.

Milk

Year	1999	2000
Western States	56	60
North-Central States	57	59

SOURCE: Department of Agriculture/*New York Times,* June 28, 2001, p. C1. Figures are approximate.

Cheese

Year	1999	2000
Western States	2.7	3.0
North-Central States	3.9	4.0

SOURCE: Department of Agriculture/*New York Times,* June 28, 2001, p. C1. Figures are approximate.

29. Use the data from both years to find a linear relationship giving the amount of milk w produced in the western states as a function of the amount of milk n produced in north-central states. Use your model to predict how much milk western states will produce if north-central states produce 50 billion pounds.

30. Use the data from both years to find a linear relationship giving the amount of cheese w produced in western states as a function of the amount of cheese n produced in north-central states. Use your model to predict how much cheese western states will produce if north-central states produce 3.4 billion pounds.

[33] Percentages of net profit are commonly called "monkey points." Few movies ever make a net profit on paper, and anyone with any clout in the business gets a share of the *gross,* not the net.

31. Use the data to model cheese production c in western states as a function of milk production m in western states. According to the model, how many pounds of cheese are produced for every 10 pounds of milk?

32. Repeat Exercise 31 for north-central states.

33. Biology The Snowtree cricket behaves in a rather interesting way: The rate at which it chirps depends linearly on the temperature. One summer evening you hear a cricket chirping at a rate of 140 chirps/minute, and you notice that the temperature is 80°F. Later in the evening, the cricket has slowed to 120 chirps/minute, and you notice that the temperature has dropped to 75°F. Express the temperature T as a linear function of the cricket's rate of chirping r. What is the temperature if the cricket is chirping at a rate of 100 chirps/minute?

34. Muscle Recovery Time Most workout enthusiasts will tell you that muscle recovery time is about 48 hours. But it is not quite as simple as that; the recovery time ought to depend on the number of sets you do involving the muscle group in question. For example, if you do no sets of biceps exercises, then the recovery time for your biceps is (of course) zero. To take a compromise position, let's assume that if you do three sets of exercises on a muscle group then its recovery time is 48 hours. Use these data to write a linear function that gives the recovery time (in hours) in terms of the number of sets affecting a particular muscle. Use this model to calculate how long it would take your biceps to recover if you did 15 sets of curls. Comment on your answer with reference to the usefulness of a linear model.

35. Profit Analysis—Aviation The operating cost of a Boeing 747-100, which seats up to 405 passengers, is estimated to be \$5132 per hour.[34] If an airline charges each passenger a fare of \$100 per hour of flight, find the hourly profit P it earns operating a 747-100 as a function of the number of passengers x (be sure to specify the domain). What is the least number of passengers it must carry in order to make a profit?

36. Profit Analysis—Aviation The operating cost of a McDonnell Douglas DC 10-10, which seats up to 295 passengers, is estimated to be \$3885 per hour.[35] If an airline charges each passenger a fare of \$100 per hour of flight, find the hourly profit P it earns operating a DC 10-10 as a function of the number of passengers x (be sure to specify the domain). What is the least number of passengers it must carry in order to make a profit?

37. Break-Even Analysis (based on a question from a CPA exam) The Oliver Company plans to market a new product. Based on its market studies, Oliver estimates that it can sell up to 5500 units in 2005. The selling price will be \$2 per unit. Variable costs are estimated to be 40% of total revenue. Fixed costs are estimated to be \$6000 for 2005. How many units should the company sell to break even?

38. Break-Even Analysis (based on a question from a CPA exam) The Metropolitan Company sells its latest product at a unit price of \$5. Variable costs are estimated to be 30% of the total revenue, and fixed costs amount to \$7000 per month. How many units should the company sell each month in order to break even, assuming that it can sell up to 5000 units per month at the planned price?

39. Break-Even Analysis (from a CPA exam) Given the following notations, write a formula for the break-even sales level.

SP = selling price per unit

FC = total fixed cost

VC = variable cost per unit

40. Break-Even Analysis (based on a question from a CPA exam) Given the following notation, give a formula for the total fixed cost.

SP = selling price per unit

VC = variable cost per unit

BE = break-even sales level in units

41. Break-Even Analysis—Organized Crime The organized crime boss and perfume king, Butch (Stinky) Rose, has daily overheads (bribes to corrupt officials, motel photographers, wages for hit men, explosives, etc.) amounting to \$20,000/day. On the other hand, he has a substantial income from his counterfeit perfume racket: He buys imitation French perfume (Chanel N° 22.5) at \$20/gram, pays an additional \$30/100 grams for transportation, and sells it via his street thugs for \$600/gram. Specify Stinky's profit function $P(x)$, where x is the quantity (in grams) of perfume he buys and sells, and use your answer to calculate how much perfume should pass through his hands each day so that he breaks even.

42. Break-Even Analysis—Disorganized Crime Lately, Butch (Stinky) Rose's counterfeit Chanel N° 22.5 racket has run into difficulties; it seems that the *authentic* Chanel N° 22.5 perfume is selling at only \$500/gram, whereas his street thugs have been selling the counterfeit perfume for \$600/gram, and his costs are \$400/gram plus \$30/gram transportation costs and commission. (The perfume's smell is easily detected by specially trained Chanel Hounds, and this necessitates elaborate packaging measures.) He therefore decides to price it at \$420/gram in order to undercut the competition. Specify Stinky's profit function $P(x)$, where x is the quantity (in grams) of perfume he buys and sells, and use your answer to calculate how much perfume should pass through his hands each day so that he breaks even. Interpret your answer.

43. Television Advertising The cost, in millions of dollars, of a 30-second television ad during the Super Bowl in the years 1990 to 2001 can be approximated by the following piecewise linear function ($t = 0$ represents 1990):[36]

$$C(t) = \begin{cases} 0.08t + 0.6 & \text{if } 0 \le t < 8 \\ 0.355t - 1.6 & \text{if } 8 \le t \le 11 \end{cases}$$

How fast and in what direction was the cost of an ad during the Super Bowl changing in 1999?

[34] In 1992. SOURCE: Air Transportation Association of America.

[35] Ibid.

[36] SOURCE: *New York Times*, January 26, 2001, p. C1.

44. Semiconductors The percentage $J(t)$ of semiconductor equipment manufactured in Japan during the period 1981–1994 can be approximated by the following function of time t in years since 1980:[37]

$$J(t) = \begin{cases} 17 + 3.1t & \text{if } 1 \le t \le 10 \\ 68 - \ \ 2t & \text{if } 10 < t \le \end{cases}$$

How fast and in what direction was the percentage of Japan's contribution to the semiconductor market changing in 1986?

45. Internet Sales The following table shows the percentage of buyers of new cars who used the Internet for research or purchase between 1997 and 2000 ($t = 0$ represents 1997):

Year, t	0	1	3
Percentage, p	15	25	55

SOURCE: J. D. Power Associates/*New York Times,* January 25, 2000, p. C1. 2000 figure is an estimate.

a. Model the percentage p as a linear function of t from 1997 ($t = 0$) to 1998 ($t = 1$).
b. Model p as a linear function of time t from 1998 to 2000.
c. Use the answers you obtained in parts (a) and (b) to model p as a piecewise-defined function of t from 1997 to 2000.
d. According to the model, what percentage of buyers of new cars used the Internet for research or purchase in 1999?

46. Internet Sales The following table shows the percentage of buyers of new cars who used the Internet for purchase (including referrals to dealers) between 1997 and 2000 ($t = 0$ represents 1997):

Year, t	0	1	3
Percentage, p	0	1	5

SOURCE: J. D. Power Associates/*New York Times,* January 25, 2000, p. C1. 2000 figure was estimated.

a. Model the percentage p as a linear function of t from 1997 ($t = 0$) to 1998.
b. Model p as a linear function of time t from 1998 to 2000.
c. Use the answers you obtained in parts (a) and (b) to model p as a piecewise-defined function of t from 1997 to 2000.
d. According to the model, what percentage of buyers of new cars used the Internet for purchase in 1999?

47. Career Choices In 1989 approximately 30,000 college-bound high school seniors intended to major in computer and information sciences. This number decreased to approximately 23,000 in 1994 and rose to 60,000 in 1999.[38] Model this number C as a piecewise linear function of the time t in years since 1989 and use your model to estimate the number of college-bound high school seniors who intended to major in computer and information sciences in 1992.

48. Career Choices In 1989 approximately 100,000 college-bound high school seniors intended to major in engineering. This number decreased to approximately 85,000 in 1994 and rose to 88,000 in 1999.[39] Model this number E as a piecewise linear function of the time t in years since 1989 and use your model to estimate the number of college-bound high school seniors who intended to major in engineering in 1995.

49. Divorce Rates A study found that the divorce rate d appears to depend on the ratio r of available men to available women.[40] When the ratio was 1.3 (130 available men per 100 available women) the divorce rate was 22%. It rose to 35% if the ratio grew to 1.6 and rose to 30% if the ratio dropped to 1.1. Model these data by expressing d as a piecewise linear function of r and extrapolate your model to estimate the divorce rate if there are the same number of available men as women.

50. Retirement In 1950 the number N of retirees was approximately 150 per 1000 people aged 20–64. In 1990 this number rose to approximately 200 and is projected to rise to 275 in 2020.[41] Model N as a piecewise linear function of the time t in years since 1950 and use your model to project the number of retirees per 1000 people aged 20–64 in 2010.

COMMUNICATION AND REASONING EXERCISES

51. If y is measured in bootlags[42] and x is measured in Martian yen and $y = mx + b$, then m is measured in ______, and b is measured in ______.

52. If the slope in a linear relationship is measured in miles per dollar, then the independent variable is measured in ______, and the dependent variable is measured in ______.

53. The velocity of an object is given by $v = 0.1t + 20$ m/sec., where t is time in seconds. The object is

(A) moving with fixed speed
(B) accelerating
(C) decelerating
(D) impossible to say from the given information

[37]SOURCE: VLSI Research/*New York Times,* October 9, 1994, sec. 3, p. 2.

[38]SOURCE: The College Board; National Science Foundation/*New York Times,* September 2, 1999, p. C1.

[39]Ibid.

[40]The cited study, by Scott J. South and associates, appeared in the *American Sociological Review* (February 1995). Figures are rounded. SOURCE: *New York Times,* February 19, 1995, p. 40.

[41]SOURCE: Social Security Administration/*New York Times,* April 4, 1999, p. WK3.

[42]An ancient Martian unit of length; 1 bootlag is the mean distance from a Martian's foreleg to its hindleg.

54. The position of an object is given by $x = 0.2t - 4$, where t is time in seconds. The object is

(A) moving with fixed speed **(B)** accelerating

(C) decelerating **(D)** impossible to say from the given information

55. If a quantity is changing linearly with time and it increases by 10 units in the first day, what can you say about its behavior in the third day?

56. The quantities Q and T are related by a linear equation of the form

$$Q = mT + b$$

When $T = 0$, Q is positive but decreases to a negative quantity when T is 10. What are the signs of m and b? Explain your answers.

57. Suppose the cost function is $C(x) = mx + b$ (with m and b positive), the revenue function is $R(x) = kx$ ($k > m$), and the number of items is increased from the break-even quantity. Does this result in a loss or a profit, or is it impossible to say? Explain your answer.

58. You have been constructing a demand equation and have obtained a (correct) expression of the form $p = mq + b$, whereas you would have preferred one of the form $q = mp + b$. Should you simply switch p and q in the answer; should you start again from scratch, using p in the role of x and q in the role of y; or should you solve your demand equation for q? Give reasons for your decision.

1.5 Linear Regression

We have seen how to find a linear model given two data points: We find the equation of the line that passes through them. However, we often have more than two data points, and they will rarely all lie on a single straight line but may often come close to doing so. The problem is to find the line coming *closest* to passing through all of the points.

Suppose, for example, we are conducting research for a cable TV company interested in expanding into China and we come across the following figures showing the growth of the cable market there:

Year, t ($t = 0$ represents 2000)	−4	−3	−2	−1	0	1	2	3
Households with Cable, y (millions)	50	55	57	60	68	72	80	83

Data are approximate, and the 2001–2003 figures are estimated. SOURCES: HSBC Securities, Bear Stearns/*New York Times,* March 23, 2001, p. C1.

A plot of these data suggests a roughly linear growth of the market (Figure 20).

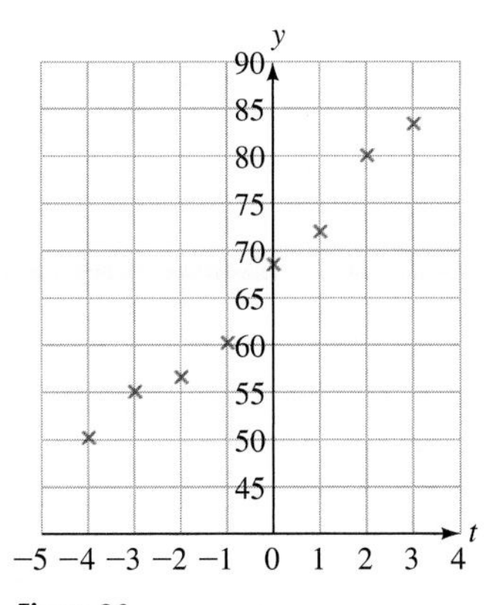

Figure 20

These points suggest a line, although they clearly do not all lie on a single straight line. Figure 21 shows the points together with several lines, some fitting better than others.

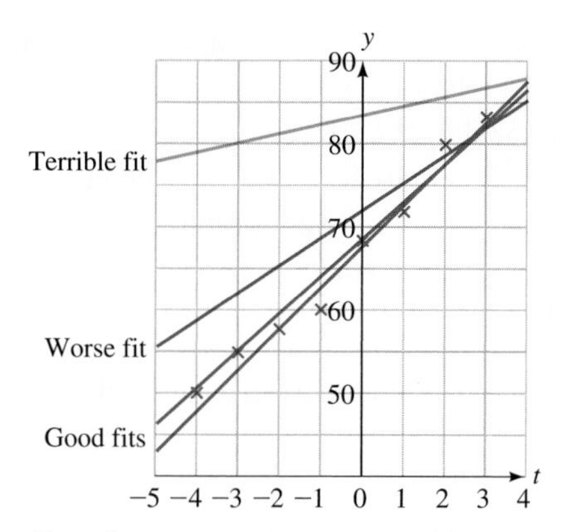

Figure 21

Question Can we precisely measure which lines fit better than others? For instance, which of the two lines labeled as "good fits" in Figure 21 models the data more accurately?

Answer We begin by considering, for each of the years 1996 through 2003, the difference between the actual number of households with cable (the **observed value**) and the number of households with cable predicted by a linear equation (the **predicted value**). The difference between the predicted value and the observed value is called the **residue:**

$$\text{residue} = \text{observed value} - \text{predicted value}$$

On the graph the residues measure the vertical distances between the (observed) data points and the line (Figure 22), and they tell us how far the linear model is from predicting the number of households with cable.

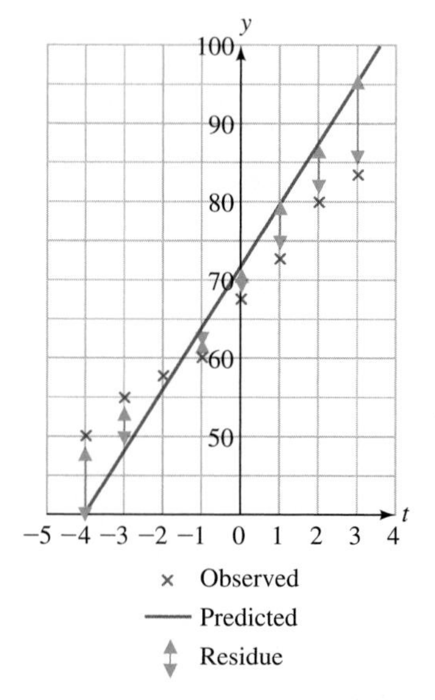

Figure 22

The more accurate our model, the smaller the residues should be. We can combine all the residues into a single measure of accuracy by adding their *squares.* (We square the residues in part to make them all positive.) The sum of the squares of the residues is called the **sum-of-squares error, SSE.**[43] Smaller values of SSE indicate more accurate models.

Here are some definitions and formulas for what we have been discussing.

Observed Values, Predicted Values, Residues, and Sum-of-Squares Error (SSE)

Observed and Predicted Values

Suppose we are given a collection of data points $(x_1, y_1), \ldots, (x_n, y_n)$. The n quantities $y_1, y_2, \ldots, y_n$ are called the **observed *y* values.** If we model these data with a linear equation

$$\hat{y} = mx + b \qquad \hat{y} \text{ stands for "estimated } y\text{" or "predicted } y\text{."}$$

then the y values we get by substituting the given x values into the equation are called the **predicted *y* values:**

$$\hat{y}_1 = mx_1 + b \qquad \text{Substitute } x_1 \text{ for } x.$$
$$\hat{y}_2 = mx_2 + b \qquad \text{Substitute } x_2 \text{ for } x.$$
$$\vdots$$
$$\hat{y}_n = mx_n + b \qquad \text{Substitute } x_n \text{ for } x.$$

Quick Example

Consider the three data points (0, 2), (2, 5), and (3, 6). The observed y values are $y_1 = 2$, $y_2 = 5$, and $y_3 = 6$. If we model these data with the equation $\hat{y} = x + 2.5$, then the predicted values are

$$\hat{y}_1 = x_1 + 2.5 = 0 + 2.5 = 2.5$$
$$\hat{y}_2 = x_2 + 2.5 = 2 + 2.5 = 4.5$$
$$\hat{y}_3 = x_3 + 2.5 = 3 + 2.5 = 5.5$$

Residues and Sum-of-Squares Error (SSE)

If we model a collection of data $(x_1, y_1), \ldots, (x_n, y_n)$ with a linear equation $\hat{y} = mx + b$, then the **residues** are the n quantities (actual value – predicted value):

$$(y_1 - \hat{y}_1), (y_2 - \hat{y}_2), \ldots, (y_n - \hat{y}_n)$$

The **sum-of-squares error (SSE)** is the sum of the squares of the residues:

$$\text{SSE} = (y_1 - \hat{y}_1)^2 + (y_2 - \hat{y}_2)^2 + \cdots + (y_n - \hat{y}_n)^2$$

Quick Example

For the data and linear approximation given above, the residues are

$$y_1 - \hat{y}_1 = 2 - 2.5 = -0.5$$
$$y_2 - \hat{y}_2 = 5 - 4.5 = 0.5$$
$$y_3 - \hat{y}_3 = 6 - 5.5 = 0.5$$
$$\text{SSE} = (-0.5)^2 + (0.5)^2 + (0.5)^2 = 0.75$$

---

[43]Why not add the absolute values of the residues instead? Here is one reason: Take any two numbers, say, 1 and 3. Their average, 2, has the property that the sum of the squares of the residues, $(\bar{x} - 1)^2 + (\bar{x} - 3)^2$, is as small as possible when $\bar{x} = 2$; other values for $\bar{x}$ would result in a larger sum. However, all values of $\bar{x}$ between 1 and 3 result in the *same* value for the sum of the absolute values of the residues, $|\bar{x} - 1| + |\bar{x} - 3| = 2$.

Example 1 • Computing the Sum-of-Squares Error

Using the preceding data on the cable TV market in China, compute SSE for the linear models $y = 8t + 72$ and $y = 5t + 68$. Which model is the better fit?

Solution We begin by creating a table showing the values of t, the observed (given) values of y, and the values predicted by the first model:

Year, t	Observed, y	Predicted, $\hat{y} = 8t + 72$
−4	50	40
−3	55	48
−2	57	56
−1	60	64
0	68	72
1	72	80
2	80	88
3	83	96

We now add two new columns for the residues and their squares:

Year, t	Observed, y	Predicted, $\hat{y} = 8t + 72$	Residue, $y - \hat{y}$	Residue², $(y - \hat{y})^2$
−4	50	40	$50 - 40 = 10$	$10^2 = 100$
−3	55	48	$55 - 48 = 7$	$7^2 = 49$
−2	57	56	$57 - 56 = 1$	$1^2 = 1$
−1	60	64	$60 - 64 = -4$	$(-4)^2 = 16$
0	68	72	$68 - 72 = -4$	$(-4)^2 = 16$
1	72	80	$72 - 80 = -8$	$(-8)^2 = 64$
2	80	88	$80 - 88 = -8$	$(-8)^2 = 64$
3	83	96	$83 - 96 = -13$	$(-13)^2 = 169$

SSE, the sum of the squares of the residues, is then the sum of the entries in the last column,

$$\text{SSE} = 479$$

Repeating the process using the second model, $y = 5t + 68$, yields SSE = 23. Thus, the second model is a better fit.

Graphing Calculator

We can use the List feature in the TI-83 to automate the computation of SSE. Press STAT `EDIT` to obtain the list screen, where you can enter the given data in the first two columns, called L_1 and L_2. (If there is already data in a column you want to use, you can clear it by highlighting the column heading—for example, L_3—using the arrow key, and pressing CLEAR ENTER.) To compute the predicted values, highlight the heading L_3 using the arrow keys and press ENTER, followed by the formula for the predicted values:

`8*L1+72` L_1 is [2nd] [1].

L_1	L_2	L_3
-4	50	
-3	55	
-2	57	
-1	60	
0	68	
1	72	
2	80	
3	83	

L_3=8*L_1+72

Pressing [ENTER] again will fill column 3 with the predicted values. Next, highlight the heading L_4, press [ENTER], and enter

(L_2−L_3)^2 Squaring the residues

L_1	L_2	L_3	L_4
-4	50	40	
-3	55	48	
-2	57	56	
-1	60	64	
0	68	72	
1	72	80	
2	80	88	
3	83	96	

L_4=(L_2−L_3)^2

Pressing [ENTER] again will fill column 4 with the squares of the residues. To compute SSE, the sum of the entries in L_4, go to the home screen and enter sum(L_4). The sum function is found by pressing [2nd] [LIST] and selecting MATH.

Excel

In Excel begin by setting up your worksheet with the observed data in two columns, t and y, and the predicted data for the first model in the third:

	A	B	C	D	E	F
1	t	y (Observed)	y (Predicted)		m	b
2	-4	50	=$E$2*A2+$F$2		8	72
3	-3	55				
4	-2	57				
5	-1	60				
6	0	68				
7	1	72				
8	2	80				
9	3	83				

Notice that, instead of using the numerical equation for the first model in column C, we used absolute references to the cells containing the slope m and the intercept b. This way, we can switch from one linear model to the next by changing only m and b in cells E2 and F2. (We have deliberately left column D empty in anticipation of the next step.)

In column D we compute the squares of the residues using the Excel formula

`=(B2−C2)^2`

Here is the completed worksheet, with SSE in cell F4:

	A	B	C	D	E	F
1	t	y (Observed)	y (Predicted)	Residue^2	m	b
2	-4	50	40	=(B2-C2)^2	8	72
3	-3	55	48			
4	-2	57	56		SSE:	=SUM(D2:D9)
5	-1	60	64			
6	0	68	72			
7	1	72	80			
8	2	80	88			
9	3	83	96			

As we have seen, SSE $= 479$. Changing m to 5 and b to 68 gives SSE for the second model, SSE $= 23$.

Before we go on . . . You can use your preferred mode of technology to plot the original data points together with the two lines.

To plot the points using the TI-83, you need to turn `PLOT1` on in the `STAT PLOT` window, obtained by pressing [2nd] [STAT PLOT]. To show the lines, enter them in the `Y=` screen as usual. To obtain a convenient window showing all the points and the lines, press [ZOOM] and choose option 9: `ZoomStat`.

In Excel use a scatter plot to graph the data in columns A through C in the worksheet discussed above. Figure 23 shows the two models with the original data points, as shown in Excel.

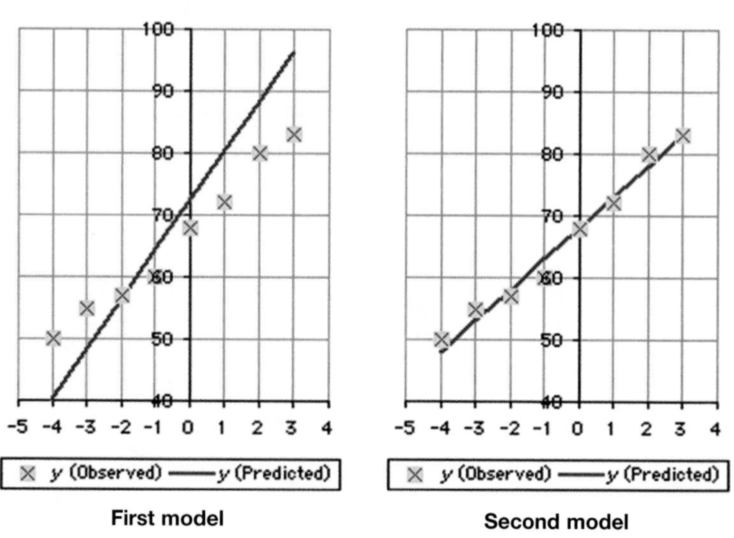

Figure 23

Question It seems clear from the figure that the second model gives a better fit. Why bother to compute SSE to tell me this?

Answer The difference between the two models we chose is so dramatic that it is clear from the graph which is the better fit. However, if we used a third model with $m = 5$ and $b = 68.1$, then its graph would be almost indistinguishable from that of the second but a better fit as measured by SSE $= 22.88$.

Among all possible lines, there ought to be one with the least possible value of SSE—that is, the greatest possible accuracy as a model. The line (and there is only one such line) that minimizes the sum of the squares of the residues is called the **regression line,** the **least-squares line,** or the **best-fit line.**

To find the regression line, we need a way to find values of m and b that give the smallest possible value of SSE. As an example, let's take the second linear model in Example 1. We said in the "Before we go on" discussion that increasing b from 68 to 68.1 had the desirable effect of decreasing SSE from 23 to 22.88. We could then decrease m to 4.9, further reducing SSE to 22.2. Imagine this as a kind of game: Alternately alter the values of m and b by small amounts until you find the lowest possible value of SSE. This works but is extremely tedious and time-consuming.

Question Is there an algebraic way to find the regression line?

Answer Yes. Here is the calculation. To justify it rigorously requires calculus of several variables or linear algebra.

Regression Line

The **regression line (least-squares line, best-fit line)** associated with the points $(x_1, y_1), (x_2, y_2), \ldots, (x_n, y_n)$ is the line that gives the minimum SSE. The regression line is

$$y = mx + b$$

where m and b are computed as follows:

$$m = \frac{n(\Sigma xy) - (\Sigma x)(\Sigma y)}{n(\Sigma x^2) - (\Sigma x)^2}$$

$$b = \frac{\Sigma y - m(\Sigma x)}{n}$$

n = number of data points

Here, Σ means "the sum of." For example,

$$\Sigma x = \text{sum of the } x \text{ values} = x_1 + x_2 + \cdots + x_n$$

$$\Sigma xy = \text{sum of products} = x_1y_1 + x_2y_2 + \cdots + x_ny_n$$

$$\Sigma x^2 = \text{sum of the squares of the } x \text{ values} = x_1^2 + x_2^2 + \cdots + x_n^2$$

On the other hand,

$$(\Sigma x)^2 = \text{square of } \Sigma x = \text{square of the sum of the } x \text{ values}$$

Example 2 • Computing a Regression Line

Find the line that best fits the following data:

x	1	2	3	4
y	1.5	1.6	2.1	3.0

Solution The best-fit line is the regression line, which we can compute by hand or by using spreadsheet technology.

Let's organize our work in the form of a table, where the original data are entered in the first two columns and the bottom row contains the column sums:

	x	y	xy	x^2
	1	1.5	1.5	1
	2	1.6	3.2	4
	3	2.1	6.3	9
	4	3.0	12.0	16
Σ (sum)	10	8.2	23.0	30

Thus, $\Sigma x = 10$, $\Sigma y = 8.2$, $\Sigma xy = 23.0$, and $\Sigma x^2 = 30$. Since there are $n = 4$ data points, we get

$$m = \frac{n(\Sigma xy) - (\Sigma x)(\Sigma y)}{n(\Sigma x^2) - (\Sigma x)^2} = \frac{4(23) - (10)(8.2)}{4(30) - 10^2} = 0.5$$

$$b = \frac{\Sigma y - m(\Sigma x)}{n} = \frac{8.2 - (0.5)(10)}{4} = 0.8$$

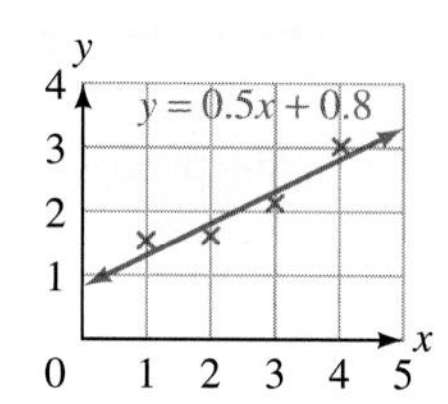

Figure 24

So, the regression line is

$$y = 0.5x + 0.8$$

The data points and the best-fit line are shown in Figure 24.

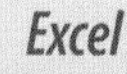

Excel

The table we used above strongly suggests a spreadsheet. The following worksheet duplicates the table and includes the formulas for m and b in row 6:

	A	B	C	D	E	F
1	x	y	xy	x^2	m	b
2	1	1.5	=A2*B2	=A2^2	=(4*C6-A6*B6)/(4*D6-A6^2)	=(B6-E2*A6)/4
3	2	1.6				
4	3	2.1				
5	4	3				
6	=SUM(A2:A5)					

Question Can we find the regression line using technology without having to use any formulas?

Answer Yes. Most computational technologies, including Excel, have built-in utilities that compute regression lines without the need for us to do any computation, as we show in Example 3.

Example 3 • Obtaining the Regression Line without Computation

In Example 1 we considered the following data about the growth of the cable TV market in China:

Year, t ($t = 0$ represents 2000)	−4	−3	−2	−1	0	1	2	3
Households with Cable, y (millions)	50	55	57	60	68	72	80	83

Find the best-fit linear model for these data and use the model to predict the number of Chinese households with cable in 2005.

Solution

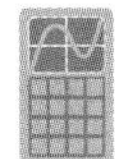

Graphing Calculator

We enter the data in the TI-83, using the List feature, putting the x coordinates (values of t) in L_1 and the y coordinates in L_2, just as in Example 1. Then we press [STAT], select `CALC`, and choose option 4: `LinReg(ax+b)`. Pressing [ENTER] will cause the equation of the regression line to be displayed in the home screen:

`y≈4.87x+68.1` Coefficients rounded to three significant digits[44]

We can use this equation as a model for the number of Chinese households with cable. To predict the number with cable in 2005, we substitute 5 for x (playing the role of t) and get $y \approx 92$ million households.

To graph the regression line without having to enter it by hand in the `Y=` screen, press [Y=], clear the contents of Y_1, press [VARS], choose option 5: `Statistics`, select `EQ`, and then choose option 1: `RegEQ`. The regression equation will then be entered under Y_1. To simultaneously show the data points, press [2nd] [STAT PLOT] and turn `PLOT1` on as in Example 1. To obtain a convenient window showing all the points and the line, press [ZOOM] and choose option 9: `ZoomStat`. The result should be something like Figure 26.

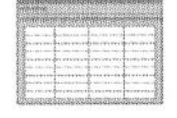

Excel

Here are two Excel shortcuts for linear regression: one graphical and one based on an Excel formula.

Using the Trendline Start with the original data and a scatter plot (Figure 25).

	A	B
1	x	y (Observed)
2	-4	50
3	-3	55
4	-2	57
5	-1	60
6	0	68
7	1	72
8	2	80
9	3	83

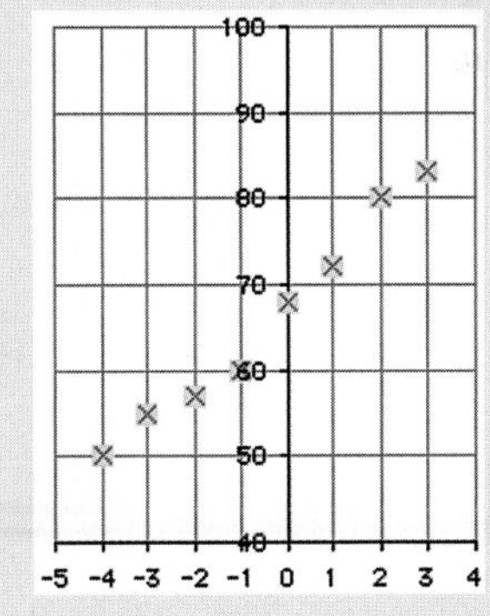

Figure 25

Click on the chart, and select `Add Trendline` from the Chart menu. Then select a `Linear` type (the default), and under `Options`, check the option `Display equation on chart`. The result should be something like Figure 26.

---

[44]The original data gives only two significant digits, so we rounded the coefficients to one more digit than that to give us a "safety margin" of one digit. Further digits are probably meaningless.

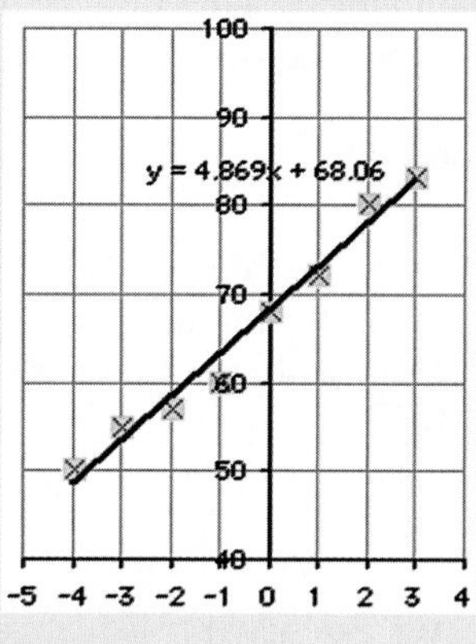

Figure 26

Using a Formula Alternatively, you can use the LINEST function (for "linear estimate"). To do this, enter your data as above and select a block of unused cells, two wide and one tall— for example, C2:D2. Then enter the formula

```
=LINEST(B2:B9,A2:A9)
```

as shown, and press Control + Shift + Enter.

	A	B	C	D
1	x	y (Observed)	m	b
2	-4	50	=LINEST(B2:B9,A2:A9)	
3	-3	55		
4	-2	57		
5	-1	60		
6	0	68		
7	1	72		
8	2	80		
9	3	83		

The result should look like this:

	A	B	C	D
1	x	y (Observed)	m	b
2	-4	50	4.86904762	68.0595238
3	-3	55		
4	-2	57		
5	-1	60		
6	0	68		
7	1	72		
8	2	80		
9	3	83		

The values of m and b appear in cells C2 and D2, as shown.

Coefficient of Correlation

Question If my data points do not all lie on one straight line, how can I measure how closely they can be approximated by a straight line?

Answer Think of SSE for a moment. It measures the sum of the squares of the deviations from the regression line and therefore itself constitutes a measurement of goodness of fit. (For instance, if SSE $= 0$, then all the points lie on a straight line.) However, SSE depends on the units we use to measure y and also on the number of data points (the more data points we use, the larger SSE tends to be). So, although we can (and do) use SSE to compare the goodness of fit of two lines to the same data, we cannot use it to compare the goodness of fit of one line to one set of data with that of another to a different set of data.

To remove this dependency, statisticians have found a related quantity that can be used to compare the goodness of fit of lines to different sets of data. This quantity, called the **coefficient of correlation** or **correlation coefficient** and usually denoted r, is between -1 and 1. The closer r is to -1 or 1, the better the fit. For an *exact* fit, we would have $r = -1$ (for a line with negative slope) or $r = 1$ (for a line with positive slope). For a bad fit, we would have r close to 0. Figure 27 shows several collections of data points with least-squares lines and the corresponding values of r.

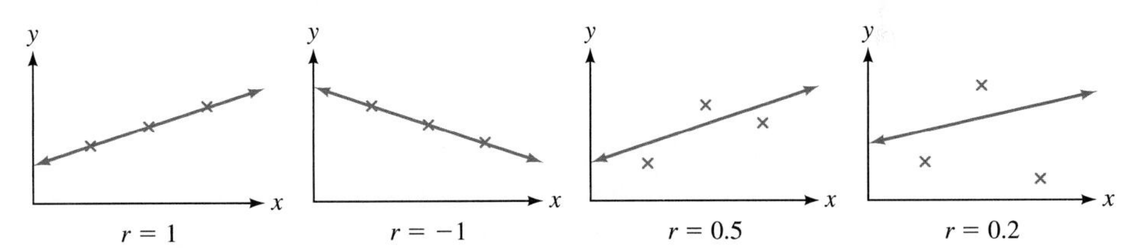

Figure 27

Correlation Coefficient

The coefficient of correlation of the line that best fits the n data points (x_1, y_1), $(x_2, y_2), \dots, (x_n, y_n)$ is

$$r = \frac{n(\Sigma xy) - (\Sigma x)(\Sigma y)}{\sqrt{n(\Sigma x^2) - (\Sigma x)^2} \cdot \sqrt{n(\Sigma y^2) - (\Sigma y)^2}}$$

Interpretation

If r is positive, the regression line has positive slope; if r is negative, the regression line has negative slope.

If $r = 1$ or -1, then all data points lie exactly on the regression line; if it is close to ± 1, then all data points are close to the regression line.

If r is close to 0, then y does not depend linearly on x.

Example 4 • Computing the Coefficient of Correlation

Find the correlation coefficient for the data in Example 2. Is the regression line a good fit?

Solution The formula for r requires Σx, Σx^2, Σxy, Σy, and Σy^2. We have all of these except for Σy^2, which we find in a new column as shown:

	x	y	xy	x^2	y^2
	1	1.5	1.5	1	2.25
	2	1.6	3.2	4	2.56
	3	2.1	6.3	9	4.41
	4	3.0	12.0	16	9.0
Σ (sum)	10	8.2	23.0	30	18.22

Substituting these values into the formula, we get

$$r = \frac{n(\Sigma xy) - (\Sigma x)(\Sigma y)}{\sqrt{n(\Sigma x^2) - (\Sigma x)^2} \cdot \sqrt{n(\Sigma y^2) - (\Sigma y)^2}}$$

$$= \frac{4(23) - (10)(8.2)}{\sqrt{4(30) - 10^2} \cdot \sqrt{4(18.22) - 8.2^2}}$$

$$\approx 0.9416$$

The fit is a fairly good one, as confirmed by Figure 24.

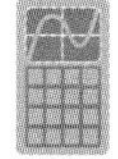

Graphing Calculator

A utility that will calculate linear regression lines will also calculate the correlation coefficient. To find the correlation coefficient using a TI-83, you need to tell the calculator to show you the coefficient at the same time that it shows you the regression line. To do this, press [CATALOG] and select `DiagnosticOn` from the list. The command will be pasted to the home screen, and you should then press [ENTER] to execute the command. Once you have done this, the `LinReg(ax+b)` command will show you not only a and b but also r and r^2.

Excel

In Excel when you add a trendline to a chart, you can select the option `Display r-squared value on chart` to show the value of r^2 on the chart (it is common to examine r^2, which takes on values between 0 and 1, instead of r). Alternatively, the LINEST function we used in Example 3 can be used to display quite a few statistics about a best-fit line, including r^2. Instead of selecting a block of cells two wide and one tall as we did in Example 3, select one two wide and *five* tall. Now enter the requisite LINEST formula with two additional arguments set to `TRUE` as shown, and press [Control]+[Shift]+[Enter].

	A	B	C	D	E
1	x	y (Observed)	m	b	
2	1	1.5	=LINEST(B2:B5,A2:A5,TRUE,TRUE)		
3	2	1.6			
4	3	2.1			
5	4	3			
6					

The result should look something like this:

	A	B	C	D
1	x	y (Observed)	m	b
2	1	1.5	0.5	0.8
3	2	1.6	0.12649111	0.34641016
4	3	2.1	0.88652482	0.28284271
5	4	3	15.625	2
6			1.25	0.16

The values of m and b appear in cells C2 and D2, as before, and the value of r^2 in cell C4. (Among the other numbers shown is SSE in cell D6. For the meanings of the remaining numbers shown, see the online help for LINEST in Excel; a good course in statistics wouldn't hurt, either.)

1.5 EXERCISES

In Exercises 1–4, compute SSE *by hand* for the given set of data and linear model.

1. $(1, 1), (2, 2), (3, 4); y = x - 1$

2. $(0, 1), (1, 1), (2, 2); y = x + 1$

3. $(0, -1), (1, 3), (4, 6), (5, 0); y = -x + 2$

4. $(2, 4), (6, 8), (8, 12), (10, 0); y = 2x - 8$

In Exercises 5–8, use technology to compute SSE for the given set of data and linear models. Indicate which linear model gives the better fit.

5. $(1, 1), (2, 2), (3, 4)$: **a.** $y = 1.5x - 1$ **b.** $y = 2x - 1.5$

6. $(0, 1), (1, 1), (2, 2)$: **a.** $y = 0.4x + 1.1$ **b.** $y = 0.5x + 0.9$

7. $(0, -1), (1, 3), (4, 6), (5, 0)$:
a. $y = 0.3x + 1.1$ **b.** $y = 0.4x + 0.9$

8. $(2, 4), (6, 8), (8, 12), (10, 0)$:
a. $y = -0.1x + 7$ **b.** $y = -0.2x + 6$

In Exercises 9–12, find the regression line associated with each set of points. Graph the data and the best-fit line. (Round all coefficients to four decimal places.)

9. $(1, 1), (2, 2), (3, 4)$ **10.** $(0, 1), (1, 1), (2, 2)$

11. $(0, -1), (1, 3), (4, 6), (5, 0)$ **12.** $(2, 4), (6, 8), (8, 12), (10, 0)$

In Exercises 13 and 14, use correlation coefficients to determine which of the given sets of data is best fit by its associated regression line and which is fit worst. Is it a perfect fit for any of the data sets?

13. a. $\{(1, 3), (2, 4), (5, 6)\}$
b. $\{(0, -1), (2, 1), (3, 4)\}$
c. $\{(4, -3), (5, 5), (0, 0)\}$

14. a. $\{(1, 3), (-2, 9), (2, 1)\}$
b. $\{(0, 1), (1, 0), (2, 1)\}$
c. $\{(0, 0), (5, -5), (2, -2.1)\}$

APPLICATIONS

15. Resources The following table shows the expenditures on water resources by the U.S. federal government, in billions of constant 1996 dollars ($t = 0$ represents 1980):

Year, t	0	5	10	15	19
Spent	7.4	5.6	5.1	4.7	4.5

SOURCE: U.S. Office of Management and Budget, *The Budget of the United States, Fiscal Year 2001* (Washington, D.C.: Government Printing Office, 2000). The numbers in this table were obtained from the Department of Energy's collection of environmental quality statistics at http://ceq.eh.doe.gov/nepa/reports/statistics/economy.html.

Find a regression line and use it to predict government expenditures on water resources in 2005. (Round regression coefficients to three significant digits and the final answer to two significant digits.)

16. Conservation The following table shows the expenditures on conservation and land management by the U.S. federal government, in billions of constant 1996 dollars ($t = 0$ represents 1980):

Year, t	0	5	10	15	19
Spent	1.8	2.0	4.1	5.4	5.4

SOURCE: U.S. Office of Management and Budget, *The Budget of the United States, Fiscal Year 2001* (Washington, D.C.: Government Printing Office, 2000). The numbers in this table were obtained from the Department of Energy's collection of environmental quality statistics at http://ceq.eh.doe.gov/nepa/reports/statistics/economy.html.

Find a regression line and use it to predict government expenditures on conservation and land management in 2005. (Round regression coefficients to three significant digits and the final answer to two significant digits.)

17. Oil Recovery In 1978 Congress conducted a study of the amount of additional oil that can be extracted from existing oil wells by "enhanced recovery techniques" (such as injecting solvents or steam into an oil well to lower the density of the oil). As the price of oil increases, the amount of oil that can be recovered economically in this manner also increases. The following table gives the study's estimates of recoverable oil based on the price per barrel:

Price/Barrel ($)	12	14	22	30
Recovery (billions of barrels)	21	30	42	49

SOURCE: U.S. Congress, Office of Technology Assessment, *Enhanced Oil Recovery Potential in the United States* (Washington, D.C.: OTA, 1978): 7. The recovery figures are based on a 10% minimum rate of return, and the prices are in constant 1976 dollars and rounded to the nearest $1.

Use a regression line to estimate the additional amount of oil that can be economically recovered if the price of oil were to drop to $10/barrel. (Round regression coefficients to three significant digits and the final answer to two significant digits.)

18. Natural Gas Reserves The following table shows the proved reserves of natural gas in the United States, in trillions of cubic feet ($t = 0$ represents 1980):

Year, t	0	5	10	15	19
Reserves	206	202	178	173	176

SOURCE: U.S. Energy Information Administration, *Annual Energy Review*, 2001, http://www.eia.doe.gov/emeu/aer/contents.html.

Use the regression line to estimate the rate at which natural gas was depleted during the given period.

19. Camera Sales The following table gives the number of cameras sold by Polaroid in the years 1994–1998:

Year	1994	1995	1996	1997	1998
Cameras Sold (millions)	6.3	5.2	5.0	5.0	4.7

SOURCES: Polaroid; Bloomberg Financial Markets/*New York Times*, March 27, 2000, p. C14.

a. Taking t to be the number of years since 1994, use linear regression to find a model of the number of cameras sold by Polaroid as a linear function of t. Include a sketch of the points and the regression line. (Round the coefficients to two decimal places.)

b. What does the slope tell you about sales of Polaroid cameras?

c. Would you say that a linear model of sales versus time is a reasonable one from which to make predictions? (Why?) Use extrapolation of your linear model to predict the number of cameras sold by Polaroid in 1999 (to the nearest 0.1 million cameras).

d. Polaroid actually sold 10 million cameras in 1999. What is the 1999 ($t = 5$) residue, and how does it compare with those of the other years?

20. Video-Game Sales The following table gives the revenue from Game Boy video games sold by Nintendo in the years 1996–1999:

Year	1996	1997	1998	1999
Game Boy Revenue ($ billions)	0.18	0.2	0.44	1.2

SOURCE: Gerard Klauer Mattison & Company/*New York Times*, March 25, 2000, p. C1.

a. Taking t to be the number of years since 1996, use linear regression to give a model of the revenue from Game Boy sales as a linear function of time t since 1996. Include a sketch of the points and the regression line. (Round the coefficients to two decimal places.)

b. What does the slope tell you about Game Boy revenues?

c. Would you say that a linear model of sales versus time is a reasonable one from which to make predictions? (Why?) Use extrapolation of your linear model to predict the revenue from Game Boy sales in 2000 (to the nearest $0.01 billion).

d. Revenues from Game Boy sales in 2000 were actually $0.98 billion. What is the 2000 ($t = 4$) residue, and how does it compare with those of the other years?

In Exercises 21–26, using a graphing calculator or Excel is recommended.

21. Big Brother In 1995 the FBI was seeking the ability to monitor 74,250 phone lines simultaneously. The following chart shows the number of phone lines monitored from 1987 through 1993:

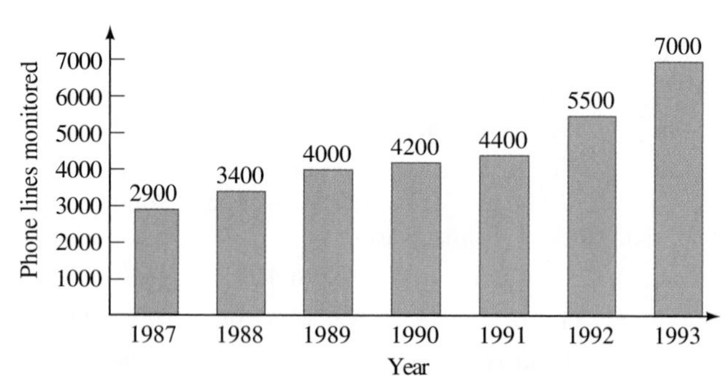

SOURCE: Electronic Privacy Information Center, Justice Department, Administrative Office of the U.S. Courts/*New York Times*, November 2, 1995, p. D5.

Use a regression model of these data to project the number of phone lines monitored by the FBI in 1995. (Round your answer to the nearest 100 phone lines.)

22. SAT Scores by Income The following chart shows 1994 U.S. verbal SAT scores as a function of parents' income level:

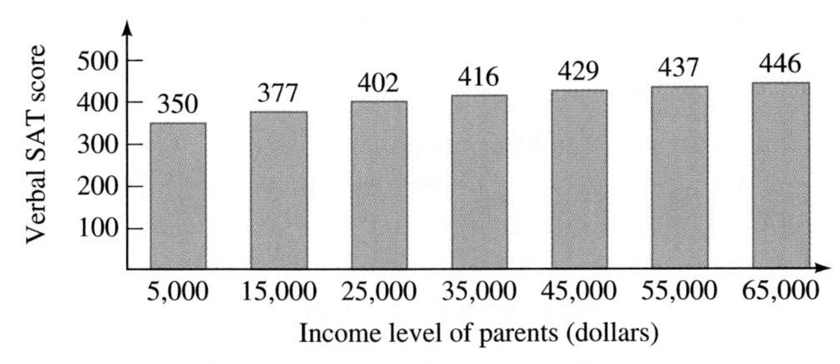

SOURCE: The College Board/*New York Times*, March 5, 1995, p. E16.

Use a linear model of these data to estimate the verbal SAT of a student whose parents have an income of $70,000.

23. **Life Expectancy** Life expectancies at birth in the United States for people born in various years is given in the following table:

Year	1920	1930	1940	1950	1960	1970	1980	1990	1998
Life Expectancy	54.1	59.7	62.9	68.2	69.7	70.8	73.7	75.4	76.7

SOURCE: Centers for Disease Control and Prevention, National Center for Health Statistics, *National Vital Statistics Report,* February 7, 2001, http://www.cdc.gov/nchs/fastats/pdf/nvsr48_18tb12.pdf.

Using a least-squares line, estimate the rate at which life expectancy was changing during the given period.

24. **Infant Mortality** The U.S. infant mortality rates (deaths per 1000 live births) for various years are given in the following table:

Year	1960	1970	1980	1985	1990	1995	1999
Mortality	26	20	12.6	10.6	9.2	7.6	7.0

SOURCES: Centers for Disease Control and Prevention, National Center for Health Statistics, *Vital Statistics of the United States,* http://www.cdc.gov/nchs/datawh/statab/pubd/hus98t23.htm; and *National Vital Statistics Report,* January 30, 2002, http://www.cdc.gov/nchs/data/lbid/nvsr50_04-t2.pdf.

Use a least-squares line to predict the infant mortality rate in the year 2010. Is this model realistic?

25. **Decreasing Homicide Rates** Homicide rates decreased in a number of major U.S. cities from 1991 to 1998. The following table shows 1991 rates (in crimes per 100,000 people) and the associated percentage drop for nine major cities:

1991 Homicide Rate	14.73	29.31	19.76	21.76	36.49	28.86	48.64	59.36	32.91
Percentage Decrease	76.4	70.6	69.3	62.8	61.3	59.3	52.4	27.6	22.3

SOURCE: Based on *FBI Uniform Crime Reports,* comp. Alfred Blumstein, Carnegie Mellon University/*New York Times,* March 4, 2000, p. A1.

a. Graph the data. Does the graph suggest a link between the 1991 homicide rate and the percentage decrease from 1991 to 1998?

b. Compute the linear regression equation for this data and compute r. Do the regression equation and correlation coefficient support your observations in part (a)?

c. What does the slope in the regression equation suggest about the relationship between homicide rate and percentage decrease?

d. Use the regression equation to estimate the percentage drop in the homicide rate for a city experiencing 40 homicides per 100,000 people in 1991.

26. **Decreasing Robbery Rates** The following data show decreasing robbery rates in major U.S. cities from 1991 to 1998:

1991 Robbery Rate	470	1340	836	395	833	1118	1095	1310	1558
Percentage Decrease	62.6	60.1	50.2	59.1	48.5	60.9	50.7	34.7	46

SOURCE: Based on *FBI Uniform Crime Reports,* comp. Alfred Blumstein, Carnegie Mellon University/*New York Times,* March 4, 2000, p. A1.

a–c. Repeat parts (a)–(c) of Exercise 25, using the data on robbery rates.

d. Use the regression equation to estimate the percentage drop in the robbery rate for a city experiencing 1000 robberies per 100,000 people in 1991.

COMMUNICATION AND REASONING EXERCISES

27. Why is the regression line associated with the two points (a, b) and (c, d) the same as the line that passes through both? (Assume that $a \neq c$.)

28. What is the smallest possible SSE if the given points happen to lie on a straight line? Why?

29. Verify that the regression line for the points $(0, 0)$, $(-a, a)$, and (a, a) has slope 0. What is the value of r^2? (Assume that $a \neq 0$.)

30. Verify that the regression line for the points $(0, a)$, $(0, -a)$, and $(a, 0)$ has slope 0. What is the value of r^2? (Assume that $a \neq 0$.)

31. If the points $(x_1, y_1), (x_2, y_2), \ldots, (x_n, y_n)$ lie on a straight line, what can you say about the regression line associated with these points?

32. If all but one of the points $(x_1, y_1), (x_2, y_2), \ldots, (x_n, y_n)$ lie on a straight line, must the regression line pass through all but one of these points?

33. Must the regression line pass through at least one of the data points? Illustrate your answer with an example.

34. Why must care be taken when using mathematical models to extrapolate?

CASE STUDY

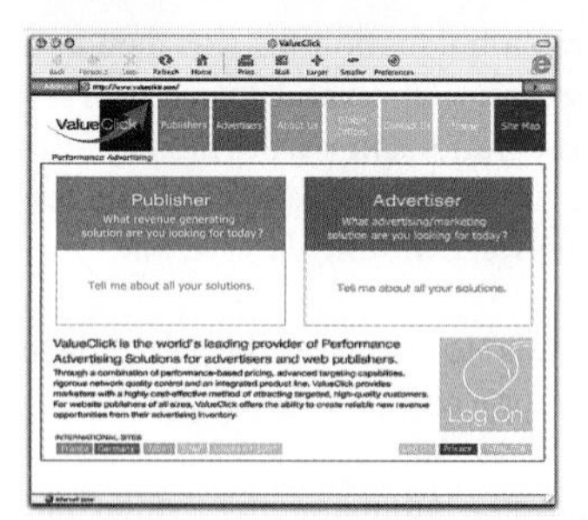

Courtesy valueclick.com

Modeling Spending on Internet Advertising

You are the new director of Impact Advertising's Internet division, which has enjoyed a steady 0.25% of the Internet advertising market. You have drawn up an ambitious proposal to expand your division in light of your anticipation that Internet advertising will continue to skyrocket. However, upper management sees things differently and, based on the following email, does not seem likely to approve the budget for your proposal.

TO: Jcheddar@impact.com (J. R. Cheddar)
CC: CVODoylePres@impact.com. (C. V. O'Doyle, CEO)
FROM: SGLombardoVP@impact.com (S. G. Lombardo, VP Financial Affairs)
SUBJECT: Your Expansion Proposal
DATE: August 3, 2001

Hi John:

Your proposal reflects exactly the kind of ambitious planning and optimism we like to see in our new upper-management personnel. Your presentation last week was most impressive and obviously reflected a great deal of hard work and preparation.

I am in full agreement with you that Internet advertising is on the increase. Indeed, our Market Research Department informs me that, based on a regression of the most recently available data, Internet advertising revenue in the U.S. will continue grow at a rate of approximately $1 billion per year. This translates into approximately $2.5 million in increased revenues per year for Impact, given our 0.25% market share. This rate of expansion is exactly what our planned 2002 budget anticipates. Your proposal, on the other hand, would require a budget of approximately *twice* the 2002 budget allocation, even though your proposal provides no hard evidence to justify this degree of financial backing.

At this stage, therefore, I'm sorry to say that I'm inclined not to approve the funding for your project, although I would be happy to discuss this further with you. I plan to present my final decision on the 2002 budget at next week's divisional meeting.

Regards, Sylvia

Refusing to admit defeat, you contact the Market Research Department and request the details of their projections on Internet advertising. They fax you the following information:

Year	1995	1996	1997	1998	1999	2000	2001
Spending on Internet Advertising ($ billions)	0	0.3	0.8	1.9	3	4.3	5.8

Figures are approximate and the 1999–2001 figures are projected. SOURCE: Nielsen NetRatings/*New York Times*, June 7, 1999, p. C17.

$$\text{Regression model: } y = 0.9857x - 0.6571 \qquad (x = \text{years since 1995})$$

Correlation coefficient: $r = 0.9781$

Now you see where the vice president got that \$1 billion figure: The slope of the regression equation is close to 1, indicating a rate of increase of just under \$1 billion per year. Also, the correlation coefficient is very high—an indication that the linear model fits the data well. In view of this strong evidence, it seems difficult to argue that revenues will increase by significantly more than the projected \$1 billion per year.

To get a better picture of what's going on, you decide to graph the data together with the regression line in your spreadsheet program. What you get is shown in Figure 28.

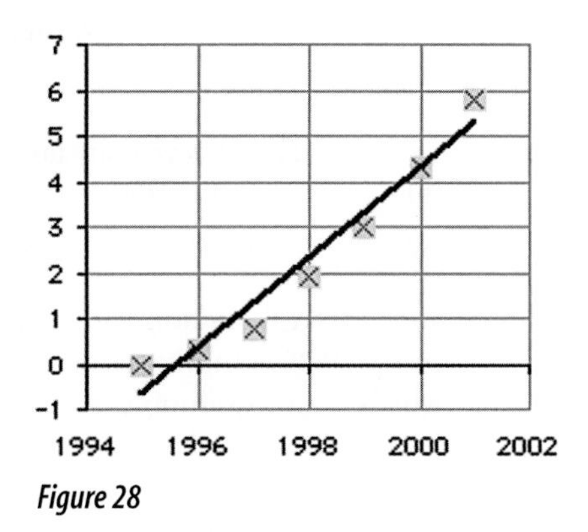

Figure 28

You immediately notice that the data points seem to suggest a curve, not a straight line; then again, perhaps the suggestion of a curve is an illusion. There are, you surmise, two possible interpretations of the data:

1. (Your first impression) As a function of time, spending on Internet advertising is nonlinear and is in fact accelerating (the rate of change is increasing), so a linear model is inappropriate.
2. (Devil's advocate) Spending on Internet advertising *is* a linear function of time; the fact that the points do not lie on the regression line is simply a consequence of such random factors as the state of the economy and the stock market performance.

You suspect that the vice president will probably opt for the second interpretation and discount the graphical evidence of accelerating growth by claiming that it is an illusion: a "statistical fluctuation." That is of course a possibility, but you wonder how likely it really is.

For the sake of comparison, you decide to try a regression based on the simplest nonlinear model you can think of—a quadratic function:

$$y = ax^2 + bx + c$$

Your spreadsheet allows you to fit such a function with a click of the mouse. The result is the following:

$$y = 0.1190x^2 + 0.2714x - 0.0619 \qquad (x = \text{number of years since 1995})$$

$$r = 0.9992$$

Figure 29 shows the graph of the regression function together with the original data.

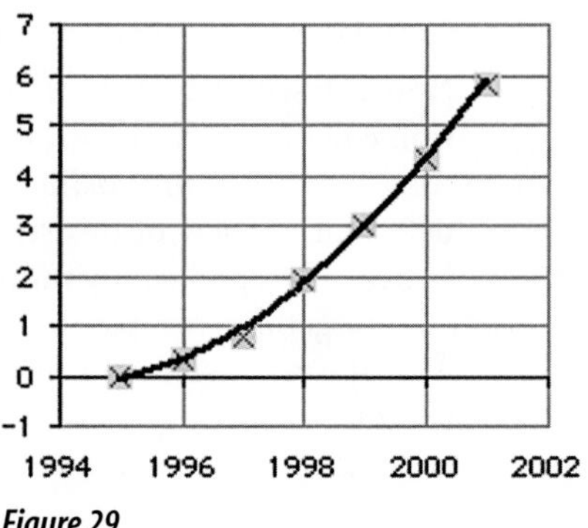

Figure 29

Aha! The fit is visually far better, and the correlation coefficient is even higher! Further, the quadratic model predicts 2002 spending as

$$y = 0.1190(7)^2 + 0.2714(7) - 0.0619 \approx \$\,7.67 \text{ billion}$$

which is \$1.87 billion above the 2001 spending figure in the table. Given Impact Advertising's 0.25% market share, this translates into an increase in revenues of \$4.7 million, which is almost double the estimate predicted by the linear model!

You quickly draft an email to Lombardo and are about to send it when you decide, as a precaution, to check with a statistician. He tells you to be cautious: The value of r will always tend to increase if you pass from a linear model to a quadratic one due to an increase in "degrees of freedom."[45] A good way to test whether a quadratic model is more appropriate than a linear one is to compute a statistic called the p value associated with the coefficient of x^2. A low p value indicates a high probability that the coefficient of x^2 cannot be zero (see below). Notice that if the coefficient of x^2 *is* zero, then you have a linear model.

You can, your friend explains, obtain the p value using Excel as follows. First, set up the data in columns, with an extra column for the values of x^2:

	A	B	C
1	x	x^2	y
2	0	0	0
3	1	1	0.3
4	2	4	0.8
5	3	9	1.9
6	4	16	3
7	5	25	4.3
8	6	36	5.8

Next, choose `Data analysis` from the `Tools` menu, and choose `Regression`. In the dialog box, give the location of the data as shown in Figure 30. Clicking `OK` then gives you a large chart of statistics. The p value you want is in the very last row of the data: $p = 0.000478$.

[45]The number of degrees of freedom in a regression model is 1 less than the number of coefficients. For a linear model, it is 1 (there are two coefficients: the slope m and the intercept b), and for a quadratic model it is 2. For a detailed discussion, consult a text on regression analysis.

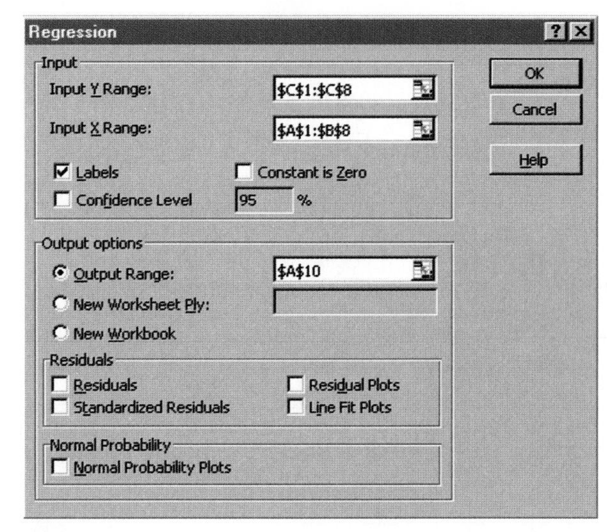

Figure 30

Question What does p measure?

Answer Roughly speaking, p is the probability, allowing for random fluctuation in the data, that you are in error in claiming that the coefficient of x^2 is not zero. In other words, you can be 99.9522% certain that you are not making an error .

Thus, you can go ahead and send your email with 99% confidence!

EXERCISES

Suppose you are given the following data:

Year	1995	1996	1997	1998	1999	2000	2001
Spending on Internet Advertising ($ billions)	0	0.3	1.5	2.6	3.4	4.3	5.0

1. Obtain a linear regression model and the correlation coefficient r. According to the model, at what rate is spending on Internet advertising increasing in the United States? How does this translate to annual revenues for Impact Advertising?
2. Use a spreadsheet or other technology to graph the data together with the best-fit line. Does the graph suggest a quadratic model (parabola)?
3. Test your impression in Exercise 2 by using technology to fit a quadratic function and graphing the resulting curve together with the data. Does the graph suggest that the quadratic model is appropriate?
4. Perform a regression analysis and find the associated p value. What does it tell you about the appropriateness of a quadratic model?

CHAPTER 1 REVIEW TEST

1. Graph the following functions and equations:
   a. $y = -2x + 5$
   b. $2x - 3y = 12$
   c. $y = \begin{cases} \frac{1}{2}x & \text{if } -1 \le x \le 1 \\ x - 1 & \text{if } 1 < x \le 3 \end{cases}$
   d. $f(x) = 4x - x^2$, with domain $[0, 4]$

2. Decide whether each of the functions specified below is based on a linear, quadratic, exponential, or absolute value model.

x	-2	0	1	2	4
$f(x)$	4	2	1	0	2
$g(x)$	-5	-3	-2	-1	1
$h(x)$	1.5	1	0.75	0.5	0
$k(x)$	0.25	1	2	4	16
$u(x)$	0	4	3	0	-12

3. Find the equation of each line.
   a. Through (3, 2) with slope -3
   b. Through $(-1, 2)$ and $(1, 0)$
   c. Through (1, 2) parallel to $x - 2y = 2$
   d. With slope $\frac{1}{2}$ crossing $3x + y = 6$ at its x intercept

OHaganBooks.com—MODELING SALES, REVENUES, AND DEMAND

4. As your online bookstore, OHaganBooks.com, has grown in popularity, you have been monitoring book sales as a function of the traffic at your site (measured in "hits" per day) and have obtained the following model:

$$n(x) = \begin{cases} 0.02x & \text{if } 0 \le x \le 1000 \\ 0.025x - 5 & \text{if } 1000 < x \le 2000 \end{cases}$$

   where $n(x)$ is the average number of books sold in a day in which there are x hits at the site.
   a. On average, how many books per day does your model predict you will sell when you have 500 hits/day? 1000 hits/day? 1500 hits/day?
   b. What does the coefficient 0.025 tell you about your book sales?
   c. According to the model, how many hits per day will be needed in order to sell an average of 30 books/day?

5. Your monthly books sales had increased quite dramatically over the past few months but now appear to be leveling off. Here are the sales figures for the past 6 months:

Month, t	1	2	3	4	5	6
Daily Book Sales, S	12.5	37.5	62.5	72.0	74.5	75.0

   a. Which of the following models best approximates the data?

   (A) $S(t) = \dfrac{300}{4 + 100(5^{-t})}$ (B) $S(t) = 13.3t + 8.0$

   (C) $S(t) = -2.3t^2 + 30.0t - 3.3$ (D) $S(t) = 7(3^{0.5t})$

   b. What do each of the above models predict for the sales in the next few months: rising, falling, or leveling off?

6. To increase business at OHaganBooks.com, you plan to place more banner ads at well-known Internet portals. So far, you have the following data on the average number of hits per day at OHaganBooks.com versus your monthly advertising expenditures:

Advertising Expenditure ($/month)	2000	5000
Web Site Traffic (hits/day)	1900	2050

   You decide to construct a linear model giving h, the average number of hits/day as a function of the advertising expenditure c.
   a. What is the model you obtain?
   b. Based on your model, how much traffic can you anticipate if you budget $6000/month for banner ads?
   c. Your goal is to eventually increase traffic at your site to an average of 2500 hits/day. Based on your model, how much do you anticipate you will need to spend on banner ads in order to accomplish this?

7. A month ago you increased expenditure on banner ads to $6000/month, and you have noticed that the traffic at OHaganBooks.com has not increased to the level predicted by the linear model in Exercise 6. Fitting a quadratic function to the data you have gives the model

$$h = -0.000005c^2 + 0.085c + 1750$$

   where h is the daily traffic (hits) at your Web site and c is the monthly advertising expenditure.
   a. According to this model, what is the current traffic at your site?
   b. Does this model give a reasonable prediction of traffic at expenditures larger than $8500/month? Why?

8. Besides selling books, you are generating additional revenue at OHaganBooks.com through your new online publishing service. Readers pay a fee to download the entire text of a novel. Author royalties and copyright fees cost you an average of $4/novel, and the monthly cost of operating and maintaining the service amounts to $900/month. You are currently charging readers $5.50/novel.
   a. What are the associated cost, revenue, and profit functions?
   b. How many novels must you sell each month in order to break even?
   c. If you lower the charge to $5.00/novel, how many books will you need to sell in order to break even?

9. To generate a profit from your online publishing service, you need to know how the demand for novels depends on the price you charge. During the first month of the service, you were charging \$10.00/novel and sold 350. Lowering the price to \$5.50/novel had the effect of increasing demand to 620 novels/month.

a. Use the given data to construct a linear demand equation.

b. Use the demand equation you constructed in part (a) to estimate the demand if you raised the price to \$15.00/novel.

c. Using the information on cost given in Exercise 8, determine which of the three prices (\$5.50, \$10.00, and \$15.00) would result in the largest profit and the size of that profit.

10. It is now several months later and you have tried selling your online novels at a variety of prices, with the following results:

Price (\$)	5.50	10.00	12.00	15.00
Demand (monthly sales)	620	350	300	100

a. Use the given data to obtain a linear regression model of demand. (Round coefficients to four decimal places.)

b. Use the demand model you constructed in part (a) to estimate the demand if you charged \$8.00/novel. (Round the answer to the nearest novel.)

ADDITIONAL ONLINE REVIEW

If you follow the path

Web site → Everything for Finite Mathematics → Chapter 1

you will find the following additional resources to help you review:

A comprehensive chapter summary (including examples and interactive features)

Additional review exercises (including interactive exercises and many with help)

Interactive section-by-section online tutorials

Section-by-section Excel tutorials

A true/false chapter quiz

Several useful utilities, including graphers and a regression tool

2

SYSTEMS OF LINEAR EQUATIONS

CASE STUDY

The Impact of Regulating Sulfur Emissions

You have been hired as a consultant to the Environmental Protection Agency (EPA). In an effort to curb the effects of acid rain on the ecosystem, the agency is considering regulations that will require a 15 million–ton reduction in sulfur emissions. You have been asked to estimate the cost of the proposed regulations to the major utility companies and also their effect on jobs in the coal-mining industry. The data you have available show the annual cost to utilities and the cost in jobs for emission reductions of up to 12 million tons. Your assignment is to use these figures to compute projections for a 15 million–ton reduction.

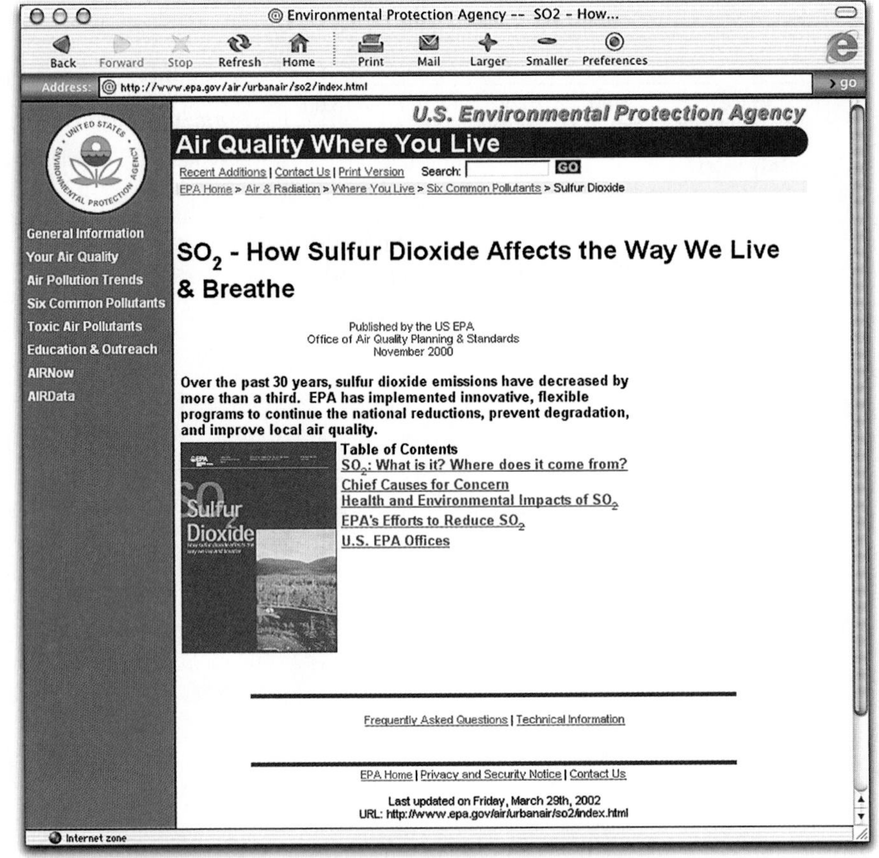

REPRODUCED FROM THE ENVIRONMENTAL PROTECTION AGENCY WEB SITE

INTERNET RESOURCES FOR THIS CHAPTER

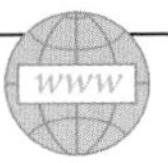

At the Web site, follow the path

Web site → Everything for Finite Math → Chapter 2

where you will find step-by-step tutorials for the main topics in this chapter, a detailed chapter summary you can print out, a true/false quiz, and a collection of chapter review questions. If you are using Excel, you can download brief Excel tutorials for every section. You will also find an online grapher, online Gauss–Jordan pivoting programs, software, graphing calculator programs, and other resources.

AND MATRICES

Introduction

In Chapter 1 we studied single functions and equations. In this chapter we seek solutions to **systems** of two or more equations. For example, suppose we need to *find two numbers whose sum is* 3 *and whose difference is* 1. In other words, we need to find two numbers x and y such that $x + y = 3$ and $x - y = 1$. The only solution turns out to be $x = 2$ and $y = 1$, a solution you might easily guess. But, how do we know that this is the only solution, and how do we find solutions systematically? When we restrict ourselves to systems of *linear* equations, there is a very elegant method for determining the number of solutions and finding them all. Moreover, as we will see, many real-world applications give rise to just such systems of linear equations.

We begin in Section 2.1 with systems of two linear equations in two unknowns and some of their applications. In Section 2.2 we study a powerful matrix method, called *row reduction,* for solving systems of linear equations in any number of unknowns. In Section 2.3 we look at more applications.

Computers have been used for many years to solve the large systems of equations that arise in the real world. You probably already have access to devices that will do the row operations used in row reduction. Many graphing calculators can do them, as can spreadsheets and various special-purpose computer programs, including utilities available at the Web site. Using such a device or program makes the calculations quicker and helps avoid arithmetic mistakes. Then there are programs (and calculators) into which you simply feed the system of equations and out pop the solutions. We can think of what we do in this chapter as looking inside the "black box" of such a program. More important, we talk about how, starting from a real-world problem, to get the system of equations to solve in the first place. This conversion no computer will yet do for us.

2.1 Systems of Two Linear Equations in Two Unknowns

Suppose you have \$3 in your pocket to spend on snacks and a drink. If x represents the amount you'll spend on snacks and y represents the amount you'll spend on a drink, you can say that $x + y = 3$. On the other hand, if for some reason you want to spend \$1 more on snacks than on your drink, you can also say that $x - y = 1$. These are simple examples of **linear equations in two unknowns.** Here's another: $3x - y = 15$. In general, a linear equation in two unknowns is an equation that can be written in the form

$$ax + by = c$$

with a, b, and c being real numbers. The number a is called the **coefficient of x,** and b is called the **coefficient of y.** In the case of $3x - y = 15$, we have $a = 3$ and $b = -1$. A

solution of an equation consists of a pair of numbers: a value for x and a value for y that satisfy the equation. For instance, $(x, y) = (5, 0)$ is a solution of $3x - y = 15$, since $3(5) - 0 = 15$. In fact, this equation has infinitely many solutions: We could solve for $y = 3x - 15$, and then for every value of x we choose, we can get the corresponding value of y, giving a solution (x, y). As we saw in Chapter 1, these solutions are the points on a straight line, the *graph* of the equation.

In this section we are concerned with pairs (x, y) that are solutions of *two* linear equations at the same time. For example, $(2, 1)$ is a solution of both of the equations $x + y = 3$ and $x - y = 1$, since substituting $x = 2$ and $y = 1$ into these equations gives $2 + 1 = 3$ (true) and $2 - 1 = 1$ (also true), respectively. So, in the simple example we began with, you could spend \$2 on snacks and \$1 on a drink.

In the following set of examples, you will see how to graphically and algebraically solve a system of two linear equations in two unknowns. Then we'll return to some more interesting applications.

Example 1 • Two Ways of Solving a System: Graphically and Algebraically

Find all solutions (x, y) of the following system of two equations:

$$x + y = 3$$

$$x - y = 1$$

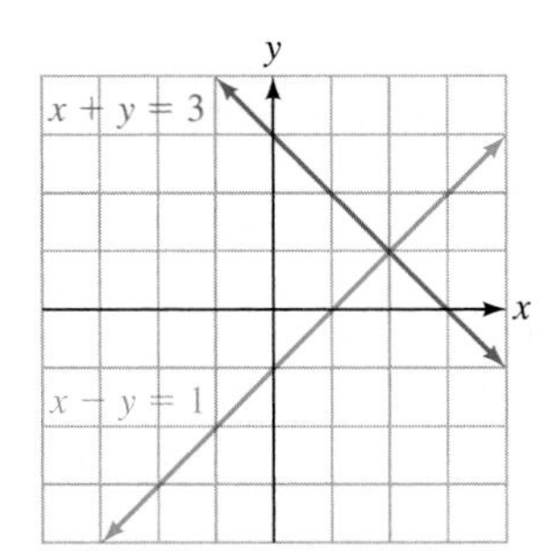

Figure 1

Solution We will see how to find the solution(s) in two ways: graphically and algebraically. Remember that a solution is a pair (x, y) that simultaneously satisfies *both* equations.

Method 1: Graphical We already know that the solutions of a single linear equation are the points on its graph, which is a straight line. For a point to represent a solution of two linear equations, it must lie simultaneously on both of the corresponding lines. In other words, it must be a point where the two lines cross, or intersect. A look at Figure 1 should convince us that the lines cross only at the point $(2, 1)$, so this is the only possible solution.

Method 2: Algebraic In the algebraic approach, we try to combine the equations in such a way as to eliminate one variable. In this case, notice that if we add the left-hand sides of the equations, the terms with y are eliminated. So we add the first equation to the second (that is, add the left-hand sides and add the right-hand sides)[1] to get

$$2x + 0 = 4$$

$$2x = 4$$

$$x = 2$$

Now that we know that x has to be 2, we can substitute back into either equation to find y. Choosing the first equation (it doesn't matter which we choose), we have

$$2 + y = 3$$

$$y = 3 - 2$$

$$= 1$$

[1]Why can we just add these equations? When we add equal amounts to both sides of an equation, the results are equal; that is, if $A = B$ and $C = D$, then $A + C = B + D$.

We have found that the only possible solution is $x = 2$ and $y = 1$, or

$$(x, y) = (2, 1)$$

Before we go on . . .

Question In our discussion of the break-even point in Section 1.4, we found the intersection of two straight lines by setting equal two linear functions of x (cost and revenue). Can we find the solution here in a similar way?

Answer Yes. We could solve our two equations for y to obtain

$$y = -x + 3$$

$$y = x - 1$$

and then equate the right-hand sides and solve to find $x = 2$. The value for y is then found by substituting $x = 2$ in either equation.

Question So why don't we solve all systems of equations this way?

Answer The elimination method extends more easily to systems with more equations and unknowns. It is the basis for the matrix method of solving systems—a method we discuss in Section 2.2. We will use it exclusively for the rest of this section.

Having found the same solution in several different ways, we suspect that it is correct. Still, we should check our solution by substituting it back into *both* of the original equations. Substituting, we get

$$2 + 1 = 3 \quad \checkmark$$

$$2 - 1 = 1 \quad \checkmark$$

The solution is correct!

Solving Graphically with a Graphing Calculator

You can use a graphing calculator to draw the graphs of the two equations on the same set of axes and to check the solution. First, solve the equations for y, obtaining $y = -x + 3$ and $y = x - 1$. On a TI-83 or similar graphing calculator, set

Y_1=-X+3

Y_2=X-1

then decide on the range of x values you want to use. As in Figure 1, let's choose the range $[-4, 4]$.[2] In the `Window` menu, set `Xmin=-4` and `Xmax=4` and then press ZOOM and select `Zoomfit` to set the y range. You can now zoom in for a more accurate view by choosing a smaller x range that includes the point of intersection, like $[1.5, 2.5]$, and using `Zoomfit` again. You can also use the Trace feature to see the coordinates of points near the point of intersection.

To check that (2, 1) is the correct solution, use the Table feature to compare the two values of y corresponding to $x = 2$. Press 2ND TABLE, set `X=2`, and compare the corresponding values of Y_1 and Y_2; they should each be 1.

---

[2]How did we come up with this interval? Trial and error. You might need to try several intervals before finding one that gives a graph showing the point of intersection clearly.

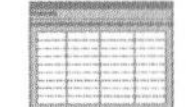

Solving Graphically with Excel

You can use Excel to draw the graphs of the two equations on the same set of axes and to check the solution. First, solve the equations for y, obtaining $y = -x + 3$ and $y = x - 1$. To graph these lines, we *could* use the Excel graphing template for smooth curves, developed in Chapter 1. However, there is no need to plot a large number of points when graphing a straight line; two points suffice. The following worksheet shows all we really need for a plot of the two lines in question:

	A	B	C
1	x	y1	y2
2	-4	=-A2+3	=A2-1
3	4		

The two values of x give the x coordinates of the two points we will use as endpoints of the lines. (We have—somewhat arbitrarily—chosen the range $[-4, 4]$ for x.) The formula for the first line, $y = -x + 3$, is in cell B2, and the formula for the second line, $y = x - 1$, is in C2. Copying these two cells as shown yields the following result:

	A	B	C
1	x	y1	y2
2	-4	7	-5
3	4	-1	3

For the graph, select all nine cells and create a scatter plot with line segments joining the data points. (Instruct Excel to insert a chart and select the `XY (Scatter)` option. In the same dialog box, select the option that shows points connected by lines. Click on `Next` to bring up a new dialog called `Data Type`, where you should make sure that the `Series in Columns` option is selected, telling the program that the x and y coordinates are arranged vertically, down columns.) Your graph should appear as shown in Figure 2.

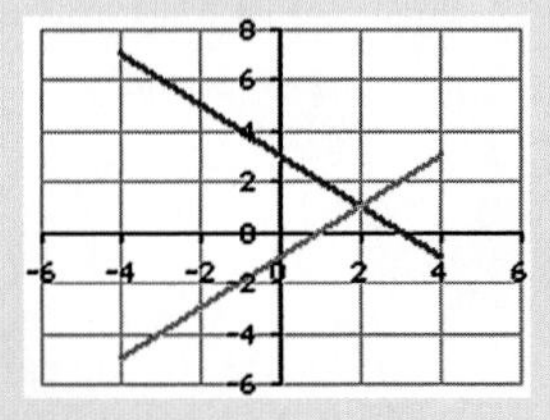

Figure 2

To zoom in, first decide on a new x range, say, $[1, 3]$. Change the value in cell A2 to 1 and the value in cell A3 to 3, and Excel will automatically update the y values and the graph.[3]

To check that $(2, 1)$ is the correct solution, enter the value 2 in cell A4 in your spreadsheet and copy the formulas in B3 and C3 down to row 4 to obtain the corresponding values of y.

[3]Excel sets the y range of the graph automatically, and the range it chooses may not always be satisfactory. It may sometimes be necessary to "narrow down" the x or y range by double-clicking on the axis in question and specifying a new range.

	A	B	C
1	x	y1	y2
2	-4	7	-5
3	4	-1	3
4	2	1	1

Since the values of y agree, we have verified that (2, 1) is a solution of both equations.

Question Is it always possible to find the *exact* answer using technology?

Answer No. Consider the system

$$2x + 3y = 1$$
$$-3x + 2y = 1$$

Its solution is $(x, y) = \left(-\frac{1}{13}, \frac{5}{13}\right)$. (Check it!) If we write this solution in decimal form, we get repeating decimals: $(-0.076\,923\,076\,923\,0\ldots, 0.384\,615\,384\,615\ldots)$. Since the graphing technology gives the coordinates of points only to several decimal places, it can show only an approximate answer.

Question How accurate is the answer shown by using the Trace feature of a graphing calculator?

Answer That depends. We can increase the accuracy up to a point by zooming in on the point of intersection of the two graphs. But there is a limit to this: Most graphing calculators are capable of giving an answer correct to about 13 decimal places. This means, for instance, that, in the eyes of the TI-83, 2.000 000 000 000 1 is exactly the same as 2 (subtracting them yields 0). It follows that if you attempt to use a window so narrow that you need approximately 13 significant digits to distinguish the left and right edges, you will run into accuracy problems. Similarly, Excel rounds all numbers to 15 significant digits (so that, in the eyes of Excel, 2.000 000 000 000 001 is exactly the same as 2).

Example 2 illustrates the points we have just made.

Example 2 • Solving a System: Algebraically versus Graphically

Solve the system

$$3x + 5y = 0$$
$$2x + 7y = 1$$

Solution

Method 1: Graphical Using the methods of Example 1, we can obtain the graph shown in Figure 3.

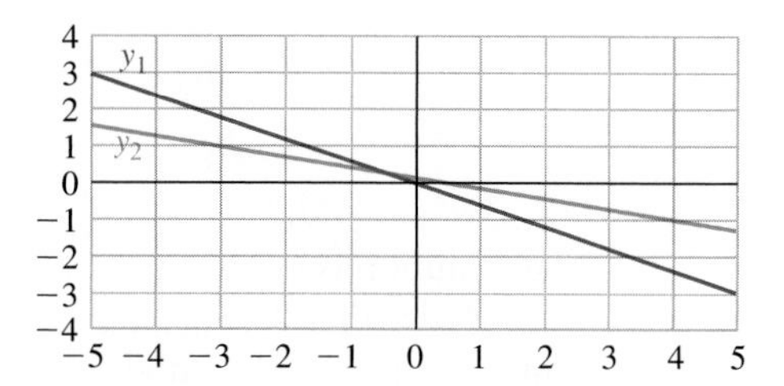

Figure 3

The lines appear to intersect slightly above and to the left of the origin, so let's zoom in for a closer view by changing the x range to $[-1, 0]$ (Figure 4).

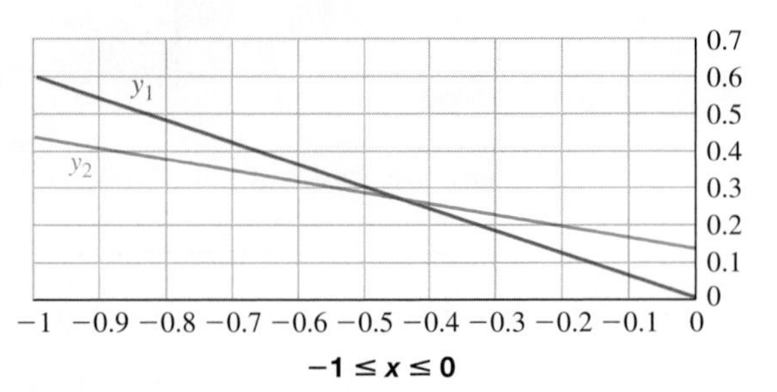

Figure 4

If we look carefully at Figure 4, we notice that the graphs intersect near $(-0.45, 0.27)$. Is the point of intersection *exactly* $(-0.45, 0.27)$? (Substitute these values into the equations to find out.) In fact, it is impossible to find the exact solution of this system graphically, but we now have a ballpark answer that we can use to help check the following algebraic solution.

Method 2: Algebraic We first notice that adding the equations is not going to eliminate either x or y. Notice, however, that if we multiply (both sides of) the first equation by 2 and the second by -3, the coefficients of x will become 6 and -6. *Then* if we add them, x will be eliminated. So we proceed as follows:

$$2(3x + 5y) = 2(0)$$
$$-3(2x + 7y) = -3(1)$$

gives

$$6x + 10y = 0$$
$$-6x - 21y = -3$$

Adding these equations we get

$$-11y = -3$$

so that

$$y = \frac{3}{11} = 0.\overline{27}$$

Substituting $y = \frac{3}{11}$ in the first equation gives

$$3x + 5\left(\frac{3}{11}\right) = 0$$
$$3x = -\frac{15}{11}$$
$$x = -\frac{5}{11} = -0.\overline{45}$$

The solution is $(x, y) = \left(-\frac{5}{11}, \frac{3}{11}\right) = (-0.\overline{45}, 0.\overline{27})$.

Notice that the algebraic method gives us the exact solution that we could not find with the graphical method. Still, we can use our graph to confirm our algebraic solution and check that we haven't made an error.

Before we go on . . .

Question In solving the above system algebraically, we multiplied (both sides of) the equations by numbers. How does that effect their graphs?

Answer Since multiplying both sides of an equation by a nonzero number has no effect on its solutions, the graph (which represents the set of all solutions) is unchanged.

We should check our answer:

$$3\left(-\frac{5}{11}\right) + 5\left(\frac{3}{11}\right) = -\frac{15}{11} + \frac{15}{11} = 0 \quad ✓$$

$$2\left(-\frac{5}{11}\right) + 7\left(\frac{3}{11}\right) = -\frac{10}{11} + \frac{21}{11} = 1 \quad ✓$$

Example 3 • An Inconsistent System

Solve the system

$$x - 3y = 5$$
$$-2x + 6y = 8$$

Solution To eliminate x we multiply the first equation by 2 and then add:

$$2x - 6y = 10$$
$$-2x + 6y = 8$$

Adding gives

$$0 = 18$$

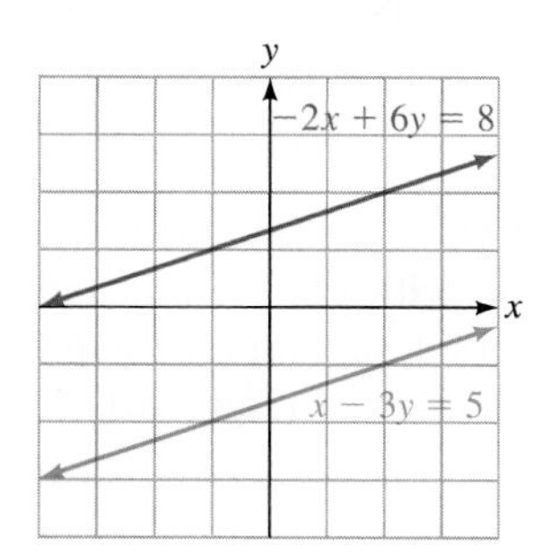

Figure 5

But this is absurd! This calculation shows that, if we had two numbers x and y that satisfied both equations, it would be true that $0 = 18$. Since 0 is *not* equal to 18, there can be no such numbers x and y. In other words, *the system has no solutions* and is called **inconsistent.**

In slope-intercept form, these lines are $y = \frac{1}{3}x - \frac{5}{3}$ and $y = \frac{1}{3}x + \frac{4}{3}$. Notice that they have the same slope. Plotting them gives the two parallel lines shown in Figure 5. Since they are parallel, they do not intersect. Because a solution must be a point of intersection, we again conclude that there is no solution.

Example 4 • A Redundant System

Solve the system

$$x + y = 2$$
$$2x + 2y = 4$$

Solution Multiplying the first equation by -2 gives

$$-2x - 2y = -4$$
$$2x + 2y = 4$$

Adding gives the not-very-enlightening result

$$0 = 0$$

Now what has happened? Looking back at the original system, we note that the second equation is really the first equation in disguise (it is the first equation multiplied by 2). Since the second equation gives us the same information as the first, we say that this is a **redundant system.** In other words, we really have only one equation in two unknowns. From Chapter 1 we know that a single linear equation in two unknowns has infinitely many solutions, one for each value of x. (Recall that to get the corresponding solution for y, we solve the equation for y and substitute the x value.) The entire set of solutions can be summarized as follows:

x is arbitrary

$y = 2 - x$ Solve the first equation for y.

This set of solutions is called the **general solution** because it includes all possible solutions. When we write the general solution this way, we say that we have a **parameterized solution** and that x is the **parameter.** We can also write the general solution as

$(x, 2 - x)$ x arbitrary

Different choices of the parameter x lead to different **particular solutions.** For instance, choosing $x = 3$ gives the particular solution

$(x, y) = (3, -1)$

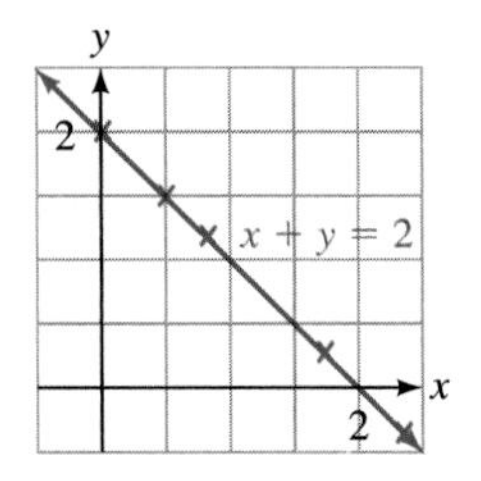

Figure 6 • The line of solutions and some particular solutions

Because there are infinitely many values of x from which to choose, there are infinitely many solutions.

We could also have solved the first equation for x instead and used y as a parameter, obtaining another form of the general solution:

$(2 - y, y)$ y arbitrary Alternate form of the general solution

What does this system of equations look like graphically? Since the two equations are really the same, their graphs are identical, each being the line with x intercept 2 and y intercept 2. The "two" lines intersect at every point, so there is a solution for each point on the common line. In other words, we have a "whole line of solutions" (Figure 6).

We summarize the three possible outcomes we have encountered.

Possible Outcomes for a System of Two Linear Equations in Two Unknowns

1. **A single (or *unique*) solution** This happens when the lines corresponding to the two equations are distinct and not parallel so that they intersect at a single point. (See Example 1.)
2. **No solution** This happens when the two lines are parallel. We say that the system is **inconsistent.** (See Example 3.)
3. **An infinite number of solutions** This occurs when the two equations represent the same straight line, and we say that such a system is **redundant,** or **dependent.** In this case we can represent the solutions by choosing one variable arbitrarily and solving for the other. (See Example 4.)

In cases 1 and 3, we say that the system of equations is **consistent** because it has at least one solution.

You should think about straight lines and convince yourself that these are the only three possibilities.

APPLICATIONS

Example 5 • Blending

Acme Babyfoods mixes two strengths of apple juice. One quart of Beginner's juice is made from 30 fluid ounces of water and 2 fluid ounces of apple juice concentrate. One quart of Advanced juice is made from 20 fluid ounces of water and 12 fluid ounces of concentrate. Every day Acme has available 30,000 fluid ounces of water and 3600 fluid ounces of concentrate. If the company wants to use all the water and concentrate, how many quarts of each type of juice should it mix?

Solution In all applications we follow the same general strategy.

1. **Identify and label the unknowns.** What are we asked to find? To answer this question, it is common to respond by saying, "The unknowns are Beginner's juice and Advanced juice." Quite frankly, this is a baffling statement. Just what is unknown about juice? We need to be more precise:

   The unknowns are (1) *the* ***number of quarts*** *of Beginner's juice and* (2) *the* ***number of quarts*** *of Advanced juice made each day.*

   So, we label the unknowns as follows: Let

   x = number of quarts of Beginner's juice made each day

   y = number of quarts of Advanced juice made each day

2. **Use the information given to set up equations in the unknowns.** This step is trickier, and the strategy varies from problem to problem. Here, the amount of juice the company can make is constrained by the fact that they have limited amounts of water and concentrate. This example shows a kind of application we will often see, and it is helpful in these problems to use a table to record the amounts of the resources used:

	Beginner's, x	Advanced, y	Available
Water (fl oz)	30	20	30,000
Concentrate (fl oz)	2	12	3,600

We can now set up an equation for each of the items listed on the left.

Water: We read across the first row. If Acme mixes x quarts of Beginner's juice, each quart using 30 fluid ounces of water, and y quarts of Advanced juice, each using 20 fluid ounces of water, it will use a total of $30x + 20y$ fluid ounces of water. But we are told that the total has to be 30,000 fluid ounces. Thus, $30x + 20y = 30{,}000$. This is our first equation.

Concentrate: We read across the second row. If Acme mixes x quarts of Beginner's juice, each using 2 fluid ounces of concentrate, and y quarts of Advanced juice, each using 12 fluid ounces of concentrate, it will use a total of $2x + 12y$ fluid ounces of concentrate. But we are told that the total has to be 3600 fluid ounces. Thus, $2x + 12y = 3600$.

Now we have two equations:

$$30x + 20y = 30{,}000$$

$$2x + 12y = 3600$$

To make the numbers easier to work with, let's divide (both sides of) the first equation by 10 and the second by 2:

$$3x + 2y = 3000$$
$$x + 6y = 1800$$

We can now eliminate x by multiplying the second equation by -3 and adding:

$$\begin{array}{r} 3x + 2y = 3000 \\ \underline{-3x - 18y = -5400} \\ -16y = -2400 \end{array}$$

So, $y = 2400/16 = 150$. Substituting this into the equation $x + 6y = 1800$ gives $x + 900 = 1800$, and so $x = 900$. The solution is $(x, y) = (900, 150)$. In other words, the company should mix 900 quarts of Beginner's juice and 150 quarts of Advanced juice.

Before we go on . . . Let's check the answer:

$$30(900) + 20(150) = 27{,}000 + 3000 = 30{,}000 \quad ✓$$
$$2(900) + 12(150) = 1800 + 1800 = 3600 \quad ✓$$

This combination of juices does use exactly 30,000 fluid ounces of water and 3600 fluid ounces of concentrate.

Example 6 • Blending

A medieval alchemist's love potion calls for a number of eyes of newt and toes of frog, the total being 20, but with twice as many newt eyes as frog toes. How many of each are required?

Solution As in the preceding example, the first step is to identify and label the unknowns. Let

x = number of newt eyes

y = number of frog toes

As for the second step—setting up the equations—a table is less appropriate here than in Example 5. Instead, we translate each phrase of the problem into an equation. The first sentence tells us that the total number of eyes and toes is 20. Thus,

$$x + y = 20$$

The end of the first sentence gives us more information, but the phrase "twice as many newt eyes as frog toes" is a little tricky: Does it mean that $2x = y$ or that $x = 2y$? We can decide which by rewording the statement using the phrases "the *number of* newt eyes," which is x, and "the *number of* frog toes," which is y. Rephrased, the statement reads:

*The **number of** newt eyes is twice the **number of** frog toes.*

(Notice how the word *twice* is forced into a different place.) With this rephrasing, we can translate directly into algebra:

$$x = 2y$$

In standard form ($ax + by = c$), this equation reads

$$x - 2y = 0$$

Thus, we have the two equations

$$x + y = 20$$
$$x - 2y = 0$$

To eliminate x, we multiply the second equation by -1 and then add:

$$x + y = 20$$
$$-x + 2y = 0$$

We'll leave it to you to finish solving the system and find that $x = 13\frac{1}{3}$ and $y = 6\frac{2}{3}$.

So, the recipe calls for exactly $13\frac{1}{3}$ eyes of newt and $6\frac{2}{3}$ toes of frog. The alchemist needed a very sharp scalpel and a very accurate balance (not to mention a very strong stomach).

We saw in Chapter 1 that the *equilibrium price* of an item (the price at which supply equals demand) and the *break-even point* (the number of items that must be sold to break even) can both be described as the intersection points of two lines. In terms of the language of this chapter, what we were really doing was solving a system of two linear equations in two unknowns, as illustrated in Example 7.

Example 7 • Equilibrium Price

The demand for refrigerators in West Podunk is given by

$$q = -\frac{p}{10} + 100$$

where q is the number of refrigerators that the citizens will buy each year if they are priced at p dollars each. The supply is

$$q = \frac{p}{20} + 25$$

where now q is the number of refrigerators the manufacturers will be willing to ship into town each year if they are priced at p dollars each. Find the equilibrium price and the number of refrigerators that will be sold at that price.

Solution Figure 7 shows the demand and supply curves. The equilibrium price occurs at the point where these two lines cross, which is where demand equals supply.

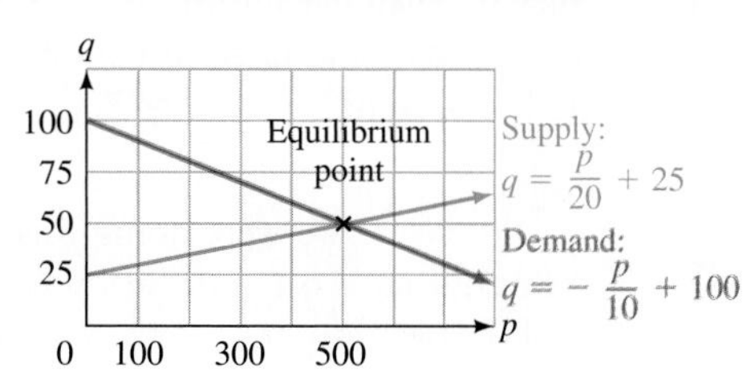

Figure 7

The graph suggests that the equilibrium price is \$500, and zooming in confirms this.

To solve this system algebraically, first write both equations in standard form:

$$\frac{p}{10} + q = 100$$

$$-\frac{p}{20} + q = 25$$

We can clear fractions and also prepare to eliminate p if we multiply the first equation by 10 and the second by 20:

$$p + 10q = 1000$$

$$-p + 20q = 500$$

and so

$$30q = 1500$$

$$q = 50$$

Substituting this value of q into either equation gives us $p = 500$.

Thus, the equilibrium price is \$500, and 50 refrigerators will be sold at this price.

Before we go on . . . We could also have solved this system of equations by setting the two expressions for q (the supply and the demand) equal to each other:

$$-\frac{p}{10} + 100 = \frac{p}{20} + 25$$

and then solving for p. (See the "Before we go on" discussion at the end of Example 1.)

Our last example shows a useful spreadsheet technique for solving a system of two linear equations without having to know explicitly what the equations are.

Example 8 • Finding a Break-Even Point Graphically

You are running a multi-user gaming Web site, www.Mudbeast.com,[4] and you charge users a fee to log on to the site. One month you charged a \$2.99 access fee, and your monthly revenue from the site amounted to \$500. After you lowered the price to \$1.89 the next month, activity increased greatly, and resulted in a monthly revenue of \$600. Monthly costs to maintain the site (commissions, server administration charges, and so on) are \$540 per month for a \$3.50 access fee and \$650 for a \$1 access fee. Graph revenue and cost as linear functions of the access fee p. What access fee should you charge to break even?

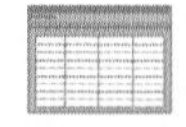

Solution If we were doing this problem algebraically, we would find equations for revenue and cost as linear functions of p and then compute the point of intersection of the two resulting graphs. (Recall that break even occurs when revenue equals cost.)

[4]Unregistered as of January 2003.

To solve the system graphically, all we need are two points on each line. We enter the coordinates of the points in the spreadsheet as follows:

	A	B	C
1	p	R	C
2	2.99	500	
3	1.89	600	
4	3.5		540
5	1		650

Graphing the data as usual (using series in columns) gives us Figure 8, where the lower line is the revenue graph.

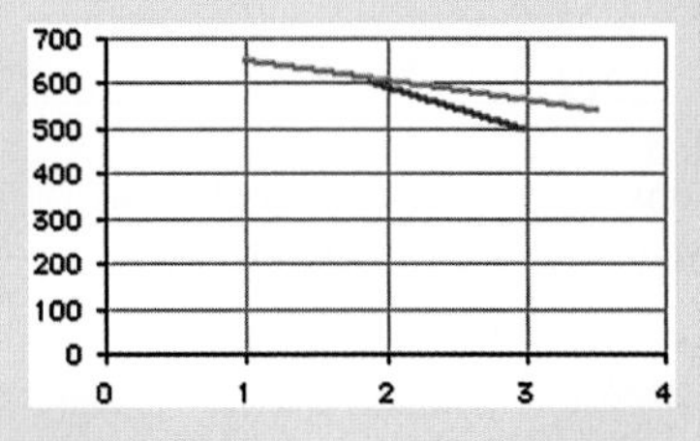

Figure 8

Question Uh-oh! The lines don't cross. How do we extend them to see the intersection point?

Answer To extend any line, we can use the Trendline feature. In Excel we select the line we wish to extend—the (lower) revenue graph in this case: Select `Add trendline` from the `Chart` menu. We will then get a dialog box, in which we select `Linear trend/regression`. Figure 9 shows the result.

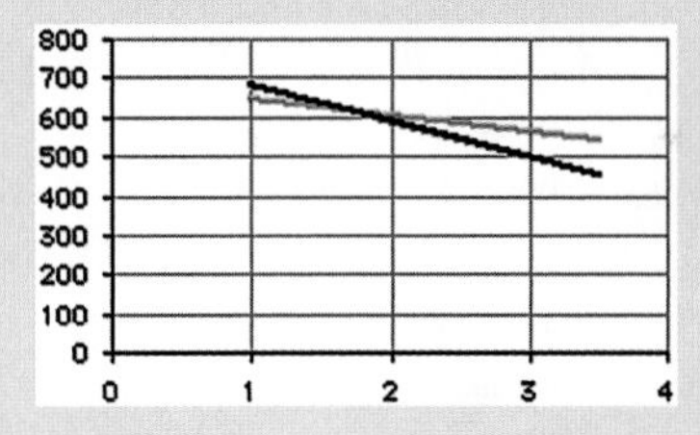

Figure 9

Since the intersection point is now visible, we can zoom in by changing the x and y ranges (double-clicking on the axis whose range we want to change and redefining the scales and units). Figure 10 was obtained this way.

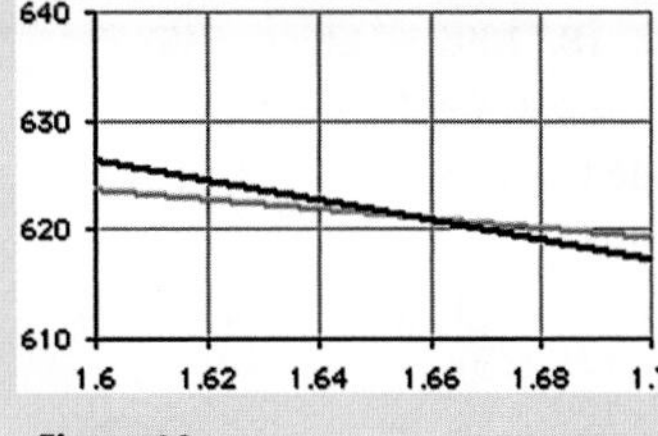

Figure 10

From the graph, we see that the break-even point occurs when $p = \$1.66$ (to the nearest 1¢). Thus, you should charge \$1.66 in order to break even. Should you charge more or less to make a profit?

Guideline: Setting Up the Equations

Question Looking through these examples, I notice that in some we can tabulate the information given and read off the equations (as in Example 5), whereas in others (like Example 6) we have to reword each sentence to turn it into an equation. How do I know what approach to use?

Answer There is no hard-and-fast rule, and indeed some applications might call for a bit of each approach. However, it is generally not hard to see when it would be useful to tabulate values: Lists of the numbers of ingredients or components generally lend themselves to tabulation, whereas phrases like "twice as many of these as those" generally require direct translation into equations (after rewording if necessary).

2.1 EXERCISES

In Exercises 1–14, find all solutions of the given system of equations and draw a graph of the system showing any solutions.

1. $x - y = 0$
$x + y = 4$

2. $x - y = 0$
$x + y = -6$

3. $x + y = 4$
$x - y = 2$

4. $2x + y = 2$
$-2x + y = 2$

5. $3x - 2y = 6$
$2x - 3y = -6$

6. $2x + 3y = 5$
$3x + 2y = 5$

7. $0.5x + 0.1y = 0.7$
$0.2x - 0.2y = 0.6$

8. $-0.3x + 0.5y = 0.1$
$0.1x - 0.1y = 0.4$

9. $\frac{x}{3} - \frac{y}{2} = 1$
$\frac{x}{4} + y = -2$

10. $-\frac{2x}{3} + \frac{y}{2} = -\frac{1}{6}$
$\frac{x}{4} - y = -\frac{3}{4}$

11. $2x + 3y = 1$
$-x - \frac{3y}{2} = -\frac{1}{2}$

12. $2x - 3y = 1$
$6x - 9y = 3$

13. $2x + 3y = 2$
$-x - \frac{3y}{2} = -\frac{1}{2}$

14. $2x - 3y = 2$
$6x - 9y = 3$

In Exercises 15–22, use technology to obtain approximate solutions graphically. All solutions should be accurate to one decimal place. (Zoom in for improved accuracy.)

15. $3.1x - 4.5y = 6$
$4.5x + 1.1y = 0$

16. $0.2x + 4.5y = 1$
$1.5x + 1.1y = 2$

17. $10.2x + 14y = 213$
$4.5x + 1.1y = 448$

18. $100x + 4.5y = 540$
$1.05x + 1.1y = 0$

19. Find the intersection of the line through (0, 1) and (4.2, 2) and the line through (2.1, 3) and (5.2, 0).

20. Find the intersection of the line through (2.1, 3) and (4, 2) and the line through (3.2, 2) and (5.1, 3).

21. Find the intersection of the line through (0, 0) and (5.5, 3) and the line through (5, 0) and (0, 6).

22. Find the intersection of the line through (4.3, 0) and (0, 5) and the line through (2.1, 2.2) and (5.2, 1).

APPLICATIONS

23. Resource Allocation You manage an ice cream factory that makes two flavors: Creamy Vanilla and Continental Mocha. Into each quart of Creamy Vanilla go 2 eggs and 3 cups of cream. Into each quart of Continental Mocha go 1 egg and 3 cups of cream. You have in stock 500 eggs and 900 cups of cream. How many quarts of each flavor should you make in order to use up all the eggs and cream?

24. Class Scheduling Enormous State University's Math Department offers two courses: Finite Math and Applied Calculus. Each section of Finite Math has 60 students, and each section of Applied Calculus has 50. The department will offer a total of 110 sections in a semester, and 6000 students would like to take a math course. How many sections of each course should the department offer in order to accommodate all students?

25. Nutrition Gerber Mixed Cereal for Baby contains, in each serving, 60 calories and 11 grams of carbohydrates. Gerber Mango Tropical Fruit Dessert contains, in each serving, 80 calories and 21 grams of carbohydrates. If you want to provide your child with 200 calories and 43 grams of carbohydrates, how many servings of each should you use?

SOURCE: Nutrition information printed on the box and the jar.

26. Nutrition Anthony Latino is mixing food for his young daughter and would like the meal to supply 1 gram of protein and 5 milligrams of iron. He is mixing together cereal, with 0.5 grams of protein and 1 milligram of iron per ounce, and fruit, with 0.2 grams of protein and 2 milligrams of iron per ounce. What mixture will provide the desired nutrition?

27. Nutrition One serving of Campbell's® Pork & Beans contains 5 grams of protein and 21 grams of carbohydrates. A typical slice of white bread provides 2 grams of protein and 11 grams of carbohydrates per slice. The U.S. RDA (Recommended Daily Allowance) is 60 grams of protein each day.

a. I am planning a meal of "beans on toast" and wish to have it supply one-half of the RDA for protein and 139 grams of carbohydrates. How should I prepare my meal?

b. Is it possible to have my meal supply the same amount of protein as in part (a) but only 100 grams of carbohydrates?

SOURCE: According to the label information on a 16-ounce can.

28. Nutrition One serving of Campbell's Pork & Beans contains 5 grams of protein and 21 grams of carbohydrates. A typical slice of "lite" rye bread contains 4 grams of protein and 12 grams of carbohydrates.

a. I'm planning a meal of "beans on toast" and wish to have it supply one-third of the RDA for protein (see the preceding exercise) and 80 grams of carbohydrates. How should I prepare my meal?

b. Is it possible to have my meal supply the same amount of protein as in part (a) but only 60 grams of carbohydrates?

SOURCE: According to the label information on a 16-ounce can.

Creatine Supplements Exercises 29 and 30 are based the following data on three body-building supplements popular in 2002. (Figures shown correspond to a single serving.)

	Creatine (g)	Carbohydrates (g)	Alpha Lipoic Acid (mg)	Cost ($)
Cell-Tech® (MuscleTech)	10	75	200	2.20
RiboForce HP® (EAS)	5	15	0	1.60
Creatine Transport® (Kaizen)	5	35	100	0.60

SOURCE: Nutritional information supplied by the manufacturers, http://www.netrition.com. Cost per serving as quoted on www.netrition.com as of February 26, 2002.

29. You are thinking of combining Cell-Tech and RiboForce HP to obtain a 10-day supply that provides exactly 80 grams of creatine and 1000 milligrams of alpha lipoic acid. How many servings of each supplement should you combine in order to meet your requirements? What will it cost?

30. You are thinking of combining Cell-Tech and Creatine Transport to obtain a 10-day supply that provides exactly 125 grams of creatine and 900 grams of carbohydrates. How many servings of each supplement should you combine in order to meet your requirements? What will it cost?

31. Investments In March 2002 Cisco (CSCO) stock rose from $15 to $16, and America Online-Time Warner (AOL) dropped from $25 to $24. If you invested a total of $6500 in these stocks at the beginning of March and sold them for $6400 at the end of March, how many shares of each stock did you buy?

Share prices are approximate. SOURCE: Quotes provided by S&P Comstock, http://money.excite.com, April 1, 2002.

32. Investments In March 2002 Dell Computer (DELL) stock started and ended the month valued at $26, and America Online-Time Warner (AOL) dropped from $25 to $24. If you invested a total of $5100 in these stocks at the beginning of March and sold them for $5000 at the end of March, how many shares of each stock did you buy?

Share prices are approximate. SOURCE: Quotes provided by S&P Comstock, http://money.excite.com, April 1, 2002.

33. Voting The U.S. House of Representatives has 435 members. If an appropriations bill passes the House with 49 more members voting in favor than against, how many voted in favor and how many voted against?

34. Voting The U.S. Senate has 100 members. For a bill to pass with a supermajority, at least twice as many senators must vote for the bill as vote against it. If all 100 senators vote, how many must vote for a bill for it to pass with a supermajority?

35. Intramural Sports The best sports dorm on campus, Lombardi House, has won a total of 12 games this semester. Some of these games were soccer games, and the others were football games. According to the rules of the university, each win in a soccer game earns the winning house 2 points, whereas each win in a football game earns them 4 points. If the total number of points Lombardi House earned was 38, how many of each type of game did they win?

36. Law Enormous State University's campus publication, *The Campus Inquirer,* ran a total of 10 exposés 5 years ago, dealing with alleged recruiting violations by the football team and with theft by the student treasurer of the film society. Each exposé dealing with recruiting violations resulted in a $4 million libel suit, and the treasurer of the film society sued the paper for $3 million as a result of each exposé concerning his alleged theft. Unfortunately for *The Campus Inquirer,* all lawsuits were successful, and the paper wound up being ordered to pay $37 million in damages. (It closed down shortly thereafter.) How many of each type of exposé did the paper run?

37. Purchasing (from the GMAT) Elena purchased brand X pens for $4.00 apiece and brand Y pens for $2.80 apiece. If Elena purchased a total of 12 of these pens for $42.00, how many brand X pens did she purchase?

38. Purchasing (based on a question from the GMAT) Earl is ordering supplies. Yellow paper costs $5.00 per ream, and white paper costs $6.50 per ream. He would like to order 100 reams total and has a budget of $560.00. How many reams of each color should he order?

39. Equilibrium Price The demand and supply functions for pet Chias are, respectively, $q = -60p + 150$ and $q = 80p - 60$, where p is the price in dollars. At what price should the Chias be marked so that there is neither a surplus nor a shortage of Chias?

40. Equilibrium Price The demand and supply functions for your college newspaper are, respectively, $q = -10{,}000p + 2000$ and $q = 4000p + 600$, where p is the price in dollars. At what price should the newspapers be sold so that there is neither a surplus nor a shortage of papers?

41. Equilibrium Price In June 2001 the retail price of a 25-kilogram bag of cornmeal was $8 in Zambia; by December the price had risen to $11. The result was that one retailer reported a drop in sales from 15 bags/day to 3 bags/day. Assume that the retailer is prepared to sell 3 bags/day at $8 and 15 bags/day at $11. Find linear demand and supply equations and then compute the retailer's equilibrium price.

The prices quoted are approximate. (Actual prices varied from retailer to retailer.) SOURCE: *New York Times,* December 24, 2001, p. A4.

42. **Equilibrium Price** At the start of December 2001, the retail price of a 25-kilogram bag of cornmeal was \$10 in Zambia; by the end of the month, the price had fallen to \$6. The result was that one retailer reported an increase in sales from 3 bags/day to 5 bags/day. Assume that the retailer is prepared to sell 18 bags/day at \$8 and 12 bags/day at \$6. Obtain linear demand and supply equations and then compute the retailer's equilibrium price.

The prices quoted are approximate. (Actual prices varied from retailer to retailer.) SOURCE: *New York Times*, December 24, 2001, p. A4.

43. **Pollution** Joe Slo, a college sophomore, neglected to wash his dirty laundry for 6 weeks. By the end of that time, his roommate had had enough and tossed Joe's dirty socks and T-shirts into the trash, counting a total of 44 items. (A pair of dirty socks counts as 1 item.) The roommate noticed that there were three times as many pairs of dirty socks as T-shirts. How many of each item did he throw out?

44. **Diet** The local sushi bar serves 1-ounce pieces of raw salmon (consisting of 50% protein) and $1\frac{1}{4}$-ounce pieces of raw tuna (40% protein). A customer's total intake of protein amounts to $1\frac{1}{2}$ ounces after consuming a total of three pieces. How many of each did the customer consume? (Fractions of pieces are permitted.)

45. **Management (from the GMAT)** A manager has \$6000 budgeted for raises for four full-time and two part-time employees. Each full-time employee receives the same raise, which is twice the raise that each part-time employee receives. What is the amount of the raise that each full-time employee receives?

46. **Publishing (from the GMAT)** There were 36,000 hardback copies of a certain novel sold before the paperback version was issued. From the time the first paperback copy was sold until the last copy of the novel was sold, nine times as many paperback copies as hardback copies were sold. If a total of 441,000 copies of the novel were sold in all, how many paperback copies were sold?

47. **Publishing** The demand per year for finite mathematics books is given by $q = -1000p + 140{,}000$, where p is the price per book in dollars. The supply is given by $q = 2000p + 20{,}000$. Find the price at which supply and demand balance.

48. **Airplane Manufacture** The demand per year for jumbo jets is given by $q = -2p + 18$, where p is the price per jet in millions of dollars. The supply is given by $q = 3p + 3$. Find the price at which supply and demand will balance.

49. **Supply and Demand (from the GRE Economics Test)** The demand curve for widgets is given by $D = 85 - 5P$, and the supply curve is given by $S = 25 + 5P$, where P is the price of widgets. When the widget market is in equilibrium, what is the quantity of widgets bought and sold?

50. **Supply and Demand (from the GRE Economics Test)** In the market for soybeans, the demand and supply functions are $Q_D = 100 - 10P$ and $Q_S = 20 + 5P$, where Q_D is quantity demanded, Q_S is quantity supplied, and P is price in dollars. If the government sets a price floor of \$7, what will be the resulting surplus or shortage?

COMMUNICATION AND REASONING EXERCISES

51. A system of three equations in two unknowns corresponds to three lines in the plane. Describe how these lines might be positioned if the system has a unique solution.

52. A system of three equations in two unknowns corresponds to three lines in the plane. Describe several ways that these lines might be positioned if the system has no solutions.

53. Both the supply and demand equations for a certain product have negative slope. Can there be an equilibrium price? Explain.

54. You are solving a system of equations with x representing the number of rocks and y representing the number of pebbles. The solution is $(200, -10)$. What do you conclude?

55. Referring to Exercise 23 but using different data, suppose that the solution of the corresponding system of equations was 198.7 gallons of vanilla and 100.89 gallons of mocha. If your factory can produce only whole numbers of gallons, would you recommend rounding the answers to the nearest whole number? Explain.

56. Referring to Exercise 23 but using different data, suppose that the general solution of the corresponding system of equations was $(200 - y, y)$, where x = number of gallons vanilla and y = number of gallons of mocha. Your factory can produce only whole numbers of gallons. There are infinitely many ways of making the ice cream mixes, right? Explain.

57. Select one: Multiplying both sides of a linear equation by a nonzero constant results in a linear equation whose graph is
(A) parallel to (B) the same as
(C) not always parallel to (D) not the same as
the graph of the original equation.

58. Select one: If the addition or subtraction of two linear equations results in the equation $3 = 3$, then the graphs of those equations are
(A) equal. (B) parallel.
(C) perpendicular. (D) none of the above.

59. Select one: If the addition or subtraction of two linear equations results in the equation $0 = 3$, then the graphs of those equations are
(A) equal. (B) parallel.
(C) perpendicular. (D) not parallel.

60. Select one: If adding two linear equations gives $x = 3$ and subtracting them gives $y = 3$, then the graphs of those equations are
(A) equal. (B) parallel.
(C) perpendicular. (D) not parallel.

61. How likely do you think it is that a "random" system of two equations in two unknowns has a unique solution? Give some justification for your answer.

62. How likely do you think it is that a "random" system of three equations in two unknowns has a unique solution? Give some justification for your answer.

63. Invent an interesting application that leads to a system of two equations in two unknowns with a unique solution.

64. Invent an interesting application that leads to a system of two equations in two unknowns with no solution.

2.2 Using Matrices to Solve Systems of Equations

In this section we describe a systematic method for solving systems of equations that makes solving large systems of equations straightforward. Although this method may seem a little cumbersome at first, it will prove *immensely* useful in this and the next several chapters.

First, notice that a linear equation (for example, $2x - y = 3$) is entirely determined by its *coefficients* (here, the numbers 2 and -1) and its *constant term* or *right-hand side* (here, 3). In other words, if we were simply given the row of numbers

$$[2 \quad -1 \quad 3]$$

we could easily reconstruct the original linear equation by multiplying the first number by x, the second by y, and inserting a plus sign and an equals sign, as follows:

$$2 \cdot x + (-1) \cdot y = 3 \quad \text{or} \quad 2x - y = 3$$

Similarly, the equation

$$-4x + 2y = 0$$

is represented by the row

$$[-4 \ 2 \ 0]$$

and the equation

$$-3y = \frac{1}{4}$$

is represented by the row by

$$\left[0 \quad -3 \quad \frac{1}{4}\right]$$

As the last example shows, the first number is always the coefficient of x, and the second is the coefficient of y. If an x or a y is missing, we write a zero for its coefficient. We call such a row the **coefficient row** of an equation.

Question What good are coefficient rows?

Answer Think about what we do when we multiply both sides of an equation by a number. For example, consider multiplying both sides of the equation $2x - y = 3$ by -2 to get $-4x + 2y = -6$. All we are really doing is multiplying the coefficients and the right-hand side by -2. This corresponds to *multiplying the row* $[2 \quad -1 \quad 3]$ *by* -2, that is, multiplying every number in the row by -2. We will see that any manipulation we want to do with equations can be done instead with rows, and this fact leads to a method of solving equations that is systematic and generalizes easily to larger systems.

Here is the same operation both in the language of equations and in the language of rows. (We refer to the equation here as *equation* 1, or simply E_1 for short, and to the row as *row* 1, or R_1.)

Equation	Row
E_1: $2x - y = 3$	$[\ 2 \quad -1 \quad 3]$ R_1
Multiply by -2: $(-2)E_1$: $-4x + 2y = -6$	$[-4 \quad 2 \quad -6]$ $(-2)R_1$

Multiplying both sides of an equation by the number a corresponds to multiplying the coefficient row by a.

Now look at what we do when we add two equations:

Equation	Row
E_1: $2x - y = 3$	$[2 \quad -1 \quad 3]$ R_1
E_2: $-x + 2y = -4$	$[-1 \quad 2 \quad -4]$ R_2
Add: $E_1 + E_2$: $x + y = -1$	$[1 \quad 1 \quad -1]$ $R_1 + R_2$

All we are really doing is *adding the corresponding entries in the rows,* or *adding the rows.* In other words,

Adding two equations corresponds to adding their coefficient rows.

In short, the manipulations of equations that we saw in Section 2.1 can be done more easily with rows, since we don't have to carry x, y, and other unnecessary notation along with us; x and y can always be inserted at the end if desired.

The manipulations we are talking about are known as **row operations.** In particular, we use three **elementary row operations.**

Elementary Row Operations*

Type 1 Replacing R_i by aR_i (where $a \neq 0$)†

In words: Multiplying or dividing a row by a nonzero number

Type 2 Replacing R_i by $aR_i \pm bR_j$ (where $a \neq 0$)

In words: Multiplying a row by a nonzero number and adding or subtracting a multiple of another row

Type 3 Switching the order of the rows

This corresponds to switching the order in which we write the equations; occasionally this will be convenient.

For types 1 and 2, we write the instruction for the row operation *next to the row we wish to replace* (see the Quick Examples below).

Quick Examples

Type 1 $\begin{bmatrix} 1 & 3 & -4 \\ 0 & 4 & 2 \end{bmatrix} \begin{matrix} \\ 3R_2 \end{matrix} \rightarrow \begin{bmatrix} 1 & 3 & -4 \\ 0 & 12 & 6 \end{bmatrix}$ Replace R_2 by $3R_2$.

Type 2 $\begin{bmatrix} 1 & 3 & -4 \\ 0 & 4 & 2 \end{bmatrix} \begin{matrix} 4R_1 - 3R_2 \\ \\ \end{matrix} \rightarrow \begin{bmatrix} 4 & 0 & -22 \\ 0 & 4 & 2 \end{bmatrix}$ Replace R_1 by $4R_1 - 3R_2$.

Type 3 $\begin{bmatrix} 1 & 3 & -4 \\ 0 & 4 & 2 \\ 1 & 2 & 3 \end{bmatrix} R_1 \leftrightarrow R_2 \rightarrow \begin{bmatrix} 0 & 4 & 2 \\ 1 & 3 & -4 \\ 1 & 2 & 3 \end{bmatrix}$ Switch R_1 and R_2.

*We are using the term *elementary row operations* a little more freely than most books. Some mathematicians insist that $a = 1$ in an operation of type 2, but the less restrictive version is very useful.

†Multiplying an equation or row by zero gives us the not very surprising result $0 = 0$. In fact, we lose any information that that equation provided, which usually means that the resulting system has more solutions than the original system.

Row Operations with a Graphing Calculator

On the TI-83 and similar calculators, start by entering the matrix into `[A]` using [MATRX] `EDIT`. You can then do row operations on `[A]` using the following instructions (`*row`, `*row+`, and `rowSwap` are found in the [MATRX] `MATH` menu):

Row Operation	TI-83 Instruction (Matrix Name Is [A].)
$R_i \to kR_i$	`*row(k,[A],i)→[A]`
Example: $R_2 \to 3R_2$	`*row(3,[A],2)→[A]`
$R_i \to R_i + kR_j$	`*row+(k,[A],j,i)→[A]`
Examples: $R_1 \to R_1 - 3R_2$	`*row+(-3,[A],2,1)→[A]`
$R_1 \to 4R_1 - 3R_2$	`*row(4,[A],1)→[A]` `*row+(-3,[A],2,1)→[A]`
Swap R_i and R_j.	`rowSwap([A],i,j)`
Example: Swap R_1 and R_2.	`rowSwap([A],1,2)`

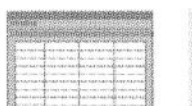

Row Operations with Excel

To Excel a row is a block of data, or an **array,** and Excel has built-in the capability to handle arrays in much the same way as it handles single cells. Consider the first Quick Example above. To get the first row of the new matrix, which is simply a copy of the first row of the old matrix, decide where you want to place the new matrix and highlight *the whole row of the new matrix,* say, `A4:C4`. Then enter the formula `=A1:C1` (the easiest way to do this is to type = and then use the mouse to select cells A1 through C1—that is, the first row of the old matrix).

	A	B	C	D
1	1	3	-4	
2	0	4	2	3R2
3				
4	=A1:C1			
5				

Enter formula for new first row.

→

	A	B	C	D
1	1	3	-4	
2	0	4	2	3R2
3				
4	1	3	-4	
5				

Press [Control]+[Shift]+[Enter].

Then press [Control]+[Shift]+[Enter] (instead of [Enter] alone) and the whole row will be copied.[5] Pressing [Control]+[Shift]+[Enter] tells Excel that your formula is an *array formula,* one that returns an array rather than a single number. Once entered, Excel will show an array formula enclosed in "curly braces." (Note that you must also use [Control]+[Shift]+[Enter] to delete any array you create: Select the block you wish to delete and press [Delete] followed by [Control]+[Shift]+[Enter].)

Similarly, to get the second row, select cells A5 through C5, where the new second row will go, enter the formula `=3*A2:C2`, and press [Control]+[Shift]+[Enter]. (Again, the easiest way to enter the formula is to type `=3*` and then select the cells `A2:C2`, using the mouse.)

	A	B	C	D
1	1	3	-4	
2	0	4	2	3R2
3				
4	1	3	-4	
5	=3*A2:C2			

Enter formula for new second row.

→

	A	B	C	D
1	1	3	-4	
2	0	4	2	3R2
3				
4	1	3	-4	
5	0	12	6	

Press [Control]+[Shift]+[Enter].

[5]On a Macintosh, [Command]+[Enter] and [Command]+[Shift]+[Enter] have the same effect as [Control]+[Shift]+[Enter].

We can perform the row operation in Quick Example 2 above in a similar way:

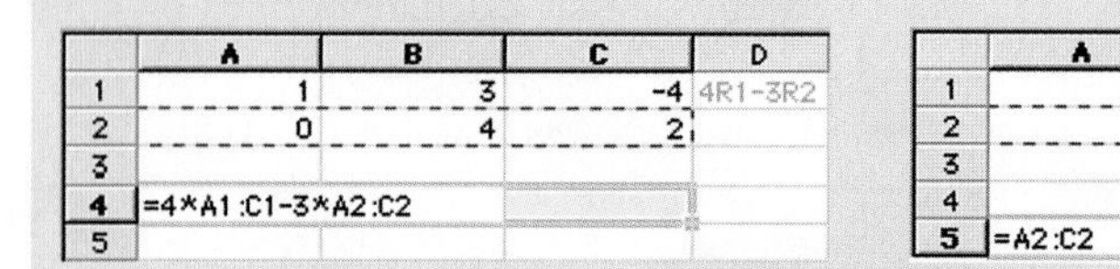

	A	B	C	D
1	1	3	-4	4R1-3R2
2	0	4	2	
3				
4	=4*A1:C1-3*A2:C2			
5				

	A	B	C	D
1	1	3	-4	4R1-3R2
2	0	4	2	
3				
4	4	0	-22	
5	=A2:C2			

(To easily enter the formula for $4R_1 - 3R_2$, type `=4*`; select the first row `A1:C1`, using the mouse; type `-3*`, and then select the second row `A2:C2`, using the mouse.)

Note One very important fact about the elementary row operations is that they do not change the solutions of the corresponding system of equations. In other words, the new system of equations that we get by applying any one of these operations will have exactly the same solutions as the original system.

Question Why?

Answer It is easy to see that any solution of the old system will be a solution of the new system; numbers that make the original equations true will also make the new equations true. This is because each of the elementary row operations corresponds to a valid operation on the original equations. That any solution of the new system is a solution of the old system follows from the fact that these row operations are *reversible*: The effects of a row operation can be reversed by applying another row operation, called its **inverse.** Here are some examples of this reversibility (try them out in the above Quick Examples):

Operation	**Inverse Operation**
Replace R_2 by $3R_2$.	Replace R_2 by $\frac{1}{3}R_2$.
Replace R_1 by $4R_1 - 3R_2$.	Replace R_1 by $\frac{1}{4}R_1 + \frac{3}{4}R_2$.
Switch R_1 and R_2.	Switch R_1 and R_2.

Our objective, then, is to use row operations to change the system we are given into one, with exactly the same set of solutions, in which it is easy to see what the solutions are.

Solving Systems of Equations Using Row Operations

Now we put rows to work for us in solving systems of equations. Let's start with a complicated looking system of equations:

$$-\frac{2x}{3} + \frac{y}{2} = -3$$

$$\frac{x}{4} - y = \frac{11}{4}$$

This system corresponds to the following two rows.

$$\begin{bmatrix} -\frac{2}{3} & \frac{1}{2} & -3 \\ \frac{1}{4} & -1 & \frac{11}{4} \end{bmatrix}$$

We call this the **augmented matrix** of the system of equations. The term *augmented* means that we have included the right-hand sides -3 and $11/4$. We will often drop the word *augmented* and simply refer to the matrix of the system. A **matrix** (plural: **matrices**) is nothing more than a rectangular array of numbers as above. (We'll be studying matrices in their own right more carefully in Chapter 3.) Now what do we do with this matrix?

Step 1

Clear the fractions and/or decimals (if any) using operations of type 1. To clear the fractions, we multiply the first row by 6 and the second row by 4. We record the operations by writing the symbolic form of an operation next to the row it will change, as follows:

$$\begin{bmatrix} -\frac{2}{3} & \frac{1}{2} & -3 \\ \frac{1}{4} & -1 & \frac{11}{4} \end{bmatrix} \begin{matrix} 6R_1 \\ 4R_2 \end{matrix}$$

By this we mean that we will replace the first row by $6R_1$ and the second by $4R_2$. Doing these operations gives

$$\begin{bmatrix} -4 & 3 & -18 \\ 1 & -4 & 11 \end{bmatrix}$$

Step 2

Designate the first nonzero entry in the first row as the ***pivot.*** In this case we designate the entry -4 in the first row as the pivot by putting a box around it:

$$\begin{bmatrix} \boxed{-4} & 3 & -18 \\ 1 & -4 & 11 \end{bmatrix} \begin{matrix} \leftarrow \text{Pivot row} \\ \\ \end{matrix}$$

↑
Pivot column

Question What is a pivot?

Answer A **pivot** is an entry in a matrix that is used to "clear a column" (see Step 3). In this procedure we will always select the first nonzero entry of a row as our pivot. In Chapter 4, when we study the simplex method, we will select our pivots differently.

Step 3

Use the pivot to clear its column using operations of type 2. By **clearing a column,** we mean changing the matrix so that the pivot is the only nonzero number in its column. The procedure of clearing a column using a designated pivot is also called **pivoting:**

$$\begin{bmatrix} \boxed{-4} & 3 & -18 \\ 0 & \# & \# \end{bmatrix} \begin{matrix} \\ \leftarrow \text{Desired row 2 (\# stands for a number not yet known)} \end{matrix}$$

↑
Cleared pivot column

We want to replace R_2 by a row of the form $aR_2 \pm bR_1$ to get a zero in column 1. Moreover—and this will be important when we discuss the simplex method in Chapter 4—*we are going to choose positive values for both a and b.*[6] We need to choose a and b so that we get the desired cancellation. We can do this quite mechanically as follows:

[6]The only place a negative sign may appear is between aR_2 and bR_1, as indicated in the formula $aR_2 \pm bR_1$.

a. Write the name of the row you need to change on the left and that of the pivot row on the right.

$$\underset{\text{Row to change}}{\overset{\boldsymbol{R_2}}{\uparrow}} \qquad \underset{\text{Pivot row}}{\overset{\boldsymbol{R_1}}{\uparrow}}$$

b. Focus on the pivot column, $\begin{bmatrix} -4 \\ 1 \end{bmatrix}$. Multiply each row by the *absolute value* of the entry currently in the other (we are not permitting a or b to be negative):

$$\underset{\text{From row 1}}{\overset{\mathbf{4}R_2}{\uparrow}} \qquad \underset{\text{From row 2}}{\overset{\mathbf{1}R_1}{\uparrow}}$$

The effect is to make the two entries in the pivot column numerically the same. Sometimes, you can accomplish this by using smaller values of a and b.

c. If the entries in the pivot column have opposite signs, insert a plus $(+)$. If they have the same sign, insert a minus $(-)$. Here, we get the instruction

$$4R_2 + 1R_1$$

d. Write the operation next to the row you want to change and then replace that row using the operation:

$$\begin{bmatrix} \boxed{-4} & 3 & -18 \\ 1 & -4 & 11 \end{bmatrix} \begin{matrix} \\ 4R_2 + 1R_1 \end{matrix} \rightarrow \begin{bmatrix} -4 & 3 & -18 \\ 0 & -13 & 26 \end{bmatrix}$$

We have cleared the pivot column and completed Step 3.

Note In general, the row operation you use should always have the following form:[7]

$$\underset{\text{Row to change}}{\overset{aR_c}{\uparrow}} \quad \pm \quad \underset{\text{Pivot row}}{\overset{bR_p}{\uparrow}}$$

with a and b both positive.

The next step is one that can be performed at any time.

Simplification Step (Optional)

If, at any stage of the process, all the numbers in a row are multiples of an integer, divide by that integer—a type 1 operation. This is an optional but extremely helpful step: It makes the numbers smaller and easier to work with. In our case the entries in R_2 are divisible by 13, so we divide that row by 13. (Alternatively, we could divide by -13. Try it.)

$$\begin{bmatrix} -4 & 3 & -18 \\ 0 & -13 & 26 \end{bmatrix} \begin{matrix} \\ \frac{1}{13}R_2 \end{matrix} \rightarrow \begin{bmatrix} -4 & 3 & -18 \\ 0 & -1 & 2 \end{bmatrix}$$

Step 4

Select the first nonzero number in the second row as the pivot and clear its column. Here we have combined two steps in one: selecting the new pivot and clearing the column (piv-

---

[7] We are deviating somewhat from the traditional procedure here. It is traditionally recommended first to divide the pivot row by the pivot, turning the pivot into a 1. This allows us to always use $a = 1$. The procedure we use here is easier for hand calculations since it avoids the use of fractions. See the end of this section for an example using the traditional procedure.

oting). The pivot is shown below, as well as the desired result when the column has been cleared:

$$\begin{bmatrix} -4 & 3 & -18 \\ 0 & \boxed{-1} & 2 \end{bmatrix} \rightarrow \begin{bmatrix} \# & 0 & \# \\ 0 & -1 & 2 \end{bmatrix} \leftarrow \text{Desired row}$$

$\uparrow$ Pivot column $\qquad$ $\uparrow$ Cleared pivot column

We now wish to get a 0 in place of the 3 in the pivot column. Let's run once again through the mechanical steps to get the row operation that accomplishes this.

a. Write the name of the row you need to change on the left and that of the pivot row on the right:

$$\underset{\text{Row to change}}{\underset{\uparrow}{\boldsymbol{R_1}}} \qquad \underset{\text{Pivot row}}{\underset{\uparrow}{\boldsymbol{R_2}}}$$

b. Focus on the pivot column, $\begin{bmatrix} 3 \\ -1 \end{bmatrix}$. Multiply each row by the absolute value of the entry currently in the other:

$$\underset{\text{From row 2}}{\underset{\uparrow}{\mathbf{1}R_1}} \qquad \underset{\text{From row 1}}{\underset{\uparrow}{\mathbf{3}R_2}}$$

c. If the entries in the pivot column have opposite signs, insert a plus (+). If they have the same sign, insert a minus (−). Here, we get the instruction

$$1R_1 + 3R_2$$

d. Write the operation next to the row you want to change and then replace that row using the operation.

$$\begin{bmatrix} -4 & 3 & -18 \\ 0 & \boxed{-1} & 2 \end{bmatrix} \begin{matrix} R_1 + 3R_2 \\ \\ \end{matrix} \rightarrow \begin{bmatrix} -4 & 0 & -12 \\ 0 & -1 & 2 \end{bmatrix}$$

Now we are essentially done, except for one last step.

Final Step

Using operations of type 1, *turn each pivot (the first nonzero entry in each row) into a* 1. We can accomplish this by dividing the first row by −4 and multiplying the second row by −1:

$$\begin{bmatrix} -4 & 0 & -12 \\ 0 & -1 & 2 \end{bmatrix} \begin{matrix} -\frac{1}{4}R_1 \\ -R_2 \end{matrix} \rightarrow \begin{bmatrix} 1 & 0 & 3 \\ 0 & 1 & -2 \end{bmatrix}$$

The matrix now has the following nice form:

$$\begin{bmatrix} \boxed{1} & 0 & \# \\ 0 & \boxed{1} & \# \end{bmatrix}$$

(This is the form we will always obtain with two equations in two unknowns when there is a unique solution.) This form is nice because, when we translate back into equations, we get

$$1x + 0y = 3$$

$$0x + 1y = -2$$

In other words,

$$x = 3 \quad \text{and} \quad y = -2$$

and so we have found the solution, which we can also write as $(x, y) = (3, -2)$.

The procedure we've just demonstrated is called **Gauss–Jordan reduction** or **row reduction.**[8]

In Example 1 we use this procedure to solve a system of linear equations in *three* unknowns: x, y, and z. Solving such a system graphically would require the graphing of planes (flat surfaces) in three dimensions. (The graph of a linear equation in three unknowns is a flat surface.) The use of row reduction makes three-dimensional graphing unnecessary.

Example 1 • Solving a System by Gauss–Jordan Reduction

Solve the system

$$\begin{aligned} x - y + 5z &= -6 \\ 3x + 3y - z &= 10 \\ x + 3y + 2z &= 5 \end{aligned}$$

Solution The augmented matrix for this system is

$$\begin{bmatrix} 1 & -1 & 5 & -6 \\ 3 & 3 & -1 & 10 \\ 1 & 3 & 2 & 5 \end{bmatrix}$$

Note that the columns correspond to x, y, z, and the right-hand side, respectively. We begin by selecting the pivot in the first row and clearing its column. Remember that clearing the column means that we turn *all* other numbers in the column into zeros. Thus, to clear the column of the first pivot, we need to change two rows, setting up the row operations in exactly the same way as above:

$$\begin{bmatrix} \boxed{1} & -1 & 5 & -6 \\ 3 & 3 & -1 & 10 \\ 1 & 3 & 2 & 5 \end{bmatrix} \begin{matrix} \\ R_2 - 3R_1 \\ R_3 - R_1 \end{matrix} \rightarrow \begin{bmatrix} 1 & -1 & 5 & -6 \\ 0 & 6 & -16 & 28 \\ 0 & 4 & -3 & 11 \end{bmatrix}$$

Notice that both row operations have the required form

$$\underset{\text{Row to change}}{\underset{\uparrow}{aR_c}} \quad \pm \quad \underset{\text{Pivot row}}{\underset{\uparrow}{bR_1}}$$

with a and b both positive.

Now we use the optional simplification step to simplify R_2:

$$\begin{bmatrix} 1 & -1 & 5 & -6 \\ 0 & 6 & -16 & 28 \\ 0 & 4 & -3 & 11 \end{bmatrix} \begin{matrix} \\ \frac{1}{2}R_2 \\ \\ \end{matrix} \rightarrow \begin{bmatrix} 1 & -1 & 5 & -6 \\ 0 & 3 & -8 & 14 \\ 0 & 4 & -3 & 11 \end{bmatrix}$$

---

[8]Carl Friedrich Gauss (1777–1855) was one of the great mathematicians, making fundamental contributions to number theory, analysis, probability and statistics, as well as many fields of science. He developed a method of solving systems of linear equations by row reduction, which was then refined by Wilhelm Jordan (1842–1899) into the form we are showing you here.

Next, we select the pivot in the second row and clear its column:

$$\begin{bmatrix} 1 & -1 & 5 & -6 \\ 0 & \boxed{3} & -8 & 14 \\ 0 & 4 & -3 & 11 \end{bmatrix} \begin{matrix} 3R_1 + R_2 \\ \\ 3R_3 - 4R_2 \end{matrix} \rightarrow \begin{bmatrix} 3 & 0 & 7 & -4 \\ 0 & 3 & -8 & 14 \\ 0 & 0 & 23 & -23 \end{bmatrix}$$

R_1 and R_3 are changed.
R_2 is the pivot row.

We simplify R_3:

$$\begin{bmatrix} 3 & 0 & 7 & -4 \\ 0 & 3 & -8 & 14 \\ 0 & 0 & 23 & -23 \end{bmatrix} \begin{matrix} \\ \\ \frac{1}{23}R_3 \end{matrix} \rightarrow \begin{bmatrix} 3 & 0 & 7 & -4 \\ 0 & 3 & -8 & 14 \\ 0 & 0 & 1 & -1 \end{bmatrix}$$

Now we select the pivot in the third row and clear its column:

$$\begin{bmatrix} 3 & 0 & 7 & -4 \\ 0 & 3 & -8 & 14 \\ 0 & 0 & \boxed{1} & -1 \end{bmatrix} \begin{matrix} R_1 - 7R_3 \\ R_2 + 8R_3 \\ \\ \end{matrix} \rightarrow \begin{bmatrix} 3 & 0 & 0 & 3 \\ 0 & 3 & 0 & 6 \\ 0 & 0 & 1 & -1 \end{bmatrix}$$

R_1 and R_2 are changed.
R_3 is the pivot row.

Finally, we turn all the pivots into 1s:

$$\begin{bmatrix} 3 & 0 & 0 & 3 \\ 0 & 3 & 0 & 6 \\ 0 & 0 & 1 & -1 \end{bmatrix} \begin{matrix} \frac{1}{3}R_1 \\ \frac{1}{3}R_2 \\ \\ \end{matrix} \rightarrow \begin{bmatrix} 1 & 0 & 0 & 1 \\ 0 & 1 & 0 & 2 \\ 0 & 0 & 1 & -1 \end{bmatrix}$$

The matrix is now reduced, so we translate back into equations to obtain the solution:

$$x = 1, y = 2, z = -1 \quad \text{or} \quad (x, y, z) = (1, 2, -1)$$

Excel

Here is the complete row reduction as it would appear in Excel:

	A	B	C	D	E
1	1	-1	5	-6	
2	3	3	-1	10	R2 - 3R1
3	1	3	2	5	R3 - R1
4					
5	=A1:D1				
6	=A2:D2-3*A1:D1				
7	=A3:D3-A1:D1				

	A	B	C	D	E
1	1	-1	5	-6	
2	3	3	-1	10	R2 - 3R1
3	1	3	2	5	R3 - R1
4					
5	1	-1	5	-6	
6	0	6	-16	28	(1/2)*R2
7	0	4	-3	11	
8					
9	=A5:D5				
10	=(1/2)*A6:D6				
11	=A7:D7				

	A	B	C	D	E
1	1	-1	5	-6	
2					
8					
9	1	-1	5	-6	3R1 + R2
10	0	3	-8	14	
11	0	4	-3	11	3R3 - 4R2
12					
13	=3*A9:D9+A10:D10				
14	=A10:D10				
15	=3*A11:D11-4*A10:D10				

	A	B	C	D	E
1	1	-1	5	-6	
2					
12					
13	3	0	7	-4	
14	0	3	-8	14	
15	0	0	23	-23	(1/23)*R3
16					
17	=A13:D13				
18	=A14:D14				
19	=(1/23)*A15:D15				

	A	B	C	D	E
1	1	-1	5	-6	
2					
16					
17	3	0	7	-4	R1 - 7R3
18	0	3	-8	14	R2 + 8R3
19	0	0	1	-1	
20					
21	=A17:D17-7*A19:D19				
22	=A18:D18+8*A19:D19				
23	=A19:D19				

	A	B	C	D	E
1	1	-1	5	-6	
2					
20					
21	3	0	0	3	(1/3)*R1
22	0	3	0	6	(1/3)*R2
23	0	0	1	-1	
24					
25	=(1/3)*A21:D21				
26	=(1/3)*A22:D22				
27	=A23:D23				

	A	B	C	D	E
1	1	-1	5	-6	
2					
24					
25	1	0	0	1	
26	0	1	0	2	
27	0	0	1	-1	

What do you notice if you change the entries in cells D1, D2, and D3?

Before we go on . . . We can check the solution by substitution:

$$1 - 2 + 5(-1) = -6 \quad \checkmark$$

$$3(1) + 3(2) - (-1) = 10 \quad \checkmark$$

$$1 + 3(2) + 2(-1) = 5 \quad \checkmark$$

Notice the form of the very last matrix in the example:

$$\begin{bmatrix} 1 & 0 & 0 & \# \\ 0 & 1 & 0 & \# \\ 0 & 0 & 1 & \# \end{bmatrix}$$

The 1s are on the **(main) diagonal** of the matrix; the goal in Gauss–Jordan reduction is to reduce our matrix to this form. If we can do so, then we can easily read the solution, as we saw in Example 1. However, as we will see in several examples in this section, it is not always possible to achieve this ideal state. After Example 6, we will give a form that is always possible to achieve.

Example 2 • Solving a System by Gauss–Jordan Reduction

Solve the system

$$2x + y + 3z = 1$$

$$4x + 2y + 4z = 4$$

$$x + 2y + z = 4$$

Solution

$$\begin{bmatrix} \boxed{2} & 1 & 3 & 1 \\ 4 & 2 & 4 & 4 \\ 1 & 2 & 1 & 4 \end{bmatrix} \begin{matrix} \\ R_2 - 2R_1 \\ 2R_3 - R_1 \end{matrix} \rightarrow \begin{bmatrix} 2 & 1 & 3 & 1 \\ 0 & 0 & -2 & 2 \\ 0 & 3 & -1 & 7 \end{bmatrix}$$

Now we have a slight problem: The number in the position where we would like to have a pivot, the second column of the second row, is a zero and thus cannot be a pivot. There are two ways out of this problem. One is to move on to the third column and pivot on the -2. Another is to switch the order of the second and third rows so that we can use the 3 as a pivot. We will do the latter.

$$\begin{bmatrix} 2 & 1 & 3 & 1 \\ 0 & 0 & -2 & 2 \\ 0 & 3 & -1 & 7 \end{bmatrix} R_2 \leftrightarrow R_3 \rightarrow \begin{bmatrix} 2 & 1 & 3 & 1 \\ 0 & \boxed{3} & -1 & 7 \\ 0 & 0 & -2 & 2 \end{bmatrix} \begin{matrix} 3R_1 - R_2 \\ \rightarrow \\ \end{matrix}$$

$$\begin{bmatrix} 6 & 0 & 10 & -4 \\ 0 & 3 & -1 & 7 \\ 0 & 0 & -2 & 2 \end{bmatrix} \begin{matrix} \\ \\ -\frac{1}{2}R_3 \end{matrix} \rightarrow \begin{bmatrix} 6 & 0 & 10 & -4 \\ 0 & 3 & -1 & 7 \\ 0 & 0 & \boxed{1} & -1 \end{bmatrix} \begin{matrix} R_1 - 10R_3 \\ R_2 + R_3 \\ \end{matrix} \rightarrow$$

$$\begin{bmatrix} 6 & 0 & 0 & 6 \\ 0 & 3 & 0 & 6 \\ 0 & 0 & 1 & -1 \end{bmatrix} \begin{matrix} \frac{1}{6}R_1 \\ \frac{1}{3}R_2 \\ \end{matrix} \rightarrow \begin{bmatrix} 1 & 0 & 0 & 1 \\ 0 & 1 & 0 & 2 \\ 0 & 0 & 1 & -1 \end{bmatrix}$$

The solution is $(x, y, z) = (1, 2, -1)$, as you can check in the original system.

Example 3 • An Inconsistent System

Solve the system

$$\begin{aligned} x + y + z &= 1 \\ 2x - y + z &= 0 \\ 4x + y + 3z &= 3 \end{aligned}$$

Solution

$$\begin{bmatrix} \boxed{1} & 1 & 1 & 1 \\ 2 & -1 & 1 & 0 \\ 4 & 1 & 3 & 3 \end{bmatrix} \begin{matrix} \\ R_2 - 2R_1 \\ R_3 - 4R_1 \end{matrix} \rightarrow \begin{bmatrix} 1 & 1 & 1 & 1 \\ 0 & \boxed{-3} & -1 & -2 \\ 0 & -3 & -1 & -1 \end{bmatrix} \begin{matrix} 3R_1 + R_2 \\ \\ R_3 - R_2 \end{matrix} \rightarrow \begin{bmatrix} 3 & 0 & 2 & 1 \\ 0 & -3 & -1 & -2 \\ 0 & 0 & 0 & 1 \end{bmatrix}$$

Stop. That last row translates into $0 = 1$, which is nonsense, and so, as in Example 3 in Section 2.1, we can say that this system has no solution. We also say, as we did for systems with only two unknowns, that a system with no solution is **inconsistent.** A system with at least one solution is **consistent.**

Before we go on . . .

Question How, exactly, does the nonsensical equation $0 = 1$ tell us that there is no solution of the system?

Answer Here is an argument similar to that in Example 3 in Section 2.1: If there *were* three numbers x, y, and z satisfying the original system of equations, then manipulating the equations according to the instructions in the row operations above leads us to conclude that $0 = 1$. Since 0 is *not* equal to 1, there can be no such numbers x, y, and z.

Example 4 • Infinitely Many Solutions

Solve the system

$$\begin{aligned} x + y + z &= 1 \\ \tfrac{1}{4}x - \tfrac{1}{2}y + \tfrac{3}{4}z &= 0 \\ x + 7y - 3z &= 3 \end{aligned}$$

Solution

$$\left[\begin{array}{ccc|c} 1 & 1 & 1 & 1 \\ \frac{1}{4} & -\frac{1}{2} & \frac{3}{4} & 0 \\ 1 & 7 & -3 & 3 \end{array}\right] \begin{matrix} \\ 4R_2 \\ \\ \end{matrix} \rightarrow \left[\begin{array}{ccc|c} \boxed{1} & 1 & 1 & 1 \\ 1 & -2 & 3 & 0 \\ 1 & 7 & -3 & 3 \end{array}\right] \begin{matrix} \\ R_2 - R_1 \\ R_3 - R_1 \end{matrix} \rightarrow$$

$$\left[\begin{array}{ccc|c} 1 & 1 & 1 & 1 \\ 0 & -3 & 2 & -1 \\ 0 & 6 & -4 & 2 \end{array}\right] \begin{matrix} \\ \\ \frac{1}{2}R_3 \end{matrix} \rightarrow \left[\begin{array}{ccc|c} 1 & 1 & 1 & 1 \\ 0 & \boxed{-3} & 2 & -1 \\ 0 & 3 & -2 & 1 \end{array}\right] \begin{matrix} 3R_1 + R_2 \\ \\ R_3 + R_2 \end{matrix} \rightarrow$$

$$\left[\begin{array}{ccc|c} 3 & 0 & 5 & 2 \\ 0 & -3 & 2 & -1 \\ 0 & 0 & 0 & 0 \end{array}\right]$$

There are no nonzero entries in the third row, so there can be no pivot in the third row. We skip to the final step and turn the pivots we did find into 1s:

$$\left[\begin{array}{ccc|c} 3 & 0 & 5 & 2 \\ 0 & -3 & 2 & -1 \\ 0 & 0 & 0 & 0 \end{array}\right] \begin{matrix} \frac{1}{3}R_1 \\ -\frac{1}{3}R_2 \\ \\ \end{matrix} \rightarrow \left[\begin{array}{ccc|c} 1 & 0 & \frac{5}{3} & \frac{2}{3} \\ 0 & 1 & -\frac{2}{3} & \frac{1}{3} \\ 0 & 0 & 0 & 0 \end{array}\right]$$

Now we translate back into equations and obtain

$$\begin{aligned} x + \tfrac{5}{3}z &= \tfrac{2}{3} \\ y - \tfrac{2}{3}z &= \tfrac{1}{3} \end{aligned}$$

But how does this help us find a solution? The thing to notice is that we can easily solve the first equation for x and the second for y, obtaining

$$\begin{aligned} x &= \tfrac{2}{3} - \tfrac{5}{3}z \\ y &= \tfrac{1}{3} + \tfrac{2}{3}z \end{aligned}$$

This is the solution! We can choose z to be any number and get corresponding values for x and y from the formulas above. This gives us infinitely many different solutions. Thus, the general solution (see Example 4 in Section 2.1) is

$$\begin{aligned} x &= \tfrac{2}{3} - \tfrac{5}{3}z \\ y &= \tfrac{1}{3} + \tfrac{2}{3}z \\ z &\text{ is arbitrary} \end{aligned}$$

General solution

We can also write the general solution as

$$\left(\tfrac{2}{3} - \tfrac{5}{3}z, \tfrac{1}{3} + \tfrac{2}{3}z, z\right) \quad z \text{ arbitrary}$$

General solution

This general solution has z as the parameter. Specific choices of values for the parameter z give particular solutions. For example, the choice $z = 6$ gives the particular solution

$$\begin{aligned} x &= \tfrac{2}{3} - \tfrac{5}{3}(6) = -\tfrac{28}{3} \\ y &= \tfrac{1}{3} + \tfrac{2}{3}(6) = \tfrac{13}{3} \qquad \text{Particular solution} \\ z &= 6 \end{aligned}$$

and the choice $z = 0$ gives the particular solution $(x, y, z) = \left(\frac{2}{3}, \frac{1}{3}, 0\right)$.

Before we go on . . . Why were there infinitely many solutions to this example? Because the third equation was really a combination of the first and second equations to begin with, so we effectively had only two equations in three unknowns.[9] Choosing a specific value for z (say, $z = 6$) has the effect of supplying the "missing" equation.

Note that, unlike the system given in Example 3, the system given in this example does have solutions and is thus *consistent.*

How we can check this solution? We can do what we always do, substitute the solution back into the original equations:

$$\left(\frac{2}{3} - \frac{5}{3}z\right) + \left(\frac{1}{3} + \frac{2}{3}z\right) + z = 1 - z + z = 1 \quad ✓$$

$$\frac{1}{4}\left(\frac{2}{3} - \frac{5}{3}z\right) - \frac{1}{2}\left(\frac{1}{3} + \frac{2}{3}z\right) + \frac{3}{4}z = \frac{1}{6} - \frac{5}{12}z - \frac{1}{6} - \frac{1}{3}z + \frac{3}{4}z = 0 \quad ✓$$

$$\left(\frac{2}{3} - \frac{5}{3}z\right) + 7\left(\frac{1}{3} + \frac{2}{3}z\right) - 3z = \frac{2}{3} - \frac{5}{3}z + \frac{7}{3} + \frac{14}{3}z - 3z = 3 \quad ✓$$

Notice that we must leave z unknown; each value of z produces a different solution.

Question How do we know when we have infinitely many solutions?

Answer When there are solutions (we have a consistent system, unlike the one in Example 3), and when the matrix we arrive at by row reduction has fewer pivots than there are unknowns. In Example 4 we had three unknowns but only two pivots.

Question How do we know which variables to use as parameters in a parameterized solution?

Answer The variables to use as parameters are those in the columns without pivots. In Example 4 there were pivots in the x and y columns but no pivot in the z column, and it was z that we used as a parameter.

Example 5 • Four Unknowns

Solve the system

$$\begin{aligned} x + 3y + 2z - w &= 6 \\ 2x + 6y + 6z + 3w &= 16 \\ x + 3y - 2z - 11w &= -2 \\ 2x + 6y + 8z + 8w &= 20 \end{aligned}$$

[9]In fact, you can check that the third equation, E_3, is equal to $3E_1 - 8E_2$. Thus, the third equation could have been left out because it conveys no more information than the first two. The process of row reduction always eliminates such a redundancy by creating a row of zeros.

Solution

$$\begin{bmatrix} \boxed{1} & 3 & 2 & -1 & 6 \\ 2 & 6 & 6 & 3 & 16 \\ 1 & 3 & -2 & -11 & -2 \\ 2 & 6 & 8 & 8 & 20 \end{bmatrix} \begin{matrix} \\ R_2 - 2R_1 \\ R_3 - R_1 \\ R_4 - 2R_1 \end{matrix} \rightarrow \begin{bmatrix} 1 & 3 & 2 & -1 & 6 \\ 0 & 0 & 2 & 5 & 4 \\ 0 & 0 & -4 & -10 & -8 \\ 0 & 0 & 4 & 10 & 8 \end{bmatrix}$$

There is no pivot available in the second column, so we move on to the third column.

$$\begin{bmatrix} 1 & 3 & 2 & -1 & 6 \\ 0 & 0 & \boxed{2} & 5 & 4 \\ 0 & 0 & -4 & -10 & -8 \\ 0 & 0 & 4 & 10 & 8 \end{bmatrix} \begin{matrix} R_1 - R_2 \\ \\ R_3 + 2R_2 \\ R_4 - 2R_2 \end{matrix} \rightarrow \begin{bmatrix} 1 & 3 & 0 & -6 & 2 \\ 0 & 0 & 2 & 5 & 4 \\ 0 & 0 & 0 & 0 & 0 \\ 0 & 0 & 0 & 0 & 0 \end{bmatrix} \begin{matrix} \\ \frac{1}{2}R_2 \\ \\ \\ \end{matrix} \rightarrow \begin{bmatrix} 1 & 3 & 0 & -6 & 2 \\ 0 & 0 & 1 & \frac{5}{2} & 2 \\ 0 & 0 & 0 & 0 & 0 \\ 0 & 0 & 0 & 0 & 0 \end{bmatrix}$$

Translating back into equations we get

$$x + 3y - 6w = 2$$

$$z + \frac{5}{2}w = 2$$

Since there are no pivots in the y or w columns, we use these two variables as parameters. We bring them over to the right-hand sides of the equations above and write the general solution as

$$x = 2 - 3y + 6w$$

y is arbitrary

$$z = 2 - \frac{5}{2}w$$

w is arbitrary

or

$$(x, y, z, w) = (2 - 3y + 6w, y, 2 - \tfrac{5}{2}w, w) \quad y, w \text{ arbitrary}$$

Before we go on . . . As in Example 4, you can check that the general solution is correct by substituting back into the original equations. (Since y and w are arbitrary, substitute nothing for them, but check that, as in Example 4, they cancel out when we substitute the formulas for x and z.)

In this example and the preceding one, you might have noticed an interesting phenomenon: If at any time in the process, two rows are equal or one is a multiple of the other, then one of those rows (eventually) becomes all zero.

Up to this point, we have always been given as many equations as there are unknowns. However, we will see in Section 2.3 that some applications lead to systems where the number of equations is not the same as the number of unknowns. As Example 6 illustrates, such systems can be handled the same way as any other.

Example 6 • Number of Equations ≠ Number of Unknowns

Solve the system

$$x + y = 1$$

$$13x - 26y = -11$$

$$26x - 13y = 2$$

Solution We proceed exactly as before and ignore the fact that there is one more equation than unknown.

$$\begin{bmatrix} \boxed{1} & 1 & 1 \\ 13 & -26 & -11 \\ 26 & -13 & 2 \end{bmatrix} \begin{matrix} \\ R_2 - 13R_1 \\ R_3 - 26R_1 \end{matrix} \rightarrow \begin{bmatrix} 1 & 1 & 1 \\ 0 & -39 & -24 \\ 0 & -39 & -24 \end{bmatrix} \begin{matrix} \\ \frac{1}{3}R_2 \\ \frac{1}{3}R_3 \end{matrix} \rightarrow$$

$$\begin{bmatrix} 1 & 1 & 1 \\ 0 & \boxed{-13} & -8 \\ 0 & -13 & -8 \end{bmatrix} \begin{matrix} 13R_1 + R_2 \\ \\ R_3 - R_2 \end{matrix} \rightarrow \begin{bmatrix} 13 & 0 & 5 \\ 0 & -13 & -8 \\ 0 & 0 & 0 \end{bmatrix} \begin{matrix} \frac{1}{13}R_1 \\ -\frac{1}{13}R_2 \\ \\ \end{matrix} \rightarrow \begin{bmatrix} 1 & 0 & \frac{5}{13} \\ 0 & 1 & \frac{8}{13} \\ 0 & 0 & 0 \end{bmatrix}$$

The solution is $(x, y) = \left(\frac{5}{13}, \frac{8}{13}\right)$.

Before we go on . . . If, instead of a row of zeros, we had obtained, say, [0 0 6] in the last row, we would immediately have concluded that the system was inconsistent.

The fact that we wound up with a row of zeros indicates that one of the equations was actually a combination of the other two; you can check that the third equation can be obtained by multiplying the first equation by 13 and adding the result to the second. Since the third equation therefore tells us nothing that we don't already know from the first two, we call the system of equations **redundant,** or **dependent** (compare Example 4 in Section 2.1).

If you now take another look at Example 5, you will find that we could have started with the following smaller system of two equations in four unknowns

$$x + 3y + 2z - w = 6$$
$$2x + 6y + 6z + 3w = 16$$

and obtained the same general solution as we did with the larger system. Verify this by solving the smaller system.

The preceding examples illustrate that we cannot always reduce a matrix to the form shown before Example 2, with pivots going all the way down the diagonal. What we *can* always do is reduce a matrix to the following form.

Reduced Row Echelon Form

A matrix is said to be in **reduced row echelon form** or to be **row-reduced** if it satisfies the following properties:

P1. The first nonzero entry in each row (called the **leading entry** of that row) is a 1.

P2. The columns of the leading entries are **clear** (that is, they contain zeros in all positions other than that of the leading entry).

P3. The leading entry in each row is to the right of the leading entry in the row above and any rows of zeros are at the bottom.

Quick Examples

$\begin{bmatrix} 1 & 0 & 0 & 2 \\ 0 & 1 & 0 & 4 \\ 0 & 0 & 1 & -3 \end{bmatrix}$, $\begin{bmatrix} 0 & 1 & -3 \\ 0 & 0 & 0 \end{bmatrix}$, and $\begin{bmatrix} 1 & 0 & 0 & -2 \\ 0 & 0 & 1 & 4 \\ 0 & 0 & 0 & 0 \end{bmatrix}$ are row-reduced.

$\begin{bmatrix} 1 & 1 & 0 & 2 \\ 0 & 1 & 0 & 4 \\ 0 & 0 & 1 & -3 \end{bmatrix}$, $\begin{bmatrix} 0 & 1 & -3 \\ 0 & 0 & 1 \end{bmatrix}$, and $\begin{bmatrix} 0 & 0 & 1 & 4 \\ 1 & 0 & 0 & -2 \\ 0 & 0 & 0 & 0 \end{bmatrix}$ are not row-reduced.

You should check in the examples we did that the final matrices were all in reduced row echelon form.

It is an interesting and useful fact, though not easy to prove, that any two people who start with the same matrix and row-reduce it will reach exactly the same row-reduced matrix, even if they use different row operations.

Web Site

The following resources are available at the Web site by following

Web site → Everything for Finite Math → Math Tools for Chapter 2

An Excel worksheet that pivots and does row operations automatically

An online Web page that pivots and does row operations automatically

A TI-83 program that pivots and does other row operations

Macintosh pivoting software available for download

The Traditional Gauss–Jordan Method (Optional)

In the version of the Gauss–Jordan method we have presented, we eliminated fractions and decimals in the first step and then worked with integer matrices, partly to make hand computation easier and partly for mathematical elegance. However, complicated fractions and decimals present no difficulty when we use technology. Example 7 illustrates the more traditional approach to Gauss–Jordan reduction used in many of the computer programs that solve the huge systems of equations that arise in practice.[10]

Example 7 • Solving a System with the Traditional Gauss–Jordan Method

Solve the following system using the traditional Gauss–Jordan method:

$$\begin{aligned} 2x + y + 3z &= 5 \\ 3x + 2y + 4z &= 7 \\ 2x + y + 5z &= 10 \end{aligned}$$

Solution We make two changes in our method. First, there is no need to get rid of decimals (since computers and calculators can handle decimals as easily as they can integers). Second, after selecting a pivot, *divide the pivot row by the pivot value, turning the pivot into a* 1. It is easier to determine the row operations that will clear the pivot column if the pivot is a 1.

If we use technology to solve this system of equations, the sequence of matrices might look like this:

$$\begin{bmatrix} \boxed{2} & 1 & 3 & 5 \\ 3 & 2 & 4 & 7 \\ 2 & 1 & 5 & 10 \end{bmatrix} \begin{matrix} \frac{1}{2}R_1 \\ \\ \\ \end{matrix} \rightarrow \begin{bmatrix} \boxed{1} & 0.5 & 1.5 & 2.5 \\ 3 & 2 & 4 & 7 \\ 2 & 1 & 5 & 10 \end{bmatrix} \begin{matrix} \\ R_2 - 3R_1 \\ R_3 - 2R_1 \end{matrix} \rightarrow \begin{bmatrix} 1 & 0.5 & 1.5 & 2.5 \\ 0 & \boxed{0.5} & -0.5 & -0.5 \\ 0 & 0 & 2 & 5 \end{bmatrix} \begin{matrix} \\ 2R_2 \\ \\ \end{matrix} \rightarrow$$

$$\begin{bmatrix} 1 & 0.5 & 1.5 & 2.5 \\ 0 & \boxed{1} & -1 & -1 \\ 0 & 0 & 2 & 5 \end{bmatrix} \begin{matrix} R_1 - 0.5R_2 \\ \\ \\ \end{matrix} \rightarrow \begin{bmatrix} 1 & 0 & 2 & 3 \\ 0 & 1 & -1 & -1 \\ 0 & 0 & \boxed{2} & 5 \end{bmatrix} \begin{matrix} \\ \\ \frac{1}{2}R_3 \end{matrix} \rightarrow \begin{bmatrix} 1 & 0 & 2 & 3 \\ 0 & 1 & -1 & -1 \\ 0 & 0 & \boxed{1} & 2.5 \end{bmatrix} \begin{matrix} R_1 - 2R_3 \\ R_2 + R_3 \\ \\ \end{matrix} \rightarrow \begin{bmatrix} 1 & 0 & 0 & -2 \\ 0 & 1 & 0 & 1.5 \\ 0 & 0 & 1 & 2.5 \end{bmatrix}$$

The solution is $(x, y, z) = (-2, 1.5, 2.5)$.

---

[10]Actually, for reasons of efficiency and accuracy, the methods used in commercial programs are close to but not exactly the method presented here. To learn more, consult a text on numerical methods.

Question That looked quite easy. Why didn't we use the traditional method from the start?

Answer It looked easy because we deliberately chose an example with simple numbers. In all but the most contrived examples, the decimals or fractions involved get very complicated very quickly.

Guideline: Getting Unstuck and Going Round in Circles

Question Help! I have been doing row operations on this matrix for half an hour. I have filled two pages, and I am getting nowhere. What do I do?

Answer Here is a way of keeping track of where you are *at any stage of the process* and also deciding what to do next.

Starting at the top row of your current matrix:

1. Scan along the row until you get to the leading entry: the first nonzero entry. If there is none—that is, the row is all zeros—go to the next row.
2. Having located the leading entry, scan up and down its *column.* If its column is not clear (that is, it contains other nonzero entries), use your leading entry as a pivot to clear its column as in the examples in this section.
3. Now go to the next row and start again at step 1.

When you have scanned all the rows and find that all the columns of the leading entries are clear, it means you are done (except possibly for reordering the rows so that the leading entries go from left to right as you read down the matrix, and zero rows are at the bottom).

Question No good. I have been following these instructions, but every time I try to clear a column, I *unclear* a column I had already cleared. What is going on?

Answer Are you using *leading entries* as pivots? Also, are you *using the pivot* to clear its column? That is, are your row operations all of the following form?

$$\underset{\text{Row to change}}{\underset{\uparrow}{aR_c}} \pm \underset{\text{Pivot row}}{\underset{\uparrow}{bR_r}}$$

The instruction next to the row you are changing should involve only that row and the pivot row, even though you might be tempted to use some other row instead.

2.2 EXERCISES

In Exercises 1–42, use Gauss–Jordan row reduction to solve the given systems of equations. We suggest doing some by hand and others using technology.

1. $\begin{aligned} x + y &= 4 \\ x - y &= 2 \end{aligned}$

2. $\begin{aligned} 2x + y &= 2 \\ -2x + y &= 2 \end{aligned}$

3. $\begin{aligned} 3x - 2y &= 6 \\ 2x - 3y &= -6 \end{aligned}$

4. $\begin{aligned} 2x + 3y &= 5 \\ 3x + 2y &= 5 \end{aligned}$

5. $\begin{aligned} 2x + 3y &= 1 \\ -x - \frac{3y}{2} &= -\frac{1}{2} \end{aligned}$

6. $\begin{aligned} 2x - 3y &= 1 \\ 6x - 9y &= 3 \end{aligned}$

7. $\begin{aligned} 2x + 3y &= 2 \\ -x - \frac{3y}{2} &= -\frac{1}{2} \end{aligned}$

8. $\begin{aligned} 2x - 3y &= 2 \\ 6x - 9y &= 3 \end{aligned}$

9. $\begin{aligned} x + y &= 1 \\ 3x - y &= 0 \\ x - 3y &= -2 \end{aligned}$

10. $\begin{aligned} x + y &= 1 \\ 3x - 2y &= -1 \\ 5x - y &= \frac{1}{5} \end{aligned}$

11. $\begin{aligned} x + y &= 0 \\ 3x - y &= 1 \\ x - y &= -1 \end{aligned}$

12. $\begin{aligned} x + 2y &= 1 \\ 3x - 2y &= -2 \\ 5x - y &= \frac{1}{5} \end{aligned}$

13. $\begin{aligned} 0.5x + 0.1y &= 1.7 \\ 0.1x - 0.1y &= 0.3 \\ x + y &= \frac{11}{3} \end{aligned}$

14. $\begin{aligned} -0.3x + 0.5y &= 0.1 \\ x - y &= 4 \\ \frac{x}{17} + \frac{y}{17} &= 1 \end{aligned}$

15. $\begin{aligned} -x + 2y - z &= 0 \\ -x - y + 2z &= 0 \\ 2x - z &= 4 \end{aligned}$

16. $\begin{aligned} x + 2y &= 4 \\ y - z &= 0 \\ x + 3y - 2z &= 5 \end{aligned}$

17.
$$\begin{aligned} x + y + 6z &= -1 \\ \frac{1}{3}x - \frac{1}{3}y + \frac{2}{3}z &= 1 \\ \frac{1}{2}x \qquad\quad + z &= 0 \end{aligned}$$

18.
$$\begin{aligned} x - \frac{1}{2}y \qquad\quad &= 0 \\ \frac{1}{3}x + \frac{1}{3}y + \frac{1}{3}z &= 2 \\ \frac{1}{2}x \qquad - \frac{1}{2}z &= -1 \end{aligned}$$

19.
$$\begin{aligned} -x + 2y - z &= 0 \\ -x - y + 2z &= 0 \\ 2x - y - z &= 0 \end{aligned}$$

20.
$$\begin{aligned} x - \frac{1}{2}y \qquad\quad &= 0 \\ \frac{1}{2}x - \qquad \frac{1}{2}z &= -1 \\ 3x - y - z &= -2 \end{aligned}$$

21.
$$\begin{aligned} x + y + 2z &= -1 \\ 2x + 2y + 2z &= 2 \\ 3x + 3y + 3z &= 2 \end{aligned}$$

22.
$$\begin{aligned} x + y - z &= -2 \\ x - y - 7z &= 0 \\ \frac{2}{7}x \qquad - \frac{8}{7}z &= 14 \end{aligned}$$

23.
$$\begin{aligned} -0.5x + 0.5y + 0.5z &= 1.5 \\ 4.2x + 2.1y + 2.1z &= 0 \\ 0.2x \qquad + 0.2z &= 0 \end{aligned}$$

24.
$$\begin{aligned} 0.25x - 0.5y \qquad\quad &= 0 \\ 0.2x + 0.2y - 0.2z &= -0.6 \\ 0.5x - 1.5y + z &= 0.5 \end{aligned}$$

25.
$$\begin{aligned} 2x - y + z &= 4 \\ 3x - y + z &= 5 \end{aligned}$$

26.
$$\begin{aligned} 3x - y - z &= 0 \\ x + y + z &= 4 \end{aligned}$$

27.
$$\begin{aligned} 0.75x - 0.75y - z &= 4 \\ x - y + 4z &= 0 \end{aligned}$$

28.
$$\begin{aligned} 2x - y + z &= 4 \\ -x + 0.5y - 0.5z &= 1.5 \end{aligned}$$

29. $3x + y - z = 12$

30. $x + y - 3z = 21$

(Yes: One equation in three unknowns!)

31.
$$\begin{aligned} x + y + 2z &= -1 \\ 2x + 2y + 2z &= 2 \\ 0.75x + 0.75y + z &= 0.25 \\ -x \qquad - 2z &= 21 \end{aligned}$$

32.
$$\begin{aligned} x + y - z &= -2 \\ x - y - 7z &= 0 \\ 0.75x - 0.5y + 0.25z &= 14 \\ x + y + z &= 4 \end{aligned}$$

33.
$$\begin{aligned} x + y + 5z \qquad &= 1 \\ y + 2z + w &= 1 \\ x + 3y + 7z + 2w &= 2 \\ x + y + 5z + w &= 1 \end{aligned}$$

34.
$$\begin{aligned} x + y \qquad + 4w &= 1 \\ 2x - 2y - 3z + 2w &= -1 \\ 4y + 6z + w &= 4 \\ 2x + 4y + 9z \qquad &= 6 \end{aligned}$$

35.
$$\begin{aligned} x + y + 5z \qquad &= 1 \\ y + 2z + w &= 1 \\ x + y + 5z + w &= 1 \\ x + 2y + 7z + 2w &= 2 \end{aligned}$$

36.
$$\begin{aligned} x + y \qquad + 4w &= 1 \\ 2x - 2y - 3z + 2w &= -1 \\ 4y + 6z + w &= 4 \\ 3x + 3y + 3z + 7w &= 4 \end{aligned}$$

37.
$$\begin{aligned} x - 2y + z - 4w &= 1 \\ x + 3y + 7z + 2w &= 2 \\ 2x + y + 8z - 2w &= 3 \end{aligned}$$

38.
$$\begin{aligned} x - 3y - 2z - w &= 1 \\ x + 3y + z + 2w &= 2 \\ 2x \qquad - z + w &= 3 \end{aligned}$$

39.
$$\begin{aligned} x + y + z + u + v &= 15 \\ y - z + u - v &= -2 \\ z + u + v &= 12 \\ u - v &= -1 \\ v &= 5 \end{aligned}$$

40.
$$\begin{aligned} x - y + z - u + v &= 1 \\ y + z + u + v &= 2 \\ z - u + v &= 1 \\ u + v &= 1 \\ v &= 1 \end{aligned}$$

41.
$$\begin{aligned} x - y + z - u + v &= 0 \\ y - z + u - v &= -2 \\ x \qquad\qquad - 2v &= -2 \\ 2x - y + z - u - 3v &= -2 \\ 4x - y + z - u - 7v &= -6 \end{aligned}$$

42.
$$\begin{aligned} x + y + z + u + v &= 15 \\ y + z + u + v &= 3 \\ x + 2y + 2z + 2u + 2v &= 18 \\ x - y - z - u - v &= 9 \\ x - 2y - 2z - 2u - 2v &= 6 \end{aligned}$$

T In Exercises 43–46, use technology to solve the systems of equations. Express all solutions as fractions.

43.
$$\begin{aligned} x + 2y - z + w &= 30 \\ 2x \qquad - z + 2w &= 30 \\ x + 3y + 3z - 4w &= 2 \\ 2x - 9y \qquad + w &= 4 \end{aligned}$$

44.
$$\begin{aligned} 4x - 2y + z + w &= 20 \\ 3y + 3z - 4w &= 2 \\ 2x + 4y \qquad - w &= 4 \\ x + 3y + 3z \qquad &= 2 \end{aligned}$$

45.
$$\begin{aligned} x + 2y + 3z + 4w + 5t &= 6 \\ 2x + 3y + 4z + 5w + t &= 5 \\ 3x + 4y + 5z + w + 2t &= 4 \\ 4x + 5y + z + 2w + 3t &= 3 \\ 5x + y + 2z + 3w + 4t &= 2 \end{aligned}$$

46.
$$\begin{aligned} x - 2y + 3z - 4w \qquad &= 0 \\ -2x + 3y - 4z \qquad + t &= 0 \\ 3x - 4y \qquad + w - 2t &= 0 \\ -4x \qquad + z - 2w + 3t &= 0 \\ y - 2z + 3w - 4t &= 1 \end{aligned}$$

T In Exercises 47–50, use technology to solve the systems of equations. Express all solutions as decimals, rounded to one decimal place.

47.
$$\begin{aligned} 1.6x + 2.4y - 3.2z &= 4.4 \\ 5.1x - 6.3y + 0.6z &= -3.2 \\ 4.2x + 3.5y + 4.9z &= 10.1 \end{aligned}$$

48.
$$\begin{aligned} 2.1x + 0.7y - 1.4z &= -2.3 \\ 3.5x - 4.2y - 4.9z &= 3.3 \\ 1.1x + 2.2y - 3.3z &= -10.2 \end{aligned}$$

49.
$$\begin{aligned} -0.2x + 0.3y + 0.4z - t &= 4.5 \\ 2.2x + 1.1y - 4.7z + 2t &= 8.3 \\ 9.2y \qquad - 1.3t &= 0 \\ 3.4x \qquad + 0.5z - 3.4t &= 0.1 \end{aligned}$$

50.
$$\begin{aligned} 1.2x - 0.3y + 0.4z - 2t &= 4.5 \\ 1.9x \qquad - 0.5z - 3.4t &= 0.2 \\ 12.1y \qquad - 1.3t &= 0 \\ 3x + 2y - 1.1z \qquad &= 9 \end{aligned}$$

COMMUNICATION AND REASONING EXERCISES

51. What is meant by a pivot? What does pivoting do?

52. Give instructions to check whether a matrix is row-reduced.

53. You are row-reducing a matrix and have chosen a -6 as a pivot in row 4. Directly above the pivot, in row 1, is a 15. What row operation can you use to clear the 15?

54. You are row-reducing a matrix and have chosen a -4 as a pivot in row 2. Directly below the pivot, in row 4, is a -6. What row operation can you use to clear the -6?

55. In the matrix of a system of linear equations, suppose two of the rows are equal. What can you say about the row-reduced form of the matrix?

56. In the matrix of a system of linear equations, suppose one of the rows is a multiple of another. What can you say about the row-reduced form of the matrix?

57. Your friend Frans tells you that the system of linear equations you are solving cannot have a unique solution because the reduced matrix has a row of zeros. Comment on his claim.

58. Your other friend Hans tells you that, since he is solving a consistent system of five linear equations in six unknowns, he will get infinitely many solutions. Comment on his claim.

59. If the reduced matrix of a consistent system of linear equations has five rows, three of which are zero, and five columns, how many parameters does the general solution contain?

60. If the reduced matrix of a consistent system of linear equations has five rows, two of which are zero, and seven columns, how many parameters does the general solution contain?

61. Suppose a system of equations has a unique solution. What must be true of the number of pivots in the reduced matrix of the system? Why?

62. Suppose a system has infinitely many solutions. What must be true of the number of pivots in the reduced matrix of the system? Why?

63. Give an example of a system of three linear equations with the general solution $x = 1$, $y = 1 + z$, z arbitrary. (Check your system by solving it.)

64. Give an example of a system of three linear equations with the general solution $x = y - 1$, y arbitrary, $z = y$. (Check your system by solving it.)

2.3 Applications of Systems of Linear Equations

In the examples and the exercises of this section, we consider scenarios that lead to systems of linear equations in three or more unknowns. Some of these applications will strike you as a little idealized or even contrived compared with the kinds of problems you might encounter in the real world.[11] One reason is that we will not have tools to handle more realistic versions of these applications until we have studied linear programming in Chapter 4.

In each example that follows, we set up the problem as a linear system and then give the solution. For practice, you should do the row reduction necessary to get the solution.

Example 1 • Blending

Arctic Juice Company makes three juice blends: PineOrange, using 2 quarts of pineapple juice and 2 quarts of orange juice per gallon; PineKiwi, using 3 quarts of pineapple juice and 1 quart of kiwi juice per gallon; and OrangeKiwi, using 3 quarts of orange juice and 1 quart of kiwi juice per gallon. Each day the company has 800 quarts of pineapple juice, 650 quarts of orange juice, and 350 quarts of kiwi juice available. How many gallons of each blend should it make each day if it wants to use up all supplies?

Solution We take the same steps to understand the problem that we took in Section 2.1. The first step is to identify and label the unknowns. Looking at the question asked in the last sentence, we see that we should label the unknowns like this:

x = number of gallons of PineOrange made each day

y = number of gallons of PineKiwi made each day

z = number of gallons of OrangeKiwi made each day

Next, we can organize the information we are given in a table:

	PineOrange, x	PineKiwi, y	OrangeKiwi, z	Total Available
Pineapple Juice (qt)	2	3	0	800
Orange Juice (qt)	2	0	3	650
Kiwi Juice (qt)	0	1	1	350

[11]See the discussion at the end of Example 1.

Notice how we have arranged the table; we have placed headings corresponding to the unknowns along the top, rather than down the side, and we have added a heading for the available totals. This gives us a table that is essentially the matrix of the system of linear equations we are looking for. (However, read the caution in the "Before we go on" section.)

Now we read across each row of the table. The fact that we want to use exactly the amount of each juice that is available leads to the following three equations:

$$\begin{aligned} 2x + 3y \quad\quad &= 800 \\ 2x \quad\quad + 3z &= 650 \\ y + \ z &= 350 \end{aligned}$$

The solution of this system is $(x, y, z) = (100, 200, 150)$, so Arctic Juice should make 100 gallons of PineOrange, 200 gallons of PineKiwi, and 150 gallons of OrangeKiwi each day.

Before we go on . . .

Caution: We do not recommend relying on the coincidence that the table we created to organize the information happened to be the matrix of the system; it is too easy to set up the table "sideways" and get the wrong matrix. You should always write down the system of equations *and be sure you know why each equation is true.* For example, the equation $2x + 3y = 800$ in this example is true because the number of quarts of pineapple juice that will be used $(2x + 3y)$ is equal to the amount available (800 quarts). By thinking of the reason for each equation, you can check that you have the correct system. If you have the wrong system of equations to begin with, solving it won't help you.

Question Just how realistic is this scenario?

Answer This is a very unrealistic scenario, for several reasons:

1. Isn't it odd that we happened to end up with exactly the same number of equations as unknowns? Real scenarios are rarely so considerate. If there had been four equations, there would in all likelihood have been no solution at all. However, we need to understand these idealized problems before we can tackle the real world.
2. Even if a real-world scenario does give the same number of equations as unknowns, there is still no guarantee that there will be a unique solution consisting of positive values. What, for instance, would we have done in this example if x had turned out to be negative?
3. The requirement that we use exactly all ingredients would be an unreasonable constraint in real life. When we discuss linear programming, we will substitute the more reasonable constraint that you use no more than is available, and we will add the more reasonable objective that you maximize profit.

Example 2 • Aircraft Purchasing

An airline is considering the purchase of aircraft to meet an estimated demand for 3200 seats. The airline has decided to buy Boeing 747s, which seat 400 passengers and are priced at \$200 million each; Boeing 777s, which seat 300 passengers and are priced at \$160 million; and Airbus A321s, which seat 200 passengers and are priced at

\$60 million.[12] Assuming that the airline wishes to buy three times as many 777s as 747s, how many of each should it order to meet the demand for seats, given a \$1540 million purchasing budget?

Solution We label the unknowns as follows:

$x =$ number of 747s ordered from Boeing

$y =$ number of 777s ordered from Boeing

$z =$ number of A321s ordered from Airbus

We must now set up the equations. We can organize some (but not all) of the given information in a table:

	Boeing 747, x	Boeing 777, y	Airbus A321, z	Total
Passengers	400	300	200	3200
Cost (\$ millions)	200	160	60	1540

Reading across, we get the equations expressing the facts that the airline needs to seat 3200 passengers and that it will spend \$1540 million:

$$400x + 300y + 200z = 3200$$
$$200x + 160y + 60z = 1540$$

There is an additional piece of information we have not yet used: The airline wishes to purchase three times as many 777s as 747s. As we said in Section 2.1, it is easiest to translate a statement like this into an equation if we first reword it using the phrase "the number of." Thus, we say: "The number of 777s ordered is three times the number of 747s ordered," or

$$y = 3x$$
$$3x - y = 0$$

We now have a system of three equations in three unknowns:

$$\begin{aligned} 400x + 300y + 200z &= 3200 \\ 200x + 160y + 60z &= 1540 \\ 3x - y \qquad\quad &= 0 \end{aligned}$$

Solving the system, we get the solution $(x, y, z) = (2, 6, 3)$. Thus, the airline should order two 747s, six 777s, and three A321s.

Example 3 • Traffic Flow

Traffic through downtown Urbanville flows through the one-way system shown in Figure 11.

[12]The prices are approximate 2001 prices; seating capacities have been rounded. Prices and seating capacities found at the companies' Web sites and in an October 2000 press release announcing Air Canada's commitment to buy 12 A321s in 2001–2002.

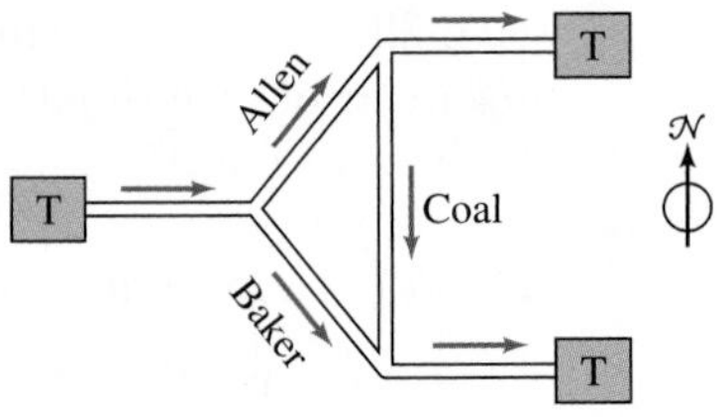

The symbol T represents traffic counters.

Figure 11

Traffic-counting devices installed in the road count 200 cars entering town from the west every hour and 100 cars leaving town on each road to the east every hour. From this information, is it possible to determine how many cars drive along Allen, Baker, and Coal Streets every hour?

Solution Our unknowns are

x = number of cars per hour on Allen Street

y = number of cars per hour on Baker Street

z = number of cars per hour on Coal Street

Assuming that, at each intersection, cars do not fall into a pit or materialize out of thin air, the number of cars entering each intersection has to equal the number exiting. For example, at the intersection of Allen and Baker Streets, 200 cars are entering and $x + y$ cars exiting, giving the equation

$$x + y = 200$$

At the intersection of Allen and Coal Streets, we get

$$x = z + 100$$

and at the intersection of Baker and Coal Streets, we get

$$y + z = 100$$

We now have the following system of equations:

$$\begin{aligned} x + y \quad\; &= 200 \\ x \quad\; - z &= 100 \\ y + z &= 100 \end{aligned}$$

This system has many solutions, which can be written in the following way:

$$x = z + 100$$

$$y = -z + 100$$

z is arbitrary

In other words, since we do not have a unique solution, it is *not* possible to determine how many cars drive along Allen, Baker, and Coal Streets every hour.

Before we go on . . . How arbitrary is z? It makes no sense for any of the variables x, y, or z to be negative in this scenario, so certainly $z \geq 0$. Further, to keep y positive, we must have $z \leq 100$. So, z is not completely arbitrary, but we can say that it has to satisfy

$0 \le z \le 100$ in order to get a realistic answer. Would it make sense for z to be fractional? What if you interpreted x, y, and z as *average* numbers of cars per hour over a long period of time?

Traffic flow is only one kind of flow in which we might be interested. Water and electricity flows are others. In each case, to analyze the flow, we use the fact that the amount entering an intersection must equal the amount leaving it.

Here is a final question to think about: If you wanted to nail down x, y, and z to see where the cars are really going, how would you do it with only one more traffic counter?

Example 4 • Transportation

A car rental company has four locations in the city: Northside, Eastside, Southside, and Westside. The Westside location has 20 more cars than it needs, and the Eastside location has 15 more cars than it needs. The Northside location needs 10 more cars than it has, and the Southside location needs 25 more cars than it has. It costs \$10 (in salary and gas) to have an employee drive a car from Westside to Northside. It costs \$5 to drive a car from Eastside to Northside. It costs \$20 to drive a car from Westside to Southside, and it costs \$10 to drive a car from Eastside to Southside. If the company will spend a total of \$475 rearranging its cars, how many cars will it drive from each of Westside and Eastside to each of Northside and Southside?

Solution Figure 12 shows a diagram of this situation. Each arrow represents a route along which the rental company can drive cars. Next to each location is written the number of extra cars the location has or the number it needs. Next to each route is written the cost of driving a car along that route.

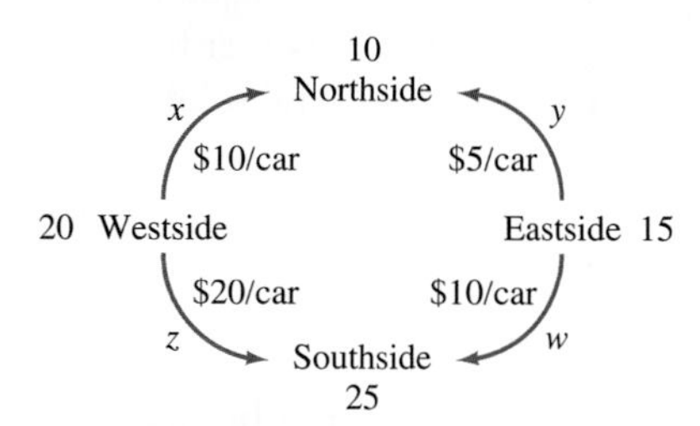

Figure 12

The unknowns are the number of cars the company will drive along each route, so we have the following four unknowns, as indicated in the figure:

x = number of cars driven from Westside to Northside

y = number of cars driven from Eastside to Northside

z = number of cars driven from Westside to Southside

w = number of cars driven from Eastside to Southside

Consider the Northside location. It needs 10 more cars, so the total number of cars being driven to Northside should be 10. This gives us the equation

$$x + y = 10$$

Similarly, the total number of cars being driven to Southside should be 25, so

$$z + w = 25$$

Considering the number of cars to be driven out of the Westside and Eastside locations, we get the following two equations as well:

$$x + z = 20$$

$$y + w = 15$$

There is one more equation that we should write down, the equation that says that the company will spend \$475:

$$10x + 5y + 20z + 10w = 475$$

Thus, we have the following system of five equations in four unknowns:

$$\begin{aligned}
x + y &= 10 \\
z + w &= 25 \\
x + z &= 20 \\
y + w &= 15 \\
10x + 5y + 20z + 10w &= 475
\end{aligned}$$

Solving this system, we find that $(x, y, z, w) = (5, 5, 15, 10)$. In words, the company will drive 5 cars from Westside to Northside, 5 from Eastside to Northside, 15 from Westside to Southside, and 10 from Eastside to Southside.

Before we go on . . . A very reasonable question to ask is, Can the company rearrange its cars for less than \$475? Even better, what is the least possible cost? In general, a question asking for the optimal cost may require the techniques of linear programming that we will discuss in Chapter 4. However, in this case we can approach the problem directly. If we remove the equation that says that the total cost is \$475 and solve the system consisting of the other four equations, we find that there are infinitely many solutions and that the general solution may be written as

$$x = w - 5$$

$$y = 15 - w$$

$$z = 25 - w$$

w is arbitrary

This allows us to write the total cost as a function of w:

$$\begin{aligned}
10x + 5y + 20z + 10w &= 10(w - 5) + 5(15 - w) + 20(25 - w) + 10w \\
&= 525 - 5w
\end{aligned}$$

So, the larger we make w, the smaller the total cost will be. The largest we can make w is 15 (why?), and if we do so we get $(x, y, z, w) = (10, 0, 10, 15)$ and a total cost of \$450.

2.3 EXERCISES

APPLICATIONS

1. **Resource Allocation** You manage an ice cream factory that makes three flavors: Creamy Vanilla, Continental Mocha, and Succulent Strawberry. Into each batch of Creamy Vanilla go 2 eggs, 1 cup of milk, and 2 cups of cream; into each batch of Continental Mocha go 1 egg, 1 cup of milk, and 2 cups of cream; into each batch of Succulent Strawberry go 1 egg, 2 cups of milk, and 1 cup of cream. You have in stock 350 eggs, 350 cups of milk, and 400 cups of cream. How many batches of each flavor should you make in order to use up all of your ingredients?

2. **Resource Allocation** You own a hamburger franchise and are planning to shut down operations for the day, but you are left with 13 bread rolls, 19 defrosted beef patties, and 15 opened cheese slices. Rather than throw them out, you decide to use them to make burgers that you will sell at a discount. Plain burgers each require 1 beef patty and 1 bread roll, double cheeseburgers each require 2 beef patties, 1 bread roll, and 2 slices of cheese, and regular cheeseburgers each require 1 beef patty, 1 bread roll, and 1 slice of cheese. How many of each should you make?

3. **Resource Allocation** Urban Community College is planning to offer courses in Finite Math, Applied Calculus, and Computer Methods. Each section of Finite Math holds 40 students and earns the college $1000 in revenue per student. Each section of Applied Calculus holds 40 students and earns the college $1500/student, and each section of Computer Methods holds 10 students and earns the college $2000/student. Assuming that the college can offer a total of six sections, wishes to accommodate 210 students, and wishes to bring in $260,000 in revenues, how many sections of each course should it offer?

4. **Resource Allocation** The Enormous State University History Department offers three courses—Ancient History, Medieval History, and Modern History—and the chairperson is trying to decide how many sections of each to offer this semester. The department is allowed to offer 45 sections total, 5000 students would like to take a course, and 60 professors are available to teach them. Sections of Ancient History hold 100 students each, sections of Medieval History hold 50 students each, and sections of Modern History hold 200 students each. Modern History sections are taught by a team of 2 professors, whereas Ancient History and Medieval History need only 1 professor/section. How many sections of each course should the chair schedule in order to offer all sections that are allowed, to accommodate all the students, and to give teaching assignments to all the professors?

5. **Purchasing** In Example 2 we saw that Boeing 747s seat 400 passengers and are priced at $200 million each, Boeing 777s seat 300 passengers and are priced at $160 million, and European Airbus A321s seat 200 passengers and are priced at $60 million. You are the purchasing manager of an airline company and have a $2100 million budget to purchase new aircraft to seat a total of 4500 passengers. Your company has a policy of supporting U.S. industries, and you have been instructed to buy twice as many U.S.-manufactured aircraft as foreign aircraft. Given the selection of three aircraft, how many of each should you order?

6. **Purchasing** Repeat Exercise 5, but with the following data: Your purchasing budget is $7200 million, you must seat a total of 17,300 passengers, and you are required to spend twice as much on U.S.-manufactured aircraft as on foreign aircraft.

7. **Feeding Schedules** Your 36-gallon tropical fish tank contains three types of carnivorous creatures: baby sharks, piranhas, and squid. You feed them three types of delicacies: goldfish, angelfish, and butterfly fish. Each baby shark can consume 1 goldfish, 2 angelfish, and 2 butterfly fish per day; each piranha can consume 1 goldfish and 3 butterfly fish per day (the piranhas are rather large as a result of their diet); each squid can consume 1 goldfish and 1 angelfish per day. After a trip to the local pet store, you were able to feed your creatures to capacity, and you noticed that 21 goldfish, 21 angelfish, and 35 butterfly fish were eaten. How many of each type of creature do you have?

8. **Resource Allocation** The Enormous State University Choral Society is planning its annual Song Festival, when it will serve three kinds of delicacies: granola treats, nutty granola treats, and nuttiest granola treats. The following table shows the ingredients required for a single serving of each delicacy, as well as the total amount of each ingredient available:

	Granola	Nutty Granola	Nuttiest Granola	Total Available
Toasted Oats (oz)	1	1	5	1,500
Almonds (oz)	4	8	8	10,000
Raisins (oz)	2	4	8	4,000

   The Song Festival planners would like to use all the ingredients. Is this possible? If so, how many servings of each kind of delicacy can they make?

9. **Supply** A bagel store orders cream cheese from three suppliers, Cheesy Cream Corp. (CCC), Super Smooth & Sons (SSS), and Bagel's Best Friend Co. (BBF). One month, the total order of cream cheese came to 100 tons (they do a booming trade). The costs were $80, $50, and $65/ton from the three suppliers, respectively, with total cost amounting to $5990. Given that the store ordered the same amount from CCC and BBF, how many tons of cream cheese were ordered from each supplier?

10. **Supply** Refer to Exercise 9. The bagel store's outlay for cream cheese the following month was $2310, when it purchased a total of 36 tons. Two more tons of cream cheese came from BBF than from SSS. How many tons of cream cheese came from each supplier?

11. **Pest Control** Conan the Great has boasted to his hordes of followers that many a notorious villain has fallen to his awesome sword: His total of 560 victims consists of evil sorcerers, warriors, and orcs. These he has slain with a total of 620 mighty thrusts of his sword—evil sorcerers and warriors each requiring 2 thrusts (to the chest) and orcs each requiring 1 thrust (to the neck). When asked about the number of warriors he has slain, he replies, "The number of warriors I, the mighty Conan, have slain is five times the number of evil sorcerers that have fallen to my sword!" How many of each type of villain has he slain?

12. **Manufacturing** Fancy French Perfume Company recently had its secret formula divulged. It turned out that it was using, as the three ingredients, rose oil, oil of fermented prunes, and alcohol. Moreover, each 22-ounce econosize bottle contained 4 more ounces of alcohol than oil of fermented prunes, and the amount of alcohol was equal to the combined volume of the other two ingredients. How much of each ingredient did it use in an econosize bottle?[13]

13. **Music CD Sales** In 2000, total revenues from sales of recorded rock, rap, and classical music amounted to $5.8 billion. Rock music brought in twice as much revenue as rap music and 900% the revenue of classical music. How much revenue was earned in each of the three categories of recorded music?

Rap includes hip hop. Revenues are based on total manufacturers' shipments at suggested retail prices and are rounded to the nearest $0.1 billion. SOURCE: Recording Industry Association of America, www.riaa.com, March 2002.

14. **Music CD Sales** In 2000, combined revenues from sales of "oldies" and soundtracks amounted to $0.23 billion. Country music brought in 15 times as much revenue as soundtracks, and revenue from the sale of oldies was 30% above that of soundtracks. How much revenue was earned in each of the three categories of recorded music?

Revenues are based on total manufacturers' shipments at suggested retail prices and are rounded to the nearest $0.1 billion. SOURCE: Recording Industry Association of America, www.riaa.com, March 2002.

T 15. **Market Share (Cereals)** We have divided U.S. cereal manufactures into four groups: Kellogg, General Mills, General Foods, and Other. Based on data from 1993 to 1998, two relationships between the domestic market shares are found to be

$$x_1 = -0.4 + 1.2x_2 + 2x_3$$

$$x_3 = 0.5x_4 + 0.08$$

where x_1, x_2, x_3, and x_4 are, respectively, the fractions of the market held by Kellogg, General Mills, General Foods, and Other. Given that the four groups account for the entire cereal market, solve the associated system of three linear equations to show how the market shares of Kellogg, General Mills, and General Foods depend on the share held by Other. (Round all answers to two decimal places.) Which of the three companies' market share is most impacted by the share held by Other?

"Other" includes Quaker and private labels. The models are based on a linear regression. SOURCE: Bloomberg Financial Markets/*New York Times*, November 28, 1998, p. C1.

T 16. **Market Share (Cars and Light Trucks)** We have divided manufacturers of cars and light trucks into four groups: Chrysler, Ford, General Motors, and Other. Based on data from 1980 to 1998, two relationships between the domestic market shares are found to be

$$x_3 = 0.66 - 2.2x_1 - 0.02x_2$$

$$x_2 = x_1 + 0.04$$

where x_1, x_2, x_3, and x_4 are, respectively, the fractions of the market held by Chrysler, Ford, General Motors, and Other. Given that the four groups account for the entire U.S. market, solve the associated system of three linear equations to show how the market shares of Chrysler, Ford, and General Motors depend on the share held by Other. (Round all answers to two decimal places.) Which of the three companies' market share is most impacted by the share held by Other?

The model is based on a linear regression. SOURCE: Ward's AutoInfoBank/*New York Times*, July 29, 1998, p. D6.

Investing in Muncipal Bonds Exercises 17 and 18 are based on the following data on three tax-exempt municipal bond funds:

	2001 Yield (%)
PNF (Pimco N.Y.)	6
CMBFX (Columbia Ore)	5
SFCOX (Safeco)	7

Yields are rounded. SOURCES: www.pimcofunds.com, www.paydenfunds.com, February 22, 2002.

17. You invested a total of $9000 in the three funds at the beginning of 2001, including an equal amount in CMBFX and SFCOX. Your interest for the year from the first two funds amounted to $400. How much did you invest in each of the three funds?

18. You invested a total of $6000 in the three funds at the beginning of 2001, including an equal amount in PNF and CMBFX. You earned a total of $360 in interest for the year. How much did you invest in each of the three funds?

Investing in Stocks Exercises 19 and 20 are based on the following data on three computer stocks:

	Price ($)	Dividend Yield (%)
IBM	100	0.5
HWP (Hewlett-Packard)	20	1.50
DELL	25	0

Stocks were trading at or near the given prices in February 2002. Dividends are rounded, based on values on February 23, 2002. SOURCE: money.excite.com, February 23, 2002.

[13]Most perfumes consist of 10–20% perfume oils dissolved in alcohol. This may or may not be reflected in this company's formula!

19. You invested a total of $12,400 in IBM, Hewlett-Packard, and Dell shares and expected to earn $56 in annual dividends. If you purchased a total of 200 shares, how many shares of each stock did you purchase?

20. You invested a total of $8250 in IBM, Hewlett-Packard, and Dell shares and expected to earn $55 in annual dividends. If you purchased a total of 200 shares, how many shares of each stock did you purchase?

21. Voting In the 75th Congress (1937–1939) the U.S. House of Representatives had 333 Democrats, 89 Republicans, and 13 members of other parties. Suppose a bill passed the House with 31 more votes in favor than against, with ten times as many Democrats voting for the bill as Republicans and with 36 more non-Democrats voting against the bill than for it. How many members of each party voted in favor of the bill?

22. Voting In the 75th Congress (1937–1939) the U.S. Senate had 75 Democrats, 17 Republicans, and 4 members of other parties. Suppose a bill passed the Senate with 16 more votes in favor than against, with three times as many Democrats voting in favor as non-Democrats voting in favor and 32 more Democrats voting in favor than Republicans voting in favor. How many members of each party voted in favor of the bill?

23. Inventory Control Big Red Bookstore wants to ship books from its warehouses in Brooklyn and Queens to its stores, one on Long Island and one in Manhattan. Its warehouse in Brooklyn has 1000 books, and its warehouse in Queens has 2000. Each store orders 1500 books. It costs $1 to ship each book from Brooklyn to Manhattan and $2 to ship each book from Queens to Manhattan. It costs $5 to ship each book from Brooklyn to Long Island and $4 to ship each book from Queens to Long Island.

a. If Big Red has a transportation budget of $9000 and is willing to spend all of it, how many books should Big Red ship from each warehouse to each store in order to fill all the orders?

b. Is there a way of doing this for less money?

24. Inventory Control Tubular Ride Boogie Board Company has manufacturing plants in Tucson, Arizona and Toronto, Ontario. You have been given the job of coordinating distribution of the latest model, the Gladiator, to outlets in Honolulu and Venice Beach. The Tucson plant, when operating at full capacity, can manufacture 620 Gladiator boards/week, whereas the Toronto plant, beset by labor disputes, can produce only 410 boards/week. The outlet in Honolulu orders 500 Gladiator boards/week, and Venice Beach orders 530 boards/week. Transportation costs are as follows:

Tucson to Honolulu: $10/board; Tucson to Venice Beach: $5/board

Toronto to Honolulu: $20/board; Toronto to Venice Beach: $10/board

a. Assuming that you wish to fill all orders and ensure full-capacity production at both plants, is it possible to meet a total transportation budget of $10,200? If so, how many boards are shipped from each manufacturing plant to each distribution outlet?

b. Is there a way of doing this for less money?

25. Tourism In the 1990s significant numbers of tourists traveled from North America and Europe to Australia and South Africa. In 1998 a total of 1,390,000 of these tourists visited Australia, and 1,140,000 of them visited South Africa. Further, 630,000 of them came from North America, and 1,900,000 of them came from Europe. (Assume that no single tourist visited both destinations or traveled from both North America and Europe.)

a. The given information is not sufficient to determine the number of tourists from each region to each destination. Why?

b. If you were given the additional information that a total of 2,530,000 tourists traveled from these two regions to these two destinations, would you now be able to determine the number of tourists from each region to each destination? If so, what are these numbers?

c. If you were given the additional information that the same number of people from Europe visited South Africa as visited Australia, would you now be able to determine the number of tourists from each region to each destination? If so, what are these numbers?

Figures are rounded to the nearest 10,000. SOURCES: South African Dept. of Environmental Affairs and Tourism; Australia Tourist Commission/*New York Times,* January 15, 2000, p. C1.

26. Tourism In the 1990s significant numbers of tourists traveled from North America and Asia to Australia and South Africa. In 1998 a total of 2,230,000 of these tourists visited Australia, and 390,000 of them visited South Africa. Also, 630,000 of these tourists came from North America, and a total of 2,620,000 tourists traveled from these two regions to these two destinations. (Assume that no single tourist visited both destinations or traveled from both North America and Asia.)

a. The given information is not sufficient to determine the number of tourists from each region to each destination. Why?

b. If you were given the additional information that a total of 1,990,000 tourists came from Asia, would you now be able to determine the number of tourists from each region to each destination? If so, what are these numbers?

c. If you were given the additional information that 200,000 tourists visited South Africa from Asia, would you now be able to determine the number of tourists from each region to each destination? If so, what are these numbers?

Figures are rounded to the nearest 10,000. SOURCES: South African Dept. of Environmental Affairs and Tourism; Australia Tourist Commission/*New York Times,* January 15, 2000, p. C1.

27. Alcohol The following table shows some data from a 2000 study on substance use among tenth-graders in the United States and Europe:

	Used Alcohol	Alcohol Free	Totals
U.S.	x	y	14,000
Europe	z	w	95,000
Totals	63,550	45,450	

"Used Alcohol" indicates consumption of alcohol at least once in the past 30 days. SOURCE: Council of Europe/University of Michigan, "Monitoring the Future"/*New York Times,* February 21, 2001, p. A10.

a. The table leads to a linear system of four equations in four unknowns. What is the system? Does it have a unique solution? What does this indicate about the given and the missing data?

b. Given that the number of U.S. tenth-graders who were alcohol free was 50% more than the number who had used alcohol, find the missing data.

28. Tobacco The following table shows some data from the same study cited in Exercise 27:

	Smoked Cigarettes	Cigarette Free	Totals
U.S.	x	y	14,000
Europe	z	w	95,000
Totals		70,210	109,000

"Smoked Cigarettes" indicates that at least one cigarette was smoked in the past 30 days. SOURCE: Council of Europe/University of Michigan, "Monitoring the Future"/*New York Times,* February 21, 2001, p. A10.

a. The table leads to a linear system of four equations in four unknowns. What is the system? Does it have a unique solution? What does this indicate about the missing data?

b. Given that 31,510 more European tenth-graders smoked cigarettes than U.S. tenth-graders, find the missing data.

29. Investments Things have not been going too well here at Accurate Accounting Inc. since we hired Todd Smiley. He has a tendency to lose important documents, especially around April, when tax returns of our business clients are due. Today, Smiley accidentally shredded Colossal Conglomerate's investment records. We must therefore reconstruct them based on the information he can gather. Todd recalls that the company earned an \$8 million return on investments totaling \$65 million last year. After a few frantic telephone calls to sources in Colossal, he learned that Colossal had made investments in four companies last year: X, Y, Z, and W. (For reasons of confidentiality, we are withholding their names.) Investments in company X earned 15% last year, investments in Y depreciated by 20% last year, investments in Z neither appreciated nor depreciated last year, and investments in W earned 20% last year. Smiley was also told that Colossal invested twice as much in company X as in company Z and three times as much in company W as in company Z. Does Smiley have sufficient information to piece together Colossal's investment portfolio before its tax return is due next week? If so, what does the investment portfolio look like?

30. Investments Things are going from bad to worse here at Accurate Accounting! Colossal Conglomerate's tax return is due tomorrow, and the accountant Todd Smiley seems to have no idea how Colossal earned a return of \$8 million on a \$65 million investment last year. It appears that, although the returns from companies X, Y, Z, and W were as listed in Exercise 29, the rest of the information there was wrong. What Smiley is now being told is that Colossal only invested in companies X, Y, and Z and that the investment in X amounted to \$30 million. His sources in Colossal still maintain that twice as much was invested in company X as in company Z. What should Smiley do?

31. Traffic Flow One-way traffic through Enormous State University is shown in the figure, where the numbers indicate daily counts of vehicles.

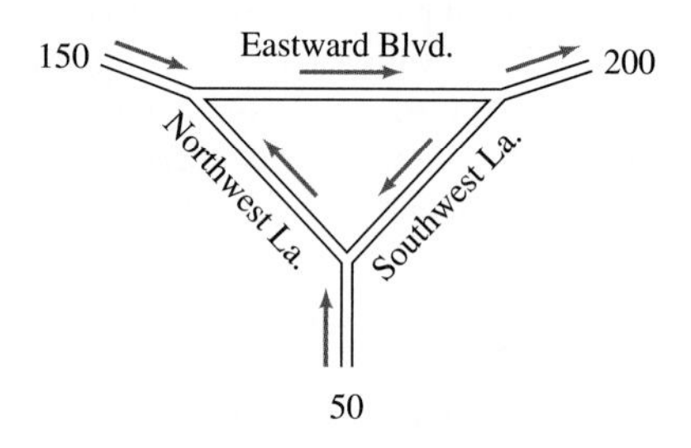

a. Is it possible to determine the daily traffic flow along each of the three streets from the information given? If your answer is yes, what is the traffic flow along each street? If your answer is no, what additional information would suffice?

b. Is a flow of 60 vehicles per day along Southwest Lane consistent with the information given?

32. Traffic Flow The traffic through downtown East Podunk flows through the one-way system shown below:

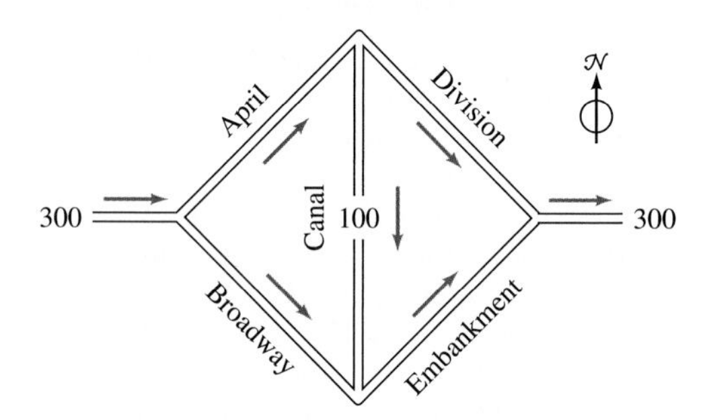

Traffic counters find that 300 vehicles enter town from the west each hour and 300 leave town toward the east each hour. Also, 100 cars drive down Canal Street each hour.

a. Write down the general solution of the associated system of linear equations. Is it possible to determine the number of vehicles on each street per hour?

b. On which street could you put another traffic counter in order to determine the flow completely?

33. Traffic Management The Outer Village Town Council has decided to convert its (rather quiet) main street, Broadway, to a one-way street but is not sure of the direction of most of the traffic. The accompanying diagram illustrates the downtown area of Outer Village, as well as the *net* traffic flow along the intersecting streets (in vehicles per day). (There are no one-way streets; a net traffic flow in a certain direction is defined as the traffic flow in that direction minus the flow in the opposite direction.)

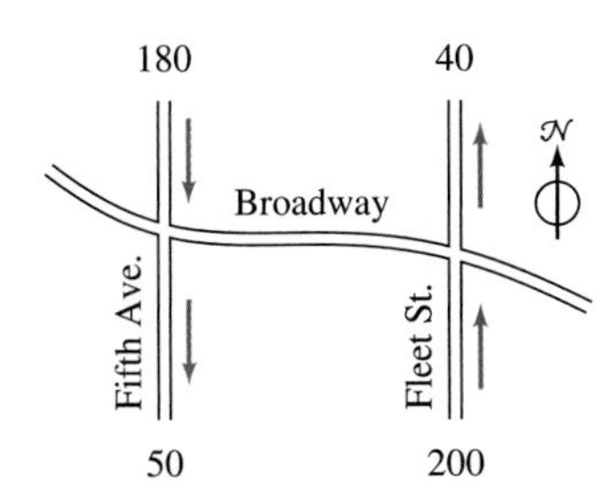

a. Is the given information sufficient to determine the net traffic flow along the three portions of Broadway shown? If your answer is yes, give the traffic flow along each stretch. If your answer is no, what additional information would suffice? (*Hint:* For the direction of net traffic flow, choose either east or west. If a corresponding value is negative, it indicates net flow in the opposite direction.)

b. Assuming that there is little traffic (less than 160 vehicles/day) east of Fleet Street, in what direction is the net flow of traffic along the remaining stretches of Broadway?

34. Electric Current Electric current measures (in amperes, or amps) the flow of electrons through wires. Like traffic flow, the current entering an intersection of wires must equal the current leaving it.[14] Here is an electrical circuit known as a Wheatstone bridge:

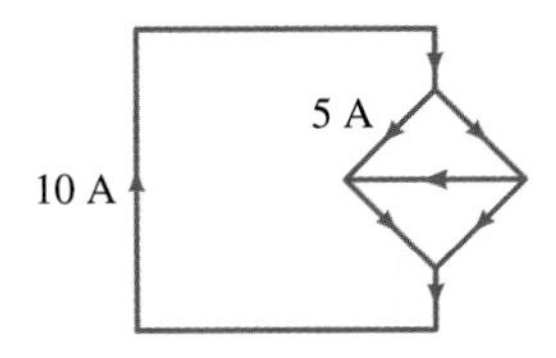

a. If the currents in two of the wires are 10 amps and 5 amps as shown, determine the currents in the unlabeled wires in terms of suitable parameters.

b. In which wire should you measure the current in order to know exactly all the currents?

35. Econometrics (from the GRE Economics Test) This and the next exercise are based on the following simplified model of the determination of the money stock.

$$M = C + D$$
$$C = 0.2D$$
$$R = 0.1D$$
$$H = R + C$$

where M = money stock
C = currency in circulation
R = bank reserves
D = deposits of the public
H = high-powered money

If the money stock were \$120 billion, what would bank reserves have to be?

36. Econometrics (from the GRE Economics Test) With the model in Exercise 35, if H were equal to \$42 billion, what would M equal?

37. Donations The Enormous State University Good Works Society recently raised funds for three worthwhile causes: the Math Professors' Benevolent Fund (MPBF), the Society of Computer Nerds (SCN), and the New York Jets (NYJ). Because the society's members are closet jocks, the society donated twice as much to the NYJ as to the MPBF and equal amounts to the first two funds (it is unable to distinguish between mathematicians and nerds). Further, for every \$1 it gave to the MPBF, it decided to keep \$1 for itself; for every \$1 it gave to the SCN, it kept \$2, and for every \$1 to the NYJ, it also kept \$2. The treasurer of the society, Johnny Treasure, was required to itemize all donations for the dean of students, but discovered to his consternation that he had lost the receipts! The only information available to him was that the society's bank account had swelled by \$4200. How much did the society donate to each cause?

38. Tenure Professor Walt is up for tenure and wishes to submit a portfolio of written student evaluations as evidence of his good teaching. He begins by grouping evaluations into four categories: good reviews; bad reviews—a typical one being "GET RID OF WALT! THE MAN CAN'T TEACH!"; mediocre reviews—such as "I suppose he's OK, given the general quality of teaching here at ______ University"; and reviews left blank. When he tallies up the piles, Walt gets a little worried: There are 280 more bad reviews than good ones and only half as many blank reviews as bad ones. The good reviews and blank reviews together total 170. On an impulse he decides to even up the piles a little by removing 280 of the bad reviews (including the memorable "GET RID OF WALT" review), and this leaves him with a total of 400 reviews of all types. How many of each category of reviews were there originally?

[14] This is known as *Kirchhoff's current law,* named after Gustav Robert Kirchhoff (1824–1887). Kirchhoff made important contributions to the fields of geometric optics, electromagnetic radiation, and electrical network theory.

CAT Scans CAT (computerized axial tomographic) scans are used to map the exact location of interior features of the human body. CAT scan technology is based on the following principles: (1) different components of the human body (water, gray matter, bone, and so on) absorb X-rays to different extents; and (2) to measure the X-ray absorption by a specific region of, say, the brain, it suffices to pass a number of line-shaped pencil beams of X-rays through the brain at different angles and measure the total absorption for each beam, which is the sum of the absorptions of the regions through which it passes. The accompanying diagram illustrates a simple example. (The number in each region shows its absorption, and the number on each X-ray beam shows the total absorption for that beam.)

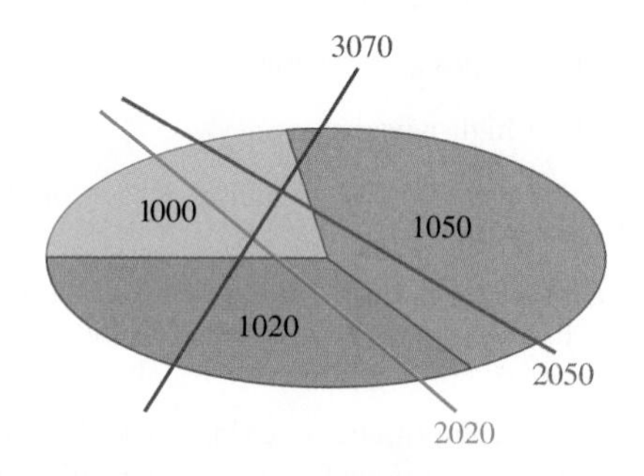

SOURCE: Based on *Geometry, New Tools for New Technologies*, by J. Malkevitch, Video Applications Library, COMAP, 1992. The absorptions are actually calibrated on a logarithmic scale. In real applications the size of the regions is very small, and very large numbers of beams must be used.

In Exercises 39–44, use the table and the given X-ray absorption diagrams to identify the composition of each of the regions marked by a letter.

Type	Air	Water	Gray matter	Tumor	Blood	Bone
Absorption	0	1000	1020	1030	1050	2000

39.

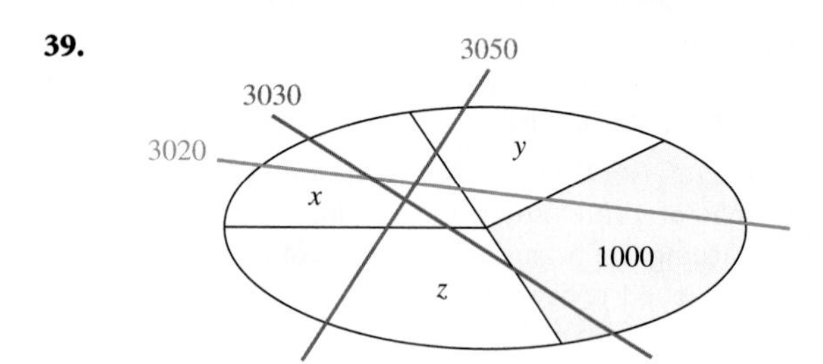

40.

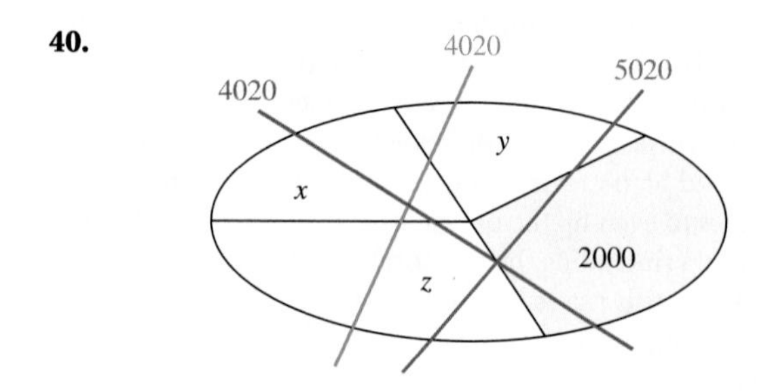

41.

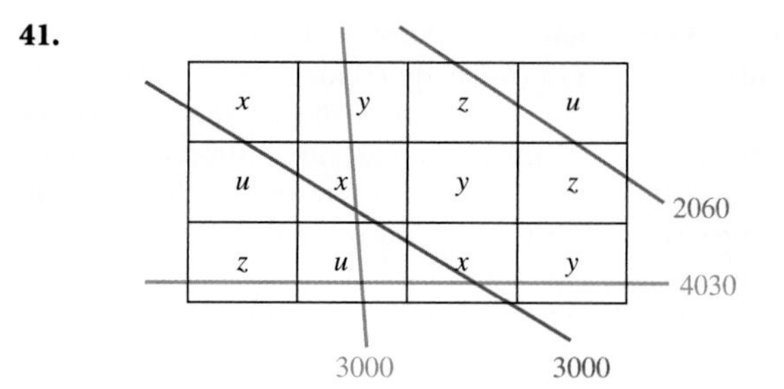

42.

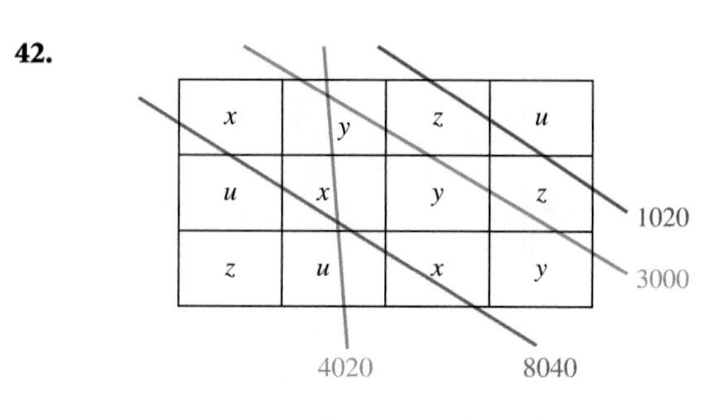

43. Identify the composition of site x.

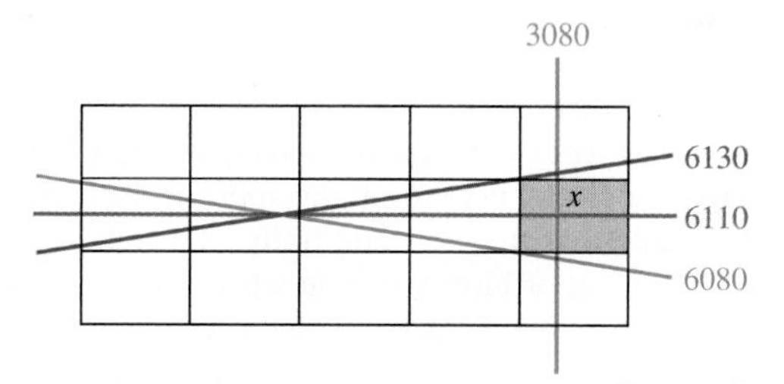

(The horizontal and slanted beams each pass through five regions.)

44. Identify the composition of site x.

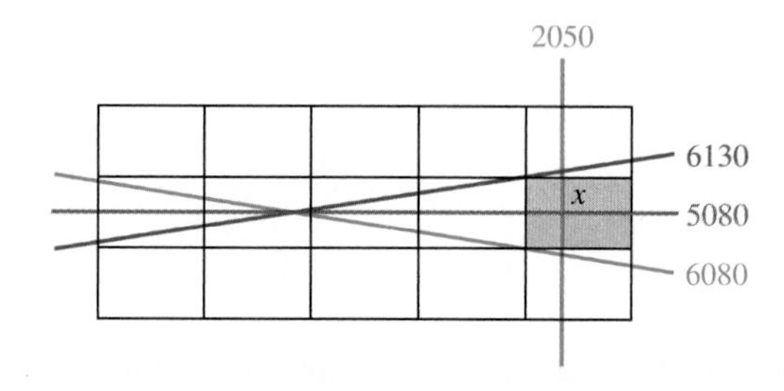

(The horizontal and slanted beams each pass through five regions.)

Airline Costs Exercises 45 and 46 are based on the following chart, which shows the amount spent by five U.S. airlines to fly one available seat 1 mile in the fourth quarter of 2000.

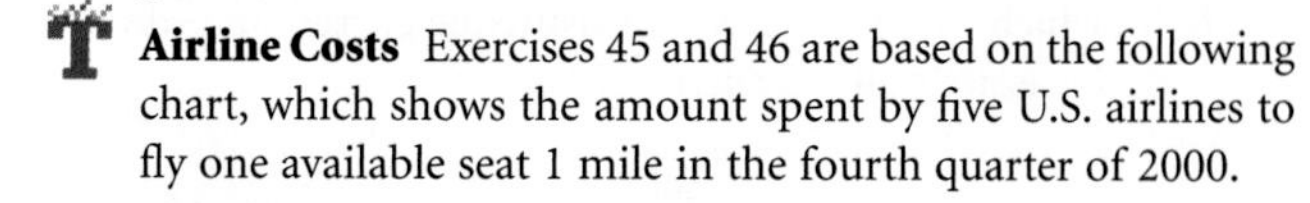

Airline	United	American	Continental	SkyWest	Southwest
Cost (¢)	11.3	11.0	9.8	20.3	7.8

SOURCE: Company financial statements, obtained from their respective Web sites. The cost per available seat-mile is a commonly used operating statistic in the airline industry.

45. On a 3000-mile New York–Los Angeles flight, suppose United, American, and Southwest flew a total of 210 empty seats, costing them a total of $65,580. If United had three times as many empty seats as American, how many empty seats did each of these three airlines carry on its flight?

46. On a 2000-mile Miami–Memphis flight, suppose Continental, SkyWest, and Southwest flew a total of 200 empty seats, costing them a total of $58,200. If SkyWest had twice as many empty seats as Southwest, how many empty seats did each of these three airlines carry on its flight?

COMMUNICATION AND REASONING EXERCISES

47. Are Exercises 1 and 2 realistic in their expectation of using up all the ingredients? What does your answer have to do with the solution(s) of the associated system of equations?

48. Suppose that you obtained a solution for Exercise 3 or 4 consisting of positive values that were not all whole numbers. What would such a solution signify about the situation in the exercise? Should you round these values to the nearest whole numbers?

In Exercises 49–54, x, y, and z represent the weights of the three ingredients X, Y, and Z in a gasoline blend. Which of the following is represented by a linear equation in x, y, and z? When it is, give a form of the equation.

49. The blend consists of 100 pounds of ingredient X.

50. The blend is free of ingredient X.

51. The blend contains 30% ingredient Y by weight.

52. The weight of ingredient X is the product of the weights of ingredients Y and Z.

53. There is at least 30% ingredient Y by weight.

54. There is twice as much ingredient X by weight as Y and Z combined.

55. Make up an entertaining word problem leading to the following system of equations:

$$\begin{aligned} 10x + 20y + 10z &= 100 \\ 5x + 15y &= 50 \\ x + y + z &= 10 \end{aligned}$$

56. Make up an entertaining word problem leading to the following system of equations:

$$\begin{aligned} 10x + 20y &= 300 \\ 10z + 20w &= 400 \\ 20x + 10z &= 400 \\ 10y + 20w &= 300 \end{aligned}$$

CASE STUDY

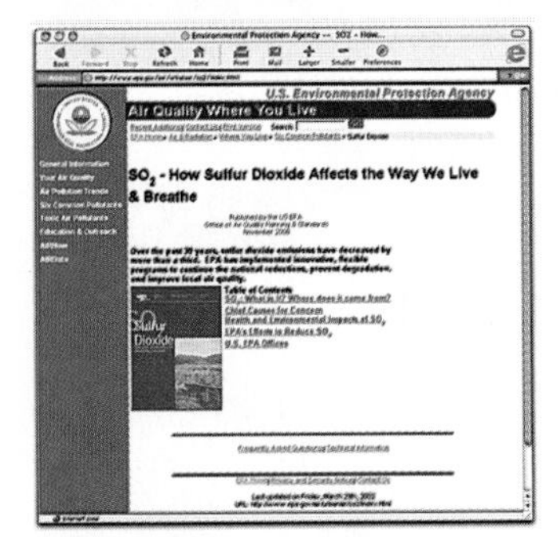

Reproduced from the Environmental Protection Agency Web site

The Impact of Regulating Sulfur Emissions

Your consulting company has been hired by the EPA. The EPA is considering regulations requiring a 15 million–ton rollback in sulfur emissions in an effort to curb the effects of acid rain on the ecosystem. The EPA would like to have an estimate of the cost to the major utility companies and the effect on jobs in the coal-mining industry. The following data are available:

Strategy	Annual Cost to Utilities ($ billions)	Cost in Jobs (number of jobs lost)
8 million–ton rollback	20.4	14,100
10 million–ton rollback	34.5	21,900
12 million–ton rollback	93.6	13,400*

SOURCE: U.S. Congress, Congressional Budget Office, *Curbing Acid Rain: Cost, Budget, and Coal Market Effects* (Washington, D.C.: Government Printing Office, 1986): xx, xxii, 23, 80.

*Job losses drop when the rollback is increased to 12 million tons because a rollback of this magnitude requires that expensive scrubbers be installed to filter emissions, even if a utility company has switched from coal to other energy sources. Once the scrubbers are installed, it is more profitable for utility companies to switch back to (cheaper) coal as a primary source of energy. A 10 million–ton reduction, on the other hand, results in a massive move away from coal—this being cheaper than installing scrubbers—and hence a dramatic job loss in the coal-mining industry.

Your assignment is to use these data to give projections of the annual cost to utilities and the cost in jobs if the regulations were to be enacted.

You decide to consider the annual cost C to utilities and the job loss J separately. After giving the situation some thought, you decide to have two equations, one giving C

in terms of the rollback tonnage t and the other giving J in terms of t. Your first inclination is to try linear equations—that is, an equation of the form

$$C = at + b \qquad (a \text{ and } b \text{ constants})$$

and a similar one for J, but you quickly discover that the data simply won't fit, no matter what the choice of the constants. The reason for this can be seen graphically by plotting C and J versus t (Figure 13). In neither case do the three points lie on a straight line. In fact, the job data are not even *close* to being linear. Thus, you will need curves to model these data.

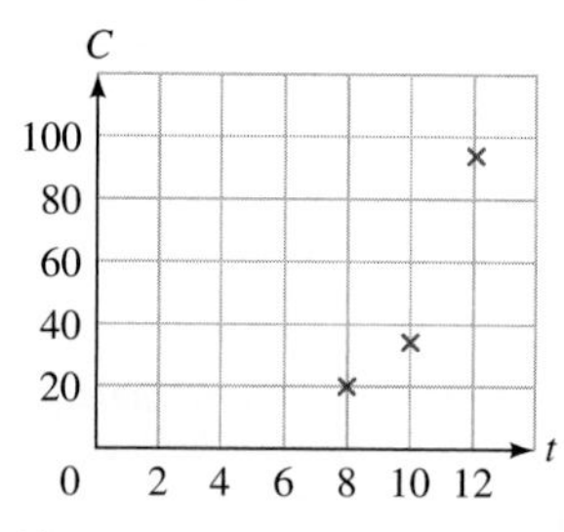

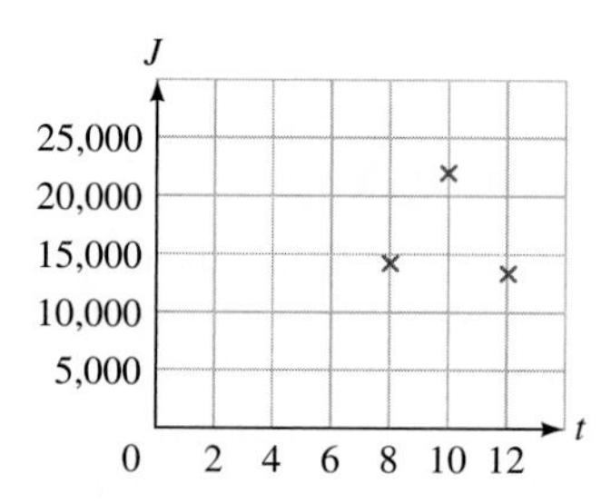

Figure 13

After giving the matter further thought, you remember something your mathematics instructor once said: The simplest curve passing through any three points not all on the same line is a parabola. Since you are looking for a simple model of the data, you decide to try a parabola. A general parabola has the equation

$$C = at^2 + bt + c$$

where a, b, and c are constants. The question now is, What are a, b, and c? You decide to try substituting the values of C and t into the general equation, and you get the following:

$$t = 8, C = 20.4 \quad \text{gives} \quad 20.4 = 64a + 8b + c$$

$$t = 10, C = 34.5 \quad \text{gives} \quad 34.5 = 100a + 10b + c$$

$$t = 12, C = 93.6 \quad \text{gives} \quad 93.6 = 144a + 12b + c$$

Now you notice that you have three linear equations in three unknowns! You solve the system:

$$a = 5.625 \qquad b = -94.2 \qquad c = 414$$

Thus, your cost equation becomes

$$C = 5.625t^2 - 94.2t + 414$$

You now substitute $t = 15$ to get $C = 266.625$. In other words, you are able to predict an annual cost to the utility industry of \$266.625 billion.

Forging ahead, you turn to the jobs equation and write

$$J = at^2 + bt + c$$

Substituting the data from the second column of the chart, you obtain

$$14{,}100 = 64a + 8b + c$$

$$21{,}900 = 100a + 10b + c$$

$$13{,}400 = 144a + 12b + c$$

You solve this system and find that

$$a = -2037.5 \qquad b = 40{,}575 \qquad c = -180{,}100$$

Thus, your jobs equation is

$$J = -2037.5t^2 + 40{,}575t - 180{,}100$$

Substituting $t = 15$, you get $J = -29{,}912.5$. Uh-oh! A negative number! Just what does this mean? Well, if there were to be $-29{,}913$ jobs lost, then there would be 29,913 new jobs *created!*[15] Figure 14 shows the parabolas superimposed on the data points.

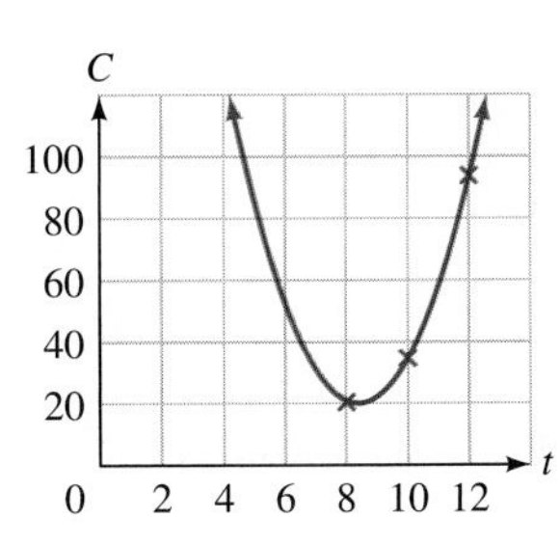

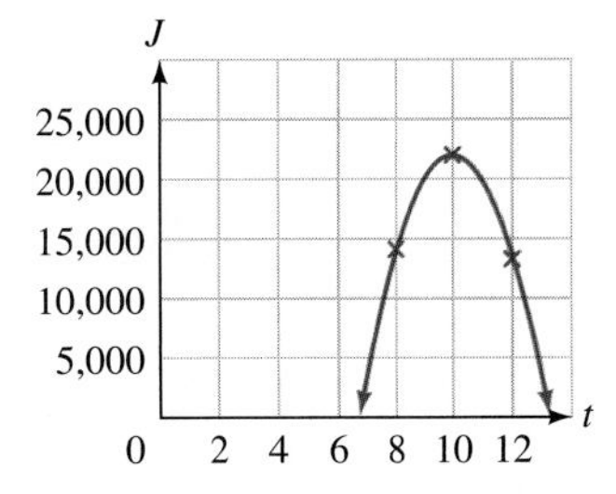

Figure 14

You submit the following projection: A 15 million–ton rollback will result in an annual cost of approximately \$266.6 billion to the utility industry, but will also result in the creation of approximately 29,900 new jobs in the coal-mining industry.

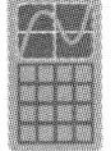

Graphing Calculator: Automating the Computation

Graphing calculators and spreadsheets usually have built-in regression features that allow you to find the equation of a curve through several points automatically. On the TI-83 press STAT, select `EDIT`, then enter the x values in list L_1 and the y values in list L_2. Next press STAT, select `CALC`, and select `QuadReg` to obtain the equation of the quadratic that fits the given data.

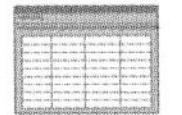

Excel: Automating the Computation

In Excel enter the data for the "Cost in Jobs" as shown:

	A	B
1	8	14100
2	10	21900
3	12	13400

Then, select the cells shown and click on the graph button on the toolbar. Select the `XY Scatter` option with points plotted (no lines or curves connecting them). On the graph, select the points on the graph (by clicking on one of them) and, from the `Chart` menu, select `Add Trendline`. In the resulting dialog box, select the polynomial option and, under `Options`, select `Forecast 3 units` (to give us the re-

[15] The logic is this: Since the stringent new emission standards would require that even more scrubbers be installed, the utility industry might as well increase its use of coal (which tends to be far cheaper than other energy sources), thus creating a boom for the coal industry.

sult for the 15 million–ton rollback) and `Display equation on chart`. Click on `OK` and a graph similar to Figure 15 will result.

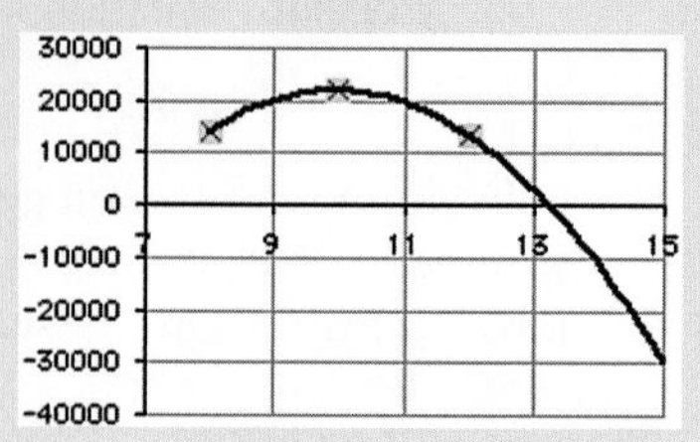

Figure 15

You can now use the graph or the equation to make the desired forecast for the 15 million–ton rollback.

Web Site: Automating the Computation

The path

Web site → Online Utilities → Simple Regression

will take you to a utility that you can use to find the equations above. Enter the x and y values and press the `y=ax^2+bx+c` button to obtain the equation of the quadratic that fits the given data.

EXERCISES

1. Repeat the computations above, using the following rounded figures:

Strategy	Annual Cost to Utilities ($ billions)	Cost in Jobs (number of jobs lost)
8 million–ton rollback	20	14,000
10 million–ton rollback	35	22,000
12 million–ton rollback	94	13,000

2. If the 8 million–ton rollback data had not been available, what projections would you have given? (Use the original data.)

3. If the 10 million–ton rollback data had not been available, what projections would you have given? (Use the original data.)

4. Find the equation of the parabola that passes through the points (1, 2), (2, 9), and (3, 19).

5. Is there a parabola that passes through the points (1, 2), (2, 9) and (3, 16)?

6. Is there a parabola that passes through the points (1, 2), (2, 9), (3, 19), and (−1, 2)?

T **7.** Use a spreadsheet to estimate the size of the rollback that results in the lowest annual cost to utilities.

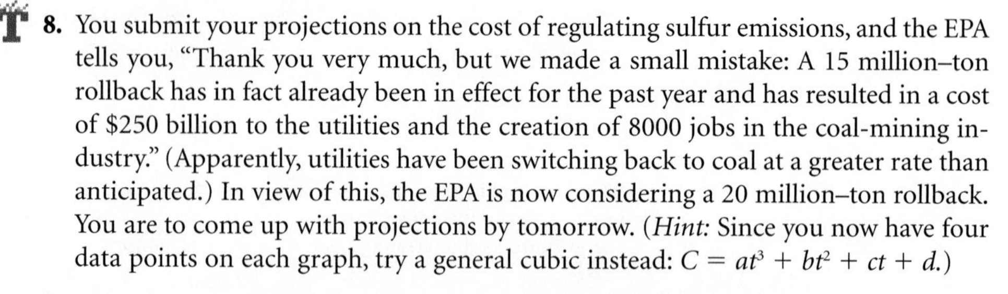

T **8.** You submit your projections on the cost of regulating sulfur emissions, and the EPA tells you, "Thank you very much, but we made a small mistake: A 15 million–ton rollback has in fact already been in effect for the past year and has resulted in a cost of $250 billion to the utilities and the creation of 8000 jobs in the coal-mining industry." (Apparently, utilities have been switching back to coal at a greater rate than anticipated.) In view of this, the EPA is now considering a 20 million–ton rollback. You are to come up with projections by tomorrow. (*Hint:* Since you now have four data points on each graph, try a general cubic instead: $C = at^3 + bt^2 + ct + d$.)

CHAPTER 2 REVIEW TEST

1. Use any method to solve the following systems of linear equations.

a. $x + 2y = 4$
$2x - y = 1$

b. $0.2x - 0.1y = 0.3$
$0.2x + 0.2y = 0.4$

c. $\frac{1}{2}x - \frac{3}{4}y = 0$
$6x - 9y = 0$

d. $2x + 3y = 2$
$-x - \frac{3y}{2} = \frac{1}{2}$

e. $x + y = 1$
$2x + y = 0.3$
$3x + 2y = \frac{13}{10}$

f. $3x + 0.5y = 0.1$
$6x + y = 0.2$
$\frac{3x}{10} - 0.05y = 0.01$

2. Solve the following systems of equations.

a. $x + 2y = -3$
$x - z = 0$
$x + 3y - 2z = -2$

b. $x - y + z = 2$
$7x + y - z = 6$
$x - \frac{1}{2}y + \frac{1}{3}z = 1$
$x + y + z = 6$

c. $x - \frac{1}{2}y + z = 0$
$\frac{1}{2}x - \frac{1}{2}z = -1$
$\frac{3}{2}x - \frac{1}{2}y + \frac{1}{2}z = -1$

d. $x + y - 2z = -1$
$-2x - 2y + 4z = 2$
$0.75x + 0.75y - 1.5z = -0.75$

e. $x = \frac{1}{2}y$
$\frac{1}{2}x = -\frac{1}{2}z + 2$
$z = -3x + y$

f. $x - y + z = 1$
$y - z + w = 1$
$x + z - w = 1$
$2x + z = 3$

3. The Fahrenheit and Celsius temperature scales are related by the equation

$$5F - 9C = 160$$

where F is the Fahrenheit temperature of an object and C is its Celsius temperature.

a. What temperature should an object be if its Fahrenheit and Celsius temperatures are the same?

b. What temperature should an object be if its Celsius temperature is half its Fahrenheit temperature?

c. Is it possible for the Fahrenheit temperature of an object to be 1.8 times its Celsius temperature? Explain.

4. Let x, y, z, and w represent the population in millions of four cities A, B, C, and D. Express each of the following as an equation in x, y, z, and w. If the equation is linear, say so and express it in the standard form $ax + by + cz + dw = k$.

a. The total population of the four cities is 10 million people.

b. City A has three times as many people as cities B and C combined.

c. City D is actually a ghost town; there are no people living in it.

d. The population of city A is the sum of the squares of the populations of the other three cities.

e. City C has 30% more people than city B.

f. City C has 30% fewer people than city B.

OHaganBooks.com—JUGGLING THE VARIABLES

5. You are the buyer for OHaganBooks.com and are considering increasing stocks of romance and horror novels at the new OHaganBooks.com warehouse in Texas. You have offers from two publishers: Duffin House and Higgins Press. Duffin offers a package of 5 horror novels and 5 romance novels for \$50, and Higgins offers a package of 5 horror and 11 romance novels for \$150.

a. How many packages should you purchase from each publisher to get exactly 4500 horror novels and 6600 romance novels?

b. If, instead, you want to spend a total of \$50,000 on books and have promised to buy twice as many packages from Duffin as from Higgins, how many packages should you purchase from each publisher?

c. You are just about to place your orders when your accountant tells you that you can now afford to spend a total of \$60,000 on romance and horror books. She also reminds you that you had signed an agreement to spend the same amount of money at both publishers and hence cannot carry out your promise in part (b). How many packages should you purchase from each publisher?

6. OHaganBooks.com has two principal competitors: JungleBooks.com and FarmerBooks.com. Combined Web-site traffic at the three sites is estimated at 10,000 hits/day. Only 10% of the hits at OHaganBooks.com result in orders, whereas JungleBooks.com and FarmerBooks.com report that 20% of the hits at their sites result in book orders. Together, the three sites process 1500 book orders/day. FarmerBooks.com appears to be the most successful of the three and gets as many book orders as the other two combined. What is the traffic (in hits per day) at each of the sites?

7. Duffin House, Higgins Press, and Sickle Publications all went public on the same day recently. John O'Hagan had the opportunity to participate in all three initial public offerings (partly because he and Marjory Duffin are good friends). He made a considerable profit when he sold all of the stock 2 days later on the open market. The following table shows the purchase price and percentage yield on the investment in each company:

	Purchase Price/Share (\$)	Yield (%)
Duffin House (DHS)	8	20
Higgins Press (HPR)	10	15
Sickle Publications (SPUB)	15	15

He invested \$20,000 in a total of 2000 shares and made a \$3400 profit from the transactions. How many shares in each company did he purchase?

8. During his lunch break, John O'Hagan decides to devote some time to assisting his son Billy Sean, who is having a terrible time coming up with a college course schedule. One reason for this is the very complicated bulletin of Suburban State University. It reads as follows:

> *All candidates for the degree of Bachelor of Arts at SSU must take a total of 124 credits from the Sciences, Fine Arts, Liberal Arts, and Mathematics,*[16] *including an equal number of Science and Fine Arts credits, and twice as many Mathematics credits as Science credits and Fine Arts credits combined, but with Liberal Arts credits exceeding Mathematics credits by exactly one-third of the number of Fine Arts credits.*

What are all the possible degree programs for Billy Sean?

9. On the same day that the sales department at Duffin House received an order for 600 packages from the OHaganBooks.com Texas headquarters, it received an additional order for 200 packages from FantasyBooks.com, based in California. Duffin House has warehouses in New York and Illinois. The Illinois warehouse is closing down and must clear all 300 packages it has in stock. Shipping costs per package of books are as follows:

New York to Texas: $20	New York to California: $50
Illinois to Texas: $30	Illinois to California: $40

Is it possible to fill both orders and clear the Illinois warehouse with a shipping budget of $22,000? If so, how many packages should be sent from each warehouse to each online bookstore?

10. All book orders received at the Order Department at OHaganBooks.com are transmitted through a small computer network to the Shipping Department. The following diagram shows the network (which uses two intermediate computers as routers), together with some of the average daily traffic measured in book orders.

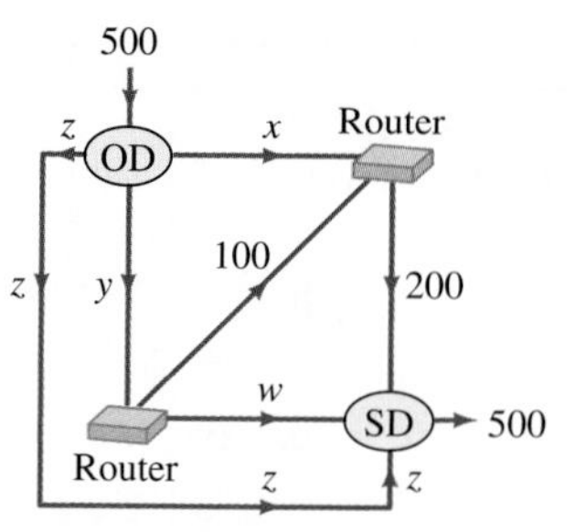

OD = Order Department
SD = Shipping Department

a. Set up a system of linear equations in which the unknowns give the average traffic along the paths labeled x, y, z, and w, and find the general solution.
b. What is the minimum volume of traffic along y?
c. What is the maximum volume of traffic along w?
d. If there is no traffic along z, find the volume of traffic along all the paths.
e. If there is the same volume of traffic along y and z, what is the volume of traffic along w?

ADDITIONAL ONLINE REVIEW

If you follow the path
Web site → Everything for Finite Math → Chapter 2
you will find the following additional resources to help you review.

A comprehensive chapter summary (including examples and interactive features)

Additional review exercises (including interactive exercises and many with help)

Interactive online tutorials for all sections

Interactive Excel tutorials for all sections

A true/false chapter quiz

Utilities and other resources for matrix operations

---

[16]Strictly speaking, mathematics is not a science; it is the Queen of the Sciences, although we like to think of it as the Mother of All Sciences.

3

MATRIX ALGEBRA AND

CASE STUDY

The Japanese Economy

A senator walks into your cubicle in the Congressional Budget Office. "Look here," she says, "I don't see why the Japanese trade representative is getting so upset with my proposal to cut down on our use of Japanese finance and insurance. He claims that it'll hurt Japan's mining operations. But just look at Japan's input–output table. The finance sector doesn't use any input from the mining sector. How can our cutting down demand for finance and insurance hurt mining?" How should you respond?

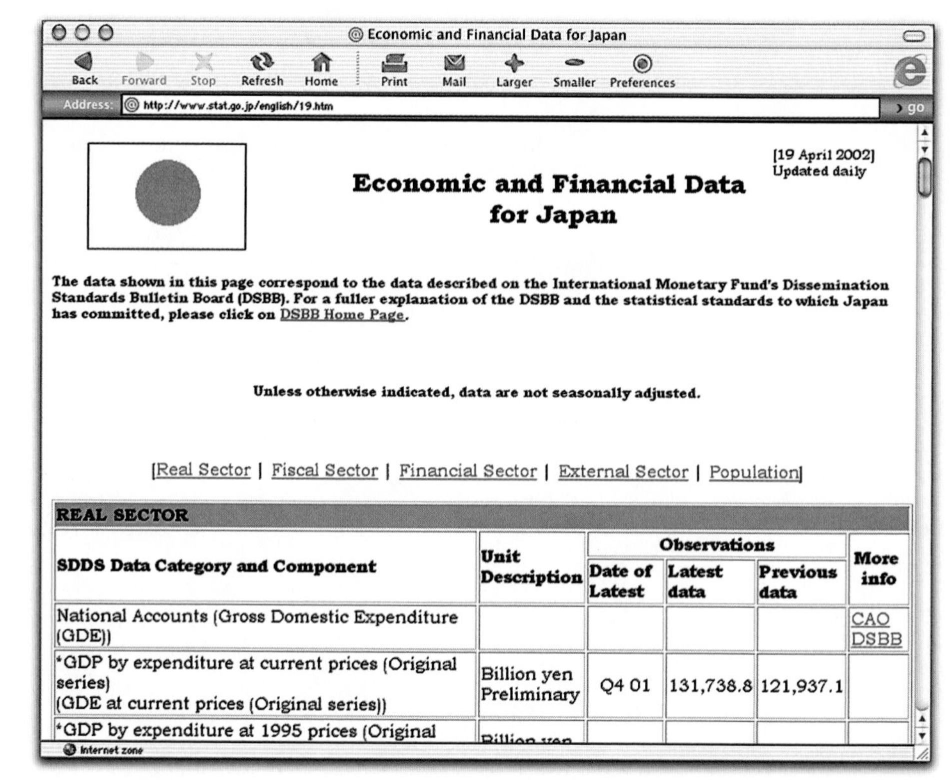

Reproduced from the Statistical Standards Department Web site, division of the Statistics Bureau, Ministry of Public Management, Japan

INTERNET RESOURCES FOR THIS CHAPTER

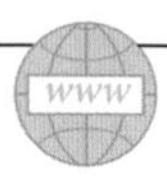

At the Web site, follow the path

Web site → Everything for Finite Math → Chapter 3

where you will find matrix algebra utilities, section-by-section interactive tutorials, a detailed chapter summary you can print out, a true/false quiz, and a collection of chapter review questions. You will also find an online matrix algebra utility and Excel spreadsheet tutorials for each section.

APPLICATIONS

Introduction

We used matrices in Chapter 2 simply to organize our work. It is time we examined them as interesting objects in their own right. There is much that we can do with matrices besides row operations: We can add, subtract, multiply, and even, in a sense, "divide" matrices. We use these operations to study input–output models in this chapter and Markov chains in a later chapter.

Many calculators, electronic spreadsheets, and other computer programs can do these matrix operations, which is a big help in doing calculations. However, we need to know how these operations are defined to see why they are useful and to understand which to use in any particular application.

3.1 Matrix Addition and Scalar Multiplication

Let's start by formally defining what a matrix is and introducing some basic terms.

Matrix, Dimension, and Entries

An $m \times n$ **matrix** A is a rectangular array of real numbers with m rows and n columns. We refer to m and n as the **dimensions** of the matrix. The numbers that appear in the matrix are called its **entries.** We customarily use capital letters A, B, C, . . . for the names of matrices.

Quick Examples

1. $A = \begin{bmatrix} 2 & 0 & 1 \\ 33 & -22 & 0 \end{bmatrix}$ is a 2×3 matrix because it has 2 rows and 3 columns.

2. $B = \begin{bmatrix} 2 & 3 \\ 10 & 44 \\ -1 & 3 \\ 8 & 3 \end{bmatrix}$ is a 4×2 matrix because it has 4 rows and 2 columns.

The entries of A are 2, 0, 1, 33, -22, and 0. The entries of B are 2, 3, 10, 44, -1, 3, 8, and 3.

Hint: Remember that the number of rows is given first and the number of columns second. An easy way to remember this is to think of the acronym RC for "Row then Column."

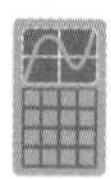

Graphing Calculator: Entering a Matrix

On the TI-83, matrices are referred to as [A], [B], and so on. To enter a matrix, press MATRX to bring up the matrix menu, select `EDIT`, select a matrix, and press ENTER. Then enter the dimensions of the matrix followed by its entries. When you want to use a matrix, press MATRX, select the matrix, and press ENTER.

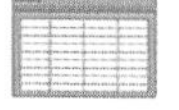

Excel: Entering a Matrix

To enter a matrix in a spreadsheet, we put its entries in any convenient block of cells. For example, the matrix A in the Quick Example above might look like this:

	A	B	C
1	2	0	1
2	33	-22	0

Web Site

The path

Web site → Online Utilities → Matrix Algebra Tool

will take you to an online computational tool that allows you to add and multiply matrices.

The path

Web site → Online Tutorials → Excel Tutorials → Matrix Addition and Scalar Multiplication

will give you a downloadable Excel tutorial that covers the use of Excel for some of the examples in this section.

Referring to the Entries of a Matrix

There is a systematic way of referring to particular entries in a matrix. If i and j are numbers, then the entry in the ith row and jth column of the matrix A is called the ***ij*th entry** of A. We usually write this entry as a_{ij} or A_{ij}. (If the matrix were called B, we would write its ijth entry as b_{ij} or B_{ij}.) Notice that this follows the "RC" convention: The row number is specified first and the column number second.

Quick Example

With $A = \begin{bmatrix} 2 & 0 & 1 \\ 33 & -22 & 0 \end{bmatrix}$,

$a_{13} = 1$ First row, third column

$a_{21} = 33$ Second row, first column

According to the labeling convention, the entries of the matrix A above are

$$A = \begin{bmatrix} a_{11} & a_{12} & a_{13} \\ a_{21} & a_{22} & a_{23} \end{bmatrix}$$

In general, the $m \times n$ matrix A has its entries labeled as follows:

$$A = \begin{bmatrix} a_{11} & a_{12} & a_{13} & \cdots & a_{1n} \\ a_{21} & a_{22} & a_{23} & \cdots & a_{2n} \\ \vdots & \vdots & \vdots & \ddots & \vdots \\ a_{m1} & a_{m2} & a_{m3} & \cdots & a_{mn} \end{bmatrix}$$

We say that two matrices A and B are **equal** if they have the same dimensions and the corresponding entries are equal. Note that a 3×4 matrix can never equal a 3×5 matrix because they do not have the same dimensions.

Example 1 • Matrix Equality

Let $A = \begin{bmatrix} 7 & 9 & x \\ 0 & -1 & y+1 \end{bmatrix}$ and $B = \begin{bmatrix} 7 & 9 & 0 \\ 0 & -1 & 11 \end{bmatrix}$. Find the values of x and y such that $A = B$.

Solution For the two matrices to be equal, we must have corresponding entries equal, so

$x = 0$ $\qquad a_{13} = b_{13}$

$y + 1 = 11, \quad \text{or} \quad y = 10$ $\qquad a_{23} = b_{23}$

Before we go on . . . Note that the matrix equation

$$\begin{bmatrix} 7 & 9 & x \\ 0 & -1 & y+1 \end{bmatrix} = \begin{bmatrix} 7 & 9 & 0 \\ 0 & -1 & 11 \end{bmatrix}$$

is really six equations in one: $7 = 7, 9 = 9, x = 0, 0 = 0, -1 = -1$, and $y + 1 = 11$. We used only the two that were interesting.

Row Matrix, Column Matrix, and Square Matrix

A matrix with a single row is called a **row matrix,** or **row vector.** A matrix with a single column is called a **column matrix,** or **column vector.** A matrix with the same number of rows as columns is called a **square matrix.**

Quick Examples

The 1×5 matrix $C = [3 \quad -4 \quad 0 \quad 1 \quad -11]$ is a row matrix.

The 4×1 matrix $D = \begin{bmatrix} 2 \\ 10 \\ -1 \\ 8 \end{bmatrix}$ is a column matrix.

The 3×3 matrix $E = \begin{bmatrix} 1 & -2 & 0 \\ 0 & 1 & 4 \\ -4 & 32 & 1 \end{bmatrix}$ is a square matrix.

Matrix Addition and Subtraction

The first matrix operations we discuss are matrix addition and subtraction. The rules for these operations are simple.

Matrix Addition and Subtraction

Two matrices can be added (or subtracted) if and only if they have the same dimensions. To add (or subtract) two matrices of the same dimensions, we add (or subtract) the corresponding entries. More formally, if A and B are $m \times n$ matrices, then $A + B$ and $A - B$ are the $m \times n$ matrices whose entries are given by

$(A + B)_{ij} = A_{ij} + B_{ij}$ — ijth entry of the sum = sum of the ijth entries

$(A - B)_{ij} = A_{ij} - B_{ij}$ — ijth entry of the difference = difference of the ijth entries

Quick Examples

1. $\begin{bmatrix} 2 & -3 \\ 1 & 0 \\ -1 & 3 \end{bmatrix} + \begin{bmatrix} 9 & -5 \\ 0 & 13 \\ -1 & 3 \end{bmatrix} = \begin{bmatrix} 11 & -8 \\ 1 & 13 \\ -2 & 6 \end{bmatrix}$ Corresponding entries added

2. $\begin{bmatrix} 2 & -3 \\ 1 & 0 \\ -1 & 3 \end{bmatrix} - \begin{bmatrix} 9 & -5 \\ 0 & 13 \\ -1 & 3 \end{bmatrix} = \begin{bmatrix} -7 & 2 \\ 1 & -13 \\ 0 & 0 \end{bmatrix}$ Corresponding entries subtracted

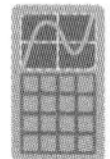

Graphing Calculator: Adding Matrices

On the TI-83, adding matrices is similar to adding numbers: The sum of the matrices [A] and [B] is [A] + [B]. Press [MATRX] and select `NAMES` to enter any of the variables [A], [B], and so on, on the home screen. If you attempt to add matrices with different dimensions, you will get an error message. If you want to save the results of your calculations for later use, store them in matrix variables. For example,

```
[A]+[B]→[C]
```

stores the sum of [A] and [B] in [C].

Excel: Adding Matrices

As we saw in Chapter 2, a matrix can be thought of as a rectangular block of data; Excel refers to such blocks of data as **arrays,** which it can handle in much the same way as it handles single cells of data. For instance, when typing a formula, just as clicking on a cell creates a reference to that cell, selecting a whole array of cells will create a reference to that array.

Arrays are referred to using an **array range** consisting of the top-left and bottom-right cell coordinates, separated by a colon. For example, the array range A1:C2 refers to the 2×3 block with top-left corner A1 and bottom-right corner C2 (see the following worksheet).

To add two matrices in Excel, first input their entries in two separate arrays in the spreadsheet:

	A	B	C	D	E	F	G
1	2	0	1		-1	2	34
2	33	-22	0		3	-2	1

Select (highlight) a block of the same size (2×3 in this case) where you would like the answer to appear. Next, enter the formula `=A1:C2+E1:G2` and then press [Control]+[Shift]+[Enter]. The easiest way to do this is as follows:

1. Type =.
2. Highlight the first matrix. — Cells A1 through C2
3. Type +.
4. Highlight the second matrix. — Cells E1 through G2
5. Press Control + Shift + Enter. — Not just Enter

	A	B	C	D	E	F	G
1	2	0	1		−1	2	34
2	33	−22	0		3	−2	1
3							
4	=A1:C2+E1:G2						
5							

Pressing Control + Shift + Enter (instead of Enter) tells Excel that your formula is an *array formula*, one that returns a matrix rather than a single number.[1] Once entered, the formula bar will show the formula you entered enclosed in "curly braces," indicating that it is an array formula. Note that you must use Control + Shift + Enter to delete any array you create: Select the block you wish to delete and press Delete followed by Control + Shift + Enter.

Example 2 • Sales

The APlus auto parts store chain has two outlets, one in Vancouver and one in Quebec. Among other things, it sells wiper blades, windshield cleaning fluid, and floor mats. The monthly sales of these items at the two stores for 2 months are given in the following tables:

January Sales

	Vancouver	Quebec
Wiper Blades	20	15
Cleaning Fluid (bottles)	10	12
Floor Mats	8	4

February Sales

	Vancouver	Quebec
Wiper Blades	23	12
Cleaning Fluid (bottles)	8	12
Floor Mats	4	5

Use matrix arithmetic to calculate the change in sales of each product in each store from January to February.

Solution The tables suggest two matrices:

$$J = \begin{bmatrix} 20 & 15 \\ 10 & 12 \\ 8 & 4 \end{bmatrix} \quad \text{and} \quad F = \begin{bmatrix} 23 & 12 \\ 8 & 12 \\ 4 & 5 \end{bmatrix}$$

[1] On a Macintosh, Command + Enter has the same effect as Control + Shift + Enter.

We want to subtract corresponding entries in these two matrices. In other words, we want to compute the difference of the two matrices:

$$F - J = \begin{bmatrix} 23 & 12 \\ 8 & 12 \\ 4 & 5 \end{bmatrix} - \begin{bmatrix} 20 & 15 \\ 10 & 12 \\ 8 & 4 \end{bmatrix} = \begin{bmatrix} 3 & -3 \\ -2 & 0 \\ -4 & 1 \end{bmatrix}$$

Thus, the change in sales of each product is the following:

	Vancouver	Quebec
Wiper Blades	3	−3
Cleaning Fluid (bottles)	−2	0
Floor Mats	−4	1

Excel: Subtracting Matrices

Here is the matrix operation as you might set it up in a worksheet:

	A	B	C	D	E	F	G
1	January Sales						
2		Vancouver	Quebec				
3	wiper blades	20	15				
4	cleaning fluid (bottles)	10	12				
5	floor mats	8	4				
6							
7	February Sales				Change in Sales		
8		Vancouver	Quebec			Vancouver	Quebec
9	wiper blades	23	12		wiper blades	=B9:C11-B3:C5	
10	cleaning fluid (bottles)	8	12		cleaning fluid (bottles)		
11	floor mats	4	5		floor mats		

Scalar Multiplication

Question When can a matrix A be added to *itself*?

Answer Always, because the expression $A + A$ is the sum of two matrices that have the same dimensions.

Question Can't we write $A + A$ as $2A$?

Answer We certainly can. Notice that when we compute $A + A$ we end up doubling every entry in A. So we can think of the expression $2A$ as telling us to *multiply every element in A by 2.*

In general, to multiply a matrix by a number, multiply every entry in the matrix by that number. For example,

$$6\begin{bmatrix} \frac{5}{2} & -3 \\ 1 & 0 \\ -1 & \frac{5}{6} \end{bmatrix} = \begin{bmatrix} 15 & -18 \\ 6 & 0 \\ -6 & 5 \end{bmatrix}$$

It is traditional when talking about matrices to call individual numbers **scalars.** For this reason we call the operation of multiplying a matrix by a number **scalar multiplication.**

Example 3 • Sales

The revenue generated by sales in the Vancouver and Quebec branches of the APlus auto parts store (see Example 2) was as follows:

January Sales in Canadian Dollars

	Vancouver	Quebec
Wiper Blades	140.00	105.00
Cleaning Fluid	30.00	36.00
Floor Mats	96.00	48.00

If the Canadian dollar was worth \$0.65US at the time, compute the revenue in U.S. dollars.

Solution We need to multiply each revenue figure by 0.65. Let A be the matrix of revenue figures in Canadian dollars:

$$A = \begin{bmatrix} 140.00 & 105.00 \\ 30.00 & 36.00 \\ 96.00 & 48.00 \end{bmatrix}$$

The revenue figures in U.S. dollars are then given by the scalar multiple

$$0.65A = 0.65\begin{bmatrix} 140.00 & 105.00 \\ 30.00 & 36.00 \\ 96.00 & 48.00 \end{bmatrix} = \begin{bmatrix} 91.00 & 68.25 \\ 19.50 & 23.40 \\ 62.40 & 31.20 \end{bmatrix}$$

In other words, in U.S. dollars, \$91 worth of wiper blades was sold in Vancouver, \$68.25 worth of wiper blades was sold in Quebec, and so on.

Graphing Calculator: Scalar Multiplication

To multiply the matrix [A] by 0.65 on the TI-83, enter

`0.65*[A]` or just `0.65[A]`

on the home screen.

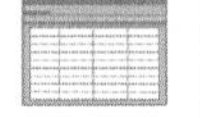

Excel: Scalar Multiplication

The following is the above matrix operation set up in Excel:

	A	B	C
1	January Sales (Can \$)		
2		Vancouver	Quebec
3	wiper blades	140	105
4	cleaning fluid (bottles)	30	36
5	floor mats	96	48
6			
7			
8	January Sales (US \$)		
9		Vancouver	Quebec
10	wiper blades	=0.65*B3:C5	
11	cleaning fluid (bottles)		
12	floor mats		

As with matrix addition, we use Control + Shift + Enter to evaluate scalar multiplication (or to evaluate any formula that uses an array).

Formally, scalar multiplication is defined as follows.

Scalar Multiplication

If A is an $m \times n$ matrix and c is a real number, then cA is the $m \times n$ matrix obtained by multiplying all the entries of A by c. (We usually use lowercase letters $c, d, e, \ldots$ to denote scalars.) Thus, the ijth entry of cA is given by

$$(cA)_{ij} = c(A_{ij})$$

Question Haven't we just moved the parentheses?

Answer There is a little more going on here than just that. In words, this rule is: To get the ijth entry of cA, multiply the ijth entry of A by c.

Example 4 • Combining Operations

Let $A = \begin{bmatrix} 2 & -1 & 0 \\ 3 & 5 & -3 \end{bmatrix}$, $B = \begin{bmatrix} 1 & 3 & -1 \\ 5 & -6 & 0 \end{bmatrix}$, and $C = \begin{bmatrix} x & y & w \\ z & t+1 & 3 \end{bmatrix}$.

Evaluate the following: $4A$, cB, and $A + 3C$.

Solution First, we find $4A$ by multiplying each entry of A by 4:

$$4A = \begin{bmatrix} 8 & -4 & 0 \\ 12 & 20 & -12 \end{bmatrix}$$

Similarly, we find cB by multiplying each entry of B by c:

$$cB = \begin{bmatrix} c & 3c & -c \\ 5c & -6c & 0 \end{bmatrix}$$

We get $A + 3C$ in two steps as follows:

$$\begin{aligned} A + 3C &= \begin{bmatrix} 2 & -1 & 0 \\ 3 & 5 & -3 \end{bmatrix} + 3\begin{bmatrix} x & y & w \\ z & t+1 & 3 \end{bmatrix} \\ &= \begin{bmatrix} 2 & -1 & 0 \\ 3 & 5 & -3 \end{bmatrix} + \begin{bmatrix} 3x & 3y & 3w \\ 3z & 3t+3 & 9 \end{bmatrix} \\ &= \begin{bmatrix} 2+3x & -1+3y & 3w \\ 3+3z & 3t+8 & 6 \end{bmatrix} \end{aligned}$$

Addition and scalar multiplication have nice properties, reminiscent of the properties of addition and multiplication of real numbers. Before we state them, we need to introduce some more notation.

If A is any matrix, then $-A$ is the matrix $(-1)A$. In other words, $-A$ is A multiplied by the scalar -1. This amounts to changing the signs of all the entries in A. For example,

$$-\begin{bmatrix} 4 & -2 & 0 \\ 6 & 10 & -6 \end{bmatrix} = \begin{bmatrix} -4 & 2 & 0 \\ -6 & -10 & 6 \end{bmatrix}$$

For any two matrices A and B, $A - B$ is the same as $A + (-B)$. (Why?)

Also, a **zero matrix** is a matrix all of whose entries are zero. For example, the 2×3 zero matrix is

$$O = \begin{bmatrix} 0 & 0 & 0 \\ 0 & 0 & 0 \end{bmatrix}$$

Now we state the most important properties of the operations that we have been talking about.

Properties of Matrix Addition and Scalar Multiplication

If A, B, and C are any $m \times n$ matrices and if O is the zero $m \times n$ matrix, then the following hold:

$A + (B + C) = (A + B) + C$	*Associative law*
$A + B = B + A$	*Commutative law*
$A + O = O + A = A$	*Additive identity law*
$A + (-A) = O = (-A) + A$	*Additive inverse law*
$c(A + B) = cA + cB$	*Distributive law*
$(c + d)A = cA + dA$	*Distributive law*
$1A = A$	*Scalar unit*
$0A = O$	*Scalar zero*

Question Aren't these properties obvious? Don't we know them already?

Answer They would be obvious if we were talking about addition and multiplication of *numbers,* but here we are talking about addition and multiplication of *matrices.* We are using + to mean something new: matrix addition. There is no reason why matrix addition has to obey *all* the properties of addition of numbers. It happens that it does obey many of them, which is why it is convenient to call it *addition* in the first place. This means that we can manipulate equations involving matrices in much the same way that we manipulate equations involving numbers. One word of caution: We haven't yet discussed how to multiply matrices, and it probably isn't what you think. It will turn out that multiplication of matrices does *not* obey all the same properties as multiplication of numbers.

Transposition

We mention one more operation on matrices.

Transposition

If A is an $m \times n$ matrix, then its **transpose** is the $n \times m$ matrix obtained by writing its rows as columns, so that the ith row of the original matrix becomes the ith column of the transpose. We denote the transpose of the matrix A by A^T.

Quick Examples

1. Let $B = \begin{bmatrix} 2 & 3 \\ 10 & 44 \\ -1 & 3 \\ 8 & 3 \end{bmatrix}$. Then $B^T = \begin{bmatrix} 2 & 10 & -1 & 8 \\ 3 & 44 & 3 & 3 \end{bmatrix}$.

4×2 matrix $\qquad$ 2×4 matrix

2. $\begin{bmatrix} -1 & 1 & 2 \end{bmatrix}^T = \begin{bmatrix} -1 \\ 1 \\ 2 \end{bmatrix}$

1 × 3 matrix   3 × 1 matrix

Properties of Transposition

If A and B are $m \times n$ matrices, then the following hold:

$$(A + B)^T = A^T + B^T$$

$$(cA)^T = c(A^T)$$

$$(A^T)^T = A$$

Question Why are the laws of transposition true?

Answer Let's consider the first one: $(A + B)^T = A^T + B^T$. The left-hand side is the transpose of $A + B$, and so is obtained by first adding A and B and then writing the rows as columns. This is the same as first writing the rows of A and B individually as columns before adding, which gives the right-hand side. Similar arguments can be used to establish the other laws of transposition.

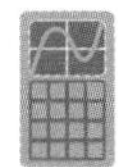

Graphing Calculator: Transposing a Matrix

On the TI-83, you can access the symbol ^{T} by pressing MATRX, selecting MATH, selecting ^{T}, and then pressing ENTER.

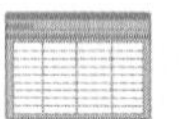

Excel: Transposing a Matrix

To transpose a matrix, first highlight the block where you would like the transpose to appear (shaded in the spreadsheet below). (Note that it should have the correct dimensions for the transpose: 3 × 2 in the case shown below.) Then type the formula `=TRANSPOSE(A1:C2)` (use the array range appropriate for the matrix you want to transpose) in the formula bar and press Control+Shift+Enter. The transpose will appear in the region you highlighted.

	A	B	C
1	2	0	1
2	33	-22	0
3			
4	=TRANSPOSE(A1:C2)		
5			
6			

3.1 EXERCISES

In Exercises 1–10, find the dimensions of the given matrix and identify the given entry.

1. $A = \begin{bmatrix} 1 & 5 & 0 & \frac{1}{4} \end{bmatrix}; a_{13}$

2. $B = \begin{bmatrix} 44 & 55 \end{bmatrix}; b_{12}$

3. $C = \begin{bmatrix} \frac{5}{2} \\ 1 \\ -2 \\ 8 \end{bmatrix}; C_{11}$

4. $D = \begin{bmatrix} 15 & -18 \\ 6 & 0 \\ -6 & 5 \\ 48 & 18 \end{bmatrix}; d_{31}$

5. $E = \begin{bmatrix} e_{11} & e_{12} & e_{13} & \cdots & e_{1q} \\ e_{21} & e_{22} & e_{23} & \cdots & e_{2q} \\ \vdots & \vdots & \vdots & \ddots & \vdots \\ e_{p1} & e_{p2} & e_{p3} & \cdots & e_{pq} \end{bmatrix}; E_{22}$

6. $A = \begin{bmatrix} 2 & -1 & 0 \\ 3 & 5 & -3 \end{bmatrix}; A_{21}$

7. $B = \begin{bmatrix} 1 & 3 \\ 5 & -6 \end{bmatrix}; b_{12}$

8. $C = \begin{bmatrix} x & y & w & e \\ z & t+1 & 3 & 0 \end{bmatrix}$; C_{23}

9. $D = [d_1 \quad d_2 \ldots d_n]$; D_{1r} (any r)

10. $E = [d \quad d \quad d \quad d]$; E_{1r} (any r)

11. Solve for x, y, z, and w:

$$\begin{bmatrix} x+y & x+z \\ y+z & w \end{bmatrix} = \begin{bmatrix} 3 & 4 \\ 5 & 4 \end{bmatrix}$$

12. Solve for x, y, z, and w:

$$\begin{bmatrix} x-y & x-z \\ y-w & w \end{bmatrix} = \begin{bmatrix} 0 & 0 \\ 0 & 6 \end{bmatrix}$$

In Exercises 13–20, let $A = \begin{bmatrix} 0 & -1 \\ 1 & 0 \\ -1 & 2 \end{bmatrix}$, $B = \begin{bmatrix} 0.25 & -1 \\ 0 & 0.5 \\ -1 & 3 \end{bmatrix}$,

and $C = \begin{bmatrix} 1 & -1 \\ 1 & 1 \\ -1 & -1 \end{bmatrix}$ and evaluate.

13. $A + B$ **14.** $A - C$ **15.** $A + B - C$ **16.** $12B$

17. $2A - C$ **18.** $2A + 0.5C$ **19.** $2A^T$ **20.** $A^T + 3C^T$

In Exercises 21–28, let $A = \begin{bmatrix} 1 & -1 & 0 \\ 0 & 2 & -1 \end{bmatrix}$, $B = \begin{bmatrix} 3 & 0 & -1 \\ 5 & -1 & 1 \end{bmatrix}$,

and $C = \begin{bmatrix} x & 1 & w \\ z & r & 4 \end{bmatrix}$ and evaluate.

21. $A + B$ **22.** $B - C$ **23.** $A - B + C$ **24.** $\frac{1}{2}B$

25. $2A - B$ **26.** $2A - 4C$ **27.** $3B^T$ **28.** $2A^T - C^T$

In Exercises 29–36, use technology. Let $A = \begin{bmatrix} 1.5 & -2.35 & 5.6 \\ 44.2 & 0 & 12.2 \end{bmatrix}$,

$B = \begin{bmatrix} 1.4 & 7.8 \\ 5.4 & 0 \\ 5.6 & 6.6 \end{bmatrix}$, and $C = \begin{bmatrix} 10 & 20 & 30 \\ -10 & -20 & -30 \end{bmatrix}$ and evaluate.

29. $A - C$ **30.** $C - A$ **31.** $1.1B$

32. $-0.2B$ **33.** $A^T + 4.2B$ **34.** $(A + 2.3C)^T$

35. $(2.1A - 2.3C)^T$ **36.** $(A - C)^T - B$

APPLICATIONS

37. Recreational Boat Sales The following table shows sales of recreational boats in the United States during the period 1999–2001. Use matrix algebra to find the sales in each category in (a) 2000 and (b) 2001.

	Motor Boats	Jet Skis	Sailboats
1999	330,000	100,000	20,000
Increase in 2000	10,000	0	0
Increase in 2001	−30,000	−10,000	10,000

Figures are approximate and represent new recreational boats sold. ("Jet Skis" includes similar vehicles, such as wave runners.) SOURCE: National Marine Manufacturers Association/*New York Times*, January 10, 2002, p. C1.

38. Recreational Boat Revenue The following table shows revenue, in billions of dollars, from sales of recreational boats in the United States during the period 1999–2001. Use matrix algebra to find the revenue in each category in (a) 2000 and (b) 2001.

	Used Boats	New Boats	Accessories
1999	$7 billion	$7 billion	$4 billion
Increase in 2000	0	0.5	1
Increase in 2001	1	0.5	0

Figures are approximate. SOURCE: National Marine Manufacturers Association/*New York Times*, January 10, 2002, p. C1.

39. Inventory The Left Coast Bookstore chain has two stores, one in San Francisco and one in Los Angeles. It stocks three kinds of book: hardcover, softcover, and plastic (for infants). At the beginning of January, the central computer showed the following books in stock:

	Hard	Soft	Plastic
San Francisco	1000	2000	5000
Los Angeles	1000	5000	2000

Suppose its sales in January were as follows: 700 hardcover books, 1300 softcover books, and 2000 plastic books sold in San Francisco, and 400 hardcover, 300 softcover, and 500 plastic books sold in Los Angeles. Write these sales figures in the form of a matrix and then show how matrix algebra can be used to compute the inventory remaining in each store at the end of January.

40. Inventory The Left Coast Bookstore chain discussed in Exercise 39 actually maintained the same sales figures for the first 6 months of the year. Each month the chain restocked the stores from its warehouse by shipping 600 hardcover, 1500 softcover, and 1500 plastic books to San Francisco and 500 hardcover, 500 softcover, and 500 plastic books to Los Angeles.

a. Use matrix operations to determine the total sales over the 6 months, broken down by store and type of book.

b. Use matrix operations to determine the inventory in each store at the end of June.

41. Inventory Microbucks Computers makes two computers, the Pomegranate II and the Pomegranate Classic, at two different factories. The Pom II requires 2 processor chips, 16 memory chips, and 20 vacuum tubes, and the Pom Classic requires 1 processor chip, 4 memory chips, and 40 vacuum tubes. At the beginning of the year, Microbucks has in stock 500 processor chips, 5000 memory chips, and 10,000 vacuum tubes at the Pom II factory and 200 processor chips, 2000 memory chips, and 20,000 vacuum tubes at the

Pom Classic factory. It manufactures 50 Pom IIs and 50 Pom Classics each month.

a. Find the company's inventory of parts after 2 months, using matrix operations.

b. When will the company run out of one of the parts?

42. Inventory Microbucks Computers, besides having the stock mentioned in Exercise 41, gets shipments of parts every month in the amounts of 100 processor chips, 1000 memory chips, and 3000 vacuum tubes at the Pom II factory and 50 processor chips, 1000 memory chips, and 2000 vacuum tubes at the Pom Classic factory.

a. What will be the company's inventory of parts after 6 months?

b. How long will it be before the company runs out of one of the parts?

Bankruptcy Filings Exercises 43–46 are based on the following chart, which shows the numbers of personal bankruptcy filings in three New York/New Jersey regions during various months of 2001–2002.

	Jan 2001	Apr 2001	Jul 2001	Oct 2001	Jan 2002
Manhattan	150	250	150	100	150
Brooklyn	300	400	300	200	250
Newark	250	400	250	200	200

Data are approximate 4-week moving averages. SOURCE: Lundquist Consulting/*New York Times*, February 10, 2002, p. L1.

43. Use matrix algebra to determine the total number of bankruptcy filings in each of the given months.

44. Use matrix algebra to determine the total number of bankruptcy filings in each of the given regions.

45. Use matrix algebra to determine in which month the difference between the number of bankruptcy filings in Brooklyn and in Newark was greatest.

46. Use matrix algebra to determine in which region the difference between the bankruptcy filings in April 2001 and January 2001 was greatest.

47. Profit Annual revenues and production costs at Luddington's Wellington Boots are shown in the following spreadsheet. Use matrix algebra to compute the profits from each sector each year.

	A	B	C	D
1	**Revenue**			
2		**2001**	**2002**	**2003**
3	**Full Boots**	$10,000	$9,000	$11,000
4	**Half Boots**	$8,000	$7,200	$8,800
5	**Sandals**	$4,000	$5,000	$6,000
6				
7	**Production Costs**			
8		**2001**	**2002**	**2003**
9	**Full Boots**	$2,000	$1,800	$2,200
10	**Half Boots**	$2,400	$1,440	$1,760
11	**Sandals**	$1,200	$1,500	$2,000

48. Revenue The following spreadsheet gives annual production costs and profits at Gauss Jordan Sneakers. Use matrix algebra to compute the revenues from each sector each year.

	A	B	C	D
1	**Production Costs**			
2		**2001**	**2002**	**2003**
3	**Gauss Grip**	$1,800	$2,200	$2,400
4	**Air Gauss**	$1,400	$1,700	$1,200
5	**Gauss Gel**	$1,500	$2,000	$1,300
6				
7	**Profit**			
8		**2001**	**2002**	**2003**
9	**Gauss Grip**	$10,000	$14,000	$16,000
10	**Air Gauss**	$8,000	$12,000	$14,000
11	**Gauss Gel**	$9,000	$14,000	$12,000

49. Tourism The following table gives the number of people (in thousands) who visited Australia and South Africa in 1998:

	To **Australia**	**South Africa**
From **North America**	440	190
Europe	950	950
Asia	1790	200

Figures are rounded to the nearest 10,000. SOURCES: South African Dept. of Environmental Affairs and Tourism; Australia Tourist Commission/*New York Times*, January 15, 2000, p. C1.

You predict that, in 2008, 20,000 fewer people from North America will visit Australia and 40,000 more will visit South Africa, 50,000 more people from Europe will visit each of Australia and South Africa, and 100,000 more people from Asia will visit South Africa, but there will be no change in the number visiting Australia.

a. Use matrix algebra to predict the number of visitors from the three regions to Australia and South Africa in 2008.

b. Let A be the 3×2 matrix whose entries are the 1998 tourism figures and let B be the 3×2 matrix whose entries are the 2008 tourism figures. Give the entries of the matrix $B - A$.

50. Tourism Referring to the 1998 tourism figures given in Exercise 49, assume that the following (fictitious) figures represent the corresponding numbers from 1988:

	To **Australia**	**South Africa**
From **North America**	500	100
Europe	900	800
Asia	1400	50

Let A be the 3×2 matrix whose entries are the 1998 tourism figures and let B be the 3×2 matrix whose entries are the 1988 tourism figures.

a. Compute the matrix $A - B$. What does this matrix represent?

b. Assuming that the changes in tourism over 1988–1998 are repeated in 1998–2008, give a formula (in terms of A and B) that predicts the number of visitors from the three regions to Australia and South Africa in 2008.

51. **Population Distribution** In 1980 the U.S. population, broken down by regions, was 49.1 million in the Northeast, 58.9 million in the Midwest, 75.4 million in the South, and 43.2 million in the West. In 1990 the population was 50.8 million in the Northeast, 59.7 million in the Midwest, 85.4 million in the South, and 52.8 million in the West. Set up the population figures for each year as a row vector and then show how to use matrix operations to find the net increase or decrease of population in each region from 1980 to 1990.

SOURCE: U.S. Census Bureau, *Statistical Abstract of the United States: 2001,* http://www.census.gov/.

52. **Population Distribution** In 1990 the U.S. population, broken down by regions, was 50.8 million in the Northeast, 59.7 million in the Midwest, 85.4 million in the South, and 52.8 million in the West. Between 1990 and 2000, the population in the Northeast grew by 2.8 million, the population in the Midwest grew by 4.7 million, the population in the South grew by 14.8 million, and the population in the West grew by 10.4 million. Set up the population figures for 1990 and the growth figures for the decade as row vectors. Assuming that the population will grow by the same numbers from 2000 to 2010 as they did from 1990 to 2000, show how to use matrix operations to find the population in each region in 2010.

SOURCE: U.S. Census Bureau, *Statistical Abstract of the United States: 2001,* http://www.census.gov/.

COMMUNICATION AND REASONING EXERCISES

53. What does it mean when we say that $(A + B)_{ij} = A_{ij} + B_{ij}$?
54. What does it mean when we say that $(cA)_{ij} = c(A_{ij})$?
55. What would a 5×5 matrix A look like if $A_{ii} = 0$ for every i?
56. What would a matrix A look like if $A_{ij} = 0$ whenever $i \neq j$?
57. Give a formula for the ijth entry of the transpose of a matrix A.
58. A matrix is **symmetric** if it is equal to its transpose. Give an example of (a) a nonzero 2×2 symmetric matrix and (b) a nonzero 3×3 symmetric matrix.
59. A matrix is **skew-symmetric,** or **antisymmetric,** if it is equal to the negative of its transpose. Give an example of (a) a nonzero 2×2 skew-symmetric matrix and (b) a nonzero 3×3 skew-symmetric matrix.
60. Referring to Exercises 58 and 59, what can be said about a matrix that is both symmetric and skew-symmetric?
61. Why is matrix addition associative?
62. Describe a scenario (possibly based on one of the preceding examples or exercises) in which you might wish to compute $A - 2B$ for certain matrices A and B.
63. Describe a scenario (possibly based on one of the preceding examples or exercises) in which you might wish to compute $A + B - C$ for certain matrices A, B, and C.

3.2 *Matrix Multiplication*

Suppose we buy 2 CDs at \$3 each and 4 Zip disks at \$5 each. We calculate our total cost by computing the products' price $\times$ quantity and adding:

$$\text{cost} = 3 \times 2 + 5 \times 4 = \$26$$

Let's instead put the prices in a row vector

$$P = [3 \quad 5] \qquad \text{The price matrix}$$

and the quantities purchased in a column vector

$$Q = \begin{bmatrix} 2 \\ 4 \end{bmatrix} \qquad \text{The quantity matrix}$$

Question Why a row and a column?

Answer It's rather a long story, but mathematicians found that it works best this way.

Since P represents the prices of the items we are purchasing and Q represents the quantities, it would be useful if the product PQ represented the total cost, a *single number* (which we can think of as a 1×1 matrix). For this to work, PQ should be calculated the same way we calculated the total cost:

$$PQ = [3 \quad 5]\begin{bmatrix} 2 \\ 4 \end{bmatrix} = [3 \times 2 + 5 \times 4] = [26]$$

Notice that we obtain the answer by multiplying each entry in P (going from left to right) by the corresponding entry in Q (going from top to bottom) and then adding the results.

The Product Row × Column

The **product** AB of a row matrix A and a column matrix B is a 1×1 matrix. The length of the row in A must match the length of the column in B for the product to be defined. To find the product, multiply each entry in A (going from left to right) by the corresponding entry in B (going from top to bottom) and then add the results.

Quick Examples

1. $[2 \quad 1]\begin{bmatrix}-3\\1\end{bmatrix} = [2 \times (-3) + 1 \times 1] = [-6 + 1] = [-5]$

2. $[2 \quad 4 \quad 1]\begin{bmatrix}2\\10\\-1\end{bmatrix} = [2 \times 2 + 4 \times 10 + 1 \times (-1)] = [4 + 40 + (-1)] = [43]$

Note It sometimes helps to visualize the process by means of a diagram:

$$[2 \quad 4 \quad 1]\begin{bmatrix}2\\10\\-1\end{bmatrix} \qquad \begin{aligned} 2 \times 2 &= 4 \\ 4 \times 10 &= 40 \\ 1 \times (-1) &= \underline{-1} \\ &\ 43 \end{aligned}$$

Product of first entries = 4
Product of second entries = 40
Product of third entries = −1
Sum of products = 43

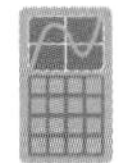

Graphing Calculator: Multiplying Matrices

On the TI-83, the format for multiplying matrices is the same as for multiplying numbers: [A][B] or [A] × [B] will give the product.

Excel: Multiplying Matrices

The formula for matrix multiplication is MMULT. To find the product AB of a row and a column in Excel, first enter the matrices A and B anywhere in the spreadsheet, select the cell where you would like the result to appear (G1 in the example below), and use MMULT as shown. (Since the product occupies only a single cell, it is not necessary to press Control + Shift + Enter.)

	A	B	C	D	E	F	G	H
1	2	4	1		2		=MMULT(A1:C1,E1:E3)	
2					10			
3					-1			

As usual, you can use the mouse to avoid typing the array ranges (see Section 3.1).

Notes

1. In the discussion so far, *the row is on the left, and the column is on the right* (RC again). For example, we *don't yet* attempt to make any sense of a product such as[2]

$$\begin{bmatrix}-3\\1\end{bmatrix}[2 \quad 1]$$

[2]We will be able to make sense of this shortly. The answer will turn out to be a 2×2 matrix.

2. The row size has to match the column size. This means that, if we have a 1×3 row on the left, then the column on the right must be 3×1 in order for the product to make sense. For example, the product

$$[a \quad b \quad c]\begin{bmatrix} x \\ y \end{bmatrix}$$

is not defined.

Example 1 • Revenue

The APlus auto parts store mentioned in Examples 2 and 3 in Section 3.1 had the following sales in its Vancouver store:

	Vancouver
Wiper Blades	20
Cleaning Fluid (bottles)	10
Floor Mats	8

The store sells wiper blades for \$7 each, cleaning fluid for \$3 per bottle, and floor mats for \$12 each. Use matrix multiplication to find the total revenue generated by sales of these items.

Solution We need to multiply each sales figure by the corresponding price and then add the resulting revenue figures. We represent the sales by a column vector, as suggested by the table:

$$Q = \begin{bmatrix} 20 \\ 10 \\ 8 \end{bmatrix}$$

We put the selling prices in a row vector:

$$P = [7 \quad 3 \quad 12]$$

We can now compute the total revenue as the product:

$$R = PQ = [7 \quad 3 \quad 12]\begin{bmatrix} 20 \\ 10 \\ 8 \end{bmatrix}$$

$$= [140 + 30 + 96] = [266]$$

So, the sale of these items generated a total revenue of \$266.

Before we go on . . . We could also have written the quantity sold as a row vector (which would be Q^T) and the prices as a column vector (which would be P^T) and then multiplied them in the opposite order (Q^TP^T). Try this.

Example 2 • Relationship with Linear Equations

a. Represent the matrix equation

$$\begin{bmatrix} 2 & -4 & 1 \end{bmatrix} \begin{bmatrix} x \\ y \\ z \end{bmatrix} = [5]$$

as an ordinary equation.

b. Represent the linear equation $3x + y - z + 2w = 8$ as a matrix equation.

Solution

a. If we perform the multiplication on the left, we get the 1×1 matrix $[2x - 4y + z]$. Thus, the equation may be rewritten as

$$[2x - 4y + z] = [5]$$

Saying that these two 1×1 matrices are equal means that their entries agree, so we get the equation

$$2x - 4y + z = 5$$

b. This is the reverse of part (a):

$$\begin{bmatrix} 3 & 1 & -1 & 2 \end{bmatrix} \begin{bmatrix} x \\ y \\ z \\ w \end{bmatrix} = [8]$$

Before we go on . . . The row matrix $[3 \quad 1 \quad -1 \quad 2]$ is the row of **coefficients** of the original equation (see Section 2.1).

Now to the general case of matrix multiplication.

The Product of Two Matrices: General Case

In general, we can take the product AB only if the number of columns of A equals the number of rows of B (so that we can multiply the rows of A by the columns of B as above). The product AB is then obtained by taking its ijth entry to be:

ijth entry of AB = row i of $A \times$ column j of B As defined above

Quick Examples

R stands for row; C stands for column.

1. $$R_1 \rightarrow \begin{bmatrix} 2 & 0 & -1 & 3 \end{bmatrix} \begin{array}{c} \begin{array}{ccc} C_1 & C_2 & C_3 \\ \downarrow & \downarrow & \downarrow \end{array} \\ \begin{bmatrix} 1 & 1 & -8 \\ 1 & -6 & 0 \\ 0 & 5 & 2 \\ -3 & 8 & 1 \end{bmatrix} \end{array} = [R_1 \times C_1 \quad R_1 \times C_2 \quad R_1 \times C_3]$$

$$= [-7 \quad 21 \quad -15]$$

2. $$\begin{array}{c} R_1 \rightarrow \\ R_2 \rightarrow \end{array} \begin{bmatrix} 1 & -1 \\ 0 & 2 \end{bmatrix} \begin{array}{c} \begin{array}{cc} C_1 & C_2 \\ \downarrow & \downarrow \end{array} \\ \begin{bmatrix} 3 & 0 \\ 5 & -1 \end{bmatrix} \end{array} = \begin{bmatrix} R_1 \times C_1 & R_1 \times C_2 \\ R_2 \times C_1 & R_2 \times C_2 \end{bmatrix} = \begin{bmatrix} -2 & 1 \\ 10 & -2 \end{bmatrix}$$

In matrix multiplication we always take

rows on the left × columns on the right

Look at the dimensions in the two products in the Quick Examples:

$$\overset{\text{Match}}{(1 \times \underset{\downarrow}{4})(\underset{\downarrow}{4} \times 3)} \rightarrow 1 \times 3 \qquad \overset{\text{Match}}{(2 \times \underset{\downarrow}{2})(\underset{\downarrow}{2} \times 2)} \rightarrow 2 \times 2$$

The fact that the number of columns in the left-hand matrix equals the number of rows in the right-hand matrix amounts to saying that the middle two numbers must match as above. If we "cancel" the middle matching numbers, we are left with the dimensions of the product.

Before continuing with examples, we state the rule for matrix multiplication formally.

Multiplication of Matrices: Formal Definition

If A is an $m \times n$ matrix and B is an $n \times k$ matrix, then the product AB is the $m \times k$ matrix whose ijth entry is the product

row i of A × column j of B

$$(AB)_{ij} = [a_{i1} \quad a_{i2} \quad a_{i3} \ldots a_{in}] \begin{bmatrix} b_{1j} \\ b_{2j} \\ b_{3j} \\ \vdots \\ b_{nj} \end{bmatrix} = a_{i1}b_{1j} + a_{i2}b_{2j} + a_{i3}b_{3j} + \cdots + a_{in}b_{nj}$$

Example 3 • Matrix Product

Calculate:

a. $\begin{bmatrix} 2 & 0 & -1 & 3 \\ 1 & -1 & 2 & -2 \end{bmatrix} \begin{bmatrix} 1 & 1 & -8 \\ 1 & 0 & 0 \\ 0 & 5 & 2 \\ -2 & 8 & -1 \end{bmatrix}$ **b.** $\begin{bmatrix} -3 \\ 1 \end{bmatrix} [2 \quad 1]$

Solution

a. Before starting the calculation, we check that the dimensions of the matrices match:

Match

$$\underset{2 \times 4}{\begin{bmatrix} 2 & 0 & -1 & 3 \\ 1 & -1 & 2 & -2 \end{bmatrix}} \underset{4 \times 3}{\begin{bmatrix} 1 & 1 & -8 \\ 1 & 0 & 0 \\ 0 & 5 & 2 \\ -2 & 8 & -1 \end{bmatrix}}$$

The product of the two matrices is defined, and the product will be a 2×3 matrix [we remove the matching 4s: $(2 \times 4)(4 \times 3) \to 2 \times 3$]. To calculate the product, we follow the prescription above:

$$\begin{matrix} R_1 \to \\ R_2 \to \end{matrix} \begin{bmatrix} 2 & 0 & -1 & 3 \\ 1 & -1 & 2 & -2 \end{bmatrix} \begin{bmatrix} 1 & 1 & -8 \\ 1 & 0 & 0 \\ 0 & 5 & 2 \\ -2 & 8 & -1 \end{bmatrix} = \begin{bmatrix} R_1 \times C_1 & R_1 \times C_2 & R_1 \times C_3 \\ R_2 \times C_1 & R_2 \times C_2 & R_2 \times C_3 \end{bmatrix}$$

(with C_1, C_2, C_3 marked by arrows above the columns of the second matrix)

$$= \begin{bmatrix} -4 & 21 & -21 \\ 4 & -5 & -2 \end{bmatrix}$$

b. The dimensions of the two matrices given are 2×1 and 1×2. Since the 1s match, the product is defined, and the result will be a 2×2 matrix:

$$\begin{matrix} R_1 \to \\ R_2 \to \end{matrix} \begin{bmatrix} -3 \\ 1 \end{bmatrix} \begin{bmatrix} 2 & 1 \end{bmatrix} = \begin{bmatrix} R_1 \times C_1 & R_1 \times C_2 \\ R_2 \times C_1 & R_2 \times C_2 \end{bmatrix} = \begin{bmatrix} -6 & -3 \\ 2 & 1 \end{bmatrix}$$

(with C_1, C_2 marked by arrows above the columns of $[2 \;\; 1]$)

Before we go on . . . In part (a) we *cannot* multiply the matrices in the opposite order—the dimensions do not match. We say simply that the product in the opposite order is *not defined.* In part (b) we *can* multiply the matrices in the opposite order, but we would get a 1×1 matrix if we did so. Thus, order is important when multiplying matrices. In general, if AB is defined, then BA need not even be defined. If BA is also defined, it may not have the same dimensions as AB. And even if AB and BA have the same dimensions, they may have different entries (see Example 4).

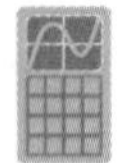

Graphing Calculator: Matrix Product

On the TI-83, if you try to multiply two matrices whose product is not defined, you will get the error `DIM MISMATCH` (dimensions mismatch).

Excel: Matrix Product

To repeat the calculation of part (a), first enter the two matrices as shown in the spreadsheet and highlight a block where you want the answer to appear. (Note that it should have the correct dimensions for the product: 2×3.)

	A	B	C	D	E	F	G	H
1	2	0	-1	3		1	1	-8
2	1	-1	2	-2		1	0	0
3						0	5	2
4						-2	8	-1
5								

Then enter the formula `=MMULT(A1:D2,F1:H4)` (using the mouse to avoid typing the array ranges if you like) and press Control+Shift+Enter. The product will appear in the region you highlighted.

If you try to multiply two matrices whose product is not defined, you will get the error `!VALUE#`.

Web Site

Use the path

Web site → Online Utilities → Matrix Algebra Tool

to get to an online computational tool that will allow you to multiply matrices.

The path

Web site → Excel Tutorials → 3.2 Matrix Multiplication

will give you a downloadable Excel tutorial covering this and other examples in this section.

Example 4 • AB versus BA

Let $A = \begin{bmatrix} 1 & -1 \\ 0 & 2 \end{bmatrix}$ and $B = \begin{bmatrix} 3 & 0 \\ 5 & -1 \end{bmatrix}$. Find AB and BA.

Solution Note first that A and B are both 2×2 matrices, so the products AB and BA are both defined and are both 2×2 matrices—unlike the case in Example 3(b). We first calculate AB:

$$AB = \begin{bmatrix} 1 & -1 \\ 0 & 2 \end{bmatrix} \begin{bmatrix} 3 & 0 \\ 5 & -1 \end{bmatrix} = \begin{bmatrix} -2 & 1 \\ 10 & -2 \end{bmatrix}$$

We might now wonder what the point is in calculating BA, since surely $AB = BA$? Well, let's calculate it anyway, if only for the sake of practice:

$$BA = \begin{bmatrix} 3 & 0 \\ 5 & -1 \end{bmatrix} \begin{bmatrix} 1 & -1 \\ 0 & 2 \end{bmatrix} = \begin{bmatrix} 3 & -3 \\ 5 & -7 \end{bmatrix}$$

Notice that BA has no resemblance to AB! Thus, we have discovered that, even for square matrices,

Matrix multiplication is not commutative.

In other words, $AB \neq BA$ in general, even when AB and BA both exist and have the same dimensions. (There are instances when $AB = BA$ for particular matrices A and B, but this is an exception, not the rule.)

Example 5 • Revenue

January sales at the APlus auto parts stores in Vancouver and Quebec are given in the following table:

	Vancouver	**Quebec**
Wiper Blades	20	15
Cleaning Fluid (bottles)	10	12
Floor Mats	8	4

The usual selling prices for these items are \$7 each for wiper blades, \$3 per bottle for cleaning fluid, and \$12 each for floor mats. The discount prices for APlusClub members are \$6 each for wiper blades, \$2 per bottle for cleaning fluid, and \$10 each for floor mats. Use matrix multiplication to compute the total revenue at each store, assuming first that all items were sold at the usual prices and then that they were all sold at the discount prices.

Solution We can do all of the requested calculations at once with a single matrix multiplication. Consider the following two labeled matrices:

$$Q = \begin{array}{r} \\ \textbf{wb} \\ \textbf{cf} \\ \textbf{fm} \end{array} \begin{array}{c} \begin{array}{cc} \textbf{V} & \textbf{Q} \end{array} \\ \begin{bmatrix} 20 & 15 \\ 10 & 12 \\ 8 & 4 \end{bmatrix} \end{array}$$

$$P = \begin{array}{r} \\ \textbf{Usual} \\ \textbf{Discount} \end{array} \begin{array}{c} \begin{array}{ccc} \textbf{wb} & \textbf{cf} & \textbf{fm} \end{array} \\ \begin{bmatrix} 7 & 3 & 12 \\ 6 & 2 & 10 \end{bmatrix} \end{array}$$

The first matrix records the quantities sold, and the second records the sales prices under the two assumptions. To compute the revenue at both stores under the two different assumptions, we calculate $R = PQ$:

$$R = PQ = \begin{bmatrix} 7 & 3 & 12 \\ 6 & 2 & 10 \end{bmatrix} \begin{bmatrix} 20 & 15 \\ 10 & 12 \\ 8 & 4 \end{bmatrix}$$

$$= \begin{bmatrix} 266 & 189 \\ 220 & 154 \end{bmatrix}$$

We can label this matrix as follows:

$$R = \begin{array}{r} \\ \textbf{Usual} \\ \textbf{Discount} \end{array} \begin{array}{c} \begin{array}{cc} \textbf{V} & \textbf{Q} \end{array} \\ \begin{bmatrix} 266 & 189 \\ 220 & 154 \end{bmatrix} \end{array}$$

In other words, if the items were sold at the usual price, then Vancouver had a revenue of \$266, Quebec had a revenue of \$189, and so on.

Before we go on . . . We were able to calculate PQ since the dimensions matched correctly: $(2 \times 3)(3 \times 2) \rightarrow 2 \times 2$. We could also have multiplied them in the opposite order and gotten a 3×3 matrix. Would the product QP be meaningful? In an application like this, not only do the dimensions have to match, but also the *labels* have to match for the result to be meaningful. The labels on the three columns of P are the parts that were sold, and these are also the labels on the three rows of Q. Therefore, we can "cancel labels" at the same time that we cancel the dimensions in the product. However, the labels on the two columns of Q do not match the labels on the two rows of P, and there is no useful interpretation of the product QP in this example.

There are very special square matrices of every size: $1 \times 1, 2 \times 2, 3 \times 3$, and so on, called the *identity matrices.*

Identity Matrix

The $n \times n$ **identity matrix** I is the matrix with 1s down the **main diagonal** (the diagonal starting at the top left) and 0s everywhere else. In symbols,

$$I_{ii} = 1 \quad \text{and} \quad I_{ij} = 0 \text{ if } i \neq j$$

Quick Examples

1. 1×1 identity matrix: $I = [1]$

2. 2×2 identity matrix: $I = \begin{bmatrix} 1 & 0 \\ 0 & 1 \end{bmatrix}$

3. 3×3 identity matrix: $I = \begin{bmatrix} 1 & 0 & 0 \\ 0 & 1 & 0 \\ 0 & 0 & 1 \end{bmatrix}$

4. 4×4 identity matrix: $I = \begin{bmatrix} 1 & 0 & 0 & 0 \\ 0 & 1 & 0 & 0 \\ 0 & 0 & 1 & 0 \\ 0 & 0 & 0 & 1 \end{bmatrix}$

Example 6 shows why I is interesting.

Example 6 • Identity Matrix

Evaluate the products AI and IA, where $A = \begin{bmatrix} a & b & c \\ d & e & f \\ g & h & i \end{bmatrix}$ and I is the 3×3 identity matrix.

Solution First notice that A is arbitrary—it could be any 3×3 matrix:

$$AI = \begin{bmatrix} a & b & c \\ d & e & f \\ g & h & i \end{bmatrix} \begin{bmatrix} 1 & 0 & 0 \\ 0 & 1 & 0 \\ 0 & 0 & 1 \end{bmatrix} = \begin{bmatrix} a & b & c \\ d & e & f \\ g & h & i \end{bmatrix}$$

and

$$IA = \begin{bmatrix} 1 & 0 & 0 \\ 0 & 1 & 0 \\ 0 & 0 & 1 \end{bmatrix} \begin{bmatrix} a & b & c \\ d & e & f \\ g & h & i \end{bmatrix} = \begin{bmatrix} a & b & c \\ d & e & f \\ g & h & i \end{bmatrix}$$

In both cases the answer is the matrix A we started with. In symbols,

$$AI = A \quad \text{and} \quad IA = A$$

no matter which 3×3 matrix A you start with. Now this should remind you of a familiar fact from arithmetic:

$$a \cdot 1 = a \quad \text{and} \quad 1 \cdot a = a$$

That is why we call the matrix I the 3×3 *identity* matrix, since it appears to play the same role for 3×3 matrices that the identity 1 does for numbers.

Before we go on . . . Try a similar calculation using 2×2 matrices: Let $A = \begin{bmatrix} a & b \\ c & d \end{bmatrix}$, let I be the 2×2 identity matrix, and check that $AI = IA = A$. In fact, the equation

$$AI = IA = A$$

works for square matrices of every dimension. It is also interesting to notice that $AI = A$ if I is the 2×2 identity matrix and A is any 3×2 matrix (try one). In fact, if I is any identity matrix, then $AI = A$ whenever the product is defined, and $IA = A$ whenever this product is defined.

Note Identity matrices are always square matrices, meaning that they have the same number of rows as columns. There is no such thing, for example, as the "2×4 identity matrix."

We can now add to the list of properties we gave for matrix arithmetic at the end of Section 3.1 by writing down properties of matrix multiplication. In stating these properties, we assume that all matrix products we write are defined—that is, that the matrices have correctly matching dimensions. The first eight properties are the ones we've already seen; the rest are new.

Properties of Matrix Addition and Multiplication

If A, B, and C are matrices, if O is a zero matrix, and if I is an identity matrix, then the following hold:

$A + (B + C) = (A + B) + C$	*Additive associative law*
$A + B = B + A$	*Additive commutative law*
$A + O = O + A = A$	*Additive identity law*
$A + (-A) = O = (-A) + A$	*Additive inverse law*
$c(A + B) = cA + cB$	*Distributive law*
$(c + d)A = cA + dA$	*Distributive law*
$1A = A$	*Scalar unit*
$0A = O$	*Scalar zero*
$A(BC) = (AB)C$	*Multiplicative associative law*
$c(AB) = (cA)B$	*Multiplicative associative law*
$c(dA) = (cd)A$	*Multiplicative associative law*
$AI = IA = A$	*Multiplicative identity law*
$A(B + C) = AB + AC$	*Distributive law*
$(A + B)C = AC + BC$	*Distributive law*
$OA = AO = O$	*Multiplication by zero matrix*

Question What about the multiplicative commutative law, $AB = BA$?

Answer It is not included because, as we have already seen, it *does not hold* in general. In other words, matrix multiplication is *not* exactly like multiplication of numbers. You have to be a little careful because it is easy to apply the commutative law without realizing it.

We should also say a bit more about transposition. Transposition and multiplication have an interesting relationship. We write down the properties of transposition again, adding one new one.

Properties of Transposition

$$(A + B)^T = A^T + B^T$$

$$(cA)^T = c(A^T)$$

$$(AB)^T = B^T A^T$$

Notice the change in order in the last one. The order is crucial.

Quick Examples

1. $\left(\begin{bmatrix} 1 & -1 \\ 0 & 2 \end{bmatrix}\begin{bmatrix} 3 & 0 \\ 5 & -1 \end{bmatrix}\right)^T = \begin{bmatrix} -2 & 1 \\ 10 & -2 \end{bmatrix}^T = \begin{bmatrix} -2 & 10 \\ 1 & -2 \end{bmatrix}$ $\quad (AB)^T$

2. $\begin{bmatrix} 3 & 0 \\ 5 & -1 \end{bmatrix}^T \begin{bmatrix} 1 & -1 \\ 0 & 2 \end{bmatrix}^T = \begin{bmatrix} 3 & 5 \\ 0 & -1 \end{bmatrix}\begin{bmatrix} 1 & 0 \\ -1 & 2 \end{bmatrix} = \begin{bmatrix} -2 & 10 \\ 1 & -2 \end{bmatrix}$ $\quad B^TA^T$

3. $\begin{bmatrix} 1 & -1 \\ 0 & 2 \end{bmatrix}^T \begin{bmatrix} 3 & 0 \\ 5 & -1 \end{bmatrix}^T = \begin{bmatrix} 1 & 0 \\ -1 & 2 \end{bmatrix}\begin{bmatrix} 3 & 5 \\ 0 & -1 \end{bmatrix} = \begin{bmatrix} 3 & 5 \\ -3 & -7 \end{bmatrix}$ $\quad A^TB^T$

In writing down these properties, what we are doing, besides showing you what you can and cannot do with matrices, is giving you a glimpse of the field of mathematics known as **abstract algebra.** Algebraists study operations like these that resemble the operations on numbers but differ in some way, such as the lack of commutativity seen here.

We end this section with more on the relationship between linear equations and matrix equations, which is one of the important applications of matrix multiplication.

Example 7 • Matrix Form of a System of Linear Equations

a. If

$$A = \begin{bmatrix} 1 & -2 & 3 \\ 2 & 0 & -1 \\ -3 & 1 & 1 \end{bmatrix} \quad X = \begin{bmatrix} x \\ y \\ z \end{bmatrix} \quad B = \begin{bmatrix} 3 \\ -1 \\ 0 \end{bmatrix}$$

rewrite the matrix equation $AX = B$ as a system of linear equations.

b. Express the following system of equations as a matrix equation of the form $AX = B$:

$$2x + y = 3$$
$$4x - y = -1$$

Solution

a. The matrix equation $AX = B$ is

$$\begin{bmatrix} 1 & -2 & 3 \\ 2 & 0 & -1 \\ -3 & 1 & 1 \end{bmatrix}\begin{bmatrix} x \\ y \\ z \end{bmatrix} = \begin{bmatrix} 3 \\ -1 \\ 0 \end{bmatrix}$$

As in Example 2(a), we first evaluate the left-hand side and then set it equal to the right-hand side:

$$\begin{bmatrix} 1 & -2 & 3 \\ 2 & 0 & -1 \\ -3 & 1 & 1 \end{bmatrix}\begin{bmatrix} x \\ y \\ z \end{bmatrix} = \begin{bmatrix} x - 2y + 3z \\ 2x - z \\ -3x + y + z \end{bmatrix}$$

$$\begin{bmatrix} x - 2y + 3z \\ 2x - z \\ -3x + y + z \end{bmatrix} = \begin{bmatrix} 3 \\ -1 \\ 0 \end{bmatrix}$$

Since these two matrices are equal, their corresponding entries must be equal:

$$\begin{aligned} x - 2y + 3z &= 3 \\ 2x \qquad - z &= -1 \\ -3x + y + z &= 0 \end{aligned}$$

In other words, the matrix equation $AX = B$ is equivalent to this system of linear equations. Notice that the coefficients of the left-hand sides of these equations are the entries of the matrix A. We call A the **coefficient matrix** of the system of equations. The entries of X are the unknowns, and the entries of B are the right-hand sides 3, -1, and 0.

b. As we saw in part (a), the coefficient matrix A has entries equal to the coefficients of the left-hand sides of the equations. Thus,

$$A = \begin{bmatrix} 2 & 1 \\ 4 & -1 \end{bmatrix}$$

X is the column matrix consisting of the unknowns, while B is the column matrix consisting of the right-hand sides of the equations, so

$$X = \begin{bmatrix} x \\ y \end{bmatrix} \quad \text{and} \quad B = \begin{bmatrix} 3 \\ -1 \end{bmatrix}$$

The system of equations can be rewritten as the matrix equation $AX = B$ with this A, X, and B.

This translation of systems of linear equations into matrix equations is really the first step in the method of solving linear equations discussed in Chapter 2. There we worked with the **augmented matrix** of the system, which is simply A with B adjoined as an extra column.

Question When we write a system of equations as $AX = B$, couldn't we solve for the unknown X by *dividing both sides by A*?

Answer If we interpret division as multiplication by the inverse (for example, $2 \div 3 = 2 \times 3^{-1}$), we will see in Section 3.3 that *certain* systems of the form $AX = B$ can be solved in this way, by multiplying both sides by A^{-1}. We first need to discuss what we mean by A^{-1} and how to calculate it.

3.2 EXERCISES

In Exercises 1–28, compute the products. (Some of these may be undefined.) Exercises marked T should be done using technology. The others should be done two ways: by hand and by using technology where possible.

1. $\begin{bmatrix} 1 & 3 & -1 \end{bmatrix}\begin{bmatrix} 9 \\ 1 \\ -1 \end{bmatrix}$

2. $\begin{bmatrix} 4 & 0 & -1 \end{bmatrix}\begin{bmatrix} -4 \\ 1 \\ 8 \end{bmatrix}$

3. $\begin{bmatrix} -1 & \frac{1}{2} \end{bmatrix}\begin{bmatrix} -\frac{1}{3} \\ 1 \end{bmatrix}$

4. $\begin{bmatrix} -1 & 1 \end{bmatrix}\begin{bmatrix} \frac{3}{4} \\ \frac{1}{4} \end{bmatrix}$

5. $\begin{bmatrix} 0 & -2 & 1 \end{bmatrix}\begin{bmatrix} x \\ y \\ z \end{bmatrix}$

6. $\begin{bmatrix} 4 & -1 & 1 \end{bmatrix}\begin{bmatrix} -x \\ x \\ y \end{bmatrix}$

7. $\begin{bmatrix} 1 & 3 & 2 \end{bmatrix}\begin{bmatrix} 1 \\ -1 \end{bmatrix}$

8. $\begin{bmatrix} 3 & 2 \end{bmatrix}\begin{bmatrix} 1 & -2 \end{bmatrix}$

9. $\begin{bmatrix} -1 & 1 \end{bmatrix}\begin{bmatrix} -3 & 1 & 4 & 3 \\ 0 & 1 & -2 & 1 \end{bmatrix}$

10. $\begin{bmatrix} 2 & -1 \end{bmatrix}\begin{bmatrix} -3 & 1 & 4 & 3 \\ 4 & 0 & 1 & 3 \end{bmatrix}$

11. $\begin{bmatrix} 1 & -1 & 2 & 3 \end{bmatrix}\begin{bmatrix} -1 & 2 & 0 \\ 2 & -1 & 0 \\ 0 & 5 & 2 \\ -1 & 8 & 1 \end{bmatrix}$

12. $\begin{bmatrix} 0 & 1 & -1 & 2 \end{bmatrix}\begin{bmatrix} 1 & -2 & 1 \\ 0 & 1 & 3 \\ 6 & 0 & 2 \\ -1 & -2 & 11 \end{bmatrix}$

13. $\begin{bmatrix} 1 & 0 & -1 \\ 1 & 1 & 2 \end{bmatrix} \begin{bmatrix} 0 & 1 & -1 \\ 1 & 0 & 1 \\ 4 & 8 & 0 \end{bmatrix}$

14. $\begin{bmatrix} 0 & 1 & -1 \\ 3 & 1 & -1 \end{bmatrix} \begin{bmatrix} 1 & 1 \\ 4 & 2 \\ 0 & 1 \end{bmatrix}$

15. $\begin{bmatrix} 1 & 0 \\ 1 & -1 \end{bmatrix} \begin{bmatrix} 0 & 1 \\ 0 & 1 \end{bmatrix}$

16. $\begin{bmatrix} 1 & -1 \\ 1 & -1 \end{bmatrix} \begin{bmatrix} 3 & -3 \\ 5 & -7 \end{bmatrix}$

17. $\begin{bmatrix} 0 & 1 \\ 0 & 1 \end{bmatrix} \begin{bmatrix} 1 & 0 \\ 1 & -1 \end{bmatrix}$

18. $\begin{bmatrix} 3 & -3 \\ 5 & -7 \end{bmatrix} \begin{bmatrix} 1 & -1 \\ 1 & -1 \end{bmatrix}$

19. $\begin{bmatrix} 1 & -1 \\ 1 & -1 \end{bmatrix} \begin{bmatrix} 2 & 3 \\ 2 & 3 \end{bmatrix}$

20. $\begin{bmatrix} 0 & 1 \\ 1 & 0 \end{bmatrix} \begin{bmatrix} 3 & -3 \\ 2 & -1 \end{bmatrix}$

21. $\begin{bmatrix} 1 & -1 \\ -1 & 1 \end{bmatrix} \begin{bmatrix} 2 & 3 \\ 2 & 3 \\ 1 & 1 \end{bmatrix}$

22. $\begin{bmatrix} 0 & 1 & -1 \\ 0 & -1 & 1 \end{bmatrix} \begin{bmatrix} 3 & -3 \\ 2 & -1 \end{bmatrix}$

23. $\begin{bmatrix} 1 & 0 & -1 \\ 2 & -2 & 1 \\ 0 & 0 & 1 \end{bmatrix} \begin{bmatrix} 1 & -1 & 4 \\ 1 & 1 & 0 \\ 0 & 4 & 1 \end{bmatrix}$

24. $\begin{bmatrix} 1 & 2 & 0 \\ 4 & -1 & 1 \\ 1 & 0 & 1 \end{bmatrix} \begin{bmatrix} 1 & 2 & -4 \\ 4 & 1 & 0 \\ 0 & -2 & 1 \end{bmatrix}$

25. $\begin{bmatrix} 1 & 0 & 1 & 0 \\ -1 & 1 & 0 & 1 \\ -2 & 0 & 1 & 4 \\ 0 & -1 & 0 & 1 \end{bmatrix} \begin{bmatrix} 1 \\ -3 \\ 2 \\ 0 \end{bmatrix}$

26. $\begin{bmatrix} 1 & 1 & -7 & 0 \\ -1 & 0 & 2 & 4 \\ -1 & 0 & -2 & 1 \\ 1 & -1 & 1 & 1 \end{bmatrix} \begin{bmatrix} 1 \\ -3 \\ 2 \\ 1 \end{bmatrix}$

27. $\begin{bmatrix} 1.1 & 2.3 & 3.4 & -1.2 \\ 3.4 & 4.4 & 2.3 & 1.1 \\ 2.3 & 0 & -2.2 & 1.1 \\ 1.2 & 1.3 & 1.1 & 1.1 \end{bmatrix} \begin{bmatrix} -2.1 & 0 & -3.3 \\ -3.4 & -4.8 & -4.2 \\ 3.4 & 5.6 & 1 \\ 1 & 2.2 & 9.8 \end{bmatrix}$

28. $\begin{bmatrix} 1.2 & 2.3 & 3.4 & 4.5 \\ 3.3 & 4.4 & 5.5 & 6.6 \\ 2.3 & -4.3 & -2.2 & 1.1 \\ 2.2 & -1.2 & -1 & 1.1 \end{bmatrix} \begin{bmatrix} 9.8 & 1 & -1.1 \\ 8.8 & 2 & -2.2 \\ 7.7 & 3 & -3.3 \\ 6.6 & 4 & -4.4 \end{bmatrix}$

29. Find[3] $A^2 = A \cdot A, A^3 = A \cdot A \cdot A, A^4$, and A^{100}, given that $A = \begin{bmatrix} 0 & 1 & 1 & 1 \\ 0 & 0 & 1 & 1 \\ 0 & 0 & 0 & 1 \\ 0 & 0 & 0 & 0 \end{bmatrix}$.

30. Repeat Exercise 29 with $A = \begin{bmatrix} 0 & 2 & 0 & -1 \\ 0 & 0 & 2 & 0 \\ 0 & 0 & 0 & 2 \\ 0 & 0 & 0 & 0 \end{bmatrix}$.

Exercises 31–38 should be done two ways: by hand and by using technology where possible. Let

$$A = \begin{bmatrix} 0 & -1 & 0 & 1 \\ 10 & 0 & 1 & 0 \end{bmatrix} \quad B = \begin{bmatrix} 0 & -1 \\ 1 & 1 \\ -1 & 3 \\ 5 & 0 \end{bmatrix} \quad C = \begin{bmatrix} 1 & -1 \\ 1 & 1 \\ 1 & 1 \\ 1 & 1 \end{bmatrix}$$

Evaluate each expression.

31. AB **32.** AC **33.** $A(B - C)$ **34.** $(B - C)A$

Let

$$A = \begin{bmatrix} 1 & -1 \\ 0 & 2 \\ 0 & -2 \end{bmatrix} \quad B = \begin{bmatrix} 3 & 0 & -1 \\ 5 & -1 & 1 \end{bmatrix} \quad C = \begin{bmatrix} x & 1 & w \\ z & r & 4 \end{bmatrix}$$

Evaluate each expression.

35. AB **36.** AC **37.** $A(B + C)$ **38.** $(B + C)A$

In Exercises 39–44, calculate (a) $P^2 = P \cdot P$, (b) $P^4 = P^2 \cdot P^2$, and (c) P^8. (Round all entries to four decimal places.) (d) Without computing it explicitly, find P^{1000}.

39. $P = \begin{bmatrix} 0.2 & 0.8 \\ 0.2 & 0.8 \end{bmatrix}$

40. $P = \begin{bmatrix} 0.1 & 0.9 \\ 0.1 & 0.9 \end{bmatrix}$

41. $P = \begin{bmatrix} 0.1 & 0.9 \\ 0 & 1 \end{bmatrix}$

42. $P = \begin{bmatrix} 1 & 0 \\ 0.8 & 0.2 \end{bmatrix}$

43. $P = \begin{bmatrix} 0.25 & 0.25 & 0.50 \\ 0.25 & 0.25 & 0.50 \\ 0.25 & 0.25 & 0.50 \end{bmatrix}$

44. $P = \begin{bmatrix} 0.3 & 0.3 & 0.4 \\ 0.3 & 0.3 & 0.4 \\ 0.3 & 0.3 & 0.4 \end{bmatrix}$

In Exercises 45–48, translate the given matrix equations into systems of linear equations.

45. $\begin{bmatrix} 2 & -1 & 4 \\ -4 & \frac{3}{4} & \frac{1}{3} \\ -3 & 0 & 0 \end{bmatrix} \begin{bmatrix} x \\ y \\ z \end{bmatrix} = \begin{bmatrix} 3 \\ -1 \\ 0 \end{bmatrix}$

46. $\begin{bmatrix} 1 & -1 & 4 \\ -\frac{1}{3} & -3 & \frac{1}{3} \\ 3 & 0 & 1 \end{bmatrix} \begin{bmatrix} x \\ y \\ z \end{bmatrix} = \begin{bmatrix} -3 \\ -1 \\ 2 \end{bmatrix}$

47. $\begin{bmatrix} 1 & -1 & 0 & 1 \\ 1 & 1 & 2 & 4 \end{bmatrix} \begin{bmatrix} x \\ y \\ z \\ w \end{bmatrix} = \begin{bmatrix} -1 \\ 2 \end{bmatrix}$

48. $\begin{bmatrix} 0 & 1 & 6 & 1 \\ 1 & -5 & 0 & 0 \end{bmatrix} \begin{bmatrix} x \\ y \\ z \\ w \end{bmatrix} = \begin{bmatrix} -2 \\ 9 \end{bmatrix}$

In Exercises 49–52, translate the given systems of equations into matrix form.

49. $\begin{aligned} x - y &= 4 \\ 2x - y &= 0 \end{aligned}$

50. $\begin{aligned} 2x + y &= 7 \\ -x &= 9 \end{aligned}$

51. $\begin{aligned} x + y - z &= 8 \\ 2x + y + z &= 4 \\ \frac{3x}{4} + \frac{z}{2} &= 1 \end{aligned}$

52. $\begin{aligned} x + y + 2z &= -2 \\ 4x + 2y - z &= -8 \\ \frac{x}{2} - \frac{y}{3} &= 4 \end{aligned}$

---

[3] $A \cdot A \cdot A$ is $A(A \cdot A)$, or the equivalent $(A \cdot A)A$ by the associative law. Similarly, $A \cdot A \cdot A \cdot A = A(A \cdot A \cdot A) = (A \cdot A \cdot A)A = (A \cdot A)(A \cdot A)$; it doesn't matter where we place the parentheses.

APPLICATIONS

53. Revenue Your T-shirt operation is doing a booming trade. Last week you sold 50 tie-dyed shirts for \$15 each, 40 Suburban State University crew shirts for \$10 each, and 30 lacrosse T-shirts for \$12 each. Use matrix operations to calculate your total revenue for the week.

54. Revenue Karen Sandberg, your competitor in Suburban State U's T-shirt market, has apparently been undercutting your prices and outperforming you in sales. Last week she sold 100 tie-dyed shirts for \$10 each, 50 (low-quality) crew shirts at \$5 apiece, and 70 lacrosse T-shirts for \$8 each. Use matrix operations to calculate her total revenue for the week.

55. Revenue Recall the Left Coast Bookstore chain from Section 3.2. In January it sold 700 hardcover books, 1300 softcover books, and 2000 plastic books in San Francisco; it sold 400 hardcover, 300 softcover, and 500 plastic books in Los Angeles. Hardcover books sell for \$30 each, softcover books sell for \$10 each, and plastic books sell for \$15 each. Write a column matrix with the price data and show how matrix multiplication (using the sales- and price-data matrices) may be used to compute the total revenue at the two stores.

56. Profit Refer to Exercise 55 and now suppose that each hardcover book costs the stores \$10, each softcover book costs \$5, and each plastic book costs \$10. Use matrix operations to compute the total *profit* at each store in January.

57. Publishing Editors' workloads were increasing during the 1990s, as the following table shows. Use matrix multiplication to estimate the total number of books published during the years 1993–1996.

	1993	1994	1995	1996
Books/Editor	3	3.5	5	5.2
Editors	16,000	15,000	12,500	13,000

"Books/Editor" refers to new titles. Data are rounded. SOURCE: R. R. Bowker, EEOC/*New York Times,* June 29, 1998, p. E1.

58. Income The following chart shows the 1997 population and per capita income in four Texas counties along the Mexico border. Use matrix multiplication to estimate the total income in the four counties shown.

	El Paso	Webb	Starr	Cameron
Population	702,000	183,000	56,000	321,000
Per Capita Income	\$14,500	\$12,200	\$7,200	\$12,500

All figures are rounded to three significant digits. SOURCES: Census Bureau: Bureau of Labor Statistics, Bureau of Economic Analysis/*New York Times,* June 23, 1998, p. A16.

59. Costs Microbucks Computers makes two computers, the Pomegranate II and the Pomegranate Classic. The Pom II requires 2 processor chips, 16 memory chips, and 20 vacuum tubes, and the Pom Classic requires 1 processor chip, 4 memory chips, and 40 vacuum tubes. Two companies can supply these parts: Motorel supplies them at \$100/processor chip, \$50/memory chip, and \$10/vacuum tube; Intola supplies them at \$150/processor chip, \$40/memory chip, and \$15/vacuum tube. Write down all data in two matrices, one showing the parts required for each model computer and the other showing the prices for each part from each supplier. Then show how matrix multiplication allows you to compute the total cost for parts for each model when parts are bought from either supplier.

60. Profits Refer to Exercise 59. It actually costs Motorel only \$25 to make each processor chip, \$10 for each memory chip, and \$5 for each vacuum tube. It costs Intola \$50 for each processor chip, \$10 for each memory chip, and \$7 for each vacuum tube. Use matrix operations to find the total profit Motorel and Intola make on each model.

61. Cheese Production The total amount of cheese, in billions of pounds, produced in the 13 western and 12 north central U.S. states in 1999 and 2000 is shown in the following table. Thinking of this table as a (labeled) 2×2 matrix P, compute the matrix product $[-1 \quad 1]P$. What does this product represent?

	1999	2000
Western States	2.7	3.0
North Central States	3.9	4.0

All figures are rounded to two significant digits. SOURCES: Census Bureau: Bureau of Labor Statistics, Bureau of Economic Analysis/*New York Times,* June 23, 1998, p. A16.

62. Milk Production The total amount of milk, in billions of pounds, produced in the 13 western and 12 north central U.S. states in 1999 and 2000 is shown in the following table. Thinking of this table as a (labeled) 2×2 matrix P, compute the matrix product $P\begin{bmatrix} -1 \\ 1 \end{bmatrix}$. What does this product represent?

	1999	2000
Western States	56	60
North Central States	57	59

Figures are approximate. SOURCE: Department of Agriculture/*New York Times,* June 28, 2001, p. C1.

Bankruptcy Filings Exercises 63–68 are based on the following chart, which shows the number of personal bankruptcy filings in three New York/New Jersey regions during various months of 2001–2002.

	Jan 2001	Jul 2001	Jan 2002
Manhattan	150	150	150
Brooklyn	300	300	250
Newark	250	250	200

Data are approximate 4-week moving averages. SOURCE: Lundquist Consulting/*New York Times,* February 10, 2002, p. L1.

63. Each month your law firm handles 10% of all bankruptcy filings in Manhattan, 5% of all filings in Brooklyn, and 20% of all filings in Newark. Use matrix multiplication to compute the total number of bankruptcy filings handled by your firm in each of the months shown.

64. Your law firm handled 10% of all bankruptcy filings in each region in January 2001, 30% of all filings in July 2001, and 20% of all filings in January 2002. Use matrix multiplication to compute the total number of bankruptcy filings handled by your firm in each of the regions shown.

65. Let A be the 3×3 matrix whose entries are the figures in the table and let $B = [1 \quad 1 \quad 0]$. What does the matrix BA represent?

66. Let A be the 3×3 matrix whose entries are the figures in the table and let $B = [1 \quad 1 \quad 0]^T$. What does the matrix AB represent?

67. Write a matrix product whose computation gives the total number by which the combined filings in Manhattan and Newark exceeded the filings in Brooklyn.

68. Write a matrix product whose computation gives the total number by which bankruptcy filings in January 2001 exceeded filings in January 2002.

69. Tourism The following table gives the number of people (in thousands) who visited Australia and South Africa in 1998.

	To **Australia**	**South Africa**
From **North America**	440	190
Europe	950	950
Asia	1790	200

Figures are rounded to the nearest 10,000. SOURCES: South African Dept. of Environmental Affairs and Tourism; Australia Tourist Commission/*New York Times*, January 15, 2000, p. C1.

You estimate that 5% of all visitors to Australia and 4% of all visitors to South Africa decide to settle there permanently. Let A be the 3×2 matrix whose entries are the 1998 tourism figures in the above table and let

$$B = \begin{bmatrix} 0.05 \\ 0.04 \end{bmatrix} \qquad C = \begin{bmatrix} 0.05 & 0 \\ 0 & 0.04 \end{bmatrix}$$

Compute the products AB and AC. What do the entries in these matrices represent?

70. Tourism Referring to the tourism figures in Exercise 69, you estimate that from 1998 to 2008 tourism from North America to each of Australia and South Africa will have increased by 20%, tourism from Europe by 30%, and tourism from Asia by 10%. Let A be the 3×2 matrix whose entries are the 1998 tourism figures and let

$$B = [1.2 \quad 1.3 \quad 1.1] \qquad C = \begin{bmatrix} 1.2 & 0 & 0 \\ 0 & 1.3 & 0 \\ 0 & 0 & 1.1 \end{bmatrix}$$

Compute the products BA and CA. What do the entries in these matrices represent?

71. Population Movement In 1999 the U.S. population, broken down by regions, was 53.9 million in the Northeast, 64.3 million in the Midwest, 100.0 million in the South, and 63.3 million in the West. The matrix P below shows the population movement during the period 1999–2000. (Thus, 98.86% of the population in the Northeast stayed there, while 0.15% of the population in the Northeast moved to the Midwest, and so on.) Set up the 1999 population figures as a row vector. Then use matrix multiplication to compute the population in each region in 2000. (Round all answers to the nearest 0.1 million.)

$$P = \begin{array}{r} \\ \textbf{From NE} \\ \textbf{From MW} \\ \textbf{From S} \\ \textbf{From W} \end{array} \begin{array}{c} \begin{array}{cccc} \textbf{To NE} & \textbf{To MW} & \textbf{To S} & \textbf{To W} \end{array} \\ \begin{bmatrix} 0.9886 & 0.0015 & 0.0075 & 0.0024 \\ 0.0011 & 0.9900 & 0.0057 & 0.0032 \\ 0.0018 & 0.0042 & 0.9897 & 0.0043 \\ 0.0017 & 0.0035 & 0.0077 & 0.9871 \end{bmatrix} \end{array}$$

This exercise ignores migration into or out of the country. The internal migration figures and 2000 population figures are accurate. SOURCE: U.S. Census Bureau, Current Population Survey, March 2000, http://www.census.gov/.

72. Population Movement Assume that the percentages given in Exercise 71 also describe the population movements from 2000 to 2001. Use two matrix multiplications to predict from the data in Exercise 71 the population in each region in 2001. (Round the *final* answers to the nearest 0.1 million but do not round the intermediate answer.)

COMMUNICATION AND REASONING EXERCISES

73. Give an example of two matrices A and B such that AB is defined but BA is not defined.

74. Give an example of two matrices A and B of different dimensions such that both AB and BA are defined.

75. Comment on the following claim: Every matrix equation represents a system of equations.

76. When is it true that both AB and BA are defined, even though neither A nor B is a square matrix?

77. Find a scenario in which it would be useful to "multiply" two row vectors according to the rule

$$[a \quad b \quad c][d \quad e \quad f] = [ad \quad be \quad cf]$$

78. Make up an application whose solution reads as follows:

$$\text{total revenue} = [10 \quad 100 \quad 30]\begin{bmatrix} 10 & 0 & 3 \\ 1 & 2 & 0 \\ 0 & 1 & 40 \end{bmatrix}$$

79. What happens in Excel if, instead of using the function MMULT you use "ordinary multiplication" as shown here?

	A	B	C	D	E	F	G
1	2	0	7		1	1	-8
2	1	-1	0		1	0	0
3	-2	1	1		0	5	2
4							
5	=A1:C3*E1:G3						
6							
7							

80. Define the *naïve product* $A \square B$ of two $m \times n$ matrices A and B by

$$(A \square B)_{ij} = A_{ij}B_{ij}$$

(This is how someone who has never seen matrix multiplication before might think to multiply matrices.) Referring to Example 1 in this section, compute and comment on the meaning of $P \square (Q^T)$.

81. Compare addition and multiplication of 1×1 matrices to the arithmetic of numbers.

82. In comparing the algebra of 1×1 matrices, as discussed so far, to the algebra of real numbers (see Exercise 81), what important difference do you find?

3.3 Matrix Inversion

Now that we've discussed matrix addition, subtraction, and multiplication, you may well be wondering about matrix *division.* In the realm of real numbers, division can be thought of as a form of multiplication: Dividing 3 by 7 is the same as multiplying 3 by $\frac{1}{7}$, the inverse of 7. In symbols, $3 \div 7 = 3 \times \frac{1}{7}$, or 3×7^{-1}. To imitate division of real numbers in the realm of matrices, we need to discuss the multiplicative **inverse,** A^{-1}, of a matrix A.

Note Since multiplication of real numbers is commutative, we can write, for example, $\frac{3}{7}$ as either 3×7^{-1} or $7^{-1} \times 3$. In the realm of matrices, multiplication is not commutative, so from now on we will *never* talk about "division" of matrices (by $\frac{B}{A}$ should we mean $A^{-1}B$ or BA^{-1}?).

Before we try to find the inverse of a matrix, we must first know exactly what we *mean* by the inverse. Recall that the inverse of a number a is the number, often written a^{-1}, with the property that $a^{-1} \cdot a = a \cdot a^{-1} = 1$. For example, the inverse of 76 is the number $76^{-1} = \frac{1}{76}$, since $\frac{1}{76} \cdot 76 = 76 \cdot \frac{1}{76} = 1$. This is the number calculated by the $\boxed{x^{-1}}$ button found on most calculators. Not all numbers have an inverse. For example—and this is the *only* example—the number 0 has no inverse, since you cannot get 1 by multiplying 0 by anything.

The inverse of a matrix is defined similarly. To make life easier, we will restrict attention to **square** matrices, matrices that have the same number of rows as columns.[4]

Inverse of a Matrix

The **inverse** of an $n \times n$ matrix A is that $n \times n$ matrix A^{-1} which, when multiplied by A on either side, yields the $n \times n$ identity matrix I. Thus,

$$AA^{-1} = A^{-1}A = I$$

[4]Nonsquare matrices *cannot* have inverses in the sense that we are talking about here. This is not a trivial fact to prove.

If A has an inverse, it is said to be **invertible**; otherwise, it is said to be **singular.**

Quick Examples

1. The inverse of the 1×1 matrix $[3]$ is $[\frac{1}{3}]$, since $[3][\frac{1}{3}] = [1] = [\frac{1}{3}][3]$.
2. The inverse of the $n \times n$ identity matrix I is I itself, since $I \times I = I$. Thus, $I^{-1} = I$.
3. The inverse of the 2×2 matrix $A = \begin{bmatrix} 1 & -1 \\ -1 & -1 \end{bmatrix}$ is $A^{-1} = \begin{bmatrix} \frac{1}{2} & -\frac{1}{2} \\ -\frac{1}{2} & -\frac{1}{2} \end{bmatrix}$, since

$$\begin{bmatrix} 1 & -1 \\ -1 & -1 \end{bmatrix} \begin{bmatrix} \frac{1}{2} & -\frac{1}{2} \\ -\frac{1}{2} & -\frac{1}{2} \end{bmatrix} = \begin{bmatrix} 1 & 0 \\ 0 & 1 \end{bmatrix} \qquad AA^{-1} = I$$

and
$$\begin{bmatrix} \frac{1}{2} & -\frac{1}{2} \\ -\frac{1}{2} & -\frac{1}{2} \end{bmatrix} \begin{bmatrix} 1 & -1 \\ -1 & -1 \end{bmatrix} = \begin{bmatrix} 1 & 0 \\ 0 & 1 \end{bmatrix} \qquad A^{-1}A = I$$

Notes

1. It is possible to show that if A and B are square matrices with $AB = I$, then it must also be true that $BA = I$. In other words, once we have checked that $AB = I$, we know that B is the inverse of A. The second check, that $BA = I$, is unnecessary.
2. If B is the inverse of A, then we can also say that A is the inverse of B. (Why?) Thus, we sometimes refer to such a pair of matrices as an **inverse pair** of matrices.

Example 1 • Singular Matrix

Can $A = \begin{bmatrix} 1 & 1 \\ 0 & 0 \end{bmatrix}$ have an inverse?

Solution No. To see why not, notice that both entries in the second row of AB will be zero no matter what B is. So AB cannot equal I no matter what B is. Hence, A is singular.

Before we go on . . . If you think about it, you can write down many similar examples of singular matrices. There is only one number with no multiplicative inverse (0), but there are many matrices having no inverses.

Finding the Inverse of a Square Matrix

Question *How* did you find the inverse of $A = \begin{bmatrix} 1 & -1 \\ -1 & -1 \end{bmatrix}$?

Answer We can think of the problem of finding A^{-1} as a problem of finding four unknowns, the four unknown entries of A^{-1}:

$$A^{-1} = \begin{bmatrix} x & y \\ z & w \end{bmatrix}$$

These unknowns must satisfy the equation $AA^{-1} = I$, or

$$\begin{bmatrix} 1 & -1 \\ -1 & -1 \end{bmatrix} \begin{bmatrix} x & y \\ z & w \end{bmatrix} = \begin{bmatrix} 1 & 0 \\ 0 & 1 \end{bmatrix}$$

If we were to try to find the first column of A^{-1}, consisting of x and z, we would have to solve

$$\begin{bmatrix} 1 & -1 \\ -1 & -1 \end{bmatrix} \begin{bmatrix} x \\ z \end{bmatrix} = \begin{bmatrix} 1 \\ 0 \end{bmatrix}$$

or

$$\begin{aligned} x - z &= 1 \\ -x - z &= 0 \end{aligned}$$

To solve this system by Gauss–Jordan reduction, we would row-reduce the augmented matrix, which is A with the column $\begin{bmatrix} 1 \\ 0 \end{bmatrix}$ adjoined:

$$\left[\begin{array}{cc|c} 1 & -1 & 1 \\ -1 & -1 & 0 \end{array}\right] \to \left[\begin{array}{cc|c} 1 & 0 & x \\ 0 & 1 & z \end{array}\right]$$

To find the second column of A^{-1}, we would similarly row-reduce the augmented matrix obtained by tacking on to A the second column of the identity matrix:

$$\left[\begin{array}{cc|c} 1 & -1 & 0 \\ -1 & -1 & 1 \end{array}\right] \to \left[\begin{array}{cc|c} 1 & 0 & y \\ 0 & 1 & w \end{array}\right]$$

The row operations used in doing these two reductions would be exactly the same. We could do both reductions simultaneously by "doubly augmenting" A, putting both columns of the identity matrix to the right of A:

$$\left[\begin{array}{cc|cc} 1 & -1 & 1 & 0 \\ -1 & -1 & 0 & 1 \end{array}\right] \to \left[\begin{array}{cc|cc} 1 & 0 & x & y \\ 0 & 1 & z & w \end{array}\right]$$

We carry out this reduction in Example 2.

Example 2 • Computing the Inverse of a Matrix

Find the inverse of each matrix.

a. $P = \begin{bmatrix} 1 & -1 \\ -1 & -1 \end{bmatrix}$ **b.** $Q = \begin{bmatrix} 1 & 0 & 1 \\ 2 & -2 & -1 \\ 3 & 0 & 0 \end{bmatrix}$

Solution

a. As described above, we put the matrix P on the left and the identity matrix I on the right to get a 2×4 matrix:

$$\underbrace{\left[\begin{array}{cc|cc} 1 & -1 & 1 & 0 \\ -1 & -1 & 0 & 1 \end{array}\right]}_{P \qquad I}$$

We now row-reduce the whole matrix:

$$\begin{bmatrix} 1 & -1 & 1 & 0 \\ -1 & -1 & 0 & 1 \end{bmatrix} \begin{matrix} \\ R_2 + R_1 \end{matrix} \to \begin{bmatrix} 1 & -1 & 1 & 0 \\ 0 & -2 & 1 & 1 \end{bmatrix} \begin{matrix} 2R_1 - R_2 \\ \\ \end{matrix} \to$$

$$\begin{bmatrix} 2 & 0 & 1 & -1 \\ 0 & -2 & 1 & 1 \end{bmatrix} \begin{matrix} \frac{1}{2}R_1 \\ -\frac{1}{2}R_2 \end{matrix} \to \left[\begin{array}{cc|cc} 1 & 0 & \frac{1}{2} & -\frac{1}{2} \\ 0 & 1 & -\frac{1}{2} & -\frac{1}{2} \end{array}\right]$$

$I \qquad P^{-1}$

We have now solved the systems of linear equations that define the entries of P^{-1}:

$$P^{-1} = \begin{bmatrix} \frac{1}{2} & -\frac{1}{2} \\ -\frac{1}{2} & -\frac{1}{2} \end{bmatrix}$$

b. The procedure to find the inverse of a 3 × 3 matrix (or larger) is just the same as for a 2 × 2 matrix. We place Q on the left and the identity matrix (now 3 × 3) on the right and reduce:

$$\underset{Q\qquad\qquad I}{\left[\begin{array}{ccc|ccc} 1 & 0 & 1 & 1 & 0 & 0 \\ 2 & -2 & -1 & 0 & 1 & 0 \\ 3 & 0 & 0 & 0 & 0 & 1 \end{array}\right]} \begin{array}{l} \\ R_2 - 2R_1 \\ R_3 - 3R_1 \end{array} \rightarrow \left[\begin{array}{ccc|ccc} 1 & 0 & 1 & 1 & 0 & 0 \\ 0 & -2 & -3 & -2 & 1 & 0 \\ 0 & 0 & -3 & -3 & 0 & 1 \end{array}\right] \begin{array}{l} 3R_1 + R_3 \\ R_2 - R_3 \\ \\ \end{array} \rightarrow$$

$$\left[\begin{array}{ccc|ccc} 3 & 0 & 0 & 0 & 0 & 1 \\ 0 & -2 & 0 & 1 & 1 & -1 \\ 0 & 0 & -3 & -3 & 0 & 1 \end{array}\right] \begin{array}{l} \frac{1}{3}R_1 \\ -\frac{1}{2}R_2 \\ -\frac{1}{3}R_3 \end{array} \rightarrow \underset{I\qquad\qquad Q^{-1}}{\left[\begin{array}{ccc|ccc} 1 & 0 & 0 & 0 & 0 & \frac{1}{3} \\ 0 & 1 & 0 & -\frac{1}{2} & -\frac{1}{2} & \frac{1}{2} \\ 0 & 0 & 1 & 1 & 0 & -\frac{1}{3} \end{array}\right]}$$

Thus,

$$Q^{-1} = \begin{bmatrix} 0 & 0 & \frac{1}{3} \\ -\frac{1}{2} & -\frac{1}{2} & \frac{1}{2} \\ 1 & 0 & -\frac{1}{3} \end{bmatrix}$$

Before we go on . . . We have already checked that P^{-1} is the inverse of P. You should also check that Q^{-1} is the inverse of Q.

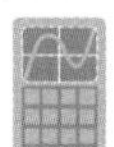

Graphing Calculator: Computing the Inverse of a Matrix

On a TI-83 you can invert the square matrix [A] by entering `[A]` [x⁻¹] [ENTER]. You can also use the calculator to help you go through the row reduction, as described in Chapter 2.

Excel: Computing the Inverse of a Matrix

The function MINVERSE computes the inverse of a matrix. Below we have entered Q in cells A1:C3. Now, choose the block where you would like the inverse to appear, highlight the whole block, enter the formula `=MINVERSE(A1:C3)`, and press [Control]+[Shift]+[Enter].

	A	B	C	D	E	F	G
1	1	0	1		=MINVERSE(		
2	2	-2	-1		A1:C3)		
3	3	0	0				

The inverse will appear in the region you highlighted. (To convert the answer to fractions, format the cells as fractions.) Although Excel appears to invert the matrix in one step, it is going through the above procedure or some variation of it to find the inverse. You can also use Excel to help you go through the row reduction, just as in Chapter 2.

If a matrix is singular, Excel will register an error by showing `#NUM!` in each cell.

Web Site

The path

Web site → Online Utilities → Matrix Algebra Tool

will take you to the online matrix algebra tool, which allows you to invert matrices directly. To invert a matrix using the row-reduction method in Example 2, use either the browser-based Pivot and Gauss–Jordan Tool or the Excel-based Excel Pivot and Gauss–Jordan Tool, also accessible from

Web site → Online Utilities

The path

Web site → Excel Tutorials → 3.3 Matrix Inversion

will give you a downloadable Excel tutorial that covers this and other examples in this section.

The method we used above can be summarized as follows:

Inverting an $n \times n$ Matrix

To determine whether an $n \times n$ matrix A is invertible and to find A^{-1} if it does exist, follow this procedure:

1. Write down the $n \times 2n$ matrix $[A \mid I]$ (this is A with the $n \times n$ identity matrix set next to it).
2. Row-reduce $[A \mid I]$.
3. If the reduced form is $[I \mid B]$ (that is, has the identity matrix in the left part), then A is invertible, and $B = A^{-1}$. If you cannot obtain I in the left part, then A is singular. (See Example 3.)

Question Surely there must be a formula for the inverse of a matrix so that we do not have to row-reduce each time?

Answer There is such a formula, but not a simple one. In fact, using the formula for anything larger than a 3×3 matrix is so inefficient that row reduction is the method of choice even for computers. However, the formula is very simple for 2×2 matrices.

Formula for the Inverse of a 2×2 Matrix

The inverse of a 2×2 matrix is

$$\begin{bmatrix} a & b \\ c & d \end{bmatrix}^{-1} = \frac{1}{ad - bc}\begin{bmatrix} d & -b \\ -c & a \end{bmatrix} \qquad (ad - bc \neq 0)$$

If the quantity $ad - bc$ is zero, then the matrix is singular (noninvertible). The quantity $ad - bc$ is called the **determinant** of the matrix $\begin{bmatrix} a & b \\ c & d \end{bmatrix}$.

Quick Examples

1. $\begin{bmatrix} 1 & 2 \\ 3 & 4 \end{bmatrix}^{-1} = \dfrac{1}{(1)(4) - (2)(3)}\begin{bmatrix} 4 & -2 \\ -3 & 1 \end{bmatrix} = -\dfrac{1}{2}\begin{bmatrix} 4 & -2 \\ -3 & 1 \end{bmatrix} = \begin{bmatrix} -2 & 1 \\ \frac{3}{2} & -\frac{1}{2} \end{bmatrix}.$
2. $\begin{bmatrix} 1 & -1 \\ 2 & -2 \end{bmatrix}$ has determinant $ad - bc = (1)(-2) - (-1)(2) = 0$ and so is singular.

Question Where does this formula for the inverse of a 2×2 matrix come from?

Answer It can be obtained using the technique of row reduction. (See the Communication and Reasoning exercises at the end of the section.)

Example 3 • Singular 3 × 3 Matrix

Find the inverse of the matrix $S = \begin{bmatrix} 1 & 1 & 2 \\ -2 & 0 & 4 \\ 3 & 1 & -2 \end{bmatrix}$, if it exists.

Solution We proceed as before:

$$\left[\begin{array}{ccc|ccc} 1 & 1 & 2 & 1 & 0 & 0 \\ -2 & 0 & 4 & 0 & 1 & 0 \\ 3 & 1 & -2 & 0 & 0 & 1 \end{array}\right] \begin{array}{l} \\ R_2 + 2R_1 \to \\ R_3 - 3R_1 \end{array} \left[\begin{array}{cccccc} 1 & 1 & 2 & 1 & 0 & 0 \\ 0 & 2 & 8 & 2 & 1 & 0 \\ 0 & -2 & -8 & -3 & 0 & 1 \end{array}\right] \begin{array}{l} 2R_1 - R_2 \\ \to \\ R_3 + R_2 \end{array} \left[\begin{array}{cccccc} 2 & 0 & -4 & 0 & -1 & 0 \\ 0 & 2 & 8 & 2 & 1 & 0 \\ 0 & 0 & 0 & -1 & 1 & 1 \end{array}\right]$$

We stopped here, even though the reduction is incomplete, because there is *no hope* of getting the identity on the left-hand side. Completing the row reduction will not change the three zeros in the bottom row. So what is wrong? Nothing. As in Example 1, we have a singular matrix. Any square matrix that, after row reduction, winds up with a row of zeros is singular.

Before we go on . . . If you try to invert a singular matrix using a spreadsheet, calculator, or computer program, you should get an error. Sometimes, instead of an error, you will get a spurious answer due to round-off errors in the device.

In practice, deciding whether a given matrix is invertible or singular is easy: Try to find its inverse. If the process works, then the matrix is invertible, and we get its inverse. If the process fails, then the matrix is not invertible; it is singular and that's that.

Using the Inverse to Solve a System of n Linear Equations in n Unknowns

Having used systems of equations and row reduction to find matrix inverses, we now use matrix inverses to solve systems of equations. Recall that, at the end of Section 3.2, we saw that a system of linear equations could be written in the form

$$AX = B$$

where A is the coefficient matrix, X is the column matrix of unknowns, and B is the column matrix of right-hand sides. Now suppose there are as many unknowns as equations, so that A is a square matrix, and suppose A is invertible. The object is to solve for the matrix X of unknowns, so we multiply both sides of the equation by the inverse A^{-1} of A, getting

$$A^{-1}AX = A^{-1}B$$

Notice that we put A^{-1} on the left on both sides of the equation. Order matters when multiplying matrices, so we have to be careful to really do the same thing to both sides of the equation. But now $A^{-1}A = I$, so we can rewrite the last equation as

$$IX = A^{-1}B$$

Also, $IX = X$ (I being the identity matrix), so we really have

$$X = A^{-1}B$$

and we have solved for X!

Actually, we have shown that, if A is invertible and $AX = B$, then the only *possible* solution is $X = A^{-1}B$. We should check that $A^{-1}B$ is actually a solution by substituting back into the original equation:

$$AX = A(A^{-1}B) = (AA^{-1})B = IB = B$$

Thus, $X = A^{-1}B$ is a solution and the only solution. Therefore, if A is invertible, $AX = B$ has exactly one solution.

On the other hand, if $AX = B$ has no solutions or has infinitely many solutions, we can conclude that A is not invertible. (Why?) We summarize as follows:

Solving the Matrix Equation $AX = B$

If A is an invertible matrix, then the matrix equation $AX = B$ has the unique solution

$$X = A^{-1}B$$

Quick Example

The system of linear equations

$$\begin{aligned} 2x \phantom{{}+y-{}} + z &= 9 \\ 2x + y - z &= 6 \\ 3x + y - z &= 9 \end{aligned}$$

can be written as $AX = B$, where

$$A = \begin{bmatrix} 2 & 0 & 1 \\ 2 & 1 & -1 \\ 3 & 1 & -1 \end{bmatrix} \qquad X = \begin{bmatrix} x \\ y \\ z \end{bmatrix} \qquad B = \begin{bmatrix} 9 \\ 6 \\ 9 \end{bmatrix}$$

The matrix A is invertible with inverse

$$A^{-1} = \begin{bmatrix} 0 & -1 & 1 \\ 1 & 5 & -4 \\ 1 & 2 & -2 \end{bmatrix}$$

You should check this.

Thus,

$$X = A^{-1}B = \begin{bmatrix} 0 & -1 & 1 \\ 1 & 5 & -4 \\ 1 & 2 & -2 \end{bmatrix} \begin{bmatrix} 9 \\ 6 \\ 9 \end{bmatrix} = \begin{bmatrix} 3 \\ 3 \\ 3 \end{bmatrix}$$

so that $(x, y, z) = (3, 3, 3)$ is the (unique) solution to the system.

Example 4 • Using an Inverse to Solve Multiple Systems

Solve the following three systems of equations.

a. $\begin{aligned} 2x \phantom{{}+y-{}} + z &= 1 \\ 2x + y - z &= 1 \\ 3x + y - z &= 1 \end{aligned}$ **b.** $\begin{aligned} 2x \phantom{{}+y-{}} + z &= 0 \\ 2x + y - z &= 1 \\ 3x + y - z &= 2 \end{aligned}$ **c.** $\begin{aligned} 2x \phantom{{}+y-{}} + z &= 0 \\ 2x + y - z &= 0 \\ 3x + y - z &= 0 \end{aligned}$

Solution We *could* go ahead and row-reduce all three augmented matrices as we did in Chapter 2, but this would require a lot of work. Notice that the coefficients are the same in all three systems. In other words, we can write the three systems in matrix form as

a. $AX = B$ **b.** $AX = C$ **c.** $AX = D$

where the matrix A is the same in all three cases:

$$A = \begin{bmatrix} 2 & 0 & 1 \\ 2 & 1 & -1 \\ 3 & 1 & -1 \end{bmatrix}$$

Now the solutions to these systems are

a. $X = A^{-1}B$ **b.** $X = A^{-1}C$ **c.** $X = A^{-1}D$

so the main work is the calculation of the single matrix A^{-1}, which we have already noted (Quick Example above) is

$$A^{-1} = \begin{bmatrix} 0 & -1 & 1 \\ 1 & 5 & -4 \\ 1 & 2 & -2 \end{bmatrix}$$

The three solutions are

a. $X = A^{-1}B = \begin{bmatrix} 0 & -1 & 1 \\ 1 & 5 & -4 \\ 1 & 2 & -2 \end{bmatrix}\begin{bmatrix} 1 \\ 1 \\ 1 \end{bmatrix} = \begin{bmatrix} 0 \\ 2 \\ 1 \end{bmatrix}$

b. $X = A^{-1}C = \begin{bmatrix} 0 & -1 & 1 \\ 1 & 5 & -4 \\ 1 & 2 & -2 \end{bmatrix}\begin{bmatrix} 0 \\ 1 \\ 2 \end{bmatrix} = \begin{bmatrix} 1 \\ -3 \\ -2 \end{bmatrix}$

c. $X = A^{-1}D = \begin{bmatrix} 0 & -1 & 1 \\ 1 & 5 & -4 \\ 1 & 2 & -2 \end{bmatrix}\begin{bmatrix} 0 \\ 0 \\ 0 \end{bmatrix} = \begin{bmatrix} 0 \\ 0 \\ 0 \end{bmatrix}$

Excel: Solving Multiple Systems

Spreadsheets like Excel instantly update calculated results every time the contents of a cell are changed. We can take advantage of this to solve the three systems of equations, using the same worksheet as follows.

First, enter the matrices A and B from the matrix equation $AX = B$. Then select a 3×1 block of cells for the matrix X. The Excel formula we can use to calculate X is

```
=MMULT(MINVERSE(A1:C3),E1:E3)
```

$A^{-1}B$

	A	B	C	D	E
1	2	0	1		1
2	2	1	-1		1
3	3	1	-1		1
4		A			B
5					
6	=MMULT(MINVERSE(A1 :C3),E1 :E3)				
7					
8					

(As usual, use the mouse to select the ranges for A and B while typing the formula, and don't forget to press Control+Shift+Enter.) Having obtained the solution to part (a), you can now simply modify the entries in column E to see the solutions for parts (b) and (c).

Before we go on . . . Your spreadsheet for part (a) may look like this:

	A	B	C	D	E
1	2	0	1		1
2	2	1	-1		1
3	3	1	-1		1
4		A			B
5					
6	1.11022E-16				
7	2				
8	1				

Question What is that strange number doing in cell A6?

Answer `E-16` represents $\times 10^{-16}$, so the entry is really

$$1.110\ 22 \times 10^{-16} = 0.000\ 000\ 000\ 000\ 000\ 111\ 022 \approx 0$$

Mathematically, it is supposed to be *exactly* zero [see the solution to part (a)], but Excel made a small error in computing the inverse of A, resulting in this spurious value. Note, however, that it is accurate (agrees with zero) to 15 decimal places! In practice, when we see numbers arise in matrix calculations that are far smaller than all the other entries, we can usually assume they are supposed to be zero.

Question We have been speaking of *the* inverse of a matrix A. Are we entitled to do this? Could an invertible matrix have two or more inverses?

Answer A matrix A cannot have more than one inverse. This is not hard to prove: If B and C were both inverses of A, then

$$
\begin{aligned}
B &= BI && \text{Property of the identity} \\
&= B(AC) && \text{Because } C \text{ is an inverse of } A \\
&= (BA)C && \text{Associative law} \\
&= IC && \text{Because } B \text{ is an inverse of } A \\
&= C && \text{Property of the identity}
\end{aligned}
$$

In other words, if B and C were both inverses of A, then B and C would have to be equal.

Guideline: Which Method to Use in Solving a System

Question Now we have two methods to solve the a system of linear equations $AX = B$: (1) Compute $X = A^{-1}B$ or (2) row-reduce the augmented matrix. Which is the best method?

Answer Each method has its advantages and disadvantages. Method 1, as we have seen, is very efficient when you must solve several systems of equations with the same coefficients, but it only works when the coefficient matrix is *square* (meaning that you have the same number of equations as unknowns) and *invertible* (meaning that there is a unique solution). The row-reduction method will work for all systems. Moreover, for all but the smallest systems, the most efficient way to find A^{-1} is to use row-reduction. In practice the two methods are essentially the same when both apply.

3.3 EXERCISES

In Exercises 1–6, determine whether the given pairs of matrices are inverse pairs.

1. $A = \begin{bmatrix} 0 & 1 \\ 1 & 0 \end{bmatrix}, B = \begin{bmatrix} 0 & 1 \\ 1 & 0 \end{bmatrix}$ **2.** $A = \begin{bmatrix} 2 & 0 \\ 0 & 3 \end{bmatrix}, B = \begin{bmatrix} \frac{1}{2} & 0 \\ 0 & \frac{1}{2} \end{bmatrix}$

3. $A = \begin{bmatrix} 2 & 1 & 1 \\ 0 & 1 & 1 \\ 0 & 0 & 1 \end{bmatrix}, B = \begin{bmatrix} \frac{1}{2} & -\frac{1}{2} & 0 \\ 0 & 1 & -1 \\ 0 & 0 & 1 \end{bmatrix}$

4. $A = \begin{bmatrix} 1 & 1 & 1 \\ 0 & 1 & 1 \\ 0 & 0 & 1 \end{bmatrix}, B = \begin{bmatrix} 1 & -1 & 0 \\ 0 & 1 & -1 \\ 0 & 0 & 1 \end{bmatrix}$

5. $A = \begin{bmatrix} a & 0 & 0 \\ 0 & b & 0 \\ 0 & 0 & 0 \end{bmatrix}, B = \begin{bmatrix} a^{-1} & 0 & 0 \\ 0 & b^{-1} & 0 \\ 0 & 0 & 0 \end{bmatrix} \quad (a, b \neq 0)$

6. $A = \begin{bmatrix} a & 0 & 0 \\ 0 & b & 0 \\ 0 & 0 & c \end{bmatrix}, B = \begin{bmatrix} a^{-1} & 0 & 0 \\ 0 & b^{-1} & 0 \\ 0 & 0 & c^{-1} \end{bmatrix} \quad (a, b, c \neq 0)$

In Exercises 7–26, use row reduction to find the inverses of the given matrices if they exist, and check your answers by multiplication.

7. $\begin{bmatrix} 1 & 1 \\ 2 & 1 \end{bmatrix}$ **8.** $\begin{bmatrix} 0 & 1 \\ 1 & 1 \end{bmatrix}$ **9.** $\begin{bmatrix} 0 & 1 \\ 1 & 0 \end{bmatrix}$ **10.** $\begin{bmatrix} 4 & 0 \\ 0 & 2 \end{bmatrix}$

11. $\begin{bmatrix} 2 & 1 \\ 1 & 1 \end{bmatrix}$ **12.** $\begin{bmatrix} 3 & 0 \\ 0 & \frac{1}{2} \end{bmatrix}$ **13.** $\begin{bmatrix} 2 & 1 \\ 4 & 2 \end{bmatrix}$ **14.** $\begin{bmatrix} 1 & 1 \\ 6 & 6 \end{bmatrix}$

15. $\begin{bmatrix} 1 & 1 & 1 \\ 0 & 1 & 1 \\ 0 & 0 & 1 \end{bmatrix}$ **16.** $\begin{bmatrix} 1 & 2 & 3 \\ 0 & 1 & 2 \\ 0 & 0 & 1 \end{bmatrix}$

17. $\begin{bmatrix} 1 & 1 & 1 \\ 1 & 0 & 2 \\ 1 & -1 & 1 \end{bmatrix}$ **18.** $\begin{bmatrix} 1 & 2 & 3 \\ 0 & 2 & 3 \\ 1 & 0 & 1 \end{bmatrix}$

19. $\begin{bmatrix} 1 & 1 & 1 \\ 1 & -1 & 0 \\ 1 & 2 & 3 \end{bmatrix}$ **20.** $\begin{bmatrix} 1 & -1 & 3 \\ 0 & 1 & 3 \\ 1 & 1 & 1 \end{bmatrix}$

21. $\begin{bmatrix} 1 & 1 & 1 \\ 1 & 0 & 1 \\ 1 & -1 & 1 \end{bmatrix}$ **22.** $\begin{bmatrix} 1 & 1 & 1 \\ 0 & 1 & 1 \\ 1 & 0 & 0 \end{bmatrix}$

23. $\begin{bmatrix} 1 & 0 & 1 & 0 \\ -1 & 1 & 0 & 1 \\ -1 & 0 & 0 & 1 \\ 0 & -1 & 0 & 1 \end{bmatrix}$ **24.** $\begin{bmatrix} 0 & 1 & 1 & 0 \\ -1 & 1 & 1 & 1 \\ -1 & 1 & 0 & 1 \\ 0 & -1 & 0 & 1 \end{bmatrix}$

25. $\begin{bmatrix} 1 & 2 & 3 & 4 \\ 0 & 1 & 2 & 3 \\ 0 & 0 & 1 & 2 \\ 0 & 0 & 0 & 1 \end{bmatrix}$ **26.** $\begin{bmatrix} 0 & 0 & 0 & 1 \\ 0 & 0 & 1 & 0 \\ 0 & 1 & 0 & 0 \\ 1 & 0 & 0 & 0 \end{bmatrix}$

In Exercises 27–34, compute the determinant of the given matrix. If the determinant is nonzero, use the formula for inverting a 2×2 matrix to calculate the inverse of the matrix.

27. $\begin{bmatrix} 1 & 1 \\ 1 & -1 \end{bmatrix}$ **28.** $\begin{bmatrix} 4 & 1 \\ 0 & 2 \end{bmatrix}$ **29.** $\begin{bmatrix} 1 & 2 \\ 3 & 4 \end{bmatrix}$ **30.** $\begin{bmatrix} 1 & 0 \\ 0 & 1 \end{bmatrix}$

31. $\begin{bmatrix} \frac{1}{6} & -\frac{1}{6} \\ 0 & \frac{1}{6} \end{bmatrix}$ **32.** $\begin{bmatrix} 2 & 1 \\ 4 & 2 \end{bmatrix}$ **33.** $\begin{bmatrix} 1 & 0 \\ \frac{3}{4} & 0 \end{bmatrix}$ **34.** $\begin{bmatrix} 1 & 1 \\ 1 & 1 \end{bmatrix}$

T In Exercises 35–42, use technology to find the inverse of the given matrix (when it exists). Round all entries in your answer to two decimal places. [Caution: Because of rounding errors, technology sometimes produces "inverses" of singular matrices. These are often recognized by their huge entries.]

35. $\begin{bmatrix} 1.1 & 1.2 \\ 1.3 & -1 \end{bmatrix}$ **36.** $\begin{bmatrix} 0.1 & -3.2 \\ 0.1 & -1.5 \end{bmatrix}$

37. $\begin{bmatrix} 3.56 & 1.23 \\ -1.01 & 0 \end{bmatrix}$ **38.** $\begin{bmatrix} 9.09 & -5.01 \\ 1.01 & 2.20 \end{bmatrix}$

39. $\begin{bmatrix} 1.1 & 3.1 & 2.4 \\ 1.7 & 2.4 & 2.3 \\ 0.6 & -0.7 & -0.1 \end{bmatrix}$ **40.** $\begin{bmatrix} 2.1 & 2.4 & 3.5 \\ 6.1 & -0.1 & 2.3 \\ -0.3 & -1.2 & 0.1 \end{bmatrix}$

41. $\begin{bmatrix} 0.01 & 0.32 & 0 & 0.04 \\ -0.01 & 0 & 0 & 0.34 \\ 0 & 0.32 & -0.23 & 0.23 \\ 0 & 0.41 & 0 & 0.01 \end{bmatrix}$

42. $\begin{bmatrix} 0.01 & 0.32 & 0 & 0.04 \\ -0.01 & 0 & 0 & 0.34 \\ 0 & 0.32 & -0.23 & 0.23 \\ 0.01 & 0.96 & -0.23 & 0.65 \end{bmatrix}$

In Exercises 43–48, use matrix inversion to solve the given systems of linear equations. (You previously solved all of these systems using row reduction in Chapter 2.)

43. $x + y = 4$
$x - y = 1$

44. $2x + y = 2$
$2x - 3y = 2$

45. $\frac{x}{3} - \frac{y}{2} = 1$
$\frac{x}{4} + y = -2$

46. $-\frac{2x}{3} + \frac{y}{2} = -\frac{1}{6}$
$\frac{x}{4} - y = -\frac{3}{4}$

47. $-x + 2y - z = 0$
$-x - y + 2z = 0$
$2x - z = 6$

48. $x + 2y = 4$
$y - z = 0$
$x + 3y - 2z = 5$

In Exercises 49 and 50, use matrix inversion to solve each collection of systems of linear equations.

49. a. $-x - 4y + 2z = 4$
$x + 2y - z = 3$
$x + y - z = 8$

b. $-x - 4y + 2z = 0$
$x + 2y - z = 3$
$x + y - z = 2$

c. $-x - 4y + 2z = 0$
$x + 2y - z = 0$
$x + y - z = 0$

50. **a.**
$$\begin{aligned} -x - 4y + 2z &= 8 \\ x \quad\quad - z &= 3 \\ x + y - z &= 8 \end{aligned}$$

b.
$$\begin{aligned} -x - 4y + 2z &= 8 \\ x \quad\quad - z &= 3 \\ x + y - z &= 2 \end{aligned}$$

c.
$$\begin{aligned} -x - 4y + 2z &= 0 \\ x \quad\quad - z &= 0 \\ x + y - z &= 0 \end{aligned}$$

APPLICATIONS

Some of the following exercises are similar—but not identical—to exercises and examples in Chapter 2. Use matrix inverses to find the solutions.

51. Nutrition A 4-ounce serving of Campbell's Pork & Beans contains 5 grams of protein and 21 grams of carbohydrates. A typical slice of "lite" rye bread contains 4 grams of protein and 12 grams of carbohydrates.

a. I'm planning a meal of "beans-on-toast," and I want it to supply 20 grams of protein and 80 grams of carbohydrates. How should I prepare my meal?

b. If I require A grams of protein and B grams of carbohydrates, give a formula that tells me how many slices of bread and how many servings of Pork & Beans to use.

SOURCE: According to the label information on a 16-ounce can.

52. Nutrition According to the nutritional information on a package of Honey Nut Cheerios® cereal, each 1-ounce serving of Cheerios contains 3 grams protein and 24 grams carbohydrates.[5] Each $\frac{1}{2}$-cup serving of enriched skim milk contains 4 grams protein and 6 grams carbohydrates.

a. I'm planning a meal of cereal and milk, and I want it to supply 26 grams of protein and 78 grams of carbohydrates. How should I prepare my meal?

b. If I require A grams of protein and B grams of carbohydrates, give a formula that tells me how many servings of milk and Cheerios to use.

53. Resource Allocation You manage an ice cream factory that makes three flavors: Creamy Vanilla, Continental Mocha, and Succulent Strawberry. Into each batch of Creamy Vanilla go 2 eggs, 1 cup of milk, and 2 cups of cream. Into each batch of Continental Mocha go 1 egg, 1 cup of milk, and 2 cups of cream. Into each batch of Succulent Strawberry go 1 egg, 2 cups of milk, and 1 cup of cream. Your stocks of eggs, milk, and cream vary from day to day. How many batches of each flavor should you make in order to use up all of your ingredients if you have the following amounts in stock?

a. 350 eggs, 350 cups of milk, and 400 cups of cream

b. 400 eggs, 500 cups of milk, and 400 cups of cream

c. A eggs, B cups of milk, and C cups of cream

54. Resource Allocation Arctic Juice Company makes three juice blends: PineOrange, using 2 quarts of pineapple juice and 2 quarts of orange juice per gallon; PineKiwi, using 3 quarts of pineapple juice and 1 quart of kiwi juice per gallon; and OrangeKiwi, using 3 quarts of orange juice and 1 quart of kiwi juice per gallon. The amount of each kind of juice the company has on hand varies from day to day. How many gallons of each blend can it make on a day with the following stocks?

a. 800 quarts of pineapple juice, 650 quarts of orange juice, 350 quarts of kiwi juice

b. 650 quarts of pineapple juice, 800 quarts of orange juice, 350 quarts of kiwi juice

c. A quarts of pineapple juice, B quarts of orange juice, C quarts of kiwi juice

Investing in Municipal Bonds Exercises 55 and 56 are based on the following data on three tax-exempt municipal bond funds:

	2001 Yield (%)
PNF (Pimco N.Y.)	6
CMBFX (Columbia Ore.)	5
SFCOX (Safeco)	7

Yields are rounded. SOURCES: www.pimcofunds.com and www.paydenfunds.com, February 22, 2002.

55. You invested a total of \$9000 in the three funds at the beginning of 2001, including an equal amount in CMBFX and SFCOX. Your interest for the year from the first two stocks amounted to \$400. How much did you invest in each of the three funds?

56. You invested a total of \$6000 in the three funds at the beginning of 2001, including an equal amount in PNF and CMBFX. You earned a total of \$360 in interest for the year. How much did you invest in each of the three funds?

Investing in Stocks Exercises 57 and 58 are based the following data on three computer stocks:

	Price (\$)	Dividend Yield (%)
IBM	100	0.5
HWP (Hewlett-Packard)	20	1.50
DELL	25	0

Stocks were trading at or near the given prices in February 2002. Earnings per share are rounded, based on values on February 23, 2002. SOURCE: http://money.excite.com, February 23, 2002.

57. You invested a total of \$12,400 in IBM, Hewlett-Packard, and Dell shares and expected to earn \$56 in annual dividends. If you purchased a total of 200 shares, how many shares of each stock did you purchase?

58. You invested a total of \$8250 in IBM, Hewlett-Packard, and Dell shares and expected to earn \$55 in annual dividends. If you purchased a total of 200 shares, how many shares of each stock did you purchase?

T In Exercises 59–62, use technology to solve.

59. Population Movement In 1999 the U.S. population, broken down by regions, was 53.9 million in the Northeast, 64.3 million in the Midwest, 100.0 million in the South, and 63.3 million in the West. The matrix P below shows the population movement during the period 1999–2000. (Thus, 98.86% of

[5]Actually, it is 23 grams carbohydrates. We made it 24 grams to simplify the calculation.

the population in the Northeast stayed there, while 0.15% of the population in the Northeast moved to the Midwest, and so on.)

$$P = \begin{array}{r} \\ \\ \textbf{From NE} \\ \textbf{From MW} \\ \textbf{From S} \\ \textbf{From W} \end{array} \begin{array}{cccc} \textbf{To} & \textbf{To} & \textbf{To} & \textbf{To} \\ \textbf{NE} & \textbf{MW} & \textbf{S} & \textbf{W} \\ \end{array}$$

$P =$	To NE	To MW	To S	To W
From NE	0.9886	0.0015	0.0075	0.0024
From MW	0.0011	0.9900	0.0057	0.0032
From S	0.0018	0.0042	0.9897	0.0043
From W	0.0017	0.0035	0.0077	0.9871

Set up the 1999 population figures as a row vector. Assuming that these percentages also describe the population movements from 1998 to 1999, show how matrix inversion and multiplication allow you to compute the population in each region in 1998. (Round all answers to the nearest 0.1 million.)

This exercise ignores migration into or out of the country. The internal migration figures and 2000 population figures are accurate. SOURCE: U.S. Census Bureau, Current Population Survey, March 2000, http://www .census.gov/.

60. Population Movement Assume that the percentages given in Exercise 59 also describe the population movements from 1997 to 1998. Show how matrix inversion and multiplication allow you to compute the population in each region in 1997. (Round all answers to the nearest 0.1 million.)

61. Rotations If a point (x, y) in the plane is rotated counterclockwise through an angle of 45°, its new coordinates (x', y') are given by

$$\begin{bmatrix} x' \\ y' \end{bmatrix} = R\begin{bmatrix} x \\ y \end{bmatrix}$$

where R is the 2×2 matrix $\begin{bmatrix} a & -a \\ a & a \end{bmatrix}$ and $a = \sqrt{1/2} \approx 0.7071$.

a. If the point (2, 3) is rotated counterclockwise through an angle of 45°, what are its (approximate) new coordinates?

b. Multiplication by what matrix would result in a counterclockwise rotation of 90°? Of 135°? (Express the matrices in terms of R. *Hint:* Think of a rotation through 90° as two successive rotations through 45°.)

c. Multiplication by what matrix would result in a *clockwise* rotation of 45°?

62. Rotations If a point (x, y) in the plane is rotated counterclockwise through an angle of 60°, its new coordinates are given by

$$\begin{bmatrix} x' \\ y' \end{bmatrix} = S\begin{bmatrix} x \\ y \end{bmatrix}$$

where S is the 2×2 matrix $\begin{bmatrix} a & -b \\ b & a \end{bmatrix}$, $a = 1/2$, and $b = \sqrt{3/4} \approx 0.8660$.

a. If the point (2, 3) is rotated counterclockwise through an angle of 60°, what are its (approximate) new coordinates?

b. Referring to Exercise 61, multiplication by what matrix would result in a counterclockwise rotation of 105°? (Express the matrix in terms of S and the matrix R from Exercise 61. *Hint:* Think of a rotation through 105° as a rotation through 60° followed by one through 45°.)

c. Multiplication by what matrix would result in a *clockwise* rotation of 60°?

Encryption Matrices are commonly used to encrypt data. Here is a simple form such an encryption can take. First, we represent each letter in the alphabet by a number. For example, if we take <space> = 0, A = 1, B = 2, and so on,

"ABORT MISSION" becomes

[1 2 15 18 20 0 13 9 19 19 9 15 14]

To encrypt this coded phrase, we use an invertible matrix of any size with integer entries. For instance, let's take A to be the 2×2 matrix $\begin{bmatrix} 1 & 2 \\ 3 & 4 \end{bmatrix}$. We can first arrange the coded sequence of numbers in the form of a matrix with two rows (using zero in the last place if we have an odd number of characters) and then multiply on the left by A:

$$\text{encrypted matrix} = \begin{bmatrix} 1 & 2 \\ 3 & 4 \end{bmatrix}\begin{bmatrix} 1 & 15 & 20 & 13 & 19 & 9 & 14 \\ 2 & 18 & 0 & 9 & 19 & 15 & 0 \end{bmatrix}$$

$$= \begin{bmatrix} 5 & 51 & 20 & 31 & 57 & 39 & 14 \\ 11 & 117 & 60 & 75 & 133 & 87 & 42 \end{bmatrix}$$

which we can also write as

[5 11 51 117 20 60 31 75 57 133 39 87 14 42]

To decipher the encoded message, multiply the encrypted matrix by A^{-1}.

63. Use the matrix A to encode the phrase "GO TO PLAN B."

64. Use the matrix A to encode the phrase "ABANDON SHIP."

65. Decode the following message, which was encrypted using the matrix A:

[33 69 54 126 11 27 20 60 29 59 65 149 41 87]

66. Decode the following message, which was encrypted using the matrix A:

[59 141 43 101 7 21 29 59 65 149 41 87]

COMMUNICATION AND REASONING EXERCISES

67. If A and B are square matrices with $AB = I$ and $BA = I$, then

(A) B is the inverse of A.

(B) A and B must be equal.

(C) A and B must both be singular.

(D) At least one of A and B is singular.

68. If A is a square matrix with $A^3 = I$, then

(A) A must be the identity matrix.

(B) A is invertible.

(C) A is singular.

(D) A is both invertible and singular.

69. What can you say about the inverse of a 2×2 matrix of the form $\begin{bmatrix} a & b \\ a & b \end{bmatrix}$?

70. If you think of numbers as 1×1 matrices, which numbers are invertible 1×1 matrices?

71. Use matrix multiplication to check that the inverse of a general 2×2 matrix is given by

$$\begin{bmatrix} a & b \\ c & d \end{bmatrix}^{-1} = \frac{1}{ad - bc} \begin{bmatrix} d & -b \\ -c & a \end{bmatrix} \qquad (ad - bc \neq 0)$$

72. Derive the formula in Exercise 71, using row reduction. (Assume that $ad - bc \neq 0$.)

73. If a square matrix A row-reduces to a matrix with a row of zeros, can it be invertible? If so, give an example of such a matrix; and if not, say why not.

74. If a square matrix A row-reduces to the identity matrix, must it be invertible? If so, say why; and if not, give an example of such a (singular) matrix.

75. Your friend has two square matrices A and B, neither of them the zero matrix, with the property that AB is the zero matrix. You immediately tell him that neither A nor B can possibly be invertible. How can you be so sure?

76. A **diagonal** matrix D has the following form:

$$D = \begin{bmatrix} d_1 & 0 & 0 & \dots & 0 \\ 0 & d_2 & 0 & \dots & 0 \\ 0 & 0 & d_3 & \dots & 0 \\ \vdots & \vdots & \vdots & \ddots & \vdots \\ 0 & 0 & 0 & \dots & d_n \end{bmatrix}$$

When is D singular?

77. If A and B are invertible, check that $B^{-1}A^{-1}$ is the inverse of AB.

78. Solve the matrix equation $A(B+CX) = D$ for X. (You may assume that all the matrices are square and invertible.)

3.4 Input–Output Models

In this section we look at input–output models, an application of matrix algebra developed by Wassily Leontief in the middle of the twentieth century. He won the Nobel Prize in economics in 1973 for this work. The application involves analyzing national and regional economies by looking at how various parts of the economy interrelate. We'll work out some of the details by looking at a simple scenario.

First, we can think of the economy of a country or a region as being composed of various **sectors,** or groups, of one or more industries. Typical sectors are the manufacturing sector, the utilities sector, and the agricultural sector. To introduce the basic concepts, we consider two specific sectors: the coal-mining sector (sector 1) and the electric utilities sector (sector 2). Both produce a commodity: The coal-mining sector produces coal, and the electric utilities sector produces electricity. We measure these products by their dollar value. By **1 unit** of a product, we mean $1 worth of that product.

Here is the scenario:

1. To produce 1 unit ($1 worth) of coal, assume that the coal-mining sector uses 50¢ worth of coal (to power mining machinery, say) and 10¢ worth of electricity.
2. To produce 1 unit ($1 worth) of electricity, assume that the electric utilities sector uses 25¢ worth of coal and 25¢ worth of electricity.

These are *internal* usage figures. In addition to this, assume that there is an *external* demand (from the rest of the economy) of 7000 units ($7000 worth) of coal and 14,000 units ($14,000 worth) of electricity over a specific time period (1 year, say). Our basic question is, How much should each of the two sectors supply in order to meet both internal and external demand?

The key to answering this question is to set up equations of the form

total supply = total demand

The unknowns, the values we are seeking, are

x_1 = total supply (in units) from sector 1 (coal)

x_2 = total supply (in units) from sector 2 (electricity)

Our equations then take the following form:

total supply from sector 1 = total demand for sector 1 products

$$\underset{\substack{\uparrow \\ \text{Coal required by sector 1}}}{x_1 = 0.50x_1} \quad + \quad \underset{\substack{\uparrow \\ \text{Coal required by sector 2}}}{0.25x_2} \quad + \quad \underset{\substack{\uparrow \\ \text{External demand for coal}}}{7000}$$

total supply from sector 2 = total demand for sector 2 products

$$\underset{\substack{\uparrow \\ \text{Electricity required by sector 1}}}{x_2 = 0.10x_1} \quad + \quad \underset{\substack{\uparrow \\ \text{Electricity required by sector 2}}}{0.25x_2} \quad + \quad \underset{\substack{\uparrow \\ \text{External demand for electricity}}}{14{,}000}$$

This is a system of two linear equations in two unknowns:

$$x_1 = 0.50x_1 + 0.25x_2 + 7000$$

$$x_2 = 0.10x_1 + 0.25x_2 + 14{,}000$$

We can rewrite this system of equations in matrix form as follows:

$$\underbrace{\begin{bmatrix} x_1 \\ x_2 \end{bmatrix}}_{\text{Production}} = \underbrace{\begin{bmatrix} 0.50 & 0.25 \\ 0.10 & 0.25 \end{bmatrix} \begin{bmatrix} x_1 \\ x_2 \end{bmatrix}}_{\text{Internal demand}} + \underbrace{\begin{bmatrix} 7{,}000 \\ 14{,}000 \end{bmatrix}}_{\text{External demand}}$$

In symbols,

$$X = AX + D$$

Here,

$$X = \begin{bmatrix} x_1 \\ x_2 \end{bmatrix}$$

is called the **production vector.** Its entries are the amounts produced by the two sectors. The matrix

$$D = \begin{bmatrix} 7{,}000 \\ 14{,}000 \end{bmatrix}$$

is called the **external demand** vector, and

$$A = \begin{bmatrix} 0.50 & 0.25 \\ 0.10 & 0.25 \end{bmatrix}$$

is called the **technology matrix.** The entries of the technology matrix have the following meanings:

a_{11} = units of sector 1 needed to produce 1 unit of sector 1

a_{12} = units of sector 1 needed to produce 1 unit of sector 2

a_{21} = units of sector 2 needed to produce 1 unit of sector 1

a_{22} = units of sector 2 needed to produce 1 unit of sector 2

You can remember this order by the slogan "In the side, out the top."

Now that we have the matrix equation

$$X = AX + D$$

we can solve it as follows. First, subtract AX from both sides:

$$X - AX = D$$

Since $X = IX$, where I is the 2×2 identity matrix, we can rewrite this as

$$IX - AX = D$$

Now factor out X:

$$(I - A)X = D$$

If we multiply both sides by the inverse of $(I - A)$ we get the solution

$$X = (I - A)^{-1}D$$

Setting Up and Solving an Input–Output Problem

The ijth entry of the **technology matrix** A is

a_{ij} = units of sector i needed to produce 1 unit of sector j
("In the side, out the top")

To meet an **external demand** of D, the economy must produce X, where X is the **production vector** and satisfies the matrix equation

$$X = AX + D$$

We can solve for the production vector using

$$X = (I - A)^{-1}D$$ Provided that $(I - A)$ is invertible

Quick Example

In the preceding scenario,

$$A = \begin{bmatrix} 0.50 & 0.25 \\ 0.10 & 0.25 \end{bmatrix} \qquad X = \begin{bmatrix} x_1 \\ x_2 \end{bmatrix} \qquad D = \begin{bmatrix} 7{,}000 \\ 14{,}000 \end{bmatrix}$$

The solution is

$$X = (I - A)^{-1}D$$

$$\begin{bmatrix} x_1 \\ x_2 \end{bmatrix} = \left(\begin{bmatrix} 1 & 0 \\ 0 & 1 \end{bmatrix} - \begin{bmatrix} 0.50 & 0.25 \\ 0.10 & 0.25 \end{bmatrix}\right)^{-1} \begin{bmatrix} 7{,}000 \\ 14{,}000 \end{bmatrix}$$

$$= \begin{bmatrix} 0.50 & -0.25 \\ -0.10 & 0.75 \end{bmatrix}^{-1} \begin{bmatrix} 7{,}000 \\ 14{,}000 \end{bmatrix}$$ Calculate $I - A$.

$$= \begin{bmatrix} \frac{15}{7} & \frac{5}{7} \\ \frac{2}{7} & \frac{10}{7} \end{bmatrix} \begin{bmatrix} 7{,}000 \\ 14{,}000 \end{bmatrix}$$ Calculate $(I - A)^{-1}$.

$$= \begin{bmatrix} 25{,}000 \\ 22{,}000 \end{bmatrix}$$

Interpreting the Solution

To meet the demand, the economy must produce \$25,000 worth of coal and \$22,000 worth of electricity.

The next example uses actual data from the U.S. economy (we have rounded the figures to make the computations less complicated).[6] It is rare to find input–output data already packaged for you as a technology matrix. Instead, the data commonly found in statistical sources comes in the form of *input–output tables,* from which we will have to construct the technology matrix.

Example 1 • Petroleum and Natural Gas

Consider two sectors of the U.S. economy: crude petroleum and natural gas (*crude*) and petroleum-refining and related industries (*refining*). According to government figures, in 1998 the crude sector used \$27,000 million worth of its own products and \$750 million worth of the products of the refining sector to produce \$87,000 million worth of goods (crude oil and natural gas).[7] The refining sector in the same year used \$59,000 million worth of the products of the crude sector and \$15,000 million worth of its own products to produce \$140,000 million worth of goods (refined oil and the like). What was the technology matrix for these two sectors? What was left over from each of these sectors for use by other parts of the economy or for export?

Solution First, for convenience, we record the given data in the form of a table, called the **input–output table** (all figures are in millions of dollars):

	To	**Crude**	**Refining**
From	**Crude**	27,000	59,000
	Refining	750	15,000
	Total Output	87,000	140,000

The entries in the top portion are arranged in the same way as those of the technology matrix: The *ij*th entry represents the number of units of sector i that went to sector j. Thus, for instance, the 59,000 million entry in the 1, 2 position represents the number of units of sector 1, crude, that were used by sector 2, refining ("In the side, out the top").

We now construct the technology matrix. The technology matrix has entries a_{ij} = units of sector i used to produce *1* unit of sector j. Thus,

a_{11} = units of crude to produce 1 unit of crude. We are told that 27,000 million units of crude were used to produce 87,000 million units of crude. Thus, to produce 1 unit of crude, $27{,}000/87{,}000 \approx 0.31$ unit of crude were used, and so $a_{11} \approx 0.31$. (We have rounded this value to two significant digits; further digits are not reliable due to rounding of the original data.)

a_{12} = units of crude to produce 1 unit of refined:
$$a_{12} = 59{,}000/140{,}000 \approx 0.42$$

a_{21} = units of refined to produce 1 unit of crude:
$$a_{21} = 750/87{,}000 \approx 0.0086$$

a_{22} = units of refined to produce 1 unit of refined:
$$a_{22} = 15{,}000/140{,}000 \approx 0.11$$

[6]In Example 3 and some of the exercises intended for the use of technology, we will give you the original data in all its gory detail!

[7]The data have been rounded to two significant digits. SOURCE: U.S. Department of Commerce, Survey of Current Business, December 2001. The Survey of Current Business and the input–output tables themselves are available at the Web site of the Department of Commerce's Bureau of Economic Analysis, http://www.bea.gov/.

This gives the technology matrix

$$A = \begin{bmatrix} 0.31 & 0.42 \\ 0.0086 & 0.11 \end{bmatrix} \qquad \text{Technology matrix}$$

In short, *we obtained the technology matrix from the input–output table by dividing the sector 1 column by the sector 1 total, and the sector 2 column by the sector 2 total.*

Now we also know the total output from each sector, so *we have already been given the production vector:*

$$X = \begin{bmatrix} 87{,}000 \\ 140{,}000 \end{bmatrix} \qquad \text{Production vector}$$

What we are asked for is the external demand vector D, the amount available for the outside economy. To find D, we use the equation

$$X = AX + D \qquad \text{Relationship of } X, A, \text{ and } D$$

where, this time, we are given A and X, and must solve for D. Solving for D gives

$$\begin{aligned} D &= X - AX \\ &= \begin{bmatrix} 87{,}000 \\ 140{,}000 \end{bmatrix} - \begin{bmatrix} 0.31 & 0.42 \\ 0.0086 & 0.11 \end{bmatrix}\begin{bmatrix} 87{,}000 \\ 140{,}000 \end{bmatrix} \\ &\approx \begin{bmatrix} 87{,}000 \\ 140{,}000 \end{bmatrix} - \begin{bmatrix} 86{,}000 \\ 16{,}000 \end{bmatrix} = \begin{bmatrix} 1{,}000 \\ 124{,}000 \end{bmatrix} \qquad \text{We rounded to two digits. (Why?)} \end{aligned}$$

The first number, \$1000 million, is the amount produced by the crude sector that is available to be used by other parts of the economy or to be exported. (In fact, since something has to happen to all that crude petroleum and natural gas, this is the amount actually used or exported, where use can include stockpiling.) The second number, \$124,000, represents the amount produced by the refining sector that is available to be used by other parts of the economy or to be exported.

Before we go on . . . We could have calculated D more simply from the input–output table. The internal use of units from the crude sector was the sum of the outputs from that sector:

$$27{,}000 + 59{,}000 = 86{,}000$$

Since 87,000 units were actually produced by the sector, that left a surplus of 87,000 − 86,000 = 1000 units for export. We could compute the surplus from the refining sector similarly. (The two calculations actually come out slightly different, because we rounded the intermediate results.) The calculation in Example 2 cannot be done as trivially, however.

Input-Output Table

The ijth entry in the top portion of the input–output table is the number of units that go from sector i to sector j. The "Total Outputs" are the total numbers of units produced by each sector. We obtain the technology matrix from the input–output table by dividing the sector 1 column by the sector 1 total, the sector 2 column by the sector 2 total, and so on.

Quick Example

Input–output table:

To		**Skateboards**	**Wood**
From	**Skateboards**	20,000*	0
	Wood	100,000	500,000
	Total Output	200,000	5,000,000

*The production of skateboards required skateboards due to the fact that skateboard workers tend to commute to work on (what else?) skateboards!

Technology Matrix:

$$A = \begin{bmatrix} \frac{20,000}{200,000} & \frac{0}{5,000,000} \\ \frac{100,000}{200,000} & \frac{500,000}{5,000,000} \end{bmatrix} = \begin{bmatrix} 0.1 & 0 \\ 0.5 & 0.1 \end{bmatrix}$$

Example 2 • Rising Demand

Suppose external demand for refined petroleum rises to \$200,000 million, but the demand for crude remains \$1000 million (as in Example 1). How do the production levels of the two sectors considered in Example 1 have to change?

Solution We are being told that now

$$D = \begin{bmatrix} 1,000 \\ 200,000 \end{bmatrix}$$

and we are asked to find X. Remember that we can calculate X from the formula

$$X = (I - A)^{-1}D$$

Now

$$I - A = \begin{bmatrix} 1 & 0 \\ 0 & 1 \end{bmatrix} - \begin{bmatrix} 0.31 & 0.42 \\ 0.0086 & 0.11 \end{bmatrix} = \begin{bmatrix} 0.69 & -0.42 \\ -0.0086 & 0.89 \end{bmatrix}$$

We take the inverse using our favorite form of technology and find that, to four significant digits,[8]

$$(I - A)^{-1} \approx \begin{bmatrix} 1.458 & 0.6880 \\ 0.01409 & 1.130 \end{bmatrix}$$

Now we can compute X:

$$X = (I - A)^{-1}D \approx \begin{bmatrix} 1.458 & 0.6880 \\ 0.01409 & 1.130 \end{bmatrix} \begin{bmatrix} 1,000 \\ 200,000 \end{bmatrix} = \begin{bmatrix} 140,000 \\ 230,000 \end{bmatrix}$$

[8] A is accurate to two digits, so we should use more than two significant digits in intermediate calculations so that additional accuracy is not lost. We must of course round the answer to two digits.

(As in Example 1, we have rounded all entries in the answer to two significant digits.) Comparing this vector to the production vector used in Example 1, we see that production in the crude sector has to increase from $87,000 million to $140,000 million, while production in the refining sector has to increase from $140,000 million to $230,000 million.

Before we go on . . . Using the matrix $(I - A)^{-1}$, we have a slightly different way of solving this problem. We are asking for the effect on production of a *change* in the final demand of 0 for crude and $200,000 – 124,000 = $76,000 million for refined products. If we multiply $(I - A)^{-1}$ by the matrix representing this *change*, we obtain

$$\begin{bmatrix}1.458 & 0.6880\\ 0.01409 & 1.130\end{bmatrix}\begin{bmatrix}0\\ 76{,}000\end{bmatrix} \approx \begin{bmatrix}53{,}000\\ 90{,}000\end{bmatrix}$$

$(I - A)^{-1} \times$ change in demand = change in production

We see the changes required in production: an increase of $53,000 million in the crude sector and an increase of $90,000 million in the refining sector.

Notice that the increase in external demand for the products of the refining sector requires the crude sector to also increase production, even though there is no increase in the *external* demand for its products. The reason is that, in order to increase production, the refining sector needs to use more crude oil, so the *internal* demand for crude oil goes up. The inverse matrix $(I - A)^{-1}$ takes these **indirect effects** into account in a nice way.

By replacing the $76,000 by $1 in the computation we just did, we see that a $1 increase in external demand for refined products will require an increase in production of $0.6880 in the crude sector, as well as an increase in production of $1.130 in the refining sector. This is how we interpret the entries in $(I - A)^{-1}$, and this is why it is useful to look at this matrix inverse rather than just solve $(I - A)X = D$ for X using, say, Gauss–Jordan reduction. Looking at $(I - A)^{-1}$, we can also find the effects of an increase of $1 in external demand for crude: an increase in production of $1.458 in the crude sector and an increase of $0.01409 in the refining sector.

Here are some questions to think about: Why are the diagonal entries of $(I - A)^{-1}$ (slightly) larger than 1? Why is the entry in the lower left so small compared to the others?

Interpreting $(I - A)^{-1}$: Indirect Effects

If A is the technology matrix for an economy, then the ijth entry of $(I - A)^{-1}$ is the change in the number of units sector i must produce in order to meet a 1-unit increase in external demand for sector j products. To meet a rising external demand, the necessary change in production for each sector is given by

$$\text{change in production} = (I - A)^{-1}D^{+}$$

where D^{+} is the change in external demand.

Quick Example

Let sector 1 be skateboards and sector 2 be wood and assume that

$$(I - A)^{-1} = \begin{bmatrix}1.1 & 0\\ 0.6 & 1.1\end{bmatrix}$$

Then

$a_{11} = 1.1 =$ number of additional units of skateboards that must be produced to meet a 1-unit increase in the demand for skateboards (Why is this number larger than 1?)

$a_{12} = 0 \quad =$ number of additional units of skateboards that must be produced to meet a 1-unit increase in the demand for wood (Why is this number 0?)

$a_{21} = 0.6 =$ number of additional units of wood that must be produced to meet a 1-unit increase in the demand for skateboards

$a_{22} = 1.1 =$ number of additional units of wood that must be produced to meet a 1-unit increase in the demand for wood

To meet an increase in external demand of 100 skateboards and 400 units of wood, the necessary change in production is

$$(I - A)^{-1}D^{+} = \begin{bmatrix} 1.1 & 0 \\ 0.6 & 1.1 \end{bmatrix} \begin{bmatrix} 100 \\ 400 \end{bmatrix} = \begin{bmatrix} 110 \\ 500 \end{bmatrix}$$

so 110 additional skateboards and 500 additional units of wood will need to be produced.

In the preceding examples, we used only two sectors of the economy. The data for Examples 1 and 2 were taken from an input–output table published by the U.S. Department of Commerce, in which the whole U.S. economy was broken down into 85 sectors. This in turn was a simplified version of a model in which the economy was broken into about 500 sectors. Obviously, computers are required to make a realistic input–output analysis possible. Many governments collect and publish input–output data as part of their national planning. The United Nations collects these data and publishes collections of national statistics. The United Nations also has a useful set of links to government statistics at the following Web site:

http://www.un.org/Depts/unsd/sd_natstat.htm

Example 3 • Kenya Economy

Consider four sectors of the economy of Kenya: (1) the traditional economy, (2) agriculture, (3) manufacture of metal products and machinery, and (4) wholesale and retail trade. The input–output table for these four sectors for 1976 looks like this (all numbers are 1000s of Kenya pounds, K£):

To	**1**	**2**	**3**	**4**
From **1**	8,600	0	0	0
2	0	19,847	24	0
3	1,463	530	15,315	660
4	814	8,529	5,773	2,888
Total Output	87,160	531,131	112,780	178,911

SOURCE: Central Bureau of Statistics of the Ministry of Economic Planning and Community Affairs, Kenya, *Input–Output Tables for Kenya 1976.*

Suppose external demand for agriculture increased by K£50,000,000 and external demand for metal products and machinery increased by K£10,000,000. How would production in these four sectors have to change to meet this rising demand?

Solution To find the change in production necessary to meet the rising demand, we need to use the formula

$$\text{change in production} = (I - A)^{-1}D^{+}$$

where A is the technology matrix and D^{+} is the change in demand:

$$D^{+} = \begin{bmatrix} 0 \\ 50{,}000 \\ 10{,}000 \\ 0 \end{bmatrix}$$

We can use technology to do the number crunching for us.

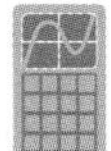

Graphing Calculator

On the TI-83 enter the entries of A (using MATRX `EDIT`) as the quotients (column entry/column total). Rounded to five decimal places, the matrix [A] is

$$A = \begin{bmatrix} 0.098\,67 & 0 & 0 & 0 \\ 0 & 0.037\,37 & 0.000\,21 & 0 \\ 0.016\,79 & 0.001\,00 & 0.135\,80 & 0.003\,69 \\ 0.009\,34 & 0.016\,06 & 0.051\,19 & 0.016\,14 \end{bmatrix}$$

Enter the matrix D^{+} as `[D]`. You can then compute the change in production with the following formula:

$$\texttt{(Identity(4)}-\texttt{[A])}^{-1}\texttt{[D]}$$

Rounded to the nearest K£1000, the answer is

$$\text{change in production} = (I - A)^{-1}D^{+} = \begin{bmatrix} 0 \\ 51{,}943 \\ 11{,}638 \\ 1{,}453 \end{bmatrix}$$

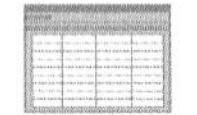

Excel

First, enter the input–output table in the spreadsheet. Next, compute the technology matrix by dividing each column by the column total. Also insert the identity matrix I in preparation for the next step.

	A	B	C	D	E	F	G	H	I
1	8600	0	0	0					
2	0	19847	24	0					
3	1463	530	15315	660					
4	814	8529	5773	2888					
5	87160	531131	112780	178911	Totals				
6		Technology Matrix					Identity Matrix		
7	=A1/A$5					1	0	0	0
8						0	1	0	0
9						0	0	1	0
10						0	0	0	1

Now, to see how each sector reacts to rising external demand, we must calculate the inverse matrix $(I - A)^{-1}$:

	A	B	C	D	E	F	G	H	I
1	8600	0	0	0					
2	0	19847	24	0					
3	1463	530	15315	660					
4	814	8529	5773	2888					
5	87160	531131	112780	178911	Totals				
6		Technology	Matrix				Identity	Matrix	
7	0.09866911	0	0	0		1	0	0	0
8	0	0.03736743	0.0002128	0		0	1	0	0
9	0.01678522	0.00099787	0.13579535	0.00368899		0	0	1	0
10	0.00933915	0.01605819	0.05118815	0.0161421		0	0	0	1
11		$(I - A)$					$(I-A)^{-1}$		
12	=F7:I10-A7:D10					=MINVERSE(A12:D15)			
13									
14									
15									

As in Example 2 in Section 3.3, we calculate the inverse $(I - A)^{-1}$ as follows: With the block highlighted as shown, type the formula `=MINVERSE(A12:D15)` in the formula bar and press Control + Shift + Enter. The inverse will then appear in the region you highlighted.

$$(I - A)^{-1} = \begin{bmatrix} 1.109\,47 & 0 & 0 & 0 \\ 0.000\,005 & 1.038\,82 & 0.000\,256 & 0.000\,001 \\ 0.021\,599 & 0.001\,272 & 1.157\,39 & 0.004\,340 \\ 0.011\,655 & 0.017\,021 & 0.060\,221 & 1.016\,63 \end{bmatrix}$$

To compute $(I - A)^{-1}D^{+}$, enter D^{+} as a column and use the MMULT operation (see Example 3 in Section 3.2). The answer is given above in the graphing calculator discussion.

Note This is one of the beauties of spreadsheet programs: Once you are done with the calculation, you can use the spreadsheet as a template for any 4×4 input–output table by just changing the entries of the input–output matrix and/or D^{+}. The rest of the computation will then be done automatically as the spreadsheet is updated. In other words, you can use it do your homework! You can download a version of this spreadsheet from

Web site → Excel Tutorials → 3.4 Input–Output Models

Web Site

The matrix algebra tool at the Web site is simpler to use than a graphing calculator: Follow the path

Web site → Online Utilities → Matrix Algebra Tool

enter A as the quotients (column entry/column total), enter `D`, and then use the format

```
(I-A)^(-1)*D
```

Interpretation of Results

Looking at $(I - A)^{-1}D^{+}$, we see that the changes in external demand will leave the traditional economy unaffected, production in agriculture will rise by K£51,943,000, production in the manufacture of metal products and machinery will rise by K£11,638,000, and activity in wholesale and retail trade will rise by K£1,453,000.

Before we go on . . . Can you see why the traditional economy was unaffected? Although it takes inputs from other parts of the economy, it is not itself an input to any other part. In other words, there is no intermediate demand for the products of the traditional economy coming from any other part of the economy, and so an increase in production in any other sector of the economy requires no increase from the traditional economy. On the other hand, the wholesale and retail trade sector does provide input to the agriculture and manufacturing sectors, so increases in those sectors do require an increase in the trade sector.

One more point: Notice how small the off-diagonal entries in $(I - A)^{-1}$ are. This says that increases in each sector have relatively small effects on the other sectors. We say that these sectors are **loosely coupled.** Regional economies, where many products are destined to be shipped out to the rest of the country, tend to show this phenomenon even more strongly. Notice in Example 2 that those two sectors are **strongly coupled,** since a rise in demand for refined products requires a comparable rise in the production of crude.

3.4 EXERCISES

1. Let A be the technology matrix $A = \begin{bmatrix} 0.2 & 0.05 \\ 0.8 & 0.01 \end{bmatrix}$, where sector 1 is paper and sector 2 is wood. Fill in the missing quantities.

a. _____ units of wood are needed to produce 1 unit of paper.

b. _____ units of paper are used in the production of 1 unit of paper.

c. The production of each unit of wood requires the use of _____ units of paper.

2. Let A be the technology matrix $A = \begin{bmatrix} 0.01 & 0.001 \\ 0.2 & 0.004 \end{bmatrix}$, where sector 1 is computer chips and sector 2 is silicon. Fill in the missing quantities.

a. _____ units of silicon are required in the production of 1 unit of silicon.

b. _____ units of computer chips are used in the production of 1 unit of silicon.

c. The production of each unit of computer chips requires the use of _____ units of silicon.

3. Each unit of television news requires 0.2 unit of television news and 0.5 unit of radio news. Each unit of radio news requires 0.1 unit of television news and no radio news. With sector 1 as television news and sector 2 as radio news, set up the technology matrix A.

4. Production of 1 unit of cologne requires no cologne and 0.5 unit of perfume. Into 1 unit of perfume go 0.1 unit of cologne and 0.3 unit of perfume. With sector 1 as cologne and sector 2 as perfume, set up the technology matrix A.

In Exercises 5–12, you are given a technology matrix A and an external demand vector D. Find the corresponding production vector X.

5. $A = \begin{bmatrix} 0.5 & 0.4 \\ 0 & 0.5 \end{bmatrix}$, $D = \begin{bmatrix} 10{,}000 \\ 20{,}000 \end{bmatrix}$

6. $A = \begin{bmatrix} 0.5 & 0.4 \\ 0 & 0.5 \end{bmatrix}$, $D = \begin{bmatrix} 20{,}000 \\ 10{,}000 \end{bmatrix}$

7. $A = \begin{bmatrix} 0.1 & 0.4 \\ 0.2 & 0.5 \end{bmatrix}$, $D = \begin{bmatrix} 25{,}000 \\ 15{,}000 \end{bmatrix}$

8. $A = \begin{bmatrix} 0.1 & 0.2 \\ 0.4 & 0.5 \end{bmatrix}$, $D = \begin{bmatrix} 24{,}000 \\ 14{,}000 \end{bmatrix}$

9. $A = \begin{bmatrix} 0.5 & 0.1 & 0 \\ 0 & 0.5 & 0.1 \\ 0 & 0 & 0.5 \end{bmatrix}$, $D = \begin{bmatrix} 1000 \\ 1000 \\ 2000 \end{bmatrix}$

10. $A = \begin{bmatrix} 0.5 & 0.1 & 0 \\ 0.1 & 0.5 & 0.1 \\ 0 & 0 & 0.5 \end{bmatrix}$, $D = \begin{bmatrix} 3000 \\ 3800 \\ 2000 \end{bmatrix}$

11. $A = \begin{bmatrix} 0.2 & 0.2 & 0 \\ 0.2 & 0.4 & 0.2 \\ 0 & 0.2 & 0.2 \end{bmatrix}$, $D = \begin{bmatrix} 16{,}000 \\ 8{,}000 \\ 8{,}000 \end{bmatrix}$

12. $A = \begin{bmatrix} 0.2 & 0.2 & 0.2 \\ 0.2 & 0.4 & 0.2 \\ 0.2 & 0.2 & 0.2 \end{bmatrix}$, $D = \begin{bmatrix} 7{,}000 \\ 14{,}000 \\ 7{,}000 \end{bmatrix}$

13. Given $A = \begin{bmatrix} 0.1 & 0.4 \\ 0.2 & 0.5 \end{bmatrix}$, find the changes in production required to meet an increase in demand of 50 units of sector 1 products and 30 units of sector 2 products.

14. Given $A = \begin{bmatrix} 0.5 & 0.4 \\ 0 & 0.5 \end{bmatrix}$, find the changes in production required to meet an increase in demand of 20 units of sector 1 products and 10 units of sector 2 products.

15. Let $(I - A)^{-1} = \begin{bmatrix} 1.5 & 0.1 & 0 \\ 0.2 & 1.2 & 0.1 \\ 0.1 & 0.7 & 1.6 \end{bmatrix}$ and assume that the external demand for the products in sector 1 increases by 1 unit. By how many units should each sector increase production? What do the columns of the matrix $(I - A)^{-1}$ tell you?

16. Let $(I - A)^{-1} = \begin{bmatrix} 1.5 & 0.1 & 0 \\ 0.1 & 1.1 & 0.1 \\ 0 & 0 & 1.3 \end{bmatrix}$ and assume that the external demand for the products in each of the sectors increases by 1 unit. By how many units should each sector increase production?

In Exercises 17 and 18, obtain the technology matrix from the given input–output table.

17.

To	**A**	**B**	**C**
From **A**	1000	2000	3000
B	0	4000	0
C	0	1000	3000
Total Output	5000	5000	6000

18.

To	**A**	**B**	**C**
From **A**	0	100	300
B	500	400	300
C	0	0	600
Total Output	1000	2000	3000

APPLICATIONS

19. **Campus Food** The two campus cafeterias, the Main Dining Room and Bits & Bytes, typically use each other's food in doing business on campus. One weekend, the input–output table was as follows:

To	**Main DR ($)**	**Bits & Bytes ($)**
From **Main DR**	10,000*	20,000
Bits & Bytes	5,000	0
Total Output	50,000	40,000

*For some reason, the Main Dining Room consumes a lot of its own food!

Given that the demand for food on campus last weekend was $45,000 from the Main Dining Room and $30,000 from Bits & Bytes, how much did the two cafeterias have to produce to meet the demand last weekend?

20. **Plagiarism** Two student groups at Enormous State University, the Choral Society and the Football Club, maintain files of term papers that they write and offer to students for research purposes. Some of these papers they use themselves in generating more papers. To avoid suspicion of plagiarism by faculty members (who seem to have astute memories), each paper is given to students or used by the clubs only once (no copies are kept). The number of papers that were used in the production of new papers last year is shown in the following input–output table:

To	**Choral Society**	**Football Club**
From **Choral Society**	20	10
Football Club	10	30
Total Output	100	200

Given that 270 Choral Society papers and 810 Football Club papers will be used by students outside of these two clubs next year, how many new papers do the two clubs need to write?

T 21. **Communication Equipment** Two sectors of the U.S. economy are (1) audio, video, and communication equipment and (2) electronic components and accessories. In 1998 the input–output table involving these two sectors was as follows (all figures are in millions of dollars):

To	**Equipment**	**Components**
From **Equipment**	6,000	500
Components	24,000	30,000
Total Output	90,000	140,000

The data have been rounded. SOURCE: U.S. Department of Commerce, Survey of Current Business, December 2001.

Determine the production levels necessary in these two sectors to meet an external demand for $80,000 million of communication equipment and $90,000 million of electronic components. Round answers to two significant digits.

T 22. **Wood and Paper** Two sectors of the U.S. economy are (1) lumber and wood products and (2) paper and allied products. In 1998 the input–output table involving these two sectors was as follows (all figures are in millions of dollars):

To	**Wood**	**Paper**
From **Wood**	36,000	7,000
Paper	100	17,000
Total Output	120,000	120,000

The data have been rounded. SOURCE: U.S. Department of Commerce, Survey of Current Business, December 2001.

If external demand for lumber and wood products rises by $10,000 million and external demand for paper and allied products rises by $20,000 million, what increase in output of these two sectors is necessary? Round answers to two significant digits.

23. **Australian Economy** Two sectors of the Australian economy are (1) textiles and (2) clothing and footwear. The 1977 input–output table involving these two sectors results in the following value for $(I - A)^{-1}$:

$$(I - A)^{-1} = \begin{bmatrix} 1.228 & 0.182 \\ 0.006 & 1.1676 \end{bmatrix}$$

Complete the following sentences:

a. ______ additional dollars worth of clothing and footwear must be produced to meet a $1 increase in the demand for textiles.

b. 0.182 additional dollar worth of ______ must be produced to meet a $1 increase in the demand for ______.

SOURCE: Australian Bureau of Statistics, *Australian National Accounts and Input–Output Tables 1978–1979.*

24. Australian Economy Two sectors of the Australian economy are (1) community services and (2) recreation services. The 1978–1979 input–output table involving these two sectors results in the following value for $(I - A)^{-1}$:

$$(I - A)^{-1} = \begin{bmatrix} 1.006\ 6 & 0.005\ 76 \\ 0.004\ 96 & 1.042\ 06 \end{bmatrix}$$

Complete the following sentences:

a. 0.00496 additional dollar worth of ______ must be produced to meet a \$1 increase in the demand for ______.

b. ______ additional dollars worth of community services must be produced to meet a \$1 increase in the demand for community services.

Exercises 25–28 require the use of technology.

25. U.S. Input–Output Table Four sectors of the U.S. economy are (1) livestock and livestock products, (2) other agricultural products, (3) forestry and fishery products, and (4) agricultural, forestry, and fishery services. In 1998 the input–output table involving these four sectors was as follows (all figures are in millions of dollars):

To	1	2	3	4
From **1**	11,937	9	109	855
2	26,649	4,285	0	4,744
3	0	0	439	61
4	5,423	10,952	3,002	216
Total Output	97,795	120,594	14,642	47,473

SOURCE: U.S. Department of Commerce, Survey of Current Business, December 2001.

Determine how these four sectors would react to an increase in demand for livestock (sector 1) of \$1000 million, how they would react to an increase in demand for other agricultural products (sector 2) of \$1000 million, and so on.

26. U.S. Input–Output Table Four sectors of the U.S. economy are (1) motor vehicles, (2) truck and bus bodies, trailers, and motor vehicle parts, (3) aircraft and parts, and (4) other transportation equipment. In 1998 the input-output table involving these four sectors was as follows (all figures in millions of dollars):

To	1	2	3	4
From **1**	75	1,092	0	1,207
2	64,858	13,081	7	1,070
3	0	0	21,782	0
4	0	0	0	1,375
Total Output	230,676	135,108	129,376	44,133

SOURCE: U.S. Department of Commerce, Survey of Current Business, December 2001.

Determine how these four sectors would react to an increase in demand for motor vehicles (sector 1) of \$1000 million, how they would react to an increase in demand for truck and bus bodies (sector 2) of \$1000 million, and so on.

27. Australian Input–Output Table Four sectors of the Australian economy are (1) agriculture, (2) forestry, fishing, and hunting, (3) meat and milk products, and (4) other food products. In 1978–1979 the input–output table involving these four sectors was as follows (all figures are in millions of Australian dollars):

To	1	2	3	4
From **1**	678.4	3.7	3341.5	1023.5
2	15.5	6.9	17.1	124.5
3	47.3	4.3	893.1	145.8
4	312.5	22.1	83.2	693.5
Total Output	9401.3	685.8	6997.3	4818.3

SOURCE: Australian Bureau of Statistics, *Australian National Accounts and Input–Output Tables 1978–1979.*

a. How much additional production by the meat and milk sector is necessary to accommodate a \$100 increase in the demand for agriculture?

b. Which sector requires the most of its own product in order to meet a \$1 increase in external demand for that product?

28. Australian Input–Output Table Four sectors of the Australian economy are (1) petroleum and coal products, (2) nonmetallic mineral products, (3) basic metals and products, and (4) fabricated metal products. In 1978–1979 the input–output table involving these four sectors was as follows (all figures are in millions of Australian dollars):

To	1	2	3	4
From **1**	174.1	30.5	120.3	14.2
2	0	190.1	55.8	12.6
3	2.1	40.2	1418.7	1242.0
4	0.1	7.3	40.4	326.0
Total Output	3278.0	2188.8	6541.7	4065.8

SOURCE: Australian Bureau of Statistics, *Australian National Accounts and Input–Output Tables 1978–1979.*

a. How much additional production by the petroleum and coal products sector is necessary to accommodate a \$1000 increase in the demand for fabricated metal products?

b. Which sector requires the most of the product of some other sector in order to meet a \$1 increase in external demand for that product?

COMMUNICATION AND REASONING EXERCISES

29. What would it mean if the technology matrix A were the zero matrix?

30. Can an external demand be met by an economy whose technology matrix A is the identity matrix? Explain.

31. What would it mean if the total output figure for a particular sector of an input–output table were equal to the sum of the figures in the row for that sector?

32. What would it mean if the total output figure for a particular sector of an input–output table were less than the sum of the figures in the row for that sector?

33. What does it mean if an entry in the matrix $(I - A)^{-1}$ is zero?

34. Why do we expect the diagonal entries in the matrix $(I - A)^{-1}$ to be slightly larger than 1?

35. Why do we expect the off-diagonal entries of $(I - A)^{-1}$ to be less than 1?

36. Why do we expect all the entries of $(I - A)^{-1}$ to be nonnegative?

CASE STUDY

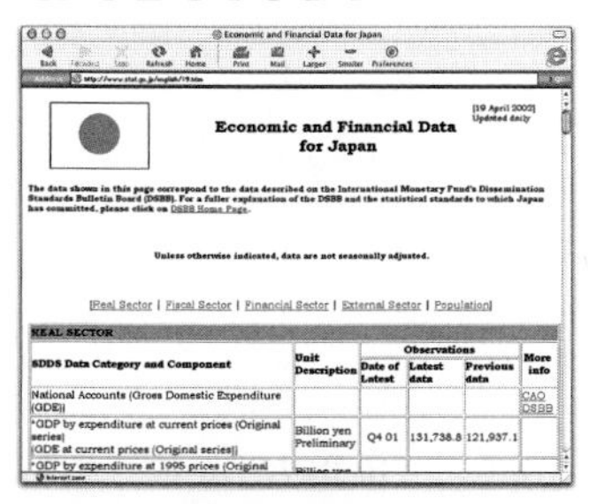

Reproduced from the Statistical Standards Department Web site, division of the Statistics Bureau, Ministry of Public Management, Japan

The Japanese Economy

A senator walks into your cubicle in the Congressional Budget Office. "Look here," she says, "I don't see why the Japanese trade representative is getting so upset with my proposal to cut down on our use of Japanese finance and insurance. He claims that it'll hurt Japan's mining operations. But just look at Japan's input–output table. The finance sector doesn't use any input from the mining sector. How can our cutting back on finance and insurance hurt mining?" Indeed, the senator is right about the input–output table, which you have hanging on your wall. Here is what it looks like (all figures are in 100 million yen):

	1	2	3	4	5	6	7	8	9	10	11	12	13
1	19,221	8	99,417	1,610	0	100	0	1	23	0	20	12,491	0
2	0	41	53,006	8,187	13,189	0	0	0	0	0	7	45	7
3	25,376	959	1,247,342	259,049	14,488	38,247	13,316	1,635	54,426	3,827	26,614	267,704	4,963
4	503	105	13,909	2,242	11,664	5,924	1,340	22,788	4,712	1,591	4,627	11,793	0
5	716	471	59,111	6,203	25,035	11,654	1,940	2,265	8,764	1,811	8,528	46,068	915
6	6,559	291	171,655	61,848	3,146	11,242	2,225	1,066	18,053	762	4,683	78,454	1,123
7	5,303	733	43,394	9,533	7,238	58,662	35,348	32,706	30,879	2,252	824	53,827	9,003
8	42	157	11,353	2,731	2,533	38,416	6,772	4,790	8,308	2,445	498	27,639	735
9	7,273	4,021	93,244	46,994	6,781	53,416	7,046	1,623	52,905	4,151	8,369	38,765	1,416
10	135	76	8,682	4,913	1,118	19,014	6,740	435	3,465	9,167	3,826	37,155	98
11	0	0	0	0	0	0	0	0	0	0	0	0	4,614
12	1,780	688	207,289	69,945	24,464	53,225	37,826	10,267	65,577	19,902	18,595	142,920	3,327
13	1,507	228	23,154	1,788	1,702	5,874	1,456	5,112	2,286	1,196	4,263	11,511	0

SOURCE: Management and Coordination Agency, Government of Japan, *1995 Input–Output Tables for Japan*, March 2000. Obtained from http://www.stat.go.jp/.

The sectors are

1. Agriculture, forestry, and fishery
2. Mining
3. Manufacturing
4. Construction
5. Electric power, gas, and water supply
6. Commerce
7. Finance and insurance
8. Real estate
9. Transport
10. Communication and broadcasting
11. Public administration
12. Services
13. Activities not elsewhere classified

The total output from each sector is given in the following table:

1	2	3	4	5	6	7	8	9	10	11	12	13
158,178	16,595	3,145,585	881,493	264,635	1,023,216	363,346	641,852	501,138	147,628	262,170	1,909,996	55,176

"If I look at just the mining and finance sectors," says the senator, "I'm looking at this input-output table."

	To	**Mining (2)**	**Finance (7)**
From	**Mining (2)**	41	0
	Finance (7)	733	35,348
	Total Output	16,595	363,346

"That gives me

$$A = \begin{bmatrix} 0.0025 & 0 \\ 0.0442 & 0.0973 \end{bmatrix}$$

and so

$$(I - A)^{-1} = \begin{bmatrix} 1.0025 & 0 \\ 0.0491 & 1.1078 \end{bmatrix}$$

That last column tells me that any change in demand for finance will have no effect on demand for mining."

Now you have to explain to the senator a point that we fudged a bit in Section 3.4. What she said assumes that changing the external demand for finance (that is, the demand from outside of these two sectors) will not change the external demand for mining. But that in fact is unlikely to be true. Changing the demand for finance will change the demand for other sectors in the economy directly (the manufacturing sector, for example), which in turn may change the demand for mining. To see these indirect effects properly, you tell the senator that she must look at the whole Japanese economy. The technology matrix A is then

$$\begin{bmatrix}
0.121\,515 & 0.000\,482 & 0.031\,605 & 0.001\,826 & 0 & 0.000\,098 & 0 & 0.000\,002 & 0.000\,046 & 0 & 0.000\,076 & 0.006\,540 & 0 \\
0 & 0.002\,471 & 0.016\,851 & 0.009\,288 & 0.049\,838 & 0 & 0 & 0 & 0 & 0 & 0.000\,027 & 0.000\,024 & 0.000\,127 \\
0.160\,427 & 0.057\,788 & 0.396\,537 & 0.293\,875 & 0.054\,747 & 0.037\,379 & 0.036\,648 & 0.002\,547 & 0.108\,605 & 0.025\,923 & 0.101\,514 & 0.140\,159 & 0.089\,949 \\
0.003\,180 & 0.006\,327 & 0.004\,422 & 0.002\,543 & 0.044\,076 & 0.005\,790 & 0.003\,688 & 0.035\,504 & 0.009\,403 & 0.010\,777 & 0.017\,649 & 0.006\,174 & 0 \\
0.004\,527 & 0.028\,382 & 0.018\,792 & 0.007\,037 & 0.094\,602 & 0.011\,390 & 0.005\,339 & 0.003\,529 & 0.017\,488 & 0.012\,267 & 0.032\,529 & 0.024\,119 & 0.016\,583 \\
0.041\,466 & 0.017\,535 & 0.054\,570 & 0.070\,163 & 0.011\,888 & 0.010\,987 & 0.006\,124 & 0.001\,661 & 0.036\,024 & 0.005\,162 & 0.017\,862 & 0.041\,075 & 0.020\,353 \\
0.033\,526 & 0.044\,170 & 0.013\,795 & 0.010\,815 & 0.027\,351 & 0.057\,331 & 0.097\,285 & 0.050\,956 & 0.061\,618 & 0.015\,255 & 0.003\,143 & 0.028\,182 & 0.163\,169 \\
0.000\,266 & 0.009\,461 & 0.003\,609 & 0.003\,098 & 0.009\,572 & 0.037\,544 & 0.018\,638 & 0.007\,463 & 0.016\,578 & 0.016\,562 & 0.001\,900 & 0.014\,471 & 0.013\,321 \\
0.045\,980 & 0.242\,302 & 0.029\,643 & 0.053\,312 & 0.025\,624 & 0.052\,204 & 0.019\,392 & 0.002\,529 & 0.105\,570 & 0.028\,118 & 0.031\,922 & 0.020\,296 & 0.025\,663 \\
0.000\,853 & 0.004\,580 & 0.002\,760 & 0.005\,573 & 0.004\,225 & 0.018\,583 & 0.018\,550 & 0.000\,678 & 0.006\,914 & 0.062\,095 & 0.014\,594 & 0.019\,453 & 0.001\,776 \\
0 & 0 & 0 & 0 & 0 & 0 & 0 & 0 & 0 & 0 & 0 & 0 & 0.083\,623 \\
0.011\,253 & 0.041\,458 & 0.065\,898 & 0.079\,348 & 0.092\,444 & 0.052\,017 & 0.104\,105 & 0.015\,996 & 0.130\,856 & 0.134\,812 & 0.070\,927 & 0.074\,827 & 0.060\,298 \\
0.009\,527 & 0.013\,739 & 0.007\,361 & 0.002\,028 & 0.006\,431 & 0.005\,741 & 0.004\,007 & 0.007\,964 & 0.004\,562 & 0.008\,101 & 0.016\,260 & 0.006\,027 & 0
\end{bmatrix}$$

and $(I - A)^{-1}$ is

$$\begin{bmatrix}
1.151\,615 & 0.008\,820 & 0.064\,373 & 0.023\,826 & 0.008\,161 & 0.004\,945 & 0.005\,439 & 0.001\,751 & 0.011\,777 & 0.005\,486 & 0.009\,378 & 0.019\,111 & 0.009\,187 \\
0.006\,998 & 1.008\,375 & 0.032\,070 & 0.020\,476 & 0.059\,556 & 0.003\,058 & 0.002\,844 & 0.001\,357 & 0.006\,714 & 0.003\,262 & 0.006\,507 & 0.007\,120 & 0.005\,696 \\
0.354\,517 & 0.202\,960 & 1.754\,282 & 0.569\,941 & 0.190\,926 & 0.114\,240 & 0.119\,857 & 0.039\,317 & 0.281\,756 & 0.113\,835 & 0.232\,270 & 0.296\,456 & 0.228\,125 \\
0.008\,384 & 0.014\,417 & 0.013\,625 & 1.010\,041 & 0.053\,425 & 0.010\,797 & 0.007\,977 & 0.037\,093 & 0.016\,950 & 0.015\,866 & 0.023\,000 & 0.012\,313 & 0.007\,258 \\
0.018\,575 & 0.046\,643 & 0.046\,090 & 0.028\,805 & 1.117\,416 & 0.020\,504 & 0.014\,828 & 0.006\,923 & 0.035\,840 & 0.023\,719 & 0.046\,783 & 0.039\,408 & 0.032\,861 \\
0.075\,374 & 0.047\,784 & 0.112\,144 & 0.115\,655 & 0.038\,606 & 1.026\,034 & 0.022\,400 & 0.009\,075 & 0.068\,981 & 0.023\,330 & 0.041\,195 & 0.067\,975 & 0.044\,748 \\
0.065\,086 & 0.085\,917 & 0.054\,440 & 0.045\,494 & 0.055\,681 & 0.080\,239 & 1.122\,213 & 0.062\,448 & 0.098\,977 & 0.034\,915 & 0.023\,922 & 0.053\,391 & 0.199\,228 \\
0.008\,625 & 0.021\,395 & 0.016\,655 & 0.015\,301 & 0.018\,399 & 0.044\,074 & 0.025\,858 & 1.010\,222 & 0.028\,489 & 0.023\,738 & 0.008\,612 & 0.022\,987 & 0.023\,259 \\
0.081\,959 & 0.291\,054 & 0.084\,629 & 0.098\,878 & 0.065\,802 & 0.071\,023 & 0.036\,062 & 0.010\,081 & 1.143\,621 & 0.046\,935 & 0.055\,252 & 0.046\,613 & 0.053\,068 \\
0.007\,101 & 0.013\,239 & 0.012\,988 & 0.015\,212 & 0.011\,790 & 0.024\,942 & 0.026\,509 & 0.003\,336 & 0.017\,217 & 1.072\,079 & 0.020\,839 & 0.027\,392 & 0.011\,983 \\
0.001\,291 & 0.001\,568 & 0.001\,402 & 0.000\,791 & 0.000\,964 & 0.000\,731 & 0.000\,602 & 0.000\,760 & 0.000\,851 & 0.000\,969 & 1.001\,677 & 0.000\,902 & 0.084\,107 \\
0.067\,590 & 0.122\,810 & 0.160\,730 & 0.160\,074 & 0.153\,207 & 0.093\,094 & 0.148\,358 & 0.034\,321 & 0.206\,047 & 0.181\,389 & 0.117\,705 & 1.129\,124 & 0.127\,110 \\
0.015\,436 & 0.018\,751 & 0.016\,768 & 0.009\,458 & 0.011\,522 & 0.008\,746 & 0.007\,199 & 0.009\,085 & 0.010\,175 & 0.011\,591 & 0.020\,055 & 0.010\,782 & 1.005\,784
\end{bmatrix}$$

You tell the senator to look at the seventh column to see the effects of a change in demand for finance. There is indeed an effect on mining: Every 1-yen increase in demand for finance produces a 0.002 844-yen increase in demand for mining. This also means that every 1-yen *decrease* in demand for finance produces a 0.002 844-yen *decrease* in demand for mining. So the Japanese trade representative is right to complain that the senator's plan will hurt their mining companies.

EXERCISES

1. What does the $(2, 7)$ entry in the matrix A tell you?
2. What does the $(2, 7)$ entry in the matrix $(I - A)^{-1}$ tell you?
3. Why are none of the entries of $(I - A)^{-1}$ negative?
4. Why are the diagonal entries of $(I - A)^{-1}$ close to and larger than 1?
5. An increase in demand for the products of which sector of the Japanese economy would have the least impact on the mining sector?
6. An increase in demand for the products of which sector of the Japanese economy would have the most impact on the mining sector?
7. Which sector of the Japanese economy has the greatest percentage of its total output available for external consumption?
8. Referring to the conclusion of this section, try to account for most of the 0.002 844 yen by looking at the technology matrix A. For example, a 1-yen increase in demand for finance produces directly a 0.036 648-yen increase in demand for manufacturing, which in turn produces a $(0.036\,648)(0.016\,851) = 0.000\,618$-yen increase in demand for mining. What other two-step effects are there? Do they account for all of the 0.002 844 yen?
9. Explain why A^2 gives you the total two-step effects of increases in each sector on the others. What about three-step effects? How do you see *all* of the direct and indirect effects?

CHAPTER 3 REVIEW TEST

1. Let

$$A = \begin{bmatrix} 1 & 2 & 3 \\ 4 & 5 & 6 \end{bmatrix} \qquad B = \begin{bmatrix} 1 & -1 \\ 0 & 1 \end{bmatrix}$$

$$C = \begin{bmatrix} -1 & 0 \\ 1 & 1 \\ 0 & 1 \end{bmatrix} \qquad D = \begin{bmatrix} -3 & -2 & -1 \\ 1 & 2 & 3 \end{bmatrix}$$

For each of the following, determine whether the expression is defined and, if it is, evaluate it.

a. $A + B$ **b.** $A - D$

c. $2A^T + C$ **d.** AB

e. A^TB **f.** A^2

g. B^2 **h.** B^3

2. For each matrix find the inverse or determine that the matrix is singular.

a. $\begin{bmatrix} 1 & -1 \\ 0 & 1 \end{bmatrix}$ **b.** $\begin{bmatrix} 1 & 2 \\ 0 & 0 \end{bmatrix}$

c. $\begin{bmatrix} 1 & 2 & 3 \\ 0 & 4 & 1 \\ 0 & 0 & 1 \end{bmatrix}$ **d.** $\begin{bmatrix} 1 & 2 & 3 & 4 \\ 1 & 3 & 4 & 2 \\ 0 & 1 & 2 & 3 \\ 0 & 0 & 1 & 2 \end{bmatrix}$

e. $\begin{bmatrix} 1 & 2 & 3 & 4 \\ 2 & 3 & 3 & 3 \\ 0 & 1 & 2 & 3 \\ 0 & 0 & 1 & 2 \end{bmatrix}$ **f.** $\begin{bmatrix} 0 & 1 & 0 & 0 \\ 1 & 0 & 0 & 0 \\ 0 & 0 & 0 & 1 \\ 0 & 0 & 1 & 0 \end{bmatrix}$

3. Write each system of linear equations as a matrix equation and solve by inverting the coefficient matrix.

a. $x + 2y = 0$
$3x + 4y = 2$

b. $x + y + z = 3$
$y + 2z = 4$
$y - z = 1$

c. $x + y + z = 2$
$x + 2y + z = 3$
$x + y + 2z = 1$

d. $x + y = 0$
$y + z = 1$
$z + w = 0$
$x - w = 3$

OHaganBooks.com—ORGANIZING THE DATA

4. It is now July 1 and online sales of romance, science fiction, and horror novels at OHaganBooks.com were disappointingly slow over the past month. The following table shows the inventory of books in stock on June 1 at the OHagan-Books.com warehouses in Texas and Nevada:

	Romance	Sci-Fi	Horror
Texas	2500	4000	3000
Nevada	1500	3000	1000

Online sales during June are listed in the following table:

	Romance	Sci-Fi	Horror
Texas	300	500	100
Nevada	100	450	200

a. Use matrix algebra to compute the inventory at each warehouse at the end of June.

b. Due to an extensive banner ad campaign, the e-commerce manager at OHaganBooks.com is hoping for a 20% increase in sales for this month. At the same time, in anticipation of increasing sales, she is thinking of ordering 6000 romance novels, 8000 science fiction novels, and 4500 horror novels, to be divided equally between the two warehouses. Use matrix algebra to compute the anticipated inventory at each warehouse at the end of this coming month, assuming the 20% sales increase is met and she goes ahead with the order.

c. It is now July 15, and the e-commerce manager realizes that she probably overestimated the anticipated increase in sales. Based on July sales to date, she now projects July sales as follows:

	Romance	Sci-Fi	Horror
Texas	280	550	100
Nevada	50	500	120

Assuming that sales continue at this level for the next few months, write a matrix equation showing the inventory N at each warehouse x months after July 1 (assuming no new books are ordered during that period).

5. It is now the end of July and OHaganBooks.com's e-commerce manager bursts into the CEO's office. "I thought you might want to know, John, that our sales figures are exactly what I projected two weeks ago [see Exercise 4(c)]. Is that good market analysis or what?"

a. OHaganBooks.com has charged an average of $5 for romance novels, $6 for science fiction novels, and $5.50 for horror novels. Use the projected sales figures from Exercise 4(c) and matrix arithmetic to compute the total revenue OHaganBooks.com earned at each warehouse in July.

b. OHaganBooks.com pays an average of $2 for romance novels, $3.50 for science fiction novels, and $1.50 for horror novels. Use this information together with the information in part (a) to compute the profit OHaganBooks.com earned from sales of these books at each warehouse in July.

6. OHaganBooks.com has two main competitors: JungleBooks.com and FarmerBooks.com, and no other competitors of any significance on the horizon. The following matrix shows the movement of customers during July.[9] (Thus, for instance, the first row tells us that 80% of OHaganBooks.com's customers remained loyal, 10% of them went to JungleBooks.com, and the remaining 10% went to FarmerBooks.com.)

	To OHagan	To Jungle	To Farmer
From OHagan	0.8	0.1	0.1
From Jungle	0.4	0.6	0
From Farmer	0.2	0	0.8

At the beginning of July, OHaganBooks.com had an estimated 2000 customers, whereas its two competitors had 4000 each.

a. Set up the July 1 customer numbers in a row matrix and use matrix arithmetic to estimate the number of customers each company has at the end of July.

b. Assuming the July trends continue in August, predict the number of customers each company will have at the end of August.

c. Name one or more important factors that the model we have used in this question does not take into account.

7. Acting on a "tip" from Marjory Duffin, John O'Hagan decided that his company should invest a significant sum in Duffin House stock. The following table shows what information he pieced together later, after some of the records had been lost:

Date	Number of Shares	Price/Share ($)
July 1	?	20
August 1	?	10
September 1	?	5
Total	5000	

Over the 3 months shown, the company invested a total of $50,000 in Duffin stock and, on August 15, was paid dividends of 10¢/share, for a total of $300.

[9]By a "customer" of one of the three e-commerce sites, we mean someone who purchases more at that site than at any of the two competitors.

a. Use matrix inversion to determine how many shares OHaganBooks.com purchased on each of the three dates shown.

b. On October 1 the shares purchased on July 1 were sold at $3/share. The remaining shares were sold 1 month later at $1/share. Use matrix algebra to determine the total loss (taking into account the dividends paid on August 15) incurred as a result of the Duffin stock fiasco.

8. Approximately $1700 worth of the books sold each month by OHaganBooks.com are printed at Bruno Mills, a combined paper mill and printing company. In addition, OHaganBooks.com uses approximately $170 worth of Bruno Mills' paper products each month. A technology matrix for the paper and book printing sectors of Bruno's facility is

$$A = \begin{bmatrix} 0.1 & 0.5 \\ 0.01 & 0.05 \end{bmatrix}$$

where sector 1 represents paper and sector 2 represents book printing.

a. Compute $(I - A)^{-1}$. What is the significance of the (1, 2) entry?

b. What is the total value of paper and books that must be produced by Bruno Mills in order to meet demand from OHaganBooks.com?

c. Currently, Bruno Mills has a monthly capacity of $500,000 of paper products and $200,000 of books. What level of external demand would cause Bruno to meet the capacity for both products?

ADDITIONAL ONLINE REVIEW

If you follow the path

Web site → Everything for Finite Math → Chapter 3

you will find the following additional resources to help you review:

A comprehensive chapter summary (including examples and interactive features)

Additional review exercises (including interactive exercises and many with help)

Interactive section-by-section tutorials

A matrix algebra utility

A true/false chapter quiz

4

LINEAR PROGRAMMING

CASE STUDY

Airline Scheduling

Fly-by-Night Airlines flies airplanes from five cities: Los Angeles, Chicago, Atlanta, New York, and Boston. Due to a strike, the airline has not had sufficient crews to fly its planes, so some planes have been stranded in Chicago and Atlanta. In fact, 15 extra planes are in Chicago, and 30 extra planes are in Atlanta. To get back on schedule, Los Angeles needs at least 10 planes, New York needs at least 20, and Boston needs at least 10. It costs \$50,000 to fly a plane from Chicago to Los Angeles, \$10,000 to fly from Chicago to New York, and \$20,000 to fly from Chicago to Boston. It costs \$70,000 to fly a plane from Atlanta to Los Angeles, \$20,000 to fly from Atlanta to New York, and \$50,000 to fly from Atlanta to Boston. To get back on schedule at the lowest cost, how should Fly-by-Night rearrange its planes?

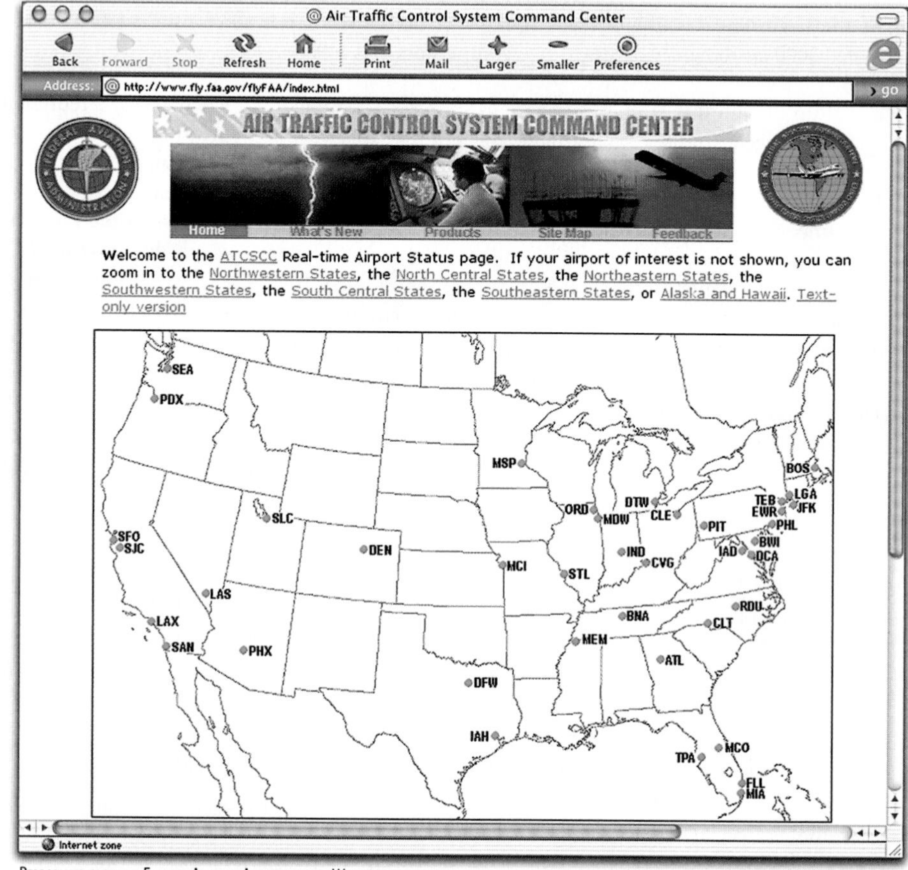

Reproduced from the Federal Aviation Administration Web site

INTERNET RESOURCES FOR THIS CHAPTER

At the Web site, follow the path

Web site → Everything for Finite Math → Chapter 4

where you will find links to step-by-step tutorials for the main topics in this chapter, a detailed chapter summary you can print out, a true/false quiz, and a collection of chapter review questions. You will also find downloadable Excel tutorials for each section and other resources, such as a matrix pivot utility, some software, and an LP resource that solves linear programming problems using the simplex method and shows all the steps.

Introduction

In this chapter we begin to look at one of the most important types of problems for business and the sciences: finding the largest or smallest possible value of some quantity (such as profit or cost) under certain constraints (for example, limited resources). We call such problems **optimization** problems because we are trying to find the best, or optimum, value. The optimization problems we look at in this chapter involve linear functions only and are known as **linear programming (LP)** problems. One of the main purposes of calculus, which you may study later, is to solve nonlinear optimization problems.

Linear programming problems involving only two unknowns can usually be solved by a graphical method, which we discuss in Sections 4.1 and 4.2. When there are three or more unknowns, we must use an algebraic method, as we had to do for systems of linear equations. The method we use is called the **simplex method.** Invented in 1947 by George Dantzig, the simplex method is still the most commonly used technique to solve LP problems in real applications, from finance to the computation of trajectories for guided missiles.

The simplex method can be used for hand calculations when the numbers are fairly small and the unknowns are few. Practical problems often involve large numbers and many unknowns, however. Problems like routing telephone calls or airplane flights, or allocating resources in a manufacturing process can involve tens of thousands of unknowns. Solving such problems by hand is obviously impractical, and so computers are regularly used. Although computer programs most often use the simplex method, faster methods are always being sought by mathematicians. One example is **Karmarkar's algorithm,** invented by N. Karmarkar in 1984. This method caused considerable excitement when first published, but it has not replaced the simplex method as the method of choice in most applications.

Calculators and spreadsheets are very useful aids in the simplex method. In practice, software packages do most of the work, so you can think of what we teach you here as a peek inside a "black box." What the software cannot do for you is convert a real situation into a mathematical problem, so the most important lessons to get out of this chapter are (1) how to recognize and set up a linear programming problem and (2) how to interpret the results.

4.1 Graphing Linear Inequalities

By the end of the next section, we will be solving LP problems with two unknowns. We use inequalities to describe the constraints in a problem, such as limitations on resources. Recall the basic notation for inequalities.

Nonstrict Inequalities	*Quick Examples*
$a \le b$ means that a **is less than or equal to** b.	$3 \le 99$, $-2 \le -2$, $0 \le 3$
$a \ge b$ means that a **is greater than or equal to** b.	$3 \ge 3$, $1.78 \ge 1.76$, $\frac{1}{3} \ge \frac{1}{4}$

There are also the inequalities $<$ and $>$, called **strict** inequalities because they do not permit equality. We do not use them in this chapter.

Following are some of the basic rules for manipulating inequalities. Although we illustrate all of them with the inequality $\le$, they apply equally well to inequalities with $\ge$ and to the strict inequalities $<$ and $>$.

Rules for Manipulating Inequalities	*Quick Examples*
1. The same quantity can be added or subtracted to both sides of an inequality: If $x \le y$, then $x + a \le y + a$ for any real number a.	$x \le y$ implies $x - 4 \le y - 4$
2. Both sides of an inequality can be multiplied or divided by a positive constant: If $x \le y$ and a is positive, then $ax \le ay$.	$x \le y$ implies $3x \le 3y$
3. Both sides of an inequality can be multiplied or divided by a negative constant if the inequality is *reversed:* If $x \le y$ and a is negative, then $ax \ge ay$.	$x \le y$ implies $-3x \ge -3y$
4. The left and right sides of an inequality can be switched if the inequality is *reversed*: If $x \le y$, then $y \ge x$; if $y \ge x$, then $x \le y$.	$3x \ge 5y$ implies $5y \le 3x$

Linear Inequalities and Solving Inequalities

An **inequality in the unknown x** is the statement that one expression involving x is less than or equal to (or greater than or equal to) another. Similarly, we can have an **inequality in x and y,** which involves expressions that contain x and y; an **inequality in x, y, and z;** and so on. A **linear inequality** in one or more unknowns is an inequality of the form

$ax \le b$ (or $ax \ge b$), *a and b real constants*

$ax + by \le c$ (or $ax + by \ge c$), *a, b, and c real constants*

$ax + by + cz \le d$, *a, b, c, and d real constants*

$ax + by + cz + dw \le e$, *a, b, c, d, and e real constants*

and so on.

Quick Examples

$2x + 8 \geq 89$	Linear inequality in x
$2x^3 \leq x^3 + y$	Nonlinear inequality in x and y
$3x - 2y \geq 8$	Linear inequality in x and y
$x^2 + y^2 \leq 19z$	Nonlinear inequality in x, y, and z
$3x - 2y + 4z \leq 0$	Linear inequality in x, y, and z

A **solution** of an inequality in the unknown x is a value for x that makes the inequality true. For example, $2x + 8 \geq 89$ has a solution $x = 50$ because $2(50) + 8 \geq 89$. Similarly, a solution of an inequality in x and y is a pair of values (x, y) making the inequality true. For example, $(5, 1)$ is a solution of $3x - 2y \geq 8$ because $3(5) - 2(1) \geq 8$. To **solve** an inequality is to find the set of *all* solutions.

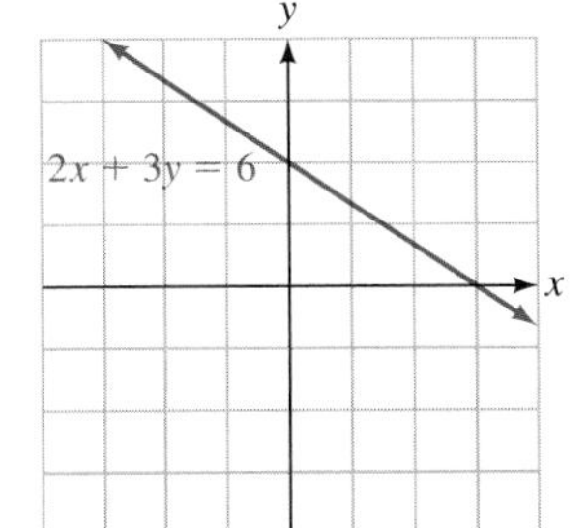

Figure 1

Our first goal is to solve linear inequalities in two variables—that is, inequalities of the form $ax + by \leq c$. As an example, let's solve

$$2x + 3y \leq 6$$

We already know how to solve the *equation* $2x + 3y = 6$. As we saw in Chapter 1, the solution of this equation may be pictured as the set of all points (x, y) on the straight-line graph of the equation. This straight line has x intercept 3 (obtained by putting $y = 0$ in the equation) and y intercept 2 (obtained by putting $x = 0$ in the equation) and is shown in Figure 1.

Notice that, if (x, y) is any point on the line, then x and y not only satisfy the *equation* $2x + 3y = 6$ but also satisfy the *inequality* $2x + 3y \leq 6$, because being equal to 6 qualifies as being less than or equal to 6.

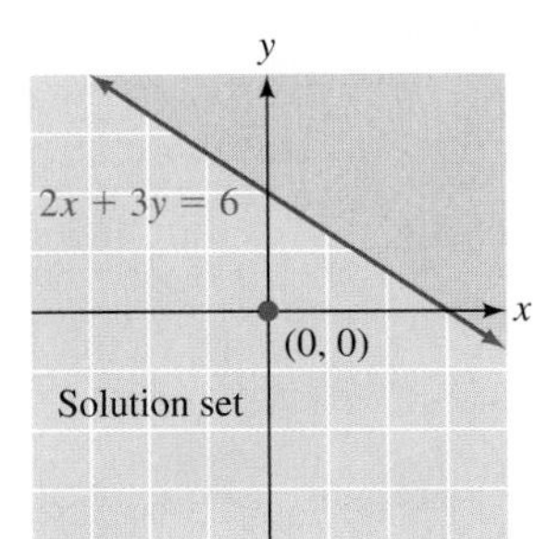

Figure 2

Question Do the points on the line give all possible solutions to the inequality?

Answer No. For example, try the origin, $(0, 0)$. Since $2(0) + 3(0) = 0 \leq 6$, the point $(0, 0)$ is another solution.

Question So how do we get all other solutions?

Answer Here is a possibly surprising fact: The solution to any linear inequality in two unknowns is represented by an entire **half plane:** the set of all points on one side of the line (including the line itself). Thus, since $(0, 0)$ is a solution of $2x + 3y \leq 6$ and is not on the line, every point on the same side of the line as $(0, 0)$ is a solution as well (the colored region below the line in Figure 2 shows part of the solution set).

Question Why is this so?

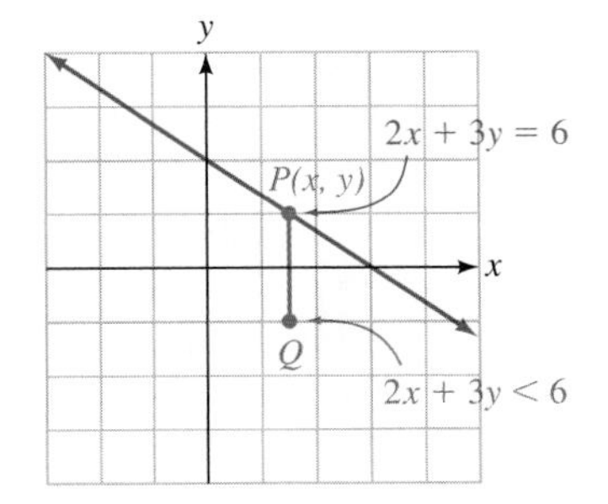

Figure 3

Answer Start with any point P on the line $2x + 3y = 6$. We already know that P is a solution of $2x + 3y \leq 6$. If we choose any point Q directly below P, the x coordinate of Q will be the same as that of P, and the y coordinate will be smaller. So the value of $2x + 3y$ at Q will be smaller than the value at P, which is 6. Thus, $2x + 3y < 6$ at Q, and so Q is another solution of the inequality. (See Figure 3.) In other words, *every point beneath the line is a solution of* $2x + 3y \leq 6$.

Question What about points on the other side of the line (the gray region above the line in Figure 2)?

Answer Any point above the line is directly above a point on the line, and so $2x + 3y > 6$ for such a point. Thus, *no point above the line is a solution of* $2x + 3y \leq 6$.

Question This seems to suggest that the solution of every inequality of the form $ax + by \leq c$ consists of the half plane below the line $ax + by = c$. Is this true?

Answer No. For instance, you can check that the solutions of $2x - 3y \leq 6$ are on or *above* the line $2x - 3y = 6$. The test-point procedure we describe below gives us an easy method for deciding whether the solution set includes the region above or below the corresponding line.

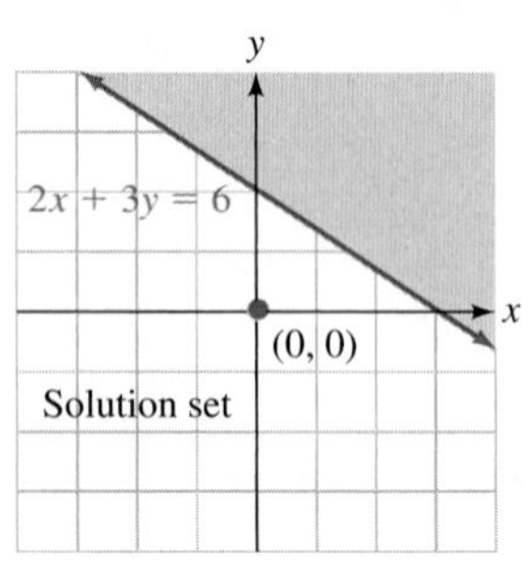

Figure 4

Now we are going to do something that will appear backward at first (but makes it simpler to sketch sets of solutions of *systems* of linear inequalities). For our standard drawing of the region of solutions of $2x + 3y \leq 6$, we are going to *shade the part that we do not want and leave the solution region blank.* Think of covering over or blocking out the unwanted points, leaving those that we do want in full view (but remember that the points on the boundary line are also points that we want). The result is Figure 4. The reason we do this should become clear in Example 2.

Sketching the Region Represented by a Linear Inequality in Two Variables

1. Sketch the straight line obtained by replacing the given inequality with an equality.
2. Choose a test point not on the line; (0, 0) is a good choice if the line does not pass through the origin.
3. If the test point satisfies the inequality, then the set of solutions is the entire region on the same side of the line as the test point. Otherwise, it is the region on the other side of the line. In either case, shade (block out) the side that does *not* contain the solutions, leaving the solution set unshaded.

Quick Example

Here are the three steps used to graph the inequality $x + 2y \geq 5$.

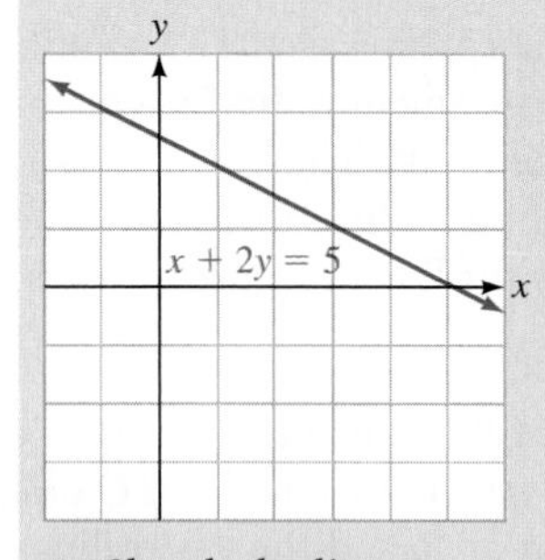

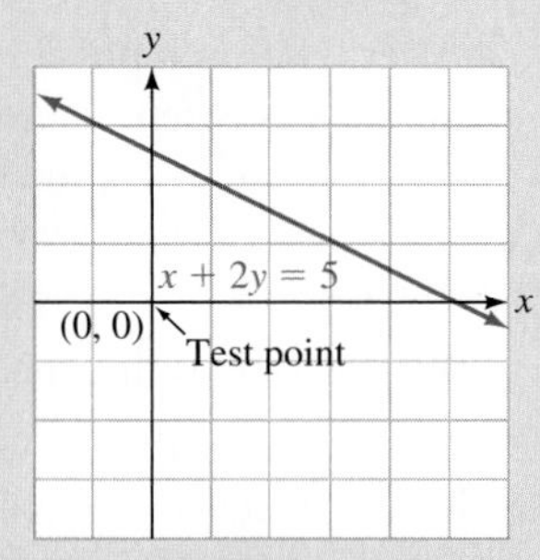

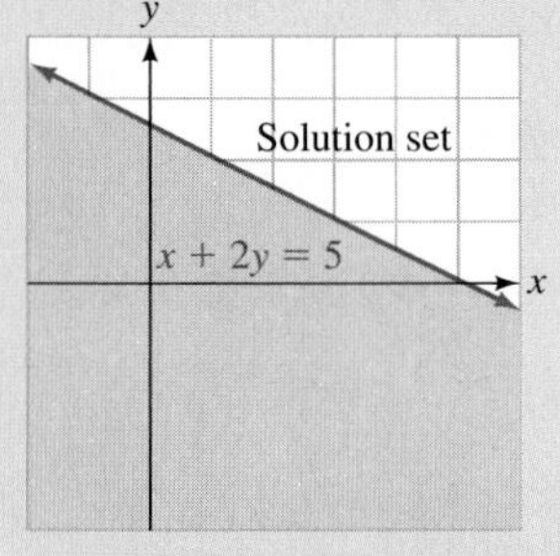

1. Sketch the line:

   $x + 2y = 5$

2. Test the point (0, 0):

   $0 + 2(0) \not\geq 5$

   Inequality is not satisfied.

3. Since the inequality is not satisfied, shade the region containing the test point.

Graphing Calculator: Graphing a Linear Inequality

Some calculators, including the TI-83, will shade one side of a graph, but you need to tell the calculator which side to shade. For instance, to obtain the solution set of $2x + 3y \leq 6$ shown in Figure 4, first solve the corresponding equation $2x + 3y = 6$ for y and use the following input:

```
▼ Y1=-(2/3)*X+2
```

The icon to the left of Y_1 tells the calculator to shade above the line. You can cycle through the various shading options by positioning the cursor to the left of Y_1 and pressing ENTER until you see the one you want.

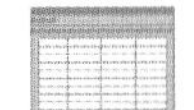

Excel: Graphing a Linear Inequality

Excel is not a particularly good tool for graphing linear inequalities because it cannot easily shade one side of a line. One solution is to use the Error Bar feature to indicate which side of the line *should* be shaded. First, create a scatter plot using two points to construct a line segment (as in Chapter 1):

	A	B
1	x	y
2	-10	= -(2/3)*A2 + 2
3	10	

Next, double-click on the line segment, and use the X-Error Bars feature to obtain a diagram similar to Figure 5, where the error bars indicate the direction of shading.

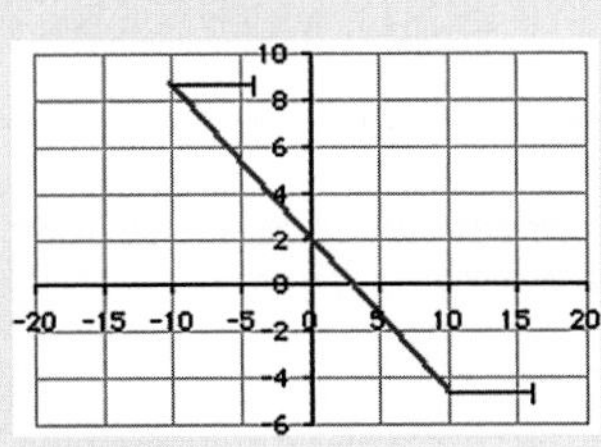

Figure 5

Example 1 • Graphing Single Inequalities

Sketch the regions determined by each of the following inequalities:

a. $3x - 2y \leq 6$ **b.** $6x \leq 12 + 4y$ **c.** $x \leq -1$ **d.** $y \geq 0$ **e.** $x \geq 3y$

Solution

a. The boundary line $3x - 2y = 6$ has x intercept 2 and y intercept -3 (Figure 6). We use $(0, 0)$ as a test point (because it is not on the line). Since $3(0) - 2(0) \leq 6$, the inequality is satisfied by the test point $(0, 0)$, and so it lies inside the solution set. The solution set is shown in Figure 6.

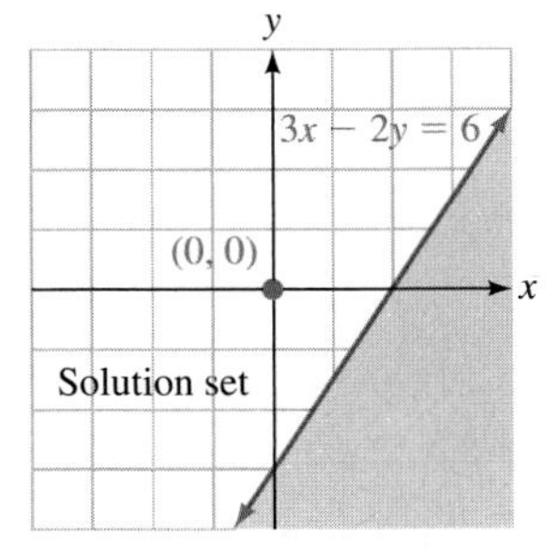

Figure 6

b. The given inequality, $6x \leq 12 + 4y$, can be rewritten in the form $ax + by \leq c$ by subtracting $4y$ from both sides:

$$6x - 4y \leq 12$$

Dividing both sides by 2 gives the inequality $3x - 2y \leq 6$, which we considered in part (a). Now, *applying the rules for manipulating inequalities does not affect the set of solutions.* Thus, the inequality $6x \leq 12 + 4y$ has the same set of solutions as $3x - 2y \leq 6$ (see Figure 6).

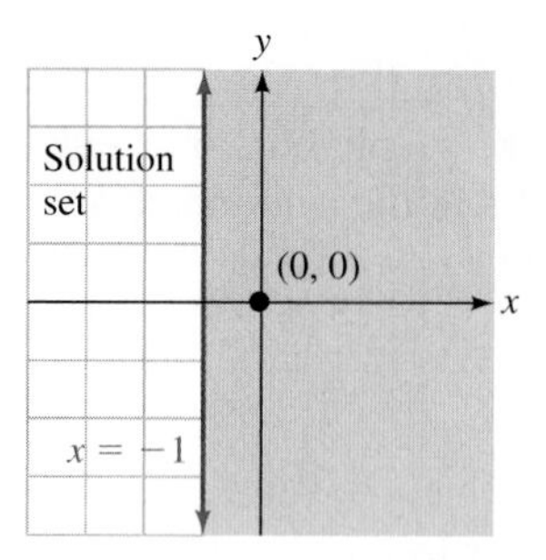

Figure 7

c. The region $x \leq -1$ has as boundary the vertical line $x = -1$. The test point (0, 0) is not in the solution set, as shown in Figure 7.

d. The region $y \geq 0$ has as boundary the horizontal line $y = 0$ (that is, the x axis). We cannot use (0, 0) for the test point because it lies on the boundary line. Instead, we choose a convenient point not on the line $y = 0$—say, (0, 1). Since $1 \geq 0$, this point is in the solution set, giving us the region shown in Figure 8.

e. The line $x \geq 3y$ has as boundary the line $x = 3y$ or, solving for y,

$$y = \frac{1}{3}x$$

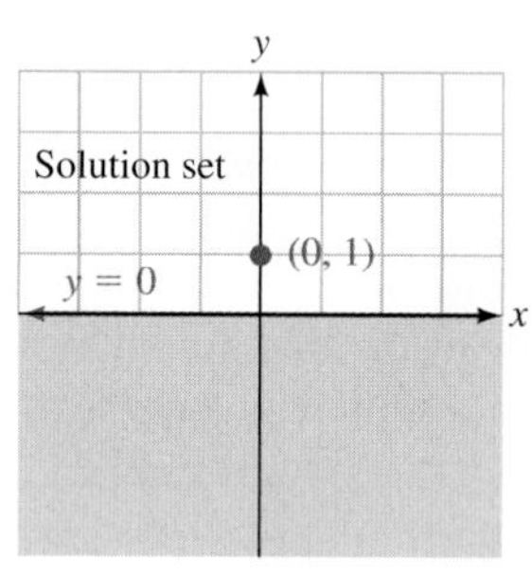

Figure 8

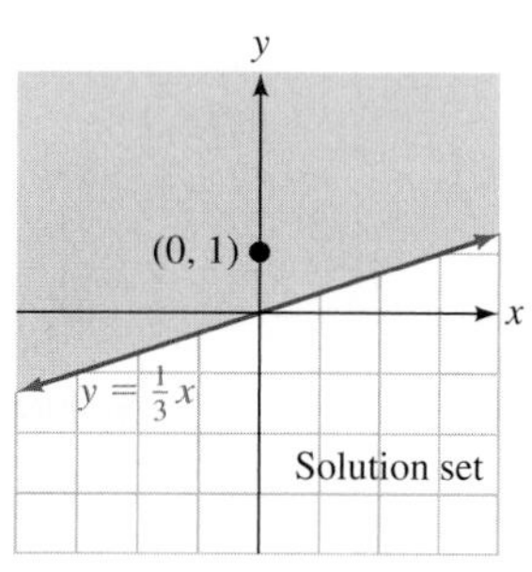

Figure 9

This line passes through the origin with slope $\frac{1}{3}$, so again we cannot use the origin as a test point. Instead, we use (0, 1). Substituting these coordinates in $x \geq 3y$ gives $0 \geq 3(1)$, which is false, so (0, 1) is not in the solution set, as shown in Figure 9.

Example 2 • Graphing Simultaneous Inequalities

Sketch the region of points that satisfy both inequalities:

$$2x - 5y \leq 10$$
$$x + 2y \leq 8$$

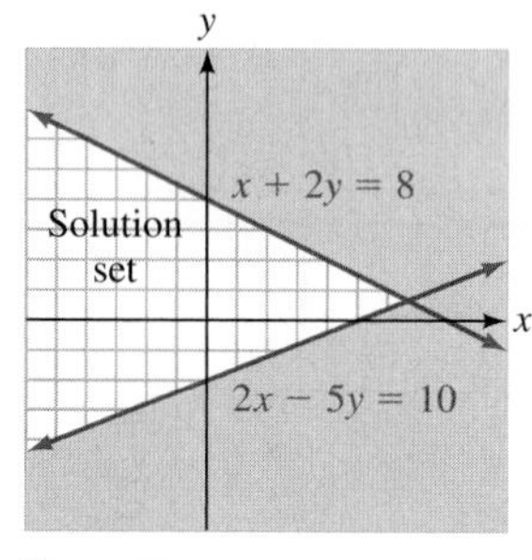

Figure 10

Solution Each inequality has a solution set that is a half plane. If a point is to satisfy *both* inequalities, it must lie in both sets of solutions. Put another way, if we cover the points that are not solutions to $2x - 5y \leq 10$ and then also cover the points that are not solutions to $x + 2y \leq 8$, the points that remain uncovered must be the points we want, those that are solutions to both inequalities. The result is shown in Figure 10, where the unshaded region is the set of solutions.[1]

Before we go on . . . As a check we can look at points in various regions in Figure 10. For example, our graph shows that (0, 0) should satisfy both inequalities, and it does:

$$2(0) - 5(0) = 0 \leq 10 \quad ✓$$
$$0 + 2(0) = 0 \leq 8 \quad ✓$$

[1] *Technology note:* Although these graphs are quite easy to do by hand, the more lines we have to graph, the more difficult it becomes to get everything in the right place; this is where graphing technology can become important. This is especially true when, for instance, three or more lines intersect in points that are very close together and hard to distinguish in hand-drawn graphs.

On the other hand, (0, 5) should fail to satisfy one of the inequalities:

$$2(0) - 5(5) = -25 \le 10 \quad ✓$$

$$0 + 2(5) = 10 > 8 \quad ✗$$

One more: (5, −1) should fail one of the inequalities:

$$2(5) - 5(-1) = 15 > 10 \quad ✗$$

$$5 + 2(-1) = 3 \le 8 \quad ✓$$

Example 3 • Corner Points

Sketch the region of solutions of the following system of inequalities and list the coordinates of all the corner points.

$$\begin{aligned} 3x - 2y &\le 6 \\ x + y &\ge -5 \\ y &\le 4 \end{aligned}$$

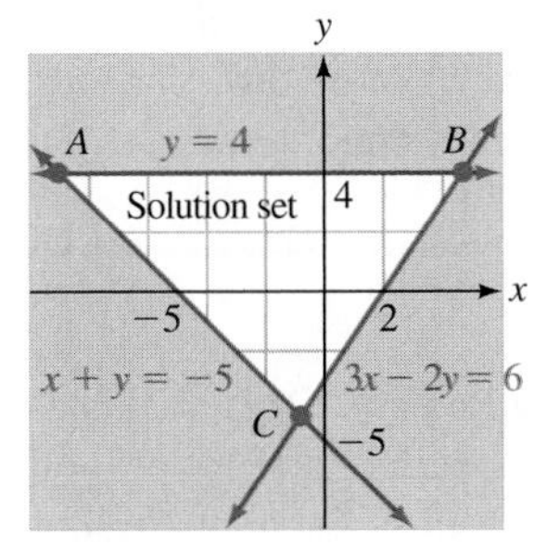

Figure 11

Solution Shading the regions that we do not want leaves us with the triangle shown in Figure 11. We label the corner points *A*, *B*, and *C* as shown. Each of these corner points lies at the intersection of two of the bounding lines. So, to find the coordinates of each corner point, we need to solve the system of equations given by the two lines. To do this systematically, we make the following table:

Point	Lines through Point	Coordinates
A	$y = 4$ $x + y = -5$	$(-9, 4)$
B	$y = 4$ $3x - 2y = 6$	$\left(\frac{14}{3}, 4\right)$
C	$x + y = -5$ $3x - 2y = 6$	$\left(-\frac{4}{5}, -\frac{21}{5}\right)$

Here, we have solved each system of equations in the middle column to get the point on the right, using the techniques of Chapter 2. You should do this for practice.[2]

Before we go on . . . As a partial check that we have drawn the correct region, let's choose any point in its interior—say, (0, 0). We can easily check that (0, 0) satisfies all three given inequalities. It follows that all of the points in the triangular region containing (0, 0) are also solutions.

[2] *Technology note:* Using the Trace feature makes it easy to locate corner points graphically. Remember to zoom in for additional accuracy when appropriate. Of course, you can also use technology to help solve the systems of equations, as we discussed in Chapter 2.

Take another look at the regions of solutions in Examples 2 and 3 (Figure 12).

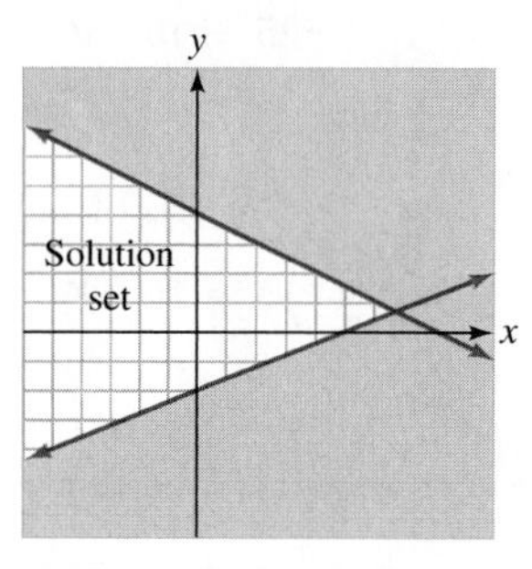

(a) Example 2 solution set

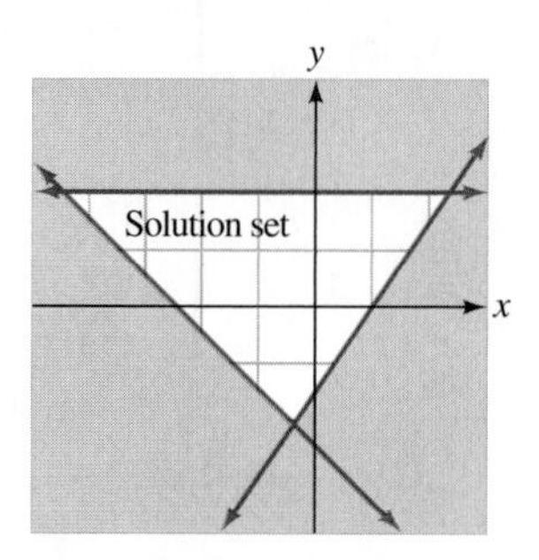

(b) Example 3 solution set

Figure 12

Notice that the solution set in Figure 12(a) extends infinitely far to the left, whereas the one in Figure 12(b) is completely enclosed by a boundary. Sets that are completely enclosed are called **bounded,** and sets that extend infinitely in one or more directions are **unbounded.** For example, all solution sets in Example 1 are unbounded.

Example 4 • Resource Allocation

Socaccio Pistachio make two types of pistachio nuts: Dazzling Red and Organic. Pistachio nuts require food color and salt, and the following table shows the amount of food color and salt required for a 1-kilogram batch of pistachios, as well as the total amount of these ingredients available each day.

	Dazzling Red	Organic	Total Available
Food Color (g)	2	1	20
Salt (g)	10	20	220

Use a graph to show the possible numbers of batches of each type of pistachio Socaccio can produce each day. This region is called the **feasible region.**

Solution As we did in Chapter 2, we start by identifying the unknowns: Let x be the number of batches of Dazzling Red manufactured each day and let y be the number of batches of Organic manufactured each day.

Now, because of our experience with systems of linear equations, we are tempted to say: for food color, $2x + y = 20$, and for salt, $10x + 20y = 220$. However, no one is saying that Socaccio has to use *all* available ingredients; the company might choose to use fewer than the total available amounts if this proves more profitable. Thus, $2x + y$ can be anything *up to a total of* 20. In other words,

$$2x + y \leq 20$$

Similarly,

$$10x + 20y \leq 220$$

There are two more restrictions not explicitly mentioned: Neither x nor y can be negative (the company cannot produce a negative number of batches of nuts). Therefore, we have the additional restrictions

$$x \geq 0 \qquad y \geq 0$$

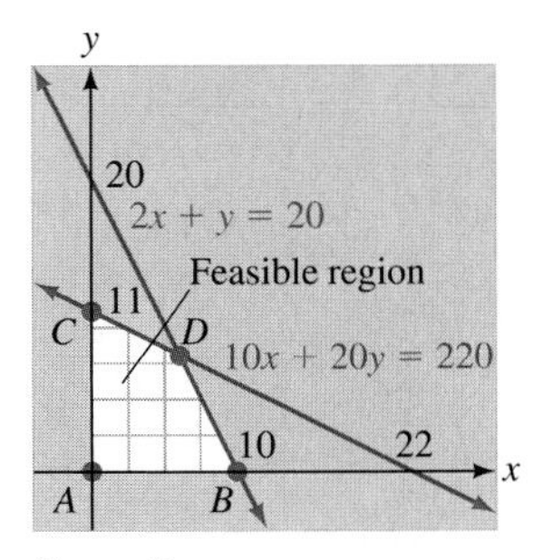

Figure 13

These two inequalities tell us that the feasible region (solution set) is restricted to the first quadrant, because in the other quadrants either x or y or both x and y are negative. So instead of shading out all other quadrants, we can simply restrict our drawing to the first quadrant.

The (bounded) feasible region shown in Figure 13 is a graphical representation of the limitations the company faces.

Before we go on . . . Every point in the feasible region represents a value for x and a value for y that do not violate any of the company's restrictions. For example, the point (5, 6) lies well inside the region, so the company can produce 5 batches of Dazzling Red nuts and 6 batches of Organic without exceeding the limitations on ingredients [that is, $2(5) + 6 = 16 < 20$, and $10(5) + 20(6) = 170 < 220$]. The corner points A, B, C, and D are significant if the company wishes to realize the greatest profit, as we will see in Section 4.2. We can find the corners as in the following table:

Point	Lines through Point	Coordinates
A		(0, 0)
B		(10, 0)
C		(0, 11)
D	$2x + y = 20$ $10x + 20y = 220$	(6, 8)

(We have not listed the lines through the first three corners because their coordinates can be read easily from the graph.) Points on the line segment DB represent use of all the food color (since the segment lies on the line $2x + y = 20$), and points on the line segment CD represent use of all the salt (since the segment lies on the line $10x + 20y = 220$). Note that the point D is the only solution that uses all of both ingredients.

Guideline: Recognizing Whether to Use a Linear Inequality or a Linear Equation

Question How do I know whether to model a situation by a linear inequality like $3x + 2y \leq 10$ or by a linear equation like $3x + 2y = 10$?

Answer Here are some key phrases to look for: *at most, up to, no more than, at least, or more,* and *exactly.* Suppose, for instance, that nuts cost 3¢, bolts cost 2¢, x is the number of nuts you can buy, and y is the number of bolts you can buy.

If you have *up to* 10¢ to spend, then $3x + 2y \leq 10$.

If you must spend *exactly* 10¢, then $3x + 2y = 10$.

If you plan to spend *at least* 10¢, then $3x + 2y \geq 10$.

The use of inequalities to model a situation is often more realistic than the use of equations; for instance, one cannot always expect to exactly fill all orders, spend the exact amount of one's budget, or keep a plant operating at exactly 100% capacity.

4.1 EXERCISES

In Exercises 1–26, sketch the region that corresponds to the given inequalities, say whether the region is bounded or unbounded, and find the coordinates of all corner points (if any).

1. $2x + y \le 10$

2. $4x - y \le 12$

3. $-x - 2y \le 8$

4. $-x + 2y \ge 4$

5. $3x + 2y \ge 5$

6. $2x - 3y \le 7$

7. $x \le 3y$

8. $y \ge 3x$

9. $\frac{3x}{4} - \frac{y}{4} \le 1$

10. $\frac{x}{3} + \frac{2y}{3} \ge 2$

11. $x \ge -5$

12. $y \le -4$

13. $4x - y \le 8$; $x + 2y \le 2$

14. $2x + y \le 4$; $x - 2y \ge 2$

15. $3x + 2y \ge 6$; $3x - 2y \le 6$; $x \ge 0$

16. $3x + 2y \le 6$; $3x - 2y \ge 6$; $-y \ge 2$

17. $x + y \ge 5$; $x \le 10$; $y \le 8$; $x \ge 0, y \ge 0$

18. $2x + 4y \ge 12$; $x \le 5$; $y \le 3$; $x \ge 0, y \ge 0$

19. $20x + 10y \le 100$; $10x + 20y \le 100$; $10x + 10y \le 60$; $x \ge 0, y \ge 0$

20. $30x + 20y \le 600$; $10x + 40y \le 400$; $20x + 30y \le 450$; $x \ge 0, y \ge 0$

21. $20x + 10y \ge 100$; $10x + 20y \ge 100$; $10x + 10y \ge 80$; $x \ge 0, y \ge 0$

22. $30x + 20y \ge 600$; $10x + 40y \ge 400$; $20x + 30y \ge 600$; $x \ge 0, y \ge 0$

23. $-3x + 2y \le 5$; $3x - 2y \le 6$; $x \le 2y$; $x \ge 0, y \ge 0$

24. $-3x + 2y \le 5$; $3x - 2y \ge 6$; $y \le \frac{x}{2}$; $x \ge 0, y \ge 0$

25. $2x - y \ge 0$; $x - 3y \le 0$; $x \ge 0, y \ge 0$

26. $-x + y \ge 0$; $4x - 3y \ge 0$; $x \ge 0, y \ge 0$

In Exercises 27–32, we suggest you use technology. Graph the regions corresponding to the inequalities and find the coordinates of all corner points (if any) to two decimal places:

27. $2.1x - 4.3y \ge 9.7$

28. $-4.3x + 4.6y \ge 7.1$

29. $-0.2x + 0.7y \ge 3.3$; $1.1x + 3.4y \ge 0$

30. $0.2x + 0.3y \ge 7.2$; $2.5x - 6.7y \le 0$

31. $4.1x - 4.3y \le 4.4$; $7.5x - 4.4y \le 5.7$; $4.3x + 8.5y \le 10$

32. $2.3x - 2.4y \le 2.5$; $4.0x - 5.1y \le 4.4$; $6.1x + 6.7y \le 9.6$

APPLICATIONS

33. Resource Allocation You manage an ice cream factory that makes two flavors: Creamy Vanilla and Continental Mocha. Into each quart of Creamy Vanilla go 2 eggs and 3 cups of cream. Into each quart of Continental Mocha go 1 egg and 3 cups of cream. You have in stock 500 eggs and 900 cups of cream. Draw the feasible region showing the number of quarts of vanilla and number of quarts of mocha that can be produced. Find the corner points of the region.

34. Resource Allocation Podunk Institute of Technology's Math Department offers two courses: Finite Math and Applied Calculus. Each section of Finite Math has 60 students, and each section of Applied Calculus has 50. The department is allowed to offer a total of up to 110 sections. Furthermore, no more than 6000 students want to take a math course (no student will take more than one math course). Draw the feasible region that shows the number of sections of each class that can be offered. Find the corner points of the region.

35. Nutrition Ruff Inc. makes dog food out of chicken and grain. Chicken has 10 grams of protein and 5 grams of fat per ounce, and grain has 2 grams of protein and 2 grams of fat per ounce. A bag of dog food must contain at least 200 grams of protein and at least 150 grams of fat. Draw the feasible region that shows the number of ounces of chicken and number of ounces of grain Ruff can mix into each bag of dog food. Find the corner points of the region.

36. Purchasing Enormous State University's Business School is buying computers. The school has two models from which to choose, the Pomegranate and the iZac. Each Pomegranate comes with 400 megabytes of memory and 80 gigabytes of disk space, while each iZac has 300 megabytes of memory and 100 gigabytes of disk space. For reasons related to its accreditation, the school would like to say that it has a total of at least 48,000 megabytes of memory and at least 12,800 gigabytes of disk space. Draw the feasible region that shows the number of each kind of computer it can buy. Find the corner points of the region.

37. Nutrition Each serving of Gerber Mixed Cereal for Baby contains 60 calories and 11 grams of carbohydrates, and each serving of Gerber Mango Tropical Fruit Dessert contains 80 calories and 21 grams of carbohydrates. You want to provide your child with at least 140 calories and at least 32 grams of carbohydrates. Draw the feasible region that shows the number of servings of cereal and number of servings of dessert that you can give your child. Find the corner points of the region.

SOURCE: Nutrition information printed on the containers.

38. Nutrition Each serving of Gerber Mixed Cereal for Baby contains 60 calories, 11 grams of carbohydrates, and no vitamin C, and each serving of Gerber Apple Banana Juice contains 60 calories, 15 grams of carbohydrates, and 120% of the U.S. Recommended Daily Allowance (RDA) of vitamin C for infants. You want to provide your child with at least 120 calories, at least 26 grams of carbohydrates, and at least 50% of the RDA of vitamin C for infants. Draw the feasible region that shows the number of servings of cereal and number of servings of juice that you can give your child. Find the corner points of the region.

SOURCE: Nutrition information printed on the containers.

39. **Municipal Bond Funds** The Pimco New York Municipal Bond Fund (PNF) and the Payden Short Duration Tax Exempt Fund (PYSDX) are tax-exempt municipal bond funds. In 2001 the Pimco fund yielded 6%, and the Payden fund yielded 5%. You would like to invest a total of up to \$80,000 and earn at least \$4200 in interest in the coming year (assuming the given yields). Draw the feasible region that shows how much money you can invest in each fund. Find the corner points of the region.

    Yields are rounded. SOURCES: www.pimcofunds.com and www.paydenfunds.com, February 22, 2002.

40. **Municipal Bond Funds** The Columbia Oregon Municipal Bond Fund (CMBFX) yielded 5% in the year ending January 31, 2002; Safeco Municipal Bond Fund (SFCOX) yielded 7% for the same period. You would like to invest a total of up to \$60,000 and earn at least \$3500 in interest. Draw the feasible region that shows how much money you can invest in each fund. Find the corner points of the region.

    Yields are rounded. SOURCES: www.safecofunds.com and www.columbiafunds.com, February 22, 2002.

41. **Investments** Your portfolio manager has suggested two high-yielding stocks: Philip Morris (MO) and R. J. Reynolds Tobacco (RJR). Each MO share costs \$50 and yields 4.5% in dividends. Each RJR share costs \$55 and yields 5% in dividends. You have up to \$12,100 to invest and would like to earn at least \$550 in dividends. Draw the feasible region that shows how many shares in each company you can buy. Find the corner points of the region. (Round each coordinate to the nearest whole number.)

    Share prices and yields are approximate as of February 2002. SOURCE: http://money.excite.com, February 23, 2002.

42. **Investments** Your friend's portfolio manager has suggested two high-yielding stocks: Consolidated Edison (ED) and Royal Bank of Scotland (RBS-K). Each ED share costs \$40 and yields 5.5% in dividends. Each RBS-K share costs \$25 and yields 7.5% in dividends. You have up to \$30,000 to invest and would like to earn at least \$1650 in dividends. Draw the feasible region that shows how many shares in each company you can buy. Find the corner points of the region. (Round each coordinate to the nearest whole number.)

    Share prices and yields are approximate as of February 2002. SOURCE: http://money.excite.com, February 23, 2002.

43. **Advertising** You are the marketing director for a company that manufactures body-building supplements, and you are planning to run ads in *Sports Illustrated* and *GQ* magazine. Based on readership data, you estimate that each one-page ad in *Sports Illustrated* will be read by 650,000 people in your target group, and each one-page ad in *GQ* will be read by 150,000. You would like your ads to be read by at least 3 million people in the target group, and to ensure the broadest possible audience, you would like to place at least three full-page ads in each magazine. Draw the feasible region that shows how many pages you can purchase in each magazine. Find the corner points of the region. (Round each coordinate to the nearest whole number.)

    The readership data for *Sports Illustrated* is based, in part, on the results of a readership survey taken in March 2000. The readership data for *GQ* is fictitious. SOURCE: Mediamark Research Inc./*New York Times,* May 29, 2000, p. C1.

44. **Advertising** You are the marketing director for a company that manufactures body-building supplements and you are planning to run ads in *Sports Illustrated* and *Muscle and Fitness.* Based on readership data, you estimate that each one-page ad in *Sports Illustrated* will be read by 650,000 people in your target group, and each one-page ad in *Muscle and Fitness* will be read by 250,000 people in your target group. You would like your ads to be read by at least 4 million people in the target group, and to ensure the broadest possible audience, you would like to place at least three full-page ads in each magazine during the year. Draw the feasible region showing how many pages you can purchase in each magazine. Find the corner points of the region. (Round each coordinate to the nearest whole number.)

    The readership data for both magazines are based on the results of a readership survey taken in March 2000. SOURCE: Mediamark Research Inc./*New York Times,* May 29, 2000, p. C1.

COMMUNICATION AND REASONING EXERCISES

45. Find a system of inequalities whose solution set is unbounded.

46. Find a system of inequalities whose solution set is empty.

47. How would you use linear inequalities to describe the triangle with corner points (0, 0), (2, 0), and (0, 1)?

48. Explain the advantage of shading the region of points that do not satisfy the given inequalities. Illustrate with an example.

In Exercises 49–52, you are mixing x grams of ingredient A and y grams of ingredient B. Choose the equation or inequality that models the given requirement.

49. There should be at least 3 more grams of ingredient A than ingredient B.
    (A) $3x - y \le 0$ **(B)** $x - 3y \ge 0$
    (C) $x - y \ge 3$ **(D)** $3x - y \ge 0$

50. The mixture should contain at least 25% of ingredient A by weight.
    (A) $4x - y \le 0$ **(B)** $x - 4y \ge 0$
    (C) $x - y \ge 4$ **(D)** $3x - y \ge 0$

51. There should be at least 3 parts (by weight) of ingredient A to 2 parts of ingredient B.
    (A) $3x - 2y \ge 0$ **(B)** $2x - 3y \ge 0$
    (C) $3x + 2y \ge 0$ **(D)** $2x + 3y \ge 0$

52. There should be no more of ingredient A (by weight) than ingredient B.
    (A) $x - y = 0$ **(B)** $x - y \le 0$
    (C) $x - y \ge 0$ **(D)** $x + y \ge y$

53. You are setting up a system of inequalities in the unknowns x and y. The inequalities represent constraints faced by Fly-by-Night Airlines, where x represents the number of first-class tickets it should issue for a specific flight and y represents the number of business-class tickets it should issue for that flight. You find that the feasible region is empty. How do you interpret this?

54. In the situation described in Exercise 53, is it possible instead for the feasible region to be unbounded? Explain your answer.

55. Create an interesting scenario that leads to the following system of inequalities:

$$\begin{aligned} 20x + 40y &\leq 1000 \\ 30x + 20y &\leq 1200 \\ x \geq 0, y &\geq 0 \end{aligned}$$

56. Create an interesting scenario that leads to the following system of inequalities:

$$\begin{aligned} 20x + 40y &\geq 1000 \\ 30x + 20y &\geq 1200 \\ x \geq 0, y &\geq 0 \end{aligned}$$

57. Describe at least one drawback to the method of finding the corner points of a feasible region by drawing its graph, when the feasible region arises from real-life constraints.

58. Draw several bounded regions described by linear inequalities. For each region you draw, find the point that gives the greatest possible value of $x + y$. What do you notice?

4.2 *Solving Linear Programming Problems Graphically*

As we saw in Example 4 in Section 4.1, in some scenarios the possibilities are restricted by a system of linear inequalities. In that example, it would be natural to ask which of the various possibilities gives the company the largest profit. This is a kind of problem known as a linear programming problem (commonly referred to as an LP problem).

Linear Programming Problems

A **linear programming problem** in two unknowns x and y is one in which we are to find the maximum or minimum value of a linear expression

$$ax + by$$

called the **objective function,** subject to a number of linear **constraints** of the form

$$cx + dy \leq e \quad \text{or} \quad cx + dy \geq e$$

The largest or smallest value of the objective function is called the **optimal value,** and a pair of values of x and y that gives the optimal value constitutes an **optimal solution.**

Quick Example

Maximize $p = x + y$ — Objective function

subject to

$$\left.\begin{aligned} x + 2y &\leq 12 \\ 2x + y &\leq 12 \\ x \geq 0, y &\geq 0 \end{aligned}\right\} \text{ Constraints}$$

See Example 1 for a method of solving this LP problem (that is, finding an optimal solution and value).

Example 1 • Solving an LP Problem

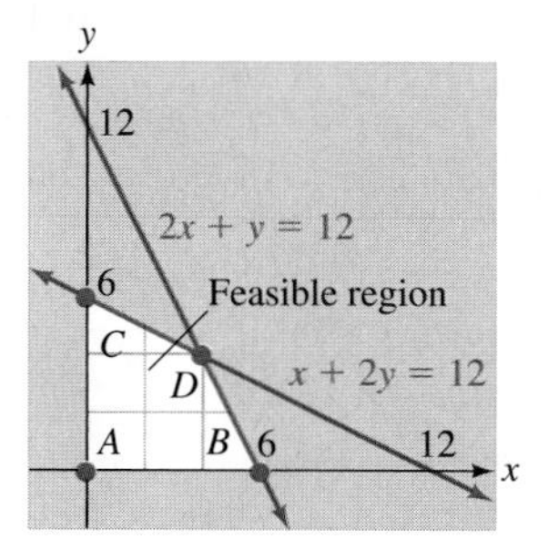

Figure 14

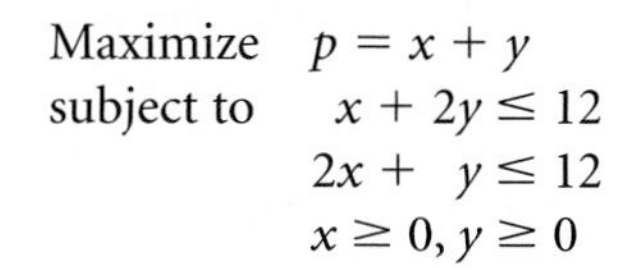

$$\begin{aligned} \text{Maximize} \quad & p = x + y \\ \text{subject to} \quad & x + 2y \le 12 \\ & 2x + y \le 12 \\ & x \ge 0, y \ge 0 \end{aligned}$$

Solution We begin by drawing the **feasible region** for the problem, which is the set of points representing solutions to the constraints. We do this using the techniques of Section 4.1, and we get Figure 14. Each **feasible point** (point in the feasible region) gives an x and a y satisfying all the constraints. The question now is, Which of these points gives the largest value of the objective function $p = x + y$? The following fact is crucial:

Fundamental Theorem of Linear Programming

- If an LP problem has optimal solutions, then at least one of these solutions occurs at a corner point.
- If there is only one optimal solution, it must occur at a corner point.
- Linear programming problems with bounded, nonempty feasible regions always have optimal solutions.

We will motivate the theorem using our example shortly, but first we use the theorem to solve the problem.

In the following table, we list the coordinates of each corner point and compute the value of the objective function at each corner:

Corner Point	Lines through Point	Coordinates	$p = x + y$
A		(0, 0)	0
B		(6, 0)	6
C		(0, 6)	6
D	$x + 2y = 12$ $2x + y = 12$	(4, 4)	8

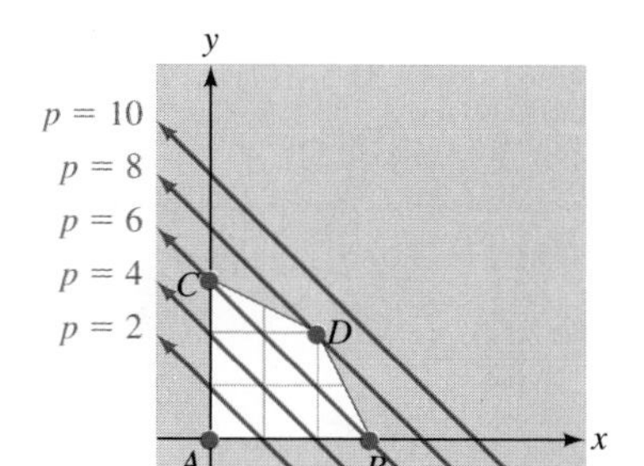

Figure 15

Since one of these points must give the optimal value, we simply pick the one that gives the largest value for p, which is D. Therefore, the optimal value of p is 8, and an optimal solution is (4, 4).

Before we go on . . . Now we owe you an explanation of why one of the corner points should be an optimal solution. The question is, Which point in the feasible region gives the largest possible value of $p = x + y$?

Consider first an easier question: Which points result in a *particular value* of p? For example, which points result in $p = 2$? These would be the points on the line $x + y = 2$, which is the line labeled $p = 2$ in Figure 15.

Now suppose we want to know which points make $p = 4$: These would be the points on the line $x + y = 4$, which is the line labeled $p = 4$ in Figure 15. Notice that this line is parallel to but higher than the line $p = 2$. (If p represented profit in an application, we would call these **isoprofit lines,** or **constant-profit lines.**) Imagine moving this line up or down in the picture. As we move the line down, we see smaller values of p, and as we move it up, we see larger values. Several more of these lines are drawn in Figure 15. Look, in particular, at the line labeled $p = 10$. This line does not meet the feasible region, meaning that no feasible point makes p as large as 10. Starting with the line $p = 2$, as we move the line up, increasing p, there will be a last line that meets the feasible region. In the figure it is clear that this is the line $p = 8$, and this meets the feasible region in only one point, which is the corner point D. Therefore, D gives the greatest value of p of all feasible points.

If we had been asked to maximize some other objective function, such as $p = x + 3y$, then the optimal solution might be different. Figure 16 shows some of the iso-profit lines for this objective function.

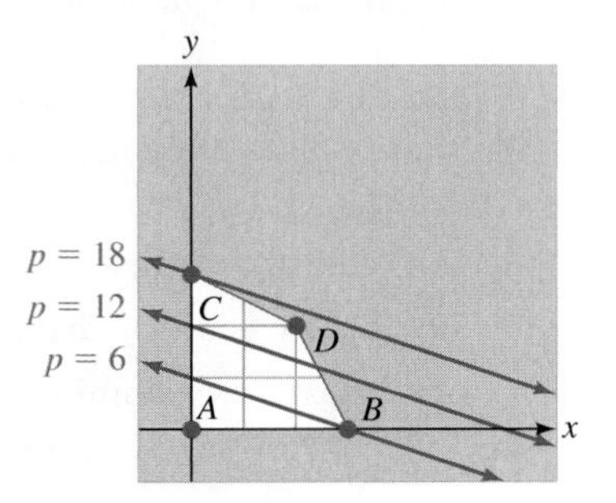

Figure 16

This time the last point hit as p increases is C, not D. This tells us that the optimal solution is $(0, 6)$, giving the optimal value $p = 18$.

This discussion should convince you that the optimal value in an LP problem will always occur at one of the corner points. By the way, it is possible for the optimal value to occur at *two* corner points and at all points along an edge connecting them. (Do you see why?) We will see this in Example 3(b).

Here is a summary of the method we have just been using.

Graphical Method for Solving Linear Programming Problems in Two Unknowns (Bounded Feasible Regions)

1. Graph the feasible region and check that it is bounded.
2. Compute the coordinates of the corner points.
3. Substitute the coordinates of the corner points into the objective function to see which gives the maximum (or minimum) value of the objective function.
4. Any such corner point is an optimal solution.

Note If the feasible region is unbounded, this method will work only if there are optimal solutions; otherwise, it will not work. We will show you a method for deciding this below.

APPLICATIONS

Example 2 • Resource Allocation

Acme Babyfoods mixes two strengths of apple juice. One quart of Beginner's juice is made from 30 fluid ounces of water and 2 fluid ounces of apple juice concentrate. One quart of Advanced juice is made from 20 fluid ounces of water and 12 fluid ounces of concentrate. Every day Acme has available 30,000 fluid ounces of water and 3600 fluid ounces of concentrate. Acme makes a profit of 20¢ on each quart of Beginner's juice and 30¢ on each quart of Advanced juice. How many quarts of each should Acme make each day to get the largest profit? How would this change if Acme made a profit of 40¢ on Beginner's juice and 20¢ on Advanced juice?

Solution Looking at the question that we are asked, we see that our unknown quantities are

x = number of quarts of Beginner's juice made each day

y = number of quarts of Advanced juice made each day

(In this context x and y are often called the **decision variables,** because we must decide what their values should be in order to get the largest profit.) We can write down the data given in the form of a table (the numbers in the first two columns are amounts per quart of juice):

	Beginner's, x	Advanced, y	Available
Water (fl oz)	30	20	30,000
Concentrate (fl oz)	2	12	3,600
Profit (¢)	20	30	

Since nothing in the problem says that Acme must use up all the water or concentrate, just that it can use no more than what is available, the first two rows of the table give us two inequalities:

$$30x + 20y \le 30{,}000$$
$$2x + 12y \le 3600$$

Dividing the first inequality by 10 and the second by 2 gives

$$3x + 2y \le 3000$$
$$x + 6y \le 1800$$

We also have that $x \ge 0$ and $y \ge 0$ because Acme can't make a negative amount of juice. To finish setting up the problem, we are asked to maximize the profit, which is

$$p = 20x + 30y \qquad \text{Expressed in cents}$$

This gives us our LP problem:

$$\begin{array}{ll} \text{Maximize} & p = 20x + 30y \\ \text{subject to} & 3x + 2y \le 3000 \\ & x + 6y \le 1800 \\ & x \ge 0, y \ge 0 \end{array}$$

The (bounded) feasible region is shown in Figure 17.

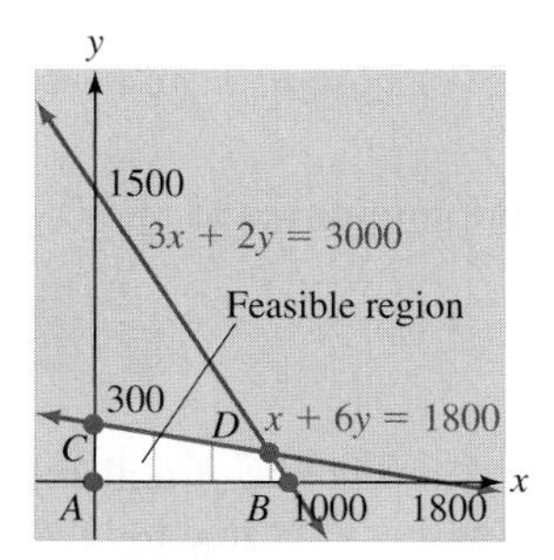

Figure 17

The corners and the values of the objective function are listed in the following table:

Point	Lines through Point	Coordinates	$p = 20x + 30y$
A		(0, 0)	0
B		(1000, 0)	20,000
C		(0, 300)	9,000
D	$3x + 2y = 3{,}000$ $x + 6y = 1{,}800$	(900, 150)	22,500

We are seeking to maximize the objective function p, so we look for corner points that give the maximum value for p. Since the maximum occurs at the point D, we conclude that the (only) optimal solution occurs at D. Thus, the company should make 900 quarts of Beginner's juice and 150 quarts of Advanced juice, for a largest possible profit of 22,500¢, or $225.

If, instead, the company made a profit of 40¢ on each quart of Beginner's juice and 20¢ on each quart of Advanced juice, then we would have $p = 40x + 20y$. This gives the following table:

Point	Lines through Point	Coordinates	$p = 40x + 20y$
A		(0, 0)	0
B		(1000, 0)	40,000
C		(0, 300)	6,000
D	$3x + 2y = 3{,}000$ $x + 6y = 1{,}800$	(900, 150)	39,000

We can see that, in this case, Acme should make 1000 quarts of Beginner's juice and no Advanced juice, for a largest possible profit of 40,000¢, or $400.

Before we go on . . . Notice that, in the first version of this problem, the company used all the water and juice concentrate. In the second version, it did not; it used all the water but not all the concentrate.

Example 3 • Investments

The Solid Trust Savings & Loan Company has set aside $25 million for loans to home buyers. Its policy is to allocate at least $10 million annually for luxury condominiums. A government housing development grant it receives requires, however, that at least one-third of its total loans be allocated to low-income housing.

a. Solid Trust's return on condominiums is 12%, and its return on low-income housing is 10%. How much should the company allocate for each type of housing to maximize its total return?

b. Redo part (a), assuming that the return is 12% on both condominiums and low-income housing.

Solution

a. We first identify the unknowns: Let x be the annual amount (in millions of dollars) allocated to luxury condominiums and let y be the annual amount allocated to low-income housing.

We now look at the constraints. The first constraint is mentioned in the first sentence: The total the company can invest is $25 million. Thus,

$$x + y \leq 25$$

(The company is not required to invest all $25 million; rather, it can invest *up to* $25 million.) Next, the company has allocated at least $10 million to condos. Rephrasing this in terms of the unknowns, we get

The amount allocated to condos is at least $10 *million.*

The phrase "is at least" means $\geq$. Thus, we obtain a second constraint:

$$x \geq 10$$

The third constraint is that at least one-third of the total financing must be for low-income housing. Rephrasing this, we say:

The amount allocated to low-income housing is at least one-third of the total.

Since the total investment will be $x + y$, we get

$$y \geq \frac{1}{3}(x + y)$$

We put this in the standard form of a linear inequality as follows:

$$3y \geq x + y \qquad \text{Multiply both sides by 3.}$$

$$-x + 2y \geq 0 \qquad \text{Subtract } x + y \text{ from both sides.}$$

There are no further constraints.

Now, what about the return on these investments? According to the data, the annual return is given by

$$p = 0.12x + 0.10y$$

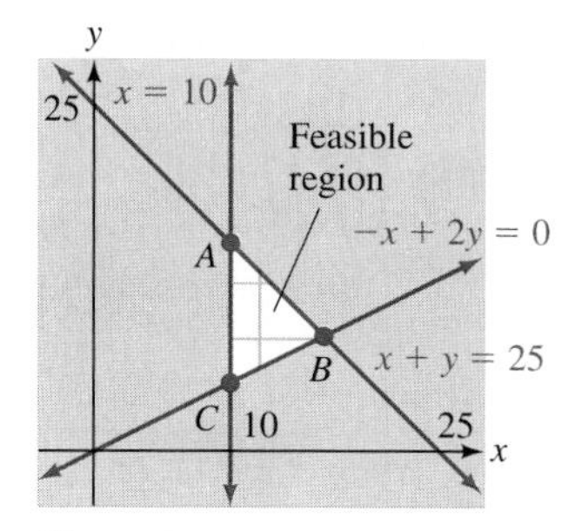

Figure 18

We want to make this quantity p as large as possible. In other words, we want to

$$\begin{aligned} \text{Maximize} \quad & p = 0.12x + 0.10y \\ \text{subject to} \quad & x + y \leq 25 \\ & x \geq 10 \\ & -x + 2y \geq 0 \\ & x \geq 0, y \geq 0 \end{aligned}$$

(Do you see why the inequalities $x \geq 0$ and $y \geq 0$ are slipped in here?) The feasible region is shown in Figure 18.

We now make a table that gives the return on investment at each corner point:

Point	Lines through Point	Coordinates	$p = 0.12x + 0.10y$
A	$x = 10$ $x + y = 25$	(10, 15)	2.7
B	$x + y = 25$ $-x + 2y = 0$	$\left(\frac{50}{3}, \frac{25}{3}\right)$	2.833
C	$x = 10$ $-x + 2y = 0$	(10, 5)	1.7

From the table we see that the values of x and y that maximize the return are $x = \frac{50}{3}$ and $y = \frac{25}{3}$, which give a total return of $2.833 million. In other words, the most profitable course of action is to invest $16.667 million in loans for condominiums and \$8.333 million in loans for low-income housing, giving a maximum annual return of \$2.833 million.

b. The LP problem is the same as for part (a) except for the objective function:

$$\begin{aligned} \text{Maximize} \quad & p = 0.12x + 0.12y \\ \text{subject to} \quad & x + y \le 25 \\ & x \ge 10 \\ & -x + 2y \ge 0 \\ & x \ge 0, y \ge 0 \end{aligned}$$

Here are the values of p at the three corners:

Point	Coordinates	$p = 0.12x + 0.12y$
A	$(10, 15)$	3
B	$\left(\frac{50}{3}, \frac{25}{3}\right)$	3
C	$(10, 5)$	1.8

Looking at the table, we see that a curious thing has happened: We get the same maximum annual return at both A and B. Thus, we could choose either option to maximize annual return. In fact, any point along the line segment AB will yield an annual return of \$3 million. For example, the point $(12, 13)$ lies on the line segment AB and also yields an annual revenue of \$3 million. This happens because the "isoreturn" lines are parallel to that edge.

Before we go on . . . What breakdowns of investments would lead to the *lowest* return for parts (a) and (b)?

Question The preceding examples all had bounded feasible regions. How do we deal with a problems whose feasible region is unbounded?

Answer If the region is unbounded, then, *provided there are optimal solutions,* the fundamental theorem of linear programming guarantees that the above method will work. The following procedure determines whether or not optimal solutions exist and finds them when they do.

Solving Linear Programming Problems in Two Unknowns (Unbounded Feasible Regions)

If the feasible region of an LP problem is unbounded, proceed as follows:

1. Draw a rectangle large enough so that all the corner points are inside the rectangle (and not on its boundary):

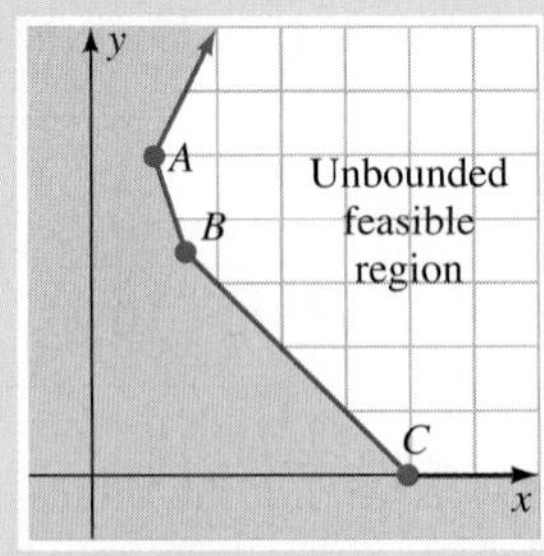

Corner points: *A*, *B*, and *C*

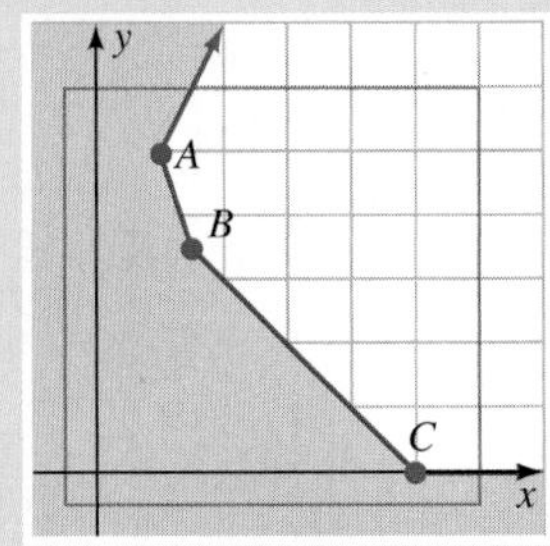

Corner points inside the rectangle

2. Shade the outside of the rectangle so as to define a new, bounded feasible region and locate the new corner points:

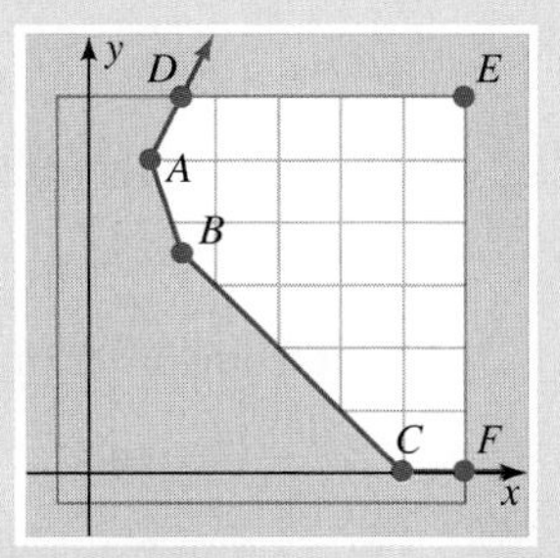

New corner points: *D, E,* and *F*

3. Obtain the optimal solutions, using this bounded feasible region.
4. If any optimal solutions occur at one of the original corner points (A, B, and C in the figure), then the LP problem has that corner point as an optimal solution. Otherwise, the LP problem has no optimal solutions. When the latter occurs, we say that the **objective function is unbounded,** because it can assume arbitrarily large (positive or negative) values.

Example 4 • Cost

You are the manager of a small store that specializes in hats, sunglasses, and other accessories. You are considering a sales promotion of a new line of hats and sunglasses. You will offer the sunglasses only to those who purchase two or more hats, so you will sell at least twice as many hats as pairs of sunglasses. Moreover, your supplier tells you that, due to seasonal demand, your order of sunglasses cannot exceed 100 pairs. To ensure that the sale items fill out the large display you have set aside, you estimate that you should order at least 210 items in all.

a. Assume that you will lose \$3 on every hat and \$2 on every pair of sunglasses sold. Given the constraints above, how many hats and pairs of sunglasses should you order to lose the least amount of money in the sales promotion?

b. Suppose instead that you lose \$1 on every hat sold but make a profit of \$5 on every pair of sunglasses sold. How many hats and pairs of sunglasses should you order to make the largest profit in the sales promotion?

c. Now suppose that you make a profit of \$1 on every hat sold but lose \$5 on every pair of sunglasses sold. How many hats and pairs of sunglasses should you order to make the largest profit in the sales promotion?

Solution

a. The unknowns are

x = number of hats you order

y = number of pairs of sunglasses you order

The objective is to minimize the total loss:

$$c = 3x + 2y$$

Now for the constraints. The requirement that you will sell at least twice as many hats as sunglasses can be rephrased as

The number of hats is at least twice the number of pairs of sunglasses

or $\quad x \geq 2y$

which, in standard form, is

$$x - 2y \geq 0$$

Next, your order of sunglasses cannot exceed 100 pairs, so

$$y \leq 100$$

Finally, you would like to sell at least 210 items in all, giving

$$x + y \geq 210$$

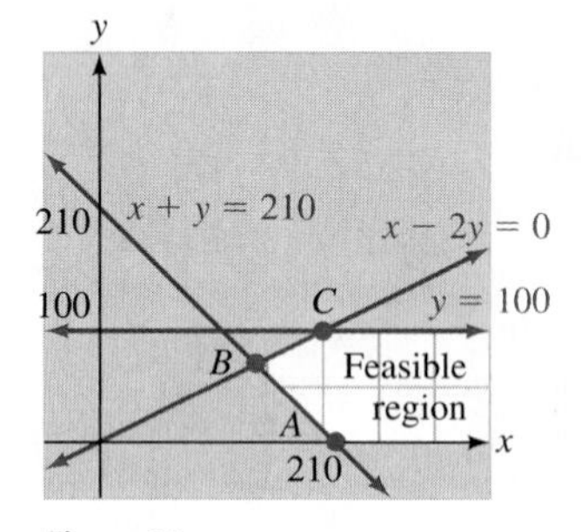

Figure 19

Thus, the LP problem is the following:

$$\begin{aligned} \text{Minimize} \quad & c = 3x + 2y \\ \text{subject to} \quad & x - 2y \geq 0 \\ & y \leq 100 \\ & x + y \geq 210 \\ & x \geq 0, y \geq 0 \end{aligned}$$

The feasible region is shown in Figure 19. This region is unbounded, so there is no guarantee that there are any optimal solutions. Following the procedure described above, we enclose the corner points in a rectangle as shown in Figure 20.

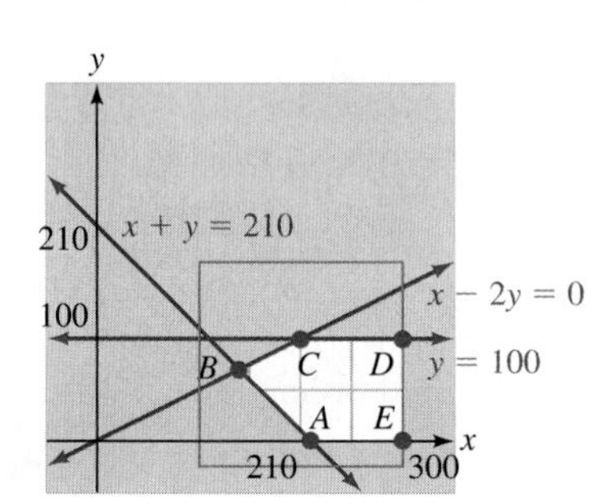

Figure 20

(There are many infinitely many possible rectangles we could have used. We chose one that gives convenient coordinates for the new corners.)

We now list all the corners of this bounded region along with the corresponding values of the objective function *c:*

Point	Lines through Point	Coordinates	$c = 3x + 2y$ (\$)
A		(210, 0)	630
B	$x + y = 210$ $x - 2y = 0$	(140, 70)	560
C	$x - 2y = 0$ $y = 100$	(200, 100)	800
D		(300, 100)	1100
E		(300, 0)	900

The corner point that gives the minimum value of the objective function c is B. Since B is one of the corner points of the original feasible region, we conclude that our linear programming problem has an optimal solution at B. Thus, the combination that gives the smallest loss is 140 hats and 70 pairs of sunglasses.

b. The LP problem is the following:

$$\begin{aligned}\text{Maximize}\quad & p = -x + 5y \\ \text{subject to}\quad & x - 2y \geq 0 \\ & y \leq 100 \\ & x + y \geq 210 \\ & x \geq 0, y \geq 0\end{aligned}$$

Since most of the work is already done for us in part (a), all we need to do is change the objective function in the table that lists the corner points:

Point	Lines through Point	Coordinates	$p = -x + 5y$ (\$)
A		$(210, 0)$	-210
B	$x + y = 210$ $x - 2y = 0$	$(140, 70)$	210
C	$x - 2y = 0$ $y = 100$	$(200, 100)$	300
D		$(300, 100)$	200
E		$(300, 0)$	-300

The corner point that gives the maximum value of the objective function p is C. Since C is one of the corner points of the original feasible region, we conclude that our LP problem has an optimal solution at C. Thus, the combination that gives the largest profit ($300) is 200 hats and 100 pairs of sunglasses.

c. The objective function is now $p = x - 5y$, which is the negative of the objective function used in part (b). Thus, the table of values of p is the same as in part (b), except that it has opposite signs in the p column. This time we find that the maximum value of p occurs at E. However, E is not a corner point of the original feasible region, so the LP problem has no optimal solution. Referring to Figure 19, we can make the objective p as large as we like by choosing a point far to the right in the unbounded feasible region. Thus, the objective function is unbounded; that is, it is possible to make an arbitrarily large profit.

Example 5 • Resource Allocation

You are composing a very avant-garde ballade for violins and bassoons. In your ballade each violinist plays a total of two notes and each bassoonist only one note. To make your ballade long enough, you decide that it should contain at least 200 instrumental notes. Furthermore, after playing the requisite two notes, each violinist will sing one soprano note, while each bassoonist will sing three soprano notes.[3] To make the ballade sufficiently interesting, you have decided on a minimum of 300 soprano notes. To give your composition a sense of balance, you wish to have no more than three times as many bassoonists as violinists. Violinists charge $200 per performance and bassoonists $400 per

[3] Whether or not these musicians are capable of singing decent soprano notes will be left to chance. You reason that a few bad notes will add character to the ballade.

performance. How many of each should your ballade call for in order to minimize personnel costs?

Solution First, the unknowns are x = number of violinists and y = number of bassoonists. The constraint on the number of instrumental notes implies that

$$2x + y \geq 200$$

because the total number is to be *at least* 200. Similarly, the constraint on the number of soprano notes is

$$x + 3y \geq 300$$

The next one is a little tricky. As usual, we reword it in terms of the quantities x and y:

The number of bassoonists should be no more than three times the number of violinists.

Thus,

$$y \leq 3x \quad \text{or} \quad 3x - y \geq 0$$

Finally, the total cost per performance will be

$$c = 200x + 400y$$

We wish to minimize total cost, so our linear programming problem is as follows:

$$\begin{aligned} \text{Minimize} \quad & c = 200x + 400y \\ \text{subject to} \quad & 2x + y \geq 200 \\ & x + 3y \geq 300 \\ & 3x - y \geq 0 \\ & x \geq 0, y \geq 0 \end{aligned}$$

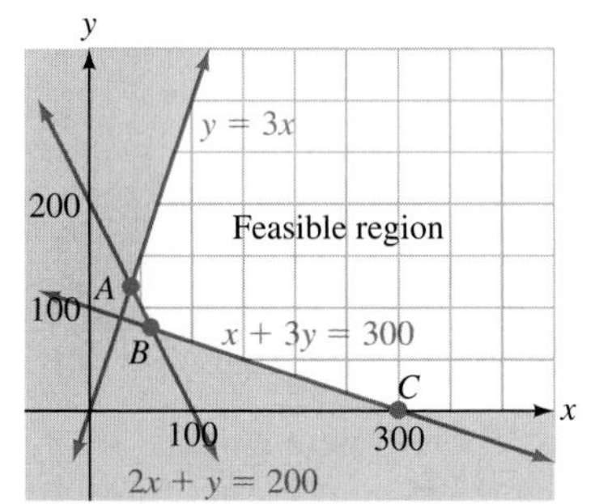

Figure 21

We get the feasible region shown in Figure 21.[4] The feasible region is unbounded, and so we add a convenient rectangle as before (Figure 22).

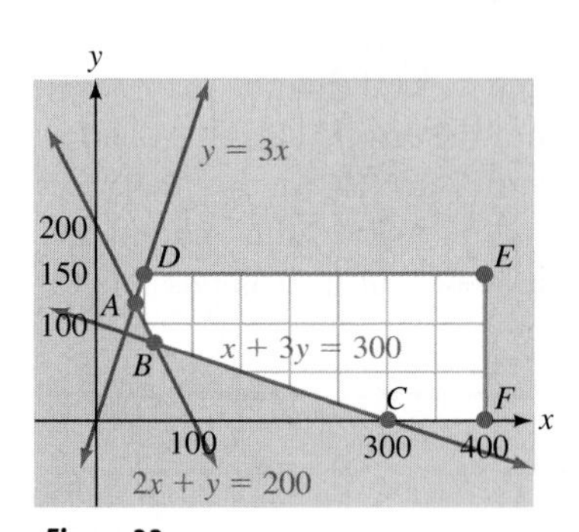

Figure 22

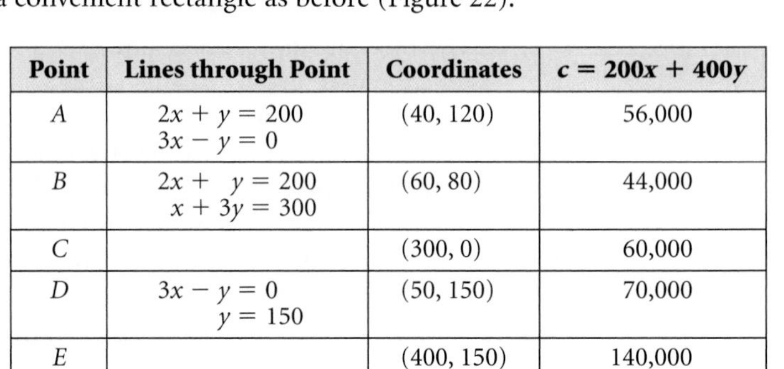

Point	Lines through Point	Coordinates	$c = 200x + 400y$
A	$2x + y = 200$ $3x - y = 0$	(40, 120)	56,000
B	$2x + y = 200$ $x + 3y = 300$	(60, 80)	44,000
C		(300, 0)	60,000
D	$3x - y = 0$ $y = 150$	(50, 150)	70,000
E		(400, 150)	140,000
F		(400, 0)	80,000

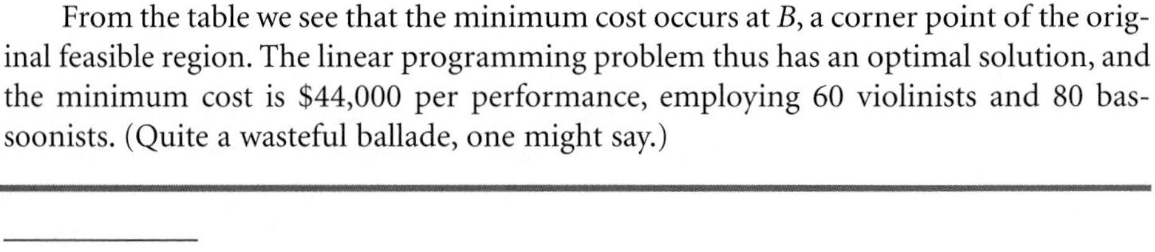

From the table we see that the minimum cost occurs at B, a corner point of the original feasible region. The linear programming problem thus has an optimal solution, and the minimum cost is \$44,000 per performance, employing 60 violinists and 80 bassoonists. (Quite a wasteful ballade, one might say.)

[4]Here is an example where graphing technology would help in determining the corner points. Unless you are very confident in the accuracy of your sketch, how do you know that the line $y = 3x$ falls to the left of the point B, for example? If it were to fall to the right, then B would not be a corner point, and the solution would be different. You could (and should) check that B satisfies the inequality $3x - y \geq 0$ so that the line falls to the left of B as shown. However, if you use a graphing calculator or computer, you can be fairly confident of the picture produced without doing further calculations.

Example 6 • Using Excel Solver to Solve a Linear Programming Problem

Excel has a built-in machine, called Solver, that solves optimization problems numerically, and it can be used to solve LP problems. (If Solver does not appear in the Tools menu, you should first install it using your Excel installation software; Solver is one of the Excel Add-Ins.) Let's now use Solver to solve the LP problem in the Example 5.

Minimize	$c = 200x + 400y$	Objective
subject to	$2x + y \geq 200$	Constraint 1
	$x + 3y \geq 300$	Constraint 2
	$3x - y \geq 0$	Constraint 3
	$x \geq 0, y \geq 0$	Constraint 4, constraint 5

Solution First, set up the problem in spreadsheet form as follows:

	A	B	C	D	E	F
1	x	y	Objective	Constraints		
2	0	0	=200*A2+400*B2	=2*A2+B2	200	Constraint 1
3	initial value of x	initial value of y		=A2+3*B2	300	Constraint 2
4				=3*A2-B2	0	Constraint 3
5				=A2	0	Constraint 4
6				=B2	0	Constraint 5

Cell A2 contains the value of x, and cell B2 contains the value of y. (We have set them both to zero in anticipation that Solver will adjust them to satisfy the constraints and give an optimal solution.) Next, select `Solver` in the `Tools` menu to bring up the Solver dialog box. Figure 23 shows the dialog box with all the fields necessary to solve the problem completed.

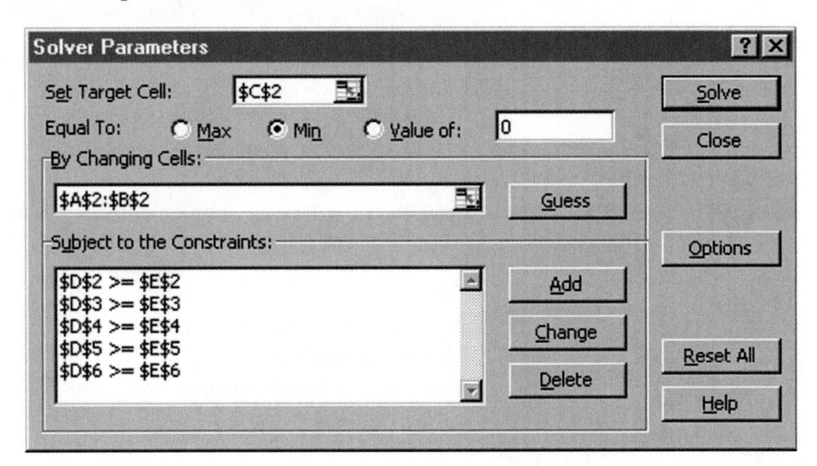

Figure 23

- The `Target Cell` refers to the cell that contains the objective function.
- `Min` is selected because we are minimizing the objective function.
- `Changing Cells` are obtained by selecting the cells that contain the current values of x and y.
- Constraints are added one at a time by clicking on the `Add` button and selecting the cells that contain the left- and right-hand sides of each inequality, as well as the type of inequality.

When you are done, click on `Solve` and the optimal solution appears in A2 and B2, with the minimum value of c appearing in cell C2.

Guideline: Recognizing a Linear Programming Problem, Setting Up Inequalities, and Dealing with Unbounded Regions

Question How do I recognize when an application leads to an LP problem as opposed to a system of linear equations?

Answer Here are some cues that suggest an LP problem:

- Key phrases suggesting inequalities rather than equalities, like *at most, up to, no more than, at least,* and *or more.*
- A quantity that is being maximized or minimized (this will be the objective). Key phrases are *maximum, minimum, most, least, largest, greatest, smallest, as large as possible,* and *as small as possible.*

Question How do I deal with tricky phrases like "there should be no more than twice as many nuts as bolts" or "at least 50% of the total should be bolts"?

Answer The easiest way to deal with phrases like this is to use the technique we discussed in Chapter 2: Reword the phrases using "the number of," as in

The number of nuts (x) is no more than twice the number of bolts (y).	$x \le 2y$
The number of bolts is at least 50% of the total.	$y \ge 0.50(x + y)$

Question Do I always have to add a rectangle to deal with unbounded regions?

Answer Under some circumstances you can tell right away whether optimal solutions exist, even when the feasible region is unbounded:

Note that the following apply only when we have the constraints $x \ge 0$ and $y \ge 0$.

1. If you are minimizing $c = ax + by$ with a and b nonnegative, then optimal solutions always exist. [Examples 4(a) and 5 are of this type.]
2. If you are maximizing $p = ax + by$ with a and b nonnegative (and not both zero), then there is no optimal solution unless the feasible region is bounded.

Do you see why statements 1 and 2 are true?

Web Site

For an online utility that does everything (solves LP problems with two unknowns and even draws the feasible region) follow

Web site → Online Utilities → Linear Programming Grapher

4.2 EXERCISES

In Exercises 1–20, solve the LP problems. If no optimal solution exists, indicate whether the feasible region is empty or the objective function is unbounded.

1. Maximize $p = x + y$
subject to $x + 2y \le 9$
$2x + y \le 9$
$x \ge 0, y \ge 0$

2. Maximize $p = x + 2y$
subject to $x + 3y \le 24$
$2x + y \le 18$
$x \ge 0, y \ge 0$

3. Minimize $c = x + y$
subject to $x + 2y \ge 6$
$2x + y \ge 6$
$x \ge 0, y \ge 0$

4. Minimize $c = x + 2y$
subject to $x + 3y \ge 30$
$2x + y \ge 30$
$x \ge 0, y \ge 0$

5. Maximize $p = 3x + y$
subject to $3x - 7y \le 0$
$7x - 3y \ge 0$
$x + y \le 10$
$x \ge 0, y \ge 0$

6. Maximize $p = x - 2y$
subject to $x + 2y \le 8$
$x - 6y \le 0$
$3x - 2y \ge 0$
$x \ge 0, y \ge 0$

7. Maximize $p = 3x + 2y$
subject to $0.2x + 0.1y \le 1$
$0.15x + 0.3y \le 1.5$
$10x + 10y \le 60$
$x \ge 0, y \ge 0$

8. Maximize $p = x + 2y$
subject to $30x + 20y \le 600$
$0.1x + 0.4y \le 4$
$0.2x + 0.3y \le 4.5$
$x \ge 0, y \ge 0$

9. Minimize $c = 0.2x + 0.3y$
subject to $0.2x + 0.1y \ge 1$
$0.15x + 0.3y \ge 1.5$
$10x + 10y \ge 80$
$x \ge 0, y \ge 0$

10. Minimize $c = 0.4x + 0.1y$
subject to $30x + 20y \ge 600$
$0.1x + 0.4y \ge 4$
$0.2x + 0.3y \ge 4.5$
$x \ge 0, y \ge 0$

11. Maximize and minimize $p = x + 2y$
subject to $x + y \ge 2$
$x + y \le 10$
$x - y \le 2$
$x - y \ge -2$

12. Maximize and minimize $p = 2x - y$
subject to $x + y \ge 2$
$x - y \le 2$
$x - y \ge -2$
$x \le 10, y \le 10$

13. Maximize $p = 2x + 3y$
subject to $0.1x + 0.2y \ge 1$
$2x + y \ge 10$
$x \ge 0, y \ge 0$

14. Maximize $p = 3x + 2y$
subject to $0.1x + 0.1y \ge 0.2$
$y \le 10$
$x \ge 0, y \ge 0$

15. Minimize $c = 2x + 4y$
subject to $0.1x + 0.1y \ge 1$
$x + 2y \ge 14$
$x \ge 0, y \ge 0$

16. Maximize $p = 2x + 3y$
subject to $-x + y \ge 10$
$x + 2y \le 12$
$x \ge 0, y \ge 0$

17. Minimize $c = 3x - 3y$
subject to $\frac{x}{4} \le y$
$y \le \frac{2x}{3}$
$x + y \ge 5$
$x + 2y \le 10$
$x \ge 0, y \ge 0$

18. Minimize $c = -x + 2y$
subject to $y \le \frac{2x}{3}$
$x \le 3y$
$y \ge 4$
$x \ge 6$
$x + y \le 16$

19. Maximize $p = x + y$
subject to $x + 2y \ge 10$
$2x + 2y \le 10$
$2x + y \ge 10$
$x \ge 0, y \ge 0$

20. Minimize $c = 3x + y$
subject to $10x + 20y \ge 100$
$0.3x + 0.1y \ge 1$
$x \ge 0, y \ge 0$

APPLICATIONS

21. Resource Allocation You manage an ice cream factory that makes two flavors: Creamy Vanilla and Continental Mocha. Into each quart of Creamy Vanilla go 2 eggs and 3 cups of cream. Into each quart of Continental Mocha go 1 egg and 3 cups of cream. You have in stock 500 eggs and 900 cups of cream. You make a profit of \$3/quart of Creamy Vanilla and \$2/quart of Continental Mocha. How many quarts of each flavor should you make in order to earn the largest profit?

22. Resource Allocation Podunk Institute of Technology's Math Department offers two courses: Finite Math and Applied Calculus. Each section of Finite Math has 60 students, and each section of Applied Calculus has 50. The department is allowed to offer a total of up to 110 sections. Furthermore, no more than 6000 students want to take a math course (no student will take more than one math course). Suppose the university makes a profit of \$100,000/section of Finite Math and \$50,000/section of Applied Calculus (the profit is the difference between what the students are charged and what the professors are paid). How many sections of each course should the department offer to make the largest profit?

23. Nutrition Ruff Inc. makes dog food out of chicken and grain. Chicken has 10 grams of protein and 5 grams of fat per ounce, and grain has 2 grams of protein and 2 grams of fat per ounce. A bag of dog food must contain at least 200 grams of protein and at least 150 grams of fat. If chicken costs 10¢/ounce and grain costs 1¢/ounce, how many ounces of each should Ruff use in each bag of dog food in order to minimize cost?

24. Purchasing Enormous State University's Business School is buying computers. The school has two models from which to choose, the Pomegranate and the iZac. Each Pomegranate comes with 400 megabytes of memory and 80 gigabytes of disk space; each iZac has 300 megabytes of memory and 100 gigabytes of disk space. For reasons related to its accreditation, the school would like to say that it has a total of at least 48,000 megabytes of memory and at least 12,800 gigabytes of disk space. If the Pomegranate and the iZac cost $2000 each, how many of each should the school buy to keep the cost as low as possible?

25. **Nutrition** Each serving of Gerber Mixed Cereal for Baby contains 60 calories and 11 grams of carbohydrates. Each serving of Gerber Mango Tropical Fruit Dessert contains 80 calories and 21 grams of carbohydrates. If the cereal costs 30¢/serving and the dessert costs 50¢/serving and you want to provide your child with at least 140 calories and at least 32 grams of carbohydrates, how can you do so at the least cost? (Fractions of servings are permitted.)

SOURCE: Nutrition information printed on the containers.

26. **Nutrition** Each serving of Gerber Mixed Cereal for Baby contains 60 calories, 10 grams of carbohydrates, and no vitamin C. Each serving of Gerber Apple Banana Juice contains 60 calories, 15 grams of carbohydrates, and 120% of the U.S. Recommended Daily Allowance (RDA) of vitamin C for infants. The cereal costs 10¢/serving and the juice costs 30¢/serving. If you want to provide your child with at least 120 calories, at least 25 grams of carbohydrates, and at least 60% of the RDA of vitamin C for infants, how can you do so at the least cost? (Fractions of servings are permitted.)

SOURCE: Nutrition information printed on the containers.

Creatine Supplements Exercises 27 and 28 are based on the following data on three body-building supplements popular in 2002 (figures shown correspond to a single serving):

	Creatine (g)	Carbohydrates (g)	Taurine (g)	Alpha Lipoic Acid (mg)	Cost ($)
Cell-Tech (MuscleTech)	10	75	2	200	2.20
RiboForce HP (EAS)	5	15	1	0	1.60
Creatine Transport (Kaizen)	5	35	1	100	0.60

SOURCE: Nutritional information supplied by the manufacturers/ www.netrition.com. Cost/serving as quoted on www.netrition.com as of February 26, 2002.

27. You are thinking of combining Cell-Tech and RiboForce HP to obtain a 10-day supply that provides at least 80 grams of creatine and at least 10 grams of taurine, but no more than 750 grams of carbohydrates and no more than 1000 milligrams of alpha lipoic acid. How many servings of each supplement should you combine to meet your specifications at the least cost?

28. You are thinking of combining Cell-Tech and Creatine Transport to obtain a 10-day supply that provides at least 80 grams of creatine and at least 10 grams of taurine, but no more than 600 grams of carbohydrates and no more than 2000 milligrams of alpha lipoic acid. How many servings of each supplement should you combine to meet your specifications at the least cost?

Investing Exercises 29 and 30 are based on the following data on four U.S. computer stocks:

	Price ($)	Dividend Yield (%)	Earnings/ Share ($)
IBM	100	0.5	4.00
HWP (Hewlett Packard)	20	1.50	0.40
DELL	25	0	0.50
CPQ (Compaq)	10	0.90	0.10

Stocks were trading at or near the given prices in February 2002. Earnings per share are rounded, based on values on February 23, 2002.
SOURCE: http://money.excite.com, February 23, 2002.

29. You are planning to invest up to $10,000 in IBM and Hewlett Packard shares. You want your investment to yield at least $100 in dividends, and you want to maximize your total earnings. How many shares of each company should you purchase?

30. Repeat Exercise 29 for a portfolio based on Dell and Compaq shares, assuming that you want your investment to yield at least $18 in dividends.

31. **Investments** Your portfolio manager has suggested two high-yielding stocks: Philip Morris (MO) and R. J. Reynolds Tobacco (RJR). Each MO share costs $50, yields 4.5% in dividends, and has a risk index of 2.0. Each RJR share costs $55, yields 5% in dividends, and has a risk index of 3.0. You have up to $12,100 to invest and would like to earn at least $550 in dividends. How many shares (to the nearest tenth of a unit) of each stock should you purchase, to meet your requirements and minimize the total risk index for your portfolio? What is the minimum total risk index?

Share prices and yields are approximate as of February 2002. SOURCE: http://money.excite.com, February 23, 2002. Risk indices are fictitious.

32. **Investments** Your friend's portfolio manager has suggested two high-yielding stocks: Consolidated Edison (ED) and Royal Bank of Scotland (RBS-K). Each ED share costs $40, yields 5.5% in dividends, and has a risk index of 1.0. Each RBS-K share costs $25, yields 7.5% in dividends, and has a risk index of 1.5. You have up to $30,000 to invest and would like to earn at least $1650 in dividends. How many shares (to the nearest tenth of a unit) of each stock should you purchase, to meet your requirements and minimize the total risk index for your portfolio?

Share prices and yields are approximate as of February 2002. SOURCE: http://money.excite.com, February 23, 2002. Risk indices are fictitious.

33. **Gasoline Stations** The following table shows 1993 statistics on gasoline sales in New York and Connecticut:

	Annual Sales (billions of gal)	Number of Gas Stations	State Tax Rates ($/gal)	Average Revenue* ($/gal)
New York	5.6	7000	0.20	1.20
Connecticut	2	2000	0.30	1.20

Figures rounded for ease of computation. SOURCE: National Petroleum News/U.S. Energy Information Administration, State Energy Offices/*New York Times*, October 31, 1994, p. B1.

*Price includes state taxes.

a. Calculate the average annual gasoline sales per gas station in each state and hence the average annual revenue of a gas station in each state from sales of gasoline.

b. Your Connecticut-based firm is planning to open a total of up to 20 gas stations in New York and Connecticut. You would like the firm to generate at least $20.4 million/year in revenues from sales of gasoline, while keeping total state taxes as low as possible. Given these requirements, how many gas stations should you open in each state?

34. Gasoline Stations Refer to the table in Exercise 33.

a. Calculate the average annual gasoline sales per gas station in each state and hence the average *after-tax* annual revenue of a gas station in each state from sales of gasoline.

b. Your Connecticut-based firm is planning to open a total of up to 30 gas stations in New York and Connecticut. You would like the firm to pay no more than $6.2 million in taxes and the firm's after-tax revenues (from sales of gasoline) to be as large as possible. Given these requirements, how many gas stations should you open in each state?

35. Resource Allocation Your salami-manufacturing plant can order up to 1000 pounds of pork and 2400 pounds of beef each day for use in manufacturing its two specialties: Count Dracula Salami and Frankenstein Sausage. Production of the Count Dracula variety requires 1 pound of pork and 3 pounds of beef for every salami, and the Frankenstein variety requires 2 pounds of pork and 2 pounds of beef for every sausage. In view of your heavy investment in advertising Count Dracula Salami, you have decided that at least one-third of the total production should be Count Dracula. On the other hand, due to the health-conscious consumer climate, your Frankenstein Sausage (sold as having less beef) is earning your company a profit of $3/sausage, while sales of the Count Dracula variety are down and it is earning your company only $1/salami. Given these restrictions, how many of each kind of sausage should you produce to maximize profits, and what is the maximum possible profit?

36. Project Design Megabuck Hospital is to build a state-subsidized nursing home catering to both homeless and high-income patients. State regulations require that every subsidized nursing home must house a minimum of 1000 homeless patients and no more than 750 high-income patients in order to qualify for state subsidies. The overall capacity of the hospital is to be 2100 patients. The board of directors, under pressure from a neighborhood group, insists that the number of homeless patients should not exceed twice the number of high-income patients. Due to the state subsidy, the hospital will make an average profit of $10,000/month for every homeless patient it houses, whereas the profit for each high-income patient is estimated at $8000/month. How many of each type of patient should it house in order to maximize profit?

37. Television Advertising In February 2002 each episode of *Becker* was typically seen in 8.3 million homes, and each episode of *The Simpsons* was seen in 7.5 million homes. Your marketing firm has been hired to promote Bald No More, Inc.'s, hair-replacement process by buying at least 30 commercial spots during episodes of *Becker* and *The Simpsons.* The cable company running *Becker* has quoted a price of $2000/spot, and the cable company showing *The Simpsons* has quoted a price of $1500/spot. Bald No More's advertising budget for TV commercials is $70,000, and it would like no more than 50% of the total number of spots to appear on *The Simpsons.* How many spots should you purchase on each show to reach the most homes most often?

Ratings are for February 18–24, 2002. SOURCE: Nielsen Media Research/www.ptd.net, February 28, 2002.

38. Television Advertising In February 2002 each episode of *Boston Public* was typically seen in 7.0 million homes, and each episode of *NYPD Blue* was seen in 7.8 million homes. Your marketing firm has been hired to promote Gauss Jordan Sneakers by buying at least 30 commercial spots during episodes of *Boston Public* and *NYPD Blue.* The cable company running *Boston Public* has quoted a price of $2000/spot, and the cable company showing *NYPD Blue* has quoted a price of $3000/spot. Gauss Jordan Sneakers' advertising budget for TV commercials is $70,000, and it would like at least 75% of the total number of spots to appear on *Boston Public.* How many spots should you purchase on each show to reach the most homes most often?

Ratings are for February 18–24, 2002. SOURCE: Nielsen Media Research/www.ptd.net, February 28, 2002.

39. Management You are the service manager for a supplier of closed-circuit television systems. Your company can provide up to 160 hours/week of technical service for your customers, although the demand for technical service far exceeds this amount. As a result, you have been asked to develop a model to allocate service technicians' time between new customers (those still covered by service contracts) and old customers (whose service contracts have expired). To ensure that new customers are satisfied with your company's service, the sales department has instituted a policy that at least 100 hours/week be allocated to servicing new customers. At the same time, your superiors have informed you that the company expects your department to generate at least $1200/week in revenues. Technical service time for new customers generates an average of $10/hour (because much of the service is still under warranty) and for old customers generates $30/hour. How many hours each week should you allocate to each type of customer to generate the most revenue?

Loosely based on a similar problem in D. R. Anderson, D. J. Sweeny, and T. A. Williams, *An Introduction to Management Science,* 6th ed. (St. Paul: West, 1991).

40. Scheduling The Scottsville Textile Mill produces several different fabrics on eight dobbie mills that operate 24 hours/day and are scheduled for 30 days in the coming month. The mill will produce only fabric 1 and fabric 2 during the coming month. Each dobbie mill can turn out 4.63 yards/hour of either fabric. Assume that there is a monthly demand of 16,000 yards of fabric 1 and 12,000 yards of fabric 2. Profits are calculated as 33¢/yard for each fabric produced on the dobbie mills.

a. Will it be possible to satisfy total demand?

b. In the event that total demand is not satisfied, Scottsville will need to purchase the fabrics from another mill to make up the shortfall. Its profits on resold fabrics ordered from another mill amount to 20¢/yard for fabric 1 and 16¢/yard for fabric 2. How many yards of each fabric should it produce to maximize profits?

Adapted from J. D. Camm, P. M. Dearing, and S. K. Tadisina, *The Calhoun Textile Mill Case*, as presented for a case study in D. R. Anderson, D. J. Sweeny, and T. A. Williams, *An Introduction to Management Science*, 6th ed. (St. Paul: West, 1991). Our exercise uses a subset of the data given in the cited study.

41. Planning My friends: I, the mighty Brutus, have decided to prepare for retirement by instructing young warriors in the arts of battle and diplomacy. For each hour spent in battle instruction, I have decided to charge 50 ducats. For each hour in diplomacy instruction, I will charge 40 ducats. Due to my advancing years, I can spend no more than 50 hours/week instructing the youths, although the great Jove knows that they are sorely in need of instruction! Due to my fondness for physical pursuits, I have decided to spend no more than one-third of the total time in diplomatic instruction. However, the present border crisis with the Gauls is a sore indication of our poor abilities as diplomats. As a result, I have decided to spend at least 10 hours/week instructing in diplomacy. Finally, to complicate things further, there is the matter of Scarlet Brew: I have estimated that each hour of battle instruction will require 10 gallons of Scarlet Brew to quench my students' thirst and that each hour of diplomacy instruction, being less physically demanding, requires half that amount. Since my harvest of red berries has far exceeded my expectations, I estimate that I'll have to use at least 400 gallons/week in order to avoid storing the fine brew at great expense. Given all these restrictions, how many hours each week should I spend in each type of instruction to maximize my income?

42. Planning Repeat Exercise 41 with the following changes: I would like to spend no more than half the total time in diplomatic instruction, and I must use at least 600 gallons of Scarlet Brew.

43. Resource Allocation One day Gillian the Magician summoned the wisest of her women. "Devoted followers," she began, "I have a quandary. As you well know, I possess great expertise in sleep spells and shock spells, but unfortunately, these are proving to be a drain on my aural energy resources; each sleep spell costs me 500 pico-shirleys of aural energy, and each shock spell requires 750 pico-shirleys. Clearly, I would like to hold my overall expenditure of aural energy to a minimum and still meet my commitments in protecting the Sisterhood from the ever-present threat of trolls. Specifically, I have estimated that each sleep spell keeps us safe for an average of 2 minutes, and every shock spell protects us for about 3 minutes. We certainly require enough protection to last 24 hours of each day, and possibly more, just to be safe. At the same time, I have noticed that each of my sleep spells can immobilize 3 trolls at once, but one of my typical shock spells (having a narrower range) can immobilize only 2 trolls at once. We are faced, my sisters, with an onslaught of 1200 trolls each day! Finally, as you are no doubt aware, the bylaws dictate that for a Magician of the Order to remain in good standing, the number of shock spells must be between one-quarter and one-third the number of shock and sleep spells combined. What do I do, oh Wise Ones?"

44. Risk Management The Grand Vizier of the Kingdom of Um is being blackmailed by numerous individuals and is having a very difficult time keeping his blackmailers from going public. He has been keeping them at bay with two kinds of payoff: gold bars from the Royal Treasury and political favors. Through bitter experience he has learned that each payoff in gold gives him peace for an average of about 1 month, whereas each political favor seems to earn him about $1\frac{1}{2}$ months of reprieve. To maintain his flawless reputation in the Court, he feels he cannot afford any revelations about his tainted past to come to light within the next year. Thus, it is imperative that his blackmailers be kept at bay for 12 months. Furthermore, he would like to keep the number of gold payoffs at no more than one-quarter of the combined number of payoffs because the outward flow of gold bars might arouse suspicion on the part of the Royal Treasurer. The Grand Vizier feels that he can do no more than seven political favors each year without arousing undue suspicion in the Court. The gold payoffs tend to deplete his travel budget. (The treasury has been subsidizing his numerous trips to the Himalayas.) He estimates that each gold bar removed from the treasury will cost him four trips. On the other hand, since the administering of political favors tends to cost him valuable travel time, he suspects that each political favor will cost him about two trips. Now, he would obviously like to keep his blackmailers silenced and lose as few trips as possible. What is he to do? How many trips will he lose in the next year?

COMMUNICATION AND REASONING EXERCISES

45. If an LP problem has a bounded, nonempty feasible region, then optimal solutions

(A) must exist. **(B)** may or may not exist.
(C) cannot exist.

46. If a linear programming problem has an unbounded, nonempty feasible region, then optimal solutions

(A) must exist. **(B)** may or may not exist.
(C) cannot exist.

47. What can you say if the optimal value occurs at two adjacent corner points?

48. Describe at least one drawback to using the graphical method to solve an LP problem arising from a real-life situation.

49. Create an LP problem in two variables that has no optimal solution.

50. Create an LP problem in two variables that has more than one optimal solution.

51. Create an interesting scenario leading to the following LP problem:

$$\begin{aligned} \text{Maximize} \quad & p = 10x + 10y \\ \text{subject to} \quad & 20x + 40y \le 1000 \\ & 30x + 20y \le 1200 \\ & x \ge 0, y \ge 0 \end{aligned}$$

52. Create an interesting scenario leading to the following LP problem:

$$\begin{aligned} \text{Minimize} \quad & c = 10x + 10y \\ \text{subject to} \quad & 20x + 40y \ge 1000 \\ & 30x + 20y \ge 1200 \\ & x \ge 0, y \ge 0 \end{aligned}$$

53. Use an example to show why there may be no optimal solutions to an LP problem if the feasible region is unbounded.

54. Use an example to illustrate that, in the event that an optimal solution does occur despite an unbounded feasible region, that solution corresponds to a corner point of the feasible region.

55. You are setting up an LP problem for Fly-by-Night Airlines with the unknowns x and y, where x represents the number of first-class tickets it should issue for a specific flight and y represents the number of business-class tickets it should issue for that flight. The problem is to maximize profit. You find that there are two different corner points that maximize the profit. How do you interpret this?

56. In the situation described in Exercise 55, you find that there are no optimal solutions. How do you interpret this?

57. Consider the following example of a *nonlinear* programming problem: Maximize $p = xy$ subject to $x \ge 0, y \ge 0, x + y \le 2$. Show that p is zero on every corner point but is greater than zero at many noncorner points.

58. Solve the nonlinear programming problem in Exercise 57.

4.3 The Simplex Method: Solving Standard Maximization Problems

The method discussed in Section 4.2 works quite well for LP problems in two unknowns, but what about three or more unknowns? Since we need an axis for each unknown, we would need to draw graphs in three dimensions (where we have x, y, and z coordinates) to deal with problems in three unknowns; we would have to draw in hyperspace to answer questions involving four or more unknowns. Given the state of technology when this book was written, we can't easily do this. So we need another method for solving LP problems that will work for any number of unknowns. One such method, called the **simplex method,** has been the method of choice since it was invented by George Dantzig in 1947.[5] To illustrate it best, we first use it to solve only so-called standard maximization problems.

Standard Maximization Problem

A **standard maximization problem in n unknowns** is an LP problem in which we are required to *maximize* (not minimize) an objective function of the form*

$$p = ax + by + cz + \cdots \text{ (}n\text{ terms)},$$

where $a, b, c, \ldots$ are numbers, subject to the constraints

$$x \ge 0, y \ge 0, z \ge 0, \ldots$$

*As in the chapter on linear equations we will seldom use the traditional subscripted variables $x_1, x_2, \ldots$. These are very useful names when you start running out of letters of the alphabet, but we should not find ourselves in that predicament.

[5] The first radically different method of solving LP problems was the ellipsoid algorithm published in 1979 by Soviet mathematician L. G. Khachiyan. In 1984 N. Karmarkar, a researcher at Bell Labs, created a much more efficient method now known as Karmarkar's algorithm. Although these methods (and others since developed) can be shown to be faster than the simplex method in the worst cases, it seems to be true that the simplex method is still the fastest in the applications that arise in practice.

and further constraints of the form

$$Ax + By + Cz + \cdots \le N$$

where $A, B, C, \ldots, N$ are numbers with N *nonnegative.* Note that the inequality here must be $\le$, *not* $=$ or $\ge$.

Quick Examples

1. Maximize $p = 2x - 3y + 3z$
subject to
$$2x + z \le 7$$
$$-x + 3y - 6z \le 6$$
$$x \ge 0, y \ge 0, z \ge 0$$
This is a standard maximization problem.

2. Maximize $p = 2x - 3y + 3z$
subject to
$$2x + z \ge 7$$
$$-x + 3y - 6z \le 6$$
$$x \ge 0, y \ge 0, z \ge 0$$
This is *not* a standard maximization problem.

The inequality $2x + z \ge 7$ cannot be written in the required form. If we reverse the inequality by multiplying both sides by -1, we get $-2x - z \le -7$, and -7 is negative.

The idea behind the simplex method is this: In any LP problem, there is a feasible region. If there are only two unknowns, we can draw the region; if there are three unknowns, it is a solid region in space; and if there are four or more unknowns, it is an abstract, higher-dimensional region. But it is a faceted region with corners (think of a diamond), and it is at one of these corners that we will find the optimal solution. Geometrically, what the simplex method does is to start at the corner where all the unknowns are zero (possible because we are talking of standard maximization problems) and then walk around the region, from corner to adjacent corner, always increasing the value of the objective function, until the best corner is found. In practice we will visit only a small number of the corners before finding the right one. Algebraically, as we are about to see, this walking around is accomplished by matrix manipulations of the same sort as those used in the chapter on systems of linear equations.

We describe the method while working through an example.

Example 1 • Introducing the Simplex Method

Maximize $p = 3x + 2y + z$
subject to
$$2x + 2y + z \le 10$$
$$x + 2y + 3z \le 15$$
$$x \ge 0, y \ge 0, z \ge 0$$

Solution

Step 1 *Convert to a system of linear equations.* The inequalities $2x + 2y + z \le 10$ and $x + 2y + 3z \le 15$ are less convenient than equations. Look at the first inequality. It says that the left-hand side, $2x + 2y + z$, must have some positive number (or zero) *added to it* if it is to equal 10. Since we don't yet know what x, y, and z are, we are not yet sure what number to add to the left-hand side. So we invent a new unknown, $s \ge 0$, called a **slack variable,** to "take up the slack," so that

$$2x + 2y + z + s = 10$$

Turning to the next inequality, $x + 2y + 3z \leq 15$, we now add a slack variable to its left-hand side, to get it up to the value of the right-hand side. We might have to add a different number than we did the last time, so we use a new slack variable, $t \geq 0$, and obtain

$$x + 2y + 3z + t = 15$$ Use a different slack variable for each constraint.

Now we write the system of equations we have (including the one that defines the objective function) in standard form:

$$\begin{aligned} 2x + 2y + z + s \qquad\qquad &= 10 \\ x + 2y + 3z \qquad + t \qquad &= 15 \\ -3x - 2y - z \qquad\qquad + p &= 0 \end{aligned}$$

Note three things: First, all the variables are neatly aligned in columns, as they were in Chapter 2. Second, in rewriting the objective function $p = 3x + 2y + z$, we have left the coefficient of p as $+1$ and brought the other variables over to the same side of the equation as p. This will be our standard procedure from now on. *Don't* write $3x + 2y + z - p = 0$ (even though it means to the same thing) because the negative coefficients will be important in later steps. Third, the above system of equations has fewer equations than unknowns and hence cannot have a unique solution.

Step 2 *Set up the initial tableau.* We represent our system of equations by the following table (which is simply the augmented matrix in disguise), called the **initial tableau:**

	x	y	z	s	t	p	
	2	2	1	1	0	0	10
	1	2	3	0	1	0	15
	−3	−2	−1	0	0	1	0

The labels along the top keep track of which columns belong to which variables.

Now notice a peculiar thing. If we rewrite the matrix using the variables s, t, and p first, we get the following matrix

$$\begin{array}{cccccc|c} s & t & p & x & y & z & \\ \end{array}$$
$$\begin{bmatrix} 1 & 0 & 0 & 2 & 2 & 1 & 10 \\ 0 & 1 & 0 & 1 & 2 & 3 & 15 \\ 0 & 0 & 1 & -3 & -2 & -1 & 0 \end{bmatrix}$$ Matrix with s, t, and p columns first

which is already in reduced form. We can therefore read off the general solution (see Section 2.2) to our system of equations as

$$s = 10 - 2x - 2y - z$$

$$t = 15 - x - 2y - 3z$$

$$p = 0 + 3x + 2y + z$$

x, y, z arbitrary

We get a whole family of solutions, one for each choice of x, y, and z. One possible choice is to set x, y, and z all equal to zero. This gives the particular solution

$$s = 10 \quad t = 15 \quad p = 0 \quad x = 0 \quad y = 0 \quad z = 0$$ Set $x = y = z = 0$ above.

This solution is called the **basic solution** associated with the tableau. The variables s and t are called the **active** variables, and x, y, and z are the **inactive** variables. (Other terms used are **basic** and **nonbasic** variables.)

We can obtain the basic solution directly from the tableau as follows:

- The active variables correspond to the cleared columns (columns with only one nonzero entry).
- The values of the active variables are calculated as shown below.
- All other variables are inactive and set equal to zero.

Inactive $x = 0$	Inactive $y = 0$	Inactive $z = 0$	Active $s = \frac{10}{1}$	Active $t = \frac{15}{1}$	Active $p = \frac{0}{1}$	
x	**y**	**z**	**s**	**t**	**p**	
2	2	1	**1**	0	0	**10**
1	2	3	0	**1**	0	**15**
−3	−2	−1	0	0	**1**	**0**

As an additional aid to recognizing which variables are active and which are inactive, we label each row with the name of the corresponding active variable. Thus, the complete initial tableau looks like this:

	x	**y**	**z**	**s**	**t**	**p**	
s	2	2	1	1	0	0	10
t	1	2	3	0	1	0	15
p	−3	−2	−1	0	0	1	0

This basic solution represents our starting position $x = y = z = 0$ in the feasible region in xyz space.

Question How do we move to another corner point?

Answer We choose a pivot in one of the first three columns of the tableau and clear its column.[6] Then we will get a different basic solution, which corresponds to another corner point. Thus, to move from corner point to corner point, all we have to do is choose suitable pivots and clear columns in the usual manner.

The next two steps give the procedure for choosing the pivot.

Step 3 *Select the pivot column* (the column that contains the pivot we are seeking).

Selecting the Pivot Column

Choose the negative number with the largest magnitude on the left-hand side of the bottom row (that is, don't consider the last number in the bottom row). Its column is the pivot column. (If there are two candidates, choose either one.) If all the numbers on the left-hand side of the bottom row are zero or positive, then we are done, and the basic solution is the optimal solution.

[6]Also see Section 2.2 for a discussion of pivots and pivoting.

Simple enough. The most negative number in the bottom row is -3, so we choose the x column as the pivot column:

	x	y	z	s	t	p	
s	**2**	2	1	1	0	0	10
t	**1**	2	3	0	1	0	15
p	**−3**	−2	−1	0	0	1	0
	↑ Pivot column						

Question Why choose the pivot column this way?

Answer The variable labeling the pivot column is going to be increased from zero to something positive. In the equation $p = 3x + 2y + z$, the fastest way to increase p is to increase x, since p would increase by 3 units for every 1-unit increase in x. (If we chose to increase y, then p would only increase by 2 units for every 1-unit increase in y, and if we increased z instead, p would grow even more slowly.) In short, choosing the pivot column this way makes it likely that we'll increase p as much as possible.

Step 4 *Select the pivot in the pivot column.*

Selecting the Pivot

1. The pivot must always be a positive number. (This rules out zeros and negative numbers, such as the -3 in the bottom row.)
2. For each positive entry b in the pivot column, compute the ratio a/b, where a is the number in the rightmost column in that row. We call this a **test ratio.**
3. Of these ratios, choose the smallest one. The corresponding number b is the pivot.

In our example the test ratio in the first row is $10/2 = 5$, and the test ratio in the second row is $15/1 = 15$. Here, 5 is the smallest, so the 2 in the upper left is our pivot:

	x	y	z	s	t	p		Test ratios
s	**[2]**	2	1	1	0	0	10	$\frac{10}{2} = 5$
t	**1**	2	3	0	1	0	15	$\frac{15}{1} = 15$
p	**−3**	−2	−1	0	0	1	0	

Question Why select the pivot this way?

Answer The rule given above guarantees that, after pivoting, all variables will be nonnegative in the basic solution. In other words, it guarantees that we will remain in the feasible region. We will explain further after finishing this example.

Step 5 *Use the pivot to clear the column in the normal manner and then relabel the pivot row with the label from the pivot column.* It is important to follow the exact prescription described in Section 2.2 for formulating the row operations:

$$aR_c \pm bR_p \qquad a \text{ and } b \text{ both positive}$$

↑ Row to change ↑ Pivot row

All entries in the last column should remain nonnegative after pivoting. Furthermore, since the x column (and no longer the s column) will be cleared, x will become an active variable. In other words, the s on the left of the pivot will be replaced by x. We call s the **departing,** or **exiting, variable** and x the **entering variable** for this step.

Entering variable ↓ (column x); Departing variable → (row s)

	x	y	z	s	t	p		
s	[2]	2	1	1	0	0	10	
t	1	2	3	0	1	0	15	$2R_2 - R_1$
p	−3	−2	−1	0	0	1	0	$2R_3 + 3R_1$

This gives

	x	y	z	s	t	p	
x	2	2	1	1	0	0	10
t	0	2	5	−1	2	0	20
p	0	2	1	3	0	2	30

This is the second tableau.

Step 6 *Go to Step 3.* But wait! According to Step 3, we are finished because there are no negative numbers in the bottom row. Thus, we can read off the answer. Remember, though, that the solution for x, the first active variable, is not just $x = 10$ but is $x = 10/2 = 5$ because the pivot has not been reduced to a 1. Similarly, $t = 20/2 = 10$ and $p = 30/2 = 15$. All other variables are zero because they are inactive. Thus, the solution is as follows: p has a maximum value of 15, and this occurs when $x = 5$, $y = 0$, and $z = 0$. (The slack variables then have the values $s = 0$ and $t = 10$.)

Question Why can we stop when there are no negative numbers in the bottom row? Why does this tableau give an optimal solution?

Answer The bottom row corresponds to the equation $2y + z + 3s + 2p = 30$, or

$$p = 15 - y - \frac{1}{2}z - \frac{3}{2}s$$

Think of this as part of the general solution to our original system of equations, with y, z, and s as the parameters. Since these variables must be nonnegative, *the largest possible value of p in any feasible solution of the system comes when all three of the parameters are zero.* Thus, the current basic solution must be an optimal solution.[7]

We owe some further explanation for Step 4. After Step 3 we knew that x would be the entering variable, and we needed to choose the departing variable. In the next basic solution, x was to have some positive value, and we wanted this value to be as large as possible (to make p as large as possible) without making any other variables negative. Look again at the equations written in Step 2:

$$s = 10 - 2x - 2y - z$$

$$t = 15 - x - 2y - 3z$$

[7]Calculators or spreadsheets could obviously be a big help in the calculations here, just as in Chapter 2. We'll say more about that after the next couple of examples.

We needed to make either s or t into an inactive variable and hence zero. Also, y and z were to remain inactive. If we had made s inactive, then we would have had $0 = 10 - 2x$, so $x = 10/2 = 5$. This would have made $t = 15 - 5 = 10$, which would be fine. On the other hand, if we had made t inactive, then we would have had $0 = 15 - x$, so $x = 15$, and this would have made $s = 10 - 2 \cdot 15 = -20$, which would *not* be fine because slack variables must be nonnegative. In other words, we had a choice of making $x = 10/2 = 5$ or $x = 15/1 = 15$, but making x larger than 5 would have made another variable negative. We were thus compelled to choose the smaller ratio, 5, and make s the departing variable. Of course, we do not have to think it through this way every time. We just use the rule stated in Step 4. (For a graphical explanation, see Example 3.)

Example 2 • Simplex Method

Find the maximum value of $p = 12x + 15y + 5z$, subject to the constraints

$$2x + 2y + z \leq 8$$

$$x + 4y - 3z \leq 12$$

$$x \geq 0, y \geq 0, z \geq 0$$

Solution Following Step 1, we introduce slack variables and rewrite the constraints and objective function in standard form:

$$\begin{aligned} 2x + 2y + z + s \qquad &= 8 \\ x + 4y - 3z \qquad + t \quad &= 12 \\ -12x - 15y - 5z \qquad + p &= 0 \end{aligned}$$

We now follow with Step 2, setting up the initial tableau:

	x	y	z	s	t	p	
s	2	2	1	1	0	0	8
t	1	4	-3	0	1	0	12
p	-12	-15	-5	0	0	1	0

For Step 3 we select the column over the negative number with the largest magnitude in the bottom row, which is the y column. For Step 4, finding the pivot, we see that the test ratios are 8/2 and 12/4, the smallest being $12/4 = 3$. So we select the pivot in the t row and clear its column:

	x	y	z	s	t	p		
s	2	2	1	1	0	0	8	$2R_1 - R_2$
t	1	[4]	-3	0	1	0	12	
p	-12	-15	-5	0	0	1	0	$4R_3 + 15R_2$

The departing variable is t, and the entering variable is y. This gives the second tableau:

	x	y	z	s	t	p	
s	3	0	5	2	-1	0	4
y	1	4	-3	0	1	0	12
p	-33	0	-65	0	15	4	180

We now go back to Step 3. Since we still have negative numbers in the bottom row, we choose the one with the largest magnitude (which is −65), and thus our pivot column is the z column. Since negative numbers can't be pivots, the only possible choice for the pivot is the 5. (We need not compute the test ratios because there would only be one from which to choose.) We now clear this column, remembering to take care of the departing and entering variables:

	x	**y**	**z**	**s**	**t**	**p**		
s	3	0	[5]	2	−1	0	4	
y	1	4	−3	0	1	0	12	$5R_2 + 3R_1$
p	−33	0	−65	0	15	4	180	$R_3 + 13R_1$

This gives

	x	**y**	**z**	**s**	**t**	**p**	
z	3	0	5	2	−1	0	4
y	14	20	0	6	2	0	72
p	6	0	0	26	2	4	232

Notice how the value of p keeps climbing: It started at zero in the first tableau, went up to $180/4 = 45$ in the second, and is currently at $232/4 = 58$. Because there are no more negative numbers in the bottom row, we are done and can write down the solution: p has a maximum value of $232/4 = 58$, and this occurs when

$$x = 0 \qquad y = \frac{72}{20} = \frac{18}{5}, \qquad z = \frac{4}{5}$$

The slack variables are both zero.

Before we go on . . . As a partial check on our answer, we can substitute these values into the objective function and the constraints:

$$58 = 12(0) + 15\left(\frac{18}{5}\right) + 5\left(\frac{4}{5}\right) \quad ✓$$

$$2(0) + 2\left(\frac{18}{5}\right) + \left(\frac{4}{5}\right) = 8 \le 8 \quad ✓$$

$$0 + 4\left(\frac{18}{5}\right) - 3\left(\frac{4}{5}\right) = 12 \le 12 \quad ✓$$

We say that this is only a partial check because it shows only that (1) our solution is feasible and (2) we have correctly calculated p. It does not show that we have the optimal solution. This check will *usually* catch any arithmetic mistakes we make, but it is not foolproof.

APPLICATIONS

In the next example (further exploits of Acme Babyfoods—compare Example 2 in Section 4.2), we show how the simplex method relates to the graphical method.

Example 3 • Resource Allocation

Acme Babyfoods makes two puddings, vanilla and chocolate. Each serving of vanilla pudding requires 2 teaspoons of sugar and 25 fluid ounces of water, and each serving of chocolate pudding requires 3 teaspoons of sugar and 15 fluid ounces of water. Acme has available each day 3600 teaspoons of sugar and 22,500 fluid ounces of water. Acme makes no more than 600 servings of vanilla pudding because that is all that it can sell each day. If Acme makes a profit of 10¢ on each serving of vanilla pudding and 7¢ on each serving of chocolate, how many servings of each should it make to maximize its profit?

Solution We first identify the unknowns. Let

$x =$ number of servings of vanilla pudding

$y =$ number of servings of chocolate pudding

The objective function is the profit $p = 10x + 7y$, which we need to maximize. For the constraints we start with the fact that Acme will make no more than 600 servings of vanilla: $x \leq 600$. We can put the remaining data in a table as follows:

	Vanilla	Chocolate	Total Available
Sugar (tsp)	2	3	3,600
Water (fl oz)	25	15	22,500

Since Acme can use no more sugar and water than is available, we get the two constraints:

$$2x + 3y \leq 3600$$

$$25x + 15y \leq 22{,}500$$ Note that all terms are divisible by 5.

Thus, our LP problem is this:

$$\begin{aligned} \text{Maximize} \quad & p = 10x + 7y \\ \text{subject to} \quad & x \leq 600 \\ & 2x + 3y \leq 3600 \\ & 5x + 3y \leq 4500 \\ & x \geq 0, y \geq 0 \end{aligned}$$

We divided $25x + 15y \leq 22{,}500$ by 5.

Next, we introduce the slack variables and set up the initial tableau:

$$\begin{aligned} x \quad + s \quad\quad\quad &= 600 \\ 2x + 3y \quad + t \quad\quad &= 3600 \\ 5x + 3y \quad\quad + u \quad &= 4500 \\ -10x - 7y \quad\quad\quad + p &= 0 \end{aligned}$$

Note that we have had to introduce a third slack variable, u. There need to be as many slack variables as there are constraints (other than those of the $x \geq 0$ variety).

Question What do the slack variables say about Acme puddings?

Answer The first slack variable, s, represents the number you must add to the number of servings of vanilla pudding actually made to obtain the maximum of 600 servings. The second slack variable, t, represents the amount of sugar that is left over once the puddings are made, and u represents the amount of water left over.

We now use the simplex method to solve the problem:

	x	y	s	t	u	p		
s	[1]	0	1	0	0	0	600	
t	2	3	0	1	0	0	3,600	$R_2 - 2R_1$
u	5	3	0	0	1	0	4,500	$R_3 - 5R_1$
p	−10	−7	0	0	0	1	0	$R_4 + 10R_1$

	x	y	s	t	u	p		
x	1	0	1	0	0	0	600	
t	0	3	−2	1	0	0	2,400	$R_2 - R_3$
u	0	[3]	−5	0	1	0	1,500	
p	0	−7	10	0	0	1	6,000	$3R_4 + 7R_3$

	x	y	s	t	u	p		
x	1	0	1	0	0	0	600	$3R_1 - R_2$
t	0	0	[3]	1	−1	0	900	
y	0	3	−5	0	1	0	1,500	$3R_3 + 5R_2$
p	0	0	−5	0	7	3	28,500	$3R_4 + 5R_2$

	x	y	s	t	u	p	
x	3	0	0	−1	1	0	900
s	0	0	3	1	−1	0	900
y	0	9	0	5	−2	0	9,000
p	0	0	0	5	16	9	90,000

Thus, the solution is as follows: The maximum value of p is $90{,}000/9 = 10{,}000¢ =$ \$100, which occurs when $x = 900/3 = 300$ and $y = 9000/9 = 1000$. (The slack variables are $s = 900/3 = 300$ and $t = u = 0$.)

Before we go on . . . Since this was a problem with only two variables, we could have solved it graphically. It is interesting to think about the relationship between the two methods. Figure 24 shows the feasible region.

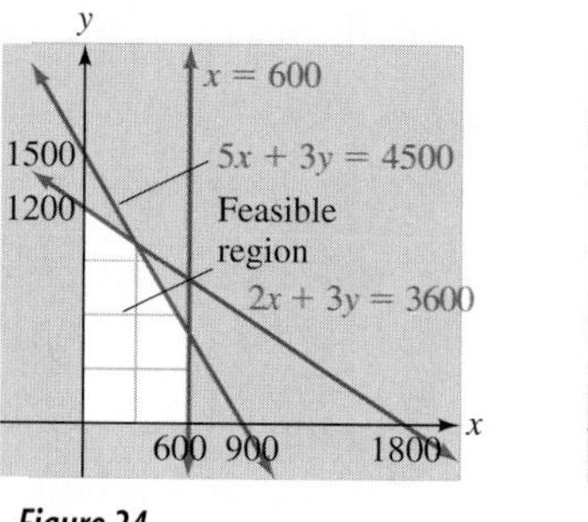

Figure 24

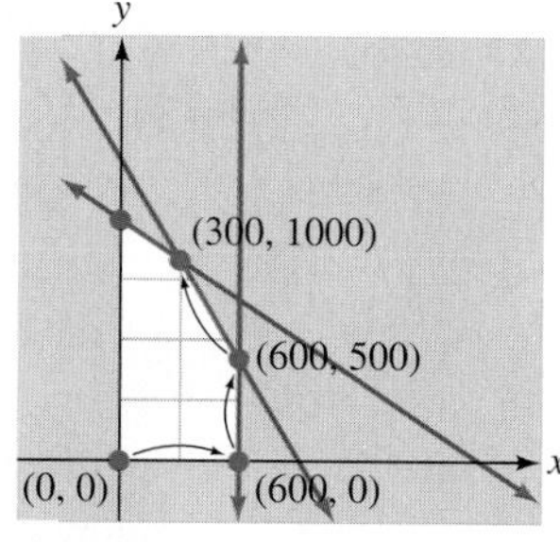

Figure 25

Each tableau in the simplex method corresponds to a corner of the feasible region, given by the corresponding basic solution. In this example the sequence of basic solutions is

$$(x, y) = (0, 0), (600, 0), (600, 500), (300, 1000)$$

This is the sequence of corners shown in Figure 25.

In general we can think of the simplex method as walking from corner to corner of the feasible region, until it locates the optimal solution. In problems with many variables and many constraints, the simplex method usually visits only a small fraction of the total number of corners.

We can also explain again, in a different way, the reason we use the test ratios when choosing the pivot. For example, when choosing the first pivot, we had to choose among the test ratios 600, 1800, and 900 (look at the first tableau). In Figure 24 you can see that those are the three x intercepts of the lines that bound the feasible region. If we had chosen 1800 or 900, we would have jumped along the x axis to a point outside of the feasible region, which we do not want to do. In general the test ratios measure the distance from the current corner to the constraint lines, and we must choose the smallest such distance to avoid crossing any of them into the unfeasible region.

It is also interesting in an application like this to think about the values of the slack variables. We said above that s is the difference between the maximum 600 servings of vanilla that might be made and the number that is actually made. In the optimal solution, $s = 300$, which says that 300 fewer servings of vanilla were made than the maximum possible. Similarly, t was the amount of sugar left over. In the optimal solution, $t = 0$, which tells us that all of the available sugar is used. Finally, $u = 0$, so all of the available water is used as well.

Using Technology

A tableau is just a matrix, and the main step in the simplex method is pivoting. Consult the technology discussion in Section 2.2 for instructions on pivoting using technology.

Web Site

For an online listing of the latest version of the PIVOT program for the TI-83, follow

Web site → Everything for Finite Math → Chapter 4

where you will find a link. If you go to

Web site → Online Utilities → Matrix and Linear Algebra Tools

you will find a number of utilities that automate the process to varying extents:

- Pivot and Gauss–Jordan Tool: Pivots and does row operations
- Excel Pivot and Gauss–Jordan Tool: Excel-based version
- Simplex Method Tool: Solves entire LP problems; shows all tableaus

For free Macintosh software that does pivoting for the simplex method, follow

Web site → Online Utilities → Miscellaneous → Mac Software Site

Note Since all of the above technology makes the handling of decimals easy, you might find the following suggestions convenient when using technology.

- There is no need to eliminate decimals and fractions; calculators can handle them with ease. Since most of the pivoting technology permits you to pivot with the click of a mouse or a single command, you need not worry about including complicated decimal expressions in your row operations. On the other hand, if your display space is limited (as in a graphing calculator), eliminating fractions and decimals does save space.
- Use the traditional Gauss–Jordan method (see the discussion at the end of Section 2.2): After selecting your pivot and prior to clearing the pivot column, *divide the pivot row by the value of the pivot, thereby turning the pivot into a* 1. Of course, this may result in decimals, so the issue of display space is again pertinent. Whether you use this approach might depend on the features in your pivoting resource: Does it automatically convert a selected pivot to 1, or does it permit you to automatically divide a row by a selected entry? (The resources outlined above all do one or the other.)

Example 4 illustrates the use of pivoting technology (technology that allows you to pivot with a single command or mouse click) to perform the simplex method using the traditional Gauss–Jordan method.

T Example 4 • Using Pivoting Technology

$$\begin{array}{ll} \text{Maximize} & p = 1.3x + 4.5y + 11.3z \\ \text{subject to} & 2.4x - 0.21y + 11.7z \le 3 \\ & 3.3x + 5.6y - 2.35z \le 5 \\ & x - y + 2.3z \le 10 \\ & x \ge 0, y \ge 0, z \ge 0 \end{array}$$

Solution One drawback of using technology is that, depending on the resource, we may not be able to label the rows or columns. We can mentally label them as we go, but we can do without the row labels anyway. First, we introduce the slack variables *s*, *t*, and *u*, as usual, and set up the initial tableau in the form of a matrix. The display will show something like the following. (This is actually based on the output from the computer program for the Macintosh available at the Web site. Present graphing calculators cannot display the whole matrix at once, but they permit us to scroll to view any part of the matrix.)

2.4	-0.21	11.7	1	0	0	0	3
3.3	5.6	-2.35	0	1	0	0	5
1	-1	2.3	0	0	1	0	10
-1.3	-4.5	-11.3	0	0	0	1	0

The most negative entry in the bottom row is -11.3, so we must select the pivot in its column (column 3). The smallest test ratio is $3/11.7 = 0.2564$, so the pivot is the entry 11.7 in position 1, 3. Next, you might want to turn the pivot into a 1 by dividing row 1 by the value of the pivot: 11.7. This gives

0.20513	-0.01795	1	0.08547	0	0	0	0.25641
3.3	5.6	-2.35	0	1	0	0	5
1	-1	2.3	0	0	1	0	10
-1.3	-4.5	-11.3	0	0	0	1	0

Now we pivot on the entry in position 1, 3, getting

0.205128	-0.017948	1	0.0854701	0	0	0	0.25641
3.78205	5.55782	0	0.200855	1	0	0	5.60256
0.528205	-0.958718	0	-0.196581	0	1	0	9.41026
1.01795	-4.70282	0	0.965812	0	0	1	2.89744

The only negative number in the bottom row is -4.70282, and we see that the pivot is the entry 5.55782 in position 2, 2. Turning the pivot into 1 (by division) now yields

0.205128	-0.017948	1	0.0854701	0	0	0	0.25641
0.680492	1	0	0.0361391	0.179927	0	0	1.00805
0.528205	-0.958718	0	-0.196581	0	1	0	9.41026
1.01795	-4.70282	0	0.965812	0	0	1	2.89744

After pivoting on the 2, 2 entry we get

0.217342	0	1	0.0861187	0.0032294	0	0	0.274503
0.680492	1	0	0.0361391	0.179927	0	0	1.00805
1.1806	0	0	-0.161934	0.172499	1	0	10.3767
4.21818	0	0	1.13577	0.846163	0	1	7.63812

Because there are no negative numbers in the bottom row, we are finished.

Question We don't have labels. How do we read off the optimal solution?

Answer Look at the columns containing one 1 and three 0s. They are the y column, the z column, the u column, and the p column. Think of the 1 that appears in each of these columns as a pivot whose column has been cleared. If we had labels, the row containing a pivot would have the same label as the column containing that pivot. We can now read off the solution as follows:

***y* column:** The pivot is in row 2, so row 2 would have been labeled with y. We scan across row 2 to the rightmost column to read off the value of y: $y = 1.00805$.

***z* column:** The pivot is in row 1, so we scan across row 1 to the rightmost column to read off the value of z: $z = 0.274503$.

***u* column:** The pivot is in row 3, so we scan across row 3 to the rightmost column to read off the value of u: $u = 10.3767$.

***p* column:** The pivot is in row 4, so we scan across row 4 to the rightmost column to read off the value of p: $p = 7.63812$.

Thus, the solution is: The maximum value of p is 7.63812, which is attained when $x = 0$, $y = 1.00805$, and $z = 0.274503$. (Since u is a slack variable, its value is not required as part of the solution of the LP problem.)

Let's end this section with a summary and some final comments.

Summary: The Simplex Method for Standard Maximization Problems

To solve a standard maximization problem using the simplex method, we take the following steps:

1. Convert to a system of equations by introducing **slack variables** to turn the constraints into equations and by rewriting the objective function in standard form.
2. Write down the initial **tableau.**
3. Select the pivot column: Choose the negative number with the largest magnitude in the left-hand side of the bottom row. Its column is the pivot column. (If there are two or more candidates, choose any of them.) If all the numbers in the left-hand side of the bottom row are zero or positive, then we are finished, and the basic solution maximizes the objective function. (See below for the basic solution.)
4. Select the pivot in the pivot column: The pivot must always be a positive number. For each positive entry b in the pivot column, compute the ratio a/b, where a is the number in the last column in that row. Of these **test ratios,** choose the smallest one. The corresponding number b is the pivot.
5. Use the pivot to clear the column in the normal manner (taking care to follow the exact prescription for formulating the row operations described in Chapter 2) and then relabel the pivot row with the label from the pivot column. The variable originally labeling the pivot row is the **departing,** or **exiting, variable,** and the variable labeling the column is the **entering variable.**

6. Go to step 3.

To get the **basic solution** corresponding to any tableau in the simplex method, set to zero all variables that do not appear as row labels. The value of a variable that does appear as a row label (an **active variable**) is the number in the rightmost column in that row divided by the number in that row in the column labeled by the same variable.

Guideline: Troubleshooting the Simplex Method

Question What if there is no candidate for the pivot in the pivot column? For example, what do we do with a tableau like the following?

	x	y	z	s	t	p	
z	0	0	5	2	0	0	4
y	−8	20	0	6	5	0	72
p	−20	0	0	26	15	4	232

Answer Here, the pivot column is the x column, but there is no suitable entry for a pivot (since zeros and negative numbers can't be pivots). This happens when the feasible region is unbounded and there is also no optimal solution. In other words, p can be made as large as we like without violating the constraints.

Question What should we do if there is a negative number in the rightmost column?

Answer A negative number will not appear above the bottom row in the rightmost column if we follow the procedure correctly. (The bottom right entry is allowed to be negative if the objective takes on negative values as in a negative profit, or loss.) Following are the most likely errors leading to this situation:

The pivot was chosen incorrectly (don't forget to choose the *smallest* test ratio). When this mistake is made, one or more of the variables will be negative in the corresponding basic solution.

The row operation instruction was written backward or performed backward (for example, instead of $R_2 - R_1$, it was $R_1 - R_2$). This mistake can be corrected by multiplying the row by -1.

An arithmetic error occurred. (We all make those annoying errors from time to time.)

Question What about zeros in the rightmost column?

Answer Zeros are permissible in the rightmost column. For example, the constraint $x - y \le 0$ will lead to a zero in the rightmost column.

Question What happens if we choose a pivot column other than the one with the most negative number in the bottom row?

Answer There is no harm in doing this as long as we choose the pivot in that column using the smallest test ratio. All it might do is slow the whole calculation by adding extra steps.

One last suggestion: If it is possible to do a simplification step (dividing a row by a positive number) *at any stage,* we should to do so. As we saw in Chapter 2, this can help prevent the numbers from getting out of hand.

4.3 EXERCISES

1. Maximize $p = 2x + y$
subject to
$x + 2y \le 6$
$-x + y \le 4$
$x + y \le 4$
$x \ge 0, y \ge 0$

2. Maximize $p = x$
subject to
$x - y \le 4$
$-x + 3y \le 4$
$x \ge 0, y \ge 0$

3. Maximize $p = x - y$
subject to
$5x - 5y \le 20$
$2x - 10y \le 40$
$x \ge 0, y \ge 0$

4. Maximize $p = 2x + 3y$
subject to
$3x + 8y \le 24$
$6x + 4y \le 30$
$x \ge 0, y \ge 0$

5. Maximize $p = 5x - 4y + 3z$
subject to
$5x + 5z \le 100$
$5y - 5z \le 50$
$5x - 5y \le 50$
$x \ge 0, y \ge 0, z \ge 0$

6. Maximize $p = 6x + y + 3z$
subject to
$3x + y \le 15$
$2x + 2y + 2z \le 20$
$x \ge 0, y \ge 0, z \ge 0$

7. Maximize $p = 7x + 5y + 6z$
subject to
$x + y - z \le 3$
$x + 2y + z \le 8$
$x + y \le 5$
$x \ge 0, y \ge 0, z \ge 0$

8. Maximize $p = 3x + 4y + 2z$
subject to
$3x + y + z \le 5$
$x + 2y + z \le 5$
$x + y + z \le 4$
$x \ge 0, y \ge 0, z \ge 0$

9. Maximize $z = 3x_1 + 7x_2 + 8x_3$
subject to
$5x_1 - x_2 + x_3 \le 1500$
$2x_1 + 2x_2 + x_3 \le 2500$
$4x_1 + 2x_2 + x_3 \le 2000$
$x_1 \ge 0, x_2 \ge 0, x_3 \ge 0$

10. Maximize $z = 3x_1 + 4x_2 + 6x_3$
subject to
$5x_1 - x_2 + x_3 \le 1500$
$2x_1 + 2x_2 + x_3 \le 2500$
$4x_1 + 2x_2 + x_3 \le 2000$
$x_1 \ge 0, x_2 \ge 0, x_3 \ge 0$

11. Maximize $p = x + y + z + w$
subject to
$x + y + z \le 3$
$y + z + w \le 4$
$x + z + w \le 5$
$x + y + w \le 6$
$x \ge 0, y \ge 0, z \ge 0, w \ge 0$

12. Maximize $p = x - y + z + w$
subject to
$x + y + z \le 3$
$y + z + w \le 3$
$x + z + w \le 4$
$x + y + w \le 4$
$x \ge 0, y \ge 0, z \ge 0, w \ge 0$

13. Maximize $p = x + y + z + w + v$
subject to
$x + y \le 1$
$y + z \le 2$
$z + w \le 3$
$w + v \le 4$
$x \ge 0, y \ge 0, z \ge 0, w \ge 0, v \ge 0$

14. Maximize $p = x + 2y + z + 2w + v$
subject to
$x + y \le 1$
$y + z \le 2$
$z + w \le 3$
$w + v \le 4$
$x \ge 0, y \ge 0, z \ge 0, w \ge 0, v \ge 0$

T In Exercises 15–20, we suggest the use of technology. Round all answers to two decimal places.

15. Maximize $p = 2.5x + 4.2y + 2z$
subject to
$0.1x + y - 2.2z \le 4.5$
$2.1x + y + z \le 8$
$x + 2.2y \le 5$
$x \ge 0, y \ge 0, z \ge 0$

16. Maximize $p = 2.1x + 4.1y + 2z$
subject to
$3.1x + 1.2y + z \le 5.5$
$x + 2.3y + z \le 5.5$
$2.1x + y + 2.3z \le 5.2$
$x \ge 0, y \ge 0, z \ge 0$

17. Maximize $p = x + 2y + 3z + w$
subject to
$x + 2y + 3z \le 3$
$y + z + 2.2w \le 4$
$x + z + 2.2w \le 5$
$x + y + 2.2w \le 6$
$x \ge 0, y \ge 0, z \ge 0, w \ge 0$

18. Maximize $p = 1.1x - 2.1y + z + w$
subject to
$x + 1.3y + z \le 3$
$1.3y + z + w \le 3$
$x + z + w \le 4.1$
$x + 1.3y + w \le 4.1$
$x \ge 0, y \ge 0, z \ge 0, w \ge 0$

19. Maximize $p = x - y + z - w + v$
subject to
$x + y \le 1.1$
$y + z \le 2.2$
$z + w \le 3.3$
$w + v \le 4.4$
$x \ge 0, y \ge 0, z \ge 0, w \ge 0, v \ge 0$

20. Maximize $p = x - 2y + z - 2w + v$
subject to $x + y \le 1.1$
$y + z \le 2.2$
$z + w \le 3.3$
$w + v \le 4.4$
$x \ge 0, y \ge 0, z \ge 0, w \ge 0, v \ge 0$

APPLICATIONS

21. Purchasing You are in charge of purchases at the student-run used-book supply program at your college, and you must decide how many introductory calculus, history, and marketing texts should be purchased from students for resale. Due to budget limitations, you cannot purchase more than 650 of these textbooks each semester. There are also shelf-space limitations: Calculus texts occupy 2 units of shelf space each, history books 1 unit each, and marketing texts 3 units each, and you can spare at most 1000 units of shelf space for the texts. If the used-book program makes a profit of \$10/calculus text, \$4/history text, and \$8/marketing text, how many of each type of text should you purchase to maximize profit? What is the maximum profit the program can make in a semester?

22. Sales The Marketing Club at your college has decided to raise funds by selling three types of T-shirt: one with a single-color "ordinary" design, one with a two-color "fancy" design, and one with a three-color "very fancy" design. The club feels that it can sell up to 300 T-shirts. Ordinary T-shirts will cost the club \$6 each, fancy T-shirts \$8 each, and very fancy T-shirts \$10 each; the club has a total purchasing budget of \$3000. It will sell ordinary T-shirts at a profit of \$4 each, fancy T-shirts at a profit of \$5 each, and very fancy T-shirts at a profit of \$4 each. How many of each kind of T-shirt should the club order to maximize profit? What is the maximum profit the club can make?

23. Resource Allocation Arctic Juice Company makes three juice blends: PineOrange, using 2 portions of pineapple juice and 2 portions of orange juice per gallon; PineKiwi, using 3 portions of pineapple juice and 1 portion of kiwi juice per gallon; and OrangeKiwi, using 3 portions of orange juice and 1 portion of kiwi juice per gallon. Each day the company has 800 portions of pineapple juice, 650 portions of orange juice, and 350 portions of kiwi juice available. Its profit on PineOrange is \$1/gallon, its profit on PineKiwi is \$2/gallon, and its profit on OrangeKiwi is \$1/gallon. How many gallons of each blend should it make each day to maximize profit? What is the largest possible profit the company can make?

24. Purchasing TransGlobal Tractor Trailers has decided to spend up to \$1.5 million on a fleet of new trucks, and it is considering three models: the Gigahaul, which has a capacity of 6000 cubic feet and is priced at \$60,000; the Megahaul, with a capacity of 5000 cubic feet and priced at \$50,000; and the Picohaul, with a capacity of 2000 cubic feet, priced at \$40,000. The anticipated annual revenues are \$500,000 for each new truck purchased (regardless of size). TransGlobal would like a total capacity of up to 130,000 cubic feet and feels that it cannot provide drivers and maintenance for more than 30 trucks. How many of each should it purchase to maximize annual revenue? What is the largest possible revenue it can make?

25. Resource Allocation The Enormous State University History Department offers three courses, Ancient History, Medieval History, and Modern History, and the department chairperson is trying to decide how many sections of each to offer this semester. They may offer up to 45 sections total and up to 5000 students would like to take a course, and there are 60 professors to teach them (no student will take more than one history course, and no professor will teach more than one section). Sections of Ancient History hold 100 students each, sections of Medieval History hold 50 students each, and sections of Modern History hold 200 students each. Modern History sections are taught by a team of 2 professors, whereas Ancient and Medieval History need only 1 professor/section. Ancient History nets the university \$10,000/section, Medieval History nets \$20,000/section, and Modern History nets \$30,000/section. How many sections of each course should the department offer to generate the largest profit? What is the largest profit possible? Will there be any unused time slots, any students who did not get into classes, or any professors without anything to teach?

26. Resource Allocation You manage an ice cream factory that makes three flavors: Creamy Vanilla, Continental Mocha, and Succulent Strawberry. Into each batch of Creamy Vanilla go 2 eggs, 1 cup of milk, and 2 cups of cream. Into each batch of Continental Mocha go 1 egg, 1 cup of milk, and 2 cups of cream. Into each batch of Succulent Strawberry go 1 egg, 2 cups of milk, and 2 cups of cream. You have in stock 200 eggs, 120 cups of milk, and 200 cups of cream. You make a profit of \$3 on each batch of Creamy Vanilla, \$2 on each batch of Continental Mocha, and \$4 on each batch of Succulent Strawberry.

a. How many batches of each flavor should you make to maximize your profit?

b. In your answer to part (a), have you used all the ingredients?

c. Due to the poor strawberry harvest this year, you cannot make more than ten batches of Succulent Strawberry. Does this affect your maximum profit?

27. Agriculture Your small farm encompasses 100 acres, and you are planning to grow tomatoes, lettuce, and carrots in the coming planting season. Fertilizer costs per acre are \$5 for tomatoes, \$4 for lettuce, and \$2 for carrots. Based on past experience, you estimate that each acre of tomatoes will require an average of 4 labor-hours/week, and tending to lettuce and carrots will each require an average of 2 labor-hours/week. You estimate a profit of \$2000 for each acre of tomatoes, \$1500 for each acre of lettuce, and \$500 for each acre of carrots. You can afford to spend no more than \$400 on fertilizer, and your farmworkers can supply 500 hours/week. How many acres of each crop should you plant to maximize total profits? In this event will you be using all 100 acres of your farm?

28. Agriculture Your farm encompasses 500 acres, and you are planning to grow soybeans, corn, and wheat in the coming planting season. Fertilizer costs per acre are \$5 for soybeans, \$2 for corn, and \$1 for wheat. You estimate that each acre of soybeans will require an average of 5 labor-hours/week, and tending to corn and wheat will each require an average of 2 labor-hours/week. Based on past yields and current market prices, you estimate a profit of \$3000 for each acre of soybeans, \$2000 for each acre of corn, and \$1000 for each acre of wheat. You can afford to spend no more than \$3000 on fertilizer, and your farmworkers can supply 3000 hours/week. How many acres of each crop should you plant to maximize total profits? In this event will you be using all the available labor?

29. Resource Allocation (Note that the following exercise is almost identical to Exercise 8 in Section 2.3, except for one important detail. Refer to your solution of that problem—if you did it—and then attempt this one.) The Enormous State University Choral Society is planning its annual Song Festival, when it will serve three kinds of delicacies: granola treats, nutty granola treats, and nuttiest granola treats. The following table shows the ingredients required (in ounces) for a single serving of each delicacy, as well as the total amount of each ingredient available:

	Granola	Nutty Granola	Nuttiest Granola	Total Available
Toasted Oats	1	1	5	1,500
Almonds	4	8	8	10,000
Raisins	2	4	8	4,000

The society makes a profit of \$6 on each serving of granola, \$8 on each serving of nutty granola, and \$3 on each serving of nuttiest granola. Assuming that they can sell all that they make, how many servings of each will maximize profits? How much of each ingredient will be left over?

30. Resource Allocation Repeat Exercise 29 but this time assume that the Choral Society makes a \$3 profit on each of its delicacies.

31. Recycling Safety-Kleen operates the world's largest oil re-refinery at Elgin, Illinois. You have been hired by the company to determine how to allocate its intake of up to 50 million gallons of used oil to its three refinery processes: A, B, and C. You are told that electricity costs for process A amount to \$150,000/million gallons treated, and for processes B and C, the costs are, respectively, \$100,000 and \$50,000/million gallons treated. Process A can recover 60% of the used oil, process B can recover 55%, and process C can recover only 50%. Assuming a revenue of \$4 million/million gallons of recovered oil and an annual electrical budget of \$3 million, how much used oil would you allocate to each process in order to maximize total revenues?

These figures are realistic: Safety-Kleen's actual 1993 capacity was 50 million gallons, its recycled oil sold for approximately \$4/gallon, its recycling process could recover approximately 55% of the used oil, and its electrical bill was \$3 million. SOURCE: "Oil Recycler Greases Rusty City's Economy," *Chicago Tribune,* May 30, 1993, sect. 7, p. 1.

32. Recycling Repeat Exercise 31 but this time assume that process C can handle only up to 20 million gallons/year.

Creatine Supplements Exercises 33 and 34 are based on the following data on four body-building supplements popular in 2002. (Figures shown correspond to a single serving.)

	Creatine (g)	Carbohydrates (g)	Taurine (g)	Alpha Lipoic Acid (mg)
Cell-Tech (MuscleTech)	10	75	2	200
RiboForce HP (EAS)	5	15	1	0
Creatine Transport (Kaizen)	5	35	1	100
Pre-Load Creatine (Optimum)	6	35	1	25

SOURCE: Nutritional information supplied by the manufacturers/www.netrition.com. Cost/serving as quoted on www.netrition.com as of February 26, 2002.

33. You are thinking of combining the first three supplements in the table to obtain a 10-day supply that gives you the maximum possible amount of creatine, but no more than 1000 milligrams of alpha lipoic acid and 225 grams of carbohydrates. How many servings of each supplement should you combine to meet your specifications, and how much creatine will you get?

34. Repeat Exercise 33 but use the last three supplements in the table instead.

Investing Exercises 35 and 36 are based the following data on four U.S. computer stocks:

	Price (\$)	Dividend Yield (%)	Earnings/ Share (\$)
IBM	100	0.5	4.00
HWP (Hewlett-Packard)	20	1.50	0.40
DELL	25	0	0.50
CPQ (Compaq)	10	0.90	0.10

Stocks were trading at or near the given prices in February 2002. Earnings per share are rounded, based on values on February 23, 2002. SOURCE: http://money.excite.com, February 23, 2002.

35. You are planning to invest up to \$10,000 in IBM, Hewlett-Packard, and Dell shares. You desire to maximize your share of the companies' earnings but (for tax reasons) want to earn no more than \$200 in dividends. Your broker suggests that, since Dell stocks pays no dividends, you should invest only in Dell. Is she right?

36. Repeat Exercise 29 for a portfolio based on Hewlett-Packard, Dell, and Compaq shares, in which you want to earn no more than \$18 in dividends.

37. Loan Planning Enormous State University's employee credit union has $5 million available for loans in the coming year. As vice president in charge of finances, you must decide how much capital to allocate to each of four different kinds of loans, as shown in the following table:

Type of Loan	Annual Rate of Return (%)
Automobile	8
Furniture	10
Signature	12
Other secured	10

SOURCE: Adapted from an exercise in D. R. Anderson, D. J. Sweeny, and T. A. Williams, *An Introduction to Management Science,* 6th ed. (St. Paul: West, 1991).

State laws and credit union policies impose the following restrictions:

- Signature loans may not exceed 10% of the total investment of funds.
- Furniture loans plus other secured loans may not exceed automobile loans.
- Other secured loans may not exceed 200% of automobile loans.

How much should you allocate to each type of loan to maximize the annual return?

T **38. Investments** You have $100,000 and are considering investing in three municipal bond funds, Pimco, Payden, and Columbia Oregon. You have the following data:

Company	Yield (%)
Pimco N.Y. (PNF)	6
Payden (PYSDX)	5
Columbia Oregon (CMBFX)	3

March 1, 2002, yields are rounded to the nearest dollar. SOURCE: Company Web sites/money.excite.com, March 3, 2002.

Your broker has made the following suggestions:

- At least 50% of your total investment should be in Columbia Oregon.
- No more than 10% of your total investment should be in Pimco.

How much should you invest in each fund to maximize your anticipated returns while following your broker's advice?

39. Portfolio Management If x dollars is invested in a company that controls, say, 30% of the market with 5 brand names, then $0.30x$ is a measure of market exposure, and $5x$ is a measure of brand-name exposure. Suppose you are a broker at a large securities firm and one of your clients would like to invest up to $100,000 in recording industry stocks. You decide to recommend a combination of stocks in four of the world's largest companies: Warner Music, Universal Music, Sony, and EMI:

	Warner Music	Universal Music	Sony	EMI
Market Share (%)	12	20	20	15
Number of Labels (brands)	8	20	10	15

The number of labels includes only major labels. Market shares are approximate and represent the period 2000–2002. SOURCES: Various, including http://www.emigroup.com, http://finance.vivendi.com/discover/financial/, http://business2.com, March 2002.

You would like your client's brand-name exposure to be as large as possible but his total market exposure to be $15,000 or less. (This would reflect an average of 15%.) Furthermore, you would like at least 20% of the investment to be in Universal because you feel that its control of the DGG and Phillips labels is advantageous for its classical music operations. How much should you advise your client to invest in each company?

40. Portfolio Management Referring to Exercise 39, suppose instead that you want your client to maximize his total market exposure but limit his brand-name exposure to 1.5 million or less (representing an average of 15 labels or fewer per company) and still invest at least 20% of the total in Universal. How much should you advise your client to invest in each company?

41. Transportation Scheduling (This exercise is almost identical to Exercise 24 in Section 2.3 but is more realistic; one cannot always expect to fill all orders exactly and keep all plants operating at 100% capacity.) Tubular Ride Boogie Board Company has manufacturing plants in Tucson, Arizona, and Toronto, Ontario. You have been given the job of coordinating distribution of the latest model, the Gladiator, to their outlets in Honolulu and Venice Beach. The Tucson plant, when operating at full capacity, can manufacture 620 Gladiator boards/week, whereas the Toronto plant, beset by labor disputes, can produce only 410/week. The outlet in Honolulu orders 500 boards/week, and Venice Beach orders 530 boards/week. Transportation costs are as follows: Tucson to Honolulu: $10/board; Tucson to Venice Beach: $5/board; Toronto to Honolulu: $20/board; Toronto to Venice Beach: $10/board. Your manager has informed you that the company's total transportation budget is $6550. You realize that it may not be possible to fill all orders, but you would like the total number of boogie boards shipped to be as large as possible. Given this, how many Gladiator boards should you order shipped from each manufacturing plant to each distribution outlet?

42. Transportation Scheduling Repeat Exercise 41 but use a transportation budget of $5050.

T **43. Transportation Scheduling** Your publishing company is about to start a promotional blitz for its new book, *Physics for the Liberal Arts.* You have 20 salespeople stationed in

Chicago and 10 in Denver. You would like to fly at most 10 into Los Angeles and at most 15 into New York. A round-trip plane flight from Chicago to Los Angeles costs \$195, from Chicago to New York costs \$182, from Denver to Los Angeles costs \$395, and from Denver to New York costs \$166. You want to spend at most \$4520 on plane flights. How many salespeople should you fly from each of Chicago and Denver to each of Los Angeles and New York to have the most salespeople on the road?

SOURCE: Prices from Travelocity, http://www.travelocity.com/, for the week of June 3, 2002, as of May 5, 2002.

44. Transportation Scheduling Repeat Exercise 43, but this time spend at most \$5320.

COMMUNICATION AND REASONING EXERCISES

45. Can the following LP problem be stated as a standard LP problem? If so, do it; if not, explain why.

$$\begin{aligned} \text{Maximize} \quad & p = 3x - 2y \\ \text{subject to} \quad & x - y + z \geq 0 \\ & x - y - z \leq 6 \end{aligned}$$

46. Can the following LP problem be stated as a standard LP problem? If so, do it; if not, explain why.

$$\begin{aligned} \text{Maximize} \quad & p = -3x - 2y \\ \text{subject to} \quad & x - y + z \geq 0 \\ & x - y - z \geq -6 \end{aligned}$$

47. Why is the simplex method useful? (After all, we do have the graphical method for solving LP problems.)

48. Are there any types of LP problems that cannot be solved with the methods of this section but that can be solved using the methods of the preceding section? Explain.

49. Your friend Janet is telling everyone that if there are only two constraints in an LP problem, then in any optimal basic solution at most two unknowns (other than the objective) will be nonzero. Is she correct? Explain.

50. Your other friend Jason is telling everyone that if there is only one constraint in a standard LP problem, then you will have to pivot at most once to obtain an optimal solution. Is he correct? Explain.

51. What is a *basic solution?*

52. In a typical simplex method tableau, there are more unknowns than equations, and we know from the chapter on systems of linear equations that this typically implies the existence of infinitely many solutions. How are the following types of solutions interpreted in the simplex method?

a. Solutions in which all the variables are positive

b. Solutions in which some variables are negative

c. Solutions in which the inactive variables are zero

53. Can the value of the objective function decrease in passing from one tableau to the next? Explain.

54. Can the value of the objective function remain unchanged in passing from one tableau to the next? Explain.

4.4 The Simplex Method: Solving General Linear Programming Problems

As we saw in Section 4.2, not all LP problems are standard maximization problems. We might have constraints like $2x + 3y \geq 4$ or perhaps $2x + 3y = 4$. Or you might have to minimize, rather than maximize, the objective function. General problems like this are almost as easy to deal with as the standard kind: There is a modification of the simplex method that works very nicely. The best way to illustrate it is by means of examples.

Example 1 • Maximizing with Mixed Constraints

$$\begin{aligned} \text{Maximize} \quad & p = 4x + 12y + 6z \\ \text{subject to} \quad & x + y + z \leq 100 \\ & 4x + 10y + 7z \leq 480 \\ & x + y + z \geq 60 \\ & x \geq 0, y \geq 0, z \geq 0 \end{aligned}$$

Solution We begin by turning the first two inequalities into equations as usual because they have the standard form. We get

$$\begin{aligned} x + y + z + s \quad\quad &= 100 \\ 4x + 10y + 7z \quad + t &= 480 \end{aligned}$$

We are tempted to use a slack variable for the third inequality, $x + y + z \geq 60$, but *adding* something positive to the left-hand side will not make it equal to the right: It will get even bigger. To make it equal to 60, we must *subtract* some nonnegative number. We will call this number u (since we have already used s and t) and refer to u as a **surplus variable** rather than a slack variable. Thus, we write

$$x + y + z - u = 60$$

Continuing with the setup, we have

$$\begin{aligned} x + y + z + s \qquad\qquad &= 100 \\ 4x + 10y + 7z \quad + t \qquad &= 480 \\ x + y + z \qquad\quad - u &= 60 \\ -4x - 12y - 6z \qquad\quad + p &= 0 \end{aligned}$$

This leads to the initial tableau:

	x	**y**	**z**	**s**	**t**	**u**	**p**	
s	1	1	1	1	0	0	0	100
t	4	10	7	0	1	0	0	480
***u**	1	1	1	0	0	−1	0	60
p	−4	−12	−6	0	0	0	1	0

Question Why have we put a star next to the third row?

Answer The basic solution corresponding to this tableau is

$$x = y = z = 0 \qquad s = 100 \qquad t = 480 \qquad u = \frac{60}{-1} = -60$$

Several things are wrong here. First, the values $x = y = z = 0$ do not satisfy the third inequality $x + y + z \geq 60$. Thus, this basic solution is *not feasible*. Second—and this is really the same problem—the surplus variable u is negative, whereas we said that it should be nonnegative. The star next to the row labeled u alerts us to the fact that the present basic solution is not feasible and that the problem is located in the starred row, where the active variable u is negative.

Whenever an active variable is negative, we star the corresponding row.

In setting up the initial tableau, we star those rows coming from $\geq$ inequalities.

The simplex method as described in Section 4.3 assumed that we began in the feasible region, but now we do not. Our first task is to get ourselves into the feasible region. In practice we can think of this as getting rid of the stars on the rows. Once we get into the feasible region, we go back to the method of the preceding section.

There are several ways to get into the feasible region. The method we have chosen is one of the simplest to state and carry out.

The Simplex Method for General Linear Programming Problems

Star all rows that give a negative value for the associated active variable (except for the objective variable, which is allowed to be negative). If there are starred rows, you will need to begin with phase I.

Phase I: Getting into the Feasible Region (Getting Rid of the Stars)
In the first starred row, find the largest positive number. Use test ratios as in Section 4.3, to find the pivot in that column (exclude the bottom row), and then pivot on that entry. Check to see which rows should now be starred. Repeat until no starred rows remain and then go on to phase II.

Phase II: Use the Simplex Method for Standard Maximization Problems
If there are any negative entries on the left side of the bottom row after phase I, use the method described in Section 4.3.

Question Why does this method work?

Answer Let's finish this example first and then discuss it.

Since there is a starred row, we need to use phase I. The largest positive number in the starred row is 1, which occurs three times. Arbitrarily select the first, which is in the first column. In that column the smallest test ratio happens to be given by the 1 in the u row, so this is our first pivot:

	x	y	z	s	t	u	p		
s	1	1	1	1	0	0	0	100	$R_1 - R_3$
t	4	10	7	0	1	0	0	480	$R_2 - 4R_3$
$^*u$	[1]	1	1	0	0	−1	0	60	
p	−4	−12	−6	0	0	0	1	0	$R_4 + 4R_3$

This gives

	x	y	z	s	t	u	p	
s	0	0	0	1	0	1	0	40
t	0	6	3	0	1	4	0	240
x	1	1	1	0	0	−1	0	60
p	0	−8	−2	0	0	−4	1	240

Question What happened to the star?

Answer It's gone! To see why, look at the basic solution given by this tableau:

$$x = 60 \qquad y = 0 \qquad z = 0 \qquad s = 40 \qquad t = 240 \qquad u = 0$$

None of the variables is negative anymore, so there are no rows to star. The basic solution is therefore feasible—it satisfies all the constraints.

Now that there are no more stars, we have completed phase I, so we proceed to phase II, which is just the method of Section 4.3:

	x	y	z	s	t	u	p		
s	0	0	0	1	0	1	0	40	
t	0	[6]	3	0	1	4	0	240	
x	1	1	1	0	0	−1	0	60	$6R_3 - R_2$
p	0	−8	−2	0	0	−4	1	240	$3R_4 + 4R_2$

	x	y	z	s	t	u	p	
s	0	0	0	1	0	1	0	40
y	0	6	3	0	1	4	0	240
x	6	0	3	0	−1	−10	0	120
p	0	0	6	0	4	4	3	1680

And we are finished. The solution is

$$p = \frac{1680}{3} = 560, \qquad x = \frac{120}{6} = 20, \qquad y = \frac{240}{6} = 40, \qquad z = 0$$

The slack and surplus variables are

$$s = 40 \qquad t = 0 \qquad u = 0$$

Question So, why does this method work?

Answer When we perform a pivot in phase I, one of two things will happen. As in Example 1, we may pivot in a starred row. In that case the negative active variable in that row will become inactive (hence, zero), and some other variable will be made active with a positive value because we are pivoting on a positive entry. Thus, at least one star will be eliminated. (We will not introduce any new stars because pivoting on the entry with the smallest test ratio will keep all nonnegative variables nonnegative.)

The second possibility is that we may pivot on some row other than a starred row. Choosing the pivot via test ratios again guarantees that no new starred rows are created. A little bit of algebra shows that the value of the negative variable in the first starred row must increase toward zero. (Choosing the *largest* positive entry in the starred row will make it a little more likely that we will increase the value of that variable as much as possible; the rationale for choosing the largest entry is the same as that for choosing the most negative entry in the bottom row during phase II.) Repeating this procedure as necessary, the value of the variable must eventually become zero or positive, assuming that there are feasible solutions to begin with.

So, one way or the other, we can eventually get rid of all of the stars.

Here is an example that begins with two starred rows.

Example 2 • More Mixed Constraints

$$\begin{aligned}
\text{Maximize} \quad & p = 2x + y \\
\text{subject to} \quad & x + y \geq 35 \\
& x + 2y \leq 60 \\
& 2x + y \geq 60 \\
& x \leq 25 \\
& x \geq 0, y \geq 0
\end{aligned}$$

Solution We introduce slack and surplus variables and write down the initial tableau:

$$\begin{aligned}
x + y - s \qquad\qquad\qquad &= 35 \\
x + 2y \quad + t \qquad\qquad &= 60 \\
2x + y \qquad\quad - u \qquad &= 60 \\
x \qquad\qquad\qquad + v &= 25 \\
-2x - y \qquad\qquad\qquad + p &= 0
\end{aligned}$$

	x	y	s	t	u	v	p	
*s	1	1	−1	0	0	0	0	35
t	1	2	0	1	0	0	0	60
*u	2	1	0	0	−1	0	0	60
v	1	0	0	0	0	1	0	25
p	−2	−1	0	0	0	0	1	0

We locate the largest positive entry in the first starred row (row 1). There are two to choose from (both 1s); let's choose the one in the x column. The entry with the smallest test ratio in that column is the 1 in the v row, so that is the entry we use as the pivot:

	x	y	s	t	u	v	p		
*s	1	1	−1	0	0	0	0	35	$R_1 - R_4$
t	1	2	0	1	0	0	0	60	$R_2 - R_4$
*u	2	1	0	0	−1	0	0	60	$R_3 - 2R_4$
v	[1]	0	0	0	0	1	0	25	
p	−2	−1	0	0	0	0	1	0	$R_5 + 2R_4$

	x	y	s	t	u	v	p	
*s	0	1	−1	0	0	−1	0	10
t	0	2	0	1	0	−1	0	35
*u	0	1	0	0	−1	−2	0	10
x	1	0	0	0	0	1	0	25
p	0	−1	0	0	0	2	1	50

Notice that both stars are still there because the basic solutions for s and u remain negative (but less so). The only positive entry in the first starred row is the 1 in the y column, and that entry also has the smallest test ratio in its column (actually, it is tied with the 1 in the u column, so we could choose either one).

	x	y	s	t	u	v	p		
*s	0	[1]	−1	0	0	−1	0	10	
t	0	2	0	1	0	−1	0	35	$R_2 - 2R_1$
*u	0	1	0	0	−1	−2	0	10	$R_3 - R_1$
x	1	0	0	0	0	1	0	25	
p	0	−1	0	0	0	2	1	50	$R_5 + R_1$

	x	y	s	t	u	v	p	
y	0	1	−1	0	0	−1	0	10
t	0	0	2	1	0	1	0	15
u	0	0	1	0	−1	−1	0	0
x	1	0	0	0	0	1	0	25
p	0	0	−1	0	0	1	1	60

The basic solution is $x = 25$, $y = 10$, $s = 0$, $t = 15$, $u = 0/(-1) = 0$, and $v = 0$. Since there are no negative variables left (even u has become 0), we are in the feasible region, so we can go on to phase II, shown next. (Filling in the instructions for the row operations is an exercise.)

	x	y	s	t	u	v	p	
y	0	1	−1	0	0	−1	0	10
t	0	0	2	1	0	1	0	15
u	0	0	[1]	0	−1	−1	0	0
x	1	0	0	0	0	1	0	25
p	0	0	−1	0	0	1	1	60

	x	y	s	t	u	v	p	
y	0	1	0	0	−1	−2	0	10
t	0	0	0	1	[2]	3	0	15
s	0	0	1	0	−1	−1	0	0
x	1	0	0	0	0	1	0	25
p	0	0	0	0	−1	0	1	60

	x	y	s	t	u	v	p	
y	0	2	0	1	0	−1	0	35
u	0	0	0	1	2	3	0	15
s	0	0	2	1	0	1	0	15
x	1	0	0	0	0	1	0	25
p	0	0	0	1	0	3	2	135

The optimal solution is

$$x = 25 \quad y = \frac{35}{2} = 17.5 \quad p = \frac{135}{2} = 67.5 \qquad (s = 7.5, t = 0, u = 7.5)$$

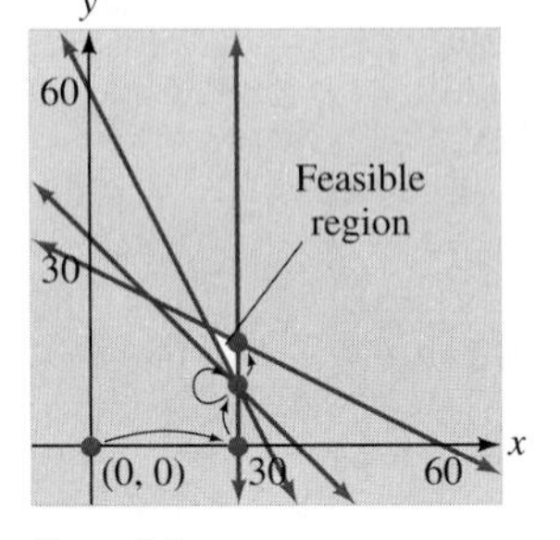

Figure 26

Before we go on . . . Since this is an example with only two unknowns, we can picture the sequence of basic solutions on the graph of the feasible region. This is shown in Figure 26. You can see that there was no way to jump from (0, 0) in the initial tableau directly into the feasible region, since the first jump must be along an axis. (Why?) Also notice that the third jump did not move at all. To which step of the simplex method does this correspond?

Now that we know how to deal with nonstandard constraints, we consider **minimization** problems, problems in which we have to minimize, rather than maximize, the objective function. The idea is to *convert a minimization problem into a maximization problem,* which we can then solve as usual.

Suppose, for instance, we want to minimize $c = 10x - 30y$ subject to some constraints. The technique is as follows: Define a new variable p by taking p to be the negative of c so that $p = -c$. Then, the larger we make p, the smaller c becomes. For example, if we can make p increase from -10 to -5, then c will decrease from 10 to 5. So, if we are looking for the smallest value of c, we might as well look for the largest value of p instead. More concisely,

Minimizing c is the same as maximizing $p = -c$.

Now, since $c = 10x - 30y$, we have $p = -10x + 30y$, and the requirement that we "minimize $c = 10x - 30y$" is now replaced by "maximize $p = -10x + 30y$."

Minimization Problems

We convert a minimization problem into a maximization problem by taking the negative of the objective function. All the constraints remain unchanged.

Quick Example

Minimization Problem

$$\begin{aligned} &\text{Minimize} && c = 10x - 30y \\ &\text{subject to} && 2x + y \le 160 \\ &&& x + 3y \ge 120 \\ &&& x \ge 0, y \ge 0 \end{aligned}$$

Maximization Problem

$$\begin{aligned} &\text{Maximize} && p = -10x + 30y \\ &\text{subject to} && 2x + y \le 160 \\ &&& x + 3y \ge 120 \\ &&& x \ge 0, y \ge 0 \end{aligned}$$

Example 3 • Purchasing

You are in charge of ordering furniture for your company's new headquarters. You need to buy at least 200 tables, 500 chairs, and 300 computer desks. Wall-to-Wall Furniture (WWF) is offering a package of 20 tables, 25 chairs, and 18 computer desks for \$2000, whereas rival Acme Furniture (AF) is offering a package of 10 tables, 50 chairs, and 24 computer desks for \$3000. How many packages should you order from each company to minimize your total cost?

Solution The unknowns here are

x = number of packages ordered from WWF

y = number of packages ordered from AF

We can put the information about the various kinds of furniture in a table:

	WWF	AF	Needed
Tables	20	10	200
Chairs	25	50	500
Computer Desks	18	24	300
Cost (\$)	2000	3000	

From this table we get the following LP problem:

$$\begin{aligned} &\text{Minimize} && c = 2000x + 3000y \\ &\text{subject to} && 20x + 10y \ge 200 \\ &&& 25x + 50y \ge 500 \\ &&& 18x + 24y \ge 300 \\ &&& x \ge 0, y \ge 0 \end{aligned}$$

Before we start solving this problem, notice that all inequalities may be simplified. The first is divisible by 10, the second by 25, and the third by 6. (However, this affects the meaning of the surplus variables; see the "Before we go on" below.) Dividing gives the following simpler problem:

$$\begin{aligned} \text{Minimize} \quad & c = 2000x + 3000y \\ \text{subject to} \quad & 2x + y \geq 20 \\ & x + 2y \geq 20 \\ & 3x + 4y \geq 50 \\ & x \geq 0, y \geq 0 \end{aligned}$$

Following the discussion that preceded this example, we convert to a maximization problem:

$$\begin{aligned} \text{Maximize} \quad & p = -2000x - 3000y \\ \text{subject to} \quad & 2x + y \geq 20 \\ & x + 2y \geq 20 \\ & 3x + 4y \geq 50 \\ & x \geq 0, y \geq 0 \end{aligned}$$

We introduce surplus variables:

$$\begin{aligned} 2x + y - s \quad &= 20 \\ x + 2y - t \quad &= 20 \\ 3x + 4y - u \quad &= 50 \\ 2000x + 3000y + p &= 0 \end{aligned}$$

The initial tableau is then

	x	y	s	t	u	p	
*s	2	1	−1	0	0	0	20
*t	1	2	0	−1	0	0	20
*u	3	4	0	0	−1	0	50
p	2000	3000	0	0	0	1	0

The largest entry in the first starred row is the 2 in the upper left, which happens to give the smallest test ratio in its column.

	x	y	s	t	u	p		
*s	[2]	1	−1	0	0	0	20	
*t	1	2	0	−1	0	0	20	$2R_2 - R_1$
*u	3	4	0	0	−1	0	50	$2R_3 - 3R_1$
p	2000	3000	0	0	0	1	0	$R_4 - 1000R_1$

	x	y	s	t	u	p		
x	2	1	−1	0	0	0	20	$3R_1 - R_2$
*t	0	[3]	1	−2	0	0	20	
*u	0	5	3	0	−2	0	40	$3R_3 - 5R_2$
p	0	2000	1000	0	0	1	−20,000	$3R_4 - 2000R_2$

	x	y	s	t	u	p		
x	6	0	−4	2	0	0	40	$5R_1 - R_3$
y	0	3	1	−2	0	0	20	$5R_2 + R_3$
*u	0	0	4	[10]	−6	0	20	
p	0	0	1000	4000	0	3	−100,000	$R_4 - 400R_3$

	x	y	s	t	u	p		
x	30	0	−24	0	6	0	180	$R_1/6$
y	0	15	9	0	−6	0	120	$R_2/3$
t	0	0	4	10	−6	0	20	$R_3/2$
p	0	0	−600	0	2400	3	−108,000	$R_4/3$

This completes phase I. We are not yet at the optimal solution, so after performing the simplifications indicated we proceed with phase II:

	x	y	s	t	u	p		
x	5	0	−4	0	1	0	30	$R_1 + 2R_3$
y	0	5	3	0	−2	0	40	$2R_2 - 3R_3$
t	0	0	[2]	5	−3	0	10	
p	0	0	−200	0	800	1	−36,000	$R_4 + 100R_3$

	x	y	s	t	u	p	
x	5	0	0	10	−5	0	50
y	0	10	0	−15	5	0	50
s	0	0	2	5	−3	0	10
p	0	0	0	500	500	1	−35,000

The optimal solution is

$$x = \frac{50}{5} = 10 \quad y = \frac{50}{10} = 5 \quad p = -35{,}000, \quad c = 35{,}000 \qquad (s = 5, t = 0, u = 0)$$

You should buy 10 packages from Wall-to-Wall Furniture and 5 from Acme Furniture, for a minimum cost of \$35,000.

Before we go on . . . The surplus variables here represent pieces of furniture over and above the minimum requirements. The order you place will give you 50 extra tables ($s = 5$, but s was introduced after we divided the first inequality by 10, so the actual surplus is $10 \times 5 = 50$), the correct number of chairs ($t = 0$), and the correct number of computer desks ($u = 0$).

The preceding LP problem is an example of a **standard minimization problem**—in a sense the opposite of a standard maximization problem: We are *minimizing* an objective function, where all the constraints have the form $Ax + By + Cz + \cdots \geq N$. We discuss standard minimization problems more fully in Section 4.5, as well as another method of solving them.

Guidelines: When to Switch to Phase II, Equality Constraints, and Troubleshooting

Question How do I know when to switch to phase II?

Answer After each step, check the basic solution for starred rows. You are not ready to proceed with phase II until all the stars are gone.

Question How do I deal with an *equality constraint,* such as $2x + 7y - z = 90$?

Answer Although we haven't given examples of equality constraints, they can be treated by the following trick: *Replace an equality by two inequalities.* For example, replace the equality $2x + 7y - z = 90$ by the two inequalities $2x + 7y - z \leq 90$ and $2x + 7y - z \geq 90$. A little thought will convince you that these two inequalities amount to the same thing as the original equality!

Question What happens if it is impossible to choose a pivot using the instructions in phase I?

Answer In that case the LP problem has no solution. In fact, the feasible region is empty. If it is impossible to choose a pivot in phase II, then the feasible region is unbounded and there is no optimal solution.

4.4 EXERCISES

1. Maximize $p = x + y$
subject to
$x + 2y \geq 6$
$-x + y \leq 4$
$2x + y \leq 8$
$x \geq 0, y \geq 0$

2. Maximize $p = 3x + 2y$
subject to
$x + 3y \geq 6$
$-x + y \leq 4$
$2x + y \leq 8$
$x \geq 0, y \geq 0$

3. Maximize $p = 12x + 10y$
subject to
$x + y \leq 25$
$x \geq 10$
$-x + 2y \geq 0$
$x \geq 0, y \geq 0$

4. Maximize $p = x + 2y$
subject to
$x + y \leq 25$
$y \geq 10$
$2x - y \geq 0$
$x \geq 0, y \geq 0$

5. Maximize $p = 2x + 5y + 3z$
subject to
$x + y + z \leq 150$
$x + y + z \geq 100$
$x \geq 0, y \geq 0, z \geq 0$

6. Maximize $p = 3x + 2y + 2z$
subject to
$x + y + 2z \leq 38$
$2x + y + z \geq 24$
$x \geq 0, y \geq 0, z \geq 0$

7. Maximize $p = 2x + 3y + z + 4w$
subject to
$x + y + z + w \leq 40$
$2x + y - z - w \geq 10$
$x + y + z + w \geq 10$
$x \geq 0, y \geq 0, z \geq 0, w \geq 0$

8. Maximize $p = 2x + 2y + z + 2w$
subject to
$x + y + z + w \leq 50$
$2x + y - z - w \geq 10$
$x + y + z + w \geq 20$
$x \geq 0, y \geq 0, z \geq 0, w \geq 0$

9. Minimize $c = 6x + 6y$
subject to
$x + 2y \geq 20$
$2x + y \geq 20$
$x \geq 0, y \geq 0$

10. Minimize $c = 3x + 2y$
subject to
$x + 2y \geq 20$
$2x + y \geq 10$
$x \geq 0, y \geq 0$

11. Minimize $c = 2x + y + 3z$
subject to
$x + y + z \geq 100$
$2x + y \geq 50$
$y + z \geq 50$
$x \geq 0, y \geq 0, z \geq 0$

12. Minimize $c = 2x + 2y + 3z$
subject to
$x + z \geq 100$
$2x + y \geq 50$
$y + z \geq 50$
$x \geq 0, y \geq 0, z \geq 0$

13. Minimize $c = 50x + 50y + 11z$
subject to
$$\begin{aligned} 2x \quad + z &\geq 3 \\ 2x + y - z &\geq 2 \\ 3x + y - z &\leq 3 \\ x \geq 0, y \geq 0, z &\geq 0 \end{aligned}$$

14. Minimize $c = 50x + 11y + 50z$
subject to
$$\begin{aligned} 3x \quad + z &\geq 8 \\ 3x + y - z &\geq 6 \\ 4x + y - z &\leq 8 \\ x \geq 0, y \geq 0, z &\geq 0 \end{aligned}$$

15. Minimize $c = x + y + z + w$
subject to
$$\begin{aligned} 5x - y + \quad w &\geq 1000 \\ z + w &\leq 2000 \\ x + y \quad &\leq 500 \\ x \geq 0, y \geq 0, z \geq 0, w &\geq 0 \end{aligned}$$

16. Minimize $c = 5x + y + z + w$
subject to
$$\begin{aligned} 5x - y + \quad w &\geq 1000 \\ z + w &\leq 2000 \\ x + y \quad &\leq 500 \\ x \geq 0, y \geq 0, z \geq 0, w &\geq 0 \end{aligned}$$

T In Exercises 17–22, we suggest the use of technology. Round all answers to two decimal places.

17. Maximize $p = 2x + 3y + 1.1z + 4w$
subject to
$$\begin{aligned} 1.2x + y + z + w &\leq 40.5 \\ 2.2x + y - z - w &\geq 10 \\ 1.2x + y + z + 1.2w &\geq 10.5 \\ x \geq 0, y \geq 0, z \geq 0, w &\geq 0 \end{aligned}$$

18. Maximize $p = 2.2x + 2y + 1.1z + 2w$
subject to
$$\begin{aligned} x + 1.5y + 1.5z + \quad w &\leq 50.5 \\ 2x + 1.5y - \quad z - \quad w &\geq 10 \\ x + 1.5y + \quad z + 1.5w &\geq 21 \\ x \geq 0, y \geq 0, z \geq 0, w &\geq 0 \end{aligned}$$

19. Minimize $c = 2.2x + y + 3.3z$
subject to
$$\begin{aligned} x + 1.5y + 1.2z &\geq 100 \\ 2x + 1.5y \quad &\geq 50 \\ 1.5y + 1.1z &\geq 50 \\ x \geq 0, y \geq 0, z &\geq 0 \end{aligned}$$

20. Minimize $c = 50.3x + 10.5y + 50.3z$
subject to
$$\begin{aligned} 3.1x \quad + 1.1z &\geq 28 \\ 3.1x + y - 1.1z &\geq 23 \\ 4.2x + y - 1.1z &\geq 28 \\ x \geq 0, y \geq 0, z &\geq 0 \end{aligned}$$

21. Minimize $c = 1.1x + y + 1.5z - w$
subject to
$$\begin{aligned} 5.12x - y + \quad w &\leq 1000 \\ z + w &\geq 2000 \\ 1.22x + y \quad &\leq 500 \\ x \geq 0, y \geq 0, z \geq 0, w &\geq 0 \end{aligned}$$

22. Minimize $c = 5.45x + y + 1.5z + w$
subject to
$$\begin{aligned} 5.12x - y + \quad w &\geq 1000 \\ z + w &\geq 2000 \\ 1.12x + y \quad &\leq 500 \\ x \geq 0, y \geq 0, z \geq 0, w &\geq 0 \end{aligned}$$

APPLICATIONS

23. Resource Allocation Succulent Citrus produces orange juice and orange concentrate. This year the company anticipates a demand of at least 10,000 quarts of orange juice and 1000 quarts of orange concentrate. Each quart of orange juice requires 10 oranges, and each quart of concentrate requires 50 oranges. The company also anticipates using at least 200,000 oranges for these products. Each quart of orange juice costs the company 50¢ to produce, and each quart of concentrate costs $2 to produce. How many quarts of each product should Succulent Citrus produce to meet the demand and minimize total costs?

24. Resource Allocation Fancy Pineapple produces pineapple juice and canned pineapple rings. This year the company anticipates a demand of at least 10,000 pints of pineapple juice and 1000 cans of pineapple rings. Each pint of pineapple juice requires 2 pineapples, and each can of pineapple rings requires 1 pineapple. The company anticipates using at least 20,000 pineapples for these products. Each pint of pineapple juice costs the company 20¢ to produce, and each can of pineapple rings costs 50¢ to produce. How many pints of pineapple juice and cans of pineapple rings should Fancy Pineapple produce to meet the demand and minimize total costs?

T **25. Nutrition** Each serving of Gerber Mixed Cereal for Baby contains 60 calories and no vitamin C. Each serving of Gerber Mango Tropical Fruit Dessert contains 80 calories and 45% of the U.S. Recommended Daily Allowance (RDA) of vitamin C for infants. Each serving of Gerber Apple Banana Juice contains 60 calories and 120% of the RDA of vitamin C for infants. The cereal costs 10¢/serving, the dessert costs 53¢/serving, and the juice costs 27¢/serving. If you want to provide your child with at least 120 calories and at least 120% of the RDA of vitamin C, how can you do so at the least cost?

SOURCE: Nutrition information supplied by Gerber.

T **26. Nutrition** Each serving of Gerber Mixed Cereal for Baby contains 60 calories, no vitamin C, and 11 grams of carbohydrates. Each serving of Gerber Mango Tropical Fruit Dessert contains 80 calories, 45% of the RDA of vitamin C for infants, and 21 grams of carbohydrates. Each serving of Gerber Apple Banana Juice contains 60 calories, 120% of the RDA of vitamin C for infants, and 15 grams of carbohydrates. Assume that the cereal costs 11¢/serving, the dessert costs 50¢/serving, and the juice costs 30¢/serving. If you want to provide your child with at least 180 calories, at least 120% of the RDA of vitamin C, and at least 37 grams of carbohydrates, how can you do so at the least cost?

SOURCE: Nutrition information supplied by Gerber.

27. Politics The political pollster Canter Inc. is preparing for a national election. It would like to poll at least 1500 Democrats and 1500 Republicans. Each mailing to the East Coast gets responses from 100 Democrats and 50 Republicans; each mailing to the Midwest gets responses from 100 Democrats and 100 Republicans; each mailing to the West Coast gets responses from 50 Democrats and 100 Republicans. Mailings to the East Coast cost $40 each to produce and mail, mailings to the Midwest cost $60 each, and mailings to the West Coast cost $50 each. How many mailings should Canter send to each area of the country to get the responses it needs at the least possible cost? What will it cost?

28. **Purchasing** Bingo's Copy Center needs to buy white paper and yellow paper. Bingo's can buy from three suppliers. Harvard Paper sells a package of 20 reams of white and 10 reams of yellow for $60, Yale Paper sells a package of 10 reams of white and 10 reams of yellow for $40, and Dartmouth Paper sells a package of 10 reams of white and 20 reams of yellow for $50. If Bingo's needs 350 reams of white and 400 reams of yellow, how many packages should it buy from each supplier to minimize the cost? What is the least possible cost?

29. **Purchasing** Cheapskate Electronics Store needs to update its inventory of stereos, TVs, and DVD players. There are three suppliers it can buy from: Nadir offers a bundle consisting of 5 stereos, 10 TVs, and 15 DVD players for $3000. Blunt offers a bundle consisting of 10 stereos, 10 TVs, and 10 DVD players for $4000. Sonny offers a bundle consisting of 15 stereos, 10 TVs, and 10 DVD players for $5000. Cheapskate Electronics needs at least 150 stereos, 200 TVs, and 150 DVD players. How can it update its inventory at the least possible cost? What is the least possible cost?

30. **Purchasing** Federal Rent-a-Car is putting together a new fleet. It is considering package offers from three car manufacturers. Fred Motors is offering 5 small cars, 5 medium cars, and 10 large cars for $500,000. Admiral Motors is offering 5 small, 10 medium, and 5 large cars for $400,000. Chrysalis is offering 10 small, 5 medium, and 5 large cars for $300,000. Federal would like to buy at least 550 small cars, at least 500 medium cars, and at least 550 large cars. How many packages should it buy from each car maker to keep the total cost as small as possible? What will be the total cost?

31. **Music CD Sales** In 2000 industry revenues from sales of recorded music amounted to $3.6 billion for rock music, $1.8 billion for rap music, and $0.4 billion for classical music. You would like the selection of music in your music store to reflect, in part, this national trend, so you have decided to stock at least twice as many rock music CDs as rap CDs. Your store has an estimated capacity of 20,000 CDs, and, as a classical music devotee, you would like to stock at least 5000 classical CDs. Rock music CDs sell for $12 on average, rap CDs for $15, and classical CDs for $12. How many of each type of CD should you stock to get the maximum retail value?

Rap includes hip hop. Revenues are based on total manufacturers' shipments at suggested retail prices and are rounded to the nearest $0.1 billion. SOURCE: Recording Industry Association of America, www.riaa.com, March 2002.

32. **Music CD Sales** Your music store's main competitor, Nuttal Hip Hop Classic Store, also wishes to stock at most 20,000 CDs, with at least half as many rap CDs as rock CDs and at least 2000 classical CDs. It anticipates an average sale price of $15/rock CD, $10/rap CD, and $10/classical CD. How many of each type of CD should it stock to get the maximum retail value, and what is the maximum retail value?

33. **Subsidies** The Miami Beach City Council has offered to subsidize hotel development in Miami Beach, and it is hoping for at least two hotels with a total capacity of at least 1400 people. Suppose you are a developer interested in taking advantage of this offer by building a small group of hotels in Miami Beach. You are thinking of three prototypes: a convention-style hotel with 500 rooms, costing $100 million; a vacation-style hotel with 200 rooms, costing $20 million; and a small motel with 50 rooms, costing $4 million. The city council will approve your plans provided that you build at least one convention-style hotel and no more than two small motels.
    a. How many of each type of hotel should you build to satisfy the city council's wishes and stipulations while minimizing your total cost?
    b. Now assume that the city council will give developers 20% of the cost of building new hotels in Miami Beach, up to $50 million.[8] Will the city's $50 million subsidy be sufficient to cover 20% of your total costs?

34. **Subsidies** Refer to Exercise 33. You are about to begin the financial arrangements for your new hotels when the city council informs you that it has changed its mind and now requires at least two vacation-style hotels and no more than four small motels.
    a. How many of each type of hotel should you build to satisfy the city council's wishes and stipulations while minimizing your total costs?
    b. Will the city's $50 million subsidy limit still be sufficient to cover 20% of your total costs?

Creatine Supplements Exercises 35 and 36 are based on the following data on four body-building supplements popular in 2002. (Figures shown correspond to a single serving.)

	Creatine (g)	Carbohydrates (g)	Taurine (g)	Alpha Lipoic Acid (mg)	Cost ($)
Cell-Tech (MuscleTech)	10	75	2	200	2.20
RiboForce HP (EAS)	5	15	1	0	1.60
Creatine Transport (Kaizen)	5	35	1	100	0.60
Pre-Load Creatine (Optimum)	6	35	1	25	0.50

SOURCE: Nutritional information supplied by the manufacturers/www.netrition.com. Cost per serving as quoted on www.netrition.com as of February 26, 2002.

[8]The Miami Beach City Council made such an offer in 1993 (*Chicago Tribune*, June 20, 1993, sect. 7, p. 8).

35. (Compare Exercise 27 in Section 4.2.) You are thinking of combining Cell-Tech, RiboForce HP, and Creatine Transport to obtain a 10-day supply that provides at least 80 grams of creatine and at least 10 grams of taurine, but no more than 750 grams of carbohydrates and 1000 milligrams of alpha lipoic acid. How many servings of each supplement should you combine, to meet your specifications for the least cost?

36. (Compare Exercise 28 in Section 4.2.) You are thinking of combining RiboForce HP, Creatine Transport, and Pre-Load Creatine to obtain a 10-day supply that provides at least 80 grams of creatine and at least 10 grams of taurine, but no more than 600 grams of carbohydrates and 2000 milligrams of alpha lipoic acid. How many servings of each supplement should you combine, to meet your specifications for the least cost?

37. Hospital Staffing As the staff director of a new hospital, you are planning to hire cardiologists, rehabilitation specialists, and infectious disease specialists. According to recent data, each cardiology case averages $12,000 in revenue; each physical rehabilitation case, $19,000; and each infectious disease case, $14,000. You judge that each specialist you employ will expand the hospital caseload by about 10 patients/week. You already have 3 cardiologists on staff, and the hospital is equipped to admit up to 200 patients/week. Based on past experience, each cardiologist and rehabilitation specialist brings in 1 government research grant per year, and each infectious disease specialist brings in 3 grants. Your board of directors would like to see a total of at least 30 grants/year and would like your weekly revenue to be as large as possible. How many of each kind of specialist should you hire?

These (rounded) figures are based on a Illinois survey of 1.3 million hospital admissions (*Chicago Tribune,* March 29, 1993, sec. 4, p. 1). SOURCE: Lutheran General Health System, Argus Associates, Inc.

38. Hospital Staffing In Exercise 37 you completely misjudged the number of patients each type of specialist would bring to the hospital each week. It turned out that each cardiologist brought in 120 new patients/*year,* each rehabilitation specialist brought in 90/year, and each infectious disease specialist brought in 70/year. It also turned out that your hospital could deal with no more than 1960 new patients per year. Repeat Exercise 37 in the light of this corrected data.

These (rounded) figures were obtained by dividing the average hospital revenue per physician by the revenue per case. SOURCE: Lutheran General Health System, Argus Associates, Inc.

39. Nutrition The Enormous State University Cafeteria, under pressure from the student council, decided to begin serving nutritious food for a change. As a goal, the cafeteria managers decided that each serving of a main course should contain at least 100 grams of protein, at most 350 grams of carbohydrate, at least 3180 milligrams of calcium, and no more than 1120 calories. The cafeteria has decided, with typical cynicism, to provide these requirements by serving hamburgers for every meal, mixed with powdered milk and eggs for nutrition. It can get hamburger at 10¢/ounce, powdered milk at 15¢/cup, and eggs at 5¢ each. Following is the chart of food values:

	Protein (g)	Carbohydrates (g)	Calcium (mg)	Calories
Hamburger (oz)	10	0	10	50
Powdered Milk (cup)	20	35	1020	100
Egg (one)	10	0	30	80

How much of each ingredient should be combined to produce hamburgers at minimum cost? Comment on the answer you get.

40. Nutrition The football team complained that it wasn't getting enough calories from the university cafeteria's hamburgers (see Exercise 39) and demanded that the cafeteria change the composition of the hamburgers to supply at least 350 g of carbohydrates per serving. Does this requirement affect the composition of the hamburgers in any way? If so, how?

41. Transportation Scheduling We return to your exploits of coordinating distribution for Tubular Ride Boogie Board Company.[9] You will recall that the company has manufacturing plants in Tucson, Arizona, and Toronto, Ontario, and you have been given the job of coordinating the distribution of their latest model, the Gladiator, to their outlets in Honolulu and Venice Beach. The Tucson plant can manufacture up to 620 boards/week, whereas the Toronto plant, beset by labor disputes, can produce no more than 410 boards/week. The outlet in Honolulu orders 500 boards/week, and Venice Beach orders 530 boards/week. Transportation costs are as follows: Tucson to Honolulu: $10/board; Tucson to Venice Beach: $5/board; Toronto to Honolulu: $20/board; Toronto to Venice Beach: $10/board. Your manager has said that you are to be sure to fill all orders and ship the boogie boards at a minimum total transportation cost. How will you do it?

42. Transportation Scheduling In the situation described in Exercise 41, you have just been notified that workers at the Toronto plant have gone on strike, resulting in a total work stoppage. You are to come up with a revised delivery schedule by tomorrow with the understanding that the Tucson plant can push production to a maximum of 1000 boards/week. What should you do?

43. Transportation Scheduling Your publishing company is about to start a promotional blitz for its new book, *Physics for the Liberal Arts.* You have 20 salespeople stationed in Chicago and 10 in Denver. You would like to fly at least 10 into Los Angeles and at least 15 into New York. A round-trip

[9] See Exercise 24 in Section 2.3 and Exercise 41 in Section 4.3. This time we will use the simplex method to solve the version of this problem that we first considered in Chapter 2.

plane flight from Chicago to Los Angeles costs \$195, from Chicago to New York costs \$182, from Denver to Los Angeles costs \$395, and from Denver to New York costs \$166. How many salespeople should you fly from each of Chicago and Denver to each of Los Angeles and New York to spend the least amount on plane flights?

SOURCE: Prices from Travelocity, at http://www.travelocity.com/, for the week of June 3, 2002, as of May 5, 2002.

T **44. Transportation Scheduling** Repeat Exercise 43 but now suppose you would like at least 15 salespeople in Los Angeles.

45. Finance Senator Porkbarrel habitually overdraws his three bank accounts, at the Congressional Integrity Bank, Citizens' Trust, and Checks R Us. There are no penalties because the overdrafts are subsidized by the taxpayer. The Senate Ethics Committee tends to let slide irregular banking activities as long as they are not flagrant. At the moment (due to Congress' preoccupation with a Supreme Court nominee), a total overdraft of up to \$10,000 will be overlooked. Porkbarrel's conscience makes him hesitate to overdraw accounts at banks whose names include expressions like "integrity" and "citizens' trust." The effect is that his overdrafts at the first two banks combined amount to no more than one-quarter of the total. On the other hand, the financial officers at Integrity Bank, aware that Senator Porkbarrel is a member of the Senate Banking Committee, "suggest" that he overdraw at least \$2500 from their bank. Find the amount he should overdraw from each bank in order to avoid investigation by the Ethics Committee and overdraw his account at Integrity by as much as his sense of guilt will allow.

46. Scheduling Since Joe Slim's brother was recently elected to the State Senate, Joe's financial advisement concern, Inside Information Inc., has been doing a booming trade, even though the financial counseling he offers is quite worthless. (None of his seasoned clients pays the slightest attention to his advice.) Slim charges different hourly rates to different categories of individuals: \$5000/hour for private citizens, \$50,000/hour for corporate executives, and \$10,000/hour for presidents of universities. Due to his taste for leisure, he feels that he can spend no more than 40 hours/week in consultation. On the other hand, Slim feels that it would be best for his intellect were he to devote at least 10 hours of consultation each week to university presidents. However, Slim always feels somewhat uncomfortable dealing with academics, so he would prefer to spend no more than half his consultation time with university presidents. Furthermore, he likes to think of himself as representing the interests of the common citizen, so he wishes to offer at least 2 more hours of his time each week to private citizens than to corporate executives and university presidents combined. Given all these restrictions, how many hours each week should he spend with each type of client in order to maximize his income?

COMMUNICATION AND REASONING EXERCISES

47. Explain the need for phase I in a nonstandard LP problem.

48. Explain the need for phase II in a nonstandard LP problem.

49. Explain briefly why we would need to use phase I in solving an LP problem with the constraint $x + 2y - z \geq 3$.

50. Which rows do we star, and why?

51. Consider the following linear programming problem:

$$\begin{array}{ll} \text{Maximize} & p = x + y \\ \text{subject to} & x - 2y \geq 0 \\ & 2x + y \leq 10 \\ & x \geq 0, y \geq 0 \end{array}$$

This problem

(A) must be solved using the techniques of Section 4.4.
(B) must be solved using the techniques of Section 4.3.
(C) can be solved using the techniques of either section.

52. Consider the following linear programming problem:

$$\begin{array}{ll} \text{Maximize} & p = x + y \\ \text{subject to} & x - 2y \geq 1 \\ & 2x + y \leq 10 \\ & x \geq 0, y \geq 0 \end{array}$$

This problem

(A) must be solved using the techniques of Section 4.4.
(B) must be solved using the techniques of Section 4.3.
(C) can be solved using the techniques of either section.

53. Find an LP problem in three variables that requires one pivot in phase I.

54. Find an LP problem in three variables that requires two pivots in phase I.

55. Find an LP problem in two or three variables with no optimal solution and show what happens when you try to solve it using the simplex method.

56. Find an LP problem in two or three variables with more than one optimal solution and investigate which solution is found by the simplex method.

4.5 *The Simplex Method and Duality (Optional)*

We mentioned **standard minimization problems** in the last section. These problems have the following form.

Standard Minimization Problem

A **standard minimization problem** is an LP problem in which we are required to *minimize* (not maximize) a linear objective function

$$c = as + bt + cu + \cdots$$

of the variables $s, t, u, \ldots$ (in this section, we will always use the letters $s, t, u, \ldots$ for the unknowns in a standard minimization problem)* subject to the constraints

$$s \geq 0, t \geq 0, u \geq 0, \ldots$$

and further constraints of the form

$$As + Bt + Cu + \cdots \geq N$$

where $A, B, C, \ldots, N$ are numbers with N nonnegative.

A **standard linear programming problem** is an LP problem that is either a standard maximization problem or a standard minimization problem. An LP problem satisfies the **nonnegative objective condition** if all the coefficients in the objective function are nonnegative.

Quick Examples: Standard Minimization and Maximization Problems

1. Minimize $c = 2s + 3t + 3u$
   subject to
   $$\begin{aligned} 2s \quad\quad + u &\geq 10 \\ s + 3t - 6u &\geq 5 \\ s \geq 0, t \geq 0, u &\geq 0 \end{aligned}$$
   This is a standard minimization problem satisfying the nonnegative objective condition.

2. Maximize $p = 2x + 3y + 3z$
   subject to
   $$\begin{aligned} 2x \quad\quad + z &\leq 7 \\ x + 3y - 6z &\leq 6 \\ x \geq 0, y \geq 0, z &\geq 0 \end{aligned}$$
   This is a standard maximization problem satisfying the nonnegative objective condition.

3. Minimize $c = 2s - 3t + 3u$
   subject to
   $$\begin{aligned} 2s \quad\quad + u &\geq 10 \\ s + 3t - 6u &\geq 5 \\ s \geq 0, t \geq 0, u &\geq 0 \end{aligned}$$
   This is a standard minimization problem that does *not* satisfy the nonnegative objective condition.

*The reasons will soon become apparent.

We saw a way of solving minimization problems in Section 4.4, but a mathematically elegant relationship between maximization and minimization problems gives us another way of solving minimization problems that satisfy the nonnegative objective condition. This relationship is called **duality.**

To describe duality, we must first represent an LP problem by a matrix. This matrix is *not* the first tableau but something simpler: Pretend you forgot all about slack variables and also forgot to change the signs of the objective function.[10] As an example, consider the following two standard problems.[11]

[10]Forgetting these things is exactly what happens to many students under test conditions!

[11]Although duality does not require the problems to be standard, it does require them to be written in so-called standard form: In the case of a maximization problem, all constraints need to be (re)written using $\leq$; for a minimization problem, all constraints need to be (re)written using $\geq$. It is least confusing to stick with standard problems, which is what we will do in this section.

Problem 1

$$\begin{array}{ll} \text{Maximize} & p = 20x + 20y + 50z \\ \text{subject to} & 2x + y + 3z \le 2000 \\ & x + 2y + 4z \le 3000 \\ & x \ge 0, y \ge 0, z \ge 0 \end{array}$$

We represent this problem by the matrix

$$\left[\begin{array}{ccc|c} 2 & 1 & 3 & 2000 \\ 1 & 2 & 4 & 3000 \\ \hline 20 & 20 & 50 & 0 \end{array}\right] \begin{array}{l} \text{Constraint 1} \\ \text{Constraint 2} \\ \text{Objective} \end{array}$$

Notice that the coefficients of the objective function go in the bottom row, and we place a zero in the bottom right corner.

Problem 2 From Example 3 in Section 4.4,

$$\begin{array}{ll} \text{Minimize} & c = 2000s + 3000t \\ \text{subject to} & 2s + t \ge 20 \\ & s + 2t \ge 20 \\ & 3s + 4t \ge 50 \\ & s \ge 0, t \ge 0 \end{array}$$

Problem 2 is represented by

$$\left[\begin{array}{cc|c} 2 & 1 & 20 \\ 1 & 2 & 20 \\ 3 & 4 & 50 \\ \hline 2000 & 3000 & 0 \end{array}\right] \begin{array}{l} \text{Constraint 1} \\ \text{Constraint 2} \\ \text{Constraint 3} \\ \text{Objective} \end{array}$$

These two problems are related: The matrix for Problem 1 is the transpose of the matrix for Problem 2. (Recall that the transpose of a matrix is obtained by writing its rows as columns; see Section 3.1.) When we have a pair of LP problems related in this way, we say that the two are *dual* LP problems.

Dual Linear Programming Problems

Two LP problems, one a maximization and one a minimization problem, are **dual** if the matrix that represents one is the transpose of the matrix that represents the other.

Finding the Dual of a Given Problem

Given an LP problem, we find its dual as follows:

1. Represent the problem as a matrix (see above).
2. Take the transpose of the matrix.
3. Write down the dual, which is the LP problem corresponding to the new matrix. If the original problem was a maximization problem, its dual will be a minimization problem, and vice versa.

The original problem is called the **primal problem,** and its dual is referred to as the **dual problem.**

Quick Example

Primal problem

Minimize $c = s + 2t$

subject to

$$\begin{aligned} 5s + 2t &\geq 60 \\ 3s + 4t &\geq 80 \\ s + t &\geq 20 \\ s \geq 0&, t \geq 0 \end{aligned} \quad \xrightarrow{\mathbf{1}} \quad \begin{bmatrix} 5 & 2 & 60 \\ 3 & 4 & 80 \\ 1 & 1 & 20 \\ 1 & 2 & 0 \end{bmatrix}$$

$$\xrightarrow{\mathbf{2}} \begin{bmatrix} 5 & 3 & 1 & 1 \\ 2 & 4 & 1 & 2 \\ 60 & 80 & 20 & 0 \end{bmatrix} \xrightarrow{\mathbf{3}}$$

Dual problem

Maximize $p = 60x + 80y + 20z$

subject to

$$\begin{aligned} 5x + 3y + z &\leq 1 \\ 2x + 4y + z &\leq 2 \\ x \geq 0, y \geq 0&, z \geq 0 \end{aligned}$$

Question Fine, now I know how to obtain the dual of ("dualize") an LP problem. What does that do for me?

Answer Quite a lot, according to the following theorem.

Fundamental Theorem of Duality

a. If an LP problem has an optimal solution, then so does its dual. Moreover, the primal problem and the dual problem have the same optimal value for their objective functions.

b. Contained in the final tableau of the simplex method applied to an LP problem is the solution to its dual problem: It is given by the bottom entries in the columns associated with the slack variables, divided by the entry under the objective variable.

This theorem gives us an alternative way of solving minimization problems that satisfy the nonnegative objective condition.[12] Let's illustrate by solving Problem 2.

Example 1 • Solving by Duality

Minimize $c = 2000s + 3000t$

subject to

$$\begin{aligned} 2s + t &\geq 20 \\ s + 2t &\geq 20 \\ 3s + 4t &\geq 50 \\ s \geq 0&, t \geq 0 \end{aligned}$$

Solution

Step 1 *Find the dual problem.* Write the primal problem in matrix form and take the transpose:

$$\begin{bmatrix} 2 & 1 & 20 \\ 1 & 2 & 20 \\ 3 & 4 & 50 \\ 2000 & 3000 & 0 \end{bmatrix} \rightarrow \begin{bmatrix} 2 & 1 & 3 & 2000 \\ 1 & 2 & 4 & 3000 \\ 20 & 20 & 50 & 0 \end{bmatrix}$$

[12]We will not prove this theorem because it is beyond the scope of this book. The proof can be found in a textbook devoted to linear programming, like *Linear Programming* by Vašek Chvátal (San Francisco: Freeman, 1983), which has a particularly well-motivated discussion.

The dual problem is

$$\begin{aligned} \text{Maximize}\quad & p = 20x + 20y + 50z \\ \text{subject to}\quad & 2x + y + 3z \le 2000 \\ & x + 2y + 4z \le 3000 \\ & x \ge 0, y \ge 0, z \ge 0 \end{aligned}$$

Step 2 *Use the simplex method to solve the dual problem.* Since we have a standard maximization problem, we do not have to worry about phase I but go straight to phase II:

	x	y	z	s	t	p	
s	2	1	[3]	1	0	0	2,000
t	1	2	4	0	1	0	3,000
p	−20	−20	−50	0	0	1	0

	x	y	z	s	t	p	
z	2	1	3	1	0	0	2,000
t	−5	[2]	0	−4	3	0	1,000
p	40	−10	0	50	0	3	100,000

	x	y	z	s	t	p	
z	9	0	6	6	−3	0	3,000
y	−5	2	0	−4	3	0	1,000
p	15	0	0	30	15	3	105,000

Note that the maximum value of the objective function is $p = 105{,}000/3 = 35{,}000$. By the theorem, this is also the optimal value of c in the primal problem!

Step 3 *Read off the solution to the primal problem by dividing the bottom entries in the columns associated with the slack variables by the entry in the p column.* Here is the final tableau again with the entries in question highlighted.

	x	y	z	s	t	p	
z	9	0	6	6	−3	0	3,000
y	−5	2	0	−4	3	0	1,000
p	15	0	0	**30**	**15**	**3**	**105,000**

The solution to the primal problem is

$$s = \frac{30}{3} = 10 \qquad t = \frac{15}{3} = 5 \qquad c = \frac{105{,}000}{3} = 35{,}000$$

(Compare this with the method we used to solve Example 3 in Section 4.4. Which method seems more efficient?)

Before we go on . . . Can you now see the reason for using the variable names $s, t, u, \ldots$ in standard minimization problems?

Question Why did we mention the nonnegative objective condition?

Answer Consider a standard minimization problem that does not satisfy the nonnegative objective condition, such as

$$\begin{aligned} \text{Minimize} \quad & c = 2s - t \\ \text{subject to} \quad & 2s + 3t \geq 2 \\ & s + 2t \geq 2 \\ & s \geq 0, t \geq 0 \end{aligned}$$

Its dual would be

$$\begin{aligned} \text{Maximize} \quad & p = 2x + 2y \\ \text{subject to} \quad & 2x + y \leq 2 \\ & 3x + 2y \leq -1 \\ & x \geq 0, y \geq 0 \end{aligned}$$

This is not a standard maximization problem because the right-hand side of the second constraint is negative. In general, if a problem does not satisfy the nonnegative condition, its dual is not standard. Therefore, to solve the dual by the simplex method will require using phase I and phase II, and we may as well just solve the primal problem that way. Thus, duality helps us solve problems only when the primal problem satisfies the nonnegative objective condition.

In many economic applications, the solution to the dual problem also gives us useful information about the primal problem, as we see in Example 2.

Example 2 • Shadow Costs

You are trying to decide how many vitamin pills to take. SuperV brand vitamin pills each contain 2 milligrams of vitamin X, 1 milligram of vitamin Y, and 1 milligram of vitamin Z. Topper brand vitamin pills each contain 1 milligram of vitamin X, 1 milligram of vitamin Y, and 2 milligrams of vitamin Z. You want to take enough pills daily to get at least 12 milligrams of vitamin X, 10 milligrams of vitamin Y, and 12 milligrams of vitamin Z. However, SuperV pills cost 4¢ each and Toppers cost 3¢ each, and you would like to minimize the total cost of your daily dosage. How many of each brand of pill should you take? How would changing your daily vitamin requirements affect your minimum cost?

Solution This is a straightforward minimization problem. The unknowns are

s = number of SuperV brand pills

t = number of Topper brand pills

The LP problem is

$$\begin{aligned} \text{Minimize} \quad & c = 4s + 3t \\ \text{subject to} \quad & 2s + t \geq 12 \\ & s + t \geq 10 \\ & s + 2t \geq 12 \\ & s \geq 0, t \geq 0 \end{aligned}$$

We solve this problem by using the simplex method on its dual, which is

$$\begin{array}{ll} \text{Maximize} & p = 12x + 10y + 12z \\ \text{subject to} & 2x + y + z \le 4 \\ & x + y + 2z \le 3 \\ & x \ge 0, y \ge 0, z \ge 0 \end{array}$$

After pivoting three times, we arrive at the final tableau:

	x	y	z	s	t	p	
x	6	0	−6	6	−6	0	6
y	0	1	3	−1	2	0	2
p	0	0	6	2	8	1	32

Therefore, the answer to the original problem is that you should take two SuperV vitamin pills and eight Toppers at a cost of 32¢ per day.

Now, the key to answering the last question, which asks you to determine how changing your daily vitamin requirements would affect your minimum cost, is to look at the solution to the dual problem. From the tableau we see that $x = 1$, $y = 2$, and $z = 0$. To see what x, y, and z might tell us about the original problem, let's look at their units. In the inequality $2x + y + z \le 4$, the coefficient 2 of x has units "mg of vitamin X/SuperV pill," and the 4 on the right-hand side has units "¢/SuperV pill." For $2x$ to have the same units as the 4 on the right-hand side, x must have units "¢/mg of vitamin X." Similarly, y must have units "¢/mg of vitamin Y" and z must have units "¢/mg of vitamin Z." One can show (although we will not do it here) that x gives the amount that would be added to the minimum cost for each increase of 1 milligram of vitamin X in our daily requirement.[13] For example, if we were to increase our requirement from 12 milligrams to 14 milligrams, an increase of 2 milligrams, the minimum cost would change by $2x = 2$¢, from 32¢ to 34¢. (Try it; you'll end up taking four SuperV pills and six Toppers.) Similarly, each increase of 1 milligram of vitamin Y in the requirements would increase the cost by $y = 2$¢. These costs are called the **marginal costs,** or the **shadow costs,** of the vitamins.

Before we go on . . . What about $z = 0$? The shadow cost of vitamin Z is 0¢/mg, meaning that you can increase your requirement of vitamin Z without changing your cost. In fact, the solution $s = 2$ and $t = 8$ provides you with 18 milligrams of vitamin Z, so you can increase the required amount of vitamin Z up to 18 milligrams without changing the solution at all.

We can also interpret the shadow costs as the effective cost to you of each milligram of each vitamin in the optimal solution. You are paying 1¢/milligram of vitamin X, 2¢/milligram of vitamin Y, and getting the vitamin Z for free. This gives a total cost of $1 \times 12 + 2 \times 10 + 0 \times 12 = 32$¢, as we know. Again, if you change your requirements slightly, these are the amounts you will pay per milligram of each vitamin.

[13] To be scrupulously correct, this works only for relatively small changes in the requirements, not necessarily for very large ones.

> ### *Guideline: When to Use Duality*
>
> ***Question*** Given a minimization problem, when should I use duality, and when should I use the two-phase method in Section 4.4?
>
> ***Answer*** If the original problem satisfies the nonnegative objective condition (none of the coefficients in the objective function are negative), then you can use duality to convert the problem to a standard maximization one, which can be solved with the one-phase method. If the original problem does not satisfy the nonnegative objective condition, then dualizing results in a nonstandard LP problem, so dualizing may not be worthwhile.
>
> ***Question*** When is it absolutely necessary to use duality?
>
> ***Answer*** Never. Duality gives us an efficient but not necessary alternative for solving standard minimization problems.

4.5 EXERCISES

In Exercises 1–8, write down (without solving) the dual LP problem.

1. Maximize $p = 2x + y$ subject to

$$\begin{aligned} x + 2y &\le 6 \\ -x + y &\le 2 \\ x \ge 0,\ y &\ge 0 \end{aligned}$$

2. Maximize $p = x + 5y$ subject to

$$\begin{aligned} x + y &\le 6 \\ -x + 3y &\le 4 \\ x \ge 0,\ y &\ge 0 \end{aligned}$$

3. Minimize $c = 2s + 2t + 3u$ subject to

$$\begin{aligned} s + t + u &\ge 100 \\ 2s + t &\ge 50 \\ s \ge 0,\ t \ge 0,\ u &\ge 0 \end{aligned}$$

4. Minimize $c = 2s + 2t + 3u$ subject to

$$\begin{aligned} s + u &\ge 100 \\ 2s + t &\ge 50 \\ s \ge 0,\ t \ge 0,\ u &\ge 0 \end{aligned}$$

5. Maximize $p = x + y + z + w$ subject to

$$\begin{aligned} x + y + z &\le 3 \\ y + z + w &\le 4 \\ x + z + w &\le 5 \\ x + y + w &\le 6 \\ x \ge 0,\ y \ge 0,\ z \ge 0,\ w &\ge 0 \end{aligned}$$

6. Maximize $p = x + y + z + w$ subject to

$$\begin{aligned} x + y + z &\le 3 \\ y + z + w &\le 3 \\ x + z + w &\le 4 \\ x + y + w &\le 4 \\ x \ge 0,\ y \ge 0,\ z \ge 0,\ w &\ge 0 \end{aligned}$$

7. Minimize $c = s + 3t + u$ subject to

$$\begin{aligned} 5s - t + v &\ge 1000 \\ u - v &\ge 2000 \\ s + t &\ge 500 \\ s \ge 0,\ t \ge 0,\ u \ge 0,\ v &\ge 0 \end{aligned}$$

8. Minimize $c = 5s + 2u + v$ subject to

$$\begin{aligned} s - t + 2v &\ge 2000 \\ u + v &\ge 3000 \\ s + t &\ge 500 \\ s \ge 0,\ t \ge 0,\ u \ge 0,\ v &\ge 0 \end{aligned}$$

In Exercises 9–22, solve the standard minimization problems using duality. (You may already have seen some of them in earlier sections, but now you will be solving them using a different method.)

9. Minimize $c = s + t$ subject to

$$\begin{aligned} s + 2t &\ge 6 \\ 2s + t &\ge 6 \\ s \ge 0,\ t &\ge 0 \end{aligned}$$

10. Minimize $c = s + 2t$ subject to

$$\begin{aligned} s + 3t &\ge 30 \\ 2s + t &\ge 30 \\ s \ge 0,\ t &\ge 0 \end{aligned}$$

11. Minimize $c = 6s + 6t$ subject to

$$\begin{aligned} s + 2t &\ge 20 \\ 2s + t &\ge 20 \\ s \ge 0,\ t &\ge 0 \end{aligned}$$

12. Minimize $c = 3s + 2t$ subject to

$$\begin{aligned} s + 2t &\ge 20 \\ 2s + t &\ge 10 \\ s \ge 0,\ t &\ge 0 \end{aligned}$$

13. Minimize $c = 0.2s + 0.3t$ subject to

$$\begin{aligned} 0.2s + 0.1t &\ge 1 \\ 0.15s + 0.3t &\ge 1.5 \\ 10s + 10t &\ge 80 \\ s \ge 0,\ t &\ge 0 \end{aligned}$$

14. Minimize $c = 0.4s + 0.1t$ subject to

$$\begin{aligned} 30s + 20t &\ge 600 \\ 0.1s + 0.4t &\ge 4 \\ 0.2s + 0.3t &\ge 4.5 \\ s \ge 0,\ t &\ge 0 \end{aligned}$$

15. Minimize $c = 2s + t$
subject to
$$\begin{aligned} 3s + t &\geq 30 \\ s + t &\geq 20 \\ s + 3t &\geq 30 \\ s \geq 0, t &\geq 0 \end{aligned}$$

16. Minimize $c = s + 2t$
subject to
$$\begin{aligned} 4s + t &\geq 100 \\ 2s + t &\geq 80 \\ s + 3t &\geq 150 \\ s \geq 0, t &\geq 0 \end{aligned}$$

17. Minimize $c = s + 2t + 3u$
subject to
$$\begin{aligned} 3s + 2t + u &\geq 60 \\ 2s + t + 3u &\geq 60 \\ s \geq 0, t \geq 0, u &\geq 0 \end{aligned}$$

18. Minimize $c = s + t + 2u$
subject to
$$\begin{aligned} s + 2t + 2u &\geq 60 \\ 2s + t + 3u &\geq 60 \\ s \geq 0, t \geq 0, u &\geq 0 \end{aligned}$$

19. Minimize $c = 2s + t + 3u$
subject to
$$\begin{aligned} s + t + u &\geq 100 \\ 2s + t &\geq 50 \\ t + u &\geq 50 \\ s \geq 0, t \geq 0, u &\geq 0 \end{aligned}$$

20. Minimize $c = 2s + 2t + 3u$
subject to
$$\begin{aligned} s + u &\geq 100 \\ 2s + t &\geq 50 \\ t + u &\geq 50 \\ s \geq 0, t \geq 0, u &\geq 0 \end{aligned}$$

21. Minimize $c = s + t + u$
subject to
$$\begin{aligned} 3s + 2t + u &\geq 60 \\ 2s + t + 3u &\geq 60 \\ s + 3t + 2u &\geq 60 \\ s \geq 0, t \geq 0, u &\geq 0 \end{aligned}$$

22. Minimize $c = s + t + 2u$
subject to
$$\begin{aligned} s + 2t + 2u &\geq 60 \\ 2s + t + 3u &\geq 60 \\ s + 3t + 6u &\geq 60 \\ s \geq 0, t \geq 0, u &\geq 0 \end{aligned}$$

APPLICATIONS

The following applications are similar to ones in preceding exercise sets. Use duality to answer them.

23. Resource Allocation Meow Inc. makes cat food out of fish and cornmeal. Fish has 8 grams of protein and 4 grams of fat per ounce, and cornmeal has 4 grams of protein and 8 grams of fat. A jumbo can of cat food must contain at least 48 grams of protein and 48 grams of fat. If fish and cornmeal both cost 5¢/ounce, how many ounces of each should Meow use in each can of cat food to minimize costs? What are the shadow costs of protein and of fat?

24. Resource Allocation Oz Inc. makes lion food out of giraffe and gazelle meat. Giraffe meat has 18 grams of protein and 36 grams of fat per pound, and gazelle meat has 36 grams of protein and 18 grams of fat per pound. A batch of lion food must contain at least 36,000 grams of protein and 54,000 grams of fat. Giraffe meat costs \$2/pound, and gazelle meat costs \$4/pound. How many pounds of each should go into each batch of lion food in order to minimize costs? What are the shadow costs of protein and fat?

25. Nutrition Ruff Inc. makes dog food out of chicken and grain. Chicken has 10 grams of protein and 5 grams of fat per ounce, and grain has 2 grams of protein and 2 grams of fat per ounce. A bag of dog food must contain at least 200 grams of protein and at least 150 grams of fat. If chicken costs 10¢/ounce and grain costs 1¢/ounce, how many ounces of each should Ruff use in each bag of dog food in order to minimize cost? What are the shadow costs of protein and fat?

26. Purchasing The Enormous State University's Business School is buying computers. The school has two models from which to choose, the Pomegranate and the iZac. Each Pomegranate comes with 400 megabytes of memory and 80 gigabytes of disk space, and each iZac has 300 megabytes of memory and 100 gigabytes of disk space. For reasons related to its accreditation, the school would like to say that it has a total of at least 48,000 megabytes of memory and at least 12,800 gigabytes of disk space. If both the Pomegranate and the iZac cost \$2000 each, how many of each should the school buy to keep the cost as low as possible? What are the shadow costs of memory and disk space?

27. Nutrition Each serving of Gerber Mixed Cereal for Baby contains 60 calories and no vitamin C. Each serving of Gerber Mango Tropical Fruit Dessert contains 80 calories and 45% of the U.S. Recommended Daily Allowance (RDA) of vitamin C for infants. Each serving of Gerber Apple Banana Juice contains 60 calories and 120% of the RDA of vitamin C for infants. Suppose the cereal costs 10¢/serving, the dessert costs 53¢/serving, and the juice costs 27¢/serving. If you want to provide your child with at least 120 calories and at least 120% of the RDA of vitamin C, how can you do so at the least cost? What are your shadow costs for calories and vitamin C?

SOURCE: Nutrition information provided by Gerber.

28. Nutrition Each serving of Gerber Mixed Cereal for Baby contains 60 calories, no vitamin C, and 11 grams of carbohydrates. Each serving of Gerber Mango Tropical Fruit Dessert contains 80 calories, 45% of the RDA of vitamin C for infants, and 21 grams of carbohydrates. Each serving of Gerber Apple Banana Juice contains 60 calories, 120% of the RDA of vitamin C for infants, and 15 grams of carbohydrates. Assume that the cereal costs 11¢/serving, the dessert costs 50¢/serving, and the juice costs 30¢/serving. If you want to provide your child with at least 180 calories, at least 120% of the RDA of vitamin C, and at least 37 grams of carbohydrates, how can you do so at the least cost? What are your shadow costs for calories, vitamin C, and carbohydrates?

SOURCE: Nutrition information provided by Gerber.

29. Politics The political pollster Canter Inc. is preparing for a national election. It would like to poll at least 1500 Democrats and 1500 Republicans. Each mailing to the East Coast gets responses from 100 Democrats and 50 Republicans; each mailing to the Midwest gets responses from 100 Democrats and 100 Republicans; each mailing to the West Coast gets responses from 50 Democrats and 100 Republicans. Mailings to the East Coast cost $40 each to produce and mail, mailings to the Midwest cost $60 each, and mailings to the West Coast cost $50 each. How many mailings should Canter send to each area of the country to get the responses it needs at the least possible cost? What will it cost? What are the shadow costs of a Democratic response and a Republican response?

30. Purchasing Bingo's Copy Center needs to buy white paper and yellow paper. Bingo's can buy from three suppliers. Harvard Paper sells a package of 20 reams of white and 10 reams of yellow for $60, Yale Paper sells a package of 10 reams of white and 10 reams of yellow for $40, and Dartmouth Paper sells a package of 10 reams of white and 20 reams of yellow for $50. If Bingo's needs 350 reams of white and 400 reams of yellow, how many packages should it buy from each supplier so as to minimize the cost? What is the lowest possible cost? What are the shadow costs of white paper and yellow paper?

31. Resource Allocation One day Gillian the Magician summoned the wisest of her women. "Devoted followers," she began, "I have a quandary. As you well know, I possess great expertise in sleep spells and shock spells, but unfortunately, these are proving to be a drain on my aural energy resources; each sleep spell costs me 500 pico-shirleys of aural energy, and each shock spell requires 750 pico-shirleys. Clearly, I would like to hold my overall expenditure of aural energy to a minimum and still meet my commitments in protecting the Sisterhood from the ever-present threat of trolls. Specifically, I have estimated that each sleep spell keeps us safe for an average of 2 minutes, and every shock spell protects us for about 3 minutes. We certainly require enough protection to last 24 hours of each day, and possibly more, just to be safe. At the same time, I have noticed that each of my sleep spells can immobilize 3 trolls at once, while one of my typical shock spells (having a narrower range) can immobilize only 2 trolls at once. We are faced, my sisters, with an onslaught of 1200 trolls per day! Finally, as you are no doubt aware, the bylaws dictate that for a Magician of the Order to remain in good standing, the number of shock spells must be between one-quarter and one-third the number of shock and sleep spells combined. What do I do, oh Wise Ones?"

32. Risk Management The Grand Vizier of the Kingdom of Um is being blackmailed by numerous individuals and is having a very difficult time keeping his blackmailers from going public. He has been keeping them at bay with two kinds of payoff: gold bars from the Royal Treasury and political favors. Through bitter experience, he has learned that each payoff in gold gives him peace for an average of about 1 month, and each political favor seems to earn him about $1\frac{1}{2}$ months of reprieve. To maintain his flawless reputation in the court, he feels he cannot afford any revelations about his tainted past to come to light within the next year. Thus, it is imperative that his blackmailers be kept at bay for 12 months. Furthermore, he would like to keep the number of gold payoffs at no more than one-quarter of the combined number of payoffs because the outward flow of gold bars might arouse suspicion on the part of the Royal Treasurer. The gold payoffs tend to deplete the Grand Vizier's travel budget. (The treasury has been subsidizing his numerous trips to the Himalayas.) He estimates that each gold bar removed from the treasury will cost him four trips. On the other hand, since the administering of political favors tends to cost him valuable travel time, he suspects that each political favor will cost him about two trips. Now, he would obviously like to keep his blackmailers silenced and lose as few trips as possible. What is he to do? How many trips will he lose in the next year?

COMMUNICATION AND REASONING EXERCISES

33. Give one possible advantage of using duality to solve a standard minimization problem.

34. To ensure that the dual of a minimization problem will result in a standard maximization problem,

(A) the primal problem should satisfy the nonnegative objective condition.

(B) the primal problem should be a standard minimization problem.

(C) the primal problem should not satisfy the nonnegative objective condition.

35. Give an example of a standard minimization problem whose dual is *not* a standard maximization problem. How would you go about solving your problem?

36. Give an example of a nonstandard minimization problem whose dual is a standard maximization problem.

37. Solve the following nonstandard minimization problem, using duality. Recall from a footnote in the text that to find the dual you must first rewrite all of the constraints using $\geq$. The Miami Beach City Council has offered to subsidize hotel development in Miami Beach and is hoping for at least two hotels with a total capacity of at least 1400 people. Suppose you are a developer interested in taking advantage of this offer by building a small group of hotels in Miami Beach. You are thinking of three prototypes: a convention-style hotel with 500 rooms costing $100 million, a vacation-style hotel with 200 rooms costing $20 million, and a small motel with 50 rooms costing $4 million. The city council will approve your plans provided you build at least one convention-style hotel and no more than two small motels. How many of each type of hotel should you build to satisfy the city council's wishes and stipulations while minimizing your total cost?

38. Given a minimization problem, when would you solve it by applying the simplex method to its dual, and when would you apply the simplex method to the minimization problem itself?

39. Create an interesting application that leads to a standard maximization problem. Solve it using the simplex method and note the solution to its dual problem. What does the solution to the dual tell you about your application?

CASE STUDY

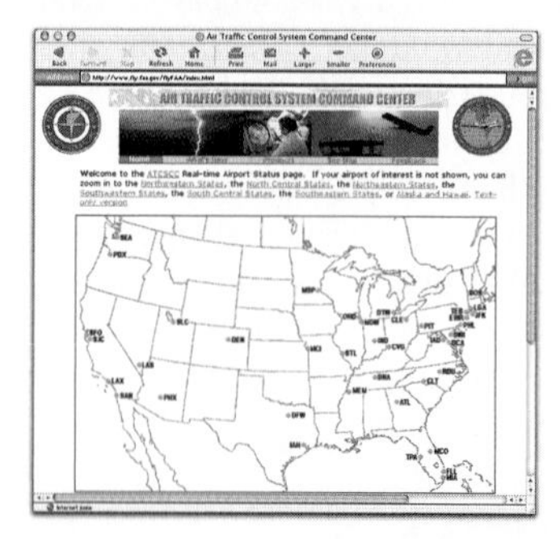

Reproduced from the Federal Aviation Administration Web site

Airline Scheduling

You are the traffic manager for Fly-by-Night Airlines, which flies airplanes from five cities: Los Angeles, Chicago, Atlanta, New York, and Boston. Due to a strike you have not had sufficient crews to fly your planes, so some of your planes have been stranded in Chicago and in Atlanta. In fact, 15 extra planes are in Chicago and 30 extra planes are in Atlanta. To get back on schedule, Los Angeles needs at least 10 planes, New York needs at least 20, and Boston needs at least 10. It costs \$50,000 to fly a plane from Chicago to Los Angeles, \$10,000 to fly from Chicago to New York, and \$20,000 to fly from Chicago to Boston. It costs \$70,000 to fly a plane from Atlanta to Los Angeles, \$20,000 to fly from Atlanta to New York, and \$50,000 to fly from Atlanta to Boston. To get back on schedule at the lowest cost, how should you rearrange your planes?

As always, you remember to start by identifying the unknowns. You need to decide how many planes you will fly from each of Chicago and Atlanta to each of Los Angeles, New York, and Boston. This gives you six unknowns:

x = number of planes to fly from Chicago to Los Angeles

y = number of planes to fly from Chicago to New York

z = number of planes to fly from Chicago to Boston

u = number of planes to fly from Atlanta to Los Angeles

v = number of planes to fly from Atlanta to New York

w = number of planes to fly from Atlanta to Boston

Los Angeles needs at least 10 planes, so you know that

$$x + u \geq 10$$

Similarly, New York and Boston give you the inequalities

$$y + v \geq 20$$

$$z + w \geq 10$$

Since Chicago has only 15 extra planes, you have

$$x + y + z \leq 15$$

Since Atlanta has 30 extra planes,

$$u + v + w \leq 30$$

Finally, you want to minimize the cost, which is given by

$$C = 50{,}000x + 10{,}000y + 20{,}000z + 70{,}000u + 20{,}000v + 50{,}000w$$

This gives you your LP problem:

$$\begin{array}{ll}
\text{Minimize} & C = 50{,}000x + 10{,}000y + 20{,}000z + 70{,}000u + 20{,}000v + 50{,}000w \\
\text{subject to} & x + u \geq 10 \\
& y + v \geq 20 \\
& z + w \geq 10 \\
& x + y + z \leq 15 \\
& u + v + w \leq 30 \\
& x \geq 0, y \geq 0, z \geq 0, u \geq 0, v \geq 0, w \geq 0
\end{array}$$

Since you have six unknowns, you know that you will have to use the simplex method, so you start converting into the proper form. You change the minimization problem into the problem of maximizing $P = -C$. At the same time, you notice that it would actually be enough to maximize

$$P = -5x - y - 2z - 7u - 2v - 5w$$

(this will keep the numbers small). You subtract the surplus variables p, q, and r from the left-hand sides of the first three inequalities, and you add the slack variables s and t to the last two. Then it's time to go to work. Remembering all the rules for pivoting and using a computer to help with the calculations, you get the following sequence of tableaus:

	x	y	z	u	v	w	p	q	r	s	t	P	
p	1	0	0	1	0	0	−1	0	0	0	0	0	10
q	0	1	0	0	1	0	0	−1	0	0	0	0	20
r	0	0	1	0	0	1	0	0	−1	0	0	0	10
s	1	1	1	0	0	0	0	0	0	1	0	0	15
t	0	0	0	1	1	1	0	0	0	0	1	0	30
P	5	1	2	7	2	5	0	0	0	0	0	1	0

	x	y	z	u	v	w	p	q	r	s	t	P	
x	1	0	0	1	0	0	−1	0	0	0	0	0	10
q	0	1	0	0	1	0	0	−1	0	0	0	0	20
r	0	0	1	0	0	1	0	0	−1	0	0	0	10
s	0	1	1	−1	0	0	1	0	0	1	0	0	5
t	0	0	0	1	1	1	0	0	0	0	1	0	30
P	0	1	2	2	2	5	5	0	0	0	0	1	−50

	x	y	z	u	v	w	p	q	r	s	t	P	
x	1	0	0	1	0	0	−1	0	0	0	0	0	10
v	0	1	0	0	1	0	0	−1	0	0	0	0	20
r	0	0	1	0	0	1	0	0	−1	0	0	0	10
s	0	1	1	−1	0	0	1	0	0	1	0	0	5
t	0	−1	0	1	0	1	0	1	0	0	1	0	10
P	0	−1	2	2	0	5	5	2	0	0	0	1	−90

	x	y	z	u	v	w	p	q	r	s	t	P	
x	1	0	0	1	0	0	−1	0	0	0	0	0	10
v	0	1	0	0	1	0	0	−1	0	0	0	0	20
w	0	0	1	0	0	1	0	0	−1	0	0	0	10
s	0	1	1	−1	0	0	1	0	0	1	0	0	5
t	0	−1	−1	1	0	0	0	1	1	0	1	0	0
P	0	−1	−3	2	0	0	5	2	5	0	0	1	−140

	x	y	z	u	v	w	p	q	r	s	t	P	
x	1	0	0	1	0	0	−1	0	0	0	0	0	10
v	0	1	0	0	1	0	0	−1	0	0	0	0	20
w	0	−1	0	$\boxed{1}$	0	1	−1	0	−1	−1	0	0	5
z	0	1	1	−1	0	0	1	0	0	1	0	0	5
t	0	0	0	0	0	0	1	1	1	1	1	0	5
P	0	2	0	−1	0	0	8	2	5	3	0	1	−125

	x	y	z	u	v	w	p	q	r	s	t	P	
x	1	1	0	0	0	−1	0	0	1	1	0	0	5
v	0	1	0	0	1	0	0	−1	0	0	0	0	20
u	0	−1	0	1	0	1	−1	0	−1	−1	0	0	5
z	0	0	1	0	0	1	0	0	−1	0	0	0	10
t	0	0	0	0	0	0	1	1	1	1	1	0	5
P	0	1	0	0	0	1	7	2	4	2	0	1	−120

Now you see what to do: You should fly 5 planes from Chicago to Los Angeles, none from Chicago to New York, and 10 from Chicago to Boston. You should fly 5 planes from Atlanta to Los Angeles, 20 from Atlanta to New York, and none from Atlanta to Boston. The total cost will be $1,200,000 to get the airline back on schedule.

EXERCISES

1. Your boss calls you on the carpet for not flying any planes from Chicago to New York, which is the cheapest run. Come up with a convincing reason for having done that. Your job may be on the line.
2. Your boss *insists* that you fly at least 5 planes from Chicago to New York. What is your best option now?
3. If Los Angeles needed 15 planes rather than 10, what would be your best option?
4. If New York needed 25 planes rather than 20, what would be your best option?
5. If Boston needed 15 planes rather than 10, what would be your best option?
6. If there were only 10 planes in Chicago, what would be your best option?
7. If there were 35 planes in Atlanta, what would be your best option?
8. Suppose the Boston airport was socked in by snow, so you only needed to send planes to Los Angeles and New York. What would be your best option?
9. Suppose the New York airport was socked in by snow, so you only needed to send planes to Los Angeles and Boston. What would be your best option?
10. A higher cost per flight from Atlanta to New York would change the answer reached in the text. By how much would the cost have to rise to change the answer?
11. The LP problem here satisfies the nonnegative objective condition. Solve it using duality. What do the values of the dual variables tell you about the situation?

CHAPTER 4 REVIEW TEST

1. Sketch the regions corresponding to the given inequalities, say whether they are bounded, and give the coordinates of all corner points.

a. $2x - 3y \le 12$

b. $x \le 2y$

c. $x + 2y \le 20$, $3x + 2y \le 30$, $x \ge 0, y \ge 0$

d. $3x + 2y \ge 6$, $2x - 3y \le 6$, $3x - 2y \ge 0$, $x \ge 0, y \ge 0$

2. Solve the following LP problems graphically.

a. Maximize $p = 2x + y$ subject to $3x + y \le 30$, $x + y \le 12$, $x + 3y \le 30$, $x \ge 0, y \ge 0$

b. Maximize $p = 2x + 3y$ subject to $x + y \ge 10$, $2x + y \ge 12$, $x + y \le 20$, $x \ge 0, y \ge 0$

c. Minimize $c = 2x + y$ subject to $3x + y \ge 30$, $x + 2y \ge 20$, $2x - y \ge 0$, $x \ge 0, y \ge 0$

d. Minimize $c = 3x + y$ subject to $3x + 2y \ge 6$, $2x - 3y \le 0$, $3x - 2y \ge 0$, $x \ge 0, y \ge 0$

3. Solve the following LP problems using the simplex method. If no optimal solution exists, indicate whether the feasible region is empty or the objective function is unbounded.

a. Maximize $p = x + y + 2z$ subject to $x + 2y + 2z \le 60$, $2x + y + 3z \le 60$, $x \ge 0, y \ge 0, z \ge 0$

b. Maximize $p = x + y + 2z$ subject to $x + 2y + 2z \le 60$, $2x + y + 3z \le 60$, $x + 3y + 6z \le 60$, $x \ge 0, y \ge 0, z \ge 0$

c. Maximize $p = x + y + 3z$ subject to $x + y + z \ge 100$, $y + z \le 80$, $x + z \le 80$, $x \ge 0, y \ge 0, z \ge 0$

d. Minimize $c = x + 2y + 3z$ subject to $3x + 2y + z \ge 60$, $2x + y + 3z \ge 60$, $x \ge 0, y \ge 0, z \ge 0$

e. Maximize $p = 2x + y$ subject to $x + 2y \ge 12$, $2x + y \le 12$, $x + y \le 5$, $x \ge 0, y \ge 0$

f. Minimize $c = x + y - z$ subject to $3x + 2y + z \ge 60$, $2x + y + 3z \ge 60$, $x + 3y + 2z \ge 60$, $x \ge 0, y \ge 0, z \ge 0$

4. Solve the following LP problems using duality.

a. Minimize $c = 2x + y$ subject to $3x + 2y \ge 60$, $2x + y \ge 60$, $x + 3y \ge 60$, $x \ge 0, y \ge 0$

b. Minimize $c = 2x + y + 2z$ subject to $3x + 2y + z \ge 100$, $2x + y + 3z \ge 200$, $x \ge 0, y \ge 0, z \ge 0$

c. Minimize $c = 2x + y$ subject to $3x + 2y \ge 10$, $2x - y \le 30$, $x + 3y \ge 60$, $x \ge 0, y \ge 0$

d. Minimize $c = 2x + y + 2z$ subject to $3x - 2y + z \ge 100$, $2x + y - 3z \le 200$, $x \ge 0, y \ge 0, z \ge 0$

5. The following is adapted from the Actuarial Exam on Operations Research.

a. You are given the following LP problem:

Minimize $c = x + 2y$ subject to $-2x + y \ge 1$, $x - 2y \ge 1$, $x \ge 0, y \ge 0$

Which of the following is true?

(A) The problem has no feasible solutions.
(B) The objective function is unbounded.
(C) The problem has optimal solutions.

b. Repeat part (a) with the following LP problem:

Maximize $p = x + y$ subject to $-2x + y \le 1$, $x - 2y \le 2$, $x \ge 0, y \ge 0$

c. Determine the optimal value of the objective function. You are given the following LP problem:

Maximize $Z = x_1 + 4x_2 + 2x_3 - 10$ subject to $4x_1 + x_2 + x_3 \le 45$, $-x_1 + x_2 + 2x_3 \le 0$, $x_1, x_2, x_3 \ge 0$

OHaganBooks.com—OPTIMIZING OPERATIONS

6. You are the buyer for OHaganBooks.com and are considering increasing stocks of romance and horror novels at the new OHaganBooks.com warehouse in Texas. You have offers from two publishers: Duffin House and Higgins Press. Duffin offers a package of 5 horror novels and 5 romance novels for $50, and Higgins offers a package of 5 horror and 10 romance novels for $80.

a. How many packages should you purchase from each publisher to obtain at least 4000 horror novels and 6000 romance novels at minimum cost? What is the minimum cost?

b. As it turns out, John O'Hagan promised Marjory Duffin that OHaganBooks.com would buy at least 20% more packages from Duffin as from Higgins, but you still want to obtain at least 4000 horror novels and 6000 romance novels at minimum cost. *Without solving the problem,* say which of the following statements are possible.

(A) The cost will stay the same.
(B) The cost will increase.
(C) The cost will decrease.
(D) It will be impossible to meet all the conditions.
(E) The cost will become unbounded.

c. If you wish to meet all the requirements in part (b) at minimum cost, how many packages should you purchase from each publisher? What is the minimum cost?

d. You are just about to place your orders when your sales manager reminds you that, in addition to all your commitments in part (b), you had also assured Sean Higgins that you would spend at least as much on purchases from Higgins as from Duffin. *Now* what should you do?

7. Duffin Press, which has become the largest publisher of books sold at the OHaganBooks.com site, prints three kinds of books: paperback, quality paperback, and hardcover. The amounts of paper, ink, and time on the presses required for each kind of book are given in this table:

	Paperback	Quality Paperback	Hardcover	Total Available
Paper (lb)	3	2	1	6,000
Ink (gal)	2	1	3	6,000
Time (min)	10	10	10	22,000
Profit ($)	1	2	3	

The table also lists the total amounts of paper, ink, and time available in a given day and the profits made on each kind of book. How many of each kind of book should Duffin print to maximize profit?

8. You are just about to place book orders from Duffin and Higgins (see Exercise 6) when Ewing Books enters the fray and offers its own package of horror and romance novels. To complicate things further, Duffin and Higgins both change their offers, you therefore decide to renege on the commitments you had made to them, and your sales manager has changed her mind about the number of books OHaganBooks.com will require. Taking all of this (and other factors too complicated to explain) into account, you arrive at the following linear programming problem:

$$\begin{aligned} \text{Minimize} \quad & c = 50x + 150y + 100z \\ \text{subject to} \quad & 5x + 10y + 5z \geq 4000 \\ & 2x + 10y + 5z \geq 6000 \\ & x - y + z \leq 0 \\ & x \geq 0, y \geq 0, z \geq 0 \end{aligned}$$

What is the solution to this LP problem?

9. During his lunch break, John O'Hagan decides to devote some time to assisting his son Billy Sean, who continues to have a terrible time planning his college course schedule. The latest Bulletin of Suburban State University claims to have added new flexibility to its course requirements, but it remains as complicated as ever. It reads as follows:

All candidates for the degree of Bachelor of Arts at SSU must take at least 120 credits from the Sciences, Fine Arts, Liberal Arts, and Mathematics combined, including at least as many science credits as fine arts credits, and at most twice as many mathematics credits as science credits, but with liberal arts credits exceeding mathematics credits by no more than one-third of the number of fine arts credits.

Science and fine arts credits cost $300 each, and liberal arts and mathematics credits cost $200 each. John would like to have Billy Sean meet all the requirements at a minimum total cost.

a. Set up the associated LP problem.

b. How many of each type of credit should Billy Sean take? What will be the total cost?

10. On the same day that the sales department at Duffin House received an order for 600 packages from the OHaganBooks.com Texas headquarters, it received an additional order for 200 packages from FantasyBooks.com, based in California. Duffin House has warehouses in New York and Illinois. The New York warehouse has 600 packages in stock, but the Illinois warehouse is closing down and has only 300 packages in stock. Shipping costs per package of books are as follows: New York to Texas, $20; New York to California, $50; Illinois to Texas, $30; Illinois to California, $40. What is the lowest total shipping cost for which Duffin House can fill the orders? How many packages should be sent from each warehouse to each online bookstore at a minimum shipping cost?

ADDITIONAL ONLINE REVIEW

If you follow the path

Web site → Everything for Finite Math → Chapter 4

you will find the following additional resources to help you review:

A comprehensive chapter summary (including examples and interactive features)

Additional review exercises (including interactive exercises and many with help)

Interactive online section tutorials

Section-by-section Excel tutorials

A true/false chapter quiz

Utilities and resources for solving linear programming problems, including a linear programming grapher, a simplex method tool, and Web- and Excel-based pivoting utilities

5

THE MATHEMATICS OF FINANCE

CASE STUDY

Saving for College

Tuition costs at colleges and universities increased at an average rate of 5.8% per year from 1991 through 2001. The 2001–2002 average cost for a private college was $23,578. How much would the parents of a newborn have to invest each month in a mutual fund expected to yield 6% per year to pay for their child's college education?

SCREEN CAPTURE: www.collegeboard.com/paying/0,3309,6-0-0-0,00.html?orig=art.

INTERNET RESOURCES FOR THIS CHAPTER

At the Web site, follow the path

Web site ⟶ Everything for Finite Mathematics ⟶ Chapter 5

where you will find a detailed chapter summary you can print out, a true/false quiz, a collection of chapter review questions, and a Time Value of Money utility. If you are using Excel, you can download brief Excel tutorials for every section.

Introduction

A knowledge of the mathematics of investments and loans is important not only for business majors but also for everyone who deals with money, which is all of us. This chapter is largely about *interest:* interest paid by an investment, interest paid on a loan, and variations on these.

We focus on three forms of investment: investments that pay simple interest, investments in which interest is compounded, and annuities. An investment that pays simple interest periodically gives interest directly to the investor, perhaps in the form of a monthly check. If instead the interest is reinvested, the interest is *compounded,* and the value of the account grows as the interest is added. An *annuity* is an investment into which periodic payments are made (increasing annuities) or from which periodic withdrawals are made (decreasing annuities or, looked at from the other side, loans).

We also look at bonds, the primary financial instrument used by companies and governments to raise money. Although bonds nominally pay simple interest, determining their worth, particularly in the secondary market, requires an annuity calculation.

5.1 Simple Interest

You deposit \$1000, called the **principal** or **present value,** into a savings account. The bank pays you 5% interest, in the form of a check, each year.

Question How much interest will I earn each year?

Answer Since the bank pays you 5% interest each year, your annual (or yearly) interest will be 5% of \$1000, or $0.05 \times 1000 = \$50$.

Generalizing this calculation, call the present value PV and the interest rate (expressed as a decimal) r. Then INT, the annual interest paid to you, is given by

$$INT = PVr$$

If the investment is made for a period of t years, then the total interest accumulated is t times this amount, which gives us the following.

Simple Interest

The **simple interest** on an investment (or loan) of PV dollars at an annual interest rate of r for a period of t years is

$$INT = PVrt$$

Quick Example

The simple interest over a period of 4 years on a \$5000 investment earning 8% per year is

$$INT = PVrt = (5000)(0.08)(4) = \$1600$$

Note on Multiletter Variables Multiletter variables like PV and INT are traditional in finance textbooks, calculators, and spreadsheets. (See later discussion of how variables like these are used in the TI-83 and Excel.) Just watch out for expressions like PVr, which is the product of two things, PV and r, not three.

Question Given my \$1000 investment at 5% simple interest, how much money will I have after 2 years?

Answer We need to add the accumulated interest to the principal to get the **future value** (FV) of your deposit.

$$FV = PV + INT = \$1000 + (1000)(0.05)(2) = \$1100$$

In general, we can compute the future value as follows:

$$FV = PV + INT = PV + PVrt = PV(1 + rt)$$

Future Value for Simple Interest

The **future value** of an investment of PV dollars at an annual simple interest rate of r for a period of t years is

$$FV = PV(1 + rt)$$

Quick Examples

1. The value, at the end of 4 years, of a \$5000 investment earning 8% simple interest per year is

$$FV = PV(1 + rt) = 5000[1 + (0.08)(4)] = \$6600$$

2. Writing the future value in Quick Example 1 as a function of time, we get

$$\begin{aligned} FV &= 5000(1 + 0.08t) \\ &= 5000 + 400t \end{aligned}$$

which is a linear function of time t. The intercept is $PV = \$5000$, and the slope is the annual interest, \$400 per year.

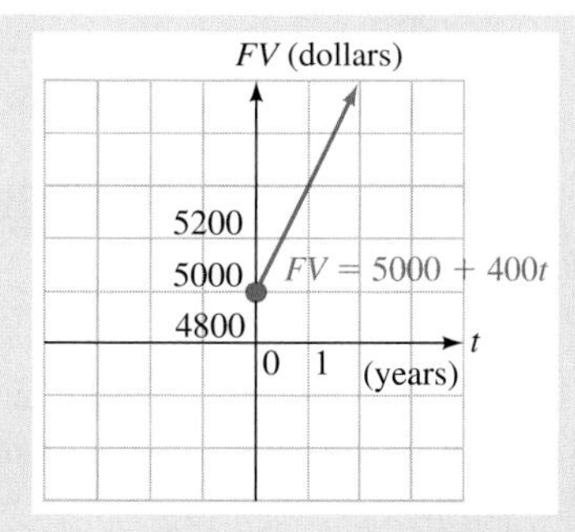

In general, *simple interest growth is a linear function of time, with intercept given by the present value and slope given by annual interest.*

Example 1 • Savings Accounts

In April 2002 the Amex Bank of Canada was paying 2.55% interest on savings accounts.[1] If the interest is paid as simple interest, find the future value of a \$2000 deposit in 6 years. What is the total interest paid over the period?

Solution We use the future value formula:

$$FV = PV(1 + rt)$$

$$= 2000[1 + (0.0255)(6)] = 2000(1.153) = \$2306$$

The total interest paid is given by the simple interest formula:

$$INT = PVrt = (2000)(0.0255)(6) = \$306$$

Before we go on . . . To find the interest paid, we could also have computed

$$INT = FV - PV = 2306 - 2000 = \$306$$

Question What is future value as a function of time?

Answer The annual interest is $PVr = (2000)(0.0255) = \$51$, so

$$FV = 2000 + 51t$$

Thus, the future value is growing linearly at a rate of \$51 per year.

Example 2 • Bridge Loans

When "trading up," homeowners sometimes have to buy a new house before they sell their old house. One way to cover the costs of the new house until they get the proceeds from selling the old house is to take out a short-term *bridge loan.* Suppose a bank charges 12% simple annual interest on such a loan. How much will be owed at the maturation of a 90-day bridge loan of \$90,000?

[1]SOURCE: Canoe Money, http://www.fan590.com/MoneyRates/savings.html, April 14, 2002.

Solution We use the future value formula

$$FV = PV(1 + rt)$$

with $t = 90/365$, the fraction of a year represented by 90 days:

$$FV = 90{,}000\left[1 + (0.12)\left(\frac{90}{365}\right)\right] = \$92{,}663.01$$

(We will always round our answers to the nearest cent after calculation. Be careful not to round intermediate results.)

Before we go on . . . Many banks count a year as 360 days for this calculation, which makes a "year" for the purposes of the loan slightly shorter than a calendar year. The effect is to increase the amount owed:

$$FV = 90{,}000\left[1 + (0.12)\left(\frac{90}{360}\right)\right] = \$92{,}700$$

One of the primary ways companies and governments raise money is by selling **bonds.** At its most straightforward, a corporate bond promises to pay simple interest, usually twice a year, for a length of time until it **matures,** at which point it returns the original investment to the investor (U.S. Treasury Notes and Bonds are similar). Things get more complicated when the selling price is negotiable, as we will see in the following sections.

Example 3 • Corporate Bonds

Megabucks Corporation is issuing 10-year bonds paying an annual rate of 6.5%. If you buy $10,000 worth of bonds, how much interest will you earn every 6 months, and how much interest will you earn over the life of the bonds?

Solution Using the simple interest formula, every 6 months you will receive

$$INT = PVrt = (10{,}000)(0.065)\left(\frac{1}{2}\right) = \$325$$

Over the 10-year life of the bonds, you will earn

$$INT = PVrt = (10{,}000)(0.065)(10) = \$6500$$

in interest. So, at the end of 10 years, when your original investment is returned to you, your $10,000 will have turned into $16,500.

U.S. Treasury bills (T-bills) are short-term investments (up to 1 year) that pay you a set amount after a period of time. A T-bill pays no interest directly; the return on your investment is the difference between what you paid for it and what it pays you when it matures.

Example 4 • Treasury Bills

A T-bill paying \$10,000 after 6 months earns 1.94% simple annual interest.[2] How much would it cost to buy?

Solution The future value of the T-bill is \$10,000; the price we pay is its present value. We know that

$$FV = \$10{,}000$$
$$r = 0.0194$$
$$t = 0.5$$

and we wish to find PV. Substituting into the future value formula, we have

$$10{,}000 = PV[1 + (0.0194)(0.5)] = PV \times 1.0097$$
$$PV = \frac{10{,}000}{1.0097} = 9903.93$$

so we will pay \$9903.93 for the T-bill.

Before we go on . . . The simplest way to find the interest earned on the T-bill is by subtraction:

$$INT = FV - PV = 10{,}000 - 9903.93 = \$96.07$$

In Example 4 we saw that we could calculate the present value, knowing the future value. It is useful to write down the general formula, which we get by solving the future value formula for PV.

Present Value for Simple Interest

The present value of an investment at an annual simple interest rate of r for a period of t years, with future value FV, is

$$PV = \frac{FV}{1 + rt}$$

Quick Example

A T-bill with a yield of 5% simple interest that pays \$50,000 after 4 months costs

$$PV = \frac{FV}{1 + rt} = \frac{50{,}000}{1 + (0.05)\left(\frac{4}{12}\right)} = \$49{,}180.33$$

[2]Figure as of April 11, 2002. Source: "Federal Reserve Statistical Release H.15: Selected Interest Rates," available on the World Wide Web through the White House's Economics Statistics Briefing Room, http://www.whitehouse.gov/fsbr/esbr.html. By the way, since interest rates are often reported to 0.01%, you will sometimes see the term *basis point,* which means 1 one-hundredth of a percent. For example, 25 basis points is the same as 0.25%, or 0.0025.

Example 5 • Treasury Bills

A T-bill paying \$10,000 after 6 months sells at a discount rate of 1.92%.[3] What does it cost? What is the annual **yield**; that is, what simple annual interest rate does it pay?

Solution To say that a 1-year T-bill sells at a **discount rate** of 1.92% is to say that its selling price is 1.92% lower than its maturity value of \$10,000. The discount rates for T-bills of other lengths are adjusted to give annual figures. Thus, the selling price for this 6-month T-bill will be 1.92%/2 = 0.96% below its maturity value. This makes the selling price, or present value,

$$PV = 10{,}000 - (10{,}000)\left(\frac{0.0192}{2}\right) \qquad \text{Maturity value} - \text{discount}$$

$$= 10{,}000(1 - 0.0096) = \$9904.00$$

To find the annual interest rate, notice that

$$FV = \$10{,}000$$

$$PV = \$9904$$

$$t = 0.5$$

and we wish to find r. Substituting in the future value formula, we get

$$FV = PV(1 + rt)$$

$$10{,}000 = 9904(1 + 0.5r)$$

so

$$1 + 0.5r = \frac{10{,}000}{9904}$$

$$r = \frac{\frac{10{,}000}{9904} - 1}{0.5} \approx 0.0194$$

The bond is paying 1.94% simple annual interest.

Before we go on . . . This is the same T-bill as in Example 4. The interest rate and the discount rate are two different ways of telling what the investment pays. One of the exercises asks you to find a formula for the interest rate in terms of the discount rate.

With T-bills it is useful to think of the return on your investment as interest. Fees on loans can also be thought of as a form of interest.

Example 6 • Tax Refunds

You are expecting a tax refund of \$800. Since it may take up to 6 weeks to get the refund, your tax preparation firm offers, for a fee of \$40, to give you an "interest-free" loan of \$800 to be paid back with the refund check. If we think of the fee as interest, what simple annual interest rate is the firm charging?

---

[3]Ibid.

Solution If we view the \$40 as interest, then the future value of the loan (the value of the loan to the firm, or the total you will pay the firm) is \$840. Thus, we have

$$FV = 840$$

$$PV = 800$$

$$t = \frac{6}{52} \qquad \text{Using 52 weeks in a year}$$

and we wish to find r. Substituting, we get

$$FV = PV(1 + rt)$$

$$840 = 800\left(1 + \frac{6r}{52}\right) = 800 + \frac{4800r}{52}$$

so

$$\frac{4800r}{52} = 840 - 800 = 40$$

$$r = \frac{40 \times 52}{4800} \approx 0.43$$

In other words, the firm is charging you 43% annual interest! Save your money and wait 6 weeks for your refund.

5.1 EXERCISES

In Exercises 1–6, compute the simple interest for the specified period and the future value at the end of the period. Round all answers to the nearest cent.

1. \$2000 is invested for 1 year at 6%/year.

2. \$1000 is invested for 10 years at 4%/year.

3. \$20,200 is invested for 6 months at 5%/year.

4. \$10,100 is invested for 3 months at 11%/year.

5. \$10,000 is borrowed for 10 months at 3%/year.

6. \$6000 is borrowed for 5 months at 9%/year.

APPLICATIONS

In Exercises 7–14, compute the specified quantity. Round all answers to the nearest month, the nearest cent, or the nearest 0.001%, as appropriate.

7. Bonds How much would you have to pay for a 6-year bond earning 4.5% simple interest whose future value is \$1000?

8. Simple Loans Your total payment on a 4-year loan, which charged 9.5% simple interest, amounted to \$30,360. How much did you originally borrow?

9. Bonds A 5-year bond costs \$1000 and will pay a total of \$250 interest over its lifetime. What is its annual interest rate?

10. Bonds A 4-year bond costs \$10,000 and will pay a total of \$2800 in interest over its lifetime. What is its annual interest rate?

11. Simple Loans A \$4000 loan, taken now, with a simple interest rate of 8%/year, will require a total repayment of \$4640. When will the loan mature?

12. Simple Loans The simple interest on a \$1000 loan at 8%/year amounted to \$640. When did the loan mature?

13. Treasury Bills At auction on January 7, 1999, 1-year T-bills were sold at a discount of 4.335%. What was the annual yield?

SOURCE: Bureau of the Public Debt's Web site: http://www.publicdebt.treas.gov/.

14. Treasury Bills At auction on December 10, 1998, 1-year T-bills were sold at a discount of 4.305%. What was the annual yield?

SOURCE: Bureau of the Public Debt's Web site: http://www.publicdebt.treas.gov/.

Corporate Income Exercises 15–22 are based on the following chart, which shows Sony Corporation's net income for fiscal years ending in March:

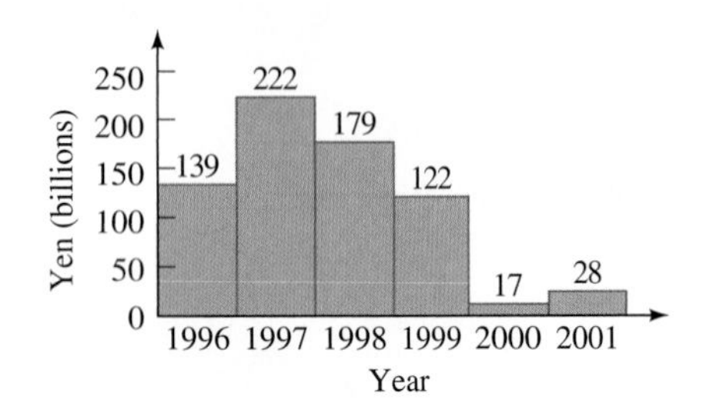

Figures are rounded to the nearest ¥1 billion. The 2001 figure is annualized based on the first 9 months of the year. Source: Sony Corporation, http://www.sony.co.jp/en/SonyInfo/, April 2002.

15. Calculate to the nearest 0.01% the percentage decrease in Sony's net income for the period 1999–2000.

16. Calculate to the nearest 0.01% the percentage increase in Sony's net income for the period 1996–1997.

17. Calculate to the nearest 0.01% (calculated on a simple interest basis) the annual percentage change in Sony's net income for the period 1996–2000.

18. Calculate to the nearest 0.01% (calculated on a simple interest basis) the annual percentage change in Sony's net income for the period 1997–2001.

19. During which 1-year period did Sony experience the largest percent decrease in net income? What was the percent decrease?

20. During which 1-year period did Sony experience the largest percent increase in net income? What was the percent increase?

21. Did Sony's net income undergo simple interest contraction in the years 1997–1999? (Give a reason for your answer.)

22. If Sony's net income experienced a simple interest expansion from 2000 to 2002, what was its 2002 net income?

Population Exercises 23–28 are based on the following graph, which shows the population of San Diego County from 1950 to 2000:

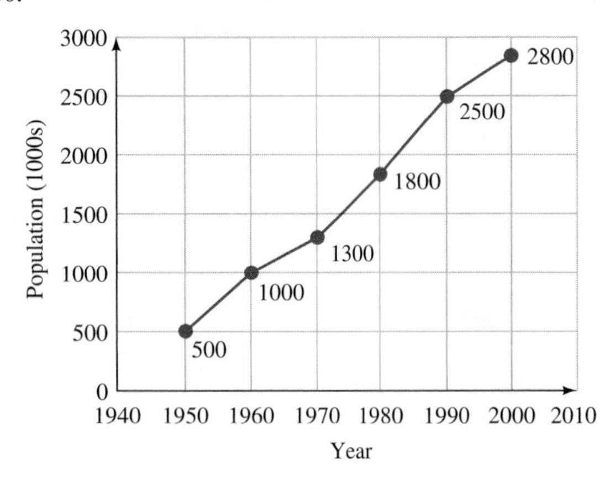

Source: U.S. Census Bureau/*New York Times*, April 2, 2002, p. F3.

23. At what annual (simple interest) rate did the population of San Diego County increase from 1950 to 2000?

24. At what annual (simple interest) rate did the population of San Diego County increase from 1950 to 1990?

25. If you used your answer to Exercise 23 as the annual (simple interest) rate at which the population was growing since 1950, what would you predict the San Diego County population to be in 2010?

26. If you used your answer to Exercise 24 as the annual (simple interest) rate at which the population was growing since 1950, what would you predict the San Diego County population to be in 2010?

27. Use your answer to Exercise 23 to give a linear model for the population of San Diego County from 1950 to 2000. Draw the graph of your model over that period of time.

28. Use your answer to Exercise 24 to give a linear model for the population of San Diego County from 1950 to 2000. Draw the graph of your model over that period of time.

COMMUNICATION AND REASONING EXERCISES

29. One or more of the following graphs represent the future value of an investment earning simple interest. Which one(s)? Give the reason for your choice(s).

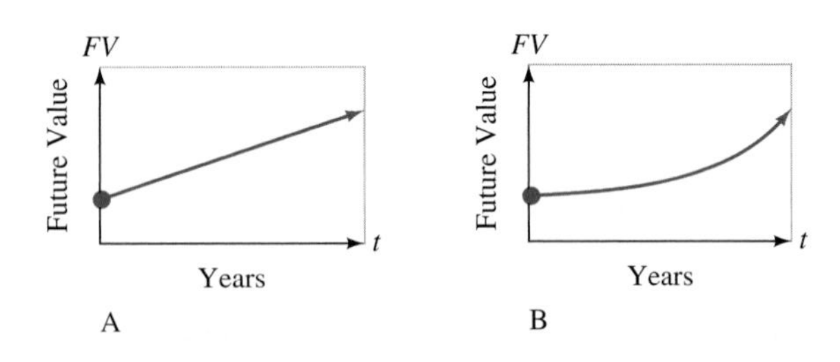

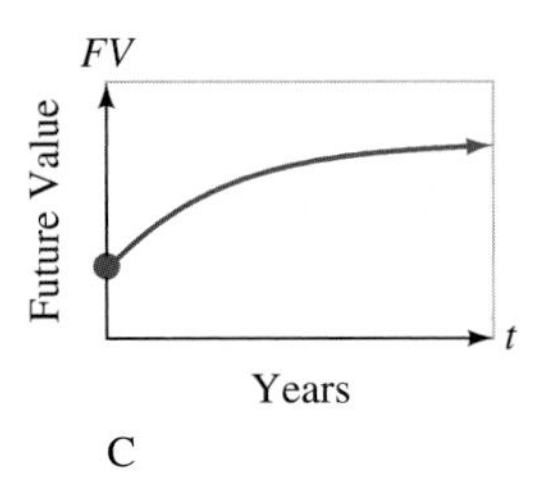

30. Given that $FV = 5t + 400$, for what interest rate is this the equation of future value (in dollars) as a function of time t (in years)?

31. Interpreting the News You hear the following on your local radio station's business news: "The economy last year grew by 1%. This was the second year in a row in which the economy showed a 1% growth." Since the rate of growth was the same 2 years in a row, this represents simple interest growth, right? Explain your answer.

32. Interpreting the News You hear the following on your local radio station's business news: "The economy last year grew by 1%. This was the second year in a row in which the economy showed a 1% growth." This means that, in dollar terms, the economy grew more last year than the year before. Why?

33. Explain why simple interest is not the appropriate way to measure interest on a savings account that pays interest directly into your account.

34. Suppose a 1-year T-bill sells at a discount rate d. Find a formula, in terms of d, for the simple annual interest rate r the bill will pay.

5.2 Compound Interest

You deposit $1000 into a savings account. The bank pays you 5% interest, which it deposits into your account, or **reinvests,** at the end of each year.

Question At the end of 5 years, how much money will I have accumulated?

Answer Let's compute the amount you have at the end of each year. At the end of the first year, the bank will pay you simple interest of 5% on your $1000, which gives you

$$PV(1 + rt) = 1000(1 + 0.05) = \$1050$$

At the end of the second year, the bank will pay you another 5% interest, but this time computed on the total in your account, which is $1050. Thus, you will have a total of

$$1050(1 + 0.05) = \$1102.50$$

If you were being paid simple interest on your original $1000, you would have only $1100 at the end of the second year. The extra $2.50 is the interest earned on the $50 interest added to your account at the end of the first year. Having interest earn interest is called **compounding** the interest. We could continue like this until the end of the fifth year, but notice what we are doing: Each year we are multiplying by $1 + 0.05$. So, at the end of 5 years, you will have

$$1000(1 + 0.05)^5 \approx \$1276.28$$

It is interesting to compare this to the amount you would have if the bank paid you simple interest:

$$1000(1 + 0.05 \times 5) = \$1250.00$$

The extra $26.28 is again the effect of compounding the interest.

Banks often pay interest more often than once a year. Paying interest quarterly (four times per year) or monthly is common.

Question If my bank pays interest monthly, how much will my $1000 deposit be worth after 5 years?

Answer The bank will not pay you 5% interest every month, but will give you 1/12 of that, or 5/12% interest each month.[4] Thus, instead of multiplying by $1 + 0.05$ every year, we should multiply by $1 + 0.05/12$ each month. Since there are $5 \times 12 = 60$ months in 5 years, the total amount you will have at the end of 5 years is

$$1000\left(1 + \frac{0.05}{12}\right)^{60} \approx \$1283.36$$

---

[4]This is approximate. They will actually give you 31/365 of the 5% at the end of January and so on, but it's simpler and reasonably accurate to call it 1/12.

Compare this to the $1276.28 you would get if the bank paid the interest every year. You earn an extra $7.08 if the interest is paid monthly because interest gets into your account and starts earning interest earlier. The amount of time between interest payments is called the **compounding period.**

The preceding example generalizes easily to give the general formula for future value when interest is compounded.

Future Value for Compound Interest

The future value of an investment of PV dollars earning interest at an annual rate of r compounded (reinvested) m times per year for a period of t years is

$$FV = PV\left(1 + \frac{r}{m}\right)^{mt}$$
$$= PV(1 + i)^n$$

where $i = r/m$ is the interest paid each compounding period and $n = mt$ is the total number of compounding periods.

Quick Examples

1. To find the future value after 5 years of a $10,000 investment earning 6% interest, with interest reinvested every month, we set $PV = 10{,}000$, $r = 0.06$, $m = 12$, and $t = 5$. Thus,

$$FV = PV\left(1 + \frac{r}{m}\right)^{mt} = 10{,}000\left(1 + \frac{0.06}{12}\right)^{60} \approx \$13{,}488.50$$

2. Writing the future value in Quick Example 1 as a function of time, we get

$$FV = 10{,}000\left(1 + \frac{0.06}{12}\right)^{12t} = 10{,}000(1.005)^{12t}$$

which is an **exponential** function of time t.

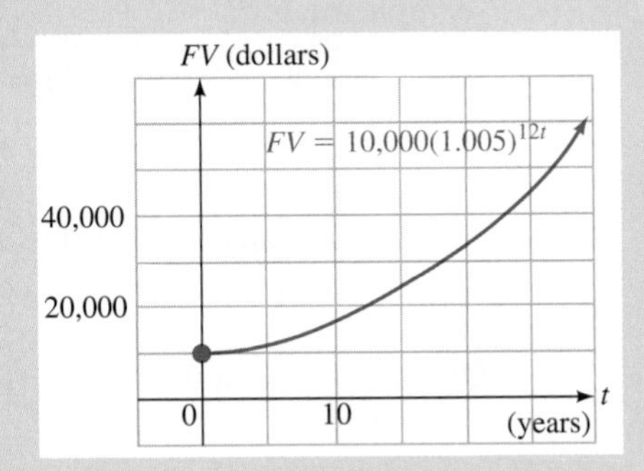

In general, *compound interest growth is an exponential function of time.*

Example 1 • Savings Accounts

In April 2002 the Amex Bank of Canada was paying 2.55% interest on savings accounts.[5] If the interest is compounded quarterly, find the future value of a $2000 deposit in 6 years. What is the total interest paid over the period?

[5] SOURCE: Canoe Money, http://www.fan590.com/MoneyRates/savings.html, April 14, 2002.

Solution We use the future value formula with $m = 4$:

$$FV = PV\left(1 + \frac{r}{m}\right)^{mt}$$

$$= 2000\left(1 + \frac{0.0255}{4}\right)^{(4)(6)} \approx \$2329.52$$

The total interest paid is

$$INT = FV - PV = 2329.52 - 2000 = \$329.52$$

Graphing Calculator: Computing Compound Interest

We could calculate the future value using the TI-83 by entering

```
2000(1+0.0255/4)^(4*6)
```

However, the TI-83 has this and other useful calculations built into its TVM (Time Value of Money) Solver. Press [2nd] [FINANCE] and then choose item 1, `TVM Solver...`, from the menu. This brings up the TVM Solver window (Figure 1).

`N=24`	Number of compounding periods
`I%=2.55`	Annual interest rate (percent, not decimal)
`PV=⁻2000`	Negative of present value (see below)
`PMT=0`	Payment per period (0 for this section)
`FV=0`	Future value (will be computed)
`P/Y=4`	Payments per year (see below)
`C/Y=4`	Compounding periods per year (see below)
`PMT: END BEGIN`	Not used in this section

Figure 1

Figure 1 shows the values you should enter for this example, together with some explanation. Notice that the present value *PV* is entered as a negative number; in general, when using the TVM Solver, any amount of money you give to someone else (such as the $2000 you deposit in the bank) will be a negative number, whereas any amount of money someone gives to you (such as the future value of your deposit, which the bank will give back to you) will be a positive number. *PMT* is not used in this example (it will be used in Section 5.3) and should be zero. *FV* is the future value, which we compute in a moment; it doesn't matter what you enter now. *P/Y* and *C/Y* stand for payments per year and compounding periods per year, respectively: They should both be set to the number of compounding periods per year for compound interest problems (setting *P/Y* automatically sets *C/Y* to the same value). Finally, `PMT: END` or `BEGIN` is not used in this example and it doesn't matter which you select.

Note that *I%* is the *annual* interest rate, corresponding to r, not i, in the compound interest formula.

To compute the future value, use the up or down arrow to put the cursor on the *FV* line and then press [ALPHA] [SOLVE]. The result should look like Figure 2.

```
  N=24
  I%=2.55
  PV=⁻2000
  PMT=0
■ FV=2329.518412
  P/Y=4
  C/Y=4
  PMT: END BEGIN
```

Figure 2

Web Site: Computing Compound Interest

If you follow the path

Web site → Online Utilities→ Time Value of Money Utility

you will find a utility very similar to the graphing calculator TVM Solver and easy instructions for its use. You can use this utility in place of the TI-83 TVM Solver.

Excel: Computing Compound Interest

As with a graphing calculator, you can either compute compound interest directly or using built-in financial functions. The following worksheet has more than we need for this example but will be useful for other examples in this and the next section.

	A	B	C	D
1		Entered	Calculated	
2	Rate	2.55%		
3	Years	6		
4	Payment	$0.00		
5	Present Value	-$2,000.00		
6	Future Value		=FV(B2/B7,B3*B7,B4,B5)	
7	Periods per year	4		

In this worksheet we have formatted the shaded cells as currency (with two decimal places). For this example the payment amount in B4 should be zero (we will use it in Section 5.3). Enter the other numbers as shown. As with the TVM Solver in the TI-83, money that you pay to others (such as the $2000 you deposit in the bank) should be entered as negative, whereas money that is paid to you is positive. The formula entered in C6 uses the built-in FV function to calculate the future value based on the entries in column B. This formula has the following format:

$FV(i, n, PMT, PV)$

where i = interest per period — We use B2/B7 for the interest.

n = number of periods — We use B3*B7 for the number of periods.

PMT = payment per period — The payment is zero (cell B4).

PV = present value — The present value is in cell B5.

Instead of using the built-in FV function, we could use

```
=-B5*(1+B2/B7)^(B3*B7)
```

based on the future value formula for compound interest. After calculation, the result, $2329.52, will appear in cell C6 (if you have formatted that cell as currency). If you change the values in column B, the future value in column C will be automatically recalculated.

As we did for simple interest, we can solve the future value formula for the present value PV and obtain the following formula.

Present Value for Compound Interest

The present value of an investment earning interest at an annual rate of r compounded m times per year for a period of t years, with future value FV, is

$$PV = \frac{FV}{\left(1 + \frac{r}{m}\right)^{mt}}$$

$$= \frac{FV}{(1+i)^n} = FV(1+i)^{-n}$$

where $i = r/m$ is the interest paid each compounding period and $n = mt$ is the total number of compounding periods.

Quick Example

To find the amount we need to invest in an investment earning 12% interest per year, compounded annually, so that we will have \$1 million in 20 years, use $FV =$ \$1,000,000, $r = 0.12$, $m = 1$, and $t = 20$:

$$PV = \frac{FV}{\left(1 + \frac{r}{m}\right)^{mt}} = \frac{1{,}000{,}000}{(1+0.12)^{20}} \approx \$103{,}666.77$$

In Section 5.1 we mentioned that a bond pays interest until it reaches maturity, at which point it pays you back an amount called its **maturity value,** or **par value.** (At the simplest, this is the price the investor paid for the bond originally, but see Section 5.3.) The two parts, the interest and the maturity value, can be separated and sold and traded by themselves. A **zero-coupon bond** is a form of corporate bond that pays no interest during its life but, like U.S. Treasury bills, promises to pay you the maturity value when it reaches maturity. Zero-coupon bonds are often created by removing, or *stripping,* the interest coupons from an ordinary bond and so are also known as **strips.** Zero-coupon bonds sell for less than their maturity value, and the return on the investment is the difference between what the investor pays and the maturity value. Although no interest is actually paid, we measure the return on investment by thinking of the interest rate that would make the selling price (the present value) grow to become the maturity value (the future value).[6]

Example 2 • Zero-Coupon Bonds

Megabucks Corporation is issuing 10-year zero-coupon bonds. How much would you pay for bonds with a maturity value of \$10,000 if you wish to get a return of 6.5% compounded annually?[7]

[6]The IRS refers to this kind of interest as **original issue discount (OID)** and taxes it as if it were interest actually paid to you each year.

[7]The return investors look for depends on a number of factors, including risk (the chance that the company will go bankrupt and you will lose your investment)—the higher the risk, the higher the return. U.S. Treasuries are considered risk free because the federal government has never defaulted on its debts. On the other hand, so-called junk bonds are high-risk investments (below investment grade) and have correspondingly high yields.

Solution As we said earlier, we think of a zero-coupon bond as if it were an account earning (compound) interest. We are asked to calculate the amount you will pay for the bond—the present value PV. We have

$$FV = \$10{,}000$$
$$r = 0.065$$
$$t = 10$$
$$m = 1$$

We can now use the present value formula:

$$PV = \frac{FV}{\left(1 + \frac{r}{m}\right)^{mt}}$$
$$= \frac{10{,}000}{\left(1 + \frac{0.065}{1}\right)^{10\times1}} \approx 5327.26$$

Thus, you should pay \$5327.26 to get a return of 6.5% annually.

Graphing Calculator: Computing Present Value

To compute the present value using a TI-83, enter the numbers shown in Figure 3 in the TVM Solver window, except for the value of PV.

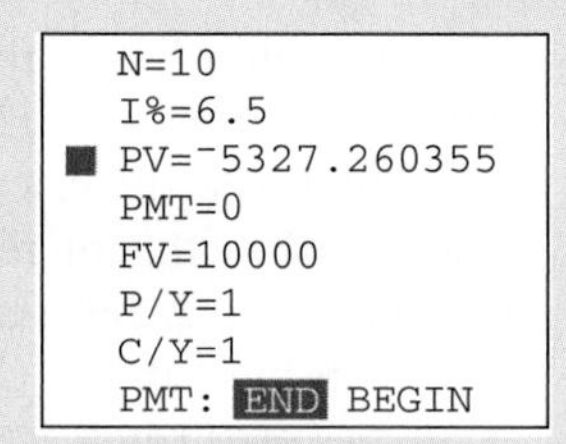

Figure 3

Use the arrow keys to put the cursor on the PV line and then press ALPHA SOLVE. Why is the present value given as negative?

Web Site: Computing Present Value

Follow the path

Web site → Online Utilities→ Time Value of Money Utility

for a utility that can be used in place of the TVM Solver.

Excel: Computing Present Value

You can compute present value, using the PV worksheet function. The following worksheet is similar to the one in Example 1, except that we have entered a formula for computing the present value from the entered values.

	A	B	C	D
1		Entered	Calculated	
2	Rate	6.50%		
3	Years	10		
4	Payment	$0.00		
5	Present Value		=PV(B2/B7,B3*B7,B4,B6)	
6	Future Value	$10,000.00		
7	Periods per year	1		

After calculation, the result will be −\$5327.26. Why is it negative?

Before we go on . . . Particularly in financial applications, you will hear the word *discounted* in place of *compounded* when discussing present value. Thus, the result of this example might be phrased, "The present value of $10,000 to be received 10 years from now, with an interest rate of 6.5% discounted annually, is $5327.26."

Question We have been discussing investments that pay various interest rates compounded at different numbers of times per year. How do we compare them?

Answer Consider the following (pleasant) scenario: You have just won $1 million in the lottery and are deciding what to do with it during the next year before you move to the South Pacific. Bank Ten offers 10% interest compounded annually, and Bank Nine offers 9.8% compounded monthly. These are the values of these investments after 1 year:

Bank Ten: $FV = 1(1 + 0.10)^1 = \$1.1$ million

Bank Nine: $FV = 1\left(1 + \frac{0.098}{12}\right)^{12} = \$1.1025$ million

Notice that, although Bank Ten pays 10% interest, or $0.1 million over the course of the year, Bank Nine will pay a total of $1.1025 million, as if it were paying 10.25% interest compounded annually. We call 10.25% the **effective** interest rate of the investment (also referred to as the **annual percentage yield,** or APY in the banking industry); the stated 9.8% is called the **nominal** interest rate. In the case of Bank Ten, the effective interest rate and the nominal rate are the same, 10%, because the interest is compounded only once per year. Obviously, you would prefer the higher effective interest rate, so you deposit your money in Bank Nine.

In general, to best compare two different investments, it is wisest to compare their *effective*—rather than nominal—interest rates.

Notice that we got 10.25% by computing

$$\left(1 + \frac{0.098}{12}\right)^{12} = 1.1025$$

and then subtracting 1. Generalizing, we get the following formula.

Effective Interest Rate

The effective interest rate r_{eff} of an investment paying a nominal interest rate of r_{nom} compounded m times per year is

$$r_{\text{eff}} = \left(1 + \frac{r_{\text{nom}}}{m}\right)^m - 1$$

To compare rates of investments with different compounding periods, always compare the effective interest rates rather than the nominal rates.

Quick Example

To calculate the effective interest rate of an investment that pays 8% per year, with interest reinvested monthly, set $r_{\text{nom}} = 0.08$ and $m = 12$, to obtain

$$r_{\text{eff}} = \left(1 + \frac{0.08}{12}\right)^{12} - 1 \approx 0.0830, \text{ or } 8.30\%$$

Example 3 • How Long to Invest

You have \$5000 to invest at 6% interest compounded monthly. How long will it take for your investment to grow to \$6000?

Solution If we use the future value formula, we have the values

$$FV = 6000$$
$$PV = 5000$$
$$r = 0.06$$
$$m = 12$$

Substituting, we get

$$6000 = 5000\left(1 + \frac{0.06}{12}\right)^{12t}$$

If you are familiar with logarithms, you can solve explicitly for t as follows:

$$\left(1 + \frac{0.06}{12}\right)^{12t} = \frac{6000}{5000} = 1.2$$

$$\log\left(1 + \frac{0.06}{12}\right)^{12t} = \log 1.2$$

$$12t \log\left(1 + \frac{0.06}{12}\right) = \log 1.2$$

$$t = \frac{\log 1.2}{12 \log\left(1 + \frac{0.06}{12}\right)} \approx 3.046 \approx 3 \text{ years}$$

Graphing calculators and spreadsheets give us alternative methods of solution.

Graphing Calculator: Computing Length of Time to Invest

You can use the TVM Solver in the TI-83 (or the online utility at the Web site) to find the length of time to invest. Enter the numbers shown in Figure 4 in the TVM Solver window, except for the value of N.

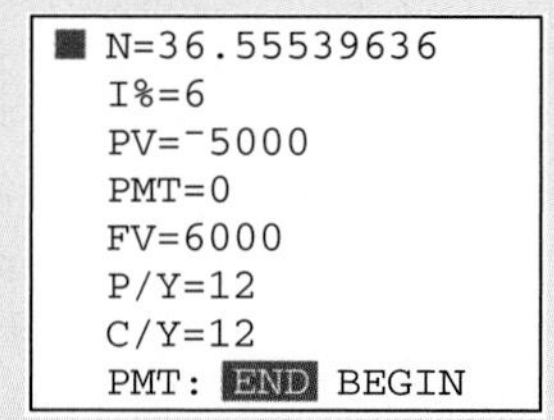

Figure 4

(Recall that $I\%$ is the annual interest rate, corresponding to r in the formula.) Use the arrow keys to put the cursor on the N line and then press [ALPHA] [SOLVE] to solve for the value $N = 36.555\,396\,36$. Thus, you will need to invest your money for about 36.5 months (compounding periods), or just over 3 years, before it grows to \$6000.

Excel: Computing Length of Time to Invest

You can compute the requisite length of an investment in Excel using the NPER worksheet function. The following worksheet shows the calculation:

	A	B	C	D
1		**Entered**	**Calculated**	
2	**Rate**	6.00%		
3	**Years**		=NPER(B2/B7,B4,B5,B6)/B7	
4	**Payment**	$0.00		
5	**Present Value**	-$5,000.00		
6	**Future Value**	$6,000.00		
7	**Periods per year**	12		

The NPER function computes the number of compounding periods, so we divide by B7, the number of periods per year, to calculate the number of years, which appears as 3.046. Once again, we see that you need to invest your money for just over 3 years for it to grow to $6000.

Compound interest shows up in other contexts as well—for example, in considering the effects of inflation. Suppose inflation is running at 5% per year; then prices will increase by 5% each year. So if PV represents the price now, the price 1 year from now will be 5% higher, or $PV(1 + 0.05)$: The price a year from then will be 5% higher still, or $PV(1 + 0.05)^2$. Thus, the effects of inflation are compounded just as reinvested interest is.

Example 4 • Inflation

Inflation in East Avalon is 5% per year. TruVision television sets cost $200 today. How much will a comparable set cost 2 years from now? How long will it take prices to double?

Solution To find the price of a television set 2 years from now, we compute the future value of $200 at an inflation rate of 5% compounded yearly:

$$FV = 200(1 + 0.05)^2 = \$220.50$$

To find the time it takes prices to double, we use one of the techniques from Example 3 to solve the following equation for t:

$$400 = 200(1 + 0.05)^t$$

We get $t = 14.2$. So, at an inflation rate of 5%, prices double in just over 14 years.

Before we go on . . . Does the time it takes a price to double depend on the starting price? For example, would it take a $300 television set any more or less time to double in price than a $200 set? Set up the equation and look at it carefully to see why the starting price does not affect the "doubling time."

Example 5 • Constant Dollars

Inflation in North Avalon is 6% per year. Which is really more expensive, a car costing \$20,000 today or one costing \$22,000 in three years?

Solution We cannot compare the two costs directly because inflation makes \$1 today worth more (it buys more) than \$1 three years from now. We need the two prices expressed in comparable terms, so we convert to **constant dollars.** We take the car costing \$22,000 in three years and ask what it would cost in today's dollars. We convert the future value of \$22,000 to its present value:

$$
\begin{aligned}
PV &= FV(1 + i)^{-n} \\
&= 22{,}000(1 + 0.06)^{-3} \\
&\approx \$18{,}471.62
\end{aligned}
$$

Thus, the car costing \$22,000 in three years actually costs less than the one costing \$20,000 now.

Before we go on . . . In the presence of inflation, the only way to compare prices at different times is to convert all prices to constant dollars. We pick some fixed time and compute future or present values as appropriate to determine what things would have cost at that time.

Guideline: Recognizing When to Use Compound Interest and the Meaning of Present Value

Question How do I distinguish a problem that calls for compound interest from one that calls for simple interest?

Answer Study the scenario to ascertain whether the interest is being withdrawn as it is earned or reinvested (deposited back into the account). If the interest is being withdrawn, the problem is calling for simple interest, since the interest is not itself earning interest. If it is being reinvested, the problem is calling for compound interest.

Question How do I distinguish present value from future value in a problem?

Answer The present value always refers to the value of an investment before any interest is included (or, in the case of a depreciating investment, before any depreciation takes place). As an example, the future value of a bond is its maturity value. The value of \$1 in constant 1990 dollars is its present value (even though 1990 is in the past).

5.2 EXERCISES

In Exercises 1–8, calculate, to the nearest cent, the future value of an investment of $10,000 at the stated interest rate and after the stated amount of time.

1. 3%/year, compounded annually, after 10 years

2. 4%/year, compounded annually, after 8 years

3. 2.5%/year, compounded quarterly (4 times per year), after 5 years

4. 1.5%/year, compounded weekly (52 times per year), after 5 years

5. 6.5%/year, compounded daily (assume 365 days per year), after 10 years

6. 11.2%/year, compounded monthly, after 12 years

7. 0.2%/month, compounded monthly, after 10 years

8. 0.45%/month, compounded monthly, after 20 years

In Exercises 9–14, calculate the present value of an investment that will be worth $1000 at the stated interest rate and after the stated amount of time.

9. 10 years, at 5%/year, compounded annually

10. 5 years, at 6%/year, compounded annually

11. 5 years, at 4.2%/year, compounded weekly (assume 52 weeks per year)

12. 10 years, at 5.3%/year, compounded quarterly

13. 4 years, depreciating 5% each year

14. 5 years, depreciating 4% each year

In Exercises 15–20, find the effective annual interest rates of the given annual interest rates. Round your answers to the nearest 0.01%.

15. 5% compounded quarterly

16. 5% compounded monthly

17. 10% compounded monthly

18. 10% compounded daily (assume 365 days per year)

19. 10% compounded hourly (assume 365 days per year)

20. 10% compounded every minute (assume 365 days per year)

APPLICATIONS

21. Savings You deposit $1000 in an account at the Lifelong Trust Savings and Loan that pays 6% interest compounded quarterly. By how much will your deposit have grown after 4 years?

22. Investments You invest $10,000 in Rapid Growth Funds, which appreciate by 2%/year, with yields reinvested quarterly. By how much will your investment have grown after 5 years?

23. Depreciation During the year ending April 2002, the S&P 500 index depreciated by approximately 6%. Assuming this trend were to continue, how much would a $3000 investment in an S&P index fund be worth in 3 years?

Values were compared in mid-April. SOURCE: *New York Times,* April 17, 2002.

24. Depreciation During the year ending April 2002, the MSCI World Index depreciated by approximately 3.4%. Assuming that this trend were to continue, how much would a $3000 investment in an MSCI index fund be worth in 3 years?

Values were compared in mid-April. SOURCE: *New York Times,* April 17, 2002.

25. Bonds You want to buy a 10-year zero-coupon bond with a maturity value of $5000 and a yield of 5.5% annually. How much will you pay?

26. Bonds You want to buy a 15-year zero-coupon bond with a maturity value of $10,000 and a yield of 6.25% annually. How much will you pay?

27. Investments When I was considering what to do with my $10,000 Lottery winnings, my broker suggested I invest half of it in gold, whose value was growing by 10%/year, and the other half in certificates of deposit (CDs), which were yielding 5%/year, compounded every 6 months. Assuming that these rates are sustained, how much will my investment be worth in 10 years?

28. Investments When I was considering what to do with the $10,000 proceeds from my sale of technology stock, my broker suggested I invest half of it in municipal bonds, whose value was growing by 6%/year, and the other half in CDs, which were yielding 3%/year, compounded every 2 months. Assuming that these interest rates are sustained, how much will my investment be worth in 10 years?

29. Depreciation During a prolonged recession, property values on Long Island depreciated by 2% every 6 months. If my house cost $200,000 originally, how much was it worth 5 years later?

30. Depreciation Stocks in the health industry depreciated by 5.1% in the first 9 months of 1993. Assuming that this trend were to continue, how much would a $40,000 investment be worth in 9 years? (*Hint:* Nine years corresponds to 12 nine-month periods.)

SOURCE: *New York Times,* October 9, 1993, p. 37.

31. Retirement Planning I want to be earning an annual salary of $100,000 when I retire in 15 years. I have been offered a job that guarantees an annual salary increase of 4%/year, and the starting salary is negotiable. What salary should I request in order to meet my goal?

32. Retirement Planning I want to be earning an annual salary of $80,000 when I retire in 10 years. I have been offered a job that guarantees an annual salary increase of 5%/year, and the starting salary is negotiable. What salary should I request in order to meet my goal?

33. Present Value Determine the amount of money, to the nearest dollar, you must invest now at 6%/year, compounded annually, so that you will be a millionaire in 30 years.

34. Present Value Determine the amount of money, to the nearest dollar, you must invest now at 7%/year, compounded annually, so that you will be a millionaire in 40 years.

35. **Stocks** Six years ago, I invested some money in Dracubunny Toy stock, acting on the advice of a "friend." As things turned out, the value of the stock decreased by 5% every 4 months, and I discovered yesterday (to my horror) that my investment was worth only $297.91. How much did I originally invest?

36. **Sales** My recent marketing idea, the *Miracle Algae Growing Kit,* has been remarkably successful, with monthly sales growing by 6% every 6 months over the past 5 years. Assuming that I sold 100 kits the first month, what is the present rate of sales?

37. **Inflation** Inflation has been running 2%/year. A car now costs $30,000. How much would it have cost 5 years ago?

38. **Inflation** (Compare Exercise 37.) Inflation has been running 1% every 6 months. A car now costs $30,000. How much would it have cost 5 years ago?

39. **Inflation** Housing prices have been rising 6%/year. A house now costs $200,000. What would it have cost 10 years ago?

40. **Inflation** (Compare Exercise 39.) Housing prices have been rising 0.5% each month. A house now costs $200,000. What would it have cost 10 years ago?

41. **Constant Dollars** Inflation is running 3%/year when you deposit $1000 in an account earning 5%/year, compounded annually. In *constant dollars* how much money will you have 2 years from now? (*Hint:* First calculate the value of your account in 2 years' time and then find its present value based on the inflation rate.)

42. **Constant Dollars** Inflation is running 1%/month when you deposit $10,000 in an account earning 8% compounded monthly. In *constant dollars* how much money will you have 2 years from now? (See the hint for Exercise 41.)

43. **Investments** You are offered two investments. One promises to earn 12% compounded annually. The other will earn 11.9% compounded monthly. Which is the better investment?

44. **Investments** You are offered three investments. The first promises to earn 15% compounded annually, the second will earn 14.5% compounded quarterly, and the third will earn 14% compounded monthly. Which is the best investment?

45. **History** Legend has it that the Manhattan Indians sold Manhattan Island to the Dutch in 1626 for $24. In 2001 the total value of Manhattan real estate was estimated to be $136,106 million. Suppose the Manhattan Indians had taken that $24 and invested it at 6.2% compounded annually (a relatively conservative investment goal). Could the Manhattan Indians have bought back the island in 2001?

SOURCE: Queens County Overall Economic Development Corporation, http://www.queensny.org/outlook/Outlook_Sp_7.pdf, 2001.

46. **History** Repeat Exercise 45, assuming that the Manhattan Indians had invested the $24 at 6.1% interest compounded annually.

Inflation Exercises 47 through 54 are based on the following table, which shows the 2002 annual inflation rates in several Latin American countries. Assume that the rates shown continue indefinitely.

	Argentina	Brazil	Colombia	Chile	Ecuador	Mexico	Uruguay
Currency	Peso	Real	Peso	Peso	Sucre	Peso	Peso
Inflation Rate (%)	8	8	6	3	15	5	4

Consumer price indices as of March 2002. SOURCE: Latin Focus, http://www.latin-focus.com, March 2002.

47. If an item in Brazil now costs 100 reals, what do you expect it to cost 5 years from now? (Answer to the nearest real.)

48. If an item in Argentina now costs 1000 pesos, what do you expect it to cost 5 years from now? (Answer to the nearest peso.)

49. If an item in Chile will cost 1000 pesos in 10 years, what does it cost now? (Answer to the nearest peso.)

50. If an item in Mexico will cost 20,000 pesos in 10 years, what does it cost now? (Answer to the nearest peso.)

51. You wish to invest 1000 pesos in Colombia at an annual interest rate of 8%, compounded twice a year. Find the value of your investment in 10 years, expressing the answer in constant pesos. (Answer to the nearest peso.)

52. You wish to invest 1000 pesos in Uruguay at an annual interest rate of 8%, compounded twice a year. Find the value of your investment in 10 years, expressing the answer in constant pesos. (Answer to the nearest peso.)

53. Which is the better investment: an investment in Chile yielding interest of 4.3%/year, compounded annually, or an investment in Ecuador yielding interest of 16.2%/year, compounded every six months? Support your answer with figures that show the future value of an investment of one unit of currency in constant units.

54. Which is the better investment: an investment in Argentina yielding interest of 10%/year, compounded annually, or an investment in Uruguay, yielding interest of 5%/year, compounded every 6 months? Support your answer with figures that show the future value of an investment of one unit of currency in constant units.

55. **Bonds** Once purchased, bonds can be sold in the secondary market. The value of a bond depends on the prevailing interest rates, which vary over time. Suppose that in January 1982 you bought a 30-year zero-coupon U.S. Treasury bond with a maturity value of $100,000 and a yield of 15% annually.
    a. How much did you pay for the bond?

b. In January 1999 your bond had 13 years remaining until maturity. Rates on U.S. Treasury bonds of comparable length were about 4.75%. If you sold your bond to an investor looking for a return of 4.75% annually, how much money would you have received?

c. Using your answers to parts (a) and (b), what was the annual yield on your 17-year investment?

56. **Bonds** Suppose that in January 1999 you bought a 30-year zero-coupon U.S. Treasury bond with a maturity value of \$100,000 and a yield of 5% annually.

a. How much did you pay for the bond?

b. Suppose that 15 years later interest rates have risen again, to 12%. If you sell your bond to an investor looking for a return of 12%, how much money will you receive?

c. Using your answers to parts (a) and (b), what will be the annual yield on your 15-year investment?

T In Exercises 57–60, use technology to graph the future value *FV* as a function of the number of years *t*.

57. \$500 invested at 10% interest/year, compounded annually; $0 \leq t \leq 10$

58. \$600 invested at 15% interest/year, compounded annually; $0 \leq t \leq 10$

59. \$500 invested at 10% interest/year, compounded daily; $0 \leq t \leq 10$

60. \$600 invested at 15% interest/year, compounded daily; $0 \leq t \leq 10$

T In Exercises 61–66, use technology to solve.

61. **Compound Interest Table** Complete the following table that shows the value of a \$1000 investment earning 5% interest/year after the specified time periods. Round all answers to the nearest dollar (but don't round intermediate results).

Years	1	2	3	4	5	6	7
Value (\$)	1050						

62. **Compound Interest Table** Complete the following table that shows the value of a \$1000 investment earning 6% interest/year after the specified time periods. Round all answers to the nearest dollar (but don't round intermediate results).

Years	1	2	3	4	5	6	7
Value (\$)	1060						

63. **Competing Investments** I have just purchased \$5000 worth of municipal funds that are expected to yield 5.4% interest/year, compounded every 6 months. My friend has just purchased \$6000 worth of CDs that will earn 4.8% interest/year, compounded every 6 months. Determine when, to the nearest year, the value of my investment will be the same as hers and what this value will be. (*Hint:* You can either graph the values of both investments or make tables of the values of both investments.)

64. **Investments** Determine when, to the nearest year, \$3000 invested at 5% interest/year, compounded daily, will be worth \$10,000.

65. **Epidemics** At the start of 1985, the incidence of AIDS was doubling every 6 months, and 40,000 cases had been reported in the United States. Assuming this trend were to continue, determine when, to the nearest tenth of a year, the number of cases would have reached 1 million.

66. **Depreciation** My investment in Genetic Splicing Inc. is now worth \$4354 and is depreciating by 5% every 6 months. For some reason I am reluctant to sell the stocks and swallow my losses. Determine when, to the nearest year, my investment will drop below \$50.

COMMUNICATION AND REASONING EXERCISES

67. Why is the graph of the future value of a compound interest investment as a function of time not a straight line (assuming a nonzero rate of interest)?

68. An investment that earns 10% (compound interest) every year is the same as an investment that earns 5% (compound interest) every 6 months, right? Explain your answer.

69. If a bacteria culture is currently 0.01 gram and increases in size by 10%/day, then its growth is linear, right? Explain your answer.

70. At what point is the future value of a compound interest investment the same as the future value of a simple interest investment at the same annual rate of interest?

71. If two equal investments have the same effective interest rate and you graph the future value as a function of time for each of them, are the graphs necessarily the same? Explain your answer.

72. For what kind of compound interest investments is the effective rate the same as the nominal rate? Explain your answer.

73. For what kind of compound interest investments is the effective rate greater than the nominal rate? When is it smaller? Explain your answer.

74. If an investment appreciates by 10%/year for 5 years (compounded annually) and then depreciates by 10%/year (compounded annually) for 5 more years, will it have the same value as it had originally? Explain your answer.

75. You can choose between two investments in zero-coupon bonds: one maturing in 10 years and the other maturing in 15 years. If you knew the rate of inflation, how would you decide which is the better investment?

76. If you knew the various inflation rates for the years 1995–2001, how would you convert \$100 in 2002 dollars to 1995 dollars?

T 77. On the same set of axes, graph the future value of a \$100 investment earning 10%/year as a function of time over a 20-year period, compounded once a year, 10 times a year, 100 times a year, 1000 times a year, and 10,000 times a year. What do you notice?

T 78. By graphing the future value of a \$100 investment that is depreciating by 1%/year, convince yourself that eventually the future value will be less than \$1.

5.3 Annuities, Loans, and Bonds

A typical defined-contribution pension plan works as follows: Every month while you work, you and your employer deposit a certain amount of money in an account.[8] This money earns interest (or dividends) from the time it is deposited. When you retire the account continues to earn interest, but you may then start withdrawing money at a rate calculated to reduce the account to zero after some number of years. While you are working, the account is an example of an **increasing annuity,** an account into which periodic payments are made. After you retire the fund is an example of a **decreasing annuity,** an account from which periodic withdrawals are made.

Increasing Annuities

Suppose you make a payment of \$100 at the end of every month into an account earning 3.6% interest per year, compounded monthly. This means that your investment is earning 3.6%/12 = 0.3% per month. We write $i = 0.036/12 = 0.003$. What will be the value of the investment at the end of 2 years (24 months)?

Think of the deposits separately. Each earns interest from the time it is deposited, and the total accumulated after 2 years is the sum of these deposits and the interest they earn. In other words, the accumulated value is the sum of the future values of the deposits, taking into account how long each deposit sits in the account. Figure 5 shows a timeline with the deposits and the contribution of each to the final value.

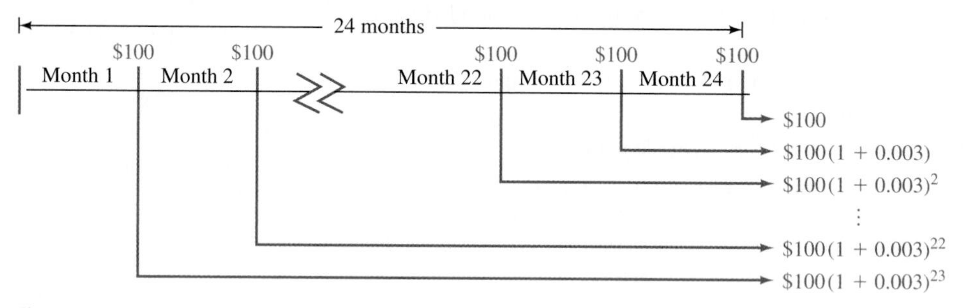

Figure 5

For example, the very last deposit (at the end of month 24) has no time to earn interest, so it contributes only \$100. The very first deposit, which earns interest for 23 months, by the future value formula for compound interest contributes $\$100(1 + 0.003)^{23}$ to the total. Adding together all the future values gives us a total future value:

$$FV = 100 + 100(1 + 0.003) + 100(1 + 0.003)^2 + \cdots + 100(1 + 0.003)^{23}$$

$$= 100[1 + (1 + 0.003) + (1 + 0.003)^2 + \cdots + (1 + 0.003)^{23}]$$

Fortunately, this sort of sum is well known (to mathematicians, anyway),[9] and there is a convenient formula for its value:[10]

$$1 + x + x^2 + \cdots + x^{n-1} = \frac{x^n - 1}{x - 1}$$

[8]Defined-contribution pensions are increasingly common, replacing the defined-benefit pensions that were once the norm. In a defined-benefit pension, the size of your pension is guaranteed; it is typically a percentage of your final working salary. In a defined-contribution plan, the size of your pension depends on how well your investments do.

[9]It is called a **geometric series.**

[10]The quickest way to convince yourself that this formula is correct is to multiply $(x - 1)(1 + x + x^2 + \cdots + x^{n-1})$ and see that you get $x^n - 1$. You should also try substituting some numbers. For example, $1 + 3 + 3^2 = 13 = (3^3 - 1)/(3 - 1)$.

In our case, with $x = 1 + 0.003$, this formula allows us to write

$$FV = 100\frac{(1+0.003)^{24}-1}{(1+0.003)-1} = 100\frac{(1.003)^{24}-1}{0.003} = \$2484.65$$

Question What if there were already some money—say, \$1000—in the account at the beginning of the 2 years?

Answer Then that amount would also earn interest. At the end of 2 years of earning 3.6% interest compounded monthly, the present value of \$1000 would contribute

$$1000(1 + 0.003)^{24} \approx \$1074.54$$

to the future value. Thus, the total in the account after 2 years would be

$$\begin{aligned} 1000(1+0.003)^{24} + 100\frac{(1+0.003)^{24}-1}{0.003} &\approx \$1074.54 + \$2484.65 \\ &= \$3559.19 \end{aligned}$$

It is now easy to generalize this calculation.

Future Value for an Increasing Annuity

An **increasing annuity** is an interest-bearing account into which periodic payments are made.* If you make a payment of PMT at the end of each compounding period into an investment with a present value of PV, paying interest at an annual rate of r compounded m times per year, then the future value after t years will be

$$FV = PV(1+i)^n + PMT\frac{(1+i)^n - 1}{i}$$

where $i = r/m$ is the interest paid each period and $n = mt$ is the total number of periods.

*An annuity in which payments are made at the end of each period is called an **ordinary annuity.** If the payments are made at the beginning of each period it is called an **annuity due.** In an annuity due, each payment earns interest for one more period than in an ordinary annuity, the effect of which is to multiply the future value of each payment by $1 + i$. The future value formula is then

$$FV = PV(1+i)^n + PMT(1+i)\frac{(1+i)^n - 1}{i}$$

Calculators and spreadsheets can do the calculations for either type of annuity. We will stick with ordinary annuities in the text.

Example 1 • Retirement Accounts

Your retirement account has \$5000 in it and earns 5% interest per year compounded monthly. Every month for the next 10 years, you will deposit \$100 into the account. How much money will there be in the account at the end of those 10 years?

Solution This is an increasing annuity with $PV = \$5000$, $PMT = \$100$, $r = 0.05$, $m = 12$, and $t = 10$. We compute $i = 0.05/12$ and $n = 12 \times 10 = 120$:

$$FV = PV(1+i)^n + PMT\,\frac{(1+i)^n - 1}{i}$$

$$= 5000\left(1 + \frac{0.05}{12}\right)^{120} + 100\,\frac{\left(1 + \frac{0.05}{12}\right)^{120} - 1}{\frac{0.05}{12}}$$

$$\approx 8{,}235.05 + 15{,}528.23$$

$$= \$23{,}763.28$$

Graphing Calculator: Computing the Future Value of an Annuity

Once more, we can use the TVM Solver in the TI-83 to calculate future values. Enter the values shown in Figure 6, except for the value of FV.

```
N=120
I%=5
PV=⁻5000
PMT=⁻100
■ FV=23763.27543
P/Y=12
C/Y=12
PMT: END BEGIN
```

Figure 6

Following the TI-83's usual convention, we set PV to the *negative* of the present value, since this is money you paid into the account. Likewise, we set PMT to -100 since you are paying \$100 each month; we set the number of payment and compounding periods to 12 per year. Finally, we set the payments to be made at the end of each period.[11]

To compute the future value, use the up or down arrow to put the cursor on the FV line and then press [ALPHA] [SOLVE].

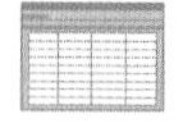

Excel: Computing the Future Value of an Annuity

We can use exactly the same worksheet that we used in Example 1 in Section 5.2.[12] In fact, we included the Payment row in that worksheet just for this purpose. On entering all the necessary values, the worksheet will look like this.

	A	B	C
1		Entered	Calculated
2	Rate	5.00%	
3	Years	10	
4	Payment	-\$100.00	
5	Present Value	-\$5,000.00	
6	Future Value		\$23,763.28
7	Periods per year	12	

[11]For an annuity due, in which payments are made at the beginning of each period, select `PMT:BEGIN`.

[12]For an annuity due, we would set the optional last argument of FV to 1 so that the formula in C6 would be `=FV(B2/B7,B3*B7,B4,B5,1)`. The other financial functions have a similar optional last argument.

Example 2 • Education Funds

Tony and Maria have just had a son, Jose Phillipe. They establish an account to save money for his college education, in which they would like to have \$100,000 after 17 years. If the account pays 4% interest per year, compounded quarterly, and they make deposits at the end of every quarter, how large must each deposit be for them to reach their goal?

Solution This is another increasing annuity. In this case $FV = \$100{,}000$, $PV = 0$, PMT is unknown, $t = 17$, $m = 4$, and $r = 0.04$. Therefore, $i = 0.04/4 = 0.01$ and $n = 4 \times 17 = 68$. From the future value formula, we get

$$100{,}000 = PMT\frac{(1 + 0.01)^{68} - 1}{0.01} = PMT\frac{1.01^{68} - 1}{0.01}$$

To solve for PMT, we multiply both sides by the reciprocal of the fraction on the right:

$$PMT = 100{,}000\frac{0.01}{1.01^{68} - 1} \approx 1033.89$$

So, Tony and Maria must deposit \$1033.89 every quarter in order to meet their goal.

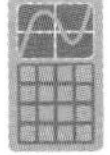

Graphing Calculator: Computing the Payments on an Annuity

In the TVM Solver in the TI-83, enter the values shown in Figure 7, except for the value of PMT, and then solve for PMT.

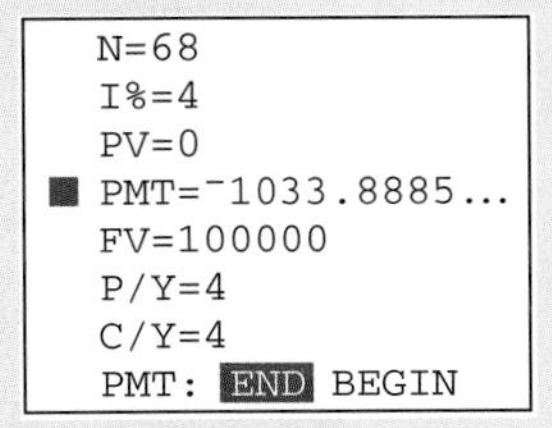

Figure 7

Why is PMT negative?

Web Site: Computing the Payments of an Annuity

Follow the path

Web site → Online Utilities→ Time Value of Money Utility

for a utility that can be used in place of the TVM Solver.

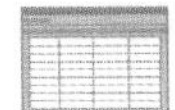

Excel: Computing the Payments on an Annuity

Use the following worksheet, in which the PMT worksheet function is used to calculate the required payments:

	A	B	C	D
1		**Entered**	**Calculated**	
2	**Rate**	4.00%		
3	**Years**	17		
4	**Payment**		=PMT(B2/B7,B3*B7,B5,B6)	
5	**Present Value**	\$0.00		
6	**Future Value**	\$100,000.00		
7	**Periods per year**	4		

After calculation, cell C4 will show the payment: −\$1033.89.

Before we go on . . . When an increasing annuity is used in this way to accumulate money to pay off an anticipated debt, it is often called a **sinking fund.** The money accumulated will be used in the future to "sink" the debt.

Decreasing Annuities

Suppose we deposit an amount PV now in an account earning 3.6% interest per year, compounded monthly. Starting 1 month from now, the bank will send us monthly payments of \$100. What must PV be so that the account will be drawn down to \$0 in exactly 2 years?

As before, we write $i = r/m = 0.036/12 = 0.003$, and we will write $PMT = 100$. The first payment of \$100 will be made 1 month from now, so its present value is

$$\frac{PMT}{(1+i)^n} = \frac{100}{1+0.003} = 100(1+0.003)^{-1} \approx \$99.70$$

In other words, that much of the original PV goes toward funding the first payment. The second payment, 2 months from now, has a present value of

$$\frac{PMT}{(1+i)^n} = \frac{100}{(1+0.003)^2} = 100(1+0.003)^{-2} \approx \$99.40$$

That much of the original PV funds the second payment. This continues for 2 years, at which point we receive the last payment, which has a present value of

$$\frac{PMT}{(1+i)^n} = \frac{100}{(1+0.003)^{24}} = 100(1+0.003)^{-24} \approx \$93.06$$

and that exhausts the account. Figure 8 shows a timeline with the payments and the present value of each.

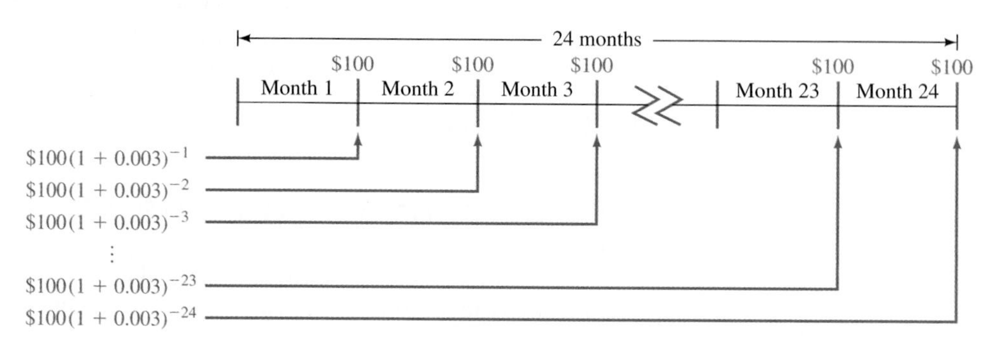

Figure 8

Since PV must be the sum of these present values, we get

$$\begin{aligned} PV &= 100(1+0.003)^{-1} + 100(1+0.003)^{-2} + \cdots + 100(1+0.003)^{-24} \\ &= 100[(1+0.003)^{-1} + (1+0.003)^{-2} + \cdots + (1+0.003)^{-24}] \end{aligned}$$

We can again find a simpler formula for this sum:

$$x^{-1} + x^{-2} + \cdots + x^{-n} = \frac{1}{x^n}(x^{n-1} + x^{n-2} + \cdots + 1) = \frac{1}{x^n} \cdot \frac{x^n - 1}{x - 1} = \frac{1 - x^{-n}}{x - 1}$$

So, in our case,

$$PV = 100\frac{1 - (1 + 0.003)^{-24}}{(1 + 0.003) - 1}$$

$$= 100\frac{1 - (1.003)^{-24}}{0.003} \approx 2312.29$$

If we deposit \$2312.29 initially and the bank sends us \$100 each month for 2 years, our account will be exhausted at the end of that time.

Question What if I want \$1000 in the account after 2 years?

Answer Then you will have to put more money in the account in the first place. The extra required is the amount that will have a future value of \$1000 in 2 years—that is, the present value, which is

$$1000(1 + 0.003)^{-24} \approx \$930.63$$

Thus, you will need to deposit

$$PV = 1000(1 + 0.003)^{-24} + 100\frac{1 - (1 + 0.003)^{-24}}{0.003}$$

$$\approx 930.63 + 2312.29$$

$$= \$3242.92$$

Generalizing, we get the following formula.

Present Value for a Decreasing Annuity

A **decreasing annuity** is an interest-bearing account from which periodic withdrawals are made.* If we invest PV dollars at an annual rate of r compounded m times per year, we receive a payment of PMT at the end of each compounding period, and the investment has value FV after t years, then

$$PV = FV(1 + i)^{-n} + PMT\frac{1 - (1 + i)^{-n}}{i}$$

where $i = r/m$ is the interest paid each period and $n = mt$ is the total number of periods.

---

*Again, if the payments are made at the beginning of each period, we have an annuity due rather than an ordinary annuity. In the case of an annuity due, each payment has 1 month less time to earn interest, so the part of the present value that funds each payment must be larger by a factor of $1 + i$. The present value formula is then

$$PV = FV(1 + i)^{-n} + PMT(1 + i)\frac{1 - (1 + i)^{-n}}{i}$$

Example 3 • Trust Funds

You wish to establish a trust fund from which your niece can withdraw \$2000 every 6 months for 15 years, at which time she will receive the remaining money in the trust, which you would like to be \$10,000. The trust will be invested at 7% interest/year compounded every 6 months. How large should the trust be?

Solution This is a decreasing annuity with $FV = 10{,}000$, $PMT = 2000$, $r = 0.07$, $m = 2$, and $t = 15$, so $i = 0.07/2 = 0.035$ and $n = 2 \times 15 = 30$. Substituting gives

$$PV = 10{,}000(1 + 0.035)^{-30} + 2000\frac{1 - (1 + 0.035)^{-30}}{0.035}$$

$$\approx \$40{,}346.87$$

The trust should start with \$40,346.87.

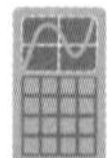

Graphing Calculator: Computing the Present Value of an Annuity

Enter the values shown in Figure 9 in the TI-83 TVM Solver window, except for the value of *PV*.

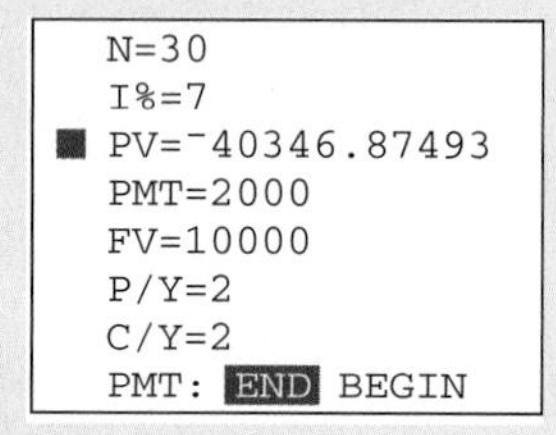

Figure 9

The payment and future value are positive because you (or your niece) will be receiving these amounts from the investment. To compute the present value, solve for *PV*.

Web Site: Computing the Present Value of an Annuity

Again, the online utility at

Web site → Online Utilities→ Time Value of Money Utility

can be used in place of the TMV Solver.

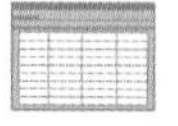

Excel: Computing the Present Value of an Annuity

You can use the same worksheet as in Example 2 in Section 5.2. After you enter all the necessary values, the worksheet will look like this:

	A	B	C
1		Entered	Calculated
2	Rate	7.00%	
3	Years	15	
4	Payment	$2,000.00	
5	Present Value		-$40,346.87
6	Future Value	$10,000.00	
7	Periods per year	2	

Example 4 • Education Funds

Tony and Maria (see Example 2), having accumulated \$100,000 for Jose Phillipe's college education, would now like to make quarterly withdrawals over the next 4 years. How much money can they withdraw each quarter in order to draw down the account to zero at the end of the 4 years? (Recall that the account pays 4% compounded quarterly.)

Solution Tony and Maria's account is now acting as a decreasing annuity with a present value of \$100,000 and a future value of 0. So, $PV = \$100{,}000$, $FV = 0$, $r = 0.04$, $m = 4$, and $t = 4$, giving $i = 0.01$ and $n = 16$. PMT is the unknown, and we have

$$100{,}000 = PMT\frac{1 - 1.01^{-16}}{0.01}$$

Solving for PMT gives

$$PMT = 100{,}000\frac{0.01}{1 - 1.01^{-16}} \approx 6794.46$$

So, if they withdraw \$6794.46 each quarter, their account balance will drop to zero at the end of the 4 years.

Graphing Calculator: Computing the Payment on a Decreasing Annuity

Enter the values shown in Figure 10 in the TI-83 TVM Solver window, except for the value of PMT.

```
N=16
I%=4
PV=⁻100000
■ PMT=6794.459682
FV=0
P/Y=4
C/Y=4
PMT: END BEGIN
```

Figure 10

The present value is negative because Tony and Maria do not possess it; the bank does. To compute the payment, solve for PMT.

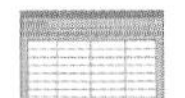

Excel: Computing the Payment on a Decreasing Annuity

You can use the same worksheet as in Example 2. After you enter all the necessary values, the worksheet will look like this:

	A	B	C
1		Entered	Calculated
2	Rate	4.00%	
3	Years	4	
4	Payment		\$6,794.46
5	Present Value	-\$100,000.00	
6	Future Value	\$0.00	
7	Periods per year	4	

Before we go on . . . If the account paid no interest, Tony and Maria could withdraw only 100,000/16 = \$6250.00 dollars each quarter.

Example 5 • Saving for Retirement

Jane Q. Employee has just started her new job with Big Conglomerate Inc. and is already looking forward to retirement. BCI offers her as a pension plan an annuity that is guaranteed to earn 6% annual interest compounded monthly. She plans to work for 40 years before retiring and would then like to be able to draw an income of $7000 per month for 20 years. How much do she and BCI together have to deposit each month into the fund to accomplish this?

Solution Here we have the situation we described at the beginning of the section: an increasing annuity accumulating money to be used later as a decreasing annuity. We know the desired payment out of the decreasing annuity, so we work backward. The first thing we need to do is find out the present value of the decreasing annuity. We use the decreasing annuity formula with $FV = 0$ (assuming that she will draw the account down to zero), $PMT = 7000$, $i = r/m = 0.06/12 = 0.005$, and $n = mt = 12 \times 20 = 240$:

$$PV = FV(1+i)^{-n} + PMT\frac{1-(1+i)^{-n}}{i}$$

$$= 7000\,\frac{1-(1+0.005)^{-240}}{0.005}$$

$$\approx \$977{,}065.40$$

This is the total that must be accumulated in the increasing annuity during the 40 years she plans to work. In other words, this is the *future* value FV of the increasing annuity. (Thus, the present value in the first step of our calculation is the future value in the second step.) To determine the payments necessary to accumulate this amount, we use the increasing annuity formula with $FV = 977{,}065.40$, $PV = 0$ (since she is a new employee), $i = 0.005$ (again), and $n = mt = 12 \times 40 = 480$:

$$FV = PMT\frac{(1+i)^n - 1}{i}$$

$$977{,}065.40 = PMT\frac{1.005^{480} - 1}{0.005}$$

Thus,

$$PMT = 977{,}065.40\,\frac{0.005}{1.005^{480} - 1}$$

$$\approx 490.62$$

So, if she and BCI collectively deposit $490.62 each month into her retirement fund, she can retire with the income she desires.

Installment Loans

In a typical installment loan, such as a car loan or a home mortgage, we borrow an amount of money and then pay it back with interest by making fixed payments (usually every month) over some number of years. From the point of view of the lender, this is a decreasing annuity with a future value of zero. Thus, loan calculations are identical to decreasing annuity calculations.

Example 6 • Home Mortgages[13]

Marc and Mira are buying a house and have taken out a 30-year, \$90,000 mortgage at 8% interest per year. What will be their monthly payments?

Solution A mortgage is essentially a decreasing annuity with a future value of zero. In this case the present value is $PV = \$90{,}000$, $r = 0.08$, $m = 12$, and $t = 30$. Thus,

$$90{,}000 = PMT \frac{1 - \left(1 + \frac{0.08}{12}\right)^{-12\times 30}}{\frac{0.08}{12}}$$

$$PMT = 90{,}000 \frac{\frac{0.08}{12}}{1 - \left(1 + \frac{0.08}{12}\right)^{-12\times 30}}$$

$$\approx \$660.39$$

Example 7 • Amortization Schedules

Continuing Example 6: Mortgage interest is tax deductible, so it is important to know how much of a year's mortgage payments represents interest. How much interest will Marc and Mira pay in the first year of their mortgage?

Solution Let's calculate how much of each month's payment is interest and how much goes to reducing the outstanding principal. At the end of the first month, Marc and Mira must pay 1 month's interest on \$90,000, which is

$$\$90{,}000 \times \frac{0.08}{12} = \$600$$

The remainder of their first monthly payment, $\$660.39 - 600 = \$60.39$ goes to reducing the principal. In the second month, the outstanding principal is $\$90{,}000 - 60.39 = \$89{,}939.61$, and part of their second monthly payment will be for the interest on this amount, which is

$$\$89{,}939.61 \times \frac{0.08}{12} \approx \$599.60$$

[13]The word *mortgage* comes from the French for "dead pledge."

The remaining $660.39 − $599.60 = $60.79 goes to further reduce the principal. If we continue this calculation for the 12 months of the first year, we get the beginning of the mortgage's **amortization schedule:**

Month	Outstanding Principal ($)	Payment on Principal ($)	Interest Payment ($)
0	90,000.00		
1	89,939.61	60.39	600.00
2	89,878.82	60.79	599.60
3	89,817.62	61.20	599.19
4	89,756.01	61.61	598.78
5	89,693.99	62.02	598.37
6	89,631.56	62.43	597.96
7	89,568.71	62.85	597.54
8	89,505.44	63.27	597.12
9	89,441.75	63.69	596.70
10	89,377.64	64.11	596.28
11	89,313.10	64.54	595.85
12	89,248.13	64.97	595.42
Total		751.87	7172.81

As we can see from the totals at the bottom of the columns, Marc and Mira will pay a total of $7172.81 in interest in the first year.

Graphing Calculator: Computing an Amortization Schedule

The TI-83 has built-in functions to compute the values in an amortization schedule. First, use the TVM Solver to find the monthly payment, as in Figure 11.

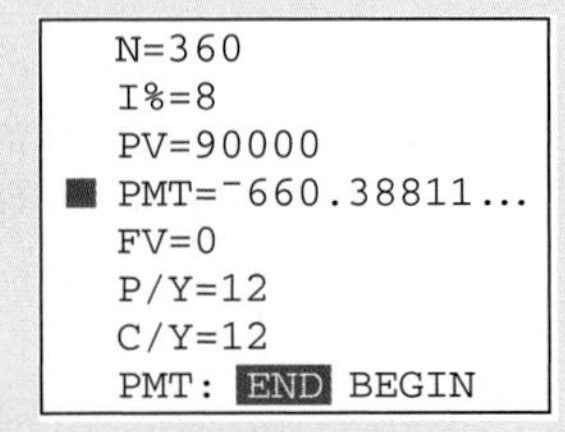

Figure 11

Three functions correspond to the last three columns of the amortization schedule: `bal`, `ΣPrn`, and `ΣInt` (found in the menu accessed through [2nd] [FINANCE]). They all require that the values of *I%*, *PV*, and *PMT* be entered or calculated ahead of time. Use `bal(n,2)` to find the balance of the principal outstanding after *n* payments. For example,

```
bal(12,2)
```

will return the value `89248.13`, the balance remaining at the end of 1 year. (The second argument, 2, tells the calculator to round all intermediate calculations to two decimal places—that is, the nearest cent—as would the mortgage lender.) Use

`ΣPrn(m,n,2)` to compute the sum of the payments on the principal from payment *m* through payment *n*. For example,

`ΣPrn(1,12,2)`

will return `-751.87`, the total paid on the principal in the first year. Similarly, use `ΣInt(m,n,2)` to compute the sum of the interest payments from payment *m* through payment *n*. For example,

`ΣInt(1,12,2)`

will return `-7172.81`, the total paid in interest in the first year.

To construct an amortization schedule, make sure that `FUNC` is selected in the `MODE` window; then enter the functions in the `Y=` window, as shown in Figure 12.

```
  Plot1 Plot2 Plot3
\ Y1=bal(X,2)
\ Y2=-ΣPrn(X,X,2)
\ Y3=-ΣInt(X,X,2)
\ Y4=
\ Y5=
\ Y6=
\ Y7=
```

Figure 12

Press 2nd TBLSET and enter the values shown in Figure 13.

```
TABLE SETUP
  TblStart=0
  ΔTbl=1
Indpnt: Auto Ask
Depend: Auto Ask
```

Figure 13

If you now press 2nd TABLE, you will get the table shown in Figure 14.

X	Y_1	Y_2
0	90000	ERROR
1	89940	60.39
2	89879	60.79
3	89818	61.2
4	89756	61.61
5	89694	62.02
6	89632	62.43
Y_1=89939.61		

Figure 14

The column labeled `X` gives the month, the column labeled Y_1 gives the outstanding principal, the column labeled Y_2 gives the payment on the principal for each month, and the column labeled Y_3 (use the right arrow button to make it visible) gives the interest payment for each month. To see later months, use the down arrow. As you can see, some values will be rounded in the table, but by selecting a value (as the outstanding principal at the end of the first month is selected in the figure) you can see its exact value at the bottom of the screen.

Excel: Computing an Amortization Schedule

You can also construct an amortization schedule using Excel. Begin with the following worksheet:

	A	B	C	D	E	F
1	Month	Outstanding Principal	Payment on Principal	Interest Payment		
2	0	$90,000.00			Rate	8%
3	1				Years	30
4	2				Payment	

In cell F4 enter the formula for the monthly payment:

```
=DOLLAR(-PMT(F2/12,F3*12,B2))
```

The function DOLLAR rounds the payment to the nearest cent, as the bank would. Calculate the interest owed at the end of the first month using the formula

```
=DOLLAR(B2*$F$2/12)
```

in cell D3. The payment on the principal is the remaining part of the payment, so enter

```
=$F$4-D3
```

in cell C3. Finally, calculate the outstanding principal by subtracting the payment on the principal from the previous outstanding principal, by entering

```
=B2-C3
```

in cell B3. We can now copy the formulas in cells B3, C3, and D3 into the cells below them to continue the table.

	A	B	C	D	E	F
1	Month	Outstanding Principal	Payment on Principal	Interest Payment		
2	0	$90,000.00			Rate	8%
3	1	$89,939.61	$60.39	$600.00	Years	30
4	2				Payment	$660.39
5						
13						
14	12					

The result should be something like the following:

	A	B	C	D	E	F
1	Month	Outstanding Principal	Payment on Principal	Interest Payment		
2	0	$90,000.00			Rate	8%
3	1	$89,939.61	$60.39	$600.00	Years	30
4	2	$89,878.82	$60.79	$599.60	Payment	$660.39
5						
13						
14	12	$89,248.13	$64.97	$595.42		

Before we go on . . . Excel has built-in functions that compute the interest payment (IPMT) or the payment on the principle (PPMT) in a given period. We could also have used the built-in future value function (FV) to calculate the outstanding principal each month. The main problem with using these functions is that, in a sense, they are too accurate. They do not take into account the fact that payments and interest are rounded to the nearest cent. Over time this rounding causes the actual value of the outstanding principal to differ from what the FV function would tell us. In fact, because the payment is rounded slightly upward (to $660.39 from 660.38811 . . .), the principal is reduced slightly faster than necessary, and a last payment of $660.39 would be $2.95 larger than needed to clear out the debt. The lender would reduce the last payment by $2.95 for this reason; Marc and Mira will pay only $657.44 for their final payment. This is common: The last payment on an installment loan is usually slightly larger or smaller than the others, to compensate for the rounding of the monthly payment amount.

Bonds

Suppose a corporation offers a 10-year bond paying 6.5% interest with payments every 6 months. As we saw in Section 5.1, this means that if we pay $10,000 for bonds with a maturity value of $10,000, we will receive 6.5/2 = 3.25% of $10,000, or $325, every 6 months for 10 years, at which time the corporation will give us the original $10,000 back. But bonds are rarely sold at their maturity value. Rather, they are auctioned off and sold at a price the bond market determines they are worth.

For example, suppose bond traders are looking for a return or **yield** of 7% rather than 6.5% (sometimes called the **coupon interest rate** to distinguish it from the rate of return). How much would they be willing to pay for the above bonds with a maturity value of $10,000? Think of the bonds as an investment that will pay the owner $325 every 6 months for 10 years, at which time it will pay $10,000. This is exactly the behavior of a decreasing annuity. So we ask, How much would an investor pay for a decreasing annuity earning 7% interest compounded semiannually, with semiannual payments of $325 and a future value of $10,000? We use the present value formula:

$$\begin{aligned} PV &= FV(1+i)^{-n} + PMT\frac{1-(1+i)^{-n}}{i} \\ &= 10{,}000\left(1+\frac{0.07}{2}\right)^{-20} + 325\,\frac{1-\left(1+\frac{0.07}{2}\right)^{-20}}{\frac{0.07}{2}} \\ &\approx 9644.69 \end{aligned}$$

Thus, an investor looking for a return of 7% would be willing to pay $9644.69 per $10,000 maturity value for these bonds.

Example 8 • Bonds

Suppose bond traders are looking for only a 6% yield on their investment. How much would they pay per $10,000 for the above bonds?

Solution We redo the calculation with $r = 0.06$:

$$PV = 10{,}000\left(1+\frac{0.06}{2}\right)^{-20} + 325\,\frac{1-\left(1+\frac{0.06}{2}\right)^{-20}}{\frac{0.06}{2}} \approx \$10{,}371.94$$

Before we go on . . . Notice how the selling price of the bonds behaves as the desired yield changes. As desired yield goes up, the price of the bonds goes down; as desired yield goes down, the price of the bonds goes up. When the desired yield equals the coupon interest rate, the selling price will equal the maturity value. Therefore, when the yield is higher than the coupon interest rate, the price of the bond will be below its maturity value; when the yield is lower than the coupon interest rate, the price will be above the maturity value.

As we've mentioned before, the desired yield depends on many factors, but it generally moves up and down with prevailing interest rates. And interest rates have gone up and down cyclically since interest was invented. The effect on the value of bonds can be quite dramatic (see Exercises 55 and 56 in Section 5.2).

Example 9 • Rate of Return on a Bond

Suppose a 5%, 20-year bond sells for \$9800 per \$10,000 maturity value. What rate of return will investors get?

Solution Assuming the usual semiannual payments, we know the following:

$$PV = 9800$$

$$FV = 10{,}000$$

$$PMT = \frac{0.05}{2} \times 10{,}000 = 250$$

$$n = 20 \times 2 = 40$$

What we do not know is r or i, the annual or semiannual interest rate.

The present value formula,

$$PV = FV(1 + i)^{-n} + PMT\frac{1 - (1 + i)^{-n}}{i}$$

$$9800 = 10{,}000(1 + i)^{-40} + 250\frac{1 - (1 + i)^{-40}}{i}$$

cannot be solved for i directly. The best we can do is a sort of trial-and-error approach, substituting a few values for i in the right-hand side of the above equation to get an estimate:

i	0.01	0.02	0.03
$10{,}000(1 + i)^{-40} + 250\frac{1 - (1 + i)^{-40}}{i}$	14,925	11,368	8844

Since we want the value to be 9800, we see that the correct answer is somewhere between $i = 0.02$ and $i = 0.03$. Let's try the value midway between 0.02 and 0.03—namely, $i = 0.025$ (the average of the two values):

i	0.02	0.025	0.03
$10{,}000(1 + i)^{-40} + 250\frac{1 - (1 + i)^{-40}}{i}$	11,368	10,000	8844

Now we know that the correct value of i is somewhere between 0.025 and 0.03, so we choose for our next estimate of i the number midway between them: 0.0275. We could continue in this fashion to obtain i as accurately as we like. In fact, $i \approx 0.02581$, corresponding to an annual rate of return of approximately 5.162%.

Alternatively, we can use a graphing calculator or spreadsheet that has this trial-and-error approach built in.

Graphing Calculator: Finding the Interest Rate

We can use the TVM Solver in the TI-83 to find the interest rate just as we use it to find any other one of the variables. Enter the values shown in Figure 15 in the TVM Solver window, except for the value of *I*%, and then solve for *I*%. (Recall that *I*% is the annual interest rate, corresponding to r in the formula.)

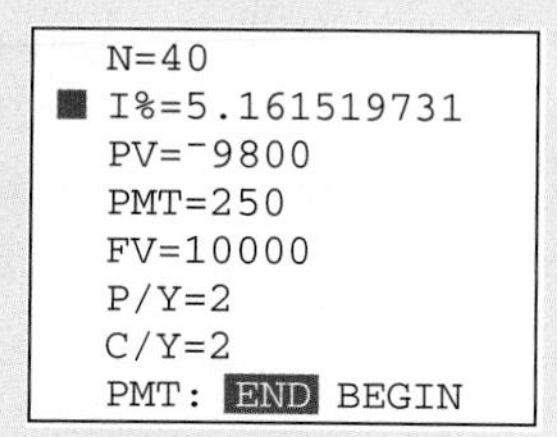

Figure 15

Thus, at $9800 per $10,000 maturity value, these bonds yield 5.162% interest compounded semiannually. The corresponding value for i is

$$i = \frac{r}{2} \approx \frac{0.05162}{2} = 0.02581$$

Web Site: Finding the Interest Rate

As usual, the online utility at

Web site → Online Utilities→ Time Value of Money Utility

can be used in place of the TMV Solver.

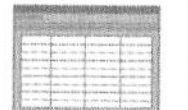

Excel: Finding the Interest Rate

Use the following worksheet, in which the RATE worksheet function is used to calculate the interest rate:

	A	B	C	D
1		**Entered**	**Calculated**	
2	**Rate**		=RATE(B3*B7,B4,B5,B6)*B7	
3	**Years**	20		
4	**Payment**	$250.00		
5	**Present Value**	-$9,800.00		
6	**Future Value**	$10,000.00		
7	**Periods per year**	2		

After calculation, cell C2 will show the desired interest rate: 5.16%.

Guideline: Which Formula to Use and the Technology Sign Convention

Question We have retirement accounts, trust funds, loans, bonds, and so on. Some are increasing annuities, and others are decreasing annuities. Which are which?

Answer In general, remember that an increasing annuity is an interest-bearing fund into which payments are made, and a decreasing annuity is an interest-bearing fund from which money is withdrawn. Here are a list of some of the accounts we have discussed in this section:

- *Retirement Accounts* A retirement account is an increasing annuity when payments are being made into the account (prior to retirement) and a decreasing annuity when an income is being withdrawn (after retirement).
- *Education Funds* These are similar to retirement accounts.
- *Trust Funds* A trust fund is a decreasing annuity if periodic withdrawals are made.
- *Installment Loans* We think of an installment loan as an investment a bank makes in the lender. In this way the lender's payments can be viewed as the bank's withdrawals, and so a loan is a decreasing annuity.
- *Bonds* Bonds pay regular interest (at a fixed coupon interest rate) until they mature, at which time they pay the maturity value. A bond trader thinks of a bond as a fictional decreasing annuity earning interest at the market rate and determines its present value accordingly.

Question In the mathematical formulas we use, all quantities (PV, FV, PMT) are positive, and yet we treat them as negative when we use technology. Are there alternative mathematical formulas that use the same sign conventions as Excel and graphing calculators?

Answer If you use the technology sign conventions, you can use the following single formula for both increasing and decreasing annuities (see Exercises 59–61):

$$FV + PV(1 + i)^n + PMT\frac{(1 + i)^n - 1}{i} = 0$$

To use it, substitute the known quantities and solve for the unknown quantity.

5.3 EXERCISES

In Exercises 1–6, find the amount accumulated in the increasing annuities. (Assume end-of-period deposits and compounding at the same intervals as deposits.)

1. \$100 deposited monthly for 10 years at 5%/year
2. \$150 deposited monthly for 20 years at 3%/year
3. \$1000 deposited quarterly for 20 years at 7%/year
4. \$2000 deposited quarterly for 10 years at 7%/year
5. \$100 deposited monthly for 10 years at 5%/year in an account containing \$5000 at the start
6. \$150 deposited monthly for 20 years at 3%/year in an account containing \$10,000 at the start

In Exercises 7–12, find the periodic payments necessary to accumulate the amounts in an increasing annuity. (Assume end-of-period deposits and compounding at the same intervals as deposits.)

7. \$10,000 in a fund paying 5%/year, with monthly payments for 5 years
8. \$20,000 in a fund paying 3%/year, with monthly payments for 10 years
9. \$75,000 in a fund paying 6%/year, with quarterly payments for 20 years
10. \$100,000 in a fund paying 7%/year, with quarterly payments for 20 years

11. \$20,000 in a fund paying 5%/year, with monthly payments for 5 years, if the fund contains \$10,000 at the start

12. \$30,000 in a fund paying 3%/year, with monthly payments for 10 years, if the fund contains \$10,000 at the start

In Exercises 13–18, find the present value of the decreasing annuity necessary to fund the withdrawals. (Assume end-of-period deposits and compounding at the same intervals as deposits.)

13. \$500/month for 20 years, if the annuity earns 3%/year

14. \$1000/month for 15 years, if the annuity earns 5%/year

15. \$1500/quarter for 20 years, if the annuity earns 6%/year

16. \$2000/quarter for 20 years, if the annuity earns 4%/year

17. \$500/month for 20 years, if the annuity earns 3%/year and if there is to be \$10,000 left in the annuity at the end of the 20 years

18. \$1000/month for 15 years, if the annuity earns 5%/year and if there is to be \$20,000 left in the annuity at the end of the 15 years

In Exercises 19–24, find the periodic withdrawals for the annuities. (Assume end-of-period deposits and compounding at the same intervals as deposits.)

19. \$100,000 at 3%, paid out monthly for 20 years

20. \$150,000 at 5%, paid out monthly for 15 years

21. \$75,000 at 4%, paid out quarterly for 20 years

22. \$200,000 at 6%, paid out quarterly for 15 years

23. \$100,000 at 3%, paid out monthly for 20 years, leaving \$10,000 in the account at the end of the 20 years

24. \$150,000 at 5%, paid out monthly for 15 years, leaving \$20,000 in the account at the end of the 15 years

In Exercises 25–28, determine the periodic payments on the loans.

25. \$10,000 borrowed at 9% for 4 years, with monthly payments

26. \$20,000 borrowed at 8% for 5 years, with monthly payments

27. \$100,000 borrowed at 5% for 20 years, with quarterly payments

28. \$1,000,000 borrowed at 4% for 10 years, with quarterly payments

In Exercises 29–32, determine the selling price, per \$1000 maturity value, of the bonds. (Assume twice-yearly interest payments.)

These are actual U.S. Treasury notes and bonds auctioned in 2001 and 2002. SOURCE: Bureau of the Public Debt's Web site, at http://www.publicdebt.treas.gov/.

29. 10 year, 4.875% bond, with a yield of 4.880%

30. 30 year, 5.375% bond, with a yield of 5.460%

31. 2 year, 3.625% bond, with a yield of 3.705%

32. 5 year, 4.375% bond, with a yield of 4.475%

In Exercises 33–36, use technology to determine the yield on the bonds. (Assume twice-yearly interest payments.)

These are actual U.S. Treasury notes and bonds auctioned in 2001 and 2002. SOURCE: Bureau of the Public Debt's Web site, at http://www.publicdebt.treas.gov/.

33. 5-year, 3.5% bond, selling for \$994.69/\$1000 maturity value

34. 10-year, 3.375% bond, selling for \$991.20/\$1000 maturity value

35. 2-year, 3% bond, selling for \$998.86/\$1000 maturity value

36. 5-year, 4.625% bond, selling for \$998.45/\$1000 maturity value

APPLICATIONS

37. Pensions Your pension plan is an annuity with a guaranteed return of 3% interest/year (compounded monthly). You would like to retire with a pension of \$5000/month for 20 years. If you work 40 years before retiring, how much must you and your employer deposit each month into the fund?

38. Pensions Meg's pension plan is an annuity with a guaranteed return of 5% interest/year (compounded quarterly). She would like to retire with a pension of \$12,000/quarter for 25 years. If she works 45 years before retiring, how much money must she and her employer deposit each quarter?

39. Pensions Your pension plan is an annuity with a guaranteed return of 4% interest/year (compounded quarterly). You can afford to put \$1200/quarter into the fund, and you will work for 40 years before retiring. After you retire you will be paid a quarterly pension based on a 25-year payout. How much will you receive each quarter?

40. Pensions Jennifer's pension plan is an annuity with a guaranteed return of 5% interest/year (compounded monthly). She can afford to put \$300/month into the fund, and she will work for 45 years before retiring. If her pension is then paid out monthly based on a 20-year payout, how much will she receive each month?

41. Car Loans While shopping for a car loan, you get the following offers: Solid Savings & Loan is willing to loan you \$10,000 at 9% interest for 4 years. Fifth Federal Bank & Trust will loan you the \$10,000 at 7% interest for 3 years. Both require monthly payments. You can afford to pay \$250/month. Which loan, if either, can you take?

42. Business Loans You need to take out a loan of \$20,000 to start up your T-shirt business. You have two possibilities: One bank is offering a 10% loan for 5 years, and another is offering a 9% loan for 4 years. Which will have the lower monthly payments? On which will you end up paying more interest total?

Exercises 43–48 require the use of the finance functions on the TI-83 or amortization tables in Excel.

43. Mortgages You take out a 15-year mortgage for \$50,000, at 8% interest, to be paid off monthly. Construct an amortization table showing how much you will pay in interest each year and how much goes toward paying off the principal.

44. Mortgages You take out a 30-year mortgage for \$95,000 at 9.75% interest, to be paid off monthly. If you sell your house after 15 years, how much will you still owe on the mortgage?

T **45. Mortgages** The following is a popular kind of mortgage. You take out a \$75,000, 30-year mortgage. For the first 5 years, the interest rate is held at 5%, but for the remaining 25 years, it rises to 9.5%. The payments for the first 5 years are calculated as if the 5% rate were going to remain in effect for all 30 years, and then the payments for the last 25 years are calculated to amortize the debt remaining at the end of the fifth year. What are your monthly payments for the first 5 years, and what are they for the last 25 years?

T **46. Adjustable Rate Mortgages** You take out an adjustable rate mortgage for \$100,000 for 20 years. For the first 5 years, the rate is 4%. It then rises to 7% for the next 10 years and then 9% for the last 5 years. What are your monthly payments in the first 5 years, the next 10 years, and the last 5 years? (Assume that each time the rate changes, the payments are recalculated to amortize the remaining debt if the interest rate were to remain constant for the remaining life of the mortgage.)

T **47. Refinancing** Your original mortgage was a \$96,000, 30-year, 9.75% mortgage.[14] After 4 years you refinance the remaining principal for 30 years at 6.875%. What was your original monthly payment? What is your new monthly payment? How much will you save in interest over the course of the loan by refinancing?

T **48. Refinancing** Kara and Michael take out a \$120,000, 30-year, 10% mortgage. After 3 years they refinance the remaining principal with a 15-year, 6.5% loan. What was their original monthly payment? What is their new monthly payment? How much did they save in interest over the course of the loan by refinancing?

T **49. Savings** You wish to accumulate \$100,000 through monthly payments of \$500. If you can earn interest at an annual rate of 4% interest compounded monthly, how long (to the nearest year) will it take to accomplish your goal?

T **50. Retirement** Alonzo plans to retire as soon as he has accumulated \$250,000 through quarterly payments of \$2500. If Alonzo invests this money at 5.4% interest compounded quarterly, when (to the nearest year) can he retire?

T **51. Loans** You have a \$2000 credit card debt, and you plan to pay it off through monthly payments of \$50. If you are being charged 15% interest/year, how long (to the nearest 0.5 year) will it take you to pay your debt?

T **52. Loans** You owe \$2000 on your credit card, which charges you 15% interest. Determine, to the nearest cent, the minimum monthly payment that will allow you to eventually pay your debt.

T **53. Savings** You are depositing \$100/month in an account that pays 4.5% interest/year (compounded monthly), while your friend Lucinda is depositing \$75/month in an account that earns 6.5% interest/year (compounded monthly). When, to the nearest year, will her balance exceed yours?

T **54. Car Leasing** You can lease a \$15,000 car for \$300/month. For how long (to the nearest year) should you lease the car so that your monthly payments are lower than purchasing it with an 8% interest/year loan?

COMMUNICATION AND REASONING EXERCISES

55. Your cousin Simon claims that you have wasted your time studying annuities: If you wish to retire on an income of \$1000/month for 20 years, you need to save \$1000/month for 20 years. Explain why he is wrong.

56. Your other cousin Trevor claims that you will earn more interest by accumulating \$10,000 through smaller payments than through larger payments made over a shorter period. Is he correct? Give a reason for your answer.

57. A real estate broker tells you that doubling the period of a mortgage halves the monthly payments. Is he correct? Support your answer by means of an example.

58. Another real estate broker tells you that doubling the size of a mortgage doubles the monthly payments. Is she correct? Support your answer by means of an example.

59. In the formula for the future value of an increasing annuity, we have taken *FV*, *PV*, and *PMT* all to be positive numbers. If we adopt the convention used by calculators and spreadsheets that payments you make are negative whereas payments to you are positive, show that the future value equation can be written as

$$FV + PV(1 + i)^n + PMT\frac{(1 + i)^n - 1}{i} = 0$$

60. In the formula for the present value of a decreasing annuity, we have taken *FV*, *PV*, and *PMT* all to be positive numbers. If we adopt the convention used by calculators and spreadsheets that payments you make are negative whereas payments to you are positive, show that the present value equation can be written as

$$FV(1 + i)^{-n} + PV + PMT\frac{1 - (1 + i)^{-n}}{i} = 0$$

61. Show that the equations in Exercises 59 and 60 are really the same equation.

62. (Needs logarithms) Show that the formula in Exercise 59 can be solved for the number n of payments to give

$$n = \frac{\ln\left(\dfrac{PMT/i - FV}{PMT/i + PV}\right)}{\ln(1 + i)}$$

[14] A somewhat simplified version of the experience of one of the authors.

CASE STUDY

www.collegeboard.com/paying/0,3309,6-0-0-0,00.html?orig=art. Copyright © 2003 by College Entrance Examination Board. Reproduced with permission. All rights reserved.

Saving for College

Tuition costs at public and private 4-year colleges and universities increased at an average rate of about 5.8% per year from 1991 through 2001. The 2001–2002 costs for tuition, fees, and room and board were \$9008 for a public in-state college and \$23,578 for a private college.[15]

Mr. and Mrs. Wong have an appointment tomorrow with you, their investment counselor, to discuss a plan to save for their newborn child's college education. You have already recommended that they invest a fixed amount each month in mutual funds expected to yield 6% per year (with earnings reinvested each month) until their child begins college at age 18. They have indicated that they are not sure when they will start making the monthly payments, however, and they don't of course know which type of college or university their child will attend.

You decide that you should create an Excel worksheet that will compute the monthly payment *PMT* for various possibilities for the age of the child at the time the Wongs begin making payments and the current cost of tuition at the college. The figures you have are for 2001, and it is now 2006, so you update your figures using the future value formula for compound interest, assuming that the 5.8% rate of increase continued:

$$9008(1 + 0.058)^5 = \$11{,}941 \quad \text{and} \quad 23{,}578(1 + 0.058)^5 = \$31{,}256$$

You begin to create your worksheet by entering various possible ages and the two tuition figures you just computed:

	A	B	C	D	E	F
1		**Tuition**				
2	**Age**	\$11,941.00	\$31,256.00			
3	0					
4	1					
5	2					
6	3					
7	4					
8	5					
9	6					
10	7					
11	8					
12	9					
13	10					

Ultimately, you would like to compute in cells B3 through C13 the monthly payments corresponding to the various possible combinations of age and tuition. You decide that it would also be good to enter the rate of inflation of tuition costs and the assumed return on the mutual fund in their own cells.

	A	B	C	D	E	F
1		**Tuition**		**Inflation**	5.8%	
2	**Age**	\$11,941.00	\$31,256.00	**Investment**	6%	
3	0					
4	1					
5	2					
6	3					
7	4					
8	5					
9	6					
10	7					
11	8					
12	9					
13	10					

---

[15]Figures are enrollment-weighted averages. SOURCE: "Trends in College Pricing 2001," The College Board, available on-line at http://www.collegeboard.com, 2001.

You realize that you need to do a two-stage calculation: The investment begins as an increasing annuity that grows until the child begins college and then changes to a decreasing annuity that pays the tuition each semester. As you learned to do when you studied the mathematics of finance, you begin by determining how much money you will need to accumulate to fund the second stage of the college fund. However, this second stage is not a simple decreasing annuity because the amount you need to withdraw for tuition will continue to rise over the course of the 4 years the child is in college. You begin by computing the eight semiannual tuition payments that will be required. The first occurs 18 years hence and pays for half a year, so will be half of the future value of the current tuition, with the assumed annual inflation rate. The second will pay for the second half of that year, so it will be identical. We can compute the future value using the FV worksheet function or directly. The next six payments also occur in pairs. You leave room above the tuition payments to compute the required accumulation:

	A	B	C	D	E	F
1		**Tuition**		**Inflation**	5.8%	
2	**Age**	$11,941.00	$31,256.00	**Investment**	6%	
3	0			**Goal**		
4	1			**Tuition**	=B$2*(1+$E$1)^18/2	
5	2			**Payments**	=B$2*(1+$E$1)^18/2	
6	3				=B$2*(1+$E$1)^19/2	
7	4				=B$2*(1+$E$1)^19/2	
8	5				=B$2*(1+$E$1)^20/2	
9	6				=B$2*(1+$E$1)^20/2	
10	7				=B$2*(1+$E$1)^21/2	
11	8				=B$2*(1+$E$1)^21/2	
12	9					
13	10					

Copying these formulas into column F gives you two columns of figures—the one in column E being the tuition payments that will be required for a public college and the one in column F being the payments that will be required for a private college:

	A	B	C	D	E	F
1		**Tuition**		**Inflation**	5.8%	
2	**Age**	$11,941.00	$31,256.00	**Investment**	6%	
3	0			**Goal**		
4	1			**Tuition**	$16,472.24	$43,116.69
5	2			**Payments**	$16,472.24	$43,116.69
6	3				$17,427.63	$45,617.46
7	4				$17,427.63	$45,617.46
8	5				$18,438.44	$48,263.27
9	6				$18,438.44	$48,263.27
10	7				$19,507.86	$51,062.54
11	8				$19,507.86	$51,062.54
12	9					
13	10					

Now, enough money needs to be accumulated in the account to fund each of these payments. Since the account continues to earn interest, the amount required to fund each payment should be discounted back to the beginning of the child's college career. Thus, for column E (public college) you want a formula to compute

$$E4 + E5(1+i)^{-1} + E6(1+i)^{-2} + \cdots + E11(1+i)^{-7}$$

where i is the semiannual interest rate for the mutual fund. Fortunately, Excel has a worksheet function just for this purpose, called NPV (Net Present Value). It takes an interest rate and a series of payments and computes the sum of the present values of those payments, assuming that the payments are made regularly at the end of each period. Since the first payment (in E4) will be made right away, you need to apply NPV to only

E5 through E11 and add in E4. One more thing you need to do is determine the semi-annual interest rate for the mutual fund. This is similar to the effective rate calculation: Since the fund earns 6%/12 each month, over 6 months it will earn interest at a rate of

$$\left(1 + \frac{0.06}{12}\right)^6 - 1$$

Thus, you enter the formula

```
=E4+NPV((1+$E$2/12)^6-1,E5:E11)
```

into cell E3 and copy it into F3 as well, getting the following result:

	A	B	C	D	E	F
1		Tuition		Inflation	5.8%	
2	Age	$11,941.00	$31,256.00	Investment	6%	
3	0			Goal	$129,162.31	$338,087.02
4	1			Tuition	$16,472.24	$43,116.69
5	2			Payments	$16,472.24	$43,116.69
6	3				$17,427.63	$45,617.46
7	4				$17,427.63	$45,617.46
8	5				$18,438.44	$48,263.27
9	6				$18,438.44	$48,263.27
10	7				$19,507.86	$51,062.54
11	8				$19,507.86	$51,062.54
12	9					
13	10					

You have now completed the calculation of the amount that needs to be accumulated in the account to pay for college. What remains is to compute the monthly payment needed to reach this goal, but this is a straightforward increasing annuity calculation. You enter the following formula in cell B3 and copy it to all the cells in the rectangle B3:C13:

	A	B	C	D	E	F
1		Tuition		Inflation	5.8%	
2	Age	$11,941.00	$31,256.00	Investment	6%	
3	0	=PMT($E$2/12,18*12-$A3*12,0,-E$3)			$129,162.31	$338,087.02
4	1			Tuition	$16,472.24	$43,116.69
5	2			Payments	$16,472.24	$43,116.69
6	3				$17,427.63	$45,617.46
7	4				$17,427.63	$45,617.46
8	5				$18,438.44	$48,263.27
9	6				$18,438.44	$48,263.27
10	7				$19,507.86	$51,062.54
11	8				$19,507.86	$51,062.54
12	9					
13	10					

Finally, this gives you the computations you were seeking:

	A	B	C	D	E	F
1		Tuition		Inflation	5.8%	
2	Age	$11,941.00	$31,256.00	Investment	6%	
3	0	$333.45	$872.81	Goal	$129,162.31	$338,087.02
4	1	$365.66	$957.13	Tuition	$16,472.24	$43,116.69
5	2	$402.26	$1,052.93	Payments	$16,472.24	$43,116.69
6	3	$444.13	$1,162.54		$17,427.63	$45,617.46
7	4	$492.41	$1,288.91		$17,427.63	$45,617.46
8	5	$548.58	$1,435.93		$18,438.44	$48,263.27
9	6	$614.62	$1,608.79		$18,438.44	$48,263.27
10	7	$693.22	$1,814.52		$19,507.86	$51,062.54
11	8	$788.15	$2,063.02		$19,507.86	$51,062.54
12	9	$904.88	$2,368.55			
13	10	$1,051.57	$2,752.51			

Thus, for example, if the Wongs start saving for a private college next year, when their child is 1 year old, they will need to invest $957.13 per month to save enough money to pay fully for private college.

EXERCISES

In Exercises 1–3, use technology to solve.

1. Graph the monthly payments needed to save for a 4-year public college as a function of the age of the child when payments start, from birth through age 17.

2. Graph the monthly payments needed to save for a 4-year private college as a function of the age of the child when payments start, from birth through age 17.

3. Redo the calculations for the case where college costs are increasing by only 4%/year and the Wongs can invest their money at 10%/year.

4. Which strategy results in a lower monthly payment: waiting a year longer to begin college or starting to save for college a year earlier?

CHAPTER 5 REVIEW TEST

1. You invest $5000 in a stock fund that pays a dividend of 4.75% every year.
   a. If you always take the dividend in cash, how much will you earn in dividends in 5 years?
   b. If you always take the dividend in cash, does it matter how often the fund pays dividends?
   c. How many years do you need to wait to accumulate at least $4000 in dividends?
   d. If, instead of taking the dividend in cash, you reinvested it in the fund, how much will your investment be worth in 5 years?
   e. How much more will your investment earn in 5 years if the fund pays one-fourth of the yearly dividend every quarter and you reinvest all dividends as they are paid?

2. a. How much do you have to pay for a 20-year bond that earns 4% simple interest with a total return of $10,000?
   b. A 20-year bond has a maturity value of $10,000 and will pay a total of $9000 simple interest over its lifetime. What is its annual interest rate?
   c. A $25,000 loan, taken now, with a simple interest rate of 8%/year, will cost a total of $33,000. When will the loan mature?
   d. A U.S. Treasury bill paying $10,000 after 2 months sells at a discount rate of 6.3%. What does it cost and what simple annual interest rate does it pay?

3. It is now January 2003, and you invest $10,000 in a mutual fund.
   a. If the fund depreciates by 4.3%/year, what will be its value at the end of 5 years?
   b. If, instead, you invest in a fund that pays dividends quarterly at an annual rate of 6.2%/year and you reinvest all dividends, how much will your investment be worth at the end of the 5 years?
   c. If inflation is running at 2%/year, what is the future value of the investment in part (b) in constant (2003) dollars?
   d. When will the investment from part (b) be worth $15,000? (Remember that the fund pays dividends only each quarter.)

4. a. If you deposit $150 each month in a savings account paying 6% interest/year, compounded monthly, how much will you have in the account after 3 years? How much of this is interest that you earned?
   b. You would like to accumulate $20,000 over the next 7 years, and you have an account that pays 5% interest compounded monthly. How much money must you deposit each month to reach your goal?
   c. Curt is buying a $12,000 car. He will make a down payment of $4000 and finance the balance with a 4-year, 7% interest loan. What will be his monthly payments?
   d. What are the monthly payments for a 30-year, $120,000 mortgage at 8% interest? What is the total interest paid during the first year of the mortgage? What is the total interest paid over the life of the mortgage?

OHaganBooks.com—INVESTING IN THE FUTURE

5. Total online revenues at OHaganBooks.com during 1999, its first year of operation, amounted to $150,000.
   a. Since December 1999, revenues have increased by a steady 20% each year. Track OHaganBooks.com's revenues for the subsequent 5 years, assuming that this rate of growth continues. During which year is the revenue predicted to surpass $300,000?
   b. Unfortunately, the picture for net income has not been so bright: The company lost $20,000 in the fourth quarter of 1999. However, the quarterly loss has been decreasing by 3.75% per quarter. How much did the company lose during the third quarter of 2001?

c. To finance anticipated expansion, you are considering making a public offering of OHaganBooks.com shares at $3/share. You are not sure how many shares to offer, but you would like the total value of the shares to reach a market value of at least $500,000 6 months after the initial offering. Your financial adviser, Sally McCormack, tells you that you can expect the value of the stock to double in the first day of trading, and then you should expect an appreciation rate of around 8%/month for the first 6 months. How many shares should you sell?

6. Unfortunately, the stock market takes a dive just as you are concluding plans for your stock offering in Exercise 5, so you decide to postpone the offering until the market shows renewed vigor. In the meantime there is an urgent need to finance the continuing losses at OHaganBooks.com, so you decide to seek a $250,000 loan.

a. The two best deals you are able to find are Industrial Bank, which offers a 10-year, 9.5% interest loan, and Expansion Loans, offering an 8-year, 6.5% interest loan. What are the monthly payments for each bank?

b. You have estimated that OHaganBooks.com can afford to pay only $3000/month to service the loan. What, to the nearest dollar, is the largest amount the company can borrow from Expansion Loans?

T **c.** What interest rate would Expansion Loans have to offer in order to meet the company's loan requirements at a price it can afford?

7. OHaganBooks.com has just introduced a retirement package for the employees. Under the annuity plan operated by Sleepy Hollow Retirement, the monthly contribution by the company on behalf of each employee is $800. Each employee can then supplement that amount through payroll deductions.

a. The current rate of return of Sleepy Hollow's retirement fund is 7.3%, and James Callahan, your site developer, plans to retire in 10 years. He contributes $1000/month to the plan (in addition to the company contribution of $800). Currently, there is $50,000 in his retirement annuity. How much (to the nearest dollar) will it be worth when he retires?

b. How much of that retirement fund was from the company contribution? (The company did not contribute toward the $50,000 that James now has.)

c. James actually wants to retire with $500,000. How much should he contribute each month to the annuity?

d. On second thought, James wants to be in a position to draw at least $5000/month for 30 years after his retirement. He feels he can invest the proceeds of his retirement annuity at 8.7% interest/year in perpetuity. Given the information in part (a), how much will he need to contribute to the plan starting now?

8. Actually, James Callahan is quite pleased with himself; 1 year ago he purchased a $50,000 government bond paying 7.2% interest/year (with interest paid every 6 months) and maturing in 10 years, and interest rates have come down since then.

a. The current interest rate on 10-year government bonds is 6.3%. If he were to auction the bond at the current interest rate, how much would he get?

b. If he holds on to the bond for 6 more months and the interest rate drops to 6%, how much will the bond be worth then?

T **c.** James suspects that interest rates will come down further during the next 6 months. If he hopes to auction the bond for $54,000 in 6 months' time, what will the interest rate need to be at that time?

WWW ADDITIONAL ONLINE REVIEW

If you follow the path

Web site → Everything for Finite Math → Chapter 5

you will find the following additional resources to help you review:

A comprehensive chapter summary (including examples and interactive features)

Additional review exercises (including interactive exercises and many with help)

An online Time Value of Money (TVM) utility

A true/false chapter quiz

6

SETS AND COUNTING

CASE STUDY

Designing a Puzzle

As product design manager for Cerebral Toys, you are constantly on the lookout for ideas for intellectually stimulating yet inexpensive toys. Your design team recently came up with an idea for a puzzle consisting of a number of plastic cubes. Each cube will have two faces colored red, two white, and two blue, and there will be exactly two cubes with each possible configuration of colors. The goal of the puzzle is to seek out the matching pairs, thereby enhancing a child's geometric intuition and three-dimensional manipulation skills. If the kit is to include every possible configuration of colors, how many cubes will the kit contain?

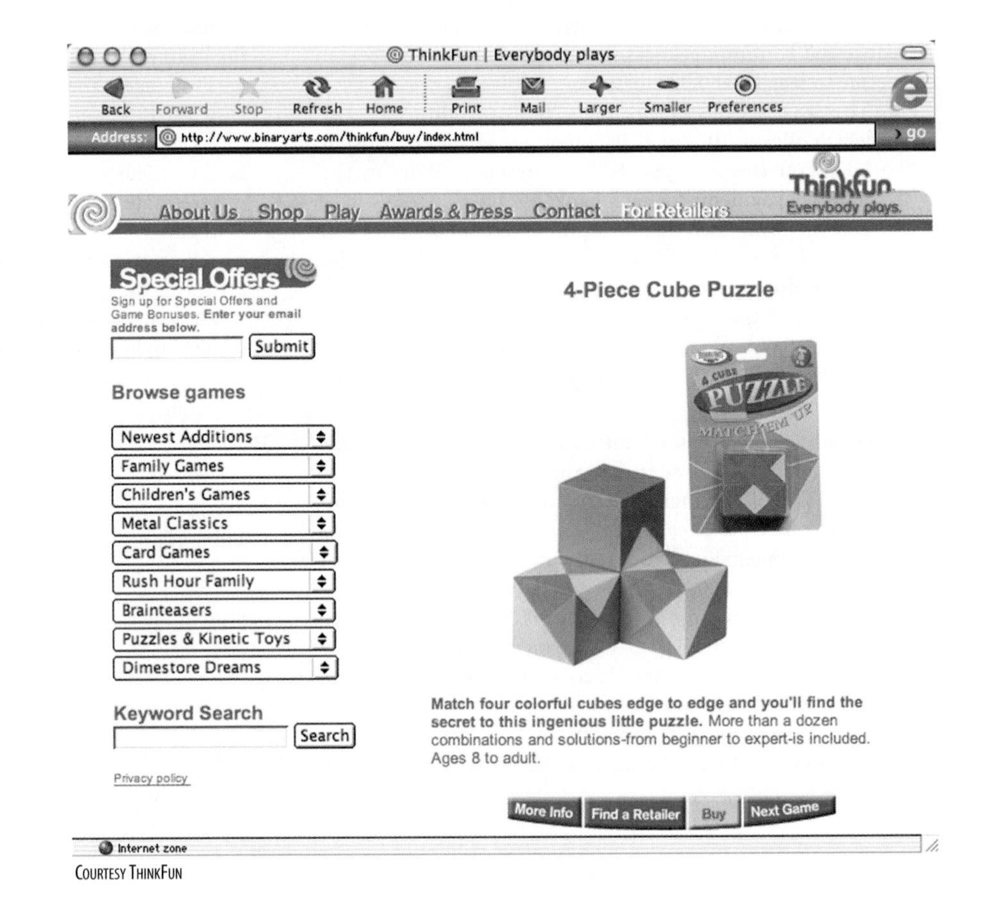

COURTESY THINKFUN

INTERNET RESOURCES FOR THIS CHAPTER

At the Web site, follow the path

Web site → Everything for Finite Math → Chapter 6

where you will find a detailed chapter summary you can print out, a true/false quiz, and a collection of sample test questions.

Introduction

The theory of sets is the foundation for most of mathematics. It also has direct applications—for example, searching computer databases. We use set theory extensively in Chapter 7, and so much of this chapter revolves around the idea of a **set of outcomes** of a procedure, such as rolling a pair of dice or choosing names from a database. Also important in probability is the theory of **counting** the number of elements in a set, which is called **combinatorics.**

To show that counting elements is not a trivial proposition, consider the following example. The betting game Lotto (used in many state lotteries) has you pick six numbers from some range—say, 1–55. If your six numbers match the six numbers chosen in the "official drawing," you win the top prize. What is your chance of winning the top prize? How many Lotto tickets would you need to buy to guarantee that you will win? By the end of this chapter, we will be able to answer these questions.

6.1 Sets and Set Operations

We assume that you're already somewhat familiar with sets. In this section we review the basic ideas of set theory, with examples in applications that recur throughout the rest of the chapter.

Sets

Sets and Elements

A **set** is a collection of items, referred to as the **elements** of the set.

Quick Examples

We usually use a capital letter to name a set and braces to enclose the elements of a set.	$W = \{\text{Amazon, eBay, Apple}\}$ $N = \{1, 2, 3, \ldots\}$
$x \in A$ means that x is an element of the set A. If x is not an element of A, we write $x \notin A$.	Amazon $\in W$ (W as above) Microsoft $\notin W$ $2 \in N$
$B = A$ means that A and B have the same elements. The order in which the elements are listed does not matter.	$\{5, -9, 1, 3\} = \{-9, 1, 3, 5\}$ $\{1, 2, 3, 4\} \neq \{1, 2, 3, 6\}$
$B \subseteq A$ means that B is a **subset** of A; every element of B is also an element of A.	$\{\text{eBay, Apple}\} \subseteq W$ $\{1, 2, 3, 4\} \subseteq \{1, 2, 3, 4\}$

$B \subset A$ means that B is a **proper subset** of A: $B \subseteq A$, but $B \neq A$.	$\{\text{eBay, Apple}\} \subset W$ $\{1, 2, 3, 4\} \not\subset \{1, 2, 3, 4\}$ $\{1, 2, 3\} \subset N$
$\varnothing$ is the **empty set,** the set containing no elements. It is a subset of every set.	$\varnothing \subseteq W$ $\varnothing \subset W$
A **finite** set has finitely many elements. An **infinite** set does not have finitely many elements.	$W = \{\text{Amazon, eBay, Apple}\}$ is a finite set. $N = \{1, 2, 3, \ldots\}$ is an infinite set.

Example 1 • Sets of Outcomes

a. If we toss a coin and observe which side faces up, there are two possible outcomes, heads (H) and tails (T). The **set of outcomes** of tossing a coin once can be written

$$S = \{\text{H, T}\}$$

b. Suppose we roll a die that has faces numbered 1 through 6 as usual and observe which number faces up. The set of outcomes *could* be represented as

$$S = \{⚀, ⚁, ⚂, ⚃, ⚄, ⚅\}$$

However, we can much more easily write

$$S = \{1, 2, 3, 4, 5, 6\}$$

Example 2 • Two Dice: Distinguishable and Indistinguishable

a. Suppose we have two dice that we can distinguish in some way—say, one is tan and one is purple. If we roll both dice, what is the set of outcomes?

b. Describe the set of outcomes if the dice are indistinguishable.

Solution

a. A systematic way of laying out the set of outcomes for a distinguishable pair of dice is shown in Figure 1.

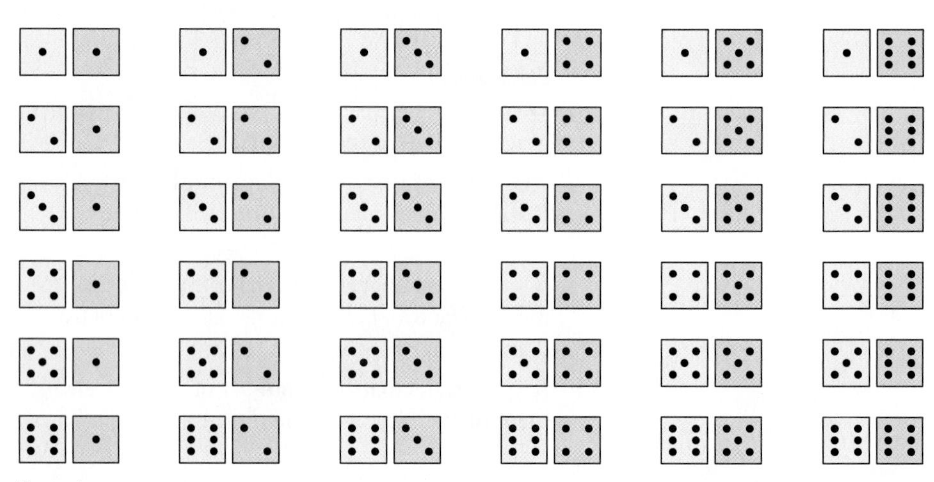

Figure 1

In the first row, all tan dice show a 1; in the second row, a 2; in the third row, a 3; and so on. Similarly, in the first column, all purple dice show a 1; in the second column, a 2; and so on. The diagonal pairs (top left to bottom right) show all the "doubles." Using the picture as a guide, we can write the set of 36 outcomes as follows:

$$S = \left\{ \begin{array}{l} (1,1),(1,2),(1,3),(1,4),(1,5),(1,6), \\ (2,1),(2,2),(2,3),(2,4),(2,5),(2,6), \\ (3,1),(3,2),(3,3),(3,4),(3,5),(3,6), \\ (4,1),(4,2),(4,3),(4,4),(4,5),(4,6), \\ (5,1),(5,2),(5,3),(5,4),(5,5),(5,6), \\ (6,1),(6,2),(6,3),(6,4),(6,5),(6,6) \end{array} \right\}$$

Distinguishable dice

Notice that S is also the set of outcomes if we roll a single die twice.

b. If the dice are truly indistinguishable, we will have no way of knowing which die is which once they are rolled. Think of placing two identical dice in a closed box and then shaking the box. When we look inside afterward, there is no way to tell which die is which. (If we make a small marking on one of the dice or somehow keep track of it as it bounces around, we are *distinguishing* the dice.) We regard two dice as **indistinguishable** if we make no attempt to distinguish them. Referring to the set of outcomes in part (a), we have to say that, for example, (1, 3) and (3, 1) represent the same outcome (one die shows a 3 and the other a 1). Since the set of outcomes should contain each outcome only once, we can remove (3, 1). Following this approach gives the following smaller set of outcomes:

$$S = \left\{ \begin{array}{r} (1,1),(1,2),(1,3),(1,4),(1,5),(1,6), \\ (2,2),(2,3),(2,4),(2,5),(2,6), \\ (3,3),(3,4),(3,5),(3,6), \\ (4,4),(4,5),(4,6), \\ (5,5),(5,6), \\ (6,6) \end{array} \right\}$$

Indistinguishable dice

Example 3 • Set-Builder Notation

Let $B = \{0, 2, 4, 6, 8\}$. B is the set of all nonnegative even integers less than 10. If we don't want to list the individual elements of B, we can instead use *set-builder notation,* and write

$$B = \{n \mid n \text{ is a nonnegative even integer less than } 10\}$$

This is read "B is the set of all n such that n is a nonnegative even integer less than 10." Here is the correspondence between the words and the symbols:

B is | the set of | all n | such that | n is a nonnegative even integer less than 10

$$B = \{n \mid n \text{ is a nonnegative even integer less than } 10\}$$

Before we go on . . . Note that the **nonnegative** integers *include* zero, whereas the **positive** integers *exclude* zero.

Venn Diagrams

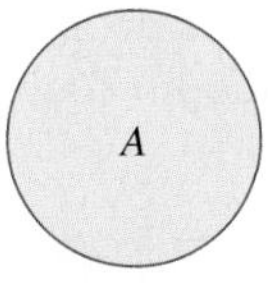

We can visualize sets and relations between sets using **Venn diagrams.** In a Venn diagram, we represent a set as a region, often a disk (Figure 2).

A

Figure 2

The elements of A are the points inside the region. The following Venn diagrams illustrate the relations we've discussed so far.

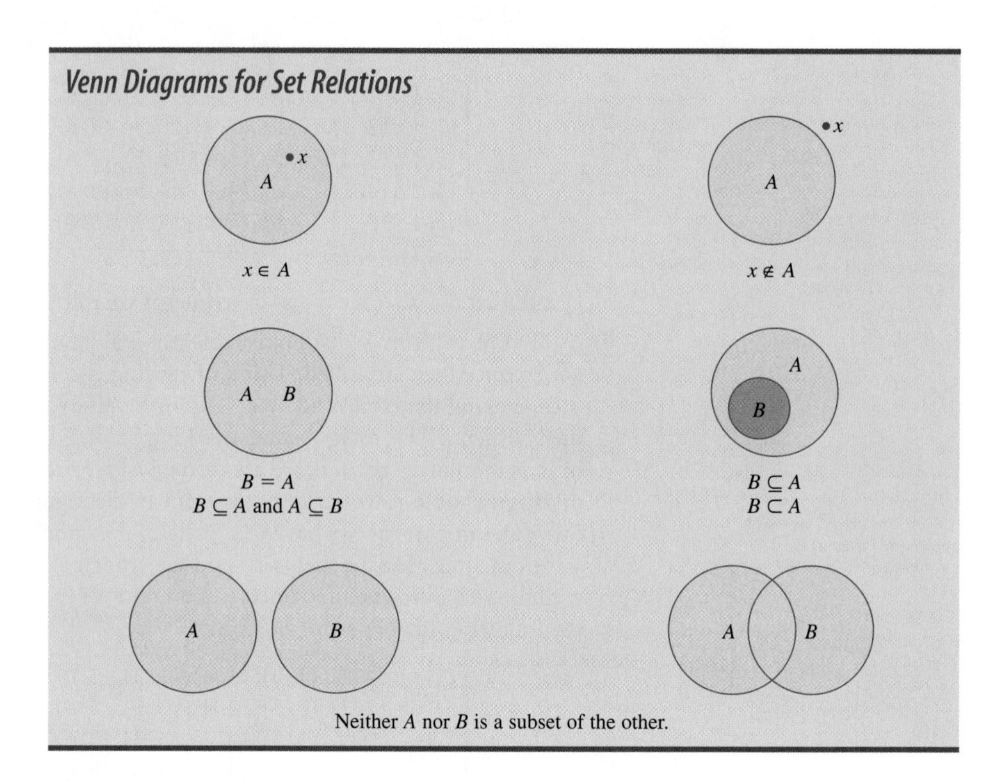

Note Although the diagram for $B \subseteq A$ suggests a proper subset, it is customary to use the same diagram for both subsets and proper subsets.

Example 4 • Customer Interests

NobelBooks.com (a fierce competitor of OHaganBooks.com) maintains a database of customers and the types of books they have purchased. In the company's database is the set of customers

$$S = \{\text{Einstein, Bohr, Millikan, Heisenberg, Schrödinger, Dirac}\}$$

A search of the database for customers who have purchased cookbooks yields the subset

$$A = \{\text{Einstein, Bohr, Heisenberg, Dirac}\}$$

Another search, this time for customers who have purchased mysteries, yields the subset

$$B = \{\text{Bohr, Heisenberg, Schrödinger}\}$$

NobelBooks.com wants to promote a new combination mystery/cookbook and wants to target two subsets of customers: those who have purchased either cookbooks or mysteries (or both) and, for additional promotions, those who have purchased both cookbooks and mysteries. Name the customers in each of these subsets.

Solution We can picture the database and the two subsets using the Venn diagram in Figure 3.

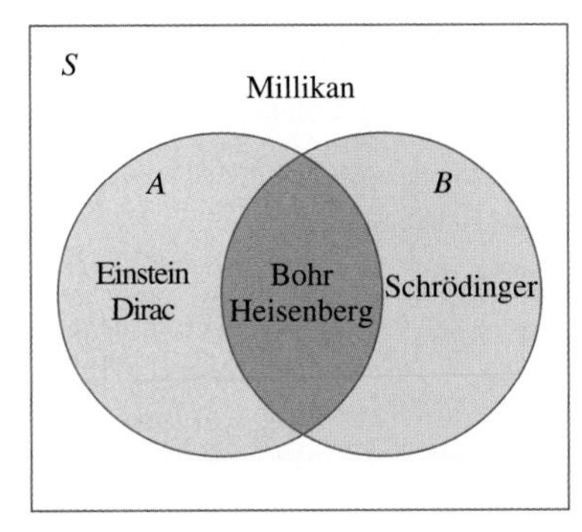

Figure 3

The set of customers who have purchased either cookbooks or mysteries (or both) consists of the customers who are in A or B or both: Einstein, Bohr, Heisenberg, Schrödinger, and Dirac. The set of customers who have purchased both cookbooks and mysteries consists of the customers in the overlap of A and B, Bohr and Heisenberg.

Set Operations

Set Operations

$A \cup B$ is the **union** of A and B, the set of all elements that are either in A or in B (or in both):

$$A \cup B = \{x \mid x \in A \text{ or } x \in B\}$$

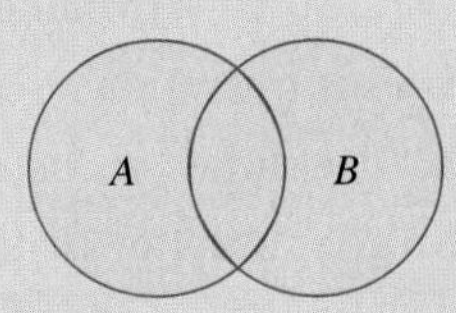

$A \cap B$ is the **intersection** of A and B, the set of all elements that are common to A and B:

$$A \cap B = \{x \mid x \in A \text{ and } x \in B\}$$

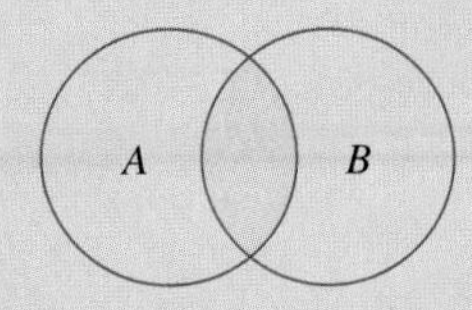

If $A \subseteq S$, then A' is the **complement** of A in S, the set of all elements of S not in A:

$$A' = \{x \in S \mid x \notin A\}$$

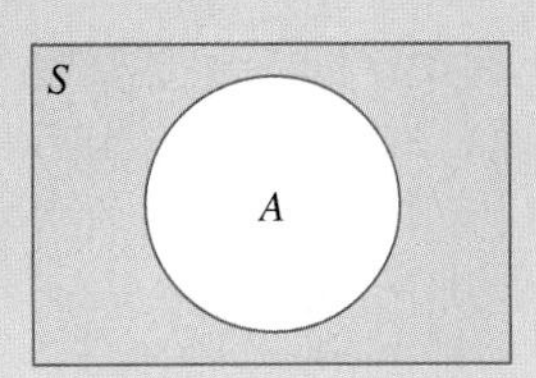

Logical Equivalents

Union For an element to be in $A \cup B$, it must be in A **or** in B.

Intersection For an element to be in $A \cap B$, it must be in A **and** in B.

Complement For an element to be in A', it must be in S but **not** in A.

Quick Examples

If $S = \{a, b, c, d, e, f, g\}$, $A = \{a, b, c, d\}$, and $B = \{c, d, e, f\}$, then

$A \cup B = \{a, b, c, d, e, f\}$
$A \cap B = \{c, d\}$
$A' = \{e, f, g\}$

Note Mathematicians always use *or* in its *inclusive* sense: one thing or another *or both*.

Example 5 • Customer Interests

NobelBooks.com maintains a database of customers and the types of books they have purchased. In the company's database is the set of customers

$S = \{\text{Einstein, Bohr, Millikan, Heisenberg, Schrödinger, Dirac}\}$

A search of the database for customers who have purchased cookbooks yields the subset

$A = \{\text{Einstein, Bohr, Heisenberg, Dirac}\}$

Another search, this time for customers who have purchased mysteries, yields the subset

$B = \{\text{Bohr, Heisenberg, Schrödinger}\}$

A third search, for customers who had registered with the site but not used their first-time customer discount yields the subset

$C = \{\text{Milliken}\}$

Use set operations to describe the following subsets:

a. The subset of customers who have purchased either cookbooks or mysteries

b. The subset of customers who have purchased both cookbooks and mysteries

c. The subset of customers who have not purchased cookbooks

d. The subset of customers who have purchased cookbooks but not used their first-time customer discount

Solution Figure 4 shows two alternative Venn diagram representations of the database.

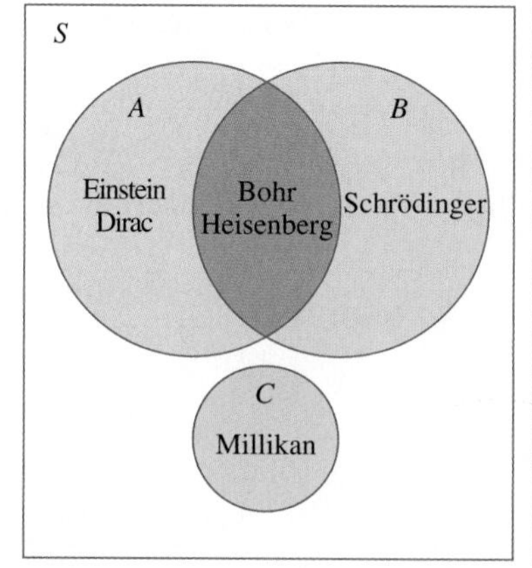

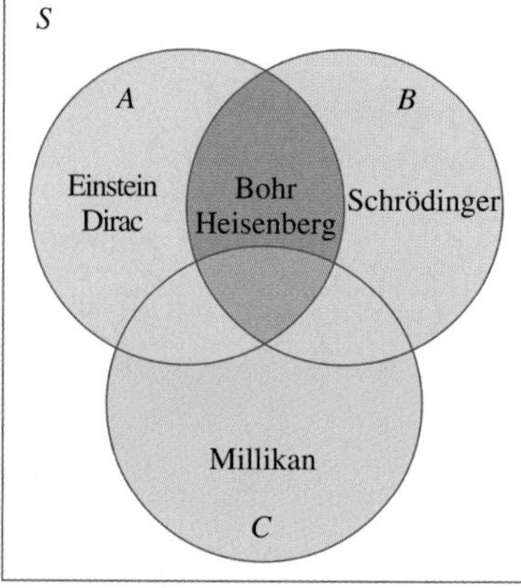

Figure 4

a. The subset of customers who have bought either cookbooks *or* mysteries is

$$A \cup B = \{\text{Einstein, Bohr, Heisenberg, Schrödinger, Dirac}\}$$

b. The subset of customers who have bought both cookbooks *and* mysteries is

$$A \cap B = \{\text{Bohr, Heisenberg}\}$$

c. The subset of customers who have *not* bought cookbooks is

$$A' = \{\text{Millikan, Schrödinger}\}$$

d. The subset of customers who have bought cookbooks but not used their first-time purchase discount is the empty set

$$A \cap C = \varnothing$$

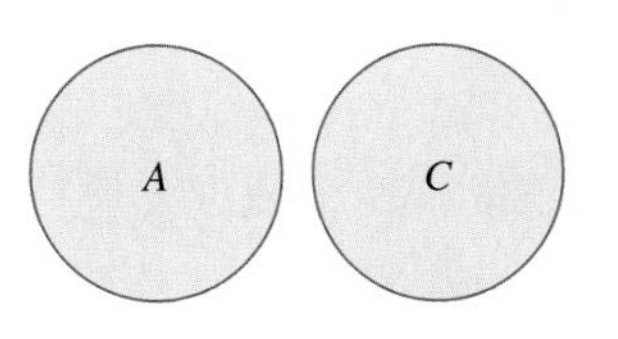

Figure 5

When an intersection is empty, we say that the two sets are **disjoint.** In a Venn diagram disjoint sets are drawn as regions that don't overlap, as in Figure 5. This is one of those times when it is useful to consider the empty set to be a valid set. If we did not, then we would have to say that $A \cap C$ was defined only when A and C had something in common, and having to deal with this special case would quickly get tiresome.

Before we go on . . . Computer databases and the Web can be searched using Boolean searches. These are search requests using *and, or,* and *not.* Using *and* gives the intersection of separate searches, using *or* gives the union, and using *not* gives the complement. In the next section we'll see an example of how the Web search engine AltaVista® allows such searches.

Question When we take the complement A' we seem to need another set, S, because

$$A' = \{x \in S \mid x \notin A\}$$

Why can't we just forget about S and take A' to be the set of all things not in A?

Answer This would amount to taking S to be the set of *all things.* Although this is tempting, talking about entities such as the "set of all things" leads to paradoxes.[1] Instead, we think of S as being the *set of all objects under consideration,* or the *universe of discourse,* and call S the **universal set** for the discussion. In Example 5, S is the set of all customers in the NobelBooks.com database. When we search the Web, S is the set of all Web pages. When talking about integers, we take S to be the set of all integers. In other words, our choice of universal set depends on the context.

Cartesian Product

There is one more set operation we need to discuss.

[1]The most famous such paradox is called "Russell's Paradox," after the mathematical logician (and philosopher and pacifist) Bertrand Russell. It goes like this: If there were a set of all things, then there would also be a (smaller) set of all sets. Call it S. Now, since S is the *set* of *all* sets, it must contain itself as a member. In other words, $S \in S$. Let P be the subset of S consisting of all sets that are *not* members of themselves. Now we pose the following question: Is P a member of itself? If it is, then, since it is the set of all sets that are *not* members of themselves, it is not. On the other hand, if it is *not* a member of itself, then it qualifies as an element of P. In other words, it *is* a member of itself! Since neither can be true, something is wrong. What is wrong is the assumption that there is such a thing as the set of all sets or the set of all things.

Cartesian Product

The **Cartesian product** of two sets, A and B, is the set of all ordered pairs (a, b) with $a \in A$ and $b \in B$:

$$A \times B = \{(a, b) \mid a \in A \text{ and } b \in B\}$$

In words, $A \times B$ is the set of all ordered pairs whose first component is in A and whose second component is in B.

Quick Examples

1. If $A = \{\text{a, b}\}$ and $B = \{1, 2, 3\}$, then

$$A \times B = \{(\text{a}, 1), (\text{a}, 2), (\text{a}, 3), (\text{b}, 1), (\text{b}, 2), (\text{b}, 3)\}$$

2. If $S = \{\text{H, T}\}$, then

$$S \times S = \{(\text{H, H}), (\text{H, T}), (\text{T, H}), (\text{T, T})\}$$

In other words, if S is the set of outcomes of tossing a coin once, then $S \times S$ is the set of outcomes of tossing a coin twice.

3. If $S = \{1, 2, 3, 4, 5, 6\}$, then

$$S \times S = \left\{\begin{matrix} (1, 1), (1, 2), (1, 3), (1, 4), (1, 5), (1, 6), \\ (2, 1), (2, 2), (2, 3), (2, 4), (2, 5), (2, 6), \\ (3, 1), (3, 2), (3, 3), (3, 4), (3, 5), (3, 6), \\ (4, 1), (4, 2), (4, 3), (4, 4), (4, 5), (4, 6), \\ (5, 1), (5, 2), (5, 3), (5, 4), (5, 5), (5, 6), \\ (6, 1), (6, 2), (6, 3), (6, 4), (6, 5), (6, 6) \end{matrix}\right\}$$

In other words, if S is the set of outcomes of a rolling a die once, then $S \times S$ is the set of outcomes of rolling a die twice (or rolling two distinguishable dice).

4. If $A = \{\text{red, yellow}\}$ and $B = \{\text{Mustang, Firebird}\}$, then

$$A \times B = \{(\text{red, Mustang}), (\text{red, Firebird}), (\text{yellow, Mustang}), (\text{yellow, Firebird})\}$$

which we might also write as

$$A \times B = \{\text{red Mustang, red Firebird, yellow Mustang, yellow Firebird}\}$$

Example 6 • Representing Cartesian Products

The manager of an automobile dealership has collected data on the number of pre-owned Acura, Infiniti, Lexus, and Mercedes cars the dealership has from the 2000, 2001, and 2002 model years. In entering this information on a spreadsheet, the manager would like to have each spreadsheet cell represent a particular year and make. Describe this set of cells.

Solution Since each cell represents a year and a make, we can think of the cell as a pair (year, make), as in (2000, Acura). Thus, the set of cells can be thought of as a Cartesian product:

$$Y = \{2000, 2001, 2002\} \qquad \text{Year of car}$$

$$M = \{\text{Acura, Infiniti, Lexus, Mercedes}\} \qquad \text{Make of car}$$

$$Y \times M = \left\{\begin{matrix} (2000, \text{Acura}), (2000, \text{Infiniti}), (2000, \text{Lexus}), (2000, \text{Mercedes}), \\ (2001, \text{Acura}), (2001, \text{Infiniti}), (2001, \text{Lexus}), (2001, \text{Mercedes}), \\ (2002, \text{Acura}), (2002, \text{Infiniti}), (2002, \text{Lexus}), (2002, \text{Mercedes}) \end{matrix}\right\} \qquad \text{Cell}$$

Thus, the manager might arrange the spreadsheet as follows:

	A	B	C	D	E
1		Acura	Infiniti	Lexus	Mercedes
2	2000	(2000 Ac)	(2000 Inf)	(2000 Lex)	(2000 Merc)
3	2001	(2001 Ac)	(2001 Inf)	(2001 Lex)	(2001 Merc)
4	2002	(2002 Ac)	(2002 Inf)	(2002 Lex)	(2002 Merc)

The highlighting shows the 12 cells to be filled in, representing the numbers of cars of each year and make. For example, in cell B2 should go the number of 2000 Acuras the dealership has. (This arrangement is consistent with the matrix notation in Chapter 2. We could also have used the elements of Y as column labels along the top and the elements of M as row labels down the side.)

Before we go on . . . We can also represent the Cartesian product $Y \times M$ as a set of points in the xy plane (Cartesian plane) as shown in Figure 6.

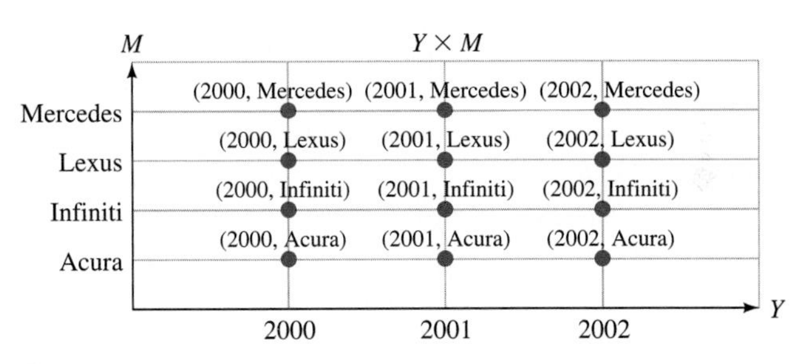

Figure 6

Guideline: The Many Meanings of And

Question Suppose A is the set of actors and B is the set of all baseball players. Then the set of all actors *and* baseball players is $A \cap B$, right?

Answer Wrong. The fact that the word *and* appears in the description of a set does not always mean that the set is an intersection; the word *and* can mean different things in different contexts. $A \cap B$ refers to the set of elements that are in both A and B, hence to actors who are also baseball players. On the other hand, the set of all actors and baseball players is the set of people who are either actors or baseball players (or both), which is $A \cup B$. We can use the word *and* to describe both sets:

$A \cap B =$ {people who are both actors *and* baseball players}

$A \cup B =$ {people who are actors *or* baseball players} $=$ {all actors *and* baseball players}

6.1 EXERCISES

In Exercises 1–16, list the elements in each of the sets.

1. The set F consisting of the four seasons
2. The set A consisting of the authors of this book
3. The set I of all positive integers no greater than 6
4. The set N of all negative integers greater than -3
5. $A = \{n \mid n \text{ is a positive integer and } 0 \leq n \leq 3\}$
6. $A = \{n \mid n \text{ is a positive integer and } 0 < n < 8\}$
7. $B = \{n \mid n \text{ is an even positive integer and } 0 \leq n \leq 8\}$
8. $B = \{n \mid n \text{ is an odd positive integer and } 0 \leq n \leq 8\}$
9. The set of all outcomes of tossing a pair of (a) distinguishable coins and (b) indistinguishable coins
10. The set of outcomes of tossing three (a) distinguishable coins and (b) indistinguishable coins
11. The set of all outcomes of rolling two distinguishable dice such that the numbers add to 6
12. The set of all outcomes of rolling two distinguishable dice such that the numbers add to 8
13. The set of all outcomes of rolling two indistinguishable dice such that the numbers add to 6
14. The set of all outcomes of rolling two indistinguishable dice such that the numbers add to 8
15. The set of all outcomes of rolling two distinguishable dice such that the numbers add to 13
16. The set of all outcomes of rolling two distinguishable dice such that the numbers add to 1

In Exercises 17–20, draw a Venn diagram that illustrates the relationships among the given sets.

17. S = {eBay, Google, Amazon, OHaganBooks, Hotmail}, A = {Amazon, OHaganBooks}, B = {eBay, Amazon}, C = {Amazon, Hotmail}
18. S = {Apple, Dell, Gateway, Pomegranate, Compaq}, A = {Gateway, Pomegranate, Compaq}, B = {Dell, Gateway, Pomegranate, Compaq}, C = {Apple, Dell, Compaq}
19. S = {eBay, Google, Amazon, OHaganBooks, Hotmail}, A = {Amazon, Hotmail}, B = {eBay, Google, Amazon, Hotmail}, C = {Amazon, Hotmail}
20. S = {Apple, Dell, Gateway, Pomegranate, Compaq}, A = {Apple, Dell, Pomegranate, Compaq}, B = {Pomegranate}, C = {Pomegranate}

In Exercises 21–34, find each set. Let A = {June, Janet, Jill, Justin, Jeffrey, Jello}, B = {Janet, Jello, Justin}, and C = {Sally, Solly, Molly, Jolly, Jello}.

21. $A \cup B$
22. $A \cup C$
23. $A \cup \varnothing$
24. $B \cup \varnothing$
25. $A \cup (B \cup C)$
26. $(A \cup B) \cup C$
27. $C \cap B$
28. $C \cap A$
29. $A \cap \varnothing$
30. $\varnothing \cap B$
31. $(A \cap B) \cap C$
32. $A \cap (B \cap C)$
33. $(A \cap B) \cup C$
34. $A \cup (B \cap C)$

In Exercises 35–42, A = {small, medium, large}, B = {blue, green}, and C = {triangle, square}.

35. List the elements of $A \times C$.
36. List the elements of $B \times C$.
37. List the elements of $A \times B$.
38. The elements of $A \times B \times C$ are the ordered triples (a, b, c), with $a \in A$, $b \in B$, and $c \in C$. List all the elements of $A \times B \times C$.
39. Represent $B \times C$ as cells in a spreadsheet (as in Example 6).
40. Represent $A \times C$ as cells in a spreadsheet (as in Example 6).
41. Represent $A \times B$ as cells in a spreadsheet (as in Example 6).
42. Represent $A \times A$ as cells in a spreadsheet (as in Example 6).

In Exercises 43–46, $A = \{H, T\}$ is the set of outcomes when a coin is tossed, and $B = \{1, 2, 3, 4, 5, 6\}$ is the set of outcomes when a die is rolled. Write each set in terms of A and/or B and list its elements.

43. The set of outcomes when a die is rolled and then a coin tossed
44. The set of outcomes when a coin is tossed twice
45. The set of outcomes when a coin is tossed three times
46. The set of outcomes when a coin is tossed twice and then a die is rolled

In Exercises 47–52, S is the set of outcomes when two distinguishable dice are rolled, E is the subset of outcomes in which at least one die shows an even number, and F is the subset of outcomes in which at least one die shows an odd number. List the elements in each subset.

47. E'
48. F'
49. $(E \cup F)'$
50. $(E \cap F)'$
51. $E' \cup F'$
52. $E' \cap F'$

In Exercises 53–60, use Venn diagrams to persuade yourself of the following identities for subsets A, B, and C of S.

53. $(A \cup B)' = A' \cap B'$ (De Morgan's law)
54. $(A \cap B)' = A' \cup B'$ (De Morgan's law)
55. $(A \cap B) \cap C = A \cap (B \cap C)$ (Associative law)
56. $(A \cup B) \cup C = A \cup (B \cup C)$ (Associative law)
57. $A \cup (B \cap C) = (A \cup B) \cap (A \cup C)$ (Distributive law)
58. $A \cap (B \cup C) = (A \cap B) \cup (A \cap C)$ (Distributive law)
59. $S' = \varnothing$
60. $\varnothing' = S$

APPLICATIONS

Databases A freelance computer consultant keeps a database of her clients, which contains the names

$S = \{\text{Acme, Brothers, Crafts, Dion, Effigy, Floyd, Global, Hilbert}\}$

The following clients owe her money:

$$A = \{\text{Acme, Crafts, Effigy, Global}\}$$

The following clients have done at least $10,000 worth of business with her:

$$B = \{\text{Acme, Brothers, Crafts, Dion}\}$$

The following clients have employed her in the last year:

$$C = \{\text{Acme, Crafts, Dion, Effigy, Global, Hilbert}\}$$

In Exercises 61–68, a subset of clients is described that the consultant could find using her database. Write each subset in terms of A, B, and C and list the clients in that subset.

61. The clients who owe her money and have done at least $10,000 worth of business with her

62. The clients who owe her money or have done at least $10,000 worth of business with her

63. The clients who have done at least $10,000 worth of business with her or have employed her in the last year

64. The clients who have done at least $10,000 worth of business with her and have employed her in the last year

65. The clients who do not owe her money and have employed her in the last year

66. The clients who do not owe her money or have employed her in the last year

67. The clients who owe her money, have not done at least $10,000 worth of business with her, and have not employed her in the last year

68. The clients who either do not owe her money, have done at least $10,000 worth of business with her, or have employed her in the last year

69. Boat Sales The amount spent in recreational boats and accessories in the United States increased during the period 1998–2001. You are given data on revenues from sales of used boats, new boats, and new boating accessories for each of the years 1998 through 2001. How would you represent these data in a spreadsheet? The cells in your spreadsheet represent elements of which set?

SOURCE: National Marine Manufacturers Association/*New York Times,* January 10, 2002, p. C1.

70. Health Care Spending Spending in most categories of health care in the United States increased dramatically in the last 30 years of the 1900s. You are given data showing total spending on prescription drugs, nursing homes, hospital care, and professional services for each of the last three decades of the 1900s. How would you represent these data in a spreadsheet? The cells in your spreadsheet represent elements of which set?

SOURCE: Department of Health and Human Services/*New York Times,* January 8, 2002, p. A14.

COMMUNICATION AND REASONING EXERCISES

71. Explain, illustrating by means of an example, why $(A \cap B) \cup C \neq A \cap (B \cup C)$.

72. Explain, making reference to operations on sets, why the statement "He plays soccer or rugby and cricket" is ambiguous.

73. You are searching online for techno music that is neither European nor Dutch. In set notation, for which set of music files are you searching?

(A) Techno ∩ (European ∩ Dutch)′

(B) Techno ∩ (European ∪ Dutch)′

(C) Techno ∪ (European ∩ Dutch)′

(D) Techno ∪ (European ∪ Dutch)′

74. You would like to see either a World War II movie or one that is based on a comic book character but does not feature aliens. Which set of movies are you interested in seeing?

(A) WWII ∩ (Comix ∩ Aliens′)

(B) WWII ∩ (Comix ∪ Aliens′)

(C) WWII ∪ (Comix ∩ Aliens′)

(D) WWII ∪ (Comix ∪ Aliens′)

75. Explain the meaning of a universal set and give two different universal sets that could be used in a discussion about sets of positive integers.

76. Is the set of outcomes when two indistinguishable dice are rolled (Example 2) a Cartesian product of two sets? If so, which two sets; if not, why not?

77. Design a database scenario that leads to the following statement: To keep the factory operating at maximum capacity, the plant manager should select the suppliers in $A \cap (B \cup C')$.

78. Design a database scenario that leads to the following statement: To keep her customers happy, the bookstore owner should stock periodicals in $A \cup (B \cap C')$.

79. Rewrite in set notation: She prefers movies that are not violent, are shorter than 2 hours, and have neither a tragic ending nor an unexpected ending.

80. Rewrite in set notation: He will cater for any event as long as there are no more than 1000 people, it lasts for at least 3 hours, and it is within a 50-mile radius of Toronto.

6.2 Cardinality

Cardinality

If A is a finite set, then its **cardinality** is

$$n(A) = \text{number of elements in } A$$

Quick Examples

1. Let $S = \{a, b, c\}$. Then $n(S) = 3$.
2. Let S be the set of outcomes when two distinguishable dice are rolled. Then $n(S) = 36$ (see Example 2 in Section 6.1).
3. $n(\varnothing) = 0$ because the empty set has no elements.

Counting the elements in a small, simple set is straightforward. To count the elements in a large, complicated set, we try to describe the set as built of simpler sets, using the set operations. We then need to know how to calculate the number of elements in, for example, a union, based on the number of elements in the simpler sets whose union we are taking.

Question How can we calculate $n(A \cup B)$ if we know $n(A)$ and $n(B)$?

Answer Our first guess might be that $n(A \cup B)$ is $n(A) + n(B)$. But consider a simple example. Let

$$A = \{a, b, c\} \quad \text{and} \quad B = \{b, c, d\}$$

Then $A \cup B = \{a, b, c, d\}$, so $n(A \cup B) = 4$, but $n(A) + n(B) = 3 + 3 = 6$. Why does $n(A) + n(B)$ give the wrong answer? Because the elements b and c are counted twice, once for being in A and again for being in B. To correct for this overcounting, we need to subtract the number of elements that get counted twice, which is the number of elements that A and B have in common, or $n(A \cap B) = 2$ in this case. This argument leads to the following general formula.

Cardinality of a Union

If A and B are finite sets, then

$$n(A \cup B) = n(A) + n(B) - n(A \cap B)$$

In particular, if A and B are disjoint (meaning that $A \cap B = \varnothing$), then

$$n(A \cup B) = n(A) + n(B)$$

(When A and B are disjoint we say that $A \cup B$ is a **disjoint union.**)

Quick Examples

1. If $A = \{a, b, c, d\}$ and $B = \{b, c, d, e, f\}$, then

$$n(A \cup B) = n(A) + n(B) - n(A \cap B) = 4 + 5 - 3 = 6$$

In fact, $A \cup B = \{a, b, c, d, e, f\}$.

2. If $A = \{a, b, c\}$ and $B = \{d, e, f\}$, then $A \cap B = \varnothing$, so

$$n(A \cup B) = n(A) + n(B) = 3 + 3 = 6$$

Example 1 • Web Searches

In May 2002 a search using the Web search engine AltaVista for the phrase "NASA space station" yielded 800 Web sites containing that phrase. A search for the phrase "NASA mars mission" yielded 201 sites. A search for sites containing both phrases yielded only one site. How many Web sites contained either "NASA space station" or "NASA mars mission" or both?

Solution Let A be the set of sites containing "NASA space station" and let B be the set of sites containing "NASA mars mission". We are told that

$$n(A) = 800$$

$$n(B) = 201$$

$$n(A \cap B) = 1$$ "NASA space station" AND "NASA mars mission"

The formula for the cardinality of the union tells us that

$$n(A \cup B) = n(A) + n(B) - n(A \cap B) = 800 + 201 - 1 = 1000$$

So, 1000 sites in the AltaVista database contained one or both of the phrases "NASA space station" and "NASA mars mission". (This was subsequently confirmed by doing a search for "NASA space station" or "NASA mars mission".)

Before we go on . . . Each search engine has a different way of specifying a search for a union or an intersection. At AltaVista or Google, a search for "NASA space station" OR "NASA mars mission" finds sites containing either phrase (or both), so finds $A \cup B$. Many search engines also support the *and* operator, so you can find $A \cap B$ by searching for "NASA space station" AND "NASA mars mission".

Although the formula $n(A \cup B) = n(A) + n(B) - n(A \cap B)$ always holds, you may sometimes find that, in an actual search, the numbers don't add up. AltaVista explains that the number reported may be only estimates of the actual number of sites in their database containing the search words.[2]

Question What about $n(A \cap B)$?

Answer The formula for the cardinality of a union can also be thought of as a formula for the cardinality of an intersection. We can solve for $n(A \cap B)$ to get

$$n(A \cap B) = n(A) + n(B) - n(A \cup B)$$

In fact, we can think of this formula as an equation relating four quantities. If we know any three of them, we can use the equation to find the fourth (see Example 2).

Question What about complements?

Answer We can get a formula for the cardinality of a complement as follows: If S is our universal set and $A \subseteq S$, then S is the disjoint union of A and its complement. That is,

$$S = A \cup A' \quad \text{and} \quad A \cap A' = \varnothing$$

Applying the cardinality formula for a disjoint union, we get

$$n(S) = n(A) + n(A')$$

---

[2]Email correspondence January 2002.

If we solve for $n(A')$ or for $n(A)$, we get the following formulas.

Cardinality of a Complement

If S is a finite universal set and A is a subset of S, then

$$n(A') = n(S) - n(A) \quad \text{and} \quad n(A) = n(S) - n(A')$$

Quick Example

If $S = \{a, b, c, d, e, f\}$ and $A = \{a, b, c, d\}$, then

$$n(A') = n(S) - n(A) = 6 - 4 = 2$$

In fact, $A' = \{e, f\}$.

Example 2 • Cookbooks

In December 2001 a search at Amazon.com found 7924 cookbooks.[3] Of these, 1218 were American cookbooks, 304 were vegetarian cookbooks, and 1513 were either American or vegetarian. How many cookbooks were not American vegetarian cookbooks?

Solution Let S be the set of all 7924 cookbooks, let A be the set of American cookbooks, and let B be the set of vegetarian cookbooks. We wish to find the size of the complement of the set of American vegetarian cookbooks—that is, $n[(A \cap B)']$. Using the formula for the cardinality of a complement, we have

$$n[(A \cap B)'] = n(S) - n(A \cap B) = 7924 - n(A \cap B)$$

To find $n(A \cap B)$, we use the formula for the cardinality of a union:

$$n(A \cup B) = n(A) + n(B) - n(A \cap B)$$

Substituting the values we were given, we find

$$1513 = 1218 + 304 - n(A \cap B)$$

which we can solve to get

$$n(A \cap B) = 1218 + 304 - 1513 = 9$$

Therefore,

$$n[(A \cap B)'] = 7924 - n(A \cap B) = 7924 - 9 = 7915$$

So, 7915 cookbooks were not American vegetarian cookbooks.

[3]Precisely, it found that many books mentioned the word *cookbook,* not all of which were food cookbooks.

Example 3 • Sales of Recreational Boats

The following table shows sales of recreational boats in the United States during the period 1999–2001:

	Motor Boats	Jet Skis	Sailboats	Total
1999	330,000	100,000	20,000	450,000
2000	340,000	100,000	20,000	460,000
2001	310,000	90,000	30,000	430,000
Total	980,000	290,000	70,000	1,340,000

Figures are approximate and represent new recreational boats sold. (Jet skis includes similar vehicles, such as wave runners.) SOURCE: National Marine Manufacturers Association/*New York Times,* January 10, 2002, p. C1.

Let S be the set of *all* these recreational boats, let M be the set of all motor boats purchased during the stated period, let K be the set of recreational boats purchased in 2001, and let L be the set of sailboats purchased during the stated period. Describe the following sets and compute their cardinality:

a. $M \cap K'$ **b.** $(M \cap K)'$ **c.** $K \cup L$

Solution

a. K' is the set of all recreational boats that were purchased in 1999 or 2000. Thus, $M \cap K'$ is the set of motor boats that were purchased in 1999 or 2000. Referring to the leftmost column of the table, we find a total of $330{,}000 + 340{,}000 = 670{,}000$ boats in this set.

b. $M \cap K$ is the set of motor boats that were purchased in 2001, and $n(M \cap K) = 310{,}000$. The complement, $(M \cap K)'$, is the set of all boats that are either not motor boats or were purchased in 1999 or 2000, and it has cardinality

$$n(M \cap K)' = n(S) - n(M \cap K)$$
$$= 1{,}340{,}000 - 310{,}000 = 1{,}030{,}000$$

c. $K \cup L$ is the set of recreational boats that were either purchased in 2001 or were sailboats. The following table shows the relevant portion shaded:

	Motor Boats	Jet Skis	Sailboats	Total
1999	330,000	100,000	20,000	450,000
2000	340,000	100,000	20,000	460,000
2001	310,000	90,000	30,000	430,000
Total	980,000	290,000	70,000	1,340,000

Thus,

$$n(K \cup L) = n(K) + n(L) - n(K \cap L)$$
$$= 430{,}000 + 70{,}000 - 30{,}000 = 470{,}000$$

Before we go on . . . Note that, to obtain the number of elements in $K \cup L$ we did not simply add the totals 430,000 and 70,000; doing so would amount to asserting that $n(K \cup L) = n(K) + n(L)$, which is not correct unless K and L are disjoint.

Question What about the cardinality of a union of three or more sets, like $n(A \cup B \cup C)$?

Answer We can think of $A \cup B \cup C$ as a union of two sets, $(A \cup B)$ and C, and then analyze each piece, using the techniques we already have. Alternatively, there are formulas for the cardinalities of unions of any number of sets, but these formulas get more and more complicated as the number of sets grows. In many applications, like Example 4, we can use Venn diagrams instead.

Example 4 • Reading Lists

A survey of 300 college students found that 100 had read *War and Peace,* 120 had read *Crime and Punishment,* and 100 had read *The Brothers Karamazov.* It also found that 40 had read only *War and Peace,* 70 had read *War and Peace* but not *The Brothers Karamazov,* and 80 had read *The Brothers Karamazov* but not *Crime and Punishment.* Only 10 had read all three novels. How many had read none of these three novels?

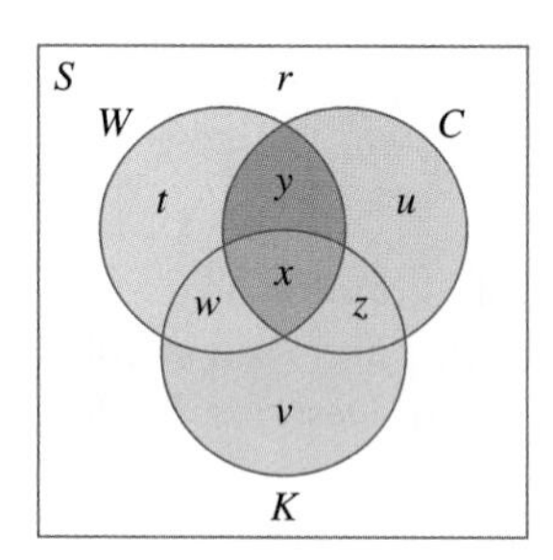

Figure 7

Solution There are four sets mentioned in the problem: the universe S consisting of the 300 students surveyed, the set W of students who had read *War and Peace,* the set C of students who had read *Crime and Punishment,* and the set K of students who had read *The Brothers Karamazov.* Figure 7 shows a Venn diagram representing these sets.

We have put labels in the various regions of the diagram to represent the number of students in each region. For example, x represents the number of students in $W \cap C \cap K$, which is the number of students who have read all three novels. We are told that this number is 10, so

$$x = 10$$

(You should draw the diagram for yourself and fill in the numbers as we go along.) We are also told that 40 students had read only *War and Peace,* so

$$t = 40$$

We are given none of the remaining regions directly. However, since 70 had read *War and Peace* but not *The Brothers Karamazov,* we see that t and y must add up to 70. Since we already know that $t = 40$, it follows that $y = 30$. Further, since a total of 100 students had read *War and Peace,* we have

$$x + y + t + w = 100$$

Substituting the known values of x, y, and t gives

$$10 + 30 + 40 + w = 100$$

so $w = 20$. Since 80 students had read *The Brothers Karamazov* but not *Crime and Punishment,* we see that $v + w = 80$, so $v = 60$ (since we know $w = 20$). We can now calculate z using the fact that a total of 100 students had read *The Brothers Karamazov:*

$$60 + 20 + 10 + z = 100$$

giving $z = 10$. Similarly, we can now get u using the fact that 120 students had read *Crime and Punishment:*

$$10 + 30 + 10 + u = 120$$

giving $u = 70$. Of the 300 students surveyed, we've now found $x + y + z + w + t + u + v = 240$. This leaves

$$r = 60$$

who had read none of the three novels.

Question We've covered all the operations except Cartesian product. What is $n(A \times B)$?

Answer Consider the following simple example:

$$A = \{\text{H, T}\}$$
$$B = \{1, 2, 3, 4, 5, 6\}$$

so that

$$A \times B = \{\text{H1, H2, H3, H4, H5, H6, T1, T2, T3, T4, T5, T6}\}$$

As we saw in Example 6 in Section 6.1, the elements of $A \times B$ can be arranged in a table or spreadsheet with $n(A) = 2$ rows and $n(B) = 6$ elements in each row:

	A	B	C	D	E	F	G
1		1	2	3	4	5	6
2	H	H1	H2	H3	H4	H5	H6
3	T	T1	T2	T3	T4	T5	T6

In a region with two rows and six columns, there are $2 \times 6 = 12$ cells. So,

$$n(A \times B) = n(A)n(B)$$

in this case. There is nothing particularly special about this example, however, and that formula holds true in general.

Cardinality of a Cartesian Product

If A and B are finite sets, then

$$n(A \times B) = n(A)n(B)$$

Quick Example

If $A = \{\text{a, b, c}\}$ and $B = \{\text{x, y, z, w}\}$, then

$$n(A \times B) = n(A)n(B) = 3 \times 4 = 12$$

Example 5 • Coin Tosses

a. If we toss a coin twice and observe the sequence of heads and tails, how many possible outcomes are there?

b. If we toss a coin three times, how many possible outcomes are there?

c. If we toss a coin ten times, how many possible outcomes are there?

Solution

a. Let $A = \{H, T\}$ be the set of possible outcomes when a coin is tossed once. The set of outcomes when a coin is tossed twice is $A \times A$, which has

$$n(A \times A) = n(A)n(A) = 2 \times 2 = 4$$

possible outcomes.

b. When a coin is tossed three times, we can think of the set of outcomes as the product of the set of outcomes for the first two tosses, which is $A \times A$, and the set of outcomes for the third toss, which is just A. The set of outcomes for the three tosses is then $(A \times A) \times A$, which we usually write as $A \times A \times A$ or A^3. The number of outcomes is

$$n[(A \times A) \times A] = n(A \times A)n(A) = (2 \times 2) \times 2 = 8$$

c. Considering the result of part (b), we can easily see that the set of outcomes here is A^{10}, the Cartesian product of ten copies of A, or the set of ordered sequences of ten H's and T's. It's also easy to see that

$$n(A^{10}) = n(A)^{10} = 2^{10} = 1024$$

Before we go on . . . We can start to see the power of these formulas for cardinality. We can calculate that 1024 outcomes are possible when we toss a coin ten times without writing out all 1024 possibilities and counting them.

6.2 EXERCISES

In Exercises 1–6, $A = \{\text{Dirk, Johan, Frans, Sarie}\}$, $B = \{\text{Frans, Sarie, Tina, Klaas, Henrika}\}$, and $C = \{\text{Hans, Frans}\}$. Find the numbers indicated.

1. $n(A) + n(B)$ **2.** $n(A) + n(C)$

3. $n(A \cup B)$ **4.** $n(A \cup C)$

5. $n[A \cup (B \cap C)]$ **6.** $n[A \cap (B \cup C)]$

7. Verify that $n(A \cup B) = n(A) + n(B) - n(A \cap B)$ with A and B as above.

8. Verify that $n(A \cup C) = n(A) + n(C) - n(A \cap C)$ with A and C as above.

In Exercises 9–14, $A = \{H, T\}$, $B = \{1, 2, 3, 4, 5, 6\}$, and $C = \{\text{red, green, blue}\}$. Find the numbers indicated.

9. $n(A \times A)$

10. $n(B \times B)$

11. $n(B \times C)$

12. $n(A \times C)$

13. $n(A \times B \times B)$

14. $n(A \times B \times C)$

15. If $n(A) = 43$, $n(B) = 20$, and $n(A \cap B) = 3$, find $n(A \cup B)$.

16. If $n(A) = 60$, $n(B) = 20$, and $n(A \cap B) = 1$, find $n(A \cup B)$.

17. If $n(A \cup B) = 100$ and $n(A) = n(B) = 60$, find $n(A \cap B)$.

18. If $n(A) = 100$, $n(A \cup B) = 150$, and $n(A \cap B) = 40$, find $n(B)$.

In Exercises 19–24, $S = \{\text{Barnsley, Manchester United, Southend, Sheffield United, Liverpool, Maroka Swallows, Witbank Aces, Royal Tigers, Dundee United, Lyon}\}$ (a universal set), $A = \{\text{Southend, Liverpool, Maroka Swallows, Royal Tigers}\}$, and $B = \{\text{Barnsley, Manchester United, Southend}\}$. Find the numbers indicated.

19. $n(A')$ **20.** $n(B')$

21. $n[(A \cap B)']$ **22.** $n[(A \cup B)']$

23. $n(A' \cap B')$ **24.** $n(A' \cup B')$

25. With S, A, and B as above, verify that $n[(A \cap B)'] = n(A') + n(B') - n[(A \cup B)']$.

26. With S, A, and B as above, verify that $n(A' \cap B') + n(A \cup B) = n(S)$.

In Exercises 27–30, use the given information to complete the solution of each partially solved Venn diagram.

27. $n(A) = 20$, $n(B) = 20$, $n(C) = 28$, $n(B \cap C) = 8$, $n(S) = 50$

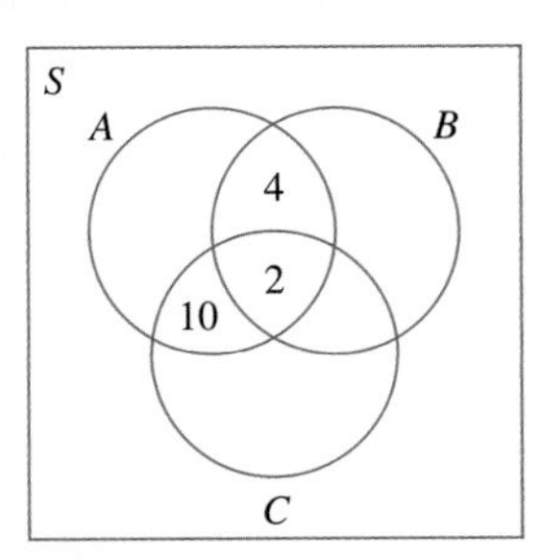

28. $n(A) = 16, n(B) = 11, n(C) = 30, n(S) = 40$

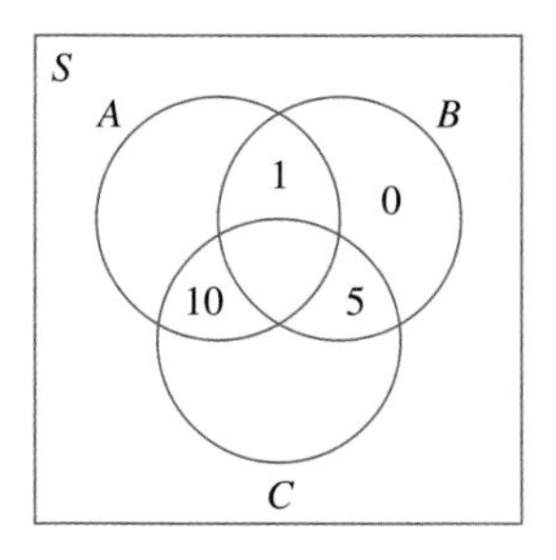

29. $n(A) = 10, n(B) = 19, n(S) = 140$

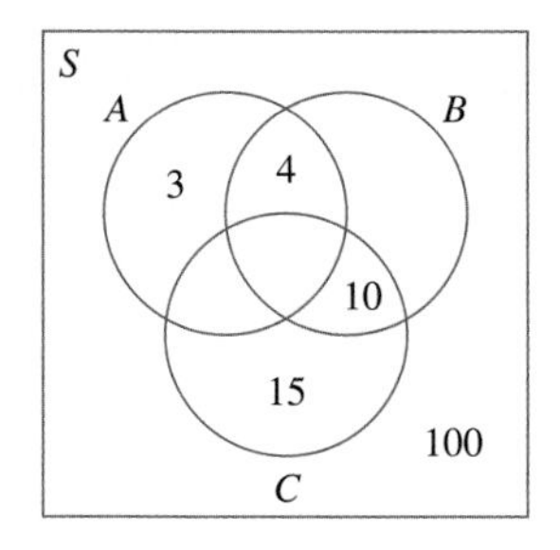

30. $n(A \cup B) = 30, n(B \cup C) = 30, n(A \cup C) = 35$

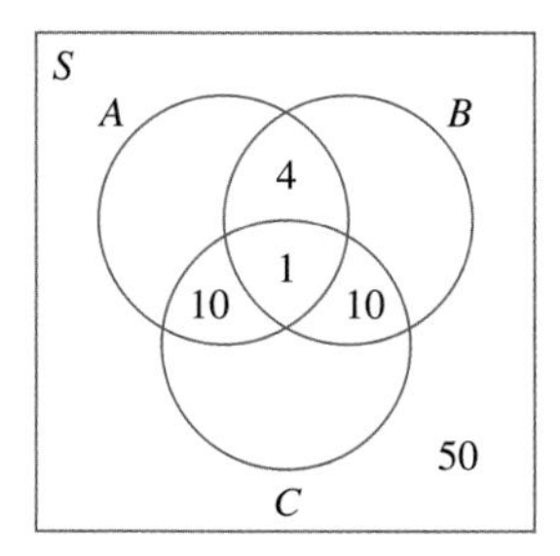

APPLICATIONS

31. Web Searches In December 2001 a search using the Web search engine AltaVista for "costenoble" yielded 1160 Web sites containing that word. A search for "waner" yielded 6407 sites. A search for sites containing both words yielded 499 sites. How many Web sites contained either "costenoble" or "waner" or both?

32. Web Searches In January 2002 a search using the Web search engine AltaVista for "seabold" yielded 951 Web sites containing that word. A search for "wu" yielded 923,607 sites. A search for sites containing both words yielded 14 sites. How many Web sites contained either "seabold" or "wu" or both?

33. Amusement On a particularly boring transatlantic flight, one of the authors amused himself by counting the heads of the people in the seats in front of him. He noticed that all 37 of them either had black hair or had a whole row of seats to themselves (or both). Of this total, 33 had black hair and 6 were fortunate enough to have a whole row to themselves. How many black-haired people had whole rows to themselves?

34. Restaurant Menus Your favorite restaurant offers a total of 14 desserts, of which 8 have ice cream as a main ingredient and 9 have fruit as a main ingredient. Assuming that all of them have either ice cream or fruit or both as a main ingredient, how many have both?

35. Ice Cream When Baskin-Robbins® was founded in 1945, it made 31 different flavors of ice cream. If you had a choice of having your ice cream in a cone, a cup, or a sundae, how many different desserts could you have?

SOURCE: Company Web site, http://www.baskinrobbins.com.

36. Ice Cream At the beginning of 2002 Baskin-Robbins claimed to have "nearly 1000 different ice cream flavors." Assuming that you could choose from 1000 different flavors, that you could have your ice cream in a cone, a cup, or a sundae, and that you could choose from a dozen different toppings, how many different desserts could you have?

SOURCE: Company Web site, http://www.baskinrobbins.com.

Publishing Exercises 37–42 are based on the following table, which shows the results of a survey of authors by a fictitious publishing company:

	New Authors	Established Authors	Total
Successful	5	25	30
Unsuccessful	15	55	70
Total	20	80	100

Consider the following subsets of the set S of all authors represented in the table: C, the set of successful authors; U, the set of unsuccessful authors; N, the set of new authors; and E, the set of established authors.

37. Describe the sets $C \cap N$ and $C \cup N$ in words. Use the table to compute $n(C)$, $n(N)$, $n(C \cap N)$, and $n(C \cup N)$. Verify that $n(C \cup N) = n(C) + n(N) - n(C \cap N)$.

38. Describe the sets $N \cap U$ and $N \cup U$ in words. Use the table to compute $n(N)$, $n(U)$, $n(N \cap U)$, and $n(N \cup U)$. Verify that $n(N \cup U) = n(N) + n(U) - n(N \cap U)$.

39. Describe the set $C \cap N'$ in words and find the number of elements it contains.

40. Describe the set $U \cup E'$ in words and find the number of elements it contains.

41. What percentage of established authors are successful? What percentage of successful authors are established?

42. What percentage of new authors are unsuccessful? What percentage of unsuccessful authors are new?

Recreational Boat Sales Exercises 43–48 are based on the following table, which shows the amount spent, in billions of dollars, on recreational boats and accessories in the United States during the period 1999–2001. (Let S be the set of all dollars represented in the table.)

	Used Boats, U	New Boats, N	Accessories, R	Total
1999, A	7	7	4	18
2000, B	7	7.5	5	19.5
2001, C	8	8	5	21
Total	22	22.5	14	58.5

SOURCE: National Marine Manufacturers Association/*New York Times*, January 10, 2002, p. C1. Figures are approximate.

In Exercises 43–48, use symbols to describe each set and compute its cardinality.

43. The set of dollars spent on new boats in 2001

44. The set of dollars spent on accessories or spent in 2001

45. The set of dollars spent in 2001 excluding new boats

46. The set of dollars spent on new boats excluding 2001

47. The set of dollars spent in 1999 on new and used boats

48. The set of dollars spent in years other than 1999 on new boats and accessories

Exercises 49–54 are based on the following table, which shows the performance of a selection of 100 stocks after 1 year. (Let S be the set of all stocks represented in the table.)

	Pharmaceutical, P	Electronic, E	Internet, I	Total
Increased, V	10	5	15	30
Unchanged,* N	30	0	10	40
Decreased, D	10	5	15	30
Total	50	10	40	100

*If a stock stayed within 20% of its original value, it is classified as "unchanged."

49. Use symbols to describe the set of non-Internet stocks that increased. How many elements are in this set?

50. Use symbols to describe the set of Internet stocks that did not increase. How many elements are in this set?

51. Compute $n(P' \cup N)$. What does this number represent?

52. Compute $n(P \cup N')$. What does this number represent?

53. Calculate $\frac{n(V \cap I)}{n(I)}$. What does the answer represent?

54. Calculate $\frac{n(D \cap I)}{n(D)}$. What does the answer represent?

55. Medicine In a study of Tibetan children, a total of 1556 children were examined. Of these, 1024 had rickets. Of the 243 urban children in the study, 93 had rickets.

a. How many children living in nonurban areas had rickets?

b. How many children living in nonurban areas did not have rickets?

SOURCE: N. S. Harris et al.,"Nutritional and Health Status of Tibetan Children Living at High Altitudes," *New England Journal of Medicine* 344(5), (2001):341–347.

56. Medicine In a study of Tibetan children, a total of 1556 children were examined. Of these, 615 had caries (cavities). Of the 1313 children living in nonurban areas, 504 had caries.

a. How many children living in urban areas had caries?

b. How many children living in urban areas did not have caries?

SOURCE: N. S. Harris et al.,"Nutritional and Health Status of Tibetan Children Living at High Altitudes," *New England Journal of Medicine* 344(5), (2001):341–347.

57. Entertainment According to a survey of 100 people regarding their movie attendance in the last year, 40 had seen a science fiction movie, 55 had seen an adventure movie, and 35 had seen a horror movie. Moreover, 25 had seen a science fiction movie and an adventure movie, 5 had seen an adventure movie and a horror movie, and 15 had seen a science fiction movie and a horror movie. Only 5 people had seen a movie from all three categories.

a. Use the given information to set up a Venn diagram and solve it.

b. Complete the following sentence: The survey suggests that ______% of science fiction movie fans are also horror movie fans.

58. Athletics Of the 4700 students at Medium Suburban College, 50 play collegiate soccer, 60 play collegiate lacrosse, and 96 play collegiate football. Only 4 students play both collegiate soccer and lacrosse, 6 play collegiate soccer and football, and 16 play collegiate lacrosse and football. No students play all three sports.

a. Use the given information to set up a Venn diagram and solve it.

b. Complete the following sentence: ______% of the college soccer players also play one of the other two sports at the collegiate level.

59. Entertainment In a survey of 100 Enormous State University students, 21 enjoyed classical music, 22 enjoyed rock music, and 27 enjoyed house music.[4] Only 5 of the students enjoyed both classical and rock. How many of those that enjoyed rock did not enjoy classical music?

60. Entertainment Referring to Exercise 59, you are also told that 5 students enjoyed all three kinds of music, and 53 enjoyed music in none of these categories. How many students enjoyed both classical and rock but disliked house music?

[4]House music is a form of techno dance music that came out of the Chicago clubs in the mid-1980s.

COMMUNICATION AND REASONING EXERCISES

61. When is $n(A \cup B) \neq n(A) + n(B)$?

62. When is $n(A \times B) = n(A)$?

63. When is $n(A \cup B) = n(A)$?

64. When is $n(A \cap B) = n(A)$?

65. Why is the Cartesian product referred to as a *product*? (*Hint:* Think about cardinality.)

66. Refer to your answer to Exercise 65. What set operation could one use to represent the *sum* of two disjoint sets A and B? Why?

67. Formulate an interesting application whose answer is $n(A \cap B) = 20$.

68. Formulate an interesting application whose answer is $n(A \times B) = 120$.

69. Use a Venn diagram or some other method to obtain a formula for $n(A \cup B \cup C)$ in terms of $n(A)$, $n(B)$, $n(C)$, $n(A \cap B)$, $n(A \cap C)$, $n(B \cap C)$, and $n(A \cap B \cap C)$.

70. Suppose A and B are nonempty sets with $A \subset B$ and $n(A)$ at least 2. Arrange the following numbers from smallest to largest (if two numbers are equal, say so): $n(A)$, $n(A \times B)$, $n(A \cap B)$, $n(A \cup B)$, $n(B \times A)$, $n(B)$, $n(B \times B)$.

6.3 The Addition and Multiplication Principles

Let's start with a really simple example. You walk into an ice cream store and find that you can choose between ice cream, of which there are 15 flavors, and frozen yogurt, of which there are 5 flavors. How many different selections can you make? Clearly, you have $15 + 5 = 20$ different dessert choices. Mathematically, this is an example of the formula for the cardinality of a disjoint union: If we let A be the set of ice creams you can choose and B the set of frozen yogurts, then $A \cap B = \varnothing$ and we want $n(A \cup B)$. But, the formula for the cardinality of a disjoint union is $n(A \cup B) = n(A) + n(B)$, which gives $15 + 5 = 20$ in this case.

This example illustrates a very useful general principle.

Addition Principle

When choosing among r disjoint alternatives, if

alternative 1 has n_1 possible outcomes

alternative 2 has n_2 possible outcomes

⋮

alternative r has n_r possible outcomes

then you have a total of $n_1 + n_2 + \cdots + n_r$ possible outcomes.

Quick Example

At a restaurant you can choose among 8 chicken dishes, 10 beef dishes, 4 seafood dishes, and 12 vegetarian dishes. This gives a total of $8 + 10 + 4 + 12 = 34$ different dishes from which to choose.

Here is another simple example. In that ice cream store, not only can you choose from 15 flavors of ice cream, but you can also choose from 3 different sizes of cone. How many different ice cream cones can you select? If we let A again be the set of ice cream flavors and now let C be the set of cone sizes, we want to pick a flavor *and* a size; that is, we want to pick an element of $A \times C$, the Cartesian product. To find the number of choices we have, we use the formula for the cardinality of a Cartesian product: $n(A \times C) = n(A)n(C)$. In this case we get $15 \times 3 = 45$ different ice cream cones we can select.

This example illustrates another general principle.

Multiplication Principle

When making a sequence of choices with r steps, if

step 1 has n_1 possible outcomes

step 2 has n_2 possible outcomes

⋮

step r has n_r possible outcomes

then you have a total of $n_1 \times n_2 \times \cdots \times n_r$ possible outcomes.

Quick Example

At a restaurant you can choose among 5 appetizers, 34 main dishes, and 10 desserts. This gives a total of $5 \times 34 \times 10 = 1700$ different meals (each including one appetizer, one main dish, and one dessert) from which you can choose.

Things get more interesting when we have to use the addition and multiplication principles in tandem.

Example 1 • Desserts

You walk into an ice cream store and find that you can choose between ice cream, of which there are 15 flavors, and frozen yogurt, of which there are 5 flavors. In addition, you can choose among 3 different sizes of cones for your ice cream or 2 different sizes of cups for your yogurt. How many different desserts can you choose from?

Solution It helps to think about a definite procedure for deciding which dessert you will choose. Here is one we can use:

Alternative 1 An Ice Cream Cone

Step 1 Choose a flavor.
Step 2 Choose a size.

Alternative 2 A Cup of Frozen Yogurt

Step 1 Choose a flavor.
Step 2 Choose a size.

That is, we can choose between alternative 1 and alternative 2. If we choose alternative 1, we have a sequence of two choices to make: flavor and size. The same is true of alternative 2. We shall call a procedure like this a **decision algorithm.**[5] Once we have a decision algorithm, we can use the addition and multiplication principles to count the number of possible outcomes.

Alternative 1 An Ice Cream Cone

Step 1 Choose a flavor: 15 choices.
Step 2 Choose a size: 3 choices.
There are $15 \times 3 = 45$ possible choices in alternative 1. Multiplication principle

[5]An algorithm is a procedure with definite rules for what to do at every step.

Alternative 2 A Cup of Frozen Yogurt

Step 1 Choose a flavor: 5 choices.

Step 2 Choose a size: 2 choices.

There are $5 \times 2 = 10$ possible choices in alternative 2. Multiplication principle

So, there are $45 + 10 = 55$ possible choices total. Addition principle

Before we go on . . . Decision algorithms are closely related to **decision trees.** To simplify the picture, suppose we had fewer choices—say, only 2 choices of ice cream flavor (vanilla and chocolate) and 2 choices of yogurt flavor (banana and raspberry). This gives us a total of $2 \times 3 + 2 \times 2 = 10$ possible desserts. We can illustrate the decisions we need to make when choosing what to buy in the diagram in Figure 8, called a decision tree.

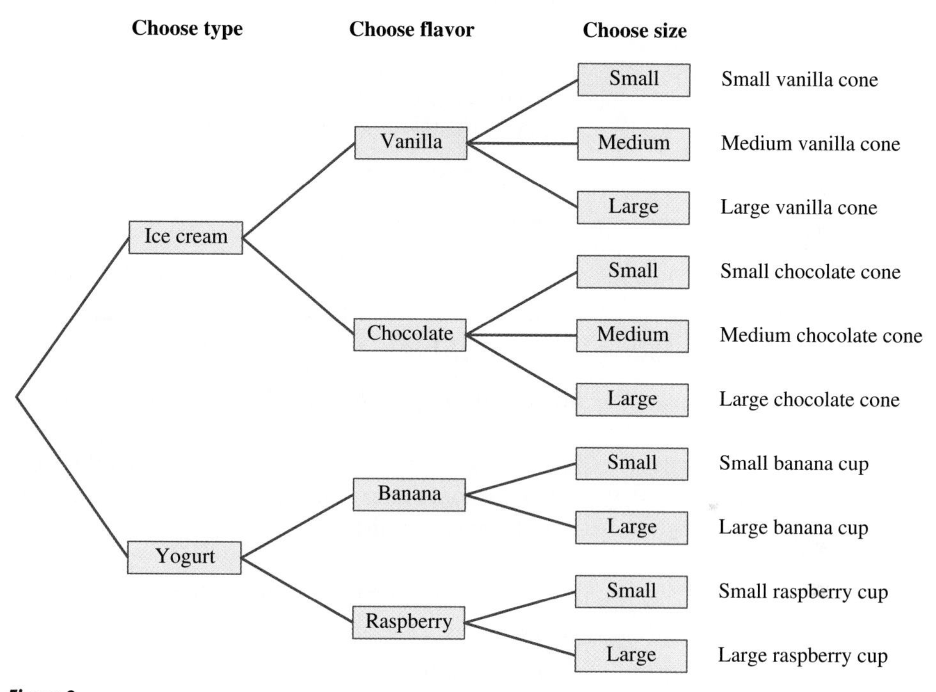

Figure 8

We do not use decision trees much in this chapter because they're not practical for counting large sets. Similar diagrams will be very useful, however, in Chapter 7, "Probability."

Example 2 • Exams

An exam has the following structure. It is broken into two parts: Part A and Part B, both of which you are required to do. In Part A you can choose between answering 10 true/false questions or answering 4 multiple-choice questions, each of which has 5 answers to choose from. In Part B you can choose between answering 8 true/false questions and answering 5 multiple-choice questions, each of which has 4 answers to choose from. How many different collections of answers are possible?

Solution While deciding what answers to write down, we use the following decision algorithm:

Step 1 Do Part A.

Alternative 1 Answer the 10 true/false questions.

Steps 1–10: Choose true or false for each question: 2 choices each.

There are $2 \times 2 \times \cdots \times 2 = 2^{10} = 1024$ choices in alternative 1.

Alternative 2 Answer the 4 multiple-choice questions.

Steps 1–4: Choose one answer for each question: 5 choices each.

There are $5 \times 5 \times 5 \times 5 = 5^4 = 625$ choices in alternative 2.

$1024 + 625 = 1649$ choices in Step 1.

Step 2 Do Part B.

Alternative 1 Answer the 8 true/false questions: 2 choices each.

There are $2^8 = 256$ choices in alternative 1.

Alternative 2 Answer the 5 multiple-choice questions: 4 choices each.

There are $4^5 = 1024$ choices in alternative 2.

$256 + 1024 = 1280$ choices in Step 2

There are $1649 \times 1280 = 2{,}110{,}720$ different collections of answers possible.

Example 3 illustrates the need to select your decision algorithm with care.

Example 3 • *Scrabble*®

You are playing *Scrabble* and have the following letters to work with: k, e, r, e. Since you are losing the game, you would like to use all your letters to make a single word, but you can't think of any four-letter words using all these letters. In desperation, you decide to list *all* the four-letter sequences possible to see if there are any valid words among them. How large is your list?

Solution It may first occur to you to try the following decision algorithm:

Step 1 Select the first letter: 4 choices.
Step 2 Select the second letter: 3 choices.
Step 3 Select the third letter: 2 choices.
Step 4 Select the last letter: 1 choice.

This gives $4 \times 3 \times 2 \times 1 = 24$ choices. However, something is wrong with the algorithm.

Question This seems ok. What could be wrong here?

Answer We didn't take into account the fact that there are two e's;[6] different decisions in Steps 1–4 can produce the same sequence. Suppose, for example, that we select the first e in Step 1, the second e in Step 2, and then the k and the r. This would produce the sequence eekr. If we select the *second* e in Step 1, the *first* e in Step 2, and then the k and r, we will obtain the *same* sequence: eekr. In other words, the decision algorithm produces two copies of the sequence eekr in the associated decision tree. (In fact, it produces two copies of each possible sequence of the letters.)

For a decision algorithm to be valid, each sequence of choices must produce a different result.

---

[6]Consider the following extreme case: If all four letters were e, then there would be only a single sequence: eeee, not the 24 predicted by the decision algorithm.

Since our original algorithm is not valid, we need a new one. Here is a strategy that works nicely for this example. Imagine, as before, that we are going to construct a sequence of four letters. This time we are going to imagine that we have a sequence of four empty slots: □□□□: Instead of selecting letters to fill the slots from left to right, we are going to select *slots* in which to place each letter. Remember that we have to use these letters: k, e, r, e. We proceed as follows, leaving the e's until last.

Step 1 Select an empty slot for the k: 4 choices (for example, □□[k]□).

Step 2 Select an empty slot for the r: 3 choices (for example, [r]□[k]□).

Step 3 Place the e's in the remaining two slots: 1 choice!

Thus, the multiplication principle yields $4 \times 3 \times 1 = 12$ choices.

Before we go on . . . Try constructing a decision tree for this example, and you will see that each sequence of four letters is produced exactly once when we use the correct (second) decision algorithm.

Guidelines: Creating and Testing a Decision Algorithm

Question How do I set up a decision algorithm to count how many items there are in a given scenario?

Answer Pretend that you are *constructing* such an item (for example, pretend that you are assembling an ice cream cone) and come up with a step-by-step procedure for doing so, listing the decisions you should make at each stage.

Question Once I have my decision algorithm, how do I check whether it is valid?

Answer Ask yourself the following question: Is it possible to get the same item (the exact same ice cream cone, say) by making different decisions when applying the algorithm? If the answer is yes, then your decision algorithm is invalid. Otherwise, it is valid.

6.3 EXERCISES

1. An experiment requires a choice among three initial setups. The first setup can result in two possible outcomes, the second in three possible outcomes, and the third in five possible outcomes. What is the total number of outcomes possible?

2. A surgical procedure requires choosing among four alternative methodologies. The first methodology can result in four possible outcomes, the second in three possible outcomes, and the remaining two can each result in two possible outcomes. What is the total number of outcomes possible?

3. An experiment requires a sequence of three steps. The first step can result in two possible outcomes, the second in three possible outcomes, and the third in five possible outcomes. What is the total number of outcomes possible?

4. A surgical procedure requires four steps. The first step can result in four possible outcomes, the second in three possible outcomes, and the remaining two can each result in two possible outcomes. What is the total number of outcomes possible?

In Exercises 5–12, find how many outcomes are possible for the decision algorithms.

5. Alternative 1
Step 1: one outcome
Step 2: two outcomes

Alternative 2
Step 1: two outcomes
Step 2: two outcomes
Step 3: one outcome

6. Alternative 1
Step 1: one outcome
Step 2: two outcomes
Step 3: two outcomes

Alternative 2
Step 1: two outcomes
Step 2: two outcomes

7. Step 1
Alternative 1: one outcome
Alternative 2: two outcomes

Step 2
Alternative 1: two outcomes
Alternative 2: two outcomes
Alternative 3: one outcome

8. Step 1
Alternative 1: one outcome
Alternative 2: two outcomes
Alternative 3: two outcomes

Step 2
Alternative 1: two outcomes
Alternative 2: two outcomes

9. Alternative 1
   Step 1
      Alternative 1: three outcomes
      Alternative 2: one outcome
   Step 2: two outcomes

   Alternative 2: five outcomes

10. Alternative 1: two outcomes

    Alternative 2
    Step 1
       Alternative 1: four outcomes
       Alternative 2: one outcome
    Step 2: two outcomes

11. Step 1
    Alternative 1
       Step 1: three outcomes
       Step 2: one outcome
    Alternative 2: two outcomes

    Step 2: five outcomes

12. Step 1: two outcomes

    Step 2
    Alternative 1
       Step 1: four outcomes
       Step 2: one outcome
    Alternative 2: two outcomes

13. How many different four-letter sequences can be formed from the letters a, a, a, b?

14. How many different five-letter sequences can be formed from the letters a, a, a, b, c?

APPLICATIONS

15. **Binary Codes** A binary digit, or bit, is either 0 or 1. A nybble is a 4-bit sequence. How many different nybbles are possible?

16. **Ternary Codes** A ternary digit is either 0, 1, or 2. How many sequences of six ternary digits are possible?

17. **Ternary Codes** A ternary digit is either 0, 1, or 2. How many sequences of six ternary digits containing a single 1 and a single 2 are possible?

18. **Binary Codes** A binary digit, or bit, is either 0 or 1. A nybble is a 4-bit sequence. How many different nybbles containing a single 1 are possible?

19. **Reward** While selecting candy for students in his class, Professor Murphy must choose between gummy candy and licorice nibs. Gummy candy packets come in three sizes, and packets of licorice nibs come in two sizes. If he chooses gummy candy, he must select gummy bears, gummy worms, or gummy dinos. If he chooses licorice nibs, he must choose between red and black. How many choices does he have?

20. **Productivity** Professor Oger must choose between an extra writing assigment and an extra reading assignment for the upcoming spring break. For the writing assignment, there are two essay topics to choose from and three different mandatory lengths (30 pages, 35 pages, or 40 pages). The reading topic would consist of one scholarly biography combined with one volume of essays. There are five biographies and two volumes of essays to choose from. How many choices does she have?

21. **Zip Disks** Zip disks come in two sizes (100MB and 250MB), packaged singly, in boxes of five, or in boxes of ten. When purchasing singly, you can choose from five colors; when purchasing in boxes of five or ten, you have two choices, black or an assortment of colors. If you are purchasing Zip disks, how many possibilities do you have to choose from?

22. **Radar Detectors** Radar detectors either are powered by their own battery or plug into the cigarette lighter socket. All radar detectors come in two models: no-frills and fancy. In addition, detectors powered by their own batteries detect either radar or laser, or both, whereas the plug-in types come in models that detect either radar or laser, but not both. How many different radar detectors can you buy?

23. **Multiple-Choice Tests** Professor Easy's final examination has ten true/false questions followed by two multiple-choice questions. In each of the multiple-choice questions, you must select the correct answer from a list of five. How many answer sheets are possible?

24. **Multiple-Choice Tests** Professor Tough's final examination has 20 true/false questions followed by three multiple-choice questions. In each of the multiple-choice questions you must select the correct answer from a list of six. How many answer sheets are possible?

25. **Tests** A test requires that you answer either Part A or Part B. Part A consists of eight true/false questions, and Part B consists of five multiple-choice questions with one correct answer out of five. How many different completed answer sheets are possible?

26. **Tests** A test requires that you answer first Part A and then either Part B or Part C. Part A consists of four true/false questions, Part B consists of four multiple-choice questions with one correct answer out of five, and Part C consists of three questions with one correct answer out of six. How many different completed answer sheets are possible?

27. **Stock Portfolios** Your broker has suggested that you diversify your investments by splitting your portfolio among mutual funds, municipal bond funds, stocks, and precious metals. She suggests four good mutual funds, three municipal bond funds, eight stocks, and three precious metals (gold, silver, and platinum).
    a. Assuming that your portfolio is to contain one of each type of investment, how many different portfolios are possible?
    b. Assuming that your portfolio is to contain three mutual funds, two municipal bond funds, one stock, and two precious metals, how many different portfolios are possible?

28. **Menus** The local diner offers a meal combination consisting of an appetizer, a soup, a main course, and a dessert. There are five appetizers, two soups, four main courses, and five desserts. Your diet restricts you to choosing between a dessert and an appetizer. (You cannot have both.) Given this restriction, how many three-course meals are possible?

29. **Computer Codes** A computer byte consists of 8 bits, each bit being either a 0 or a 1. If characters are represented using a code that uses a byte for each character, how many different characters can be represented?

30. **Computer Codes** Some written languages, like Chinese and Japanese, use tens of thousands of different characters. If a language uses roughly 50,000 characters, a computer code for this language would have to use how many bytes per character? (See Exercise 29.)

31. **Symmetries of a Five-Pointed Star** A five-pointed star will appear unchanged if it is rotated through any one of the angles 0°, 72°, 144°, 216°, or 288°. It will also appear unchanged if it is flipped about the axis shown in the figure. A *symmetry* of the five-pointed star consists of either a rotation or a rotation followed by a flip. How many different symmetries are there altogether?

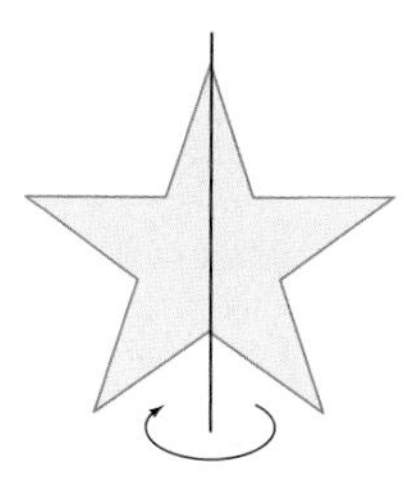

32. **Symmetries of a Six-Pointed Star** A six-pointed star will appear unchanged if it is rotated through any one of the angles 0°, 60°, 120°, 180°, 240°, or 300°. It will also appear unchanged if it is flipped about the axis shown in the figure. A *symmetry* of the six-pointed star consists of either a rotation or a rotation followed by a flip. How many different symmetries are there altogether?

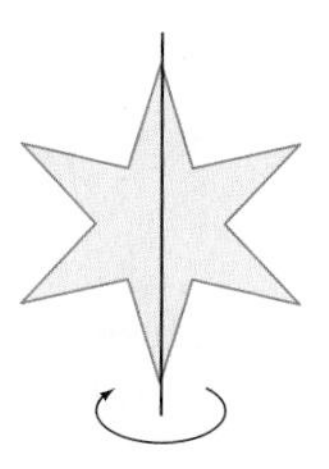

33. **Variables in BASIC** A variable name in the programming language BASIC can be either a letter or a letter followed by a decimal digit—that is, one of the numbers 0, 1, . . . , 9. How many different variables are possible?

SOURCE: F. S. Roberts, *Applied Combinatorics* (Upper Saddle River, N.J.: Prentice-Hall, 1984).

34. **Employee IDs** A company assigns to each of its employees an ID code that consists of one, two, or three letters followed by a digit from 0 through 9. How many employee codes does the company have available?

35. **Tournaments** How many ways are there of filling in the blanks for the following (fictitious) soccer tournament?

North Carolina
Central Connecticut
Virginia
Syracuse

36. **Tournaments** How many ways are there of filling in the blanks for a (fictitious) soccer tournament involving the four teams San Diego State, De Paul, Colgate, and Hofstra?

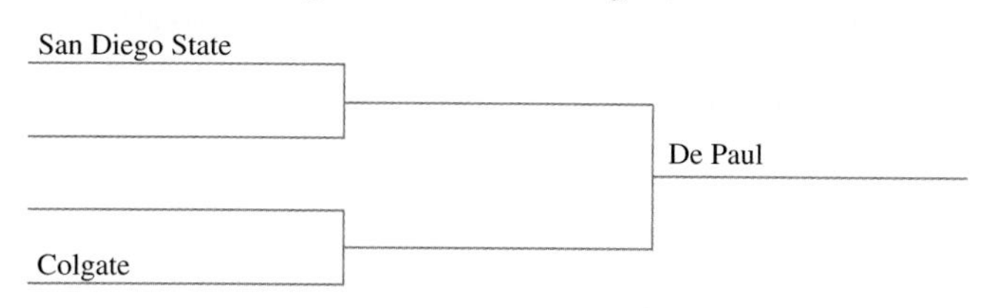

37. **Telephone Numbers** A telephone number consists of a sequence of seven digits not starting with 0 or 1.
    a. How many telephone numbers are possible?
    b. How many of them begin with either 463, 460, or 400?
    c. How many telephone numbers are possible if no two adjacent digits are the same? (For example, 235-9350 is permitted, but not 223-6789.)

38. **Social Security Numbers** A Social Security number is a sequence of nine digits.
    a. How many Social Security numbers are possible?
    b. How many of them begin with either 023 or 003?
    c. How many Social Security numbers are possible if no two adjacent digits are the same? (For example, 235-93-2345 is permitted, but not 126-67-8189.)

39. **DNA Chains** DNA (deoxyribonucleic acid) is the basic building block of reproduction in living things. A DNA chain is a sequence of chemicals called *bases*. There are four possible bases: thymine (T), cytosine (C), adenine (A), and guanine (G).
    a. How many three-element DNA chains are possible?
    b. How many n-element DNA chains are possible?
    c. A human DNA chain has 2.1×10^{10} elements. How many human DNA chains are possible?

40. **Credit Card Numbers** Each customer of Mobil Credit Corporation is given a nine-digit number for computer identification purposes.
    a. If each digit can be any number from 0 to 9, are there enough different account numbers for 10 million credit card holders?
    b. Would there be if the digits were only 0 or 1?

SOURCE: Taken from an exercise in F. S. Roberts, *Applied Combinatorics* (Upper Saddle River, N.J.: Prentice-Hall, 1984).

41. **HTML** Colors in HTML (the language in which many Web pages are written) can be represented by six-digit hexadecimal codes: sequences of six integers ranging from 0 to 15 (represented as 0, . . . , 9, A, B, . . . , F).
    a. How many different colors can be represented?
    b. Some monitors can only display colors encoded with pairs of repeating digits (such as 44DD88). How many colors can these monitors display?
    c. Grayscale shades are represented by sequences *xyxyxy*, consisting of a repeated pair of digits. How many grayscale shades are possible?
    d. The pure colors are pure red, *xy*0000; pure green, 00*xy*00; and pure blue, 0000*xy*. (*xy* = *FF* gives the brightest pure color, and *xy* = 00 gives the darkest: black). How many pure colors are possible?

42. Telephone Numbers In the past a local telephone number in the United States consisted of a sequence of two letters followed by five digits. Three letters were associated with each number from 2 to 9 (just as in the standard telephone layout shown in the figure) so that each telephone number corresponded to a sequence of seven digits. How many different sequences of seven digits were possible?

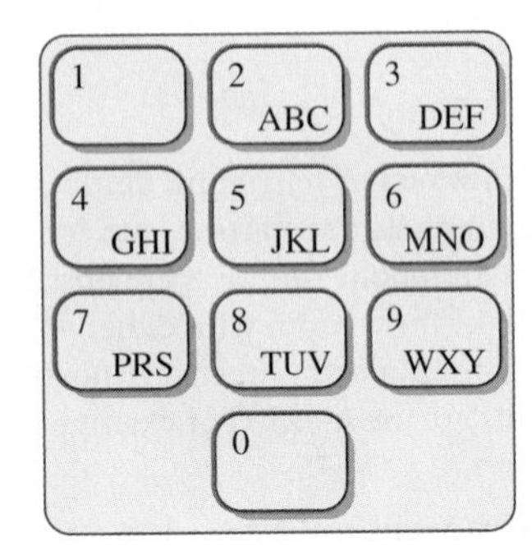

43. Romeo and Juliet Here is a list of the main characters in Shakespeare's *Romeo and Juliet.* The first seven characters are men and the last four are women.

Escalus, *prince of Verona*
Paris, *kinsman to the prince*
Romeo, *of Montague Household*
Mercutio, *friend of Romeo*
Benvolio, *friend of Romeo*
Tybalt, *nephew to Lady Capulet*
Friar Lawrence, *a Franciscan*
Lady Montague, *of Montague Household*
Lady Capulet, *of Capulet Household*
Juliet, *of Capulet Household*
Juliet's nurse

A total of ten male and eight female actors are available to play these roles. How many possible casts are there? (All roles are to be played by actors of the correct gender.)

44. Swan Lake The Enormous State University's Accounting Society has decided to produce a version of the ballet *Swan Lake,* in which all the female roles (including all of the swans) will be danced by men, and vice versa. Here are the main characters:

Prince Siegfried
Prince Siegfried's mother
Princess Odette, *the White Swan*
The Evil Duke Rotbart
Odile, *the Black Swan*
Cygnet 1, *young swan*
Cygnet 2, *young swan*
Cygnet 3, *young swan*

The ESU Accounting Society has on hand a total of 4 female dancers and 12 male dancers who are to be considered for the main roles. How many possible casts are there?

45. License Plates Many U.S. license plates display a sequence of three letters followed by three digits.

a. How many such license plates are possible?
b. To avoid confusion of letters with digits, some states do not issue standard plates with the last letter an I, O, or Q. How many license plates are still possible?
c. Assuming that the letter combinations VET, MDZ, and DPZ are reserved for disabled veterans, medical practitioners, and disabled persons, respectively, how many license plates are possible also taking the restriction in part (b) into account?

46. License Plates License plates in Montana have a sequence consisting of (1) a digit from 1 to 9, (2) a letter, (3) a dot, (4) a letter, and (5) a four-digit number.

a. How many different license plates are possible?
b. How many different license plates are available for citizens if numbers that end with 0 are reserved for official state vehicles?

SOURCE: The License Plates of the World Web site, http://servo.oit.gatech.edu/~mk5/.

47. Mazes

a. How many four-letter sequences are possible that contain only the letters R and D, with D occurring only once?
b. Use part (a) to calculate the number of possible routes from Start to Finish in the maze shown in the figure, where each move is either to the right or down.

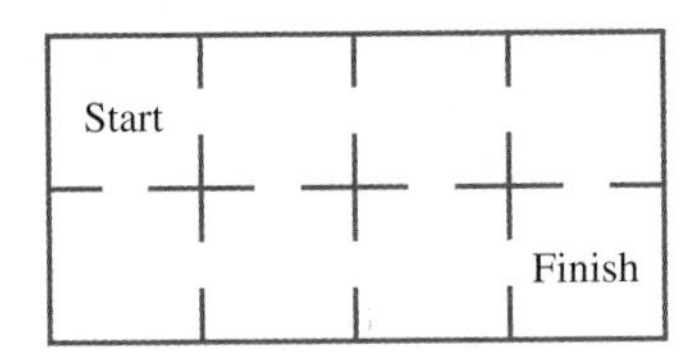

c. Comment on what would happen if we also allowed left and up moves.

48. Mazes

a. How many six-letter sequences are possible that contain only the letters R and D, with D occurring only once?
b. Use part (a) to calculate the number of possible routes from Start to Finish in the maze shown in the figure, where each move is either to the right or down.

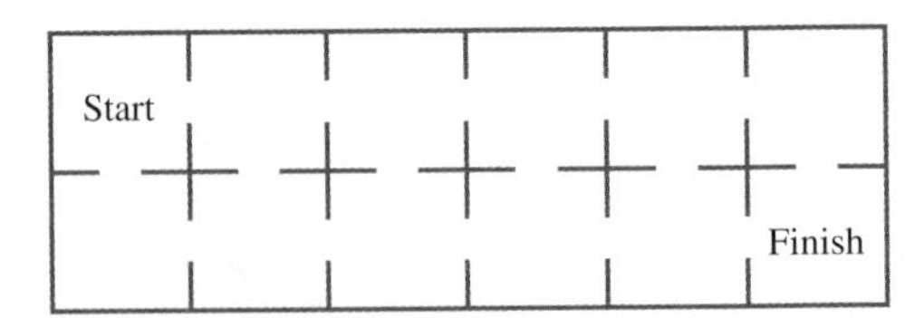

c. Comment on what would happen if we also allowed left and up moves.

49. Car Engines In a six-cylinder V6 engine, the even-numbered cylinders are on the left, and the odd-numbered cylinders are on the right. A good firing order is a sequence of the numbers 1 through 6 in which right and left sides alternate.

a. How many possible good firing sequences are there?
b. How many good firing sequences are there that start with a cylinder on the left?

SOURCE: Adapted from an exercise in D. I. A. Cohen, *Basic Techniques of Combinatorial Theory* (New York: Wiley, 1978).

50. Car Engines Repeat Exercise 49 for an eight-cylinder V8 engine.

51. Minimalist Art You are exhibiting your collection of minimalist paintings. Art critics have raved about your paintings, each of which consists of ten vertical colored lines set against a white background. You have used the following rule to produce your paintings: Every second line, starting with the first, is to be either blue or gray, while the remaining five lines are to be either all light blue, all pink, or all purple. Your collection is complete: Every possible combination that satisfies the rules occurs. How many paintings are you exhibiting?

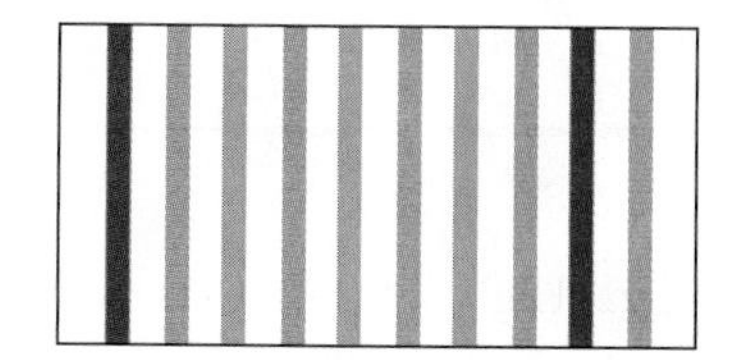

52. Combination Locks Dripping wet after your shower at the gym, you have clean forgotten the combination of your lock. It is a standard combination lock, that uses a three-number combination with each number in the range 0 through 39. All you remember is that the second number is either 27 or 37, and the third number ends in a 5. In desperation you decide to go through all possible combinations using the information you remember. Assuming that it takes about 10 seconds to try each combination, what is the longest possible time you may have to stand dripping in front of your locker?

53. Product Design Your company has patented an electronic digital padlock that a user can program with his or her own four-digit code. (Each digit can be 0 through 9.) The padlock is designed to open if either the correct code is keyed in or—and this is helpful for forgetful people—if exactly one of the digits is incorrect.

a. How many incorrect codes will open a programmed padlock?

b. How many codes will open a programmed padlock?

54. Product Design Your company has patented an electronic digital padlock that has a telephone-style keypad. Each digit from 2 through 9 corresponds to three letters of the alphabet (see the figure for Exercise 42). How many different four-letter sequences correspond to a single four-digit sequence using digits in the range 2 through 9?

55. Morse Code In Morse code each letter of the alphabet is encoded by a different sequence of dots and dashes. Different letters may have sequences of different lengths. How long should the longest sequence be in order to allow for every possible letter of the alphabet?

56. Numbers How many odd numbers between 10 and 99 have distinct digits?

57. Calendars *The World Almanac* features a "perpetual calendar," a collection of 14 possible calendars. Why does this suffice to be sure there is a calendar for every conceivable year?

SOURCE: *The World Almanac and Book of Facts 1992* (New York: Pharos Books, 1992).

58. Calendars How many possible calendars are there that have February 12 falling on a Sunday, Monday, or Tuesday?

59. Building Blocks Use a decision algorithm to show that a rectangular solid with dimensions $m \times n \times r$ can be constructed with $m \cdot n \cdot r$ cubical $1 \times 1 \times 1$ blocks (see the figure).

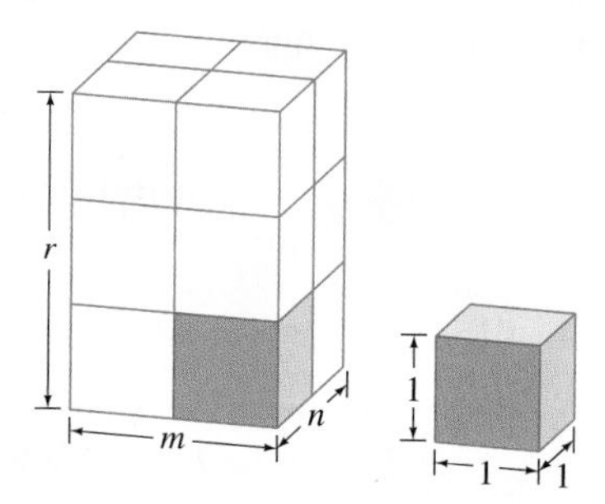

60. Matrices (Some knowledge of matrices is assumed for this exercise.) Use a decision algorithm to show that an $m \times n$ matrix must have $m \cdot n$ entries.

61. Programming in BASIC (Some programming knowledge assumed for this exercise.) How many iterations will be carried out in the following routine?

```
For i = 1 to 10
    For j = 2 to 20
        For k = 1 to 10
            Print i, j, k
        Next k
    Next j
Next i
```

62. Programming in JAVASCRIPT (Some programming knowledge assumed for this exercise.) How many iterations will be carried out in the following routine?

```
for (i = 1; i <= 2; i++) {
    for (j = 1; j <= 2; j++) {
        for (k = 1; k <= 2; k++)
            sum += i+j+k;
        }
    }
```

COMMUNICATION AND REASONING EXERCISES

63. Complete the following sentence: The multiplication principle is based on the cardinality of the _______ of two sets.

64. Complete the following sentence: The addition principle is based on the cardinality of the _______ of two disjoint sets.

65. You are packing for a short trip and want to take two of the ten shirts you have hanging in your closet. Critique the following decision algorithm and calculation of how many different ways you can choose two shirts to pack. Step 1: Choose one shirt, 10 choices. Step 2: Choose another shirt, 9 choices. Hence, there are 90 possible choices of two shirts.

66. You are designing an advertising logo that consists of a tower of five squares. Three are yellow, one is blue, and one is green. Critique the following decision algorithm and calculation of the number of different five-square sequences. Step 1: Choose the first square, 5 choices. Step 2: Choose the second square, 4 choices. Step 3: Choose the third square, 3 choices. Step 4: Choose the fourth square, 2 choices. Step 5: Choose the last square, 1 choice. Hence, there are 120 possible five-square sequences.

67. Construct a decision algorithm that gives the correct number of five-square sequences in Exercise 66.

68. Find an interesting application that requires a decision algorithm with two steps in which each step has two alternatives.

6.4 Permutations and Combinations

Certain classes of counting problems come up frequently, and it is useful to develop formulas to deal with them.

Example 1 • Casting

Ms. Birkitt, the English teacher at Brakpan Girls High School, wants to stage a production of R. B. Sheridan's play, *The School for Scandal.* The casting is going well until she is left with five unfilled characters and five seniors who are yet to be assigned roles. The characters are Lady Sneerwell, Lady Teazle, Mrs. Candour, Maria, and Snake; the unassigned seniors are April, May, June, Julia, and Augusta. How many possible assignments are there?

Solution To decide on a specific assignment, we use the following algorithm:

Step 1 Choose a senior to play Lady Sneerwell: 5 choices.

Step 2 Choose one of the remaining seniors to play Lady Teazle: 4 choices.

Step 3 Choose one of the now remaining seniors to play Mrs. Candour: 3 choices.

Step 4 Choose one of the now remaining seniors to play Maria: 2 choices.

Step 5 Choose the remaining senior to play Snake: 1 choice.

Thus, there are $5 \times 4 \times 3 \times 2 \times 1 = 120$ possible assignments of seniors to roles.

What this example has in common with many others is that we start with a set—here the set of seniors—and we want to know how many ways we can put the elements of that set in order in a list. In this example, an ordered list of the five seniors—say,

1. May
2. Augusta
3. June
4. Julia
5. April

corresponds to a particular casting:

Cast

Lady Sneerwell	May
Lady Teazle	Augusta
Mrs. Candour	June
Maria	Julia
Snake	April

We call an ordered list of items a **permutation** of those items.

Question If we have n items, how many permutations of those items are possible?

Answer We can use a decision algorithm similar to the one we used in Example 1 to select a permutation.

Step 1 Select the first item: n choices.

Step 2 Select the second item: $n - 1$ choices.

Step 3 Select the third item: $n - 2$ choices.

⋮

Step $n - 1$: Select the next-to-last item: 2 choices.

Step n Select the last item: 1 choice.

Thus, there are $n \times (n - 1) \times (n - 2) \times \cdots \times 2 \times 1$ possible permutations. We call this number ***n* factorial,** which we write as $n!$.

Permutations

A **permutation of *n* items** is an ordered list of those items. The number of possible permutations of n items is given by ***n* factorial,** which is

$$n! = n \times (n - 1) \times (n - 2) \times \cdots \times 2 \times 1$$

Quick Examples

1. The number of permutations of five items is $5! = 5 \times 4 \times 3 \times 2 \times 1 = 120$.
2. The number of ways four CDs can be played in sequence is $4! = 4 \times 3 \times 2 \times 1 = 24$.
3. The number of ways three cars can be matched with three drivers is $3! = 6$.

Using Technology

To compute 12! on the TI-83, enter

```
12!
```

To find !, press MATH and select PRB.

To compute 12! in Excel, enter

```
=FACT(12)
```

in any vacant cell.

Sometimes, instead of constructing an ordered list of *all* the items of a set, we might want to construct a list of *some* of the items, as in the next example.

Example 2 • Corporations

At the end of 2001, the ten largest companies (by market capitalization) listed on the New York Stock Exchange were, in alphabetical order, American International Group, AOL Time Warner, Citigroup, Exxon Mobil Corp., General Electric Co., IBM, Johnson & Johnson, Merck & Co., Pfizer, and Wal-Mart Stores.[7] You would like to apply to six of these companies for a job, and you would like to list them in order of job preference. How many such ordered lists are possible?

[7]SOURCE: New York Stock Exchange Web site, http://www.nyse.com/.

Solution We want to count ordered lists, but we can't use the permutation formula because we don't want all ten companies in the list, just six of them. So we fall back to a decision algorithm.

Step 1 Choose the first company: 10 choices.

Step 2 Choose the second company: 9 choices.

Step 3 Choose the third one: 8 choices.

Step 4 Choose the fourth one: 7 choices.

Step 5 Choose the fifth one: 6 choices.

Step 6 Choose the sixth one: 5 choices.

Thus, there are $10 \times 9 \times 8 \times 7 \times 6 \times 5 = 151{,}200$ possible lists of six. We call this number the **number of permutations of six items chosen from ten,** or the **number of permutations of ten items taken six at a time.**

Before we go on . . . We wrote the answer as the product $10 \times 9 \times 8 \times 7 \times 6 \times 5$. But it is useful to notice that we can write this number in a more compact way:

$$10 \times 9 \times 8 \times 7 \times 6 \times 5 = \frac{10 \times 9 \times 8 \times 7 \times 6 \times 5 \times 4 \times 3 \times 2 \times 1}{4 \times 3 \times 2 \times 1} = \frac{10!}{4!} = \frac{10!}{(10-6)!}$$

Permutations of n Items Taken r at a Time

A **permutation of *n* items taken *r* at a time** is an ordered list of r items chosen from a set of n items. The number of permutations of n items taken r at a time is given by

$$P(n, r) = n \times (n-1) \times (n-2) \times \cdots \times (n-r+1)$$

(there are r terms multiplied together). We can also write

$$P(n, r) = \frac{n!}{(n-r)!}$$

Quick Example

The number of permutations of six items taken two at a time is

$$P(6, 2) = 6 \times 5 = 30$$

which we could also calculate as

$$P(6, 2) = \frac{6!}{(6-2)!} = \frac{6!}{4!} = \frac{720}{24} = 30$$

Using Technology

To compute $P(6, 2)$ on the TI-83, enter

`6 nPr 2` To find nPr, press MATH and select PRB.

To compute $P(6, 2)$ in Excel, enter

`=PERMUT(6,2).`

in any vacant cell

Example 3 • Corporations

Suppose we simply want to apply to only two of the ten companies listed in Example 2, without regard to order. How many possible choices do we have? What if we want to apply to six, without regard to order?

Solution Our first guess might be $P(10, 2) = 10 \times 9 = 90$. However, that is the number of *ordered lists* of two companies. We said that we don't care which is first and which second. For example, we consider the list

1. Merck **2.** Pfizer

to be the same as

1. Pfizer **2.** Merck

Since every set of two companies occurs twice in the 90 lists, once in one order and again in the reverse order, we would count every set of two twice. Thus, there are $90/2 = 45$ possible choices of two companies.

Now, if we wish to pick six companies, again we might start with $P(10, 6) = 151{,}200$. But now every set of six companies appears as many times as there are different orders in which they could be listed. Six things can be listed in $6! = 720$ different orders, so the number of ways of choosing six companies is $151{,}200/720 = 210$.

In Example 3 we were concerned with counting not the number of ordered lists but the number of *unordered sets* of companies. For ordered lists we use the word **permutation**; for unordered sets we use the word **combination.**

Permutations and Combinations

A **permutation** of n items taken r at a time is an *ordered list* of r items chosen from n. A **combination** of n items taken r at a time is an *unordered set* of r items chosen from n.

Note Since lists are usually understood to be ordered, when we refer to a list of items, we will always mean an *ordered* list. Similarly, since sets are understood to be unordered, when we refer to a set of items we will always mean an *unordered* set. In short:

Lists are ordered. Sets are unordered.

Quick Examples

There are six permutations of the three letters a, b, c taken two at a time:

1. a, b **2.** b, a **3.** a, c **4.** c, a **5.** b, c **6.** c, b

There are six lists containing two of the letters a, b, c.

There are three combinations of the three letters a, b, c taken two at a time:

1. {a, b} **2.** {a, c} **3.** {b, c}

There are three sets containing two of the letters a, b, c.

Question How do we count the number of possible combinations of n items taken r at a time?

Answer We generalize the calculation done in Example 3. The number of permutations is $P(n, r)$, but each set of r items occurs $r!$ many times because this is the number of ways in which those r items can be ordered. So, the number of combinations is $P(n, r)/r!$.

Combinations of n Items Taken r at a Time

The number of **combinations of n items taken r at a time** is given by

$$C(n, r) = \frac{P(n, r)}{r!} = \frac{n \times (n-1) \times (n-2) \times \cdots \times (n-r+1)}{r!}$$

We can also write

$$C(n, r) = \frac{n!}{r!(n-r)!}$$

Quick Examples

1. The number of combinations of six items taken two at a time is

$$C(6, 2) = \frac{6 \times 5}{2 \times 1} = 15$$

which we can also calculate as

$$C(6, 2) = \frac{6!}{2!(6-2)!} = \frac{6!}{2!4!} = \frac{720}{2 \times 24} = 15$$

2. The number of sets of four marbles chosen from six is

$$C(6, 4) = \frac{6 \times 5 \times 4 \times 3}{4 \times 3 \times 2 \times 1} = 15$$

T

Using Technology

To compute $C(6, 2)$ on the TI-83, enter

```
6 nCr 2
```

To find nCr, press MATH and select PRB.

To compute $C(6, 2)$ in Excel, enter

```
=COMBIN(6,2)
```

in any vacant cell.

Note There are other common notations for $C(n, r)$. Calculators often have ${}_nC_r$. In mathematics we often write $\binom{n}{r}$, which is also known as a **binomial coefficient.** Since $C(n, r)$ is the number of ways of choosing a set of r items from n, it is often read "n choose r."

Example 4 • Calculating Combinations

Calculate **a.** $C(11, 3)$ and **b.** $C(11, 8)$.

Solution The easiest way to calculate $C(n, r)$ by hand is to use the formula

$$C(n, r) = \frac{n \times (n-1) \times (n-2) \times \cdots \times (n-r+1)}{r!}$$

$$= \frac{n \times (n-1) \times (n-2) \times \cdots \times (n-r+1)}{r \times (r-1) \times (r-2) \times \cdots \times 1}$$

Both the numerator and the denominator have r factors, so we can begin with n/r and then continue multiplying by decreasing numbers on the top and the bottom until we hit 1 in the denominator. When calculating, it helps to cancel common factors from the numerator and denominator before doing the multiplication in either one.

a. $C(11, 3) = \dfrac{11 \times 10 \times 9}{3 \times 2 \times 1} = \dfrac{11 \times \overset{5}{\cancel{10}} \times \overset{3}{\cancel{9}}}{\cancel{3} \times \cancel{2} \times 1} = 165$

b. $C(11, 8) = \dfrac{11 \times 10 \times 9 \times \cancel{8} \times \cancel{7} \times \cancel{6} \times \cancel{5} \times \cancel{4}}{\cancel{8} \times \cancel{7} \times \cancel{6} \times \cancel{5} \times \cancel{4} \times 3 \times 2 \times 1} = \dfrac{11 \times \overset{5}{\cancel{10}} \times \overset{3}{\cancel{9}}}{\cancel{3} \times \cancel{2} \times 1} = 165$

Before we go on . . . It is no coincidence that the answers for parts (a) and (b) are the same. Consider what each represents. $C(11, 3)$ is the number of ways of choosing 3 items from 11—for example, electing 3 trustees from a slate of 11. Electing those 3 is the same as choosing the 8 who do not get elected. Thus, there are exactly as many ways to choose 3 items from 11 as there are ways to choose 8 items from 11. So, $C(11, 3) = C(11, 8)$. In general,

$$C(n, r) = C(n, n - r)$$

We can also see this equality by using the formula

$$C(n, r) = \frac{n!}{r!(n-r)!}$$

If we substitute $n - r$ for r, we get exactly the same formula.

Use the equality $C(n, r) = C(n, n - r)$ to make your calculations easier. Choose the one with the smaller denominator to begin with.

Example 5 • Calculating Combinations

Calculate **a.** $C(11, 11)$ and **b.** $C(11, 0)$.

Solution

a. $C(11, 11) = \dfrac{11 \times 10 \times 9 \times 8 \times 7 \times 6 \times 5 \times 4 \times 3 \times 2 \times 1}{11 \times 10 \times 9 \times 8 \times 7 \times 6 \times 5 \times 4 \times 3 \times 2 \times 1} = 1$

b. What do we do with that zero? What does it mean to multiply zero numbers together? We know from above that $C(11, 0) = C(11, 11)$, so we must have $C(11, 0) = 1$. How does this fit with the formulas? Go back to the calculation of $C(11, 11)$:

$$1 = C(11, 11) = \frac{11!}{11!(11-11)!} = \frac{11!}{11!0!}$$

This equality is true only if we agree that $0! = 1$, which we do. Then

$$C(11, 0) = \frac{11!}{0!11!} = 1$$

Before we go on . . . There is nothing special about the number 11 in these calculations. In general,

$$C(n, n) = C(n, 0) = 1$$

After all, there is only one way to choose n items out of n: Choose them all. Similarly, there is only one way to choose zero items out of n: Choose none of them.

Now for a few more complicated examples that illustrate the applications of the counting techniques we've discussed.

Example 6 • Lotto

In the betting game Lotto, used in many state lotteries, you choose six different numbers in the range 1–55 (the upper number varies). The order in which you choose them is irrelevant. If your six numbers match the six numbers chosen in the "official drawing," you win the top prize. If Lotto tickets cost \$1 for two sets of numbers and you decide to buy tickets that cover every possible combination, thereby guaranteeing that you will win the top prize, how much money will you have to spend?

Solution We first need to know how many sets of numbers are possible. Since order does not matter, we are asking for the number of combinations of 55 numbers taken 6 at a time. This is

$$C(55, 6) = \frac{55 \times 54 \times 53 \times 52 \times 51 \times 50}{6 \times 5 \times 4 \times 3 \times 2 \times 1} = 28{,}989{,}675$$

Since \$1 buys you two of these, you need to spend \$28,989,675/2 = \$14,494,838 (rounding up to the nearest dollar) to be assured of a win!

Before we go on . . . This calculation shows that you should not bother buying all these tickets if the winning prize is less than about \$14.5 million. Even if the prize is higher, you need to account for the fact that many people will play and the prize may end up split among several winners, not to mention the impracticality of filling out millions of betting slips.

Example 7 • Marbles

A bag contains 3 red, 3 blue, 3 green, and 2 yellow marbles (all distinguishable from one another).

a. How many sets of 4 marbles are possible?

b. How many sets of 4 are there such that each one is a different color?

c. How many sets of 4 are there in which at least 2 are red?

d. How many sets of 4 are there in which none are red, but at least 1 is green?

Solution

a. We simply need to find the number of ways of choosing 4 marbles out of 11, which is

$$C(11, 4) = 330 \text{ possible sets of 4 marbles}$$

b. We use a decision algorithm for choosing such a set of marbles:

Step 1 Choose 1 red one from the 3 red ones: $C(3, 1) = 3$ choices.
Step 2 Choose 1 blue one from the 3 blue ones: $C(3, 1) = 3$ choices.
Step 3 Choose 1 green one from the 3 green ones: $C(3, 1) = 3$ choices.
Step 4 Choose 1 yellow one from the 2 yellow ones: $C(2, 1) = 2$ choices.

This gives a total of $3 \times 3 \times 3 \times 2 = 54$ possible sets.

c. We need another decision algorithm. To say that at least 2 must be red means that either 2 are red or 3 are red (with a total of 3 red ones). In other words, we have two *alternatives.*

Alternative 1 Exactly 2 red marbles
Step 1 Choose 2 red ones: $C(3, 2) = 3$ choices.
Step 2 Choose 2 nonred ones. There are 8 of these, so we get $C(8, 2) = 28$ possible choices.
Thus, the total number of choices for alternative 1 is $3 \times 28 = 84$.

Alternative 2 Exactly 3 red marbles
Step 1 Choose the 3 red ones: $C(3, 3) = 1$ choice.
Step 2 Choose 1 nonred one: $C(8, 1) = 8$ choices.
Thus, the total number of choices for alternative 2 is $1 \times 8 = 8$.

By the addition principle, we get a total of $84 + 8 = 92$ sets.

d. The phrase "at least 1 green" tells us that we again have some alternatives.

Alternative 1 1 green marble
Step 1 Choose 1 green marble from the 3: $C(3, 1) = 3$ choices.
Step 2 Choose 3 nongreen, nonred marbles: $C(5, 3) = 10$ choices.
Thus, the total number of choices for alternative 1 is $3 \times 10 = 30$.

Alternative 2 2 green marbles
Step 1 Choose 2 green marbles from the 3: $C(3, 2) = 3$ choices.
Step 2 Choose 2 nongreen, nonred marbles: $C(5, 2) = 10$ choices.
Thus, the total number of choices for alternative 2 is $3 \times 10 = 30$.

Alternative 3 3 green marbles
Step 1 Choose 3 green marbles from the 3: $C(3, 3) = 1$ choice.
Step 2 Choose 1 nongreen, nonred marble: $C(5, 1) = 5$ choices.
Thus, the total number of choices for alternative 3 is $1 \times 5 = 5$.

The addition principle now tells us that the number of such sets is $30 + 30 + 5 = 65$.

Before we go on . . . Here is an easier way to answer part (d). First, the total number of sets having no red marbles is $C(8, 4) = 70$. Next, of those, the number containing *no* green marbles is $C(5, 4) = 5$. This leaves $70 - 5 = 65$ sets that contain no red marbles but having at least 1 green marble. (We have really used here the formula for the cardinality of the complement of a set.)

The last example concerns poker hands. For those unfamiliar with playing cards, here is a short description. A standard deck consists of 52 playing cards. Each card is in one of 13 denominations: Ace, 2, 3, 4, 5, 6, 7, 8, 9, 10, Jack (J), Queen (Q), and King (K); and in one of four suits: hearts (♥), diamonds (♦), clubs (♣), and spades (♠). For instance, the Jack of spades, J♠, refers to the denomination of Jack in the suit of spades. The entire deck of cards is

A♥	2♥	3♥	4♥	5♥	6♥	7♥	8♥	9♥	10♥	J♥	Q♥	K♥
A♦	2♦	3♦	4♦	5♦	6♦	7♦	8♦	9♦	10♦	J♦	Q♦	K♦
A♣	2♣	3♣	4♣	5♣	6♣	7♣	8♣	9♣	10♣	J♣	Q♣	K♣
A♠	2♠	3♠	4♠	5♠	6♠	7♠	8♠	9♠	10♠	J♠	Q♠	K♠

Example 8 • Poker Hands

In the card game poker, a hand consists of a set of 5 cards from a standard deck of 52. A *full house* is a hand consisting of 3 cards of one denomination ("three of a kind"—for example, three 10s) and 2 of another ("two of a kind"—for example, two Queens). Here is an example of a full house: 10♣ , 10♦, 10♠, Q♥, Q♣.

a. How many different poker hands are there?

b. How many different full houses are there that contain three 10s and two Queens?

c. How many different full houses are there altogether?

Solution

a. Since the order of the cards doesn't matter, we simply need to know the number of ways of choosing a set of 5 cards out of 52, which is

$$C(52, 5) = 2{,}598{,}960 \text{ hands}$$

b. Here is a decision algorithm for choosing a full house with three 10s and two Queens:

Step 1 Choose three 10s. Since there are four 10s from which to choose, we have $C(4, 3) = 4$ choices.

Step 2 Choose 2 Queens: $C(4, 2) = 6$ choices.

So, there are $4 \times 6 = 24$ possible full houses with three 10s and two Queens.

c. Here is a decision algorithm for choosing a full house:

Step 1 Choose a denomination for the three of a kind: 13 choices.

Step 2 Choose 3 cards of that denomination. Since there are 4 cards of each denomination (one for each suit), we get $C(4, 3) = 4$ choices.

Step 3 Choose a different denomination for the two of a kind. There are only 12 denominations left, so we have 12 choices.

Step 4 Choose 2 of that denomination: $C(4, 2) = 6$ choices.

So, by the multiplication principle, there are a total of $13 \times 4 \times 12 \times 6 = 3744$ possible full houses.

Guideline: Recognizing When to Use Permutations or Combinations

Question How can I tell whether a given application calls for permutations or combinations?

Answer Decide whether the application calls for ordered lists (as in situations where order is implied) or for unordered sets (as in situations where order is not relevant). Ordered lists are permutations, whereas unordered sets are combinations.

6.4 EXERCISES

In Exercises 1–16, evaluate each number.

1. 6! **2.** 7!

3. 8!/6! **4.** 10!/8!

5. $P(6, 4)$ **6.** $P(8, 3)$

7. $P(6, 4)/4!$ **8.** $P(8, 3)/3!$

9. $C(3, 2)$ **10.** $C(4, 3)$

11. $C(10, 8)$ **12.** $C(11, 9)$

13. $C(20, 1)$ **14.** $C(30, 1)$

15. $C(100, 98)$ **16.** $C(100, 97)$

17. How many ordered lists are there of four items chosen from six?

18. How many ordered sequences are possible that contain three objects chosen from seven?

19. How many unordered sets are possible that contain three objects chosen from seven?

20. How many unordered sets are there of four items chosen from six?

21. How many five-letter sequences are possible that use the letters b, o, g, e, y once each?

22. How many six-letter sequences are possible that use the letters q, u, a, k, e, s once each?

23. How many three-letter sequences are possible that use the letters q, u, a, k, e, s at most once each?

24. How many three-letter sequences are possible that use the letters b, o, g, e, y at most once each?

25. How many three-letter (unordered) sets are possible that use the letters q, u, a, k, e, s at most once each?

26. How many three-letter (unordered) sets are possible that use the letters b, o, g, e, y at most once each?

27. How many six-letter sequences are possible that use the letters a, u, a, a, u, k? (*Hint:* Use the decision algorithm discussed in Example 3 of Section 6.3.)

28. How many six-letter sequences are possible that use the letters f, f, a, a, f, f? (See the hint for Exercise 27.)

APPLICATIONS

29. Itineraries Your international diplomacy trip requires stops in Thailand, Singapore, Hong Kong, and Bali. How many possible itineraries are there?

30. Itineraries Referring to Exercise 29, how many possible itineraries are there in which the last stop is Thailand?

Marbles For Exercises 31–44, a bag contains three red, two green, one lavender, two yellow, and two orange marbles.

31. How many possible sets of four marbles are there?

32. How many possible sets of three marbles are there?

33. How many sets of four marbles include all the red ones?

34. How many sets of three marbles include all the yellow ones?

35. How many sets of four marbles include none of the red ones?

36. How many sets of three marbles include none of the yellow ones?

37. How many sets of four marbles include one of each color other than lavender?

38. How many sets of five marbles include one of each color?

39. How many sets of five marbles include at least two red ones?

40. How many sets of five marbles include at least one yellow one?

41. How many sets of five marbles include at most one of the yellow ones?

42. How many sets of five marbles include at most one of the red ones?

43. How many sets of five marbles include either the lavender one or exactly one yellow one, but not both colors?

44. How many sets of five marbles include at least one yellow one, but no green ones?

Poker Hands A poker hand consists of 5 cards from a standard deck of 52. (See the chart preceding Example 8.) In Exercises 45–50, find the number of different poker hands of the specified type.

45. Two pairs (two of one denomination, two of another denomination, and one of a third)

46. Three of a kind (three of one denomination, one of another denomination, and one of a third)

47. Two of a kind (two of one denomination and three of different denominations)

48. Four of a kind (all four of one denomination and one of another)

49. Straight (five cards of consecutive denominations: A, 2, 3, 4, 5 up through 10, J, Q, K, A, not all of the same suit) (Note that the Ace counts either as a 1 or as the denomination above King.)

50. Flush (five cards all of the same suit, but not consecutive denominations)

Dice If a die is rolled 30 times, there are 6^{30} different sequences possible. In Exercises 51–54, find how many of these sequences satisfy certain conditions. (*Hint:* Use the decision algorithm discussed in Example 3 of Section 6.3.)

51. What fraction of these sequences have exactly five 1s?

52. What fraction of these sequences have exactly five 1s and five 2s?

53. What fraction of these sequences have exactly 15 even numbers?

54. What fraction of these sequences have exactly ten numbers less than or equal to 2?

55. Traveling Salesperson Suppose you are a salesperson who must visit 23 cities: Dallas, Tampa, Orlando, Fairbanks, Seattle, Detroit, Chicago, Houston, Arlington, Grand Rapids, Urbana, San Diego, Aspen, Little Rock, Tuscaloosa, Honolulu, New York, Ithaca, Charlottesville, Lynchville, Raleigh, Anchorage, and Los Angeles. Leave all your answers in factorial form.

a. How many possible itineraries are there that visit each city exactly once?

b. Repeat part (a) in the event that the first five stops have already been determined.

c. Repeat part (a) in the event that your itinerary must include the sequence Anchorage, Fairbanks, Seattle, Chicago, and Detroit, in that order.

56. Traveling Salesperson Refer to Exercise 55 (and leave all your answers in factorial form).

a. How many possible itineraries are there that start and end at Detroit and visit every other city exactly once?

b. How many possible itineraries are there that start and end at Detroit and visit Chicago twice and every other city once?

c. Repeat part (a) in the event that your itinerary must include the sequence Anchorage, Fairbanks, Seattle, Chicago, and New York, in that order.

57. Committees The Judicial Committee of the Student Senate is to consist of one chief investigator (Party Party), two assistant investigators (Study Party), two rabble-rousers, and three do-nothing members. The committee is to be selected from a pool of 20 senators, half of whom were elected on the Party Party ticket and half on the Study Party ticket, including freshman Senator Boondoggle (Study Party). Boondoggle is hoping desperately to serve on the committee, preferably as a do-nothing member. Unfortunately, Boondoggle's roommate and bitterest enemy, Senator Porkbarrel (Party Party), is also in the pool of candidates. After giving the matter some thought, Boondoggle finally decides that he will refuse to serve unless (a) he is a do-nothing member and (b) Porkbarrel is not also serving on the committee. How many possible committees are there that would make Boondoggle happy? [Leave your answer as a product of terms of the form $C(n, r)$.]

58. Committees Refer to Exercise 57. Senator Porkbarrel is furious about having been dropped from the Judicial Committee of the Student Senate and has decided to retaliate by forming his own committee, the Committee to Judge the Judicial Committee. It is to have the following members: one supreme investigator (Porkbarrel), two semisupreme investigators (either party), and two demisemisupreme investigators (Party Party). There are a total of eight members of the Party Party (including himself) and nine members of the Study Party (excluding Boondoggle) he is prepared to consider as committee members. How many committees are possible? [Leave your answer as a product of terms of the form $C(n, r)$.]

In Exercises 59–64, calculate how many different sequences can be formed that use the letters of each given word. [Decide where, for example, all the s's will go, rather than what will go in each position. Leave your answer as a product of terms of the form $C(n, r)$.]

59. Mississippi

60. Mesopotamia

61. Megalomania

62. Schizophrenia

63. Casablanca

64. Desmorelda

65. Tests A test requires that you answer Part A, Part B, or Part C. Part A consists of eight true/false questions, Part B consists of five multiple-choice questions with one correct answer out of five, and Part C requires you to match five questions with five different answers. (Each question matches exactly one correct answer.) How many different completed answer sheets are possible?

66. Tests A test requires that you answer first Part A and then either Part B or Part C. Part A consists of four true/false questions, Part B consists of four multiple-choice questions with one correct answer out of five, and Part C requires you to match six questions with six different answers. (Each question matches exactly one correct answer.) How many different completed answer sheets are possible?

67. (From the GMAT) Ben and Ann are among seven contestants from which four semifinalists are to be selected. Of the different possible selections, how many contain neither Ben nor Ann?

(A) 5 **(B)** 6 **(C)** 7 **(D)** 14 **(E)** 21

68. (Based on a Question from the GMAT) Ben and Ann are among seven contestants from which four semifinalists are to be selected. Of the different possible selections, how many contain Ben but not Ann?

(A) 5 **(B)** 8 **(C)** 9 **(D)** 10 **(E)** 20

69. (From the GMAT) If ten persons meet at a reunion and each person shakes hands exactly once with each of the others, what is the total number of handshakes?

(A) $10 \cdot 9 \cdot 8 \cdot 7 \cdot 6 \cdot 5 \cdot 4 \cdot 3 \cdot 2 \cdot 1$

(B) $10 \cdot 10$

(C) $10 \cdot 9$

(D) 45

(E) 36

70. (Based on a Question from the GMAT) If 12 businesspeople have a meeting and each pair exchanges business cards, how many business cards, total, get exchanged?

(A) $12 \cdot 11 \cdot 10 \cdot 9 \cdot 8 \cdot 7 \cdot 6 \cdot 5 \cdot 4 \cdot 3 \cdot 2 \cdot 1$

(B) $12 \cdot 12$

(C) $12 \cdot 11$

(D) 66

(E) 72

71. Product Design Honest Lock Company plans to introduce what it refers to as the "true combination lock." The lock will open if the correct set of three numbers from 0 to 39 is entered in any order.

a. How many different combinations of three different numbers are possible?

b. If it is allowed that a number appear twice (but not three times), how many more possibilities are created?

c. If it is allowed that any or all of the numbers may be the same, how many total possibilities are there?

72. Product Design A vending machine company is planning production of a new vending machine where the customer will be required to enter a sequence of one or two digits 0–9 followed by one of the letters A, B, C, or D. How many different products could the machine offer for sale?

73. Theory of Linear Programming (Some familiarity with linear programming is assumed for this exercise.) Suppose you have a linear programming problem with two unknowns and 20 constraints. You know that graphing the feasible region would take a lot of work, but then you recall that corner points are obtained by solving a system of two equations in two unknowns obtained from two of the constraints. You then decide that it might instead be easier to locate all possible corner points by solving all possible combinations of two equations and then checking whether each solution is a feasible point.

a. How many systems of two equations in two unknowns will you be required to solve?

b. Generalize this to n constraints.

74. More Theory of Linear Programming (Some familiarity with linear programming is assumed for this exercise.) Before the advent of the simplex method for solving linear programming problems, the following method was used: Suppose you have a linear programming problem with three unknowns and 20 constraints. You locate corner points as follows. Selecting three of the constraints, you turn them into equations (by replacing the inequalities with equalities), solve the resulting system of three equations in three unknowns, and then check to see whether the solution is feasible.

a. How many systems of three equations in three unknowns will you be required to solve?

b. Generalize this to n constraints.

COMMUNICATION AND REASONING EXERCISES

75. If you were hard pressed to study for an exam on counting and had only enough time to study one topic, would you choose the formula for the number of permutations or the multiplication principle? Give reasons for your choice.

76. Which of the following represent permutations?

a. An arrangement of books on a shelf

b. A group of ten people in a bus

c. A committee of 5 senators chosen from 100

d. A presidential cabinet of 5 portfolios chosen from 20

77. You are tutoring your friend for a test on sets and counting, and she asks, "How do I know what formula to use for a given problem?" What is a good way to respond?

78. A textbook has the following exercise. "Three students from a class of 50 are selected to take part in a play. How many casts are possible?" Comment on this exercise.

79. Complete the following sentences. If a counting procedure has five alternatives, each of which has four steps of two choices each, then there are ______ outcomes. On the other hand, if there are five steps, each of which has four alternatives of two choices each, then there are ______ outcomes.

80. Explain why the coefficient of a^2b^4 in $(a+b)^6$ is $C(6, 2)$ (this is a consequence of the **binomial theorem**). [*Hint:* In the product $(a + b)(a + b) \cdots (a + b)$ (six times), in how many different ways can you pick two a's and four b's to multiply together?]

CASE STUDY

Courtesy ThinkFun

Designing a Puzzle

As product design manager for Cerebral Toys, you are constantly on the lookout for ideas for intellectually stimulating yet inexpensive toys. You recently received the following memo from Felix Frost, the developmental psychologist on your design team.

> TO: Felicia
> FROM: Felix
> SUBJECT: Crazy Cubes
>
> We've hit on an excellent idea for a new educational puzzle (which we are calling "Crazy Cubes" until marketing comes up with a better name). Basically, Crazy Cubes will consist of a set of plastic cubes. Two faces of each cube will be colored red, two will be colored blue, and two white, and there will be exactly two cubes with each possible configuration of colors. The goal of the puzzle is to seek out the matching pairs, thereby enhancing a child's geometric intuition and three-dimensional manipulation skills. The kit will include every possible configuration of colors. We are, however, a little stumped on the following question: How many cubes will the kit contain? In other words, how many possible ways can one color the faces of a cube so that two faces are red, two are blue, and two are white?

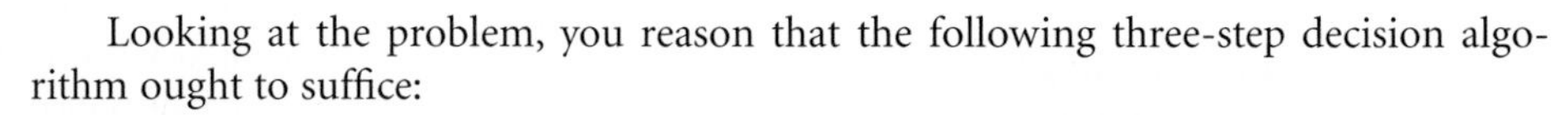

Looking at the problem, you reason that the following three-step decision algorithm ought to suffice:

Step 1 Choose a pair of faces to color red: $C(6, 2) = 15$ choices.

Step 2 Choose a pair of faces to color blue: $C(4, 2) = 6$ choices.

Step 3 Choose a pair of faces to color white: $C(2, 2) = 1$ choice.

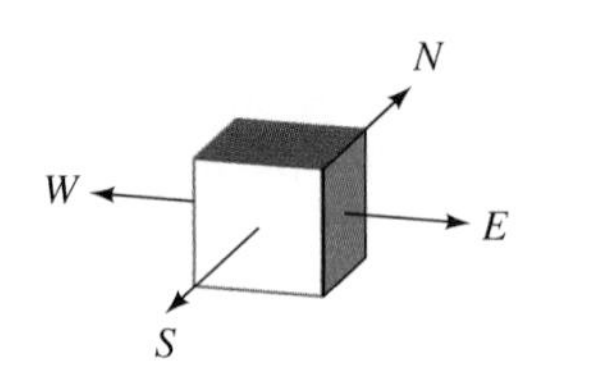

Figure 9

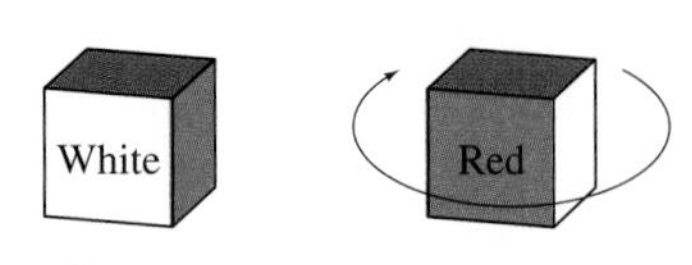

Figure 10

This algorithm appears to give a total of $15 \times 6 \times 1 = 90$ possible cubes. However, before sending your reply to Felix, you realize that something is wrong because different choices result in the same cube. To describe some of these choices, imagine a cube oriented so that four of its faces are facing the four compass directions (Figure 9). Consider choice 1 with the top and bottom faces blue, north and south faces white, and east and west faces red and choice 2 with the top and bottom faces blue, north and south faces red, and east and west faces white. These cubes are actually the same, as you see by rotating the second cube 90° (Figure 10).

You therefore decide that you need a more sophisticated decision algorithm. Here is one that works:

Alternative 1 Faces with the same color opposite each other. Place one of the blue faces down. Then the top face is also blue. The cube must look like the one drawn in Figure 10. Thus there is only 1 choice here.

Alternative 2 Red faces opposite each other and the other colors on adjacent pairs of faces. Again there is only 1 choice, as you can see by putting the red faces on the top and bottom and then rotating.

Alternative 3 White faces opposite each other and the other colors on adjacent pairs of faces: 1 possibility.

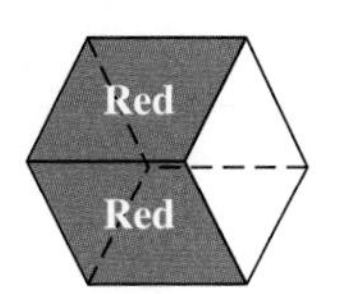

Figure 11

Alternative 4 Blue faces opposite each other and the other colors on adjacent pairs of faces: 1 possibility.

Alternative 5 Faces with the same color adjacent to each other. Look at the cube so that the edge common to the two red faces is facing you and horizontal (Figure 11). Then the faces on the left and right must be of different colors because they are opposite each other. Assume that the face on the right is white. (If it's blue, then rotate the die with the red edge still facing you to move it there, as in Figure 12.)

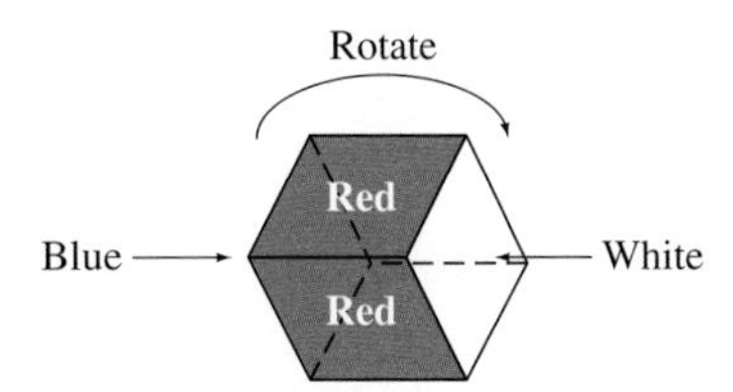

Figure 12

This leaves two choices for the other white face, on the upper or the lower of the two back faces. This alternative gives two choices.

It follows that there are $1 + 1 + 1 + 1 + 2 = 6$ choices. Since the Crazy Cubes kit will feature two of each cube, the kit will require 12 different cubes.[8]

EXERCISES

In all the following exercises there are three colors to choose from: red, white, and blue.

1. In order to enlarge the kit, Felix suggests including two each of two-colored cubes (using two of the colors red, white, and blue) with three faces one color and three another. How many additional cubes will be required?
2. If Felix now suggests adding two each of cubes with two faces one color, one face another color, and three faces the third color, how many additional cubes will be required?
3. Felix changes his mind and suggests the kit use tetrahedral blocks with two colors instead (see the figure).

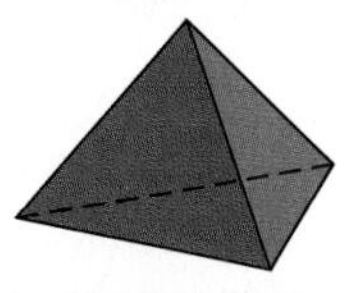

A tetrahedron

How many of these would be required?

4. Once Felix finds the answer to Exercise 3, he decides to go back to the cube idea, but this time insists that all possible combinations of up to three colors should be included. (For instance, some cubes will be all one color, others will be two colors.) How many cubes should the kit contain?

[8]There is a beautiful way of calculating this and similar numbers, called Polya enumeration, but it requires a discussion of topics well outside the scope of this book. Take this as a hint that counting techniques can use some of the most sophisticated mathematics.

CHAPTER 6 REVIEW TEST

1. List the elements of each set.

a. The set N of all negative integers greater than or equal to -3.

b. The set of all outcomes of tossing a coin five times.

c. The set of all outcomes of tossing two distinguishable dice such that the numbers are different.

d. The sets $(A \cap B) \cup C$ and $A \cap (B \cup C)$, where $A = \{1, 2, 3, 4, 5\}$, $B = \{3, 4, 5\}$, and $C = \{1, 2, 5, 6, 7\}$

e. The sets $A \cup B'$ and $A \times B'$, where $A = \{a, b\}$, $B = \{b, c\}$, and $S = \{a, b, c, d\}$.

2. Write each of the indicated sets in terms of the given sets A and B.

a. *S:* the set of all customers; *A:* the set of all customers who owe money; *B:* the set of all customers who owe at least \$1000. The set of all customers who owe money but owe less than \$1000.

b. *A:* the set of outcomes when a day in August is selected; *B:* the set of outcomes when a time of day is selected. The set of outcomes when a day in August and a time of that day are selected.

c. *S:* the set of outcomes when two dice are rolled; *E:* those outcomes in which at most one die shows an even number; *F:* those outcomes in which the sum of the numbers is 7. The set of outcomes in which both dice show an even number or sum to 7.

d. *S:* the set of all integers; *P:* the set of all positive integers; *E:* the set of all even integers; *Q:* the set of all integers that are perfect squares ($Q = \{0, 1, 4, 9, 16, 25, \ldots\}$). The set of all integers that are not positive odd perfect squares.

3. Give a formula for the cardinality rule or rules needed to answer each question and then give the solution.

a. There are 32 students in categories A and B combined; 24 are in A and 24 are in B. How many are in both A and B?

b. You have read 150 of the 400 novels in your home, but your sister Roslyn has read 200, of which only 50 are novels you have read as well. How many have neither of you read?

c. The Apple iMac comes in three models, each with five colors from which to choose. How many combinations are possible?

d. You roll two dice, one red and one green. Losing combinations are doubles (both dice show the same number) and outcomes in which the green die shows an odd number and the red die shows an even number. The other combinations are winning ones. How many winning combinations are there?

4. Recall that a poker hand consists of 5 cards from a standard deck of 52. In each case find the number of different poker hands of the specified type. Leave your answer in terms of combinations.

a. A full house with either two Kings and three Queens or two Queens and three Kings

b. Two of a kind with no Aces

c. Three of a kind with no Aces

d. Straight flush (5 cards of the same suit with consecutive denominations: A, 2, 3, 4, 5 up through 10, J, Q, K, A)

5. A bag contains four red, two green, one transparent, three yellow, and two orange marbles.

a. How many possible sets of five marbles are there?

b. How many sets of five marbles include all the red ones?

c. How many sets of five marbles do not include all the red ones?

d. How many sets of five marbles include at least two yellow ones?

e. How many sets of five marbles include at most one of the red ones but no yellow ones?

OHaganBooks.com—ORGANIZING INFORMATION

6. OHaganBooks.com currently operates three warehouses: in Washington, in California, and in Texas. Book inventories are shown in the following table:

	Science Fiction	Horror	Romance	Other	Total
Washington	10,000	12,000	12,000	30,000	64,000
California	8,000	12,000	6,000	16,000	42,000
Texas	15,000	15,000	20,000	44,000	94,000
Total	33,000	39,000	38,000	90,000	200,000

Take the first letter of each category to represent the corresponding set of books; for instance, S is the set of science fiction books in stock, W is the set of books in the Washington warehouse, and so on. Describe the following sets in words and compute their cardinalities.

a. $S \cup T$

b. $H \cap C$

c. $C \cup S'$

d. $(R \cap T) \cup H$

e. $R \cap (T \cup H)$

f. $(S \cap W) \cup (H \cap C')$

7. OHaganBooks.com has two main competitors: JungleBooks.com and FarmerBooks.com. At the beginning of August, OHaganBooks.com had 3500 customers. Of these, a total of 2000 customers were shared with JungleBooks.com and 1500 with FarmerBooks.com. Furthermore, 1000 customers were shared with both.

a. How many of all these customers are exclusive OHaganBooks.com customers?

b. JungleBooks.com has a total of 3600 customers, FarmerBooks.com has 3400, and they share 1100 customers between them. How many of their customers are not customers of OHaganBooks.com?

c. Which of the three companies has the largest number of exclusive customers?

d. OHaganBooks.com is interested in merging with one of its two competitors. Which merger would give it the largest combined customer base, and how large would that be?

e. Refer to part (c). Which merger would give OHaganBooks.com the largest *exclusive* customer base, and how large would that be?

8. As the customer base at OHaganBooks.com grows, software manager Ruth Nabarro is thinking of introducing identity codes for all the online customers.

a. If she uses three-letter codes, how many different customers can be identified?

b. If she uses codes with three different letters, how many different customers can be identified?

c. It appears that Nabarro has finally settled on codes consisting of two letters followed by two digits. For technical reasons, the letters must be different, and the first digit cannot be a zero. How many different customers can be identified?

d. The CEO sends Nabarro the following memo:

> To: Ruth Nabarro, Software Manager
> From: John O'Hagan, CEO
> Subject: Customer Identity Codes
> I have read your proposal for the customer ID codes. However, due to our ambitious expansion plans, I would like our system software to allow for at least 500,000 customers. Please adjust your proposal accordingly.

Nabarro is determined to have a sequence of letters followed by some digits, and, for reasons too complicated to explain, there cannot be more than two letters, the letters must be different, all the digits must be different, and the first digit cannot be a zero. What is the form of the shortest code she can use to satisfy the CEO, and how many different customers can be identified?

9. After an exhausting day at the office (you are CEO of OHaganBooks.com), you return home and find yourself having to assist your son Billy Sean, who continues to have a terrible time planning his first-year college course schedule. The latest Bulletin of Suburban State University reads as follows:

> All candidates for the degree of Bachelor of Arts at SSU must take, in their first year, at least ten courses in the Sciences, Fine Arts, Liberal Arts, and Mathematics combined, of which at least two must be in each of the Sciences and Fine Arts, and exactly three must be in each of the Liberal Arts and Mathematics.

a. If the Bulletin lists exactly five first-year-level science courses and six first-year-level courses in each of the other categories, how many course combinations are possible that meet the minimum requirements?

b. Reading through the course descriptions in the bulletin a second time, you notice that Calculus I (listed as one of the mathematics courses) is a required course for many of the other courses, and you decide that it would be best if Billy Sean included Calculus I. Further, two of the Fine Arts courses cannot both be taken in the first year. How many course combinations are possible that meet the minimum requirements and include Calculus I?

c. To complicate things further, in addition to the requirement in part (b), Physics II has Physics I as a prerequisite (both are listed as first-year science courses, but it is not necessary to take both). How many course combinations are possible that include Calculus I and meet the minimum requirements?

ADDITIONAL ONLINE REVIEW

If you follow the path

Web site → Everything for Finite Math → Chapter 6

you will find the following additional resources to help you review:

A comprehensive chapter summary (including examples and interactive features)

Additional review exercises (including interactive exercises and many with help)

A true/false chapter quiz

7

PROBABILITY

CASE STUDY

The Monty Hall Problem

On the game show *Let's Make a Deal,* you are shown three doors, A, B, and C, and behind one of them is the Big Prize. After you select one of them—say, door A—to make things more interesting the host (Monty Hall) opens one of the other doors—say, door B—revealing that the Big Prize is not there. He then offers you the opportunity to change your selection to the remaining door, door C. Should you switch or stick with your original guess? Does it make any difference?

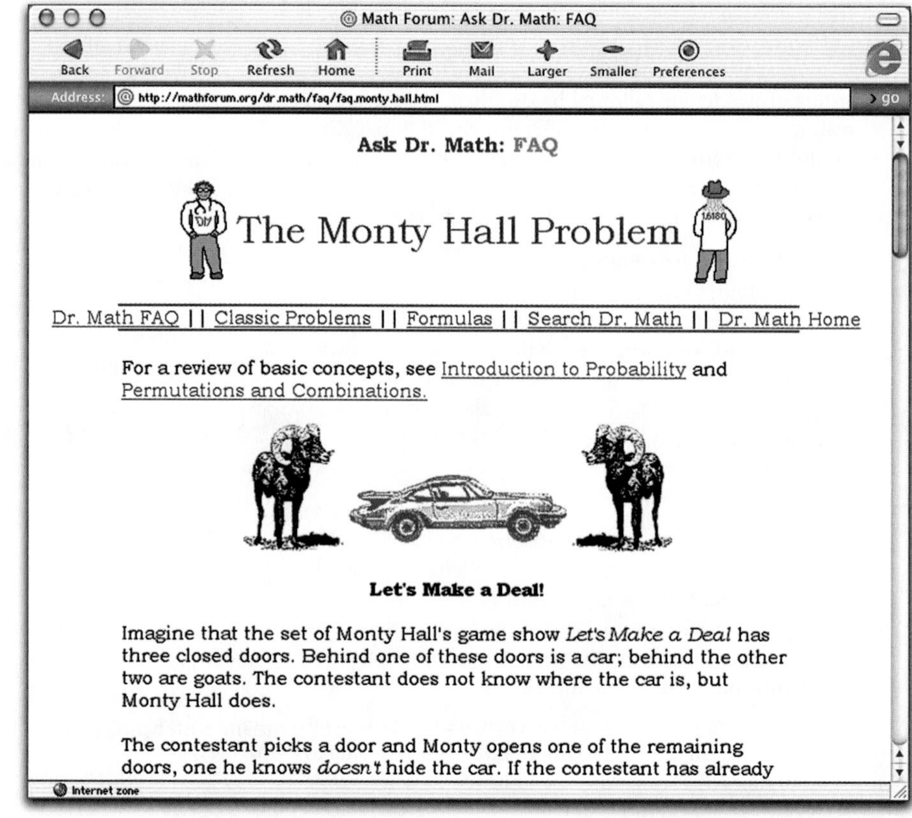

INTERNET RESOURCES FOR THIS CHAPTER 

At the Web site, follow the path

Web site → Everything for Finite Math → Chapter 7

where you will find step-by-step tutorials for the main topics in this chapter, a detailed chapter summary you can print out, a true/false quiz, and a collection of chapter review questions.

Introduction

What is the probability of winning the lottery twice? What are the chances that a college athlete whose drug test is positive for steroid use is actually using steroids? You are playing poker and have been dealt two Jacks. What is the likelihood that one of the next three cards you are dealt will also be a Jack? These are all questions about probability.

Understanding probability is important in many fields, ranging from risk management in business through hypothesis testing in psychology to quantum mechanics in physics. The goal of this chapter is to familiarize you with the basic concepts of probability theory and to give you a working knowledge that you can apply in a variety of situations.

In the first two sections, the emphasis is on translating real-life situations into the language of sample spaces, events, and probability. Once we have mastered the language of probability, we spend the rest of the chapter studying some of its theory and applications.

7.1 Sample Spaces and Events

Sample Spaces

When we can't decide between two alternatives, it's common to toss a coin to make the decision. If we toss the coin and observe which side faces up, there are two possible results: heads (H) and tails (T). These are the *only* possible results, ignoring the (remote) possibility that the coin lands on its edge. The act of tossing the coin is an example of an **experiment.** The two possible results, H and T, are possible **outcomes** of the experiment, and the set $S = \{H, T\}$ of all possible outcomes is the **sample space** for the experiment.

Experiments, Outcomes, and Sample Spaces

An **experiment** is an occurrence with a result, or **outcome,** that is uncertain. The set of all possible outcomes is called the **sample space** for the experiment.

Quick Examples

1. *Experiment:* Flip a coin and observe the side facing up.
   Outcomes: H, T
   Sample space: $S = \{H, T\}$
2. *Experiment:* Select a student in your class.
   Outcomes: The students in your class
   Sample space: The set of students in your class

3. *Experiment:* Select a student in your class and observe the color of his or her hair.
   Outcomes: red, black, brown, blond, green, . . .
   Sample space: {red, black, brown, blond, green, . . . }
4. *Experiment:* Cast a die and observe the number facing up.
   Outcomes: 1, 2, 3, 4, 5, 6
   Sample space: $S = \{1, 2, 3, 4, 5, 6\}$
5. *Experiment:* Cast two distinguishable dice and observe the numbers facing up.
   Outcomes: (1, 1), (1, 2), . . . , (6, 6) (36 outcomes)

$$\textit{Sample space: } S = \left\{\begin{matrix} (1,1), & (1,2), & (1,3), & (1,4), & (1,5), & (1,6), \\ (2,1), & (2,2), & (2,3), & (2,4), & (2,5), & (2,6), \\ (3,1), & (3,2), & (3,3), & (3,4), & (3,5), & (3,6), \\ (4,1), & (4,2), & (4,3), & (4,4), & (4,5), & (4,6), \\ (5,1), & (5,2), & (5,3), & (5,4), & (5,5), & (5,6), \\ (6,1), & (6,2), & (6,3), & (6,4), & (6,5), & (6,6) \end{matrix}\right\}$$

$n(S) = 36$

6. *Experiment:* Cast two indistinguishable dice and observe the numbers facing up.
   Outcomes: (1, 1), (1, 2), . . . , (6, 6) (21 outcomes)

$$\textit{Sample space: } S = \left\{\begin{matrix} (1,1), & (1,2), & (1,3), & (1,4), & (1,5), & (1,6), \\ & (2,2), & (2,3), & (2,4), & (2,5), & (2,6), \\ & & (3,3), & (3,4), & (3,5), & (3,6), \\ & & & (4,4), & (4,5), & (4,6), \\ & & & & (5,5), & (5,6), \\ & & & & & (6,6) \end{matrix}\right\}$$

$n(S) = 21$

7. *Experiment:* Cast two dice and observe the *sum* of the numbers facing up.
   Outcomes: 2, 3, 4, 5, 6, 7, 8, 9, 10, 11, 12
   Sample space: $S = \{2, 3, 4, 5, 6, 7, 8, 9, 10, 11, 12\}$
8. *Experiment:* Choose two cars (without regard to order) at random from a fleet of ten.
   Outcomes: Collections of two cars chosen from ten
   Sample space: The set of all collections of two cars chosen from ten

$n(S) = C(10, 2) = 45$

Example 1 • School and Work

In a survey conducted by the Bureau of Labor Statistics, the high school graduating class of 2000 was divided into those who went on to college and those who did not.[1] Those who went on to college were further divided into those who went to 2-year colleges and those who went to 4-year colleges. All graduates were also asked whether they were working or not. Find the sample space for the experiment "Select a member of the high school graduating class of 2000 and classify his or her subsequent school and work activity."

[1]SOURCE: "College Enrollment and Work Activity of High School Graduates," U.S. Bureau of Labor Statistics, http://www.bls.gov/news.release/hsgec.toc.htm, April 2001.

Solution The tree in Figure 1 shows the various possibilities.

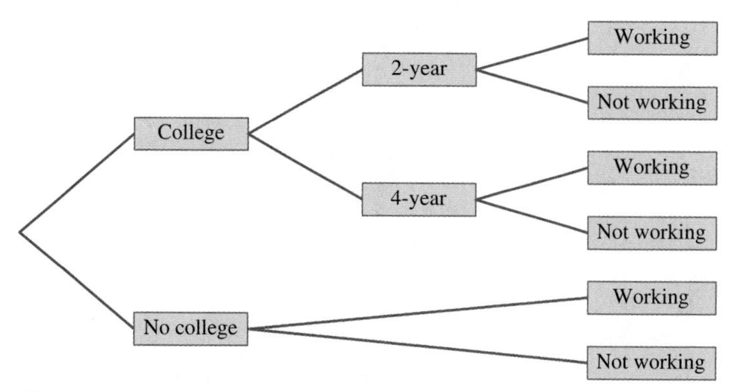

Figure 1

The sample space is

$S = \{$2-year college and working, 2-year college and not working, 4-year college and working, 4-year college and not working, no college and working, no college and not working$\}$

Events

In Example 1 suppose we are interested in the event that a 2000 high school graduate was working. In mathematical language we are interested in the *subset* of the sample space consisting of all outcomes in which the graduate was working.

Events

Given a sample space S, an **event** E is a subset of S. The outcomes in E are called the **favorable** outcomes. We say that E **occurs** in a particular experiment if the outcome of that experiment is one of the elements of E—that is, if the outcome of the experiment is favorable.

Quick Examples

1. *Experiment:* Roll a die and observe the number facing up.

$S = \{1, 2, 3, 4, 5, 6\}$

Event: E: The number observed is odd.

$E = \{1, 3, 5\}$

2. *Experiment:* Roll two distinguishable dice and observe the numbers facing up.

$S = \{(1, 1), (1, 2), \ldots, (6, 6)\}$

Event: F: The dice show the same number.

$F = \{(1, 1), (2, 2), (3, 3), (4, 4), (5, 5), (6, 6)\}$

3. *Experiment:* Roll two distinguishable dice and observe the numbers facing up.

$S = \{(1, 1), (1, 2), \ldots, (6, 6)\}$

Event: G: The sum of the numbers is 1.

$G = \varnothing$ There are *no* favorable outcomes.

4. *Experiment:* Select a city beginning with "J."
   Event: E: The city is Johannesburg.

   $E = \{\text{Johannesburg}\}$ An event can consist of a single outcome.

5. *Experiment:* Roll a die and observe the number facing up.
   Event: E: The number observed is either even or odd.

   $E = S = \{1, 2, 3, 4, 5, 6\}$ An event can consist of all possible outcomes.

6. *Experiment:* Select a student in your class.
   Event: E: The student has red hair.

   $E = \{\text{red-haired students in your class}\}$

7. *Experiment:* Draw a hand of 2 cards from a deck of 52.
   Event: H: Both cards are diamonds.

   H is the set of all hands of 2 cards chosen from 52 such that both cards are diamonds.

Example 2 • Dice

We roll a red die and a green die and observe the numbers facing up. Describe the following events as subsets of the sample space:

a. E: The sum of the numbers showing is 6.

b. F: The sum of the numbers showing is 2.

Solution Here (again) is the sample space for the experiment of throwing two dice:

$$S = \begin{Bmatrix} (1,1), & (1,2), & (1,3), & (1,4), & (1,5), & (1,6), \\ (2,1), & (2,2), & (2,3), & (2,4), & (2,5), & (2,6), \\ (3,1), & (3,2), & (3,3), & (3,4), & (3,5), & (3,6), \\ (4,1), & (4,2), & (4,3), & (4,4), & (4,5), & (4,6), \\ (5,1), & (5,2), & (5,3), & (5,4), & (5,5), & (5,6), \\ (6,1), & (6,2), & (6,3), & (6,4), & (6,5), & (6,6) \end{Bmatrix}$$

a. In mathematical language, E is the subset of S that consists of all those outcomes in which the sum of the numbers showing is 6. Here is the sample space once again, with the outcomes in question shown in color:

$$S = \begin{Bmatrix} (1,1), & (1,2), & (1,3), & (1,4), & (1,5), & (1,6), \\ (2,1), & (2,2), & (2,3), & (2,4), & (2,5), & (2,6), \\ (3,1), & (3,2), & (3,3), & (3,4), & (3,5), & (3,6), \\ (4,1), & (4,2), & (4,3), & (4,4), & (4,5), & (4,6), \\ (5,1), & (5,2), & (5,3), & (5,4), & (5,5), & (5,6), \\ (6,1), & (6,2), & (6,3), & (6,4), & (6,5), & (6,6) \end{Bmatrix}$$

Thus, $E = \{(1, 5), (2, 4), (3, 3), (4, 2), (5, 1)\}$.

b. The only outcome in which the numbers showing add to 2 is (1, 1). Thus, $F = \{(1, 1)\}$.

Example 3 • School and Work

Let S be the sample space of Example 1. List the elements in the following events:

a. The event E that a 2000 high school graduate was working

b. The event F that a 2000 high school graduate was not going to a 2-year college

Solution

a. We had this sample space:

S = {2-year college and working, 2-year college and not working, 4-year college and working, 4-year college and not working, no college and working, no college and not working}

We are asked for the event that a graduate was working. Whenever we encounter a phrase involving "the event that," we mentally translate this into mathematical language by changing the wording:

Replace the phrase "the event that" by the phrase "the subset of the sample space consisting of all outcomes in which"

Thus, we are interested in the subset of the sample space consisting of all outcomes in which the graduate was working. This gives

E = {2-year college and working, 4-year college and working, no college and working}

b. We are looking for the event that a graduate was not going to a 2-year college. Translating this into mathematical language, we want the subset of the sample space consisting of all outcomes in which the graduate was not going to a 2-year college:

F = {4-year college and working, 4-year college and not working, no college and working, no college and not working}

Complement, Union, and Intersection of Events

Events may often be described in terms of other events, using set operations such as complement, union, and intersection.

Complement, Union, and Intersection of Events

Complement of an Event

The **complement** of an event E is the set of outcomes not in E. Thus, the complement of E represents the event that E *does not occur.*

Quick Examples

1. You take four shots at the goal during a soccer game and record the number of times you score. Describe the event that you score at least twice, and also its complement.

$S = \{0, 1, 2, 3, 4\}$ Set of outcomes

$E = \{2, 3, 4\}$ Event that you score at least twice

$E' = \{0, 1\}$ Event that you do *not* score at least twice

2. You roll a red die and a green die and observe the two numbers facing up. Describe the event that the sum of the numbers is not 6.

$$S = \{(1, 1), (1, 2), \ldots, (6, 6)\}$$

$$F = \{(1, 5), (2, 4), (3, 3), (4, 2), (5, 1)\} \qquad \text{Sum of numbers is 6.}$$

$$F' = \left\{\begin{matrix} (1,1), & (1,2), & (1,3), & (1,4), & & (1,6), \\ (2,1), & (2,2), & (2,3), & & (2,5), & (2,6), \\ (3,1), & (3,2), & & (3,4), & (3,5), & (3,6), \\ (4,1), & & (4,3), & (4,4), & (4,5), & (4,6), \\ & (5,2), & (5,3), & (5,4), & (5,5), & (5,6), \\ (6,1), & (6,2), & (6,3), & (6,4), & (6,5), & (6,6) \end{matrix}\right\} \qquad \text{Sum of numbers is } \textit{not } 6.$$

Union of Events

The **union** of the events E and F is the set of all outcomes in E or F (or both). Thus, $E \cup F$ represents the event that E occurs *or* F occurs (or both).*

Quick Example

Roll a die.

E: The outcome is a 5; $E = \{5\}$.

F: The outcome is an even number; $F = \{2, 4, 6\}$.

$E \cup F$: The outcome is either a 5 *or* an even number; $E \cup F = \{2, 4, 5, 6\}$.

Intersection of Events

The **intersection** of the events E and F is the set of all outcomes common to E and F. Thus, $E \cap F$ represents the event that both E *and* F occur.

Quick Example

Roll two dice: one red and one green.

E: The red die is 2.

F: The green die is odd.

$E \cap F$: The red die is 2, and the green die is odd;
$E \cap F = \{(2, 1), (2, 3), (2, 5)\}$.

*As in Chapter 6, when we use the word *or,* we agree to mean one or the other *or both.* This is called the **inclusive or,** and mathematicians have agreed to take this as the meaning of *or* to avoid confusion.

Example 4 • Weather

Let R be the event that it will rain tomorrow, let P be the event that it will be pleasant, let C be the event that it will be cold, and let H be the event that it will be hot.

a. Express in words: $R \cap P'$, $R \cup (P \cap C)$.

b. Express in symbols: Tomorrow will be either a pleasant day or a cold and rainy day; it will not, however, be hot.

Solution The key here is to remember that intersection corresponds to *and* and union to *or*.

a. $R \cap P'$ is the event that it will rain *and* it will not be pleasant.
$R \cup (P \cap C)$ is the event that either it will rain or it will be pleasant and cold.

b. If we rephrase the given statement using *and* and *or* we get "Tomorrow will be either a pleasant day or a cold and rainy day, and it will not be hot":

$[P \cup (C \cap R)] \cap H'$ Pleasant, or cold and rainy, and not hot

The nuances of the English language play an important role in this formulation. For instance, the effect of the pause (comma) after "rainy day" suggests placing the preceding clause $P \cup (C \cap R)$ in parentheses. In addition, the phrase "cold and rainy" suggests that C and R should be grouped together in their own parentheses.

Example 5 • Sales of Recreational Boats

The following table shows sales of recreational boats in the United States during the period 1999–2001:

	Motor Boats	Jet Skis	Sailboats	Total
1999	330,000	100,000	20,000	450,000
2000	340,000	100,000	20,000	460,000
2001	310,000	90,000	30,000	430,000
Total	980,000	290,000	70,000	1,340,000

SOURCE: National Marine Manufacturers Association/*New York Times*, January 10, 2002, p. C1. Figures are approximate, and represent new recreational boats sold. (Jet skis includes similar vehicles, such as wave runners.)

Consider the experiment in which a recreational boat is selected at random from those in the table. Let E be the event that the boat was a motor boat, let F be the event that the boat was purchased in 2001, and let G be the event that the boat was a sailboat. Describe the following events: (a) $E \cap F$, (b) G', and (c) $E \cup F'$. How many outcomes are there in each event?

Solution Before we answer the questions, note that S is the set of all recreational boats represented in the table, so S has a total of 1,340,000 outcomes.

a. $E \cap F$ is the event that the boat selected was a motor boat *and* was purchased in 2001. Referring to the leftmost column of the table, we find 310,000 outcomes in this event.

b. G' is the event that the boat selected was not a sailboat. So, G' is the event that the boat selected was a motor boat or a jet ski, and $n(G') = 980,000 + 290,000 = 1,270,000$.

c. $E \cup F'$ is the event that either the boat selected was a motor boat *or* it was *not* purchased in 2001. The following table shows the relevant outcomes shaded:

	Motor Boats	Jet Skis	Sailboats	Total
1999	330,000	100,000	20,000	450,000
2000	340,000	100,000	20,000	460,000
2001	310,000	90,000	30,000	430,000
Total	980,000	290,000	70,000	1,340,000

Rather than add these seven numbers together, we can use the formula for the cardinality of a union to shorten the calculation:

$$n(E \cup F') = n(E) + n(F') - n(E \cap F')$$
$$= 980{,}000 + (450{,}000 + 460{,}000) - (330{,}000 + 340{,}000)$$
$$= 1{,}220{,}000 \text{ boats}$$

Before we go on . . . We could shorten the calculation in part (c) even further using De Morgan's law to write $n(E \cup F') = n((E' \cap F)') = 1{,}340{,}000 - (90{,}000 + 30{,}000)$, a calculation suggested by looking at the table.

The case where $E \cap F$ is empty is interesting, and we give it a name.

Mutually Exclusive Events

If E and F are events, then E and F are said to be **disjoint** or **mutually exclusive** if $E \cap F$ is empty. (Hence, they have no outcomes in common.)

Interpretation

It is impossible for mutually exclusive events to occur simultaneously.

Quick Examples

In each of the following examples, E and F are mutually exclusive events.

1. Roll a die and observe the number facing up. E: The outcome is even; F: The outcome is odd. $E = \{2, 4, 6\}$ $F = \{1, 3, 5\}$
2. Toss a coin three times and record the sequence of heads and tails. E: All three tosses land the same way up; F: One toss shows heads, and the other two show tails. $E = \{\text{HHH, TTT}\}$, $F = \{\text{HTT, THT, TTH}\}$
3. Observe tomorrow's weather. E: It is raining; F: There is not a cloud in the sky.

Guideline: Specifying the Sample Space

Question How do I determine the sample space in a given application?

Answer Strictly speaking, an experiment should include a description of what kinds of objects are in the sample space, as in

1. Cast a die and observe the number facing up.

   Sample space: the possible numbers facing up, $\{1, 2, 3, 4, 5, 6\}$

2. Choose a person at random and record her Social Security number and whether she is blonde.

   Sample space: pairs (9-digit number, Y/N)

However, in many of the scenarios discussed in this chapter and the next, an experiment is specified more vaguely, as in "Select a student in your class." In cases like this, the nature of the sample space should be determined from the context. For example, if the discussion is about grade-point averages and gender, the sample space can be taken to consist of pairs (grade-point average, M/F).

7.1 EXERCISES

In Exercises 1–18, describe the sample space S of the experiment and list the elements of the given event. (Assume that the coins are distinguishable and that what is observed are the faces or numbers that face up.)

1. Two coins are tossed; the result is at most one tail.

2. Two coins are tossed; the result is one or more heads.

3. Three coins are tossed; the result is at most one head.

4. Three coins are tossed, the result is more tails than heads.

5. Two distinguishable dice are rolled; the numbers add to 5.

6. Two distinguishable dice are rolled; the numbers add to 9.

7. Two indistinguishable dice are rolled; the numbers add to 4.

8. Two indistinguishable dice are rolled; one of the numbers is even, and the other is odd.

9. Two indistinguishable dice are rolled; both numbers are prime.[2]

10. Two indistinguishable dice are rolled; neither number is prime.

11. A letter is chosen at random from those in the word *Mozart;* the letter is a vowel.

12. A letter is chosen at random from those in the word *Mozart;* the letter is neither *a* nor *m*.

13. A sequence of two different letters is randomly chosen from those of the word *sore;* the first letter is a vowel.

14. A sequence of two different letters is randomly chosen from those of the word *hear;* the second letter is not a vowel.

15. A sequence of two different digits is randomly chosen from the digits 0–4; the first digit is larger than the second.

16. A sequence of two different digits is randomly chosen from the digits 0–4; the first digit is twice the second.

17. You are considering purchasing either a domestic car, an imported car, a van, an antique car, or an antique truck; you do not buy a car.

18. You are deciding whether to enroll for Psychology 1, Psychology 2, Economics 1, General Economics, or Math for Poets; you decide to avoid economics.

19. A packet of gummy candy contains four strawberry, four lime, two black currant, and two orange gums. April May sticks her hand in and selects four at random. Complete the following sentences:

a. The sample space is the set of ______.

b. April is particularly fond of the combination of two strawberry and two black currant. The event that April will get the combination she desires is the set of ______.

20. A bag contains three red, two blue, and four yellow marbles, and Alexandra pulls out three of them at random. Complete the following sentences:

a. The sample space is the set of ______.

b. The event that Alexandra gets one of each color is the set of ______.

21. President George W. Bush's cabinet consists of the Secretaries of State, Treasury, Defense, Interior, Agriculture, Commerce, Labor, Health and Human Services, Housing and Urban Development, Transportation, Energy, Education, Veterans Affairs, and the Attorney General. Assuming that President Bush had 20 candidates, including Colin Powell, to fill these posts (and wished to assign no one to more than one post), complete the following sentences:

a. The sample space is the set of ______.

b. The event that Colin Powell is the Secretary of State is the set of ______.

SOURCE: The Whitehouse Web site, http://www.whitehouse.gov/.

22. A poker hand consists of a set of 5 cards chosen from a standard deck of 52 playing cards. You are dealt a poker hand. Complete the following sentences:

a. The sample space is the set of ______.

b. The event "a full house" is the set of ______. (A full house is 3 cards of one denomination and 2 of another.)

Suppose two dice (one red, one green) are rolled. Consider the following events. A: The red die shows 1; B: The numbers add to 4; C: At least one of the numbers is 1; and D: The numbers do not add to 11. In Exercises 23–30, express the given event in symbols and say how many elements it contains.

23. The red die shows 1, and the numbers add to 4.

24. The red die shows 1, but the numbers do *not* add to 11.

25. The numbers do not add to 4.

26. The numbers add to 11.

27. The numbers do not add to 4, but they do add to 11.

28. Either the numbers add to 11, or the red die shows a 1.

29. At least one of the numbers is 1, or the numbers add to 4.

30. Either the numbers add to 4, or they add to 11, or at least one of them is 1.

Let W be the event that you will use the Web site tonight, let I be the event that your math grade will improve, and let E be the event that you will use the Web site every night. In Exercises 31–36, express the given event in symbols.

31. You will use the Web site tonight, and your math grade will improve.

32. You will use the Web site tonight, or your math grade will *not* improve.

33. Either you will use the Web site every night or your math grade will *not* improve.

34. Your math grade will not improve, even though you use the Web site every night.

35. Either your math grade will improve or you will use the Web site tonight but not every night.

36. You will either use the Web site tonight with no grade improvement or every night with grade improvement.

---

[2]A positive integer is **prime** if it is neither 1 nor a product of smaller integers.

APPLICATIONS

Housing Prices Exercises 37–42 are based on the following graphic that shows the percent increase in housing prices from September 30, 2000, to September 30, 2001, in each of nine regions (U.S. census divisions):

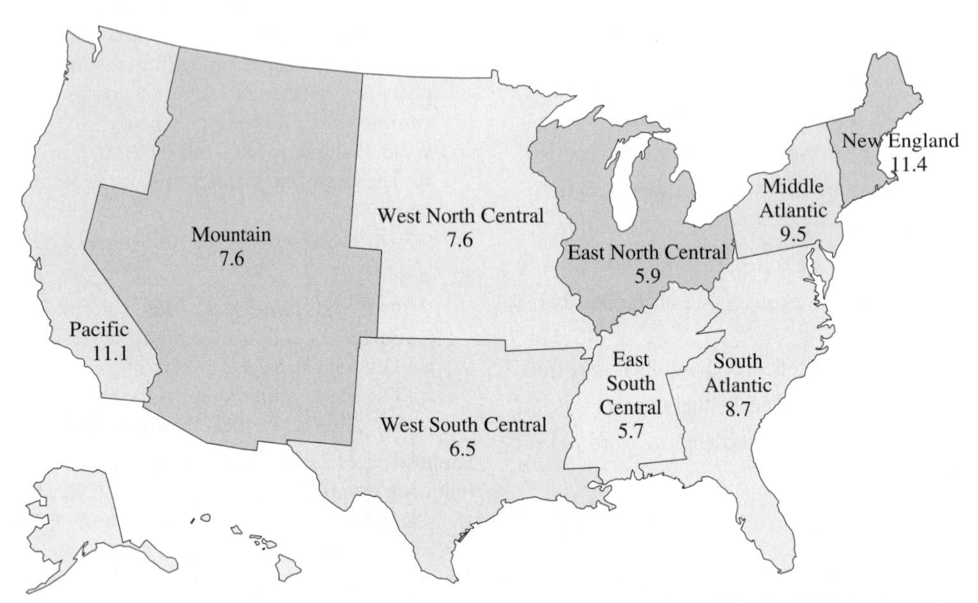

SOURCE: Third Quarter 2001 House Price Index, released November 30, 2001, by the Office of Federal Housing Enterprise Oversight; available online at http://www.ofheo.gov/house/3q01hpi.pdf.

37. You are choosing a region of the country to move to. Describe the event E that the region you choose saw an increase in housing prices of 9% or more.

38. You are choosing a region of the country to move to. Describe the event F that the region you choose saw an increase in housing prices of less than 9%.

39. You are choosing a region of the country to move to. Let E be the event that the region you choose saw an increase in housing prices of 9% or more and let F be the event that the region you choose is on the East Coast. Describe the events $E \cup F$ and $E \cap F$ both in words and by listing the outcomes of each.

40. You are choosing a region of the country to move to. Let E be the event that the region you choose saw an increase in housing prices of less than 9% and let F be the event that the region you choose is not on the East Coast. Describe the events $E \cup F$ and $E \cap F$ both in words and by listing the outcomes of each.

41. You are choosing a region of the country to move to. Which of the following pairs of events are mutually exclusive?

a. E: You choose a region from among the three with the highest percent increase in housing prices. F: You choose a region that is not on the East Coast or West Coast.

b. E: You choose a region from among the three with the highest percent increase in housing prices. F: You choose a region that is not on the West Coast.

42. You are choosing a region of the country to move to. Which of the following pairs of events are mutually exclusive?

a. E: You choose a region from among the three with the lowest percent increase in housing prices. F: You choose a region from among the central divisions.

b. E: You choose a region from among the three with the lowest percent increase in housing prices. F: You choose a region from among the noncentral divisions.

Publishing Exercises 43–50 are based on the following table, which shows the results of a survey of authors by a (fictitious) publishing company:

	New Authors	Established Authors	Total
Successful	5	25	30
Unsuccessful	15	55	70
Total	20	80	100

Consider the following events. S: An author is successful; U: An author is unsuccessful; N: An author is new; and E: An author is established.

43. Describe the events $S \cap N$ and $S \cup N$ in words. Use the table to compute $n(S \cap N)$ and $n(S \cup N)$.

44. Describe the events $N \cap U$ and $N \cup U$ in words. Use the table to compute $n(N \cap U)$ and $n(N \cup U)$.

45. Which of the following pairs of events are mutually exclusive: N and E, N and S, S and E?

46. Which of the following pairs of events are mutually exclusive: U and E, U and S, S and N?

47. Describe the event $S \cap N'$ in words and find the number of elements it contains.

48. Describe the event $U \cup E'$ in words and find the number of elements it contains.

49. What percentage of established authors are successful? What percentage of successful authors are established?

50. What percentage of new authors are unsuccessful? What percentage of unsuccessful authors are new?

Exercises 51–58 are based on the following table that shows the performance of a selection of 100 stocks after 1 year. (Take S to be the set of all stocks represented in the table.)

	Companies			
	Pharmaceutical, *P*	**Electronic, *E***	**Internet, *I***	**Total**
Increased, *V*	10	5	15	30
Unchanged,* *N*	30	0	10	40
Decreased, *D*	10	5	15	30
Total	50	10	40	100

*If a stock stayed within 20% of its original value, it is classified as unchanged.

51. Use symbols to describe the event that a stock's value increased, but it was not an Internet stock. How many elements are in this event?

52. Use symbols to describe the event that an Internet stock did not increase. How many elements are in this event?

53. Compute $n(P' \cup N)$. What does this number represent?

54. Compute $n(P \cup N')$. What does this number represent?

55. Find all pairs of mutually exclusive events.

56. Find all pairs of events that are *not* mutually exclusive.

57. Calculate $\dfrac{n(V \cap I)}{n(I)}$. What does the answer represent?

58. Calculate $\dfrac{n(D \cap I)}{n(D)}$. What does the answer represent?

Animal Psychology Exercises 59–64 concern the following chart that shows the way in which a dog moves its facial muscles when torn between the drives of fight and flight. The fight drive increases from left to right; the flight drive increases from top to bottom. (Notice that an increase in the fight drive causes its upper lip to lift, and an increase in the flight drive draws its ears downward.)

SOURCE: Konrad Lorenz, *On Aggression* (Fakenham, Norfolk: University Paperback Edition, Cox & Wyman Limited, 1967).

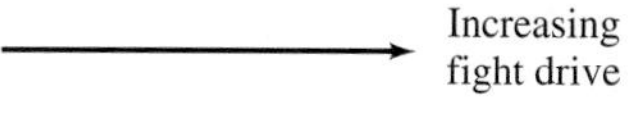

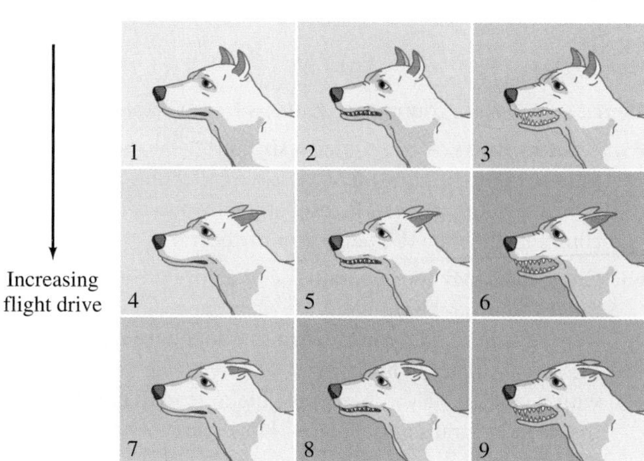

59. Let E be the event that the dog's flight drive is the strongest, let F be the event that the dog's flight drive is weakest, let G be the event that the dog's fight drive is the strongest, and let H be the event that the dog's fight drive is weakest. Describe the following events in terms of E, F, G, and H, using the symbols $\cap$, $\cup$, and $'$.

a. The dog's flight drive is not strongest, and its fight drive is weakest.

b. The dog's flight drive is strongest, or its fight drive is weakest.

c. Neither the dog's flight drive nor fight drive are strongest.

60. Let E be the event that the dog's flight drive is the strongest, let F be the event that the dog's flight drive is weakest, let G be the event that the dog's fight drive is the strongest, and let H be the event that the dog's fight drive is weakest. Describe the following events in terms of E, F, G, and H, using the symbols $\cap$, $\cup$, and $'$.

a. The dog's flight drive is weakest, and its fight drive is not weakest.

b. The dog's flight drive is not strongest, or its fight drive is weakest.

c. Either the dog's flight drive or its fight drive fails to be strongest.

61. Describe the following events explicitly (as subsets of the sample space as numbered in the diagram):

a. The dog's fight and flight drives are both strongest.

b. The dog's fight drive is strongest, but its flight drive is neither weakest nor strongest.

62. Describe the following events explicitly (as subsets of the sample space as numbered in the diagram):

a. Neither the dog's fight drive nor its flight drive is strongest.

b. The dog's fight drive is weakest, but its flight drive is neither weakest nor strongest.

63. Describe the following events in words:

a. $\{1, 4, 7\}$

b. $\{1, 9\}$

c. $\{3, 6, 7, 8, 9\}$

64. Describe the following events in words:

a. $\{7, 8, 9\}$

b. $\{3, 7\}$

c. $\{1, 2, 3, 4, 7\}$

Exercises 65–72 use counting arguments from Chapter 6.

65. Gummy Bears A bag contains six gummy bears. Noel picks four at random. How many possible outcomes are there? If one of the gummy bears is raspberry, how many of these outcomes include the raspberry gummy bear?

66. Chocolates My couch potato friend enjoys sitting in front of the TV and grabbing handfuls of 5 chocolates at random from his snack jar. Unbeknownst to him, I have replaced 1 of the 20 chocolates in his jar with a cashew. (He hates cashews with a passion.) How many possible outcomes are there the first time he grabs 5 chocolates? How many of these include the cashew?

67. Horse Races The seven contenders in the fifth horse race at Aqueduct on February 18, 2002, were Pipe Bomb, Expect a Ship, All That Magic, Electoral College, Celera, Cliff Glider, and Inca Halo. You are interested in the first three places (winner, second place and third place) for the race.

a. Find the cardinality $n(S)$ of the sample space S of all possible finishes of the race. (A finish for the race consists of a first-, second-, and third-place winner.)

b. Let E be the event that Electoral College is in second or third place and let F be the event that Celera is the winner. Express the event $E \cap F$ in words and find its cardinality.

SOURCE: *Newsday*, February 18, 2002, p. A36.

68. Intramurals The following five teams will be participating in Urban University's hockey intramural tournament: the Independent Wildcats, the Phi Chi Bulldogs, the Gate Crashers, the Slide Rule Nerds, and the City Slickers. Prizes will be awarded for the winner and runner-up.

a. Find the cardinality $n(S)$ of the sample space S of all possible outcomes of the tournament. (An outcome of the tournament consists of a winner and a runner-up.)

b. Let E be the event that the City Slickers are runners-up and let F be the event that the Independent Wildcats are neither the winners nor runners-up. Express the event $E \cup F$ in words and find its cardinality.

In Exercises 69–72, Suzy randomly picks three marbles from a bag of eight marbles (four red ones, two green ones, and two yellow ones).

69. How many outcomes are there in the sample space?

70. How many outcomes are there in the event that Suzy grabs three red marbles?

71. How many outcomes are there in the event that Suzy grabs one marble of each color?

72. How many outcomes are there in the event that Suzy's marbles are not all the same color?

COMMUNICATION AND REASONING EXERCISES

73. Complete the following sentence. An event is a ______.

74. Complete the following sentence. Two events E and F are mutually exclusive if their intersection is ______.

75. If E and F are events, then $(E \cap F)'$ is the event that ______.

76. If E and F are events, then $(E' \cap F')$ is the event that ______.

77. True or false: Every set S is the sample space for some experiment. Explain.

78. True or false: Every sample space S is a finite set. Explain.

79. Describe an experiment in which a die is cast and the set of outcomes is $\{0, 1\}$.

80. Describe an experiment in which two coins are flipped and the set of outcomes is $\{0, 1, 2\}$.

81. Two distinguishable dice are rolled. Could there be two mutually exclusive events that both contain outcomes in which the numbers facing up add to 7?

82. Describe an experiment in which two dice are rolled and describe two mutually exclusive events that both contain outcomes in which both dice show a 1.

7.2 Estimated Probability

Suppose you have a coin that you think is not fair and you would like to measure the likelihood that heads will come up when it is tossed. You could estimate this likelihood by tossing the coin a large number of times and counting the number of times heads come up. Suppose, for instance, that in 100 tosses of the coin, heads come up 58 times. The fraction of times that heads come up, $58/100 = .58$, is the **estimated probability,** or **relative frequency,** of heads coming up when the coin is tossed. In other words, saying that the estimated probability of heads coming up is .58 is the same as saying that heads came up 58% of the time in your series of experiments.

Now let's think about this example in terms of sample spaces and events. First of all, there is an experiment that has been repeated $N = 100$ times: Toss the coin and observe the side facing up. The sample space for this experiment is $S = \{H, T\}$. Also, there is an event E in which we are interested: the event that heads come up, which is $E = \{H\}$. The number of times E has occurred, or the **frequency** of E, is $fr(E) = 58$. The estimated probability of the event E is then

$$P(E) = \frac{fr(E)}{N} \qquad \frac{\text{Frequency of event } E}{\text{Number of repetitions } N}$$

$$= \frac{58}{100} = .58$$

Notes

1. The estimated probability gives us an *estimate* of the likelihood that heads will come up when that particular coin is tossed.
2. The larger the number of times the experiment is performed, the more accurate an estimate we expect the estimated probability to be.
3. One always speaks of the probability *of an event E*.

Estimated Probability

When an experiment is performed a number of times, the **estimated probability** or **relative frequency** of an event E is the fraction of times that the event E occurs. If the experiment is performed N times and the event E occurs $fr(E)$ times, then the estimated probability is given by

$$P(E) = \frac{fr(E)}{N} \qquad \text{Fraction of times that } E \text{ occurs}$$

The number $fr(E)$ is called the **frequency** of E. The number N, the number of times that the experiment is performed, is called the number of **trials** or the **sample size.** If E consists of a single outcome s, then we refer to $P(E)$ as the estimated probability, or relative frequency, of the outcome s, and we write $P(s)$.

The collection of the estimated probabilities of *all* the outcomes is the **estimated probability distribution,** or **relative frequency distribution.**

Note In this text, we almost always use the term *estimated probability* rather than *relative frequency.*

Quick Examples

1. *Experiment:* Roll a pair of dice and add the numbers that face up.
   Event: E: The sum is 5.

   If the experiment is repeated 100 times and E occurs on 10 of the rolls, then the estimated probability of E is

$$P(E) = \frac{fr(E)}{N} = \frac{10}{100} = .10$$

2. If 10 rolls of a single die resulted in the outcomes 2, 1, 4, 4, 5, 6, 1, 2, 2, 1, then the associated estimated probability distribution is shown in the following table:

Outcome	1	2	3	4	5	6
Probability	.3	.3	0	.2	.1	.1

Example 1 • School and Work

In a survey of 55,120 members of the high school graduating class of 2000, the Bureau of Labor Statistics found that 11,780 had gone on to a 2-year college, 23,120 had gone on to a 4-year college, and the rest had not gone on to college.[3] What is the estimated probability that a member of the high school graduating class of 2000 went on to college?

Solution The experiment consisted of asking a member of the high school graduating class of 2000 what kind of college that person attended, if any. The sample space is $S = \{$2-year college, 4-year college, no college$\}$, and we are interested in the event $E = \{$2-year college, 4-year college$\}$. The sample size is $N = 55{,}120$, and the frequency of E is $fr(E) = 11{,}780 + 23{,}120 = 34{,}900$. Thus, the estimated probability of E is

$$P(E) = \frac{fr(E)}{N}$$

$$= \frac{34{,}900}{55{,}120} \approx .633$$

Before we go on . . . You might ask how accurate this estimate of .633 is or how well it reflects the activity of *all* the estimated 2,756,000 members of the high school graduating class of 2000. The field of statistics provides the tools needed to say to what extent the estimated probability can be trusted.

Note Probabilities are often expressed as percentages. We can say that there was about a 63% chance that a member of the 2000 high school graduating class went on to college. Similarly, if the weather service announces that there is a 30% chance of rain tomorrow, it is saying that the probability that it will rain tomorrow is .30.

Example 2 • Sales of Recreational Boats

The following table shows the different types of recreational boats in a sample of 430 sold in the United States in 2001:

Type	Motor Boat	Jet Ski	Sailboat	Total
Frequency	310	90	30	430

The proportions are based on 2001 sales data. SOURCE: National Marine Manufacturers Association/*New York Times*, January 10, 2002, p. C1.

Consider the experiment in which a recreational boat is selected at random and its type is observed.

a. Find the estimated probability distribution.

b. Find the estimated probability that a recreational boat is not a motor boat.

[3]SOURCE: This example is based on figures in "College Enrollment and Work Activity of High School Graduates," U.S. Bureau of Labor Statistics, http://www.bls.gov/news.release/hsgec.toc.htm, April 2001. This document did not give the actual number of people in the sample.

Solution

a. The following table shows the estimated probability of each outcome, which we find by dividing each frequency by $N = 430$:

Type	Motor Boat	Jet Ski	Sailboat
Probability	$\frac{310}{430} \approx .721$	$\frac{90}{430} \approx .209$	$\frac{30}{430} \approx .070$

b. $E = \{\text{jet ski, sailboat}\}$; thus,

$$P(E) = \frac{90 + 30}{430} = \frac{120}{430} \approx .279$$

Excel: Calculating Probabilities

Calculating estimated probabilities is easy using Excel. We first enter the given data like this:

	A	B	C	D	E
1	**Type**	**Motor Boat**	**Jet Ski**	**Sailboat**	**Total**
2	**Frequency**	310	90	30	430

We have Excel calculate N for us by entering the formula

```
=SUM(B2:D2)
```

in cell E2. We can then calculate the estimated probabilities like this:

	A	B	C	D	E
1	**Type**	**Motor Boat**	**Jet Ski**	**Sailboat**	**Total**
2	**Frequency**	310	90	30	430
3	**Probability**	=B2/$E$2			

The worksheet with the probabilities calculated and rounded to three decimal points looks like the following:

	A	B	C	D	E
1	**Type**	**Motor Boat**	**Jet Ski**	**Sailboat**	**Total**
2	**Frequency**	310	90	30	430
3	**Probability**	0.721	0.209	0.070	

Following are some important properties of estimated probability that we can observe in Example 2.

Some Properties of Estimated Probability

Let $S = \{s_1, s_2, \ldots, s_n\}$ be a sample space and let $P(s_i)$ be the estimated probability of the event $\{s_i\}$. Then,

1. $0 \leq P(s_i) \leq 1$

2 $P(s_1) + P(s_2) + \cdots + P(s_n) = 1$

3. If $E = \{e_1, e_2, \ldots, e_r\}$, then $P(E) = P(e_1) + P(e_2) + \cdots + P(e_r)$

In words:

1. The estimated probability of each outcome is a number between 0 and 1 (inclusive).
2. The estimated probabilities of all the outcomes add up to 1.
3. The estimated probability of an event E is the sum of the estimated probabilities of the individual outcomes in E.

Estimated Probability and Increasing Sample Size

A fair coin is one that is as likely to come up heads as it is to come up tails. In other words, we expect heads to come up 50% of the time if we toss such a coin many times. Put more precisely, we expect the estimated probability to approach .5 as the number of trials gets larger. Figure 2 shows how the estimated probability behaved for one sequence of coin tosses. For each N, we have plotted what fraction of times the coin came up heads in the first N tosses.

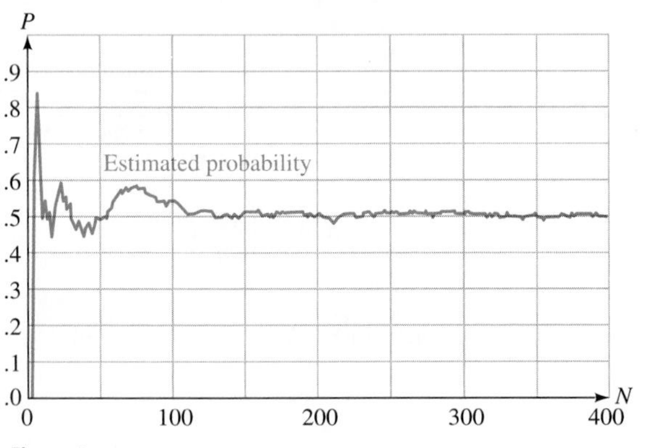

Figure 2

Notice that the estimated probability graph meanders as N increases, sometimes getting closer to .5 and sometimes drifting away again. However, the graph tends to meander within smaller and smaller distances of .5 as N increases.[4]

In general, this is how estimated probability behaves; as N gets large, the estimated probability approaches some fixed value.

Using Technology to Simulate Experiments

Most calculators, spreadsheets, and the like are equipped with *random number generators;* that is, they have the ability to produce a random number, usually a decimal between 0 and 1, when you press the appropriate key or enter the appropriate formula (usually called something like RAND or RAN#). Each time you press the key(s), you get a new random number between 0 and 1. Here is a sequence of 50 random numbers (rounded to three decimal places) produced by a word processor:

.136	.405	.835	.754	.285	.801	.654	.029	.601	.250
.595	.852	.816	.508	.288	.037	.268	.462	.639	.747
.579	.184	.427	.751	.705	.501	.501	.877	.508	.964
.852	.654	.411	.541	.039	.354	.197	.312	.532	.077
.278	.153	.168	.451	.712	.865	.558	.926	.880	.334

Example 3 shows how we can use such data to simulate tossing of coins or rolling of dice. As we explained, a **fair coin** is equally likely to land with heads or tails up. Sim-

[4]This can be made precise by the concept of *limit* used in calculus.

ilarly, a **fair die** is equally likely to land with any of the six numbers facing up. Since we expect heads to come up approximately 50% of the time if we toss a fair coin a very large number of times, we say that the **theoretical probability** of heads coming up in a toss of a fair coin is .5. Similarly, the theoretical probability of throwing a 3 with a fair die is 1/6. Theoretical probability is discussed in detail in Section 7.3.

Example 3 • Simulating Experiments with Random Numbers

Use a simulated experiment to check the following:

a. The estimated probability of heads coming up in a toss of a fair coin approaches 1/2 as the number of trials gets large.

b. The estimated probability of heads coming up in two consecutive tosses of a fair coin approaches 1/4 as the number of trials gets large.[5]

Solution

a. Let's use 1 to represent heads and 0 to represent tails. We need to generate a list of **random binary digits** (0 or 1). One way to do this—and a method that works for most forms of technology—is to generate a random number between 0 and 1 as above and then round it to the nearest whole number, which will be either 0 or 1.

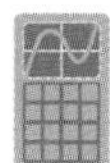

Graphing Calculator: Generating Random Numbers

To round the number X to the nearest whole number on the TI-83, follow [MATH] →NUM, select `round`, and enter `round(X,0)`. This instruction rounds X to zero decimal places—that is, to the nearest whole number. Since we wish to round the random number `rand`, we need to enter

`round(rand,0)` To obtain `rand`, follow [MATH] →PRB.

The result will be either 0 or 1. Each time you press [ENTER], you will now get another 0 or 1. The TI-83 can also generate a random integer directly (without the need for rounding) through the instruction

`randInt(0,1)` To obtain `randInt`, follow [MATH] →PRB.

In general, the command `randInt(m, n)` generates a random integer in the range $[m, n]$. The following sequence of 100 random binary digits was produced using technology:[6]

```
0 1 0 0 1 1 0 1 0 0
0 1 0 0 0 0 0 0 1 0
1 1 0 0 0 1 0 0 1 1
1 1 1 0 1 0 0 0 1 0
1 1 1 1 1 1 1 0 0 1
1 0 1 1 1 0 0 1 1 0
0 1 0 1 1 1 0 1 1 1
1 0 0 0 0 0 0 1 1 1
1 1 1 1 0 0 1 1 1 0
1 1 1 0 1 1 0 1 0 0
```

---

[5]Since the set of outcomes of a pair of coin tosses is {HH, HT, TH, TT}, we expect HH to come up once in every four trials, on average.

[6]The instruction `randInt(0,1,100)→L1` will generate a list of 100 random 0s and 1s and store it in L_1, where it can be summed with `Sum(L1)` (under [2nd] [LIST] →MATH).

If we use only the first row of data (corresponding to the first ten tosses) we find

$$P(\mathrm{H}) = \frac{fr(1)}{N} = \frac{4}{10} = .4$$

Using the first two rows ($N = 20$) gives

$$P(\mathrm{H}) = \frac{fr(1)}{N} = \frac{6}{20} = .3$$

Using all ten rows ($N = 100$) gives

$$P(\mathrm{H}) = \frac{fr(1)}{N} = \frac{54}{100} = .54$$

This is somewhat closer to the theoretical probability of 1/2 and supports our intuitive notion that the larger the number of trials, the more closely the estimated probability should approximate the theoretical value.[7]

Excel: Generating Random Numbers

In Excel, the formula `RAND()` is the same as `rand` on the TI-83, giving a random number between 0 and 1.[8] Furthermore, the rounding instruction is identical to that on the TI-83. Therefore, to obtain a random binary digit in any cell just enter the following formula.

```
=ROUND(RAND(),0)
```

Excel can also generate a random integer directly (without the need for rounding) through the formula

```
=RANDBETWEEN(0,1)
```

To obtain a whole array of random numbers like that above, just drag this formula into the cells you wish to use.

b. We need to generate pairs of random binary digits and then check whether they are both 1s. Although the TI-83 will generate a pair of random digits if you enter `round(rand(2),0)`, it would be a lot more convenient if the calculator (or spreadsheet) could tell you right away whether both digits are 1s (corresponding to two consecutive heads in a coin toss). Here is a simple way of accomplishing this. Notice that if we *add* the two random digits, we obtain either 0, 1, or 2, telling us the number of heads that result from the two consecutive throws. Therefore, all we need to do is add the pairs of random digits and then count the number of times 2 comes up. Formulas we can use are

TI-83: `randInt(0,1)+randInt(0,1)`

Excel: `=RANDBETWEEN(0,1)+RANDBETWEEN(0,1)`

What would be even *more* convenient is if the result of the calculation would be either 0 or 1, with 1 signifying success (two consecutive heads) and 0 signifying failure. Then, we could simply add up all the results to obtain the number of times two heads occurred. To do this, we first divide the result of the calculation above by 2 (obtain-

---

[7]Do not expect this to happen every time. Compare, for example, $P(\mathrm{H})$ for the first five rows and for all ten rows.

[8]The parentheses after `RAND` are necessary, even though the function takes no arguments.

ing 0, .5, or 1, where now 1 signifies success) and then round *down* to an integer using a function called "int":

TI-83: `int(0.5*(randInt(0,1)+randInt(0,1)))`

Excel: `=INT(0.5*(RANDBETWEEN(0,1)+RANDBETWEEN(0,1)))`

Following is the result of 100 such pairs of coin tosses, with 1 signifying success (two heads) and 0 signifying failure (all other outcomes). The last column records the number of successes in each row and the total number at the end:

```
1  1  0  0  0  0  0  0  0  0   2
0  1  0  0  0  0  0  1  0  1   3
0  1  0  0  1  1  0  0  0  1   4
0  0  0  0  0  0  0  0  1  0   1
0  1  0  0  1  0  0  1  0  0   3
1  0  1  0  0  0  0  0  0  0   2
0  0  0  0  0  0  0  0  0  1   1
0  1  1  1  1  0  0  0  0  1   5
1  1  0  1  0  0  1  1  0  0   5
0  0  0  0  0  0  0  0  1  0   1
                              27
```

Now, as in part (a), we can compute estimated probabilities, with D standing for the outcome "two heads":

First 10 trials: $P(D) = \dfrac{fr(1)}{N} = \dfrac{2}{10} = .2$

First 20 trials: $P(D) = \dfrac{fr(1)}{N} = \dfrac{5}{20} = .25$

First 50 trials: $P(D) = \dfrac{fr(1)}{N} = \dfrac{13}{50} = .26$

100 trials: $P(D) = \dfrac{fr(1)}{N} = \dfrac{27}{100} = .27$

Question What is happening with the data? The probabilities seem to be getting *less* accurate as N increases!

Answer Quite by chance, exactly 5 of the first 20 trials resulted in success, which matches the theoretical probability. Figure 3 shows an Excel plot of estimated probability versus N (for N a multiple of 10). Notice that, as N increases, the graph seems to meander within smaller distances of .25.

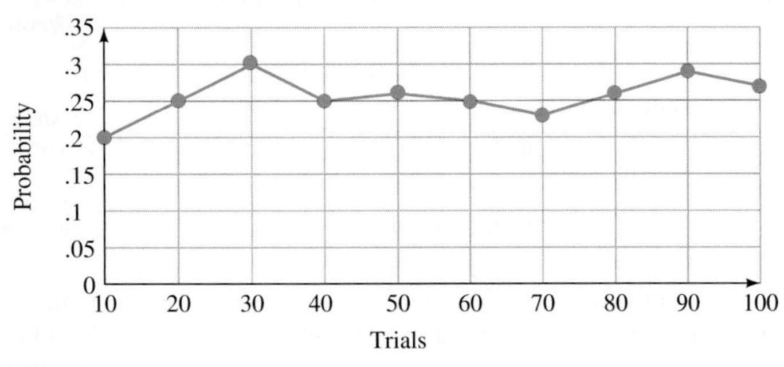

Figure 3

Before we go on . . .

Question The above techniques work fine for simulating coin tosses. What about rolls of a fair die, where we want outcomes between 1 and 6?

Answer We can simulate a roll of a die by generating a random integer in the range 1 through 6. The following formula accomplishes this:

TI-83: `1+int(5.99999*rand)`

Excel: `=1+INT(5.99999*RAND())`

(We used 5.99999 instead of 6 to avoid the outcome 7.)

7.2 EXERCISES

In Exercises 1–4, calculate the estimated probability $P(E)$ using the given information.

1. $N = 100, fr(E) = 40$ **2.** $N = 500, fr(E) = 300$

3. Eight hundred adults are polled, and 640 of them support universal health care coverage. E is the event that an adult supports universal health coverage.

4. Eight hundred adults are polled, and 640 of them support universal health care coverage. E is the event that an adult does not support universal health coverage.

Exercises 5–10 are based on the following table, which shows the frequency of outcomes when two distinguishable coins were tossed 4000 times and the uppermost faces were observed.

Outcome	HH	HT	TH	TT
Frequency	1100	950	1200	750

5. Determine the estimated probability distribution.

6. What is the estimated probability that heads comes up at least once?

7. What is the estimated probability that the second coin lands with heads up?

8. What is the estimated probability that the first coin lands with heads up?

9. Would you judge the second coin to be fair? Give a reason for your answer.

10. Would you judge the first coin to be fair? Give a reason for your answer.

Exercises 11–14 require the use of a calculator or computer with a random number generator.

11. Simulate 100 tosses of a fair coin and compute the estimated probability that heads comes up.

12. Simulate 100 throws of a fair die and calculate the estimated probability that the result is a 6.

13. Simulate 50 tosses of two coins and compute the estimated probability that the outcome is one head and one tail (in any order).

14. Simulate 100 throws of two fair dice and calculate the estimated probability that the result is a double 6.

APPLICATIONS

15. Fast-Food Stores In 2000 the top 100 chain restaurants in the United States owned a total of approximately 130,000 outlets. Of these, the three largest (in sales) were McDonald's with approximately 13,000 outlets, Burger King with approximately 8300 outlets, and Wendy's with approximately 5100 outlets. Find the estimated probability that a randomly selected outlet from among the top 100 chains was (a) a McDonald's outlet and (b) not a Burger King outlet.

SOURCE: "Technomic 2001 Top 100 Report," Technomic, Inc.; information obtained from their Web site, www.technomic.com.

16. Fast-Food Revenues In 2000 the top 100 chain restaurants in the United States earned a total of approximately $123 billion in sales. The three largest earners were McDonald's with approximately $20 billion, Burger King with approximately $8.5 billion, and Wendy's with approximately $5.8 billion. Find the estimated probability that a randomly selected dollar earned by one of the top 100 chain restaurants was earned by (a) McDonald's and (b) either Burger King or Wendy's.

SOURCE: "Technomic 2001 Top 100 Report," Technomic, Inc.; information obtained from their Web site, www.technomic.com.

17. Motor Vehicle Safety The following table shows crashworthiness ratings for 10 small SUVs (3 = good, 2 = acceptable, 1 = marginal, 0 = poor):

Frontal Crash Test Rating	3	2	1	0
Frequency	1	4	4	1

SOURCES: Stacy C. Davis and Lorena F. Truett, Oak Ridge National Laboratory, "An Analysis of the Impact of Sport Utility Vehicles in the United States," August 2000/ Insurance Institute for Highway Safety, http://www-cta.ornl.gov/Publications/Final SUV report.pdf.

a. Find the estimated probability distribution for the experiment of choosing a small SUV at random and determining its frontal crash rating.

b. What is the estimated probability that a randomly selected small SUV will have a crash test rating of acceptable or better?

18. Motor Vehicle Safety The following table shows crashworthiness ratings for 16 small cars (3 = good, 2 = acceptable, 1 = marginal, 0 = poor):

Frontal Crash Test Rating	3	2	1	0
Frequency	1	11	2	2

SOURCES: Stacy C. Davis and Lorena F. Truett, Oak Ridge National Laboratory: "An Analysis of the Impact of Sport Utility Vehicles in the United States," August 2000/Insurance Institute for Highway Safety, http://www-cta.ornl.gov/Publications/Final SUV report.pdf.

a. Find the estimated probability distribution for the experiment of choosing a small car at random and determining its frontal crash rating.

b. What is the estimated probability that a randomly selected small car will have a crash test rating of marginal or worse?

19. Internet Use The following pie chart shows the number of Internet users and nonusers found in a survey taken in the United States in August 2000, among adults with and without a college degree:

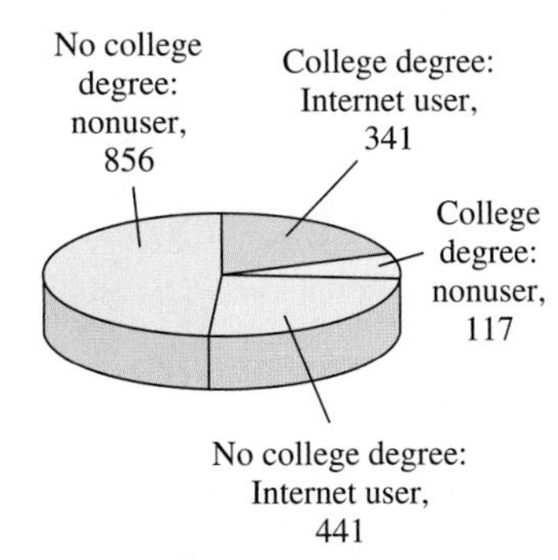

SOURCE: Based on data in "Falling through the Net: Toward Digital Inclusion. A Report on Americans' Access to Technology Tools," U.S. Department of Commerce, http://www.ntia.doc.gov/ntiahome/fttn00/contents00.html, October 2000. The survey results are fictitious, but the proportions reflect those in the report.

a. Determine the estimated probability of each of the outcomes for the experiment of choosing an adult at random and determining his or her college education and Internet use. (Round probabilities to two decimal places.)

b. What is the estimated probability that an adult uses the Internet?

20. Internet Use Repeat Exercise 19, using the following pie chart that shows the results of a survey taken in September 2001:

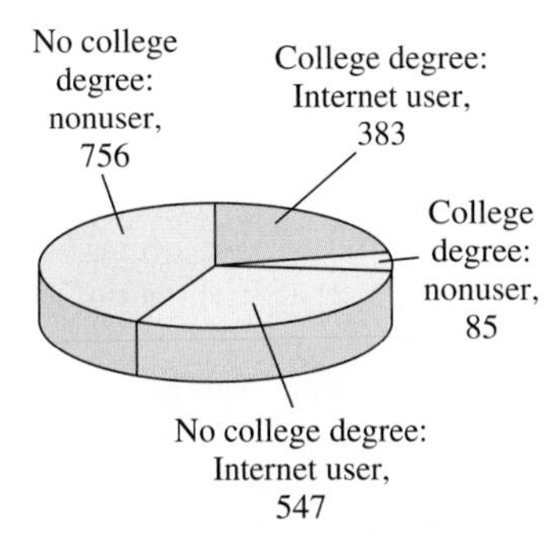

SOURCE: Based on "A Nation Online: How Americans Are Expanding Their Use of the Internet," U.S. Department of Commerce, http://www.ntia.doc.gov/ntiahome/dn/index.html, February 2002. Again, the survey results are fictitious but the proportions are correct.

Publishing Exercises 21–30 are based on the following table, which shows the results of a survey of 100 authors by a (fictitious) publishing company:

	New Authors	**Established Authors**	**Total**
Successful	5	25	30
Unsuccessful	15	55	70
Total	20	80	100

Compute the estimated probabilities of the given events if an author as specified is chosen at random.

21. An author is established and successful.

22. An author is unsuccessful and new.

23. An author is a new author.

24. An author is successful.

25. An author is unsuccessful.

26. An author is established.

27. A successful author is established.

28. An unsuccessful author is established.

29. An established author is successful.

30. A new author is unsuccessful.

31. Public Health A random sampling of chicken in supermarkets revealed that approximately 80% was contaminated with the organism *Campylobacter.* Of the contaminated chicken, 20% had the strain resistant to antibiotics. Construct an estimated probability distribution showing the following outcomes when chicken is purchased at a supermarket: *U*: The chicken is not infected with *Campylobacter; C*: The chicken is infected with nonresistant *Campylobacter; R*: The chicken is infected with resistant *Campylobacter.*

Campylobacter is one of the leading causes of food poisoning in humans. Thoroughly cooking the meat kills the bacteria. SOURCE: The *New York Times,* October 20, 1997, p. A1. Publication of this article first brought *Campylobacter* to the attention of a wide audience.

32. Public Health A random sampling of turkey in supermarkets found 58% to be contaminated with *Campylobacter* and 84% of those to be resistant to antibiotics. Construct an estimated probability distribution showing the following outcomes when turkey is purchased at a supermarket: *U*: The turkey is not infected with *Campylobacter; C*: The turkey is infected with nonresistant *Campylobacter; R*: The turkey is infected with resistant *Campylobacter.*

SOURCE: The *New York Times*, October 20, 1997, p. A1.

33. Organic Produce A 2001 Department of Agriculture study of more than 94,000 samples from more than 20 crops showed that 73% of conventionally grown foods had residues from at least one pesticide. Moreover, conventionally grown foods were six times as likely to contain multiple pesticides as organic foods. Of the organic foods tested, 23% had pesticide residues, which includes 10% with multiple pesticide residues. Compute two probability distributions: one for conventional produce and one for organic produce, showing the estimated probabilities that a randomly selected product has no pesticide residues, has residues from a single pesticide, and has residues from multiple pesticides.

The 10% figure is an estimate. SOURCE: *New York Times*, May 8, 2002, p. A29.

34. Organic Produce Repeat Exercise 33, using the following information for produce from California: 31% of conventional food and 6.5% of organic food had residues from at least one pesticide. Assume that, as in Exercise 33, conventionally grown foods were six times as likely to contain multiple pesticides as organic foods. Also assume that 3% of the organic food has residues from multiple pesticides.

35. Stock Index The following table shows the week-end closing values of the Dow Jones Industrial Average over the 10-week period beginning January 7, 2002:

Week	1	2	3	4	5	6	7	8	9	10
Dow	9990	9770	9840	9910	9740	9900	9970	10,370	10,570	10,610

Values are rounded to the nearest 10 points. SOURCE: Yahoo! Finance, http://finance.yahoo.com/.

Use these data to construct the estimated probability distribution using the following three outcomes. Low: The index is at or below 9900; Middle: The index is above 9900 but not above 10,100; High: The index is above 10,100.

36. Stock Prices The following table shows the week-end closing values of the Dow Jones Industrial Average over the 10-week period beginning January 1, 1993:

Week	1	2	3	4	5	6	7	8	9	10
Dow	3250	3260	3250	3310	3460	3390	3310	3380	3400	3420

SOURCE: *New York Times*, July 18, 1993, p. C7. Averages are rounded to the nearest 10 points.

Use these data to construct the estimated probability distribution using the following three outcomes. Low: The index is at or below 3300; Middle: The index is above 3300 but not above 3400; High: The index is above 3400.

37. Steroids Testing A pharmaceutical company is running trials on a new test for anabolic steroids. The company uses the test on 400 athletes known to be using steroids and 200 athletes known not to be using steroids. Of those using steroids, the new test is positive for 390 and negative for 10. Of those not using steroids, the test is positive for 10 and negative for 190. What is the estimated probability of a *false negative* result (the probability that an athlete using steroids will test negative)? What is the estimated probability of a *false positive* result (the probability that an athlete not using steroids will test positive)?

38. Lie Detectors A manufacturer of lie detectors is testing its newest design. It asks 300 subjects to lie deliberately and another 500 to tell the truth. Of those who lied, the lie detector caught 200. Of those who told the truth, the lie detector accused 200 of lying. What is the estimated probability of the machine wrongly letting a liar go, and what is the probability that it will falsely accuse someone who is telling the truth?

T **39. Public Health** Refer to Exercise 31. Simulate the experiment of selecting chicken at a supermarket and determining the following outcomes. *U*: The chicken is not infected with *Campylobacter; C*: The chicken is infected with nonresistant *Campylobacter; R*: The chicken is infected with resistant *Campylobacter.* (*Hint:* Generate integers in the range 1–100. The outcome is determined by the range. For instance, if the number is in the range 1–20, regard the outcome as *U*.)

T **40. Public Health** Repeat Exercise 39 but use turkeys and the data given in Exercise 32.

COMMUNICATION AND REASONING EXERCISES

41. Complete the following sentence. The estimated probability of an event *E* is defined to be ______.

42. Interpret the popularity rating of the student council president as an estimated probability by specifying an appropriate experiment and also what is observed.

43. How would you measure the estimated probability that the weather service accurately predicts the next day's temperature?

44. Suppose you toss a coin 100 times and get 70 heads. If you continue tossing the coin, the estimated probability of heads overall should approach 50% if the coin is fair. Will you have to get more tails than heads in subsequent tosses to "correct" for the 70 heads you got in the first 100 tosses?

45. Tony has had a "losing streak" at the casino—the chances of winning the game he is playing are 40%, but he has lost five times in a row. Tony argues that, since he should have won two times, the game must obviously be "rigged." Comment on his reasoning.

46. Maria is on a winning streak at the casino. She has already won four times in a row and concludes that her chances of winning a fifth time are good. Comment on her reasoning.

7.3 Theoretical Probability

It is understandable if you are a little uncomfortable with estimated probability because it does not always confirm what you intuitively feel to be true. For instance, if you toss a fair coin 100 times and heads happen to come up 62 times, the experiment seems to suggest that the probability of heads is .62, even though you *know* that the "actual" probability is .50 (because the coin is fair).

Question So what do we mean by the "actual" probability?

Answer First, some terminology: We will refer to the "actual" probability as the **theoretical probability.** We can define it in two ways: one more intuitive (but not very precise) and one more precise (but not very intuitive).

Theoretical Probability

Intuitive Definition of Theoretical Probability

The **theoretical probability,** or **probability,** $P(E)$ of an event E is the fraction of times we *expect* E to occur if we repeat the same experiment over and over.

More Precise Definition of Theoretical Probability

The **theoretical probability,** or **probability,** $P(E)$ of an event E is the *limiting value* of the estimated probability as the number of trials gets larger and larger. That is, the estimated probability approaches the theoretical probability as the number of trials gets larger and larger.* Thus,

Estimated probability is an approximation, or estimate, of theoretical probability. The larger the number of trials, the more accurate we expect this approximation to be.

If E consists of a single outcome s, we refer to $P(E)$ as the probability of the outcome s, and write $P(s)$ for $P(E)$. The collection of the probabilities of all the outcomes is the **probability distribution.**

Determining Theoretical Probability

Theoretical probability is determined *analytically*—that is, by using our knowledge about the nature of the experiment rather than through actual experimentation. The best we can obtain through actual experimentation is an *estimate* of the theoretical probability (hence the term *estimated probability*).

Notes

- Another distinction between estimated and theoretical probability is that the estimated probability of an event E is the fraction of times E *actually occurs* in a repeated experiment, whereas the theoretical probability is the fraction of times we *expect* E to occur in the long term.
- We use the same letter P for theoretical probability as we used for estimated probability. Whether we are talking about estimated or theoretical probability will either be stated or be clear from the context. In the same vein, we will often simply speak of "probability" in reference to either estimated or theoretical probability.

*See Figures 2 and 3 in Section 7.2. In calculus the idea of a "limiting value" can be made very precise, and we say that $P(E)$ is the limit of the estimated probability of E as the number of trials $N \to +\infty$.

Quick Examples

1. Toss a fair coin and observe the side that faces up. Since we expect that heads is as likely to come up as tails, we conclude that the theoretical probability distribution is

$$P(\mathrm{H}) = \frac{1}{2} \qquad P(\mathrm{T}) = \frac{1}{2}$$

2. Roll a fair die. Since we expect to roll a 1 one-sixth of the time,

$$P(1) = \frac{1}{6}$$

Similarly, $P(2) = \frac{1}{6}, P(3) = \frac{1}{6}, \ldots, P(6) = \frac{1}{6}$.

3. Roll a pair of fair dice (recall that there are a total of 36 outcomes if the dice are distinguishable). If E is the event that the sum of the numbers that face up is 5, then $E = \{(1, 4), (2, 3), (3, 2), (4, 1)\}$. Since all 36 outcomes are equally likely,

$$P(E) = \frac{4}{36} = \frac{1}{9}$$

Question How do we calculate theoretical probability "using our knowledge about the nature of the experiment"?

Answer The simple answer is that there is no easy way to calculate theoretical probability in general, and sometimes there is no way at all. For instance, the probability that it will snow in Toronto on November 1 is impossible to determine theoretically, so we must rely on estimated probabilities for weather prediction. However, notice that in the preceding Quick Examples all outcomes were equally likely and for an event E, all we did was compute the ratio

$$\frac{\text{number of favorable outcomes}}{\text{total number of outcomes}} = \frac{n(E)}{n(S)}$$

Computing Theoretical Probability: Equally Likely Outcomes

In an experiment in which all outcomes are equally likely, the theoretical probability of an event E is given by

$$P(E) = \frac{\text{number of favorable outcomes}}{\text{total number of outcomes}} = \frac{n(E)}{n(S)}$$

Note Remember that this formula will work *only* when the outcomes are equally likely. If, for example, a die is *weighted,* then the outcomes may not be equally likely, and the formula above will not apply.

Quick Examples

1. Toss a fair coin three times. The probability that we throw exactly two heads is

$$P(E) = \frac{n(E)}{n(S)} = \frac{3}{8} \quad \text{There are eight outcomes, and } E = \{\text{HHT, HTH, THH}\}.$$

2. Roll a pair of fair dice. The probability that we roll a double (both dice show the same number) is

$$P(E) = \frac{n(E)}{n(S)} = \frac{6}{36} = \frac{1}{6} \qquad E = \{(1, 1), (2, 2), (3, 3), (4, 4), (5, 5), (6, 6)\}$$

3. Randomly choose a person from a class of 40, in which 6 have red hair. If E is the event that a randomly selected person in the class has red hair, then

$$P(E) = \frac{n(E)}{n(S)}\ \frac{6}{40} = .15$$

Example 1 • Employment

In December 2001 approximately 134 million people were employed in the United States.[9] Of these, 42 million were employed in managerial or professional specialties, and 3 million were employed in farming, forestry, or fishing. Determine the following:

a. The probability that a randomly chosen working person was employed in a managerial or professional specialty

b. The probability that a randomly chosen working person was not employed in farming, forestry, or fishing.

Solution

a. The experiment suggested by the question is to randomly choose a person working at the end of 2001 and determine how that person was employed. We are interested in the event E that the person was employed in a managerial or professional specialty. So,

S = set of people employed at the end of 2001; $n(S) = 134$ million

E = set of people in managerial or professional specialties; $n(E) = 42$ million

Are the outcomes equally likely in this experiment? Yes, because we are as likely to choose one person as another. Thus,

$$P(E) = \frac{n(E)}{n(S)} = \frac{42}{134} \approx .31$$

b. Let the event F consist of those working people who were not employed in farming, forestry, or fishing. Since only 3 million people were so employed, the remaining $134 - 3 = 131$ million were not. Hence,

$$P(F) = \frac{n(F)}{n(S)} = \frac{131}{134} \approx .98$$

In other words, approximately 98% of all working people were employed in some way other than in farming, forestry, or fishing.

[9]SOURCE: U.S. Bureau of Labor Statistics press release, http://www.bls.gov/, January 4, 2002.

Before we go on . . .

Question In Example 1 in Section 7.2, we had a similar example about people, but we called the probabilities calculated there estimated. Here they are theoretical. What is the difference?

Answer In that example the data were based on the results of a survey, or sample, of only 55,120 members of the high school graduating class of 2000 (out of a total of about 2.76 million) and were therefore incomplete. (A statistician would say that we were given *sample statistics.*) It follows that any inference we draw from the 55,120 surveyed, such as the probability that a student went on to college, is uncertain, and this is the cue that tells us that we are working with estimated probability. Think of the survey as an experiment (for instance, select a graduating senior) repeated 55,120 times—exactly the setting for estimated probability.

In this example, on the other hand, the data do not describe how a sample of *some* working people were employed, but they describe how *all 134 million people* who were working were employed. (The statistician would say that we were given *population statistics* in this case, since the data describe the entire population of workers.)

Question We saw that estimated probability has the following properties for a sample space $S = \{s_1, s_2, \ldots, s_n\}$:

1. $0 \le P(s_i) \le 1$
2. $P(s_1) + P(s_2) + \cdots + P(s_n) = 1$
3. We can obtain the probability of an event E by adding up the probabilities of the outcomes in E.

Do these properties hold for theoretical probability as well?

Answer Yes. Since the probability of an outcome is the fraction of times we expect it to occur, probabilities are always numbers between 0 and 1, inclusive. To get the fraction of times an event will occur, we clearly need to add the fractions of times each outcome in the event will occur. And, since the event $E = S$ *must* occur every time, $P(S) = P(s_1) + \cdots + P(s_n) = 1$.

Example 2 • Indistinguishable Dice

We recall from Section 7.1 that the sample space when we roll a pair of indistinguishable dice is

$$S = \left\{ \begin{array}{cccccc} (1, 1), & (1, 2), & (1, 3), & (1, 4), & (1, 5), & (1, 6), \\ & (2, 2), & (2, 3), & (2, 4), & (2, 5), & (2, 6), \\ & & (3, 3), & (3, 4), & (3, 5), & (3, 6), \\ & & & (4, 4), & (4, 5), & (4, 6), \\ & & & & (5, 5), & (5, 6), \\ & & & & & (6, 6) \end{array} \right\}$$

Compute the probabilities of all the outcomes.

Solution Since there are 21 outcomes, it is tempting to say that the probability of each outcome is 1/21. However, the outcomes are not all equally likely. For instance, the outcome (2, 3) is twice as likely as (2, 2), since (2, 3) can occur in two ways (it corresponds to the event $\{(2, 3), (3, 2)\}$ for distinguishable dice). For purposes of calculating probability, it is easiest to use calculations for distinguishable dice.[10] Here are some examples:

Outcome (indistinguishable dice)	(1, 1)	(1, 2)	(2, 2)	(1, 3)	(2, 3)	(3, 3)
Corresponding Event (distinguishable dice)	$\{(1, 1)\}$	$\{(1, 2), (2, 1)\}$	$\{(2, 2)\}$	$\{(1, 3), (3, 1)\}$	$\{(2, 3), (3, 2)\}$	$\{(3, 3)\}$
Probability	$\frac{1}{36}$	$\frac{2}{36} = \frac{1}{18}$	$\frac{1}{36}$	$\frac{2}{36} = \frac{1}{18}$	$\frac{2}{36} = \frac{1}{18}$	$\frac{1}{36}$

If we continue this process for all 21 outcomes, we will find that they add to 1.

Example 3 • Weighted Dice

To impress your friends with your die-rolling skills, you have surreptitiously weighted your die in such a way that 6 is three times as likely to come up as any one of the other numbers. Find the probability distribution for a roll of the die and use it to calculate the probability of an even number coming up.

Solution We haven't a clue yet as to what the individual probabilities are—these are our *unknowns.* All we know is that they must add to 1. Let's label our unknowns (there appear to be two of them):

x = probability of rolling a 6

y = probability of rolling any one of the other numbers

We are first told that "6 is three times as likely to come up as any other number." If we rephrase this in terms of our unknown probabilities, we get "the probability of rolling a 6 is three times the probability of rolling any other number." In symbols,

$$x = 3y$$

We must also use a piece of information not given to us, but one we know must be true: The sum of the probabilities of all the outcomes is 1:

$$x + y + y + y + y + y = 1$$

$$x + 5y = 1$$

We now have two linear equations in two unknowns, and we solve for x and y. Substituting the first equation ($x = 3y$) in the second ($x + 5y = 1$) gives

$$8y = 1$$

$$y = \frac{1}{8}$$

To get x we substitute the value of y back into either equation and find

$$x = \frac{3}{8}$$

[10]Note that any pair of real dice can be distinguished in principle because they possess slight differences, although we may regard them as indistinguishable by not attempting to distinguish them. Thus, the probabilities of events must be the same as for the corresponding events for distinguishable dice.

Thus, the probability distribution is the one shown in the following table:

Outcome	1	2	3	4	5	6
Probability	$\frac{1}{8}$	$\frac{1}{8}$	$\frac{1}{8}$	$\frac{1}{8}$	$\frac{1}{8}$	$\frac{3}{8}$

We can use the distribution to calculate the probability of an even number coming up by adding the probabilities of the favorable outcomes:

$$P(\{2, 4, 6\}) = \frac{1}{8} + \frac{1}{8} + \frac{3}{8} = \frac{5}{8}$$

Thus, there is a $5/8 = .625$ chance that an even number will come up.

Before we go on . . . We should check that the distribution satisfies the requirements: 6 is indeed three times as likely to come up as any other number. Also, the probabilities we calculated do add up to 1:

$$\frac{1}{8} + \frac{1}{8} + \frac{1}{8} + \frac{1}{8} + \frac{1}{8} + \frac{3}{8} = 1 \quad ✓$$

Guideline: Distinguishing Theoretical Probability from Estimated Probability

Question How do I know whether a given probability is estimated or theoretical?

Answer Ask yourself this: Has the probability been arrived at experimentally, by performing a number of trials and counting the number of times the event occurred? If so, the probability is estimated. If, on the other hand, the probability was computed by analyzing the experiment under consideration rather than by performing actual trials of the experiment, it is theoretical.

Question Twenty-two out of every 100 homes have broadband Internet service. Thus, the probability that a house has broadband service is .22. Is this probability estimated or theoretical?

Answer That depends on how the ratio 22 out of 100 was found. If it is based on a poll of *all* homes, then the probability is theoretical. If it is based on a survey of only a *sample* of homes, it is estimated (see "Before we go on" in Example 1).

7.3 EXERCISES

In Exercises 1–6, calculate the (theoretical) probability $P(E)$ using the given information, assuming that all outcomes are equally likely.

1. $n(S) = 20, n(E) = 5$

2. $n(S) = 8, n(E) = 4$

3. $n(S) = 10, n(E) = 10$

4. $n(S) = 10, n(E) = 0$

5. $S = \{a, b, c, d\}, E = \{a, b, d\}$

6. $S = \{1, 3, 5, 7, 9\}, E = \{3, 7\}$

In Exercises 7–16, an experiment is given together with an event. Find the probability of each event, assuming that the coins and dice are distinguishable and fair and that what is observed are the faces or numbers uppermost. (Compare with Exercises 1–10 in Section 7.1.)

7. Two coins are tossed; the result is at most one tail.

8. Two coins are tossed; the result is one or more heads.

9. Three coins are tossed; the result is at most one head.

10. Three coins are tossed; the result is more tails than heads.

11. Two dice are rolled; the numbers add to 5.

12. Two dice are rolled; the numbers add to 9.

13. Two dice are rolled; the numbers add to 1.

14. Two dice are rolled; one of the numbers is even, and the other is odd.

15. Two dice are rolled; both numbers are prime.[11]

16. Two dice are rolled; neither number is prime.

17. If two indistinguishable dice are rolled, what is the probability of the event {(4, 4), (2, 3)}? What is the corresponding event for a pair of distinguishable dice?

18. If two indistinguishable dice are rolled, what is the probability of the event {(5, 5), (2, 5), (3, 5)}? What is the corresponding event for a pair of distinguishable dice?

19. A die is weighted in such a way that 2, 4, and 6 are twice as likely to come up as 1, 3, and 5. Find the probability distribution. What is the probability of rolling a 1, 2, or 3?

20. Another die is weighted in such a way that each of 1 and 2 is three times as likely to come up as each of the other numbers. Find the probability distribution. What is the probability of rolling an even number?

21. A tetrahedral die has 4 faces, numbered 1–4. If the die is weighted in such a way that each number is twice as likely to land facing down as the next number (1 twice as likely as 2, 2 twice as likely as 3, and so on), what is the probability distribution for the face landing down?

22. A dodecahedral die has 12 faces, numbered 1–12. If the die is weighted in such a way that 2 is twice as likely to land facing up as 1, 3 is three times as likely to land facing up as 1, and so on, what is the probability distribution for the face landing up?

23. Complete the following theoretical probability distribution table and then calculate the stated probabilities.

Outcome	a	b	c	d	e
Probability	.1	.05	.6	.05	

a. $P(\{a, c, e\})$

b. $P(E \cup F)$, where $E = \{a, c, e\}$ and $F = \{b, c, e\}$

c. $P(E')$, where E is as in part (b)

d. $P(E \cap F)$, where E and F are as in part (b)

24. Repeat the preceding exercise using the following table:

Outcome	a	b	c	d	e
Probability	.1		.65	.1	.05

[11]A positive integer is prime if it is neither 1 nor a product of smaller integers.

APPLICATIONS

Employment (Compare Example 1) In December 2001 approximately 134 million people were employed in the United States. The following table shows the categories of occupations of these workers. In Exercises 25–30, use the data to compute the probabilities of the given events, rounded to the nearest 1%.

Occupation	**Workers (millions)**
Managerial and Professional Specialty	42
Technical, Sales, and Administrative Support	39
Service Occupations	18
Precision Production, Craft, and Repair	15
Operators, Fabricators, and Laborers	17
Farming, Forestry, and Fishing	3

SOURCE: U.S. Bureau of Labor Statistics press release, http://www.bls.gov/, January 4, 2002.

25. A worker was employed in a service occupation.

26. A worker was employed in precision production, craft, or repair.

27. A worker was employed in either technical, sales, or administrative support or was an operator, fabricator, or laborer.

28. A worker was employed in either a service occupation or farming, forestry, or fishing.

29. A worker was employed in neither a managerial or professional specialty nor a service occupation.

30. A worker was employed in neither a service occupation nor farming, forestry, or fishing.

31. Ethnic Diversity The following pie chart shows the ethnic makeup of California schools in the 2000–2001 academic year:

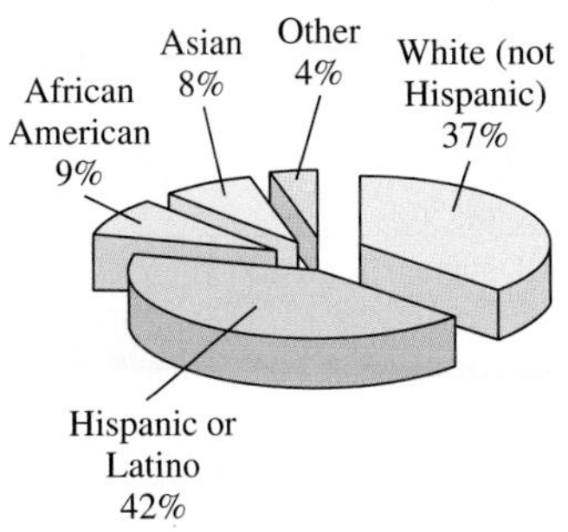

SOURCE: CBEDS data collection, Educational Demographics, http://www.cde.ca.gov/resrc/factbook/ethnicpies.htm, October 2000.

Write down the probability distribution showing the probability that a randomly selected California student in 2000–2001 belonged to one of the ethnic groups named. What is the probability that a student is neither White nor Asian?

32. Ethnic Diversity The following pie chart shows the ethnic makeup of California schools in the 1981–1982 academic year:

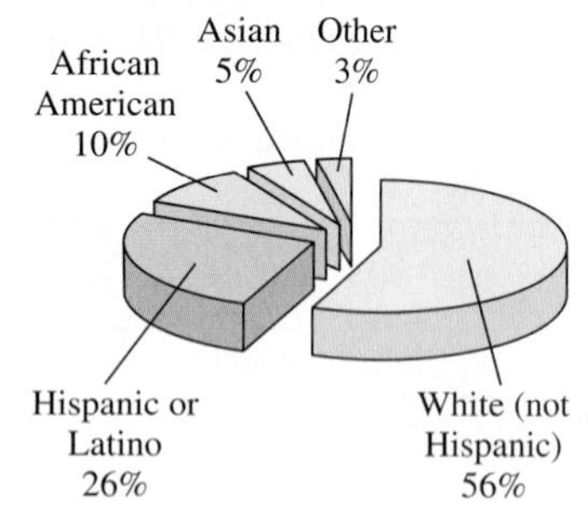

SOURCE: CBEDS data collection, Educational Demographics, http://www.cde.ca.gov/resrc/factbook/ethnicpies.htm, October 2000.

Write down the probability distribution showing the probability that a randomly selected California student in 1981–1982 belonged to one of the ethnic groups named. What is the probability that a student is neither Hispanic, Latino, nor African American?

33. Music Downloading The following table shows some results of a 2001 study on how recorded music purchases were affected by downloading of music files by experienced file sharers:

Outcome	Decreased music purchases	Increased music purchases	No change
Probability	.34	.52	.14

SOURCE: *New York Times,* May 6, 2002, p. C6.

Assuming that this probability distribution applied to a community of 2000 experienced file sharers at a university, how many of them did not increase their music purchases?

34. Music Downloading The following table shows some results of a 2001 study on how recorded music purchases were affected by downloading of music files by experienced file sharers with broadband Internet access and CD burners:

Outcome	Decreased music purchases	Increased music purchases	No change
Probability	.37	.47	.16

SOURCE: *New York Times,* May 6, 2002, p. C6.

Assuming that this probability distribution applied to a community of 400 experienced file sharers with broadband Internet access and CD burners at a university residence, how many of them did not decrease music purchases?

35. Internet Investments The following excerpt is from an article in the *New York Times:* July, 1999.

> While statistics are not available for Web entrepreneurs who fail, the venture capitalists that finance such Internet start-up companies have a rule of thumb. For every 10 ventures that receive financing—and there are plenty who do not—2 will be stock market successes, which means spectacular profits for early investors; 3 will be sold to other concerns, which translates into more modest profits; and the rest will fail.[12]

a. What is a sample space for the scenario?
b. Write down the associated probability distribution.
c. What is the probability that a start-up venture that receives financing will realize profits for early investors?

36. Internet Investments The following excerpt is from an article in the *New York Times:*

> Right now, the market for Web stocks is sizzling. Of the 126 initial public offerings of Internet stocks priced this year, 73 are trading above the price they closed on their first day of trading. . . . Still, 53 of the offerings have failed to live up to their fabulous first-day billings, and 17 [of these] are below the initial offering price.[13]

Assume that, on the first day of trading, all stocks closed higher than their initial offering price.
a. What is a sample space for the scenario?
b. Write down the associated probability distribution. (Round your answers to two decimal places.)
c. What is the probability that an Internet stock purchased during the period reported ended either below its initial offering price or above the price it closed on its first day of trading?

Student Admissions Exercises 37–44 are based on the following table that shows the profile, by Math SAT I scores, of admitted students at UCLA for the fall 2002 semester:

	200–399	400–499	500–599	600–699	700–800	Total
Admitted	4	358	1355	3361	5035	10,113
Not Admitted	764	3229	8642	11,953	4741	29,329
Total Applicants	768	3587	9997	15,314	9776	39,442

SOURCE: University of California Web site, http://www.admissions.ucla.edu/Prospect/Adm_fr/Frosh_Prof.htm, May 2002.

Determine the theoretical probabilities of the following events. (Round your answers to the nearest .001.)

37. An applicant was admitted.
38. An applicant had a Math SAT below 400.
39. An applicant with a Math SAT of 700 or above was admitted.
40. An applicant with a Math SAT below 400 was admitted.
41. An admitted student had a Math SAT of 700 or above.
42. An admitted student had a Math SAT below 400.

[12]"Not All Hit It Rich in the Internet Gold Rush," *New York Times,* July 20, 1999, p. A1. SOURCE: Comm-Scan/*New York Times.*
[13]Ibid.

43. A rejected applicant had a Math SAT below 600.

44. A rejected applicant had a Math SAT of at least 600.

45. Computer Sales In 1999 (1 year after the iMac was first launched by Apple), a retail or mail-order purchase of a personal computer was approximately seven times as likely to be a non-Apple PC as an Apple PC. What is the probability that a randomly chosen personal computer purchase was an Apple?

SOURCE: PC Data/*New York Times,* April 26, 1999, p. C1. Figure is approximate.

46. Computer Sales At the start of 1998, shortly before the iMac was first launched, a retail or mail-order purchase of a personal computer was approximately 17 times as likely to be a non-Apple PC as an Apple PC. What is the probability that a randomly chosen personal computer purchase was not an Apple?

SOURCE: PC Data/*New York Times,* April 26, 1999, p. C1. Figure is approximate.

47. Auto Sales In 1999 automobile sales in Europe equaled combined sales in NAFTA (North American Free Trade Agreement) countries and Asia. Further, sales in Europe were 70% more than sales in NAFTA countries.

a. Write down the associated probability distribution.

b. A total of 34 million automobiles were sold in these three regions. How many were sold in Europe?

SOURCE: Economist Intelligence Unit (EIU), http://www.autoindustry.co.uk/statistics/sales/world.html, March 15, 2002.

48. Auto Sales In 2000 automobile sales in Western Europe were three times combined sales in Japan and other Asian countries. Further, sales in Japan were four times sales in other Asian countries.

a. Write down the associated probability distribution.

b. A total of 20 million automobiles were sold in these three regions. How many were sold in Japan?

SOURCE: Economist Intelligence Unit (EIU), http://www.autoindustry.co.uk/statistics/sales/world.html, March 15, 2002.

Gambling Exercises 49–54 detail some of the nefarious dicing practices of the Win Some/Lose Some Casino. In each case find the probabilities of all possible outcomes and the probability that an odd number or an odd sum faces up.

49. Some of the dice are cleverly weighted so that each of 2, 3, 4, and 5 is twice as likely to come up as 1, and 1 and 6 are equally likely.

50. Other dice are weighted so that each of 2, 3, 4, and 5 is half as likely to come up as 1, and 1 and 6 are equally likely.

51. Some pairs of dice are magnetized so that each pair of mismatching numbers is twice as likely to come up as each pair of matching numbers.

52. Other pairs of dice are so strongly magnetized that mismatching numbers never come up.

53. Some dice are constructed in such a way that deuce (2) is five times as likely to come up as 4 and three times as likely to come up as each of 1, 3, 5, and 6.

54. Other dice are constructed in such a way that deuce (2) is six times as likely to come up as 4 and four times as likely to come up as each of 1, 3, 5, and 6.

Exercises 55 and 56 use counting techniques based on permutations and/or combinations.

55. Marbles A bag contains six marbles. Suzy chooses four at random. How many possible outcomes are there? If one of the marbles is red, how many of these outcomes include the red marble? What is the probability that Suzy grabs the red marble?

56. Chocolates My couch potato friend enjoys sitting in front of the TV and grabbing handfuls of 5 chocolates at random from his snack jar. Unbeknownst to him, I have replaced 1 of the 20 chocolates in his jar with a cashew. (He hates cashews with a passion.) How many possible outcomes are there the first time he grabs 5 chocolates? How many of these include the cashew? What is the probability that he will grab the cashew?

COMMUNICATION AND REASONING EXERCISES

57. Complete the following sentences. _____ probability is an estimate of _____ probability. This estimate improves as the _____.

58. The theoretical probability of an event E is the number of outcomes in E divided by the total number of outcomes, right? Explain.

59. Ruth tells you that when you roll a pair of fair dice, the theoretical probability of obtaining a pair of matching numbers is 1/6. To test this claim, you roll a pair of fair dice 20 times and never once get a pair of matching numbers. This proves that either Ruth is wrong or the dice are not fair, right? Explain.

60. Design an experiment based on rolling a fair die for which there are at least three outcomes with different probabilities.

61. It is said that lightning never strikes twice in the same spot. Assuming this to be the case, what is the theoretical probability that lightning will strike your favorite dining spot during a thunderstorm? Explain.

62. Your friend tells you that, once lightning has struck a particular spot, it is bound to do so again within the next five thunderstorms. Last night, lightning struck the first-hole green at the golf course. What can be said about the theoretical probability that it will strike there again? Explain.

63. If the weather service says that there is a 50% chance of rain today, does this suggest estimated probability or theoretical probability? Explain your answer.

64. The theory of quantum mechanics predicts that an electron in an atom can be in one of many energy states, according to a probability distribution that depends on the particular atom, and can be calculated analytically. Does this suggest estimated probability or theoretical probability? Explain your answer.

65. A certain event has theoretical probability equal to zero. This means it will never occur, right? Explain.

66. A certain experiment is performed a large number of times, and the event E has estimated probability equal to zero. This means that it has theoretical probability zero, right? Explain.

7.4 Probability and Counting Techniques (Optional)

We saw in Section 7.3 that when all outcomes in a sample space are equally likely we can use the following formula.

Computing Theoretical Probability: Equally Likely Outcomes

In an experiment in which all outcomes are equally likely, the theoretical probability of an event E is given by

$$P(E) = \frac{\text{number of favorable outcomes}}{\text{total number of outcomes}} = \frac{n(E)}{n(S)}$$

This formula is simple, but calculating $n(E)$ and $n(S)$ may not be. In this section we look at some examples in which we need to use the counting techniques discussed in Chapter 6.

Example 1 • Marbles

A bag contains four red and two green marbles. Upon seeing the bag, Suzy (who has compulsive marble-grabbing tendencies) sticks her hand in and grabs three at random. Find the probability that she will get both green marbles.

Solution According to the formula, we need to know these numbers:

- The number of elements in the sample space S
- The number of elements in the event E

What is the sample space? The sample space is the set of all possible outcomes, and each outcome consists of a set of three marbles (in Suzy's hand). So, the set of outcomes is the set of all sets of three marbles chosen from a total of six marbles (four red and two green). So,

$$n(S) = C(6, 3) = 20$$

Now what about E? This is the event that Suzy gets both green marbles. We must *rephrase this as a subset of S* in order to deal with it: "E is the collection of sets of three marbles such that one is red and two are green." Thus, $n(E)$ is the *number* of such sets, which we determine using a decision algorithm.

Step 1 Choose a red marble: $C(4, 1) = 4$ possible outcomes.

Step 2 Choose the two green marbles: $C(2, 2) = 1$ possible outcome.

We get $n(E) = 4 \times 1 = 4$. Now,

$$P(E) = \frac{n(E)}{n(S)} = \frac{4}{20} = \frac{1}{5}$$

Thus, there is a 1 in 5 chance of Suzy's getting both green marbles.

Example 2 • Investment Lottery

To "spice up" your investment portfolio, you decided to ignore your broker's cautious advice and select three stocks at random from the six most active stocks listed on the New York Stock Exchange on February 1, 2002:

	Close	Change
Tyco Intl	\$35.63	+0.48
AOL Time Warner	25.99	−0.32
Taiwan Semi	17.02	+0.05
QWEST Comms Intl	10.00	−0.50
Ford Motor Co.	14.90	−0.40
Waste Mgmt	25.13	−3.69

SOURCE: Yahoo! Finance, http://finance.yahoo.com, February 2, 2002.

Find the probabilities of the following events:

a. Your portfolio included Tyco and AOL Time Warner.

b. At most two of the stocks in your portfolio declined in value.

Solution The sample space is the set of all collections of three stocks chosen from the six. So,

$$n(S) = C(6, 3) = 20$$

a. The event E of interest is the event that your portfolio includes Tyco and AOL Time Warner. Thus, E is the set of all groups of three stocks that include Tyco and AOL Time Warner. Since there is only one more stock left to choose,

$$n(E) = C(4, 1) = 4$$

We now have

$$P(E) = \frac{n(E)}{n(S)} = \frac{4}{20} = \frac{1}{5} = .2$$

b. Let F be the event that at most two of the stocks in your portfolio declined in value. To calculate $n(F)$, we use the following decision algorithm:

Alternative 1 None of the stocks declined in value.
There are no possibilities for this alternative, since only two of the listed stocks did not decline in value.

Alternative 2 One of the stocks declined in value.
Step 1 Choose one stock that declined in value: $C(4, 1) = 4$ possibilities.
Step 2 Choose two stocks that did not decline in value: $C(2, 2) = 1$ possibility.
This gives $4 \times 1 = 4$ possibilities for this alternative.

Alternative 3 Two of the stocks declined in value.
Step 1 Choose two stocks that declined in value: $C(4, 2) = 6$ possibilities.
Step 2 Choose one stock that did not decline in value: $C(2, 1) = 2$ possibilities.
This gives $6 \times 2 = 12$ possibilities for this alternative.

So, we have a total of $4 + 12 = 16$ possible portfolios. Thus,

$$n(F) = 16$$

$$P(F) = \frac{n(F)}{n(S)} = \frac{16}{20} = .8$$

Example 3 • Poker Hands

You are dealt 5 cards from a well-shuffled standard deck of 52 cards. Find the probability that you have a full house. (Recall that a full house consists of 3 cards of one denomination and 2 of another.)

Solution The sample space S is the set of all possible 5-card hands dealt from a deck of 52. So,

$$n(S) = C(52, 5) = 2{,}598{,}960$$

If the deck is thoroughly shuffled, then each of these 5-card hands is equally likely. Now consider the event E, the set of all possible 5-card hands that constitute a full house. To calculate $n(E)$, we use a decision algorithm, which we show in the following compact form:

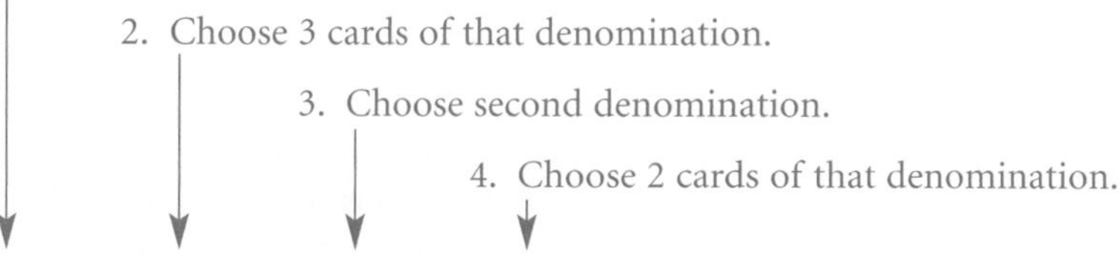
1. Choose first denomination.
2. Choose 3 cards of that denomination.
3. Choose second denomination.
4. Choose 2 cards of that denomination.

$$n(E) = C(13, 1) \times C(4, 3) \times C(12, 1) \times C(4, 2) = 3744$$

Thus,

$$P(E) = \frac{n(E)}{n(S)} = \frac{3744}{2{,}598{,}960} \approx .00144$$

In other words, there is approximately a .144% chance that you will be dealt a full house.

Example 4 • More Poker Hands

You are playing poker, and you have been dealt the following hand:

J♠, J♦, J♥, 2♣, 10♠

You decide to exchange the last two cards. The exchange works as follows: The two cards are discarded (not replaced in the deck), and you are dealt two new cards.

a. Find the probability that you end up with a full house.

b. Find the probability that you end up with four Jacks.

c. What is the probability that you end up with either a full house or four Jacks?

Solution

a. To get a full house, you must be dealt two of a kind. The sample space S is the set of all pairs of cards selected from what remains of the original deck of 52. You were dealt five cards originally, so there are $52 - 5 = 47$ cards left in the deck. Thus, $n(S) = C(47, 2) = 1081$. The event E is the set of all pairs of cards that constitute two of a kind. Note that you cannot get two Jacks because only one is left in the deck. Also, only three 2s and three 10s are left in the deck. We have

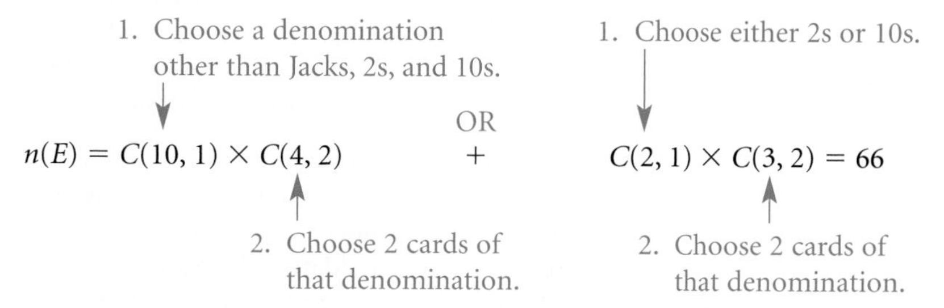

$$n(E) = C(10, 1) \times C(4, 2) + C(2, 1) \times C(3, 2) = 66$$

So,

$$P(E) = \frac{n(E)}{n(S)} = \frac{66}{1081} \approx .0611$$

b. We have the same sample space as in part (a). Let F be the set of all pairs of cards that include the missing Jack of clubs. So,

1. Choose the Jack of clubs

2. Choose 1 card from the remaining 46.

$$n(F) = C(1, 1) \times C(46, 1) = 46$$

Thus,

$$P(F) = \frac{n(F)}{n(S)} = \frac{46}{1081} \approx .0426$$

c. We are asked to calculate the probability of the event $E \cup F$. From the addition principle, we have

$$n(E \cup F) = n(E) + n(F) - n(E \cap F)$$

Since $E \cap F$ means "E and F," $E \cap F$ is the event that the pair of cards you are dealt are two of a kind and include the Jack of clubs. But this is impossible because only one Jack is left. Thus, $E \cap F = \varnothing$, and so $n(E \cap F) = 0$. This gives us

$$P(E \cup F) = \frac{n(E \cup F)}{n(S)} = \frac{n(E) + n(F)}{n(S)} = \frac{66 + 46}{1081} = \frac{112}{1081} \approx .1036$$

In other words, there is slightly better than a 1 in 10 chance that you will wind up with either a full house or four of a kind, given the original hand.

Before we go on . . . The answer to part (c) is the sum of the answers to parts (a) and (b), but you may have noticed that .1036 is not quite the sum of .0611 and .0426. The difference is due to the fact that these numbers are rounded.

Example 5 • Committees

The University Senate bylaws at Hofstra University state the following:

> The Student Affairs Committee shall consist of one elected faculty senator, one faculty senator-at-large, one elected Student Senator, five student senators-at-large (including one from the graduate school), two delegates from the Student Government Association, and the President of the Student Government Association or his/her designate. It shall be chaired by the elected student senator on the Committee and it shall be advised by the Dean of Students or his/her designate.[14]

You are an undergraduate student and, even though not an elected student senator, would very much like to serve on the Student Affairs Committee. The senators-at-large as well as the student government delegates are chosen by means of a random drawing from a list of candidates. There are already 13 undergraduate candidates for the position of senator-at-large and 6 candidates for student government delegates, and you have been offered a position on the Student Government Association (SGA), should you wish to join it. (This would make you ineligible for a senator-at-large position.) What should you do?

Solution You have two options. Option 1 is to include your name on the list of candidates for the senator-at-large position. Option 2 is to join the SGA and add your name to its list of candidates. Let's look at the two options separately.

Option 1: Add Your Name to the Senator-at-Large List

This will result in a list of 14 undergraduates for 4 undergraduate positions. The sample space is the set of all possible outcomes of the random drawing. Each outcome consists of a set of 4 lucky students chosen from 14:

$$n(S) = C(14, 4) = 1001$$

We are interested in the probability that you are among the chosen 4. Thus, E is the set of sets of 4 that include you:

1. Choose yourself.
2. Choose 3 from the remaining 13.

$$n(E) = C(1, 1) \times C(13, 3) = 286$$

So,

$$P(E) = \frac{n(E)}{n(S)} = \frac{286}{1001} = \frac{2}{7} \approx .2857$$

Option 2: Join the SGA and Add Your Name to Its List

This results in a list of 7 candidates from which 2 are selected. For this case, the sample space consists of all sets of 2 chosen from 7, so

$$n(S) = C(7, 2) = 21$$

---

[14]Hofstra University Senate Bylaws (as of 2002).

and

1. Choose yourself.

2. Choose 1 from the remaining 6.

$$n(E) = C(1, 1) \times C(6, 1) = 6$$

So,

$$P(E) = \frac{n(E)}{n(S)} = \frac{6}{21} = \frac{2}{7} \approx .2857$$

In other words, the probability of being selected is exactly the same for option 1 as it is for option 2! You can choose either option, and you will have slightly less than a 29% chance of being selected.

7.4 EXERCISES

Recall from Example 1 that, whenever Suzy sees a bag of marbles, she grabs a handful at random. In Exercises 1–10, she has seen a bag containing four red, three green, two white, and one purple marbles. She grabs five of them. Find the probabilities of the following events, expressing each as a fraction in lowest terms.

1. She has all the red ones.

2. She has none of the red ones.

3. She has at least one white one.

4. She has at least one green one.

5. She has two red ones and one of each of the other colors.

6. She has two green ones and one of each of the other colors.

7. She has at most one green one.

8. She has no more than one white one.

9. She does not have all the red ones.

10. She does not have all the green ones.

Investments Exercises 11–16 are based on the following table, which shows a selection of well-known companies listed in the Standard & Poors 500, together with their earnings per share (EPS) for the fourth quarter (Q4) of 2001:

Company	Q4 2001 EPS ($)
AOL Time Warner	0.33
Walt Disney	0.13
Apple Computer	0.18
Compaq Computer	0.06
Dell Computer	0.17
Hewlett-Packard	0.19
IBM	1.33

SOURCE: Earnings data obtained from www.earnings.com, a service of Thomson Financial Solutions.

11. Assuming that you had selected two of these companies' stocks at random, what is the probability that both the stocks in your selection had earnings per share of $0.30 or more?

12. Assuming that you had selected two of these companies' stocks at random, what is the probability that both the stocks in your selection had earnings per share of less than $0.30?

13. If you selected four of these stocks at random, find the probability that they included Compaq Computer, but not Dell Computer.

14. If you selected four of these stocks at random, what is the probability that your selection included the company with the largest earnings per share and excluded the company with the smallest earnings per share?

15. If your portfolio included 100 shares of AOL Time Warner and you then purchased 100 shares each of any two companies on the list at random, find the probability that you have 200 shares of AOL Time Warner.

16. If your portfolio included 100 shares of IBM and you then purchased 100 shares each of any three companies on the list at random, find the probability that you have 200 shares of IBM.

Poker In Exercises 17–22, you are asked to calculate the probability of being dealt various poker hands. (Recall that a poker player is dealt 5 cards at random from a standard deck of 52 cards.) Express each of your answers as a decimal rounded to four decimal places, unless otherwise stated.

17. Two of a kind: two cards with the same denomination and three cards with other denominations (different from each other and that of the pair). *Example:* K♣, K♥, 2♠, 4♦, J♠.

18. Three of a kind: three cards with the same denomination and two cards with other denominations (different from each other and that of the three). *Example:* Q♣, Q♥, Q♠, 4♦, J♠.

19. Two pair: two cards with one denomination, two with another and one with a third. *Example:* 3♣, 3♥, Q♠, Q♥, 10♠.

20. Straight flush: five cards of the same suit with consecutive denominations but not a royal flush (a royal flush consists of the 10, J, Q, K, and A of one suit); round the answer to one significant digit. *Examples:* A♣, 2♣, 3♣, 4♣, 5♣, or 9♦, 10♦, J♦, Q♦, K♦, or A♥, 2♥, 3♥, 4♥, 5♥, but *not* 10♦, J♦, Q♦, K♦, A♦.

21. Flush: five cards of the same suit, but not a straight flush or royal flush. *Example:* A♣, 5♣, 7♣, 8♣, K♣.

22. Straight: five cards with consecutive denominations, but not all of the same suit. *Examples:* 9♦, 10♦, J♣, Q♥, K♦ and 10♥, J♦, Q♦, K♦, A♦.

23. Lotteries The Sorry State Lottery requires you to select five different numbers from 0 through 49. (Order is not important.) You are a Big Winner if the five numbers you select agree with those in the drawing, and you are a Small-Fry Winner if four of your five numbers agree with those in the drawing. What is the probability of being a Big Winner? What is the probability of being a Small-Fry Winner? What is the probability that you win something?

24. Lotto The Sad State Lottery requires you to select a sequence of three different numbers from 0 through 49. (Order is important.) You are a winner is your sequence agrees with that in the drawing, and you are a booby-prize winner if your selection of numbers is correct, but in the wrong order. What is the probability of being a winner? What is the probability of being a booby-prize winner? What is the probability that you are either a winner or a booby-prize winner?

25. Transfers Your company is considering offering 400 employees the opportunity to transfer to its new headquarters in Ottawa. As personnel manager, you decide that it would be fairest if the transfer offers are decided by means of a lottery. Assuming that your company currently employs 100 managers, 100 factory workers, and 500 miscellaneous staff, find the following probabilities, leaving the answers as formulas.

a. All managers will be offered the opportunity.

b. You will be offered the opportunity.

26. Transfers Refer to Exercise 25. After thinking about your proposed method of selecting employees for the opportunity to move to Ottawa, you decide it might be a better idea to select 50 managers, 50 factory workers, and 300 miscellaneous staff, all chosen at random. Find the probability that you will be offered the opportunity. (Leave your answer as a formula.)

27. Lotteries In the New York State daily lottery game, a sequence of three (not necessarily different) digits in the range 0–9 is selected at random. Find the probability that all three are different.

28. Lotteries Refering to Exercise 27, find the probability that two of the three digits are the same.

29. The Monkey at the Typewriter Suppose a monkey is seated at a computer keyboard and randomly strikes the 26 letter keys and the space bar. Find the probability that its first 39 characters (including spaces) will be "to be or not to be that is the question." (Leave your answer as a formula.)

30. The Cat on the Piano A standard piano keyboard has 88 different keys. Find the probability that a cat, jumping on 4 keys in sequence at random (possibly with repetition), will strike the first four notes of Beethoven's *Fifth Symphony.* (Leave your answer as a formula.)

31. Sports The following table shows the results of the Big Eight Conference for the 1988 college football season:

	Won	Lost
Nebraska (NU)	7	0
Oklahoma (OU)	6	1
Oklahoma State (OSU)	5	2
Colorado (CU)	4	3
Iowa State (ISU)	3	4
Missouri (MU)	2	5
Kansas (KU)	1	6
Kansas State (KSU)	0	7

SOURCE: "On the Probability of a Perfect Progression," *American Statistician* 45(3) (1991): 214.

This is referred to as a "perfect progression." Assuming that the "Won" scores are chosen at random in the range 0–7, find the probability that they form a perfect progression in the Big Eight Conference.[15] (Leave your answer as a formula.)

32. Sports Referring to Exercise 31, find the probability of a perfect progression with Nebraska scoring 7 wins and 0 losses. (Leave your answer as a formula.)

33. Graph Searching A graph consists of a collection of **nodes** (the dots in the figure) connected by **edges** (line segments from one node to another). A **move on a graph** is a move from one node to another along a single edge. Find the probability of going from Start to Finish in a sequence of two random moves in the graph shown. (All directions are equally likely.)

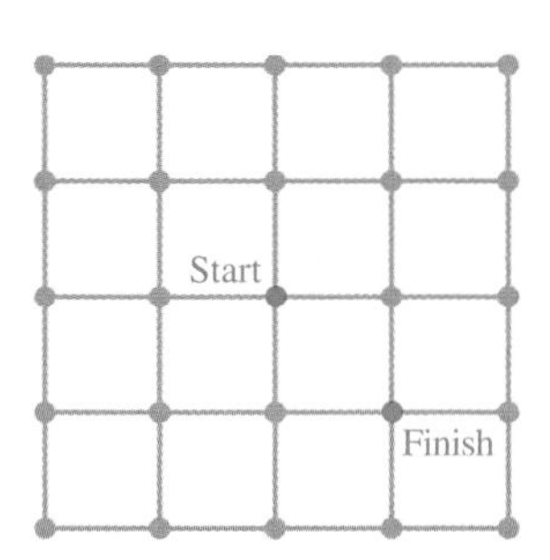

[15]Even if all the teams are equally likely to win each game, the chances of a perfect progression actually coming up are a little more difficult to estimate, because the number of wins by one team impacts directly on the number of wins by the others. For instance, it is impossible for all eight teams to show a score of 7 wins and 0 losses at the end of the season—someone must lose! It is, however, not too hard to come up with a counting argument to estimate the total number of win–lose scores actually possible.

34. **Graph Searching** Referring to Exercise 33, find the probability of going from Start to one of the Finish nodes in a sequence of two random moves in the following figure. (All directions are equally likely.)

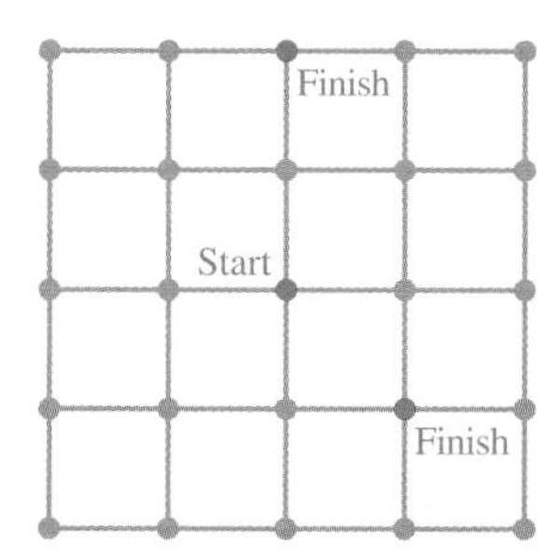

35. **Tests** A test has three parts. Part A consists of eight true/false questions, Part B consists of five multiple-choice questions with five choices each, and Part C requires you to match five questions with five different answers one-to-one. Assuming that you make random guesses in filling out your answer sheet, what is the probability that you will earn 100% on the test? (Leave your answer as a formula.)

36. **Tests** A test has three parts. Part A consists of four true/false questions, Part B consists of four multiple choice questions with five choices each, and Part C requires you to match six questions with six different answers one-to-one. Assuming that you make random choices in filling out your answer sheet, what is the probability that you will earn 100% on the test? (Leave your answer as a formula.)

37. **Tournaments** What is the probability that North Carolina will beat Central Connecticut but lose to Virginia in the following (fictitious) soccer tournament? (Assume that all outcomes are equally likely.)

North Carolina
Central Connecticut
Virginia
Syracuse

38. **Tournaments** In a (fictitious) soccer tournament involving the four teams San Diego State, De Paul, Colgate, and Hofstra, find the probability that Hofstra will play Colgate in the finals and win. (Assume that all outcomes are equally likely and that the teams not listed in the first round slots are placed at random.)

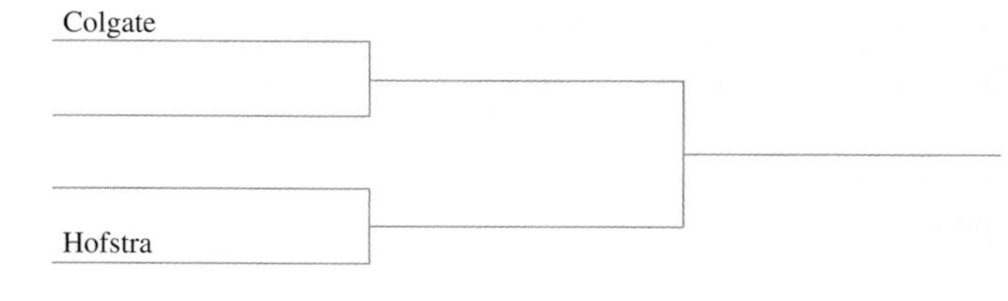

39. **Product Design** Your company has patented an electronic digital padlock that a user can program with his or her own four-digit code. (Each digit can be 0 through 9, and repetitions are allowed.) The padlock is designed to open either if the correct code is keyed in or—and this is helpful for forgetful people—if exactly one of the digits is incorrect. What is the probability that a randomly chosen sequence of four digits will open a programmed padlock?

40. **Product Design** Assume that you already know the first digit of the combination for the lock described in Exercise 39. Find the probability that a random guess of the remaining three digits will open the lock.

41. **Committees** An investigatory committee in the kingdom of Utopia consists of a chief investigator (a Royal Party member), an assistant investigator (a Birthday Party member), two at-large investigators (either party), and five ordinary members (either party). Royal Party member Larry Sifford is hoping to avoid serving on the committee, unless he is the chief investigator and Otis Taylor, a Birthday Party member, is the assistant investigator. The committee is to be selected at random from a pool of 12 candidates, including Larry Sifford and Otis Taylor, half of whom are Royal Party and half of whom are Birthday Party.
    a. How many different committees are possible?
    b. How many committees are possible in which Larry's hopes are fulfilled? (This includes the possibility that he's not on the committee at all.)
    c. What is the probability that he'll be happy with a randomly selected committee?

42. **Committees** A committee is to consist of a chair, three hagglers, and four do-nothings. The committee is formed by choosing randomly from a pool of ten people and assigning them to the various "jobs."
    a. How many different committees are possible?
    b. Norman is eager to be the chair of the committee. What is the probability that he will get his wish?
    c. Norman's girlfriend Norma is less ambitious and would be happy to hold any position on the committee, provided that Norman is also selected as a committee member. What is the probability that she will get her wish and serve on the committee?
    d. Norma does not get along with Oona (who is also in the pool of prospective members) and would be most unhappy if Oona were to chair the committee. Find the probability that all her wishes will be fulfilled: She and Norman are on the committee, and it is not chaired by Oona.

43. **(Based on a Question from the GMAT)** Tyler and Gebriella are among seven contestants from whom four semifinalists are to be selected at random. Find the probability that neither Tyler nor Gebriella is selected.

44. **(Based on a Question from the GMAT)** Tyler and Gebriella are among seven contestants from whom four semifinalists are to be selected at random. Find the probability that Tyler is selected but not Gebriella.

COMMUNICATION AND REASONING EXERCISES

45. What is wrong with the following argument? A bag contains two blue and two red marbles; two are drawn at random. Since there are four possibilities—(red, red), (blue, blue), (red, blue), and (blue, red)—the probability that both are red is 1/4.

46. What is wrong with the following argument? When we roll two indistinguishable dice, the number of possible outcomes (unordered groups of two not necessarily distinct numbers) is 21, and the number of outcomes in which both numbers are the same is 6. Hence, the probability of throwing a double is $6/21 = 2/7$.

47. Suzy grabs two marbles out of a bag of five red and four green marbles. She could do so in two ways: She could take them out one at a time so that there is a first and a second marble, or she could grab two at once so that there is no order. Does the method she uses to grab the marbles affect the probability that she gets two red marbles?

48. If Suzy grabs two marbles, one at a time, out of a bag of five red and four green marbles, find an event with a probability that depends on the order in which the two marbles are drawn.

49. Create an interesting application whose solution requires finding a probability using combinations.

50. Create an interesting application whose solution requires finding a probability using permutations.

7.5 Probability Distributions

In Sections 7.2 and 7.3, we studied estimated and theoretical probability, and we spent some time discussing the differences between them. What they have in common is the idea of a **probability distribution.** In fact, they are best understood as two different ways, appropriate in different contexts, to find probability distributions. In this section we look a little more closely at probability distributions in the abstract and some of the things that can be done with them.

Looking back over the discussions of estimated and theoretical probability distributions, we notice the following common properties:

1. The probability of each outcome in the sample space is a real number between 0 and 1 (inclusive).
2. The probabilities of all the outcomes add to 1.
3. We can compute the probability of any event E by adding the probabilities of all the outcomes in E.

We record these as the defining properties of a probability distribution, deliberately ignoring any discussion of how the probabilities were determined or even what experiment we may have in mind. Note that everything we say about probability distributions applies equally well to estimated and theoretical probability.

Sample Space, Probability, and Probability Distribution

A **finite sample space** is a finite set S. A **probability distribution** is an assignment of a number $P(s_i)$ to each outcome of a sample space $S = \{s_1, s_2, \ldots, s_n\}$ such that

1. $0 \le P(s_i) \le 1$

2. $P(s_1) + P(s_2) + \cdots + P(s_n) = 1$

$P(s_i)$ is called the **probability of s_i.** An **event** E is a subset of S. Given a probability distribution, we obtain the probability of an event E, written $P(E)$, by adding up the probabilities of the outcomes in E.

If $P(E) = 0$, we call E an **impossible event.** The event $\varnothing$ is always impossible because *something* must happen.

Quick Examples

1. All examples of estimated and theoretical probability distributions we have considered are examples of probability distributions.
2. Let's take $S = \{H, T\}$ and make the assignments $P(H) = 0.2$ and $P(T) = 0.8$. Since these numbers are between 0 and 1 and add to 1, they specify a probability distribution.
3. With $S = \{H, T\}$ again, we could also take $P(H) = 1$ and $P(T) = 0$ so that T is an impossible event.
4. The following table gives a probability distribution for the sample space

   $S = \{1, 2, 3, 4, 5, 6\}$

Outcome	1	2	3	4	5	6
Probability	.3	.3	0	.1	.2	.1

It follows that

$P(\{1, 6\}) = .3 + .1 = .4$

$P(\{2, 3\}) = .3 + 0 = .3$

$P(3) = 0$ An impossible event

The distribution in Quick Example 4 could be the probability distribution of a weighted die. Whether or not a die with these exact (theoretical) probabilities for outcomes can actually be manufactured is a question we leave for die engineers to ponder. As mathematicians, we can certainly *conceive* of such a die, and that is all the justification we need to consider its probability distribution. Mathematicians think of probability distributions in this rather abstract way; instead of thinking of the actual probability that an outcome will occur in a real-life experiment, they think of the probability of an outcome as an assignment of a real number between 0 and 1 to each of the outcomes in some sample space. (All we have to ensure is that the numbers add up to 1. We can then pass the entire probability distribution along to the engineering department for realization.)

Probability of Unions, Intersections, and Complements

So far, all we know about computing the probability of an event E is that $P(E)$ is the sum of the probabilities of the individual outcomes in E. However, it is often the case that we do not know the probabilities of the individual outcomes, but we do know something else. For instance, it may be the case that E is the union or intersection of two events A and B, where we happen to know $P(A)$ and $P(B)$.

Question Given two events A and B, how do we compute the probability of $A \cup B$?

Answer We might be tempted to say that $P(A \cup B)$ is $P(A) + P(B)$, but let's look at an example using the probability distribution in Quick Example 4. Let A be the event $\{2, 4, 5\}$ and let B be the event $\{2, 4, 6\}$. $A \cup B$ is then the event $\{2, 4, 5, 6\}$. We know that we can find the probabilities $P(A)$, $P(B)$, and $P(A \cup B)$ by adding the probabilities of all the outcomes in these events, so

$$P(A) = P(\{2, 4, 5\}) = .3 + .1 + .2 = .6$$

$$P(B) = P(\{2, 4, 6\}) = .3 + .1 + .1 = .5$$

$$P(A \cup B) = P(\{2, 4, 5, 6\}) = .3 + .1 + .2 + .1 = .7$$

Our first guess was wrong: $P(A \cup B) \neq P(A) + P(B)$. Notice, however, that the outcomes in $A \cap B$ are counted twice in computing $P(A) + P(B)$, but only once in computing $P(A \cup B)$:

$$\begin{aligned} P(A) + P(B) &= P(\{2, 4, 5\}) + P(\{2, 4, 6\}) && A \cap B = \{2, 4\} \\ &= (.3 + .1 + .2) + (.3 + .1 + .1) && P(A \cap B) \text{ counted twice} \\ &= 1.1 \end{aligned}$$

whereas

$$\begin{aligned} P(A \cup B) &= P(\{2, 4, 5, 6\}) = .3 + .1 + .2 + .1 && P(A \cap B) \text{ counted once} \\ &= .7 \end{aligned}$$

So, if we take $P(A) + P(B)$ and then subtract the surplus $P(A \cap B)$, we get $P(A \cup B)$. In symbols,

$$P(A \cup B) = P(A) + P(B) - P(A \cap B)$$

$$.7 = .6 + .5 - .4$$

One more thing: Notice that our original guess $P(A \cup B) = P(A) + P(B)$ would have worked if we had chosen A and B with no outcomes in common; that is, if $A \cap B = \emptyset$. When $A \cap B = \emptyset$, recall that we say that A and B are mutually exclusive.

Some Principles of Probability Distributions

Addition Principle

If A and B are any two events, then

$$P(A \cup B) = P(A) + P(B) - P(A \cap B)$$

Mutually Exclusive Events

If $A \cap B = \emptyset$, we say that A and B are **mutually exclusive,** and we have

$$P(A \cup B) = P(A) + P(B) \qquad \text{Because } P(A \cap B) = 0$$

This holds true also for more than two events: If $A_1, A_2, \ldots, A_n$ are mutually exclusive events (that is, the intersection of every pair of them is empty), then

$$P(A_1 \cup A_2 \cup \cdots \cup A_n) = P(A_1) + P(A_2) + \cdots + P(A_n) \qquad \text{Many mutually exclusive events}$$

Quick Examples

1. There is a 10% chance of rain (R) tomorrow, a 20% chance of high winds (W), and a 5% chance of both. The probability of either rain or high winds (or both) is

$$\begin{aligned} P(R \cup W) &= P(R) + P(W) - P(R \cap W) \\ &= .10 + .20 - .05 = .25 \end{aligned}$$

2. The probability that you will be in Cairo at 6 A.M. tomorrow (C) is .3, whereas the probability that you will be in Alexandria at 6 A.M. tomorrow (A) is .2. Thus, the probability that you will be either in Cairo or Alexandria at 6 A.M. tomorrow is

$$P(C \cup A) = P(C) + P(A) \qquad \text{A and C are mutually exclusive.}$$
$$= .3 + .2 = .5$$

3. When a pair of fair dice is rolled, the probability of the numbers that face up adding to 7 is 6/36, the probability of their adding to 8 is 5/36, and the probability of their adding to 9 is 4/36. Thus, the probability of the uppermost numbers adding to 7, 8, or 9 is

$$P(7 \cup 8 \cup 9) = P(7) + P(8) + P(9) \qquad \text{The events are mutually exclusive.*}$$
$$= \frac{6}{36} + \frac{5}{36} + \frac{4}{36} = \frac{15}{36} = \frac{5}{12}$$

*The sum of the numbers that face up cannot equal two different numbers at the same time.

Example 1 • *School and Work*

A survey conducted by the Bureau of Labor Statistics found that 63% of the high school graduating class of 2000 went on to college the following year, and 53% of the class was working.[16] Furthermore, 89% were either in college or working, or both.

a. What percentage went on to college and work at the same time?

b. What percentage went on to college but not work?

Solution We can think of the experiment of choosing a member of the high school graduating class of 2000 at random. The sample space is the set of all these graduates.

a. We are given information about two events:

A: A graduate went on to college; $P(A) = .63$

B: A graduate went on to work; $P(B) = .53$

We are asked for the probability that a graduate went on to both college and work. That is, we are asked for $P(A \cap B)$. We are told that $P(A \cup B) = .89$. How can we use the formula for $P(A \cup B)$ if we already *know* $P(A \cup B)$? We take advantage of the fact that the formula

$$P(A \cup B) = P(A) + P(B) - P(A \cap B)$$

can be used to calculate any one of the four quantities that appear in it, as long as we know the other three. Substituting the quantities we know, we get

$$.89 = .63 + .53 - P(A \cap B)$$

so

$$P(A \cap B) = .63 + .53 - .89 = .27$$

Thus, 27% of the graduates went on to college and work at the same time.

[16]"College Enrollment and Work Activity of High School Graduates," U.S. Bureau of Labor Statistics, http://www.bls.gov/news.release/hsgec.toc.htm, April 2001.

b. We are asked for the probability of a new event:

C: A graduate went on to college, but not work

C is the part of A outside of $A \cap B$, so $C \cup (A \cap B) = A$, and C and $A \cap B$ are mutually exclusive. Thus, applying the addition principle, we have

$$P(C) + P(A \cap B) = P(A)$$

From part (a), we know that $P(A \cap B) = .27$, so

$$P(C) + .27 = .63$$

$$P(C) = .36$$

In other words, 36% of the graduates went on to college, but not work.

We can use the addition principle to deduce other useful properties of a probability distribution.

More Principles of Probability Distributions

The following rules hold for any sample space S and any event A:

$P(S) = 1$ — The probability of *something* happening is 1.

$P(\varnothing) = 0$ — The probability of *nothing* happening is 0.

$P(A') = 1 - P(A)$ — The probability of A *not* happening is 1 minus the probability of A.

Note We can also write the third equation as

$$P(A) = 1 - P(A') \quad \text{or} \quad P(A) + P(A') = 1$$

Quick Examples

1. There is a 10% chance of rain (R) tomorrow. Therefore, the probability that it will *not* rain is

$$P(R') = 1 - P(R) = 1 - .10 = .90$$

2. The probability that Eric Ewing will score at least two goals is .6. Therefore, the probability that he will score at most one goal is $1 - .6 = .4$.

Question Can you persuade me that all these principles are true?

Answer Let's take them one at a time.

We know that $S = \{s_1, s_2, \dots, s_n\}$ is the set of all outcomes, and so

$$P(S) = P(\{s_1, s_2, \dots, s_n\})$$

$$= P(s_1) + P(s_2) + \cdots + P(s_n) \qquad \text{We add the probabilities of the outcomes to obtain the probability of an event.}$$

$$= 1 \qquad \text{By the definition of a probability distribution}$$

Now, note that $S \cap \varnothing = \varnothing$, so that S and $\varnothing$ are mutually exclusive. Applying the addition principle gives

$$P(S) = P(S \cup \varnothing) = P(S) + P(\varnothing)$$

Subtracting $P(S)$ from both sides gives $0 = P(\varnothing)$.

If A is any event in S, then we can write

$$S = A \cup A'$$

where A and A' are mutually exclusive. (Why?) So, by the addition principle,

$$P(S) = P(A) + P(A')$$

Since $P(S) = 1$, we get

$$1 = P(A) + P(A')$$

$$P(A') = 1 - P(A)$$

Example 2 • Employment

In December 2001 the probability that a randomly selected U.S. resident (of working age) was employed was approximately .63.[17] The probability that the resident was unemployed but actively searching for a job was .04. Calculate the probabilities of the following events:

a. A resident was unemployed.

b. A resident was unemployed and not actively searching for a job.

Solution

a. Let's write E for the event that a resident was employed. We are given that $P(E) = .63$. The event that the resident was *un*employed is the complement of E and is given by

$$P(E') = 1 - P(E) = 1 - .63 = .37$$

b. We are given the probability that a resident was unemployed and searching for a job, and we are asked to find the probability that a resident was unemployed and *not* searching for a job. The two corresponding events

J: A resident was unemployed and searching for a job; $P(J) = .04$

N: A resident was unemployed and not searching for a job

are mutually exclusive events whose union is E', the set of all unemployed residents. Hence,

$$P(E') = P(N) + P(J)$$

$$.37 = P(N) + .04$$

giving

$$P(N) = .37 - .04 = .33$$

Thus, there was a 33% chance that a U.S. resident of working age was unemployed and not searching for a job.

[17] SOURCE: U.S. Bureau of Labor Statistics press release, http://www.bls.gov/, January 4, 2002.

7.5 EXERCISES

In Exercises 1–16, use the given information to find the indicated probability.

1. $P(A) = .1$; $P(B) = .6$; $P(A \cap B) = .05$. Find $P(A \cup B)$.
2. $P(A) = .3$; $P(B) = .4$; $P(A \cap B) = .02$. Find $P(A \cup B)$.
3. $A \cap B = \varnothing$; $P(A) = .3$; $P(A \cup B) = .4$. Find $P(B)$.
4. $A \cap B = \varnothing$; $P(B) = .8$; $P(A \cup B) = .8$. Find $P(A)$.
5. $A \cap B = \varnothing$; $P(A) = .3$; $P(B) = .4$. Find $P(A \cup B)$.
6. $A \cap B = \varnothing$; $P(A) = .2$; $P(B) = .3$. Find $P(A \cup B)$.
7. $P(A \cup B) = .9$; $P(B) = .6$; $P(A \cap B) = .1$. Find $P(A)$.
8. $P(A \cup B) = 1.0$; $P(A) = .6$; $P(A \cap B) = .1$. Find $P(B)$.
9. $P(A) = .75$. Find $P(A')$.
10. $P(A) = .22$. Find $P(A')$.
11. A, B, and C are mutually exclusive. $P(A) = .3$; $P(B) = .4$; $P(C) = .3$. Find $P(A \cup B \cup C)$.
12. A, B, and C are mutually exclusive. $P(A) = .2$; $P(B) = .6$; $P(C) = .1$. Find $P(A \cup B \cup C)$.
13. A and B are mutually exclusive. $P(A) = .3$ and $P(B) = .4$. Find $P[(A \cup B)']$.
14. A and B are mutually exclusive. $P(A) = .4$ and $P(B) = .4$. Find $P[(A \cup B)']$.
15. $A \cup B = S$ and $A \cap B = \varnothing$. Find $P(A) + P(B)$.
16. $P(A \cup B) = .3$ and $P(A \cap B) = .1$. Find $P(A) + P(B)$.

In Exercises 17–24, determine whether the information shown is consistent with a probability distribution. If not, say why.

17.

Outcome	a	b	c	d	e
Probability	0	0	.65	.3	.05

18.

Outcome	a	b	c	d	e
Probability	.1	−.1	.65	.3	.05

19. $P(A) = .2$; $P(B) = .1$; $P(A \cup B) = .4$
20. $P(A) = .2$; $P(B) = .4$; $P(A \cup B) = .2$
21. $P(A) = .2$; $P(B) = .4$; $P(A \cap B) = .2$
22. $P(A) = .2$; $P(B) = .4$; $P(A \cap B) = .3$
23. $P(A) = .1$; $P(B) = 0$; $P(A \cup B) = 0$
24. $P(A) = .1$; $P(B) = 0$; $P(A \cap B) = 0$

APPLICATIONS

25. **Internet Use** The following pie chart shows the percentage of the population that uses the Internet, broken down by family income, based on a survey taken in August 2000:

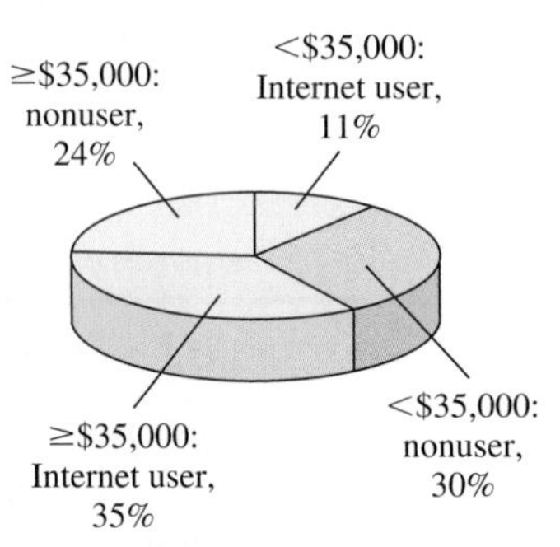

SOURCE: "Falling through the Net: Toward Digital Inclusion, A Report on Americans' Access to Technology Tools," U.S. Department of Commerce, http://www.ntia.doc.gov/ntiahome/fttn00/contents00.html, October 2000.

What is the probability that a randomly chosen person was an Internet user?

26. **Internet Use** Repeat Exercise 25, using the following pie chart that shows the results of a similar survey taken in September 2001:

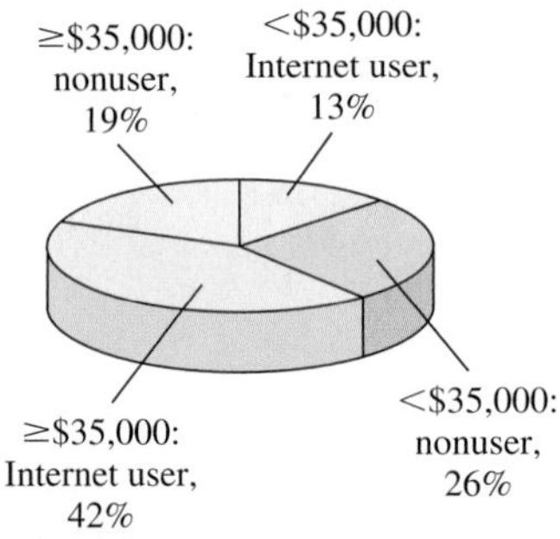

SOURCE: "A Nation Online: How Americans Are Expanding Their Use of the Internet," U.S. Department of Commerce, http://www.ntia.doc.gov/ntiahome/dn/index.html, February 2002.

27. **Astrology** The astrology software package, Turbo Kismet,[18] works by first generating random number sequences and then interpreting them numerologically. When I ran it yesterday, it informed me that there was a 1/3 probability that I would meet a tall dark stranger this month, a 2/3 probability that I would travel this month, and a 1/6 probability that I would meet a tall dark stranger and also travel this month. What is the probability that I will either meet a tall dark stranger or that I will travel this month?

[18] The name and concept were borrowed from a hilarious (as yet unpublished) novel by the science fiction writer William Orr, who also happens to be a faculty member at Hofstra University.

28. **Astrology** Another astrology software package, Java Kismet, is designed to help day traders choose stocks based on the position of the planets and constellations. When I ran it yesterday, it informed me that there was a .5 probability that Amazon.com will go up this afternoon, a .2 probability that Yahoo.com will go up this afternoon, and a .2 chance that both will go up this afternoon. What is the probability that either Amazon.com or Yahoo.com will go up this afternoon?

29. **Holiday Shopping** In 1999 the probability that a consumer would shop for holiday gifts at a discount department store was .80, and the probability that a consumer would shop for holiday gifts from catalogs was .42. Assuming that 90% of consumers shopped from one or the other, what percentage of them did both?

    SOURCE: U.S. Commerce Department; Deloitte & Touche Survey/*New York Times,* November 24, 1999, p. C1.

30. **Holiday Shopping** In 1997 the probability that a consumer would shop for holiday gifts at a discount department store was .70, and the probability that a consumer would shop for holiday gifts from catalogs was .32. Assuming that 75% of consumers shopped from one or the other, what percentage of them did both?

    SOURCE: U.S. Commerce Department; Deloitte & Touche Survey/*New York Times,* November 24, 1999, p. C1.

31. **Online Households** In 2001 the probability that a randomly selected online U.S. household (a household connected to the Internet) had cable access was .11, and the probability that an online household had DSL access was .05. What percentage of online households had neither cable nor DSL access? (Assume that the probability that an online household had both kinds of access is negligible.)

    SOURCE: *New York Times,* December 24, 2001, p. C1. The 5% figure is an estimate based on a 1999 study.

32. **Online Households** In 2001 6.1% of all U.S. households were connected to the Internet via cable, and 2.7% of them were connected to the internet through DSL. What percentage of U.S. households did not have high-speed (cable or DSL) connection to the Internet? (Assume that the percentage of households with both cable and DSL access is negligible.)

33. **Fast-Food Stores** In 2000 the top 100 chain restaurants in the United States owned a total of approximately 130,000 outlets. Of these, the three largest (in numbers of outlets) were McDonald's, Subway, and Burger King, owning among them 26% of all of the outlets. The two hamburger companies, McDonald's and Burger King, together owned approximately 16% of all outlets, and the two largest, McDonald's and Subway, together owned 19% of the outlets. What was the probability that a randomly chosen restaurant was a McDonald's?

    SOURCE: "Technomic 2001 Top 100 Report," Technomic, Inc. Information obtained from their Web site, www.technomic.com.

34. **Fast-Food Revenues** In 2000 the top 100 chain restaurants in the United States earned a total of approximately $123 billion in sales. The three largest (in numbers of outlets) were McDonald's, Subway, and Burger King, earning between them 26% of all sales. The two hamburger companies, McDonald's and Burger King, together earned approximately 23% of all sales, and the two largest (in number of outlets), McDonald's and Subway, together earned 19% of sales. What was the probability that a randomly selected dollar in sales was earned by McDonald's?

    SOURCE: "Technomic 2001 Top 100 Report," Technomic, Inc. Information obtained from their Web site, www.technomic.com.

35. **Opinion Polls** A *New York Times*/CBS News poll of 1368 people interviewed in March 1993 showed that 85% of all respondents favored a national law requiring a 7-day waiting period for handgun purchases, 13% opposed it, and 2% were undecided. Considering the poll as an experiment, find the probability that a randomly chosen resident was not opposed to such a law.

    SOURCE: *New York Times,* August 15, 1993, sect. 4, p. 4

36. **Opinion Polls** The opinion poll referred to in Exercise 35 also showed that 41% of the respondents would favor an outright ban on the sale of handguns (law enforcement officers excepted), 55% would oppose such a ban, and 4% were undecided. Considering the poll as an experiment, find the probability that a randomly chosen resident was not opposed to such a law.

37. **Greek Life** The ΤΦΦ Sorority has a tough pledging program—it requires its pledges to master the Greek alphabet forward, backward, and "sideways." During the last pledge period, two-thirds of the pledges failed to learn it backward, and three-quarters of them failed to learn it sideways; 5 of the 12 pledges failed to master it either backward or sideways. Since admission into the sisterhood requires both backward and sideways mastery, what fraction of the pledges were disqualified on this basis?

38. **Swords and Sorcery** Lance the Wizard has been informed that tomorrow there will be a 50% chance of encountering the evil Myrmidons and a 20% chance of meeting up with the dreadful Balrog. Moreover, Hugo the elf has predicted that there is a 10% chance of encountering both tomorrow. What is the probability that Lance will be lucky tomorrow and encounter neither the Myrmidons nor the Balrog?

Auto Theft Exercises 39–42 are based on the following table, which shows the probability that an owner of each model would report his or her vehicle stolen in 1997:

Brand	Jeep Wrangler	Suzuki Sidekick (two door)	Toyota Land Cruiser	Geo Tracker (two door)	Acura Integra (two door)
Probability	.0170	.0154	.0143	.0142	.0123
Brand	Mitsubishi Montero	Acura Integra (four door)	BMW 3-series (two door)	Lexus GS300	Honda Accord (two door)
Probability	.0108	.0103	.0077	.0074	.0070

Data are for insured vehicles, for 1995 to 1997 models except Wrangler, which is for 1997 models only.
SOURCE: Highway Loss Data Institute/*New York Times,* March 28, 1999, p. WK3.

39. Which of the following is true?
  (A) There is a 1.54% chance that a vehicle reported stolen in 1997 was a Suzuki Sidekick.
  (B) Of all the vehicles reported stolen by their owners in 1997, 1.43% of them were Toyota Land Cruisers.
  (C) Of all the owners of Toyota Land Cruisers, 1.43% reported their vehicles stolen in 1997.
  (D) If an owner reported his or her vehicle stolen in 1997, there was a 1.08% chance that the vehicle was a Mitsubishi Montero.

40. Which of the following is true?
  (A) There is a 98.46% chance that the owner of a Suzuki Sidekick would not report it stolen in 1997.
  (B) Of all the reports of stolen vehicles in 1997, 98.57% of them were not by owners of Toyota Land Cruisers.
  (C) If someone owned an Acura that was reported stolen in 1997, there was a 1.03% chance that it was a four-door model.
  (D) If someone reported his or her vehicle stolen in 1997, there was a 98.92% chance that it was not a Mitsubishi Montero.

41. It is now January 1, 1997, and I own a BMW 3-series and a Lexus GS300. For reasons too complicated to explain, it is simply impossible for both my cars to get stolen this year. What is the probability that one of my vehicles will get stolen this year?

42. Actually, there *is* a slight chance that both my cars will get stolen: approximately a .01% chance. How does this affect the answer to Exercise 41?

43. Judges and Juries A study revealed that the probability of a judge acquitting a randomly chosen defendant was .17, and the probability of a jury acquitting a randomly chosen defendant was .33. Further, the probability that both a judge and a jury would acquit a randomly chosen defendant was .14. Find the probability of the following events:
  a. A judge would not acquit a randomly chosen defendant.
  b. A jury would not acquit a randomly chosen defendant.
  c. Neither a judge nor a jury would acquit a randomly chosen defendant.
  d. A jury would acquit a defendant, but a judge would not acquit him or her. (*Hint:* Draw a Venn diagram.)

SOURCE: Hans Zeisel and Harry Kalven, "Parking Tickets and Missing Women: Statistics and the Law," in *Statistics: A Guide to the Unknown,* ed. J. A. Tanur et al. (San Francisco: Holden-Day, 1972), pp. 102–111, as cited in D. E. Zitarelli and R. F. Coughlin, *Finite Mathematics* (Orlando, Fla.: Saunders, 1987).

44. Judges and Juries Repeat Exercise 43, using the following data: The probability of a judge acquitting a randomly chosen defendant was .21, and the probability of a jury acquitting a randomly chosen defendant was .30. Further, the probability that both a judge and a jury would acquit a randomly chosen defendant was .15.

45. Public Health A study shows that 80% of the population has been vaccinated against the Venusian flu, but 2% of the vaccinated population gets the flu anyway. If 10% of the total population gets this flu, what percent of the population either gets the vaccine or gets the disease?

46. Public Health A study shows that 75% of the population has been vaccinated against the Martian ague, but 4% of this group gets this disease anyway. If 10% of the total population gets this disease, what is the probability that a randomly selected person has been neither vaccinated nor has contracted Martian ague?

COMMUNICATION AND REASONING EXERCISES

47. Complete the following sentence. The probability of the union of two events is the sum of the probabilities of the two events if ______.

48. If you know $P(E)$ and $P(F)$, what additional information would you need to calculate $P(E \cap F)$, and how would you calculate it?

49. Give an example of a sample space S, a probability distribution on S, and two events A and B with the property that A and B are not mutually exclusive and yet $P(A) + P(B) = P(A \cup B)$.

50. A friend of yours asserted at lunch today that, according to the weather forecast for tomorrow, there is a 52% chance of rain and a 60% chance of snow. "But that's impossible!" you blurted out, "the percentages add up to more than 100%." Explain why you were wrong.

51. Explain how the addition principle for mutually exclusive events follows from the general addition principle.

52. Explain how the property $P(A') = 1 - P(A)$ follows directly from the properties of a probability distribution.

53. Find a formula for the probability of the union of three (not necessarily mutually exclusive) events A, B, and C.

54. Four events A, B, C, and D have the following property: If any two events have an outcome in common, that outcome is common to all four events. Find a formula for the probability of their union.

7.6 Conditional Probability and Independence

Cyber Video Games has been running a television ad for its latest game, "Ultimate Hockey." As Cyber Video's director of marketing, you would like to assess the ad's effectiveness, so you ask your market research team to survey video-game players. The results of its survey of 50,000 video-game players are summarized in the following table:

	Saw Ad	Did Not See Ad	Total
Purchased Game	1,200	2,000	3,200
Did Not Purchase Game	3,800	43,000	46,800
Total	5,000	45,000	50,000

The market research team concludes in its report that the ad campaign is highly effective.

Question But wait! How could the campaign possibly have been effective? Only 1200 people who saw the ad purchased the game, while 2000 people purchased the game without seeing the ad! It looks as though potential customers are being *put off* by the ad.

Answer Let's analyze the figures a little more carefully. First, we can look at the event A that a randomly chosen game player purchased Ultimate Hockey. In the "Purchased Game" row, we see that a total of 3200 people purchased the game. So, the (estimated) probability of A is

$$P(A) = \frac{fr(A)}{N} = \frac{3200}{50{,}000} = .064$$

To test the effectiveness of the television ad, let's compare this figure with the estimated probability that *a game player who saw the ad* purchased Ultimate Hockey. This means that we restrict attention to the "Saw Ad" column. The probability we want now is the fraction

$$\frac{\text{number of people who saw the ad and bought the game}}{\text{total number of people who saw the ad}} = \frac{1200}{5000} = .24$$

In other words, 24% of those surveyed who saw the ad bought Ultimate Hockey, whereas overall, only 6.4% of those surveyed bought it. Thus, it appears that the ad campaign *was* highly successful.

Here's some terminology. In this example there were two related events of importance:

A: A video-game player purchased Ultimate Hockey

B: A video-game player saw the ad

The two probabilities we compared were the estimated probability $P(A)$ and the estimated probability that a video-game player purchased Ultimate Hockey *given that* he or she saw the ad. We call the latter probability the (estimated) **probability of *A*, given *B*,** and we write it as $P(A \mid B)$. We call $P(A \mid B)$ a **conditional probability**—it is the probability of A under the condition that B occurred. Put another way, it is the probability of A occurring if the sample space is reduced to just those outcomes in B.

Question How do we calculate conditional probabilities?

Answer In the example above we used the ratio

$$P(A \mid B) = \frac{\text{number of people who saw the ad and bought the game}}{\text{total number of people who saw the ad}}$$

The numerator is the frequency of $A \cap B$, and the denominator is the frequency of B:

$$P(A \mid B) = \frac{fr(A \cap B)}{fr(B)}$$

Now, we can write this formula in another way:

$$P(A \mid B) = \frac{fr(A \cap B)}{fr(B)} = \frac{fr(A \cap B)/N}{fr(B)/N} = \frac{P(A \cap B)}{P(B)}$$

We therefore have the following definition, which applies to general probability distributions.

Conditional Probability

If A and B are events with $P(B) \neq 0$, then the probability of A given B is

$$P(A \mid B) = \frac{P(A \cap B)}{P(B)}$$

Quick Examples

1. If there is a 50% chance of rain (R) and a 10% chance of both rain and lightning (L), then the probability of lightning, given that it rains, is

$$P(L \mid R) = \frac{P(L \cap R)}{P(R)} = \frac{.10}{.50} = .20$$

   Here are two more ways to express the result:

   - If it rains, the probability of lightning is .20.
   - Assuming that it rains, there is a 20% chance of lightning.

2. Referring to the Cyber Video data above, the probability that a video-game player did not purchase the game (NG), given that she did not see the ad (NA), is

$$P(NG \mid NA) = \frac{P(NG \cap NA)}{P(NA)} = \frac{\frac{43{,}000}{50{,}000}}{\frac{45{,}000}{50{,}000}} = \frac{43}{45} \approx .96$$

Example 1 • Dice

If you roll a fair die twice and observe the numbers that face up, find the probability that the sum of the numbers is 8, given that the first number is 3.

Solution We begin by recalling that the sample space when we roll a fair die twice is the set $S = \{(1, 1), (1, 2), \ldots, (6, 6)\}$ containing the 36 different equally likely outcomes.

The two events under consideration are

A: The sum of the numbers is 8

B: The first number is 3

We also need

$A \cap B$: The sum of the numbers is 8 and the first number is 3

But this can only happen in one way: $A \cap B = \{(3, 5)\}$. From the formula, then,

$$P(A \mid B) = \frac{P(A \cap B)}{P(B)} = \frac{\frac{1}{36}}{\frac{6}{36}} = \frac{1}{6}$$

Before we go on . . . There is another way to think about this example. When we say that the first number is 3, we are restricting the sample space to the six outcomes (3, 1), (3, 2), . . . , (3, 6), all still equally likely. Of these six, only one has a sum of 8, so the probability of the sum being 8, given that the first number is 3, is 1/6.

Notes

1. Remember that, in the expression $P(A \mid B)$, A is the event whose probability you want, given that you know the event B has occurred.

2. From the formula, notice that $P(A \mid B)$ is not defined if $P(B) = 0$. Could $P(A \mid B)$ make any sense if the event B were impossible?

Example 2 • School and Work

A survey of the high school graduating class of 2000, conducted by the Bureau of Labor Statistics, found that, if a graduate went on to college, there was a 43% chance that he or she would work at the same time.[19] On the other hand, there was a 63% chance that a randomly selected graduate would go on to college. What is the probability that a graduate went to college and work at the same time?

Solution To understand what the question asks and what information is given, it is helpful to rephrase everything using the standard wording "*the probability that* _____" and "*the probability that* _____ *given that* _____." Now we have, "The probability that a graduate worked, given that the graduate went on to college, is .43. The probability that a graduate went on to college is .63." The events in question are as follows:

W: A high school graduate went on to work

C: A high school graduate went on to college

From our rephrasing of the question, we can write:

$P(W \mid C) = .43$ and $P(C) = .63$. Find $P(W \cap C)$.

The definition

$$P(W \mid C) = \frac{P(W \cap C)}{P(C)}$$

---

[19] SOURCE: "College Enrollment and Work Activity of High School Graduates," U.S. Bureau of Labor Statistics, http://www.bls.gov/news.release/hsgec.toc.htm, April 2001.

can be used to find $P(W \cap C)$:

$$P(W \cap C) = P(W \mid C)P(C)$$
$$= (.43)(.63) \approx .27$$

So, there is a 27% chance that a member of the high school graduating class of 2000 went on to college and work at the same time.

The Multiplication Principle and Trees

In Example 2 we saw that the formula

$$P(A \mid B) = \frac{P(A \cap B)}{P(B)}$$

can be used to calculate $P(A \cap B)$ if we rewrite the formula in the following form, known as the **multiplication principle.**

Multiplication Principle

If A and B are events, then

$$P(A \cap B) = P(A \mid B)P(B)$$

Quick Example

If there is a 50% chance of rain (R) and a 20% chance of a lightning (L) if it rains, then the probability of both rain and lightning is

$$P(R \cap L) = P(L \mid R)P(R) = (.2)(.5) = .1$$

The multiplication principle is often used in conjunction with **tree diagrams.** Let's return to Cyber Video Games and its television ad campaign. Its marketing survey was concerned with the following events:

A: A video-game player purchased Ultimate Hockey

B: A video-game player saw the ad

We can illustrate the various possibilities by means of the two-stage "tree" shown in Figure 4.

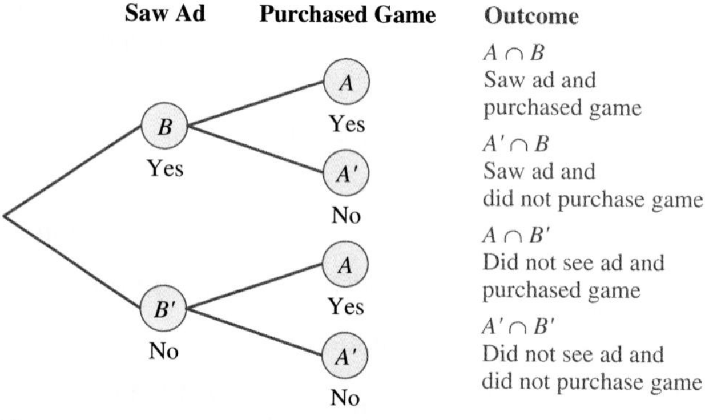

Figure 4

Consider the outcome $A \cap B$. To get there from the starting position on the left, we must first travel up to the B node. (In other words, B must occur.) Then we must travel up the branch from the B node to the A node. We are now going to associate a probability with each branch of the tree: the probability of traveling along that branch *given that we have gotten to its beginning node.* For instance, the probability of traveling up the branch from the starting position to the B node is $P(B) = 5000/50{,}000 = .1$ (see the data in the survey). The probability of going up the branch from the B node to the A node is the probability that A occurs, given that B has occurred. In other words, it is the *conditional* probability $P(A \mid B) = .24$. (We calculated this probability at the beginning of the section.) The probability of the outcome $A \cap B$ can then be computed using the multiplication principle:

$$P(A \cap B) = P(B)P(A \mid B) = (.1)(.24) = .024$$

In other words, *to obtain the probability of the outcome* $A \cap B$, *we multiply the probabilities on the branches leading to that outcome* (Figure 5).

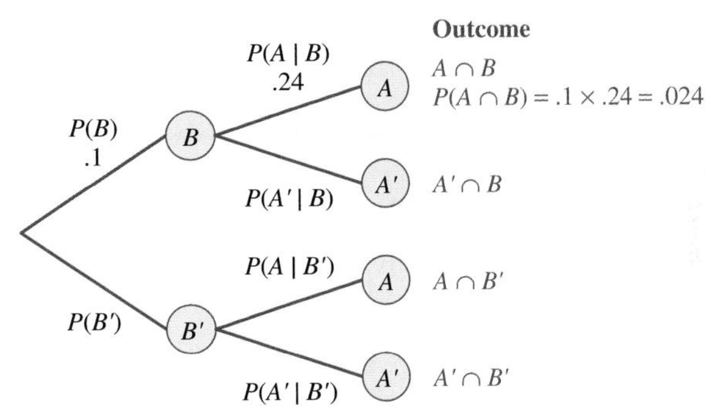

Figure 5

The same argument holds for the remaining three outcomes, and we can use the table given at the beginning of this section to calculate all the conditional probabilities shown in Figure 6.

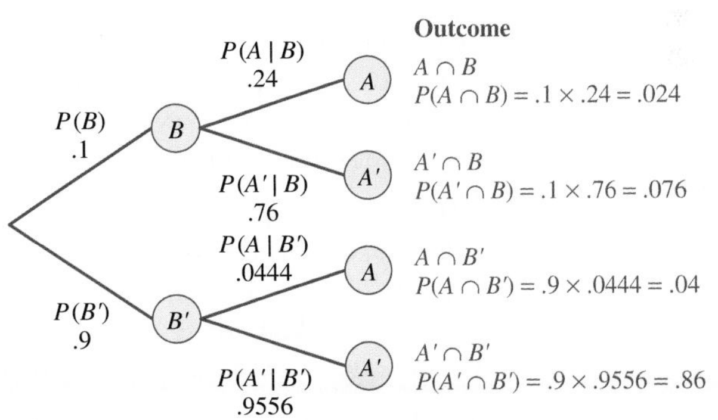

Figure 6

Note The sum of the probabilities on the branches leaving any node is always 1. (Why?) This observation often speeds things up because after we have labeled one branch we can easily label the other.

Example 3 • Unfair Coins

An experiment consists of tossing two coins. The first coin is fair, but the second coin is twice as likely to land with heads facing up as it is with tails facing up. Draw a tree diagram to illustrate all possible outcomes and use the multiplication principle to compute the probabilities of all the outcomes.

Solution A quick calculation shows that the probability distribution for the second coin is $P(\text{H}) = 2/3$ and $P(\text{T}) = 1/3$. (How did we get that?) Figure 7 shows the tree diagram and the calculations of the probabilities of the outcomes.

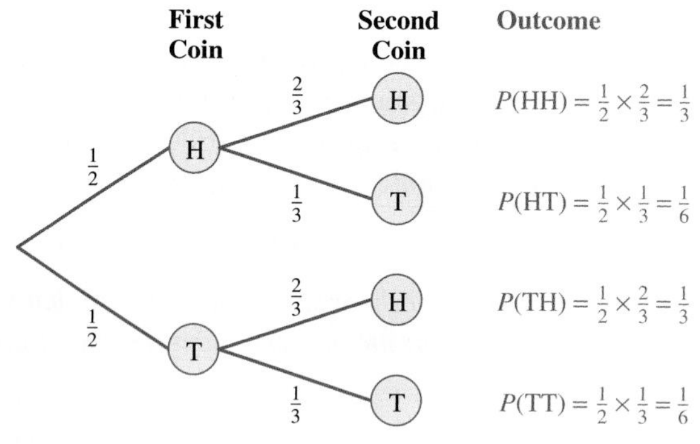

Figure 7

Independence

Let's go back once again to Cyber Video Games and its ad campaign. How did we assess the ad's effectiveness? We considered the following events:

A: A video-game player purchased Ultimate Hockey

B: A video-game player saw the ad

We used the survey data to calculate $P(A)$, the probability that a video-game player purchased Ultimate Hockey, and $P(A \mid B)$, the probability that a video-game player *who saw the ad* purchased Ultimate Hockey. When these probabilities are compared, one of three things can happen.

Case 1 $P(A \mid B) > P(A)$

This is what the survey data actually showed: A video-game player was more likely to purchase Ultimate Hockey if he or she saw the ad. This indicates that the ad is effective; seeing the ad had a positive effect on a player's decision to purchase the game.

Case 2 $P(A \mid B) < P(A)$

If this had happened, then a video-game player would have been *less* likely to purchase Ultimate Hockey if he or she saw the ad. This would have indicated that the ad had "backfired"; it had, for some reason, put off potential customers. In this case, just as in the first case, the event B would have had an effect—a negative one—on the event A.

Case 3 $P(A \mid B) = P(A)$

In this case seeing the ad would have had absolutely no effect on a potential customer's buying Ultimate Hockey. Put another way, the probability of A occurring *does not depend* on whether B occurred or not. We say in a case like this that the events A and B are **independent.**

In general, we say that two events A and B are independent if $P(A \mid B) = P(A)$. When this happens, we have

$$P(A) = P(A \mid B) = \frac{P(A \cap B)}{P(B)}$$

so

$$P(A \cap B) = P(A)P(B)$$

Conversely, if $P(A \cap B) = P(A)P(B)$, then, assuming $P(B) \neq 0$,[20] $P(A) = P(A \cap B)/P(B) = P(A \mid B)$. Thus, saying that $P(A) = P(A \mid B)$ is the same as saying that $P(A \cap B) = P(A)P(B)$. Also, we can switch A and B in this last formula and conclude that saying that $P(A \cap B) = P(A)P(B)$ is the same as saying that $P(B \mid A) = P(B)$.

Independent Events

The events A and B are **independent** if

$$P(A \cap B) = P(A)P(B)$$

Equivalent formulas are

$$P(A \mid B) = P(A) \quad \text{and} \quad P(B \mid A) = P(B)$$

If two events A and B are not independent, then they are **dependent.**

The property $P(A \cap B) = P(A)P(B)$ can be extended to three or more independent events. If, for example, A, B, and C are three mutually independent events (that is, each of them is independent of each of the other two), then, among other things,

$$P(A \cap B \cap C) = P(A)P(B)P(C)$$

Testing for Independence

To check whether two events A and B are independent, we compute $P(A)$, $P(B)$, and $P(A \cap B)$. If $P(A \cap B) = P(A)P(B)$, the events are independent; otherwise, they are dependent. Sometimes it is obvious that two events by their nature are independent, so a test is not necessary. For example, the event that a die you roll comes up 1 is clearly independent of whether or not a coin you toss comes up heads.

Quick Examples

1. Roll two distinguishable dice (one red, one green) and observe the numbers that face up.

A: The red die is even; $P(A) = \frac{18}{36} = \frac{1}{2}$

B: The dice have the same parity;* $P(B) = \frac{18}{36} = \frac{1}{2}$

$A \cap B$: Both dice are even; $P(A \cap B) = \frac{9}{36} = \frac{1}{4}$

$P(A \cap B) = P(A)P(B)$, and so A and B are independent.

2. Roll two distinguishable dice and observe the numbers that face up.

A: The sum of the numbers is 6; $P(A) = \frac{5}{36}$

B: Both numbers are odd; $P(B) = \frac{9}{36} = \frac{1}{4}$

$A \cap B$: The sum is 6, and both are odd; $P(A \cap B) = \frac{3}{36} = \frac{1}{12}$

$P(A \cap B) \neq P(A)P(B)$, and so A and B are dependent.

*Two numbers have the **same parity** if both are even or both are odd. Otherwise, they have **opposite parity.**

[20]We discuss the independence of two events only in cases where their probabilities are both nonzero.

Example 4 • Weather Prediction

According to the weather service, there is a 50% chance of rain in New York and a 30% chance of rain in Honolulu. Assuming that New York's weather is independent of Honolulu's, find the probability that it will rain in at least one of these cities.

Solution We take A to be the event that it will rain in New York and B to be the event that it will rain in Honolulu. We are asked to find the probability of $A \cup B$, the event that it will rain in at least one of the two cities. We use the addition principle:

$$P(A \cup B) = P(A) + P(B) - P(A \cap B)$$

We know that $P(A) = .5$ and $P(B) = .3$. But what about $P(A \cap B)$? Since the events A and B are independent, we can compute

$$\begin{aligned} P(A \cap B) &= P(A)P(B) \\ &= (.5)(.3) = .15 \end{aligned}$$

Thus,

$$\begin{aligned} P(A \cup B) &= P(A) + P(B) - P(A \cap B) \\ &= .5 + .3 - .15 \\ &= .65 \end{aligned}$$

So, there is a 65% chance that it will rain either in New York or in Honolulu (or in both).

Example 5 • Roulette

You are playing roulette and have decided to leave all ten of your \$1 chips on black for five consecutive rounds, hoping for a sequence of five blacks, which according to the rules will leave you with \$320. There is a 50% chance of black coming up on each spin, ignoring the complicating factor of zero or double zero. What is the probability that you will be successful?

Solution Since the roulette wheel has no memory, each spin is independent of the others. Thus, if A_1 is the event that black comes up the first time, A_2 the event that it comes up the second time, and so on, then

$$P(A_1 \cap A_2 \cap A_3 \cap A_4 \cap A_5) = P(A_1)P(A_2)P(A_3)P(A_4)P(A_5) = \left(\frac{1}{2}\right)^5 = \frac{1}{32}$$

Example 6 is a version of a well-known "brain teaser" that forces one to think carefully about conditional probability.

Example 6 • Legal Argument

A man was arrested for attempting to smuggle a bomb on board an airplane. During the subsequent trial, his lawyer claimed that, by means of a simple argument, she would prove beyond a shadow of a doubt that her client was not only innocent of any crime but also was in fact contributing to the safety of the other passengers on the flight. This was her eloquent argument: "Your Honor, first of all, my client had absolutely no intention of setting off the bomb. As the record clearly shows, the detonator was unarmed when he was apprehended. In addition—and your Honor is certainly aware of this—there is a small but definite possibility that there will be a bomb on any given flight. On the other hand, the chances of there being *two* bombs on a flight are so remote as to be negligible. There is in fact no record of this having *ever* occurred. Thus, since my client had already brought one bomb on board (with no intention of setting it off) and since we have seen that the chances of there being a second bomb on board were vanishingly remote, it follows that the flight was far safer as a result of his action! I rest my case." This argument was so elegant in its simplicity that the judge acquitted the defendant. Where is the flaw in the argument? (Think about this for a while before reading the solution.)

Solution The lawyer has cleverly confused the phrases "two bombs on board" and "a second bomb on board." To pinpoint the flaw, let B be the event that there is one bomb on board a given flight and let A be the event that there are two independent bombs on board. Let's assume for argument's sake that $P(B) = 1/1{,}000{,}000 = .000\,001$. Then the probability of the event A is

$$(.000\,001)(.000\,001) = .000\,000\,000\,001$$

This *is* vanishingly small, as the lawyer contended. It was at this point that the lawyer used a clever maneuver: She assumed in concluding her argument that the probability of having two bombs on board was the same as the probability of having a *second* bomb on board. But to say that there is a *second* bomb on board is to imply that there already is one bomb on board. This is therefore a *conditional* event: the event that there are two bombs on board, *given that there is already one bomb on board.* Thus, the probability that there is a second bomb on board is the probability that there are two bombs on board, given that there is already one bomb on board, which is

$$P(A \mid B) = \frac{P(A \cap B)}{P(B)} = \frac{.000\,000\,000\,001}{.000\,001} = .000\,001$$

In other words, it is the same as the probability of there being a single bomb on board to begin with! Thus the man's carrying the bomb onto the plane did not improve the flight's safety at all.[21]

---

[21]If we want to be picky, there was a *slight* decrease in the probability of a second bomb because there was one less seat for a potential second bomb bearer to occupy. In terms of our analysis, this is saying that the event of one passanger with a bomb and the event of a second passanger with a bomb are not completely independent.

7.6 EXERCISES

In Exercises 1–6, find the conditional probabilities of the indicated events when two fair dice (one red and one green) are rolled.

1. The sum is 5, given that the green one is not a 1.

2. The sum is 6, given that the green one is either 4 or 3.

3. The red one is 5, given that the sum is 6.

4. The red one is 4, given that the green one is 4.

5. The sum is 5, given that the dice have opposite parity.

6. The sum is 6, given that the dice have opposite parity.

Exercises 7–12 require the use of counting techniques from Chapter 6. A bag contains three red, two green, one fluorescent pink, two yellow, and two orange marbles. Suzy grabs four at random. Find the probabilities of the indicated events.

7. She gets all the red ones, given that she gets the fluorescent pink one.

8. She gets all the red ones, given that she does not get the fluorescent pink one.

9. She gets none of the red ones, given that she gets the fluorescent pink one.

10. She gets one of each color other than fluorescent pink, given that she gets the fluorescent pink one.

11. She gets one of each color other than fluorescent pink, given that she gets at least one red one.

12. She gets at least two red ones, given that she gets at least one green one.

In Exercises 13–16, supply the missing quantities.

13.

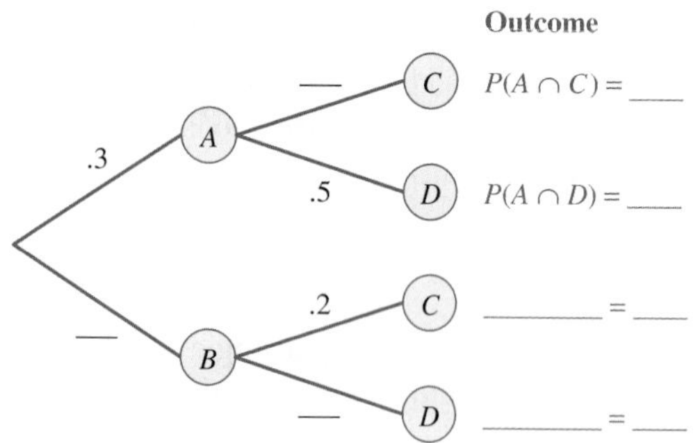

14.

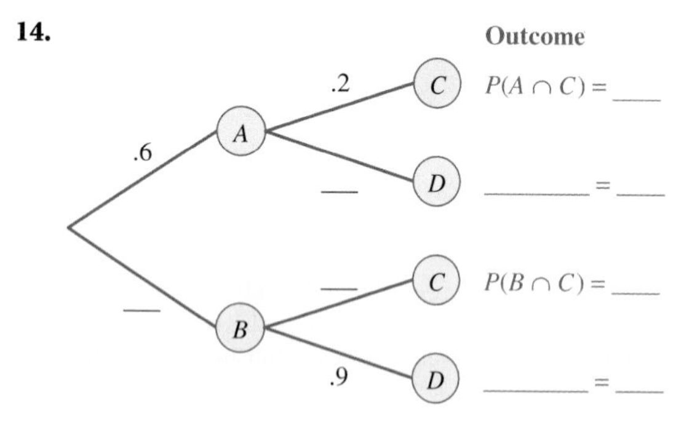

15.

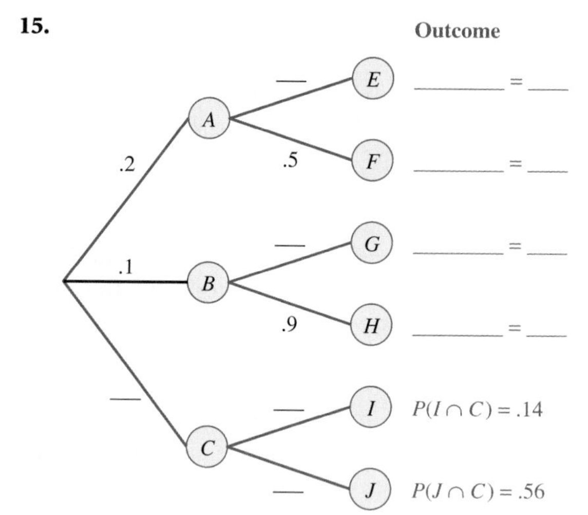

16.

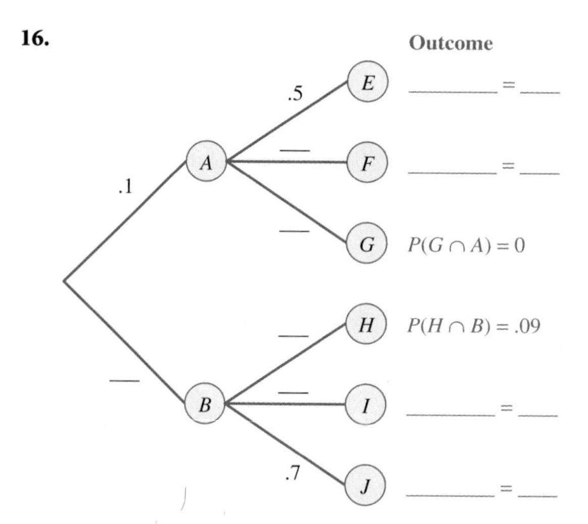

In Exercises 17–20, say whether the given pairs of events are **(A)** independent, **(B)** mutually exclusive, or **(C)** neither.

17. *A*: Your new skateboard design is a success.
B: Your new skateboard design is a failure.

18. *A*: Your new skateboard design is a success.
B: There is life in the Andromeda Galaxy.

19. *A*: Your new skateboard design is a success.
B: Your competitor's new skateboard design is a failure.

20. *A*: Your first coin flip results in heads.
B: Your second coin flip results in heads.

In Exercises 21–26, two dice (one red and one green) are rolled, and the numbers that face up are observed. Test the given pairs of events for independence.

21. *A*: The red die is 1, 2, or 3. *B*: The green die is even.

22. *A*: The red die is 1. *B*: The sum is even.

23. *A*: Exactly one die is 1. *B*: The sum is even.

24. A: Neither die is 1 or 6. B: The sum is even.

25. A: Neither die is 1. B: Exactly one die is 2.

26. A: Both dice are 1. B: Neither die is 2.

27. If a coin is tossed 11 times, find the probability of the sequence H, T, T, H, H, H, T, H, H, T, T.

28. If a die is rolled four times, find the probability of the sequence 4, 3, 2, 1.

APPLICATIONS

29. **Movies** In 1998 the probability that a randomly selected movie ticket in France was not for a French film was approximately .7, and the probability that it was for a U.S. film was .6. What is the probability that a randomly selected movie ticket was for a U.S. film, given that it was not for a French film?

SOURCE: Center National de la Cinématographie/*New York Times,* December 14, 1999, p. E1.

30. **Movies** In 1997 the probability that a randomly selected movie ticket in France was not for a U.S. film was approximately .5. Furthermore, the probability that the ticket was for a French film, given that it was not for a U.S. film, was .6. What is the probability that a randomly selected movie ticket was for a French film?

SOURCE: Center National de la Cinématographie/*New York Times,* December 14, 1999, p. E1.

31. **Road Safety** In 1999 the probability that a randomly selected vehicle would be involved in a deadly tire-related accident was approximately 3×10^{-6}, whereas the probability that a tire-related accident would prove deadly was .02. What was the probability that a vehicle would be involved in a tire-related accident?

SOURCE: *New York Times* analysis of National Traffic Safety Administration crash data/Polk Company vehicle registration data/*New York Times,* November 22, 2000, p. C5. The original data reported three tire-related deaths per million vehicles.

32. **Road Safety** In 1998 the probability that a randomly selected vehicle would be involved in a deadly tire-related accident was approximately 2.8×10^{-6}, while the probability that a tire-related accident would prove deadly was .016. What was the probability that a vehicle would be involved in a tire-related accident?

SOURCE: *New York Times* analysis of National Traffic Safety Administration crash data/Polk Company vehicle registration data/*New York Times,* November 22, 2000, p. C5. The original data reported 2.8 tire-related deaths per million vehicles.

33. **Food Safety** According to a University of Maryland study of 200 samples of ground meats, the probability that a sample was contaminated by salmonella was .20. The probability that a salmonella-contaminated sample was contaminated by a strain resistant to at least three antibiotics was .53. What was the probability that a ground meat sample was contaminated by a strain of salmonella resistant to at least three antibiotics?

SOURCE: *New York Times,* October 16, 2001, p. A12.

34. **Food Safety** According the study mentioned in Exercise 33, the probability that a ground meat sample was contaminated by salmonella was .20. The probability that a salmonella-contaminated sample was contaminated by a strain resistant to at least one antibiotic was .84. What was the probability that a ground meat sample was contaminated by a strain of salmonella resistant to at least one antibiotic?

SOURCE: *New York Times,* October 16, 2001, p. A12.

Publishing Exercises 35–42 are based on the following table, which shows the results of a survey of 100 authors by a publishing company.

	New Authors	Established Authors	Total
Successful	5	25	30
Unsuccessful	15	55	70
Total	20	80	100

Compute the following conditional probabilities:

35. An author is established, given that she is successful.

36. An author is successful, given that he is established.

37. An author is unsuccessful, given that he is a new author.

38. An author is a new author, given that she is unsuccessful.

39. An author is unsuccessful, given that she is established.

40. An author is established, given that he is unsuccessful.

41. An unsuccessful author is established.

42. An established author is successful.

43. **Internet Use** The following pie chart shows the percentage of the population that uses the Internet, broken down further by family income, based on a survey taken in August, 2000:

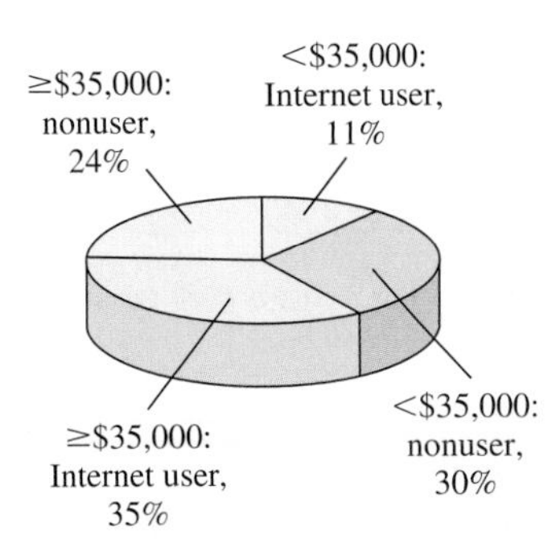

SOURCE: "Falling through the Net: Toward Digital Inclusion, A Report on Americans' Access to Technology Tools," U.S. Department of Commerce, http://www.ntia.doc.gov/ntiahome/fttn00/contents00.html, October 2000.

a. Determine the probability that a randomly chosen person was an Internet user, given that his or her family income was at least $35,000.

b. Based on the data, was a person more likely to be an Internet user if his or her family income was less than $35,000 or $35,000 or more? (Support your answer by citing the relevant conditional probabilities.)

44. Internet Use Repeat Exercise 43, using the following pie chart, which shows the results of a similar survey taken in September 2001.

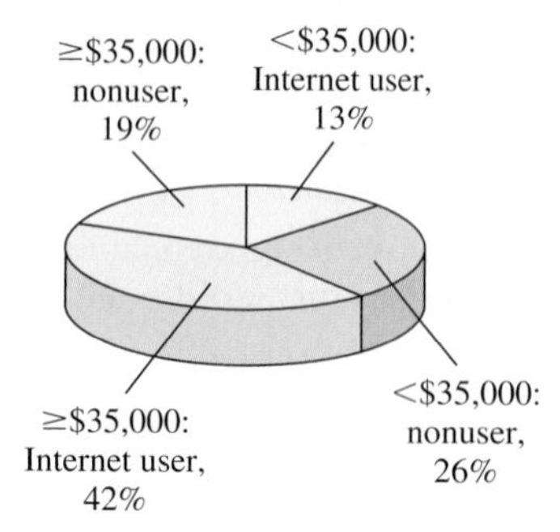

SOURCE: "A Nation Online: How Americans Are Expanding Their Use of the Internet," U.S. Department of Commerce, http://www.ntia.doc.gov/ntiahome/dn/index.html, February 2002.

Education and Employment Exercises 45–54 are based on the following table, which shows U.S. employment figures at the end of 2001, broken down by educational attainment. All numbers are in millions and represent civilians aged 16 years and over. Those classed as "Not in Labor Force" were not employed nor actively seeking employment. Round all answers to two decimal places.

	Employed	Unemployed	Not in Labor Force	Total
Less Than 4 Years High School	11.1	1.1	15.6	27.8
4 Years of High School Only	35.2	1.8	20.5	57.5
1–3 Years of College	32.2	1.3	11.9	45.4
4 or More Years of College	36.0	1.0	9.9	46.9
Total	114.5	5.2	57.9	177.6

SOURCE: U.S. Bureau of Labor Statistics press release, http://www.bls.gov/, January 4, 2002.

45. Find the probability that a person was employed, given that the person had completed at least 4 years of college.

46. Find the probability that a person was employed, given that the person had completed less than 4 years of high school.

47. Find the probability that a person had completed 4 or more years of college, given that the person was employed.

48. Find the probability that a person had completed less than 4 years of high school, given that the person was employed.

49. Find the probability that a person who had not completed 4 or more years of college was not in the labor force.

50. Find the probability that a person who had completed at least 4 years of high school was not in the labor force.

51. Find the probability that a person who had completed at least 4 years of college and was in the labor force was employed.

52. Find the probability that a person who had completed less than 4 years of high school and was in the labor force was employed.

53. Your friend claims that an unemployed person is more likely to have 1 to 3 years of college than an employed person. Respond to this claim by citing actual probabilities.

54. Your friend claims that a person not in the labor force is more likely to have only 4 years of high school than an employed person. Respond to this claim by citing actual probabilities.

55. Air-Bag Safety According to a study conducted by the Harvard School of Public Health, a child seated in the front seat who was wearing a seatbelt was 31% more likely to be killed in an accident if the car had an air bag that deployed than if it did not. Let the sample space S be the set of all accidents involving a child seated in the front seat wearing a seatbelt. Let K be the event that the child was killed and let D be the event that the air bag deployed. Fill in the missing terms and quantities: $P(\_\_\_\_ \mid \_\_\_\_) = \_\_\_\_ \times P(\_\_\_\_ \mid \_\_\_\_)$.

The study was conducted by Dr. Segul-Gomez at the Harvard School of Public Health. SOURCE: *New York Times,* December 1, 2000, p. F1.

56. Air-Bag Safety According to the study cited in Exercise 55, a child seated in the front seat not wearing a seatbelt was 84% more likely to be killed in an accident if the car had an air bag that deployed than if it did not. Let the sample space S be the set of all accidents involving a child seated in the front seat not wearing a seatbelt. Fill in the missing terms and quantities: $P(\_\_\_\_ \mid \_\_\_\_) = \_\_\_\_ \times P(\_\_\_\_ \mid \_\_\_\_)$.

The study was conducted by Dr. Segul-Gomez at the Harvard School of Public Health. SOURCE: *New York Times,* December 1, 2000, p. F1.

Auto Theft Exercises 57–62 are based on the following table, which shows the probability that an owner of the given model would report his or her vehicle stolen in 1997:

Brand	Jeep Wrangler	Suzuki Sidekick (two door)	Toyota Land Cruiser	Geo Tracker (two door)	Acura Integra (two door)
Probability	.0170	.0154	.0143	.0142	.0123
Brand	Mitsubishi Montero	Acura Integra (four door)	BMW 3-series (two door)	Lexus GS300	Honda Accord (two door)
Probability	.0108	.0103	.0077	.0074	.0070

Data are for insured vehicles, for 1995 to 1997 models except Wrangler, which is for 1997 models only. SOURCE: Highway Loss Data Institute/*New York Times,* March 28, 1999, p. WK3.

In an experiment in which a vehicle is selected, consider the following events:

R: The vehicle was reported stolen in 1997

J: The vehicle was a Jeep Wrangler

$A2$: The vehicle was an Acura Integra (two door)

$A4$: The vehicle was an Acura Integra (four door)

A: The vehicle was an Acura Integra (either two door or four door)

57. Fill in the blanks: $P(\_\_\_\_ \mid \_\_\_\_) = .0170$.

58. Fill in the blanks: $P(\_\_\_\_ \mid A4) = \_\_\_\_$.

59. Which of the following is true?
(A) There is a 1.43% chance that a vehicle reported stolen in 1997 was a Toyota Land Cruiser.
(B) Of all the vehicles reported stolen in 1997, 1.43% of them were Toyota Land Cruisers.
(C) Given that a vehicle was reported stolen in 1997, there is a .0143 probability that it was a Toyota Land Cruiser.
(D) Given that a vehicle was a Toyota Land Cruiser, there was a 1.43% chance that it was reported stolen in 1997.

60. Which of the following is true?
(A) $P(R \mid A) = .0123 + .0103 = .0226$
(B) $P(R' \mid A2) = 1 - .0123 = .9877$
(C) $P(A2 \mid A) = .0123/(.0123 + .0103) \approx .544$
(D) $P(R \mid A2') = 1 - .0123 = .9877$

61. It is now January 1997, and I own a BMW 3-series and a Lexus GS300. Since I house my vehicles in different places, the event that one of my vehicles gets stolen does not depend on the event that the other gets stolen. Compute each probability to six decimal places.
a. Both my vehicles will get stolen this year.
b. At least one of my vehicles will get stolen this year.

62. It is now December 1997, and I own a Mitsubishi Montero and a Jeep Wrangler. Since I house my vehicles in different places, the event that one of my vehicles gets stolen does not depend on the event that the other gets stolen. I just have just returned from a 1-year trip to the Swiss Alps.
a. What is the probability that my Montero, but not my Wrangler, has been stolen?
b. Which is more likely: the event that my Montero was stolen or the event that *only* my Montero was stolen?

In Exercises 63–66, draw an appropriate tree diagram and use the multiplication principle to calculate the probabilities of all outcomes.

63. Sales Each day there is a 40% chance that you will sell an automobile. You know that 30% of all the automobiles you sell are two-door models, and the rest are four-door models.

64. Product Reliability You purchase brand X floppy disks one-quarter of the time and brand Y floppy disks the rest of the time. Brand X have a 1% failure rate, and brand Y have a 3% failure rate.

65. Car Rentals Your auto rental company rents out 30 small cars, 24 luxury sedans, and 46 slightly damaged "budget" vehicles. The small cars break down 14% of the time, the luxury sedans break down 8% of the time, and the "budget" cars break down 40% of the time.

66. Travel It appears that there is only a 1 in 5 chance that you will be able to take your spring vacation to the Greek Islands. If you are lucky enough to go, you will visit either Corfu (20% chance) or Rhodes. On Rhodes there is a 20% chance of meeting a tall dark stranger; on Corfu there is no such chance.

67. Weather Prediction There is a 50% chance of rain today and a 50% chance of rain tomorrow. Assuming that the event that it rains today is independent of the event that it rains tomorrow, draw a tree diagram showing the probabilities of all outcomes. What is the probability that there will be no rain today or tomorrow?

68. Weather Prediction There is a 20% chance of snow today and a 20% chance of snow tomorrow. Assuming that the event that it snows today is independent of the event that it snows tomorrow, draw a tree diagram showing the probabilities of all outcomes. What is the probability that it will snow by the end of tomorrow?

69. Drug Tests If 90% of the athletes who test positive for steroids in fact use them and 10% of all athletes use steroids and test positive, what percentage of athletes test positive?

70. Fitness Tests If 80% of candidates for the soccer team pass the fitness test and only 20% of all athletes are soccer team candidates who pass the test, what percentage of the athletes are candidates for the soccer team?

71. Food Safety According to a University of Maryland study of 200 samples of ground meats, the probability that one of the samples was contaminated by salmonella was .20. The probability that a salmonella-contaminated sample was contaminated by a strain resistant to at least one antibiotic was .84, and the probability that a salmonella-contaminated sample was contaminated by a strain resistant to at least three antibiotics was .53. Find the probability that a ground meat sample that was contaminated by an antibiotic-resistant strain was contaminated by a strain resistant to at least three antibiotics.

SOURCE: *New York Times,* October 16, 2001, p. A12.

72. Food Safety According to a University of Maryland study of 200 samples of ground meats, the probability that a ground meat sample was contaminated by a strain of salmonella resistant to at least three antibiotics was .11. The probability that someone infected with any strain of salmonella will become seriously ill is .10. What is the probability that someone eating a randomly chosen ground meat sample will not become seriously ill with a strain of salmonella resistant to at least three antibiotics?

SOURCE: *New York Times,* October 16, 2001, p. A12.

73. Marketing A market survey shows that 40% of the population used brand X laundry detergent last year, 5% of the population gave up doing its laundry last year, and 4% of the population used brand X and then gave up doing laundry last year. Are the events of using brand X and giving up doing laundry independent? Is a user of brand X detergent more or less likely to give up doing laundry than a randomly chosen person?

74. Marketing A market survey shows that 60% of the population used brand Z computers last year, 5% of the population quit their jobs last year, and 3% of the population used brand Z computers and then quit their jobs. Are the events of using brand Z computers and quitting your job independent? Is a user of brand Z computers more or less likely to quit a job than a randomly chosen person?

75. Productivity A company wishes to enhance productivity by running a 1-week training course for its employees. Let T be the event that an employee participated in the course and let I be the event that an employee's productivity improved the week after the course was run.

a. Assuming that the course has a positive effect on productivity, how are $P(I \mid T)$ and $P(I)$ related?

b. If T and I are independent, what can one conclude about the training course?

76. Productivity Consider the events T and I in Exercise 75.

a. Assuming that everyone who improved took the course but that not everyone took the course, how are $P(T \mid I)$ and $P(T)$ related?

b. If half the employees who improved took the course and half the employees took the course, are T and I independent?

COMMUNICATION AND REASONING EXERCISES

77. You wish to ascertain the probability of an event E, but you happen to know that the event F has occurred. Is the probability you are seeking $P(E)$ or $P(E \mid F)$? Give the reason for your answer.

78. Your television ad campaign has apparently been very successful: 10,000 people who saw the ad purchased your product, whereas only 2000 people purchased the product without seeing the ad. Explain how additional data could show that your ad campaign was in fact a failure.

79. Name three events, each independent of the others, when a fair coin is tossed four times.

80. Name three pairs of independent events when a pair of distinguishable and fair dice are rolled and the numbers that face up are observed.

81. If $A \subseteq B$ and $P(B) \neq 0$, why does $P(A \mid B) = \dfrac{P(A)}{P(B)}$?

82. If $B \subseteq A$ and $P(B) \neq 0$, why does $P(A \mid B) = 1$?

83. Your best friend thinks that it is impossible for two mutually exclusive events with nonzero probabilities to be independent. Establish whether he is correct.

84. Another of your friends thinks that two mutually exclusive events with nonzero probabilities can never be dependent. Establish whether she is correct.

85. Show that, if A and B are independent, then so are A' and B' (assuming that none of these events has zero probability). (*Hint:* $A' \cap B'$ is the complement of $A \cup B$.)

86. Show that, if A and B are independent, then so are A and B' (assuming that none of these events has zero probability). (*Hint:* $P(B' \mid A) + P(B \mid A) = 1$)

7.7 Bayes' Theorem and Applications

Should schools test their athletes for drug use? A problem with drug testing is that there are always false positive results, so one can never be certain that an athlete who tests positive is in fact using drugs. Here is a typical scenario.

Example 1 • Drug Testing

Gamma Chemicals advertises its anabolic steroid detection test as being 95% effective, meaning that it will show a positive result for 95% of all anabolic steroid users. It also states that its test has a false positive rate of 6%. This means that the probability of a nonuser testing positive is .06. Estimating that about 10% of its athletes are using anabolic steroids, Enormous State University (ESU) begins testing its football players. The quarterback, Hugo V. Huge, tests positive and is promptly dropped from the team. Hugo claims that he is not using anabolic steroids. How confident can we be that he is not telling the truth?

Solution Two events are of interest here: the event T that a person tests positive and the event A that the person tested uses anabolic steroids. Here are the probabilities we are given:

$$P(T \mid A) = .95$$
$$P(T \mid A') = .06$$
$$P(A) = .10$$

We are asked to find $P(A \mid T)$, the probability that someone who tests positive is using anabolic steroids. We can use a tree diagram to calculate $P(A \mid T)$. The trick to setting up the tree diagram is to use as the first branching the events with *unconditional* probabilities we know. Since the only unconditional probability we are given is $P(A)$, we use A and A' as our first branching (Figure 8).

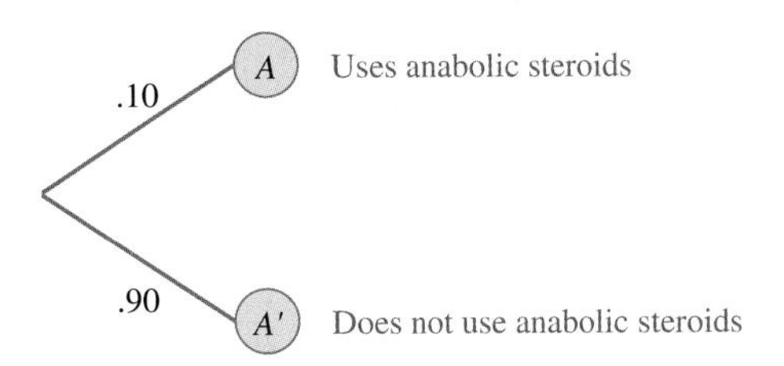

Figure 8

For the second branching, we use the outcomes of the drug test: positive (T) or negative (T'). The probabilities on these branches are conditional probabilities because they depend on whether an athlete uses steroids or not (Figure 9). (We fill in the probabilities not supplied by remembering that the sum of the probabilities on the branches leaving any node must be 1.)

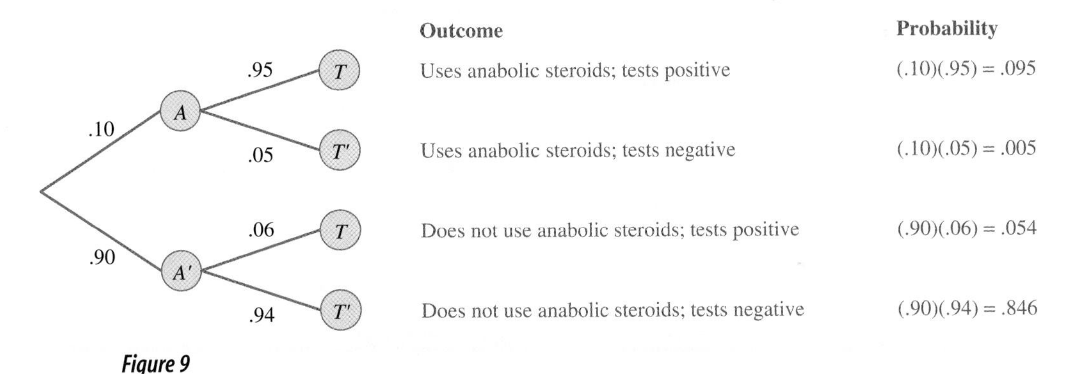

Figure 9

We can now calculate the probability we are asked to find:

$$P(A \mid T) = \frac{P(A \cap T)}{P(T)} = \frac{P(\text{uses anabolic steroids and tests positive})}{P(\text{tests positive})}$$

$$= \frac{P(\text{using } A \text{ and } T \text{ branches})}{\text{sum of } P(\text{using branches ending in } T)}$$

From the tree diagram, we see that $P(A \cap T) = .095$. To calculate $P(T)$, the probability of testing positive, notice that two outcomes on the tree diagram reflect a positive test result. The probabilities of these events are .095 and .054. Since these two events are mutually exclusive (an athlete either uses steroids or does not, but not both), the probability of a test being positive (ignoring whether or not steroids are used) is the sum of these probabilities, .149. Thus,

$$P(A \mid T) = \frac{.095}{.095 + .054} = \frac{.095}{.149} \approx .64$$

There is a 64% chance that a randomly selected athlete who tests positive, like Hugo, is using steroids. In other words, we can be 64% confident that Hugo is lying.

Before we go on . . . Note that the correct answer is 64%, *not* the 94% we might suspect from the test's false positive rating. In fact, we can't answer the question asked without knowing the percentage of athletes who actually use steroids. For instance, if *no* athletes at all use steroids, then Hugo must be telling the truth, and so the test result has no significance whatsoever. On the other hand, if *all* athletes use steroids, then Hugo is definitely lying, regardless of the outcome of the test.

False positive rates are determined by testing a large number of samples known not to contain drugs and computing estimated probabilities. False negative rates are computed similarly by testing samples known to contain drugs. However, the accuracy of the tests depends also on the skill of those administering them. Although the rate depends on the drug being tested, estimates of false positive rates for common immunoassay tests range from 10% to 30% on the high end (see, for example, the 1996 ACLU Briefing Paper, "Drug Testing in the Workplace," http://www.aclu.org/library/pbp5/html) to 3% to 5% on the low end (typical numbers quoted by test manufacturers). Because of the possibility of false positive results, positive immunoassay tests need to be confirmed by the more expensive and much more reliable gas chromatograph/mass spectrometry (GC/MS) test. See also the NCAA's drug testing policy, available at http://www.ncaa.org/sports_sciences/drugtesting (the section on Institutional Drug Testing addresses the problem of false positives).

Bayes' Theorem

The calculation we used to answer the question in Example 1 can be recast as a formula known as **Bayes' theorem.** Figure 10 shows a general form of the tree we used in Example 1.

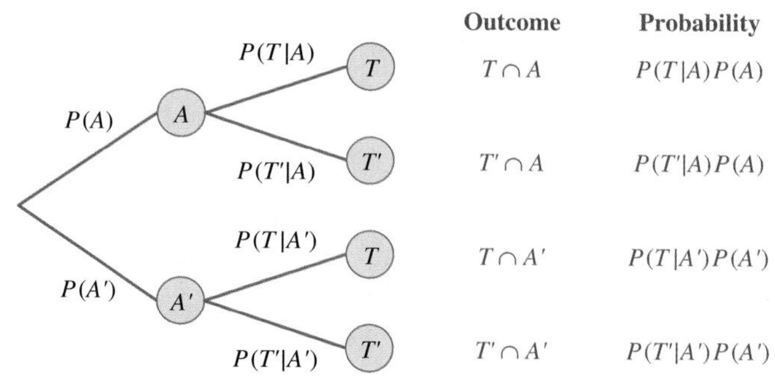

Figure 10

We calculated

$$P(A \mid T) = \frac{P(A \cap T)}{P(T)}$$

as follows. We first calculated $P(A \cap T)$, using the multiplication principle:

$$P(A \cap T) = P(T \mid A)P(A)$$

We then calculated $P(T)$ by using the addition principle for mutually exclusive events, together with the multiplication principle:

$$\begin{aligned} P(T) &= P(A \cap T) + P(A' \cap T) \\ &= P(T \mid A)P(A) + P(T \mid A')P(A') \end{aligned}$$

Substituting gives

$$P(A \mid T) = \frac{P(T \mid A)P(A)}{P(T \mid A)P(A) + P(T \mid A')P(A')}$$

This is the short form of Bayes' theorem.

Bayes' Theorem (Short Form)

If A and T are events, then

Bayes' Formula

$$P(A \mid T) = \frac{P(T \mid A)P(A)}{P(T \mid A)P(A) + P(T \mid A')P(A')}$$

Using a Tree

$$P(A \mid T) = \frac{P(\text{using } A \text{ and } T \text{ branches})}{\text{sum of } P(\text{using branches ending in } T)}$$

Quick Example

Let's calculate the probability that an athlete who tests positive is actually using steroids if only 5% of ESU athletes are using steroids. Thus,

$$P(T \mid A) = .95$$

$$P(T \mid A') = .06$$

$$P(A) = .05$$

$$P(A') = .95$$

and so

$$P(A \mid T) = \frac{P(T \mid A)P(A)}{P(T \mid A)P(A) + P(T \mid A')P(A')} = \frac{(.95)(.05)}{(.95)(.05) + (.06)(.95)} \approx .45$$

In other words, it is actually more likely that such an athlete does *not* use steroids than that he does.*

*Without knowing the results of the test, we would have said that there was a probability of $P(A) = .05$ the athlete is using steroids. The positive test result raises the probability to $P(A \mid T) = .45$, but the test gives too many false positives for us to be any more certain than that that the athlete is actually using steroids.

Remembering the Formula Although the formula looks complicated at first sight, it is not hard to remember if you notice the pattern. We want the left-hand side: $P(A \mid T)$. The numerator on the right has it the other way around, $P(T \mid A)$, multiplied by $P(A)$. This expression also appears in the denominator, added to a similar one with A replaced by A'.

Notice that we solved the original drug testing problem in two ways: using a tree and using the Bayes' theorem formula. Example 2 illustrates this dual approach.

Example 2 • Lie Detectors

Sherlock Lie Detector Company manufactures the latest in lie detectors, and the Count-Your-Pennies (CYP) store chain is eager to use them to screen their employees for theft. Sherlock's advertising claims that the test misses a lie only once in every 100 instances. On the other hand, an analysis by a consumer group reveals that 20% of people who are telling the truth fail the test anyway.[22] The local police department estimates that 1 out of every 200 employees has engaged in theft. When the CYP store first screened their employees, the test indicated Mrs. Prudence V. Good was lying when she claimed that she had never stolen from CYP. What is the probability that she had in fact stolen from the store?

Solution We are asked for the probability that Mrs. Good was lying, and in the preceding sentence we are told that the lie detector test showed her to be lying. So, we are looking for a conditional probability: the probability that she is lying, given that the lie detector test is positive. Now we can start to give names to the events:

L: A subject is lying

T: The test is positive (indicated that the subject was lying)

Using a Tree Diagram We are looking for $P(L \mid T)$. We know $P(L)$ and the conditional probabilities $P(T' \mid L)$ and $P(T \mid L')$. Figure 11 shows the tree diagram.

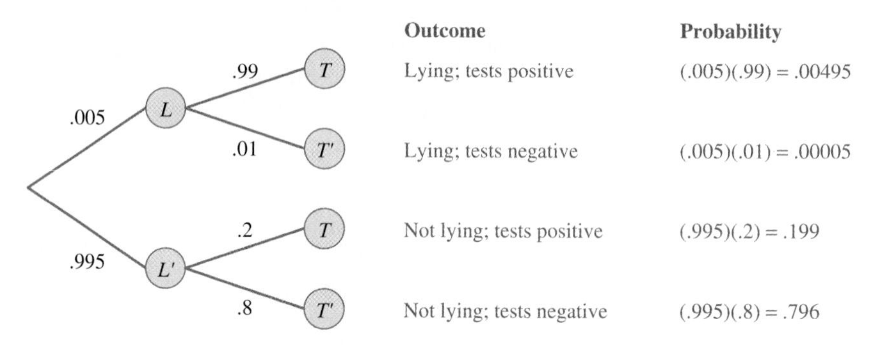

Figure 11

We see that

$$P(L \mid T) = \frac{P(\text{using } L \text{ and } T \text{ branches})}{\text{sum of } P(\text{using branches ending in } T)}$$

$$= \frac{.00495}{.00495 + .199} \approx .024$$

This means that there was only a 2.4% chance that poor Mrs. Good was lying!

[22]The reason for this is that many people show physical signs of distress when asked accusatory questions. Many people are nervous around police officers, even if they have done nothing wrong.

Using Bayes' Theorem We have

$$P(L) = .005$$

$$P(T \mid L) = .99$$

$$P(T \mid L') = .2$$

and so

$$P(L \mid T) = \frac{P(T \mid L)P(L)}{P(T \mid L)P(L) + P(T \mid L')P(L')} = \frac{(.99)(.005)}{(.99)(.005) + (.2)(.995)} \approx .024$$

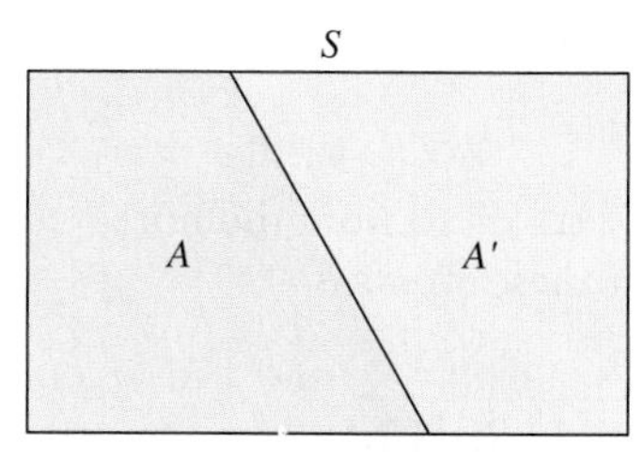

Figure 12 • A and A′ form a partition of S.

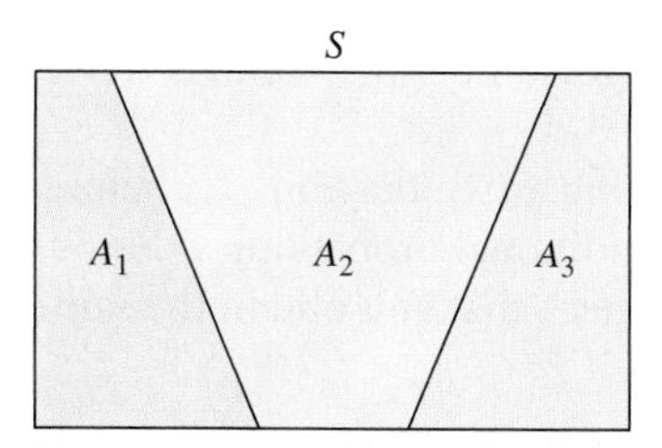

Figure 13 • A_1, A_2, and A_3 form a partition of S.

Question We have seen the "short form" of Bayes' theorem. What is the "long form?"

Answer To motivate the long form of Bayes' theorem, look again at the short formula:

$$P(A \mid T) = \frac{P(T \mid A)P(A)}{P(T \mid A)P(A) + P(T \mid A')P(A')}$$

The events A and A' form a **partition** of the sample space S; that is, their union is the whole of S and their intersection is empty (Figure 12).

For the long form of Bayes' theorem, we are given a partition of S into three or more events, as shown in the Figure 13. By saying that the events A_1, A_2, and A_3 form a partition of S, we mean that their union is the whole of S and the intersection of any two of them is empty, as in the figure. When we have a partition into three events as shown, the formula gives us $P(A_1 \mid T)$ in terms of $P(T \mid A_1)$, $P(T \mid A_2)$, $P(T \mid A_3)$ and $P(A_1)$, $P(A_2)$, $P(A_3)$.

Bayes' Theorem (Expanded Form)

If the events A_1, A_2, and A_3 form a partition of S, then

$$P(A_1 \mid T) = \frac{P(T \mid A_1)P(A_1)}{P(T \mid A_1)P(A_1) + P(T \mid A_2)P(A_2) + P(T \mid A_3)P(A_3)}$$

As for why this is true and what happens when we have a partition into *four or more* events, we will wait for the exercises. In practice, as was the case with a partition into two events, we can often compute $P(A_1 \mid T)$ by constructing a tree diagram.

Example 3 • School and Work

A survey conducted by the Bureau of Labor Statistics found that approximately 21% of the high school graduating class of 2000 went on to a 2-year college, 42% went on to a 4-year college, and the remaining 37% did not go on to college.[23] Of those who went on to a 2-year college, 61% worked at the same time, 35% of those going on to a 4-year college worked, and 70% of those who did not go on to college worked. What percentage of those working had not gone on to college?

[23] "College Enrollment and Work Activity of High School Graduates," U.S. Bureau of Labor Statistics, http://www.bls.gov/news.release/hsgec.toc.htm, April 2001.

Solution We can interpret these percentages as probabilities if we consider the experiment of choosing a member of the high school graduating class of 2000 at random. The events we are interested in are these:

R_1: A graduate went on to a 2-year college

R_2: A graduate went on to a 4-year college

R_3: A graduate did not go to college

A: A graduate went on to work

The three events R_1, R_2, and R_3 partition the sample space of all graduates into three events. We are given the following probabilities:

$$P(R_1) = .21 \qquad P(R_2) = .42 \qquad P(R_3) = .37$$
$$P(A \mid R_1) = .61 \qquad P(A \mid R_2) = .35 \qquad P(A \mid R_3) = .70$$

We are asked to find the probability that a graduate who went on to work did not go to college, so we are looking for $P(R_3 \mid A)$. Bayes' formula for these events is

$$P(R_3 \mid A) = \frac{P(A \mid R_3)P(R_3)}{P(A \mid R_1)P(R_1) + P(A \mid R_2)P(R_2) + P(A \mid R_3)P(R_3)}$$
$$= \frac{(.70)(.37)}{(.61)(.21) + (.35)(.42) + (.70)(.37)} \approx .48$$

We conclude that 48%, or nearly half, of all those working had not gone on to college.

Before we go on . . . We could also solve this problem using a tree diagram. As before, the first branching corresponds to the events with unconditional probabilities that we know: R_1, R_2, and R_3. You should complete the tree and check that you obtain the same result as above.

7.7 EXERCISES

In Exercises 1–8, use Bayes' theorem or a tree diagram to calculate the indicated probability. Round all answers to four decimal places.

1. $P(A \mid B) = .8$; $P(B) = .2$; $P(A \mid B') = .3$. Find $P(B \mid A)$.
2. $P(A \mid B) = .6$; $P(B) = .3$; $P(A \mid B') = .5$. Find $P(B \mid A)$.
3. $P(X \mid Y) = .8$; $P(Y') = .3$; $P(X \mid Y') = .5$. Find $P(Y \mid X)$.
4. $P(X \mid Y) = .6$; $P(Y') = .4$; $P(X \mid Y') = .3$. Find $P(Y \mid X)$.
5. Y_1, Y_2, and Y_3 form a partition of S. $P(X \mid Y_1) = .4$; $P(X \mid Y_2) = .5$; $P(X \mid Y_3) = .6$; $P(Y_1) = .8$; $P(Y_2) = .1$. Find $P(Y_1 \mid X)$.
6. Y_1, Y_2, and Y_3 form a partition of S. $P(X \mid Y_1) = .2$; $P(X \mid Y_2) = .3$; $P(X \mid Y_3) = .6$; $P(Y_1) = .3$; $P(Y_2) = .4$. Find $P(Y_1 \mid X)$.
7. Y_1, Y_2, and Y_3 form a partition of S. $P(X \mid Y_1) = .4$; $P(X \mid Y_2) = .5$; $P(X \mid Y_3) = .6$; $P(Y_1) = .8$; $P(Y_2) = .1$. Find $P(Y_2 \mid X)$.
8. Y_1, Y_2, and Y_3 form a partition of S. $P(X \mid Y_1) = .2$; $P(X \mid Y_2) = .3$; $P(X \mid Y_3) = .6$; $P(Y_1) = .3$; $P(Y_2) = .4$. Find $P(Y_2 \mid X)$.

APPLICATIONS

9. **Weather** It snows in Greenland an average of once every 25 days, and when it does, glaciers have a 20% chance of growing. When it does not snow in Greenland, glaciers have only a 4% chance of growing. What is the probability that it is snowing in Greenland when glaciers are growing?
10. **Weather** It rains in Spain an average of once every 10 days, and when it does, hurricanes have a 2% chance of happening in Hartford. When it does not rain in Spain, hurricanes have a 1% chance of happening in Hartford. What is the probability that it rains in Spain when hurricanes happen in Hartford?
11. **Athletic Fitness Tests** Any athlete who fails the Enormous State University's women's soccer fitness test is automatically dropped from the team. (The fitness test is traditionally given at 5 A.M. on a Sunday morning.) Last year, Mona Header failed the test but claimed that this was due to the early hour. In fact, a study by the ESU Physical Education Department suggested that 50% of athletes fit enough to play on the team would fail the soccer test, although no unfit athlete could

possibly pass the test. It also estimated that 45% of the athletes who take the test are fit enough to play soccer. Assuming these estimates are correct, what is the probability that Mona was justifiably dropped?

12. Academic Testing Professor Frank Nabarro insists that all senior physics majors take his notorious physics aptitude test. The test is so tough that anyone *not* going on to a career in physics has no hope of passing, whereas 60% of the seniors who do go on to a career in physics still fail the test. Further, 75% of all senior physics majors in fact go on to a career in physics. Assuming that you fail the test, what is the probability that you will not go on to a career in physics?

13. Road Safety In 1999 the probability that a randomly selected vehicle would be involved in a deadly accident was .00015, and 2% of deadly accidents were tire-related. Assume also that 1 out of every 10,000 randomly selected vehicles not involved in a deadly accident was involved in a tire-related accident.

a. What is the probability that a tire-related accident would prove deadly? (Round your answer to three decimal places.)

b. If we assume instead that 1 out of every 20,000 randomly selected vehicles not involved in a deadly accident was involved in a tire-related accident, how much impact does it have on the answer?

Based on data reporting tire-related deaths per million vehicles. Source: *New York Times* analysis of National Traffic Safety Administration crash data/Polk Company vehicle registration data/*New York Times,* November 22, 2000, p. C5.

14. Road Safety In 1998 the probability that a randomly selected vehicle would be involved in a deadly accident was .000175, and 1.6% of deadly accidents were tire-related. Assume also that 1 out of every 10,000 randomly selected vehicles not involved in a deadly accident was involved in a tire-related accident.

a. What is the probability that a tire-related accident would prove deadly? (Round your answer to three decimal places.)

b. If we assume instead that 1 out of every 20,000 randomly selected vehicles not involved in a deadly accident was involved in a tire-related accident, how much impact does it have on the answer?

Based on data reporting tire-related deaths per million vehicles. Source: *New York Times* analysis of National Traffic Safety Administration crash data/Polk Company vehicle registration data/*New York Times,* November 22, 2000, p. C5.

15. Market Surveys A *New York Times* survey of homeowners showed that 86% of those with swimming pools are married couples and the other 14% are single. It also showed that 15% of all homeowners had pools.

a. Assuming that 90% of all homeowners without pools are married couples, what percentage of homes owned by married couples have pools?

b. Would it best pay pool manufacturers to go after single homeowners or married homeowners? Explain.

Source: "All about Swimming Pools," *New York Times,* September 13, 1992.

16. Crime and Preschool Another *New York Times* survey of needy and disabled youths showed that 51% of those who had no preschool education were arrested or charged with a crime by the time they were 19, whereas only 31% who had preschool education wound up in this category. The survey did not specify what percentage of the youths in the survey had preschool education, so let's take a guess at that and estimate that 20% of them had attended preschool.

a. What percentage of the youths arrested or charged with a crime had no preschool education?

b. What would this figure be if 80% of the youths had attended preschool? Would youths who had preschool education be more likely to be arrested or charged with a crime than those who did not? Support your answer by quoting probabilities.

Source: "Governors Develop Plan to Help Preschool Children," *New York Times,* August 2, 1992.

17. Music Downloading According to a 2001 study on the effect of music downloading on spending on music, 11% of all Internet users had decreased their spending on music. We estimate that, in 2001, 40% of all music fans used the Internet.[24] If 20% of non-Internet users had decreased their spending on music, what percentage of those who had decreased their spending on music were Internet users?

Regardless of whether they used the Internet to download music. Source: *New York Times,* May 6, 2002, p. C6.

18. Music Downloading According to the study cited in Exercise 17, 36% of experienced file sharers with broadband access had decreased their spending on music. Let's estimate that, in 2001, 3% of all music fans were experienced file sharers with broadband access. If 20% of the other music fans had decreased their spending on music, what percentage of those who had decreased their spending on music were experienced file sharers with broadband access?

Around 15% of all online households had broadband access in 2001. Source: *New York Times,* December 24, 2001, p. C1.

19. Grade Complaints Two of the mathematics professors at Enormous State are Professor A (known for easy grading) and Professor F (known for tough grading). Last semester roughly three-quarters of Professor F's class consisted of former students of Professor A; these students apparently felt encouraged by their (utterly undeserved) high grades. (Professor F's own former students had fled in droves to Professor A's class to try to shore up their grade-point averages.) At the end of the semester, as might have been predicted, all of Professor A's former students wound up with a C− or lower. The rest of the students in the class—former students of Professor F who had decided to "stick it out"—fared better, and two-thirds of them earned higher than a C−. After discovering what had befallen them, all the students who earned C− or lower got together and decided to send a delegation to the department chair to complain that their grade-point averages had been ruined by this callous and heartless beast! The contingent was to consist of ten representatives selected at random from among them. How many of the ten would you estimate to have been former students of Professor A?

---

[24]According to the U.S. Department of Commerce, 51% of all U.S. households had computers in 2001.

20. **Weather Prediction** A local television station employs Desmorelda, "Mistress of the Zodiac," as its weather forecaster. Now, when it rains, Sagittarius is in the shadow of Jupiter one-third of the time, and it rains on 4 out of every 50 days. Sagittarius falls in Jupiter's shadow on only 1 in every 5 rainless days. The powers-that-be at the station notice a disturbing pattern to Desmorelda's weather predictions. It seems that she always predicts that it will rain when Sagittarius is in the shadow of Jupiter. What percentage of the time is she correct? Should they replace her?

21. **Employment** In a 1987 survey of married couples with earnings, 95% of all husbands were employed. Of all employed husbands, 71% of their wives were also employed. Noting that either the husband or wife in a couple with earnings had to be employed, find the probability that the husband of an employed woman was also employed.

Source: *Statistical Abstract of the United States,* 111th ed. (Washington, D.C.: U.S. Dept. of Commerce/U.S. Bureau of Labor Statistics, 1991). Figures rounded to the nearest 1%.

22. **Employment** Repeat Exercise 21 in the event that 50% of all husbands were employed.

23. **Juvenile Delinquency** According to a study at the Oregon Social Learning Center, boys who had been arrested by age 14 were 17.9 times more likely to become chronic offenders than those who had not. Use these data to estimate the percentage of chronic offenders who had been arrested by age 14 in a city where 0.1% of all boys have been arrested by age 14. (*Hint:* Use Bayes' formula rather than a tree.)

Based on a study, by Marion S. Forgatch, of 319 boys from high-crime neighborhoods in Eugene, Oregon. Source: W. Wayt Gibbs, "Seeking the Criminal Element," *Scientific American* (March 1995): 101–107.

24. **Crime** According to the same study at the Oregon Social Learning Center, chronic offenders were 14.3 times more likely to commit violent offenses than people who were not chronic offenders. In a neighborhood where 2 in every 1000 residents is a chronic offender, estimate the probability that a violent offender is also a chronic offender. (*Hint:* Use Bayes' formula rather than a tree.)

Based on a study, by Marion S. Forgatch, of 319 boys from high-crime neighborhoods in Eugene, Oregon. Source: W. Wayt Gibbs, "Seeking the Criminal Element," *Scientific American* (March 1995): 101–107.

25. **Benefits of Exercise** According to a study in the *New England Journal of Medicine,* 202 of a sample of 5990 middle-aged men had developed diabetes. It also found that men who were very active (burning about 3500 calories daily) were half as likely to develop diabetes compared with men who were sedentary. Assume that one-third of all middle-aged men are very active and the rest are classified as sedentary. What is the probability that a middle-aged man with diabetes is very active?

Source: *New York Times,* July 18, 1991.

26. **Benefits of Exercise** Repeat Exercise 25, assuming that only 1 in 10 middle-aged men is very active.

27. **University Admissions** In fall 2002 UCLA admitted 26% of its California resident applicants, 18% of its applicants from other U.S. states, and 13% of its international student applicants. Of all its applicants, 86% were California residents, 11% were from other U.S. states, and 4% were international students. What percentage of all admitted students were California residents? (Round your answer to the nearest 1%.)

Source: UCLA Web site, http://www.admissions.ucla.edu/Prospect/Adm_fr/Frosh_Prof.htm, May 2002.

28. **University Admissions** In fall 2001 UCLA admitted 29% of its California resident applicants, 19% of its applicants from other U.S. states, and 14% of its international student applicants. Of all its applicants, 85% were California residents, 11% were from other U.S. states, and 4% were international students. What percentage of all admitted students were international students? (Round your answer to the nearest 1%.)

Source: UCLA Web site, http://www.admissions.ucla.edu/Prospect/Adm_fr/Frosh_Prof.htm, May 2002.

29. **Internet Use** In 2000 86% of all Caucasians in the United States, 77% of all African Americans, 77% of all Hispanics, and 85% of residents not classified into one of these groups used the Internet for email. At that time the U.S. population was 69% Caucasian, 12% African American, and 13% Hispanic. What percentage of U.S. residents who used the Internet for email were Hispanic?

Source: NTIA and ESA, U.S. Department of Commerce, using August 2000 U.S. Bureau of the Census Current Population Survey Supplement.

30. **Internet Use** In 2000 59% of all Caucasians in the United States, 57% of all African Americans, 58% of all Hispanics, and 54% of residents not classified into one of these groups used the Internet to search for information. At that time the U.S. population was 69% Caucasian, 12% African American, and 13% Hispanic. What percentage of U.S. residents who used the Internet for information search were African American?

Source: NTIA and ESA, U.S. Department of Commerce, using August 2000 U.S. Bureau of the Census Current Population Survey Supplement.

31. **Population Migration** In 1999 the U.S. population, broken down by regions, was 53.9 million in the Northeast, 64.3 million in the Midwest, 100.0 million in the South, and 63.3 million in the West. From 1999 to 2000, 0.75% of the population in the Northeast moved to the South, 0.57% of the population in the Midwest moved to the South, 98.97% of the population in the South stayed there, and 0.77% of the population in the West moved to the South. What percentage of the population of the South in 2000 had moved there from the Northeast that year? (Round the answer to two decimal places.)

Note that this exercise ignores migration into or out of the country. The internal migration figures and 2000 population figures are accurate. Source: U.S. Census Bureau, Current Population Survey, http://www.census.gov/, March 2000.

32. **Population Migration** In 1999 the U.S. population, broken down by regions, was 53.9 million in the Northeast, 64.3 million in the Midwest, 100.0 million in the South, and 63.3 million in the West. From 1999 to 2000, 98.86% of the population in the Northeast stayed in the Northeast, 0.11% of the population in the Midwest moved to the Northeast, 0.18% of the population in the South moved to the Northeast, and

0.17% of the population in the West moved to the Northeast. What percentage of the population of the Northeast in 2000 had moved there from the South that year? (Round the answer to two decimal places.)

Note that this exercise ignores migration into or out of the country. The internal migration figures and 2000 population figures are accurate. SOURCE: U.S. Census Bureau, Current Population Survey, http://www.census.gov/, March 2000.

33. Air-Bag Safety According to a study conducted by the Harvard School of Public Health, a child seated in the front seat who was wearing a seatbelt was 31% more likely to be killed in an accident if the car had an air bag that deployed than if it did not. Air bags deployed in 25% of all accidents. For a child seated in the front seat wearing a seatbelt, what is the probability that the air bag deployed in an accident in which the child was killed? (Round your answer to two decimal places.)

The study was conducted by Dr. Segul-Gomez at the Harvard School of Public Health. SOURCE: *New York Times,* December 1, 2000, p. F1.

34. Air-Bag Safety According to the study cited in Exercise 33, a child seated in the front seat who was not wearing a seatbelt was 84% more likely to be killed in an accident if the car had an air bag that deployed than if it did not. Air bags deployed in 25% of all accidents. For a child seated in the front seat not wearing a seatbelt, what is the probability that the air bag deployed in an accident in which the child was killed? (Round your answer to two decimal places.)

The study was conducted by Dr. Segul-Gomez at the Harvard School of Public Health. SOURCE: *New York Times,* December 1, 2000, p. F1.

COMMUNICATION AND REASONING EXERCISES

35. Your friend claims that the probability of A given B is the same as the probability of B given A. How would you convince him that he is wrong?

36. Complete the following sentence. To use Bayes' formula to compute $P(E \mid F)$, you need to be given ______.

37. Give an example in which a steroid test gives a false positive only 1% of the time, and yet if an athlete tests positive, the chance that he or she has used steroids is under 10%.

38. Give an example in which a steroid test gives a false positive 30% of the time, and yet if an athlete tests positive, the chance that he or she has used steroids is over 90%.

39. Politics The following letter appeared in the *New York Times:*

To the Editor:

It stretches credulity when William Safire contends (column, Jan. 11) that 90 percent of those who agreed with his Jan. 8 column, in which he called the First Lady, Hillary Rodham Clinton, "a congenital liar," were men and 90 percent of those who disagreed were women.

Assuming these percentages hold for Democrats as well as Republicans, only 10 percent of Democratic men disagreed with him. Is Mr. Safire suggesting that 90 percent of Democratic men supported him? How naive does he take his readers to be?

A. D.
New York, Jan. 12, 1996

Comment on the letter writer's reasoning.

SOURCE: The original letter appeared in the *New York Times,* January 16, 1996, p. A16. The authors have edited the first phrase of the second paragraph slightly for clarity; the original sentence read: "Assuming the response was equally divided between Democrats and Republicans, ..."

40. Politics Refer to Exercise 39. If the letter writer's conclusion was correct, what percentage of all Democrats would have agreed with Safire's column?

41. Use a tree diagram to derive the expanded form of Bayes' theorem for a partition of the sample space S into three events R_1, R_2, and R_3.

42. Write down an expanded form of Bayes' theorem that applies to a partition of the sample space S into four events R_1, R_2, R_3, and R_4.

CASE STUDY

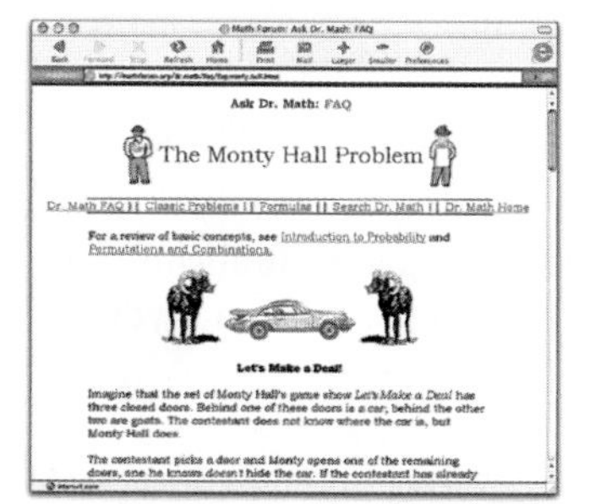

Reprinted with permission. © 1994–2003, The Math Forum @ Drexel.

The Monty Hall Problem

Here is a famous "paradox" that even mathematicians find counterintuitive. On the game show *Let's Make a Deal,* you are shown three doors, A, B, and C, and behind one of them is the Big Prize. After you select one of them—say, door A—to make things more interesting the host (Monty Hall), who knows what is behind each door, opens one of the other doors—say, door B—to reveal that the Big Prize is not there. He then offers you the opportunity to change your selection to the remaining door, door C. Should you switch or stick with your original guess? Does it make any difference?

Most people would say that the Big Prize is equally likely to be behind door A or door C, so there is no reason to switch.[25] In fact, this is wrong: The prize is more likely to be behind door C! There are several ways of seeing why this is so. Here is how you might work it out using Bayes' theorem.

Let A be the event that the Big Prize is behind door A, B the event that it is behind door B, and C the event that it is behind door C. Let F be the event that Monty has opened door B and revealed that the prize is not there. You wish to find $P(C \mid F)$ using Bayes' theorem. To use that formula, you need to find $P(F \mid A)$ and $P(A)$ and similarly for B and C. Now, $P(A) = P(B) = P(C) = 1/3$ because at the outset the prize is equally likely to be behind any of the doors. $P(F \mid A)$ is the probability that Monty will open door B if the prize is actually behind door A, and this is 1/2 because we assume that he will choose either B or C randomly in this case. On the other hand, $P(F \mid B) = 0$, since he will never open the door that hides the prize. Also, $P(F \mid C) = 1$ because if the prize is behind door C, he must open door B to keep from revealing that the prize is behind door C. Therefore,

$$P(C \mid F) = \frac{P(F \mid C)P(C)}{P(F \mid A)P(A) + P(F \mid B)P(B) + P(F \mid C)P(C)}$$

$$= \frac{1 \cdot \frac{1}{3}}{\frac{1}{2} \cdot \frac{1}{3} + 0 \cdot \frac{1}{3} + 1 \cdot \frac{1}{3}} = \frac{2}{3}$$

You conclude from this that you *should* switch to door C because it is more likely than door A to be hiding the prize.

Here is a more elementary way you might work it out. Consider the tree diagram of possibilities shown in Figure 14.

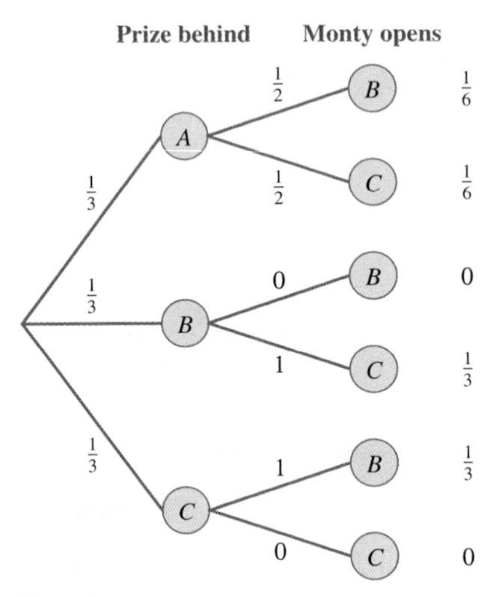

Figure 14

The top two branches of the tree give the cases in which the prize is behind door A, and there is a total probability of 1/3 for that case. The remaining two branches with nonzero probabilities give the cases in which the prize is behind the door that you did

[25]This problem caused quite a stir in late 1991 when this problem was discussed in Marilyn vos Savant's column in *Parade* magazine. Vos Savant gave the answer that you should switch. She received about 10,000 letters in response, most of them disagreeing with her, including several from mathematicians.

not choose, and there is a total probability of 2/3 for that case. Again, you conclude that you should switch your choice of doors because the one you did not choose is twice as likely as door A to be hiding the Big Prize.

EXERCISES

1. The answer you came up with, to switch to the other door, depends on the strategy Monty Hall uses in picking the door to open. Suppose he actually picks one of doors B and C at random, so that there is a chance that he will reveal the Big Prize. If he opens door B and it happens that the Prize is not there, should you switch or not?

2. What if you know that Monty's strategy is always to open door B if possible (i.e., it does not hide the Big Prize) after you choose A?
a. If he opens door B, should you switch?
b. If he opens door C, should you switch?

3. Repeat the analysis of the original game, but suppose the game uses four doors instead of three (and still only one prize).

4. Repeat the analysis of the original game, but suppose the game uses 1000 doors instead of 3 (and still only one prize).

CHAPTER 7 REVIEW TEST

1. In each case say how many elements are in the sample space *S*, list the elements of the given event *E*, and compute the probability of *E*.
a. Three coins are tossed; the result is one or more tails.
b. Four coins are tossed; the result is fewer heads than tails.
c. Two distinguishable dice are rolled; the numbers facing up add to 7.
d. A die is weighted so that each of 2, 3, 4, and 5 is half as likely to come up as either 1 or 6; however, 2 comes up.
e. Two indistinguishable dice are rolled; the numbers facing down add to 7.

2. In each case calculate the estimated probability $P(E)$, using the given information.
a. Two coins are tossed 50 times, and two heads come up 12 times. *E* is the event that at least one tail comes up.
b. Ten stocks are selected at random from a portfolio. Seven of them have increased in value since their purchase, and the rest have decreased. Eight of them are Internet stocks, and two of those have decreased in value. *E* is the event that a stock has either increased in value or is an Internet stock.
c. You have read 150 of the 400 novels in your home, but your sister Roslyn has read 200, of which only 50 are novels you have read as well. *E* is the event that a novel has been read by neither you nor your sister.
d. You roll two dice ten times. Both dice show the same number three times, and on two rolls, exactly one number is odd. *E* is the event that the sum of the numbers is even.

3. In each case compute the given theoretical probability $P(E)$.
a. There are 32 students in categories A and B combined. Some are in both, 24 are in A, and 24 are in B. *E* is the event that a randomly selected student (among the 32) is in both categories.
b. You roll two dice, one red and one green. Losing combinations are doubles (both dice showing the same number) and outcomes in which the green die shows an odd number and the red die shows an even number. The other combinations are winning ones. *E* is the event that you roll a winning combination.
c. The Apple iMac comes in three models: A, B, and C, each with five colors from which to choose, and there are equal numbers of each combination. *E* is the event that a randomly selected iMac is either orange (one of the available colors), a model A, or both.
d. The Heavy Weather Service predicts that for tomorrow there is a 50% chance of tornadoes, a 20% chance of a monsoon, and a 10% chance of both. What is the probability that we will be lucky tomorrow and encounter neither tornadoes nor a monsoon?

4. A bag contains four red, two green, one transparent, three yellow, and two orange marbles. You select five at random. Compute the probabilities of the given events.
a. You have selected all the red ones.
b. All are different colors.
c. At least one is not red.
d. At least two are yellow.
e. None are yellow and at most one is red.

5. In each case find the probability of being dealt the given type of five-card hand from a standard deck of 52 cards. (*None* of these is a recognized poker hand.) Express your answer in terms of combinations.

a. *Kings and Queens:* Each of the five cards is either a King or a Queen

b. *Five Pictures:* All picture cards (J, Q, K)

c. *Fives and Queens:* Three fives, the Queen of spades, and one other Queen

d. *Prime Full House:* A full house (three cards of one denomination, two of another) with the face value of each card a prime number (Ace $= 1$, J $= 11$, Q $= 12$, K $= 13$)

e. *Black Two Pair:* Five black cards (spades or clubs), two with one denomination, two with another, and one with a third

6. Two dice, one green and one yellow, are rolled. In each case find the conditional probability and say whether the indicated pair of events is independent.

a. The sum is 5, given that the green one is not 1 and the yellow one is 1.

b. The sum is 6, given that the green one is either 1 or 3 and the yellow one is 1.

c. The yellow one is 4, given that the green one is 4.

d. The yellow one is 5, given that the sum is 6.

e. The dice have the same parity, given that both of them are odd.

f. The sum is 7, given that the dice do not have the same parity.

OHaganBooks.com—DEALING WITH UNCERTAINTY

7. OHaganBooks.com currently operates three warehouses: in Washington, California, and Texas. Book inventories are shown in the following table:

	Science Fiction	Horror	Romance	Other	Total
Washington	10,000	12,000	12,000	30,000	64,000
California	8,000	12,000	6,000	16,000	42,000
Texas	15,000	15,000	20,000	44,000	94,000
Total	33,000	39,000	38,000	90,000	200,000

A book is selected at random. Compute the probability of each of the following events.

a. That it is either a science fiction book or stored in Texas (or both).

b. That it is a science fiction book stored in Texas.

c. That it is a science fiction book, given that it is stored in Texas.

d. That it is stored in Texas, given that it is a science fiction book.

e. That it is stored in Texas, given that it is not a science fiction book.

f. That it is not stored in Texas, given that it is a science fiction book.

8. To gauge the effectiveness of the OHaganBooks.com site, you recently commissioned a survey of online shoppers. According to the results, 2% of online shoppers visited OHaganBooks.com during a 1-week period, and 5% of them visited at least one of OHaganBooks.com's two main competitors: JungleBooks.com and FarmerBooks.com.

a. What percentage of online shoppers never visited OHaganBooks.com?

b. Assuming that visiting OHaganBooks.com was independent of visiting a competitor, what percentage of online shoppers visited either OHaganBooks.com or a competitor?

c. Under the assumption of part (b), what is the estimated probability that an online shopper will visit none of the three sites during a week?

d. Actually, the assumption in part (b) is not what was found by the survey, because an online shopper visiting a competitor was in fact more likely to visit OHaganBooks.com that a randomly selected online shopper. Let H be the event that an online shopper visits OHaganBooks.com and let C be the event that he visits a competitor. Which is greater: $P(H \cap C)$ or $P(H)P(C)$? Why?

e. What the survey found is that 25% of online shoppers who visited a competitor also visited OHaganBooks.com. Given this information, what percentage of online shoppers visited OHaganBooks.com and neither of its competitors?

9. Not all visitors to the OHaganBooks.com site actually purchase books, and not all OHaganBooks.com customers buy through the Web site (some call in orders, and others use a mail-order catalog). According to statistics gathered at the Web site, 8% of online shoppers who visit the site during the course of a single week will purchase books. However, the survey mentioned in Question 8 revealed that 2% of online shoppers visited the site during the course of a week. Another survey estimated that 0.5% of online shoppers who did not visit the site during the course of a week nonetheless purchased books at OHaganBooks.com.

a. Complete the following sentence: Online shoppers who visit the OHaganBooks.com Web site are ______ times as likely to purchase books than online shoppers who do not.

b. What is the probability that an online shopper will not visit the site during the course of a week but still purchase books?

c. What is the probability that an online shopper who purchased books during a given week also visited the site?

 ADDITIONAL ONLINE REVIEW

If you follow the path

Web site → Everything for Finite Math → Chapter 7

you will find the following additional resources to help you review:

A comprehensive chapter summary (including examples and interactive features)

Online tutorials for many sections of the chapter

Additional review exercises (including interactive exercises and many with help)

A true/false chapter quiz

8

RANDOM VARIABLES AND

CASE STUDY

Spotting Tax Fraud with Benford's Law

You are a tax-fraud specialist working for the Internal Revenue Service (IRS), and you have just been handed a portion of the tax return from Colossal Conglomerate. The IRS suspects that the portion you were handed may be fraudulent and would like your opinion. Is there any mathematical test, you wonder, that can point to a suspicious tax return based on nothing more than the numbers entered?

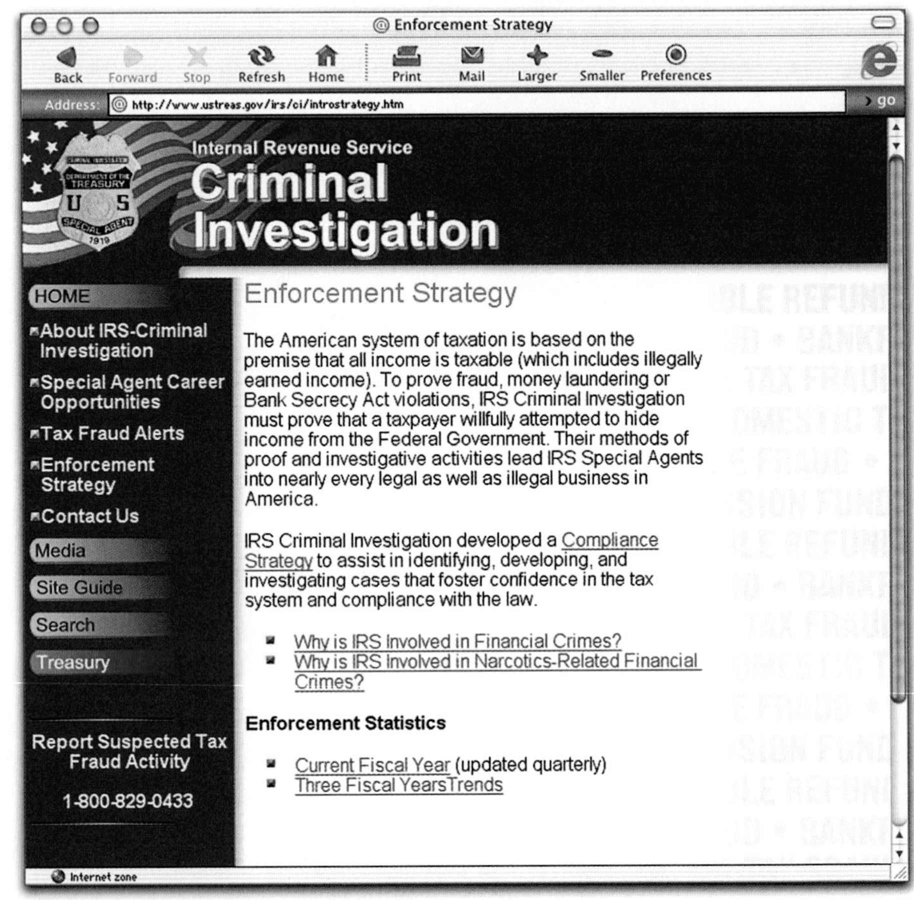

REPRODUCED FROM THE U.S. DEPARTMENT OF THE TREASURY WEB SITE

INTERNET RESOURCES FOR THIS CHAPTER

At the Web site, follow the path

Web site → Everything for Finite Math → Chapter 8

where you will find links to step-by-step tutorials for the main topics in this chapter, a detailed chapter summary you can print out, a true/false quiz, and a collection of sample test questions. You will also find downloadable Excel tutorials for each section, an online grapher, and other resources. In addition, complete text and interactive exercises have been placed on the Web site covering several optional topics:

Sampling Distributions and the Central Limit Theorem
Confidence Intervals
Hypothesis Testing

STATISTICS

Introduction

Statistics is the branch of mathematics concerned with organizing, analyzing, and interpreting numerical data. For example, given the current annual incomes of 1000 lawyers selected at random, you might wish to answer some questions: If I become a lawyer, what income am I likely to earn? Do lawyers' salaries vary widely? How widely?

To answer questions like these, it helps to begin by organizing the data in the form of tables or graphs. This is the topic of the first section of the chapter. The second section describes an important class of examples that are applicable to a wide range of situations, from tossing a coin to product testing.

Once the data are organized, the next step is to apply mathematical tools for analyzing the data and answering questions like those posed above. Numbers such as the **mean** and the **standard deviation** can be computed to reveal interesting facts about the data. These numbers can then be used to make predictions about future events.

The chapter ends with a section on one of the most important distributions in statistics, the **normal distribution.** This distribution describes many sets of data and also plays an important role in the underlying mathematical theory.

8.1 Random Variables and Distributions

Random Variables

In many experiments we can assign numerical values to the outcomes. For instance, if we roll a die, each outcome has a value from 1 through 6. If you select a lawyer and ascertain his or her annual income, the outcome is again a number. We call a rule that assigns a number to each outcome of an experiment a **random variable.**

Random Variable

A **random variable** X is a rule that assigns a number, or **value,** to each outcome in the sample space of an experiment.*

Quick Examples

1. Select a mutual fund; $X =$ the number of companies in the fund portfolio.
2. Select a computer; $X =$ the number of megabytes of memory it has.
3. Survey a group of 20 college students; $X =$ the mean SAT.

*In the language of functions (Chapter 1), a random variable is a *real-valued function* whose domain is the sample space.

Discrete and Continuous Random Variables

A **discrete** random variable can take on only specific, isolated numerical values, like the outcome of a roll of a die or the number of dollars in a randomly chosen bank account. A **continuous** random variable, on the other hand, can take on any values within a continuum or an interval, like the temperature in Central Park or the height of an athlete in centimeters. Discrete random variables that can take on only finitely many values (like the outcome of a roll of a die) are called **finite** random variables.

Quick Examples

Random Variable	Values	Type
1. Select a mutual fund; X = the number of companies in the fund portfolio.	$\{1, 2, 3, \ldots\}$	Discrete infinite
2. Take five shots at the goal during a soccer match; X = the number of times you score.	$\{0, 1, 2, 3, 4, 5\}$	Finite
3. Measure the length of an object; X = its length in centimeters.	Any positive real number	Continuous
4. Roll a die until you get a 6; X = the number of times you roll the die.	$\{1, 2, \ldots\}$	Discrete infinite
5. Bet a whole number of dollars in a race where the betting limit is \$100; X = the amount you bet.	$\{0, 1, \ldots, 100\}$	Finite
6. Bet a whole number of dollars in a race where there is no betting limit; X = the amount you bet.	$\{0, 1, \ldots, 100, 101, \ldots\}$	Discrete infinite

Note There are some borderline situations. For instance, if X is the salary of a factory worker, then X is, strictly speaking, discrete. However, the values of X are so numerous and close together that it may make sense to model X as a continuous random variable.

For the moment, we consider only finite random variables.

Example 1 • Finite Random Variable

Let X be the number of heads that come up when a coin is tossed three times. List the value of X for each possible outcome. What are the possible values of X?

Solution First, we describe X as a random variable.

X is the rule that assigns to each outcome the number of heads that come up.

We take as the outcomes of this experiment all possible sequences of three heads and tails. Then, for instance, if the outcome is HTH, the value of X is 2. An easy way to list the values of X for all the outcomes is by means of a table:

Outcome	HHH	HHT	HTH	HTT	THH	THT	TTH	TTT
Value of *X*	3	2	2	1	2	1	1	0

From the table, we also see that the possible values of X are 0, 1, 2, and 3.

Before we go on . . . Remember that X is just a rule on which we decided. We could have taken X to be a different rule, such as the number of tails or perhaps the number of heads minus the number of tails. These different rules are examples of different random variables associated with the same experiment.

Web Site

Follow the path

Web site → Everything for Finite Math → Chapter 8

At the Online Chapter Summary page, you will find an interactive simulation based on this example.

Example 2 • Stock Prices

You have purchased \$10,000 worth of stock in a biotech company whose newest arthritis drug is awaiting approval by the Food and Drug Administration. If the drug is approved this month, the value of the stock will double by the end of the month. If the drug is rejected this month, the stock's value will decline by 80%. If no decision is reached this month, its value will decline by 10%. Let X be the value of your investment at the end of this month. List the value of X for each possible outcome.

Solution There are three possible outcomes: The drug is approved this month, it is rejected this month, and no decision is reached. Once again, we express the random variable as a rule:

The random variable X is the rule that assigns to each outcome the value of your investment at the end of this month.

We can now tabulate the values of X as follows:

Outcome	Approved this month	Rejected this month	No decision
Value of *X* (\$)	20,000	2000	9000

Probability Distribution of a Random Variable

Given a random variable X, it is natural to look at certain *events*—for instance, the event that $X = 2$. By this we mean the event consisting of all outcomes that have an assigned X value of 2. Looking once again at the chart in Example 1, with X being the number of heads that face up when a coin is tossed three times, we find the following events:

The event that $X = 0$ is {TTT}.

The event that $X = 1$ is {HTT, THT, TTH}.

The event that $X = 2$ is {HHT, HTH, THH}.

The event that $X = 3$ is {HHH}.

The event that $X = 4$ is $\varnothing$. There are no outcomes with four heads.

Each event has a certain probability. For instance, the probability that $X = 2$ is 3/8 because the event in question consists of three of the eight possible (equally likely) outcomes. We abbreviate this by writing

$$P(X = 2) = \frac{3}{8}$$ The probability that $X = 2$ is $\frac{3}{8}$.

Thus, $P(X = 2)$ is the probability of the event $X = 2$, or simply the *probability that* $X = 2$. Similarly,

$$P(X = 4) = 0$$ The probability that $X = 4$ is 0.

The collection of probabilities of each of these events is called the **probability distribution** of the random variable X. Since the probabilities in a probability distribution can be estimated or theoretical, we discuss both *estimated probability distributions* (or *relative frequency distributions*) and *theoretical probability distributions* of random variables. (See the next two examples.)

Example 3 • Theoretical Probability Distribution

Let X be the number of heads that face up in three tosses of a coin. Give the probability distribution of X.

Solution X is the random variable of Example 1, so its values are 0, 1, 2, and 3. The probability distribution of X is given in the following table:

x	0	1	2	3
$P(X = x)$	$\frac{1}{8}$	$\frac{3}{8}$	$\frac{3}{8}$	$\frac{1}{8}$

Notice that the probabilities add to 1, as we might expect.

Question What is the distinction between X (uppercase) and x (lowercase) in the table?

Answer The distinction is important; X stands for the random variable in question, whereas x stands for a specific *value* (0, 1, 2, or 3) of X (so that x is always a number). For example, if $x = 2$, then $P(X = x)$ means $P(X = 2)$, the probability that X is 2. Similarly, if Y is a random variable, then $P(Y = y)$ is the probability that Y has the specific value y.

Before we go on . . . The probabilities in the table above are *theoretical* probabilities. To obtain a similar table of *estimated* probabilities, we would have to repeatedly toss a coin three times and calculate the fraction of times we got 0, 1, 2, and 3 heads.

Question This table of probabilities looks like the probability distribution associated with a sample space, as we studied in Section 7.2. What is the difference?

Answer The probability distribution of a random variable is not really new. Consider the following experiment. Toss three coins and count the number of heads. The associated probability distribution (as per Section 7.2) would be this:

Outcome	0	1	2	3
Probability	$\frac{1}{8}$	$\frac{3}{8}$	$\frac{3}{8}$	$\frac{1}{8}$

However, in this chapter we are thinking of 0, 1, 2, and 3 not as the outcomes of the experiment but as values of the random variable X. This allows us to use the more natural sample space consisting of the equally likely outcomes HHH, HHT, . . . , TTT; it also allows us to consider several different random variables on the same sample space.

Web Site

Follow the path

Web site → Everything for Finite Math → Chapter 8

At the Online Chapter Summary page, you will find a coin-tossing simulation that gives estimated probabilities for this random variable.

Example 4 • Estimated Probability Distribution

The following table shows the (fictitious) income brackets of a sample of 1000 lawyers in their first year out of law school:

Income Bracket ($)	20,000–29,999	30,000–39,999	40,000–49,999	50,000–59,999	60,000–69,999	70,000–79,999	80,000–89,999
Number of Lawyers	20	80	230	400	170	70	30

Think of the experiment of choosing a first-year lawyer at random (all being equally likely) and assign to each lawyer the number X that is the midpoint of his or her bracket. Find the probability distribution of X.

Solution Statisticians refer to the income brackets as **measurement classes.** Since the first measurement class contains incomes that are at least \$20,000 but less than \$30,000, its midpoint is \$25,000.[1] Similarly, the second measurement class has midpoint \$35,000, and so on. We can rewrite the table with the midpoints, as follows:

x ($)	25,000	35,000	45,000	55,000	65,000	75,000	85,000
Frequency	20	80	230	400	170	70	30

We have used the term *frequency* rather than *number,* although it means the same thing. This table is called a **frequency table.** It is *almost* the probability distribution for X,

[1]One might argue that the midpoint should be $(20{,}000 + 29{,}999)/2 = 24{,}999.50$, but we round this to 25,000. So, technically we are using "rounded" midpoints of the measurement classes.

except that we must replace frequencies by probabilities (we did this in calculating estimated probabilities in the preceding chapter). We start with the lowest measurement class. Since 20 of the 1000 lawyers fall in this group, we have

$$P(X = 25{,}000) = \frac{20}{1000} = .02$$

We can calculate the remaining probabilities similarly to obtain the following distribution:

x ($)	25,000	35,000	45,000	55,000	65,000	75,000	85,000
P(X = x)	.02	.08	.23	.40	.17	.07	.03

Note again the distinction between X and x: X stands for the random variable in question, whereas x stands for a specific value (25,000, 35,000, . . . , or 85,000) of X.

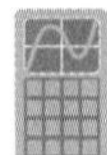

Graphing Calculator: Finding an Estimated Probability Distribution

Although the computations in this example (dividing the seven frequencies by 1000) are simple to do by hand, they could become tedious in general, so technology is helpful. In the TI-83 enter a list of frequencies as follows: press [STAT], choose EDIT, and then press [ENTER]. Clear columns L_1 and L_2 if they are not already cleared. (Select the heading of a column and press [CLEAR] [ENTER] to clear it.) Enter the values of X in the column under L_1 (pressing [ENTER] after each entry) and enter the frequencies in the column under L_2. Then, on the home screen, enter

L_2/1000→L_3

or, better yet,

L_2/sum(L_2)→L_3 — sum is found in [2nd] [STAT], under MATH.

After pressing [ENTER], go back to the [STAT] EDIT screen, and you will find the probabilities displayed in L_3 as shown:

L1	L2	L3
25000	20	0.02
35000	80	0.08
45000	230	0.23
55000	400	0.4
65000	170	0.17
75000	70	0.07
85000	30	0.03

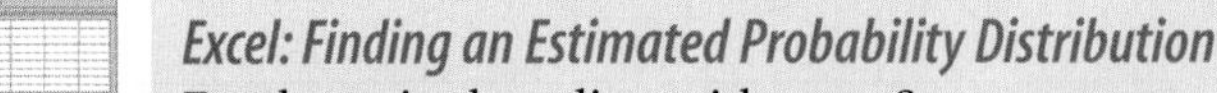

Excel: Finding an Estimated Probability Distribution

Excel manipulates lists with ease. Set up your spreadsheet as shown:

	A	B	C
1	x	Fr	P(X=x)
2	25000	20	=B2/SUM(B:B)
3	35000	80	
4	45000	230	
5	55000	400	
6	65000	170	
7	75000	70	
8	85000	30	

The formula `SUM(B:B)` gives the sum of all the numerical entries in column B. (Alternatively, you could specify `SUM(B2:B8)`.) You can now change the frequencies to see the effect on the probabilities. You can also add new values and frequencies to the list if you copy the formula in column C further down the column.

Graphing Probability Distributions: Histograms

We can use a bar graph to visualize a probability distribution. Figure 1 shows the bar graph for the probability distribution in Example 3. Such a graph is sometimes called a **histogram.**

x	0	1	2	3
$P(X = x)$	$\frac{1}{8}$	$\frac{3}{8}$	$\frac{3}{8}$	$\frac{1}{8}$

Table

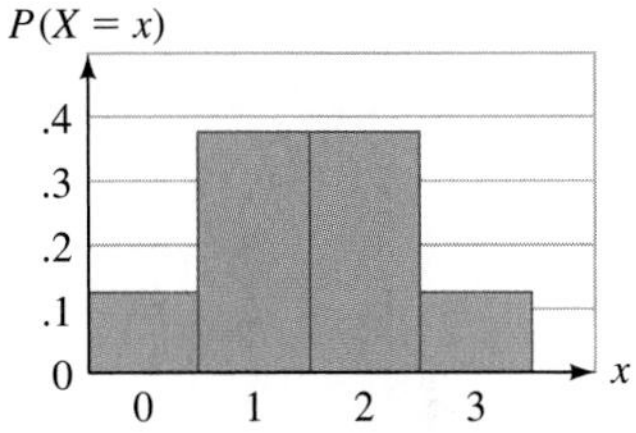

$P(X = x)$

.4 .3 .2 .1 0

0 1 2 3 x

Histogram

Figure 1

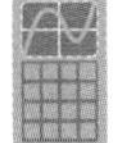

Graphing Calculator: Graphing a Probability Distribution

Most graphing calculators and spreadsheet programs include software to graph histograms. On the TI-83 press [STAT], select `EDIT`, enter the values of X in the L_1 list, and enter the probabilities in the L_2 list. To graph the data as in Figure 1, first set the [WINDOW] to $0 \le X \le 4$, $0 \le Y \le 0.5$, and Xscl $= 1$ (the width of the bars). Then turn on `STAT PLOT` ([2nd] [Y=]) and configure it by selecting the histogram icon, setting Xlist $= L_1$ and Freq $= L_2$. Then press [GRAPH].

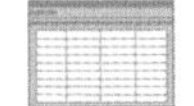

Excel: Graphing a Probability Distribution

Enter the values of X in one column and the probabilities in another, as illustrated in Example 4. Next, select *only* the column of probabilities (C2–C8 in Example 4) and then choose `Insert→Chart`. In the `Chart` dialog, select `Column` under `Standard Types` and then select `Next` and `Series`. At the bottom of the resulting dialog box, click in the blank area to the right of `Category (X) Axis Labels` and use your mouse to select the column of X values (A2–A8 in Example 4). You can then select `Finish`.

Web Site: Graphing a Probability Distribution

The Web site has an online histogram tool. Follow the path

Web site → Everything for Finite Math → Chapter 8

The Math Tools for this chapter include a histogram generator that automatically computes the probability distribution from a frequency table and also draws a histogram.

Example 5 • Probability Distribution: Greenhouse Gases

The following table shows per capita emissions of greenhouse gases for various countries, rounded to the nearest 5 metric tons.

Country	Per Capita Emissions (metric tons)	Country	Per Capita Emissions (metric tons)
Australia	30	Ireland	15
Austria	10	Italy	10
Belgium	15	Japan	10
Britain	10	Netherlands	15
Bulgaria	10	New Zealand	15
Canada	20	Norway	10
Czech Rep.	15	Poland	10
Denmark	15	Portugal	5
Estonia	10	Romania	5
Finland	15	Russian Fed.	5
France	10	Spain	10
Germany	10	Sweden	5
Greece	10	Switzerland	5
Hungary	10	Ukraine	10
Iceland	10	U.S.A.	20

Figures are measured in carbon dioxide equivalent and are based on 1998 data. SOURCE: Hal Turton and Clive Hamilton, "Comprehensive Emissions per Capita for Industrialized Countries," The Australia Institute, http://www.tai.org.au/Publications_Files/Papers&Sub_Files/Percapita.htm, 2001.

Consider the experiment in which a country is selected at random from this list and let X be the per capita greenhouse gas emissions for that country. Find the probability distribution of X and graph it with a histogram. Use the probability distribution to compute $P(X \geq 20)$ (the probability that X is 20 or more) and interpret the result.

Solution The values of X are the possible emissions figures, which we take to be 0, 5, 10, 15, 20, 25, and 30. In the table below, we first compute the frequency of each value of X by counting the number of countries that produce that per capita level of greenhouse gases. For instance, there are seven countries that have $X = 15$. Then, we divide each frequency by the sample size $N = 30$ to obtain the probabilities.

x	0	5	10	15	20	25	30
Frequency	0	5	15	7	2	0	1
$P(X = x)$	0	$\frac{5}{30}$	$\frac{15}{30}$	$\frac{7}{30}$	$\frac{2}{30}$	0	$\frac{1}{30}$

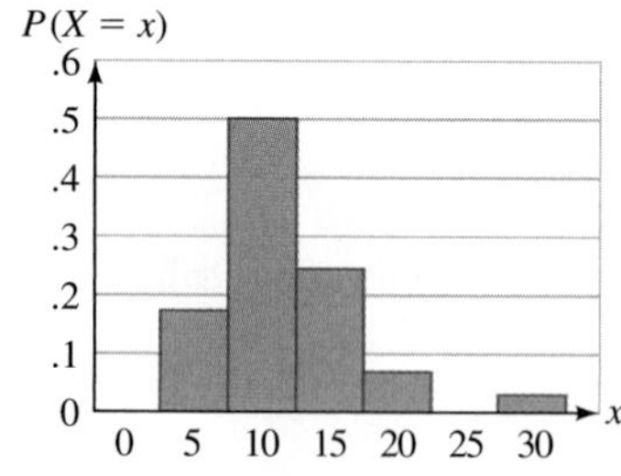

Figure 2

Figure 2 shows the resulting histogram.

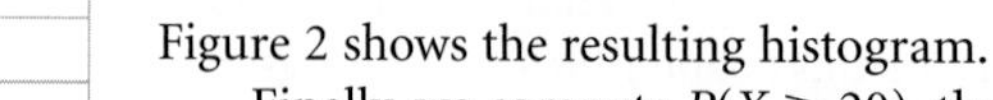

Finally, we compute $P(X \geq 20)$, the probability of all events with an X value of 20 or more. From the table we obtain

$$P(X \geq 20) = \frac{2}{30} + 0 + \frac{1}{30} = \frac{3}{30} = .1$$

Thus, there is a 10% chance that a country randomly selected from the given list produces 20 or more metric tons per capita of greenhouse gases.

Guidelines: Recognizing What to Use as a Random Variable and Deciding on Its Values

Question In an application, how, exactly, do I decide what to use as a random variable X?

Answer Be as systematic as possible: First, decide what the experiment is and its sample space. Then, based on what is asked for in the application, complete the following sentence: "X assigns ______ to each outcome." For instance, "X assigns the number of flavors to each packet of gummy bears selected," or "X assigns the average faculty salary to each college selected."

Question Once I have decided what X should be, how do I decide what values to assign it?

Answer Ask yourself: What are the conceivable values I could get for X? Then choose a collection of values that includes all of these. For instance, if X is the number of heads obtained when a coin is tossed five times, then the possible values of X are 0, 1, 2, 3, 4, and 5. If X is the average faculty salary in dollars, rounded to the nearest \$5000, then possible values of X could be 20,000, 25,000, 30,000, and so on, up to the highest salary in your data.

8.1 EXERCISES

In Exercises 1–10, classify each random variable X as finite, discrete infinite, or continuous, and indicate the values that X can take.

1. Roll two dice; $X =$ the sum of the numbers facing up.

2. Open a 500-page book on a random page; $X =$ the page number.

3. Select a stock at random; $X =$ your profit, to the nearest dollar, if you purchase one share and sell it one year later.

4. Select an electric utility company at random; $X =$ the exact amount of electricity, in gigawatt-hours, it supplies in 1 year.

5. Look at the second hand of your watch; $X =$ the time it reads in seconds.

6. Watch a soccer game; $X =$ the total number of goals scored.

7. Watch a soccer game; $X =$ the total number of goals scored, up to a maximum of 10.

8. Your class is given a mathematics exam worth 100 points; $X =$ the average score, rounded to the nearest whole number.

9. According to quantum mechanics, the energy of an electron in a hydrogen atom can assume only the values $k/1$, $k/4$, $k/9$, $k/16$, . . . for a certain constant value k; $X =$ the energy of an electron in a hydrogen atom.

10. According to classical mechanics, the energy of an electron in a hydrogen atom can assume any positive value; $X =$ the energy of an electron in a hydrogen atom.

In Exercises 11–18, (a) say what an appropriate sample space is; (b) complete the following sentence: "X is the rule that assigns to each . . ."; (c) list the values of X for all the outcomes.

11. X is the number of tails that come up when a coin is tossed twice.

12. X is the largest number of consecutive times heads comes up in a row when a coin is tossed three times.

13. X is the sum of the numbers that face up when two dice are rolled.

14. X is the value of the larger number when two dice are rolled.

15. X is the number of red marbles that Tonya has in her hand after she selects four marbles from a bag containing four red and two green marbles and then notes how many of each color there are.

16. X is the number of green marbles that Stej has in his hand after he selects four marbles from a bag containing three red and two green marbles and then notes how many of each color there are.

17. The mathematics final exam scores for the students in your study group are 89%, 85%, 95%, 63%, 92%, and 80%.

18. The capacities of the hard drives of your dormitory suite mates' computers are 10GB (gigabytes), 15GB, 20GB, 25GB, 30GB, and 35GB.

19. The random variable X has this probability distribution table:

x	2	4	6	8	10
$P(X = x)$	.1	.2			.1

a. Assuming that $P(X = 8) = P(X = 6)$, find each of the missing values.
b. Calculate $P(X \geq 6)$.

20. The random variable X has the probability distribution table shown below:

x	−2	−1	0	1	2
$P(X = x)$			.4	.1	.1

a. Calculate $P(X \geq 0)$ and $P(X < 0)$.
b. Assuming that $P(X = -2) = P(X = -1)$, find each of the missing values.

In Exercises 21–28, give the probability distribution for the indicated random variable and draw the corresponding histogram.

21. A fair die is rolled, and X is the number facing up.

22. A fair die is rolled, and X is the square of the number facing up.

23. Three fair coins are tossed, and X is the square of the number of heads showing.

24. Three fair coins are tossed, and X is the number of heads minus the number of tails.

25. A red and a green die are rolled, and X is the sum of the numbers facing up.

26. A red and a green die are rolled, and

$$X = \begin{cases} 0 \text{ if the numbers are the same} \\ 1 \text{ if the numbers are different} \end{cases}$$

27. A red and a green die are rolled, and X is the larger of the two numbers facing up.

28. A red and a green die are rolled, and X is the smaller of the two numbers facing up.

APPLICATIONS

29. SUVs—Tow Ratings The following table shows tow ratings for some popular 2000 model sport utility vehicles:

SUV	Tow Rating (lb)	SUV	Tow Rating (lb)
Mercedes Grand Marquis V8	2000	Ford Explorer V8	6000
Jeep Wrangler I6	2000	Dodge Durango V8	6000
Ford Explorer V6	3000	Dodge Ram 1500 V8	8000
Dodge Dakota V6	4000	Ford Expedition V8	8000
Mitsubishi Montero V6	5000	Hummer two-door hardtop	8000

Tow ratings vary considerably within each model. Figures cited are rounded. For more detailed information, consult http://www.rvsafety.com/towrate2k.htm.

Let X be the tow rating of a randomly chosen SUV from the list.
a. What are the values of X?
b. Compute the frequency and probability distributions of X.
c. What is the probability that an SUV is rated to tow no more than 5000 pounds?

30. Housing Prices The following table shows the average percent increase in the price of a house from 1980 to 2001 in nine regions of the United States:

Region	Price Increase (%)	Region	Price Increase (%)
New England	300	West north central	125
Pacific	225	West south central	75
Middle Atlantic	225	East north central	150
South Atlantic	150	East south central	125
Mountain	150		

Percentages are rounded to the nearest 25%. SOURCE: Third Quarter 2001 House Price Index, released November 30, 2001, by the Office of Federal Housing Enterprise Oversight. Online: http://www.ofheo.gov/house/3q01hpi.pdf.

Let X be the percent increase in the price of a house in a randomly selected region.
a. What are the values of X?
b. Compute the frequency and probability distribution of X.
c. What is the probability that, in a randomly selected region, the percent increase in the cost of a house exceeded 200%?

31. Pollen Count Allergy sufferers usually react to pollen counts higher than 30 grains/cubic meter. The following histogram shows the pollen counts in the New York metropolitan region on 23 different days in 1999:

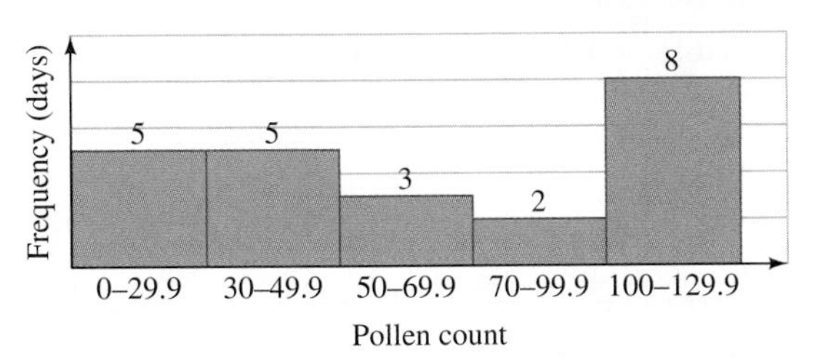

Upper limit of 129.9 was not set in original data. SOURCE: Department of Allergy and Immunology at Long Island College Hospital/*New York Times,* June 5, 1999, p. B1.

What is the associated random variable? Represent the data as an estimated probability distribution using the (rounded) midpoints of the given measurement classes.

32. Pollen Count Repeat Exercise 31, using the following histogram showing the pollen counts in the New York metropolitan region on 20 different days in 1998:

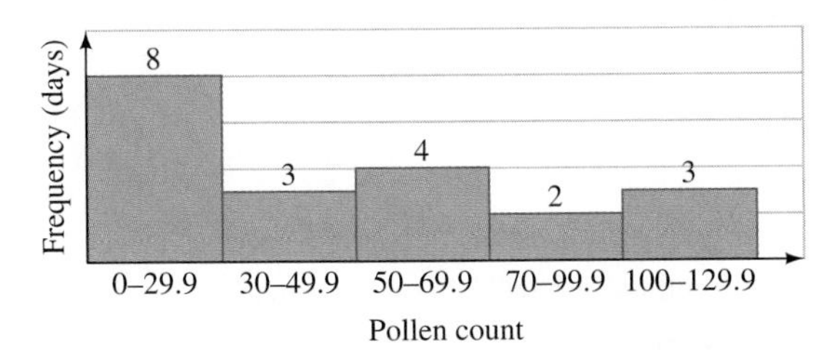

Upper limit of 129.9 was not set in original data. SOURCE: Department of Allergy and Immunology at Long Island College Hospital/*New York Times,* June 5, 1999, p. B1.

33. Grade-Point Averages The grade-point averages of the students in your mathematics class are

3.2 3.5 4.0 2.9 2.0 3.3 3.5 2.9 2.5 2.0
2.1 3.2 3.6 2.8 2.5 1.9 2.0 2.2 3.9 4.0

Use these raw data to construct a frequency table with the following measurement classes: 1.1–2.0, 2.1–3.0, and 3.1–4.0. Find the probability distribution using the (rounded) midpoint values as the values of X.

34. Test Scores Your scores for the 20 surprise math quizzes last semester were (out of 10)

4.5 9.5 10.0 3.5 8.0 9.5 7.5 6.5 7.0 8.0
8.0 8.5 7.5 7.0 8.0 9.0 10.0 8.5 7.5 8.0

Use these raw data to construct a frequency table with the following brackets: 2.1–4.0, 4.1–6.0, 6.1–8.0, and 8.1–10.0. Find the probability distribution using the (rounded) midpoint values as the values of X.

35. Car Purchases To persuade his parents to contribute to his new-car fund, Carmine has spent the last week surveying the ages of 2000 cars on campus. His findings are reflected in the following frequency table:

Age of Car (years)	0	1	2	3	4	5	6	7	8	9	10
Cars	140	350	450	650	200	120	50	10	5	15	10

Carmine's jalopy is 6 years old. He would like to make the following claim to his parents: "x% of students have cars newer than mine." Use a probability distribution to find x.

36. Car Purchases Carmine's parents, not convinced of his need for a new car, produced the following statistics showing the ages of cars owned by students on the dean's list:

Age of Car (years)	0	1	2	3	4	5	6	7	8	9	10
Cars	0	2	5	5	10	10	15	20	20	20	40

They then claimed that if he kept his 6-year old car for another year, his chances of getting on the dean's list would be increased by x%. Use a probability distribution to find x.

Highway Safety Exercises 37–46 are based on the following table, which shows crashworthiness ratings for several categories of motor vehicles. In all exercises let X be the crash-test rating of a small car, Y be the crash-test rating for a small SUV, and so on, as shown in the table.

	Number Tested	Overall Frontal Crash-Test Rating			
		3 (good)	2 (acceptable)	1 (marginal)	0 (poor)
Small Cars, X	16	1	11	2	2
Small SUVs, Y	10	1	4	4	1
Medium SUVs, Z	15	3	5	3	4
Passenger Vans, U	13	3	0	3	7
Midsize Cars, V	15	3	5	0	7
Large Cars, W	19	9	5	3	2

Ratings are by the Insurance Institute for Highway Safety. SOURCES: Stacy C. Davis and Lorena F. Truett, Oak Ridge National Laboratory, "An Analysis of the Impact of Sport Utility Vehicles in the United States," August 2000/Insurance Institute for Highway Safety. Online: http://www-cta.ornl.gov/Publications/Final SUV report.pdf http://www.highwaysafety.org/vehicle_ratings/.

37. Compute the probability distribution for X.

38. Compute the probability distribution for Y.

39. Compute $P(X \geq 2)$ and interpret the result.

40. Compute $P(Y \leq 1)$ and interpret the result.

41. Compare $P(Y \geq 2)$ and $P(Z \geq 2)$. What does the result suggest about SUVs?

42. Compare $P(V \geq 2)$ and $P(Z \geq 2)$. What does the result suggest?

43. Which of the six categories shown has the *lowest* probability of a good rating?

44. Which of the six categories shown has the *highest* probability of a poor rating?

45. You choose, at random, a small car and a small SUV. What is the probability that both will be rated at least 2?

46. You choose, at random, a small car and a midsize car. What is the probability that both will be rated at most 1?

Exercises 47 and 48 assume familiarity with counting arguments and probability (Section 7.4).

47. Camping Kent's Tents has four red tents and three green tents in stock. Karin selects four of them at random. Let X be the number of red tents she selects. Give the probability distribution and find $P(X \geq 2)$.

48. Camping Kent's Tents has five green knapsacks and four yellow ones in stock. Curt selects four of them at random. Let X be the number of green knapsacks he selects. Give the probability distribution and find $P(X \leq 2)$.

49. Income Distribution The following table shows the distribution of income in the United States in 2000, based on a sample of 1000 households:

Income Bracket ($)	0–4999	5000–9999	10,000–14,999	15,000–24,999	25,000–34,999	35,000–49,999	50,000–74,999	75,000–99,999	100,000–219,999
Households	28	61	70	134	125	155	189	104	134

Based on the actual income distribution. Upper-income limit is an estimate. SOURCE: U.S. Census Bureau, "Money Income in the United States: 2000," Current Population Reports, P60-213 (Washington, D.C.: Government Printing Office, 2001).

a. Let X be the midpoint of a bracket in which a randomly selected household falls. Find the probability distribution of X and graph its histogram.

b. Shade the area of your histogram corresponding to the probability that a U.S. household has a value of X of more than 42,500. What is this probability?

50. Income Distribution Refer to Exercise 49. The following table shows the distribution of income among Hispanic U.S. households in 2000, based on a sample of 1000 households:

Income Bracket ($)	0–4999	5000–9999	10,000–14,999	15,000–24,999	25,000–34,999	35,000–49,999	50,000–74,999	75,000–99,999	100,000–179,999
Households	33	73	83	182	147	177	174	74	57

Based on the actual income distribution. Upper-income limit is an estimate. SOURCE: U.S. Census Bureau, "Money Income in the United States: 2000," Current Population Reports, P60-213 (Washington, D.C.: Government Printing Office, 2001).

a. Let X be the midpoint of a bracket in which a randomly selected household falls. Find the probability distribution of X and graph its histogram.

b. Shade the area of your histogram corresponding to the probability that a Hispanic U.S. household has a value of X of at most 12,500. What is this probability?

51. Testing Your Calculator Use your calculator or computer to generate a sequence of 100 random digits in the range 0–9 and test the random number generator for uniformness by drawing the distribution histogram.

52. Testing Your Dice Repeat Exercise 51 but this time use a die to generate a sequence of 50 random numbers in the range 1–6.

COMMUNICATION AND REASONING EXERCISES

53. Are all infinite random variables necessarily continuous? Explain.

54. Are all continuous random variables necessarily infinite? Explain.

55. If you are unable to compute the (theoretical) probability distribution for a random variable X, how can you estimate the distribution?

56. What do you expect to happen to the probabilities in a probability distribution as you make the measurement classes smaller?

57. Give an example of a real-life situation that can be modeled by a random variable with a probability distribution whose histogram is highest on the left.

58. How wide should the bars in a histogram be so that the probability $P(a \leq X \leq b)$ equals the area of the corresponding portion of the histogram?

59. Give at least one scenario in which you might prefer to model the number of pages in a randomly selected book using a continuous random variable rather than a discrete random variable.

60. Give at least one reason why you might prefer to model a temperature using a discrete random variable rather than a continuous random variable.

8.2 *Bernoulli Trials and Binomial Random Variables*

Your electronic production plant produces video-game joysticks. Unfortunately, quality control at the plant leaves much to be desired, and 10% of the joysticks the plant produces are defective due to random errors in the electronic components. A large corporation has expressed interest in adopting your product for its new game console, and today an inspection team will be visiting to test video-game joysticks as they come off the assembly line. If the team tests five joysticks, what is the probability that none will be defective? What is the probability that more than one will be defective?

In this scenario we are interested in the following random variable: Think of the experiment as a sequence of five "trials" (in each trial the inspection team chooses one joystick at random and tests it), each with two possible outcomes: "success" (a defective joystick) and "failure" (a nondefective one).[2] If we now take X to be the number of successes (defective joysticks) the inspection team finds, we can recast the questions above as follows: Find $P(X = 0)$ and $P(X > 1)$. This is an example of a **binomial random variable.**

[2]These are customary names for the two possible outcomes and often do not indicate actual success or failure at anything.

Bernoulli Trial and Binomial Random Variable

A **Bernoulli* trial** is an experiment that has two possible outcomes, called **success** and **failure.** If the probability of success is p then the probability of failure is $q = 1 - p$.

Tossing a coin three times is an example of a **sequence of independent Bernoulli trials:** a sequence of Bernoulli trials in which the outcomes in any one trial are independent (in the sense of the preceding chapter) of those in any other trial.

A **binomial random variable** is one that counts the number of successes in a sequence of independent Bernoulli trials.

Quick Examples (of Binomial Random Variables)

1. Roll a die ten times; X is the number of times you roll a 6.
2. Provide a property with flood insurance for 20 years; X is the number of years, during the 20-year period, during which the property is flooded.[†]
3. You know that 60% of all bond funds will depreciate in value next year, and you randomly select four from a very large number of possible choices; X is the number of bond funds you hold that will depreciate next year. (X is approximately binomial.)[‡]

*Jakob Bernoulli (1654–1705) was one of the pioneers of probability theory.

[†]Assuming that the probability of flooding one year is independent of whether there was flooding in earlier years.

[‡]Since the number of bond funds is extremely large, choosing a "loser" (a fund that will depreciate next year) does not significantly deplete the pool of "losers," and so the probability that the next fund you choose will be a "loser" is hardly affected. Hence, we can think of X as being a binomial variable.

Example 1 • Probability Distribution of a Binomial Random Variable

Suppose we have a possibly unfair coin with the probability of heads p and the probability of tails $q = 1 - p$.

a. Let X be the number of heads you get in a sequence of five tosses. Find $P(X = 2)$.

b. Let X be the number of heads you get in a sequence of n tosses. Find $P(X = x)$.

Solution

a. We are looking for the probability of getting exactly two heads in a sequence of five tosses. Let's start with a simpler question.

Question What is the probability that we will get the sequence HHTTT?

Answer The probability that the first toss will come up heads is p.

The probability that the second toss will come up heads is also p.

The probability that the third toss will come up tails is q.

The probability that the fourth toss will come up tails is q.

The probability that the fifth toss will come up tails is q.

The probability that the first toss will be heads *and* the second will be heads *and* the third will be tails *and* the fourth will be tails *and* the fifth will be tails equals the probability of the *intersection* of these five events. Since these are independent events, the probability of the intersection is the product of the probabilities, which is

$$p \times p \times q \times q \times q = p^2q^3$$

HHTTT is only one of several outcomes with two heads and three tails. Two others are HTHTT and TTTHH.

Question How many such outcomes are there all together?

Answer This is the number of "words" with two H's and three T's, and we know from Chapter 6 that the answer is $C(5, 2) = 10$.

Each of these ten outcomes has the same probability: p^2q^3. (Why?) Thus, the probability of getting one of these ten outcomes is the probability of the union of all these (mutually exclusive) events, and we saw in the preceding chapter that this is just the sum of the probabilities. In other words, the probability we are after is

$$P(X = 2) = p^2q^3 + p^2q^3 + \cdots + p^2q^3 \qquad C(5, 2) \text{ times}$$
$$= C(5, 2)p^2q^3$$

The structure of this formula is as follows:

$$P(X = 2) = C(5, 2)p^2q^3$$

Number of heads (exponent 2 of p), Number of tails (exponent 3 of q)
Number of tosses (5), Number of heads (2), Probability of heads (p), Probability of tails (q)

b. There is nothing special about 2 in part (a). To get $P(X = x)$ rather than $P(X = 2)$, replace 2 with x:

$$P(X = x) = C(5, x)p^xq^{5-x}$$

Again, there is nothing special about 5. The general formula for n tosses is

$$P(X = x) = C(n, x)p^xq^{n-x}$$

Probability Distribution of Binomial Random Variable

If X is the number of successes in a sequence of n independent Bernoulli trials, then

$$P(X = x) = C(n, x)p^xq^{n-x}$$

where n = number of trials
p = probability of success
q = probability of failure $= 1 - p$

Quick Example

If you roll a fair die five times, the probability of throwing exactly two 6s is

$$P(X = 2) = C(5, 2)\left(\frac{1}{6}\right)^2\left(\frac{5}{6}\right)^3 = 10 \times \frac{1}{36} \times \frac{125}{216} \approx .1608$$

Here, we used $n = 5$ and $p = 1/6$, the probability of rolling a 6 on one roll of the die.

Example 2 • Aging

The probability that a randomly chosen person in Cape Coral, Florida, is 65 years old or older is approximately .2.[3]

a. What is the probability that, in a randomly selected sample of six Cape Coral Floridians, exactly four of them are 65 years or older?

b. If X is the number of people aged 65 years or older in a sample of six, construct the probability distribution of X and plot its histogram.

c. Compute $P(X \leq 2)$.

d. Compute $P(X \geq 2)$.

Solution

a. The experiment is a sequence of Bernoulli trials; in each trial we select a person and ascertain his or her age. If we take "success" to mean selection of a person aged 65 years or older, then the probability distribution is

$$P(X = x) = C(n, x)p^x q^{n-x}$$

where n = number of trials = 6
p = probability of success = .2
q = probability of failure = .8

So,

$$P(X = 4) = C(6, 4)(.2)^4(.8)^2 = 15 \times .0016 \times .64 = .015\ 36$$

b. We have already computed $P(X = 4)$. Here are all the calculations:

$$P(X = 0) = C(6, 0)(.2)^0(.8)^6$$
$$= 1 \times 1 \times .262\ 144 = .262\ 144$$
$$P(X = 1) = C(6, 1)(.2)^1(.8)^5$$
$$= 6 \times .2 \times .327\ 68 = .393\ 216$$
$$P(X = 2) = C(6, 2)(.2)^2(.8)^4$$
$$= 15 \times .04 \times .4096 = .245\ 76$$
$$P(X = 3) = C(6, 3)(.2)^3(.8)^3$$
$$= 20 \times .008 \times .512 = .081\ 92$$
$$P(X = 4) = C(6, 4)(.2)^4(.8)^2$$
$$= 15 \times .0016 \times .64 = .015\ 36$$
$$P(X = 5) = C(6, 5)(.2)^5(.8)^1$$
$$= 6 \times .000\ 32 \times .8 = .001\ 536$$
$$P(X = 6) = C(6, 6)(.2)^6(.8)^0$$
$$= 1 \times .000\ 064 \times 1 = .000\ 064$$

---

[3]The actual figure in 2000 was .196. SOURCE: U.S. Census Bureau, Census 2000 Summary File 1, http://www.census.gov/prod/2001pubs/c2kbr01-10.pdf.

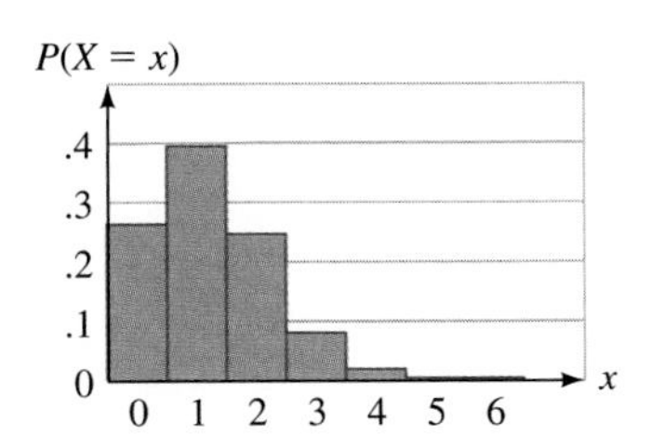

Figure 3

The probability distribution is therefore as follows:

x	0	1	2	3	4	5	6
$P(X = x)$	.262 144	.393 216	.245 76	.081 92	.015 36	.001 536	.000 064

Figure 3 shows its histogram.

c. $P(X \le 2)$, the probability that the number of people selected who are at least 65 years old is either 0, 1, or 2, is the probability of the union of these events and is thus the sum of the three probabilities:

$$P(X \le 2) = P(X = 0) + P(X = 1) + P(X = 2)$$
$$= .262\ 144 + .393\ 216 + .245\ 76$$
$$= .901\ 12$$

d. To compute $P(X \ge 2)$, we *could* compute the sum

$$P(X \ge 2) = P(X = 2) + P(X = 3) + P(X = 4) + P(X = 5) + P(X = 6)$$

but it is far easier to compute the probability of the complement of the event:

$$P(X < 2) = P(X = 0) + P(X = 1) = .262\ 144 + .393\ 216 = .655\ 36$$

and then subtract the answer from 1:

$$P(X \ge 2) = 1 - P(X < 2) = 1 - .655\ 36 = .344\ 64$$

Graphing Calculator: Computing the Distribution of a Binomial Random Variable

Question Computing the probability distribution required a lot of number crunching. Can I get my graphing calculator to do the work?

Answer Yes. Here are the instructions for the TI-83. In the `Y=` screen, enter the binomial distribution formula

```
Y1=6 nCr X*0.2^X*0.8^(6-X)
```

directly (to get `nCr`, press [MATH] and select `PRB`) and press [TABLE]. You can then replicate the table above by choosing $X = 0, 1, \ldots, 6$ (use the `TBLSET` screen to set `Indpnt` to `Ask` if you have not already done so). The TI-83 also has a built-in binomial distribution function that you can use in place of the explicit formula:

```
Y1=binompdf(6,0.2,X)
```
Press [2nd] [VARS] [7].

The TI-83 function `binompcf` (directly following `binompdf`) gives the value of the *cumulative* distribution function, $P(0 \le X \le x)$.

To graph the resulting probability distribution on your calculator, follow the instructions for graphing a histogram in Section 8.1.

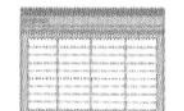

Excel: Computing the Distribution of a Binomial Random Variable

You can generate the binomial distribution as follows:

	A	B
1	x	P(X=x)
2	0	=BINOMDIST(A2,6,0.2,0)
3	1	
4	2	
5	3	
6	4	
7	5	
8	6	

The values of X are shown in column A, and the probabilities are computed in column B. The arguments of the `BINOMDIST` function are as follows:

```
BINOMDIST(x, n, p, Cumulative(0=no,1=yes))
```

Setting the last argument to 0 (as shown) gives $P(X = x)$. Setting it to 1 gives $P(X \leq x)$.

To graph the resulting probability distribution, follow the instructions for graphing a histogram in Section 8.1.

Web Site

Follow the path

Web site → Everything for Finite Math → Chapter 8 → Binomial Distribution Utility

where you can obtain the distribution, compute individual probabilities, and graph the histogram.

Guideline: Terminology and Recognizing When to Use the Binomial Distribution

Question What is the difference between Bernoulli trials and the binomial random variable?

Answer A Bernoulli trial is a type of experiment, whereas a binomial random variable is the resulting kind of random variable. More precisely, if your experiment consists of performing a sequence of n Bernoulli trials (think of throwing a dart n times at random points on a dartboard, hoping to hit the bull's-eye), then the random variable X that counts the number of successes (the number of times you actually hit the bull's-eye) is a binomial random variable.

Question How do I recognize when a situation gives a binomial random variable?

Answer Make sure that the experiment consists of a sequence of independent Bernoulli trials—that is, a repeated sequence of trials of an experiment that has two outcomes, where the outcome of each trial does not depend on the outcomes in previous trials. For instance, repeatedly throwing a dart at a dartboard hoping to hit the bull's-eye does not constitute a sequence of Bernoulli trials if you adjust your aim each time, depending on the outcome of your previous attempt. This dart-throwing experiment can be modeled by a sequence of Bernoulli trials if you make no adjustments after each attempt and your aim does not improve (or deteriorate) with time.

8.2 EXERCISES

In Exercises 1–10, you are performing five independent Bernoulli trials with $p = .1$ and $q = .9$. Calculate the probability of each of the stated outcomes.

1. Two successes

2. Three successes

3. No successes

4. No failures

5. All successes

6. All failures

7. At most two successes

8. At least four successes

9. At least three successes

10. At most three successes

In Exercises 11–18, X is a binomial variable with $n = 6$ and $p = .2$. Compute the given probabilities.

11. $P(X = 3)$

12. $P(X = 4)$

13. $P(X \leq 2)$

14. $P(X \leq 1)$

15. $P(X \geq 5)$

16. $P(X \geq 4)$

17. $P(1 \leq X \leq 3)$

18. $P(3 \leq X \leq 5)$

In Exercises 19 and 20, graph the histogram of the given binomial distribution.

19. $n = 5, p = \frac{1}{4}, q = \frac{3}{4}$

20. $n = 5, p = \frac{1}{3}, q = \frac{2}{3}$

In Exercises 21 and 22, graph the histogram of the given binomial distribution and compute the given quantity, indicating the corresponding region on the graph.

21. $n = 4, p = \frac{1}{3}, q = \frac{2}{3}; P(X \leq 2)$

22. $n = 4, p = \frac{1}{4}, q = \frac{3}{4}; P(X \leq 1)$

APPLICATIONS

23. Retirement The probability that a randomly chosen person in Britain is of pension age is approximately .25. What is the probability that, in a randomly selected sample of five people, two are of pension age?

SOURCE: Carnegie Center, Moscow/*New York Times*, March 15, 1998, p. 10.

24. Alien Retirement The probability that a randomly chosen citizen-entity of Cygnus is of pension age[4] is approximately .8. What is the probability that, in a randomly selected sample of four citizen-entities, all of them are of pension age?

25. 1990s Internet Stock Boom According to a July 1999 article in the *New York Times*, venture capitalists had this rule of thumb: The probability that an Internet start-up company will be a "stock market success," resulting in "spectacular profits for early investors" is .2. If you were a venture capitalist who invested in ten Internet start-up companies, what was the probability that at least one of them would be a stock market success? (Round your answer to four decimal places.)

"Not All Hit It Rich in the Internet Gold Rush," *New York Times*, July 20, 1999, p. A1.

26. 1990s Internet Stock Boom According to the article cited in Exercise 25, 13.5% of Internet stocks that entered the market in 1999 ended up trading below their initial offering prices. If you were an investor who purchased five Internet stocks at their initial offering prices, what was the probability that at least four of them would end up trading at or above their initial offering price? (Round your answer to four decimal places.)

27. Job Training (from the GRE Exam in Economics) In a large on-the-job training program, half of the participants are women and half are men. In a random sample of three participants, what is the probability that an investigator will draw at least one man?

28. Job Training (Based on a Question from the GRE Exam in Economics) In a large on-the-job training program, half of the participants are women and half are men. In a random sample of five participants, what is the probability that an investigator will draw at least two men?

29. Manufacturing Your manufacturing plant produces air bags, and it is known that 10% of them are defective. Five air bags are tested.

a. Find the probability that three of them are defective.

b. Find the probability that at least two of them are defective.

30. Manufacturing Compute the probability distribution of the binomial variable described in Exercise 29 and use it to compute the probability that, if five air bags are tested, at least one will be defective and at least one will not.

31. Teenage Pastimes According to a study, the probability that a randomly selected teenager watched a rented video at least once during a week was .71. What is the probability that at least eight teenagers in a group of ten watched a rented movie at least once last week?

SOURCES: Rand Youth Poll/Teen-age Research Unlimited/*New York Times*, March 14, 1998, p. D1.

32. Other Teenage Pastimes According to the study cited in Exercise 31, the probability that a randomly selected teenager studied at least once during a week was only .52. What is the probability that less than half of the students in your study group of ten have studied in the last week?

33. Triple Redundancy To ensure reliable performance of vital computer systems, aerospace engineers sometimes employ the technique of *triple redundancy*, in which three identical computers are installed in a space vehicle. If one of the three computers gives results different from the other two, it is assumed to be malfunctioning, and it is ignored. This technique will work as long as no more than one computer malfunctions. Assuming that an on-board computer is 99% reliable (that is, the probability of its failing is .01), what is the probability that at least two of the three computers will malfunction?

---

[4]12,000 bootlags, which translates to approximately 20 minutes Earth time.

34. IQ Scores Mensa is a club for people who have high IQ scores. To qualify, your IQ must be at least 132, putting you in the top 2% of the general population. If a group of ten people are chosen at random, what is the probability that at least two of them qualify for Mensa?

In Exercises 35–40, use technology to solve.

35. Standardized Tests Assume that on a standardized test of 100 questions, a person has a probability of 80% of answering each question correctly. Find the probability of answering between 75 and 85 questions, inclusive, correctly. (Assume independence, and round your answer to four decimal places.)

36. Standardized Tests Assume that on a standardized test of 100 questions, a person has a probability of 80% of answering each question correctly. Find the probability of answering at least 90 questions correctly. (Assume independence, and round your answer to four decimal places.)

37. Product Testing It is known that 43% of all the ZeroFat hamburger patties produced by your factory contain more than 10 grams of fat. Compute the probability distribution for $n = 50$ Bernoulli trials.

a. What is the most likely value for the number of burgers that contain more than 10 grams of fat?

b. Complete the following sentence: There is approximately a 71% chance that a batch of 50 ZeroFat patties contains ______ or more patties with more than 10 grams of fat.

c. Compare the graphs of the distributions for $n = 50$ trials and $n = 20$ trials. What do you notice?

38. Product Testing It is known that 65% of all the ZeroCal hamburger patties produced by your factory contain more than 1000 calories. Compute the probability distribution for $n = 50$ Bernoulli trials.

a. What is the most likely value for the number of burgers that contain more than 1000 calories?

b. Complete the following sentence: There is approximately a 73% chance that a batch of 50 ZeroCal patties contains ______ or more patties with more than 1000 calories.

c. Compare the graphs of the distributions for $n = 50$ trials and $n = 20$ trials. What do you notice?

39. Quality Control A manufacturer of light bulbs chooses bulbs at random from its assembly line for testing. If the probability of a bulb's being bad is .01, how many bulbs do they need to test before the probability of finding at least one bad one rises to more than .5? (You may have to use trial and error to solve this.)

40. Quality Control A manufacturer of light bulbs chooses bulbs at random from its assembly line for testing. If the probability of a bulb's being bad is .01, how many bulbs do they need to test before the probability of finding at least two bad ones rises to more than .5? (You may have to use trial and error to solve this.)

41. Highway Safety According to a study, a male driver in the United States will average 562 accidents per 100 million miles. Regard an n-mile trip as a sequence of n Bernoulli trials with success corresponding to having an accident during a particular mile. What is the probability that a male driver will have an accident in a 1-mile trip?

Data are based on a report by the National Highway Traffic Safety Administration released in January 1996. SOURCE: U.S. Department of Transportation/*New York Times*, April 9, 1999, p. F1.

42. Highway Safety According to the study cited in Exercise 41, a female driver in the United States will average 611 accidents per 100 million miles. Regard an n-mile trip as a sequence of n Bernoulli trials with success corresponding to having an accident during a particular mile. What is the probability that a female driver will have an accident in a 1-mile trip?

COMMUNICATION AND REASONING EXERCISES

43. A soccer player is more likely to score on his second shot if he was successful on his first. Can we model a succession of shots a player takes as a sequence of Bernoulli trials? Explain.

44. A soccer player takes repeated shots on goal. What assumption must we make if we want to model a succession of shots by a player as a sequence of Bernoulli trials?

45. Your friend just told you that "misfortunes always occur in threes." If life is just a sequence of Bernoulli trials, is this possible? Explain.

46. Suppose an experiment consists of repeatedly (every week) checking whether your graphing calculator battery has died. Is this a sequence of Bernoulli trials? Explain.

47. Why is the following not a binomial random variable? Select, without replacement, five marbles from a bag containing six red and two blue marbles, and let X be the number of red marbles you have selected.

48. By contrast with Exercise 47, why can the following be modeled by a binomial random variable? Select, without replacement, 5 electronic components from a batch of 10,000 in which 1000 are defective and let X be the number of defective components you select.

8.3 Measures of Central Tendency

Mean, Median, and Mode of a Set of Data

One day you decide to measure the popularity rating of your statistics instructor, Mr. Bravo. Ideally, you should poll every one of Mr. Bravo's students, what statisticians would refer to as the **population.** However, it would be difficult to poll all the members of the population in question (Mr. Bravo teaches more than 400 students). Instead, you decide to survey 10 of his students, chosen at random, and ask them to rate Mr. Bravo on a scale of 0–100. The survey results in the following set of data:

60 50 55 0 100 90 40 20 40 70

Statisticians call such a collection of data a **sample,** since the 10 people you polled represent only a (small) sample of Mr. Bravo's students. We should think of the individual scores 60, 50, 55, . . . as values of a random variable: Choose one of Mr. Bravo's students at random and let X be the rating the student gives to Mr. Bravo.

How do we distill a single measurement, or **statistic,** from this sample that would tell us Mr. Bravo's popularity? Perhaps the most commonly used statistic is the **average,** or **mean,** which is computed by adding the scores and dividing the sum by the number of scores in the sample:

$$\text{sample mean} = \frac{60 + 50 + 55 + 0 + 100 + 90 + 40 + 20 + 40 + 70}{10} = \frac{525}{10} = 52.5$$

We might then conclude, based on the sample, that Mr. Bravo's average popularity rating is about 52.5. The usual notation for the sample mean is $\bar{x}$, and the formula we use to compute it is

$$\bar{x} = \frac{x_1 + x_2 + \cdots + x_n}{n}$$

where $x_1, x_2, \ldots, x_n$ are the values of X in the sample.

A convenient way of writing the sum that appears in the numerator is to use **summation** or **sigma notation.** We write the sum $x_1 + x_2 + \cdots + x_n$ as

$$\sum_{i=1}^{n} x_i \qquad \text{The sum of the } x_i \text{ values from } i = 1 \text{ to } n$$

We think of i as taking on the values $1, 2, \ldots, n$ in turn, making x_i equal $x_1, x_2, \ldots, x_n$ in turn, and we then add up these values.

Sample and Mean

A **sample** is a sequence of values of a random variable X. (The process of collecting such a sequence is sometimes called **sampling** X.) The **sample mean** is the average of the values, or **scores,** in the sample. To compute the sample mean, we use the following formula:

$$\bar{x} = \frac{x_1 + x_2 + \cdots + x_n}{n} = \frac{\sum_{i=1}^{n} x_i}{n}$$

or simply

$$\bar{x} = \frac{\sum_i x_i}{n} \qquad \textstyle\sum_i \text{ stands for "sum over all } i\text{."}$$

Here, n is the **sample size** (number of scores), and $x_1, x_2, \ldots, x_n$ are the individual scores.

If the sample $x_1, x_2, \ldots, x_n$ consists of all the values of X from the entire population* (for instance, the ratings given Mr. Bravo by *all* his students), we refer to the mean as the **population mean** and write it as μ (Greek mu) instead of $\bar{x}$.

Quick Examples

1. The mean of the sample 1, 2, 3, 4, 5 is $\bar{x} = 3$.
2. The mean of the sample $-1, 0, 2$ is $\bar{x} = \dfrac{-1 + 0 + 2}{3} = \dfrac{1}{3}$.
3. The mean of the population $-3, -3, 0, 0, 1$ is $\mu = \dfrac{-3 - 3 + 0 + 0 + 1}{5} = -1$.

Sample Mean versus Population Mean

Measuring a population mean can be difficult or even impossible. For instance, computing the mean household income for the United States would entail recording the income of every single household. Instead of attempting to do this, statisticians usually use sample means. The larger the sample used, the more accurately we expect the sample mean to approximate the population mean. Estimating how accurately a sample mean of a given size approximates the population mean is possible, but we will not go into that in this book.

*When we talk about *populations,* the understanding is that the underlying experiment consists of selecting a member of a given population and ascertaining the value of X.

The mean $\bar{x}$ is an attempt to measure where the center of the sample is. It is therefore called a **measure of central tendency.** There are two other common measures of central tendency: the middle score, or **median,** and the most frequent score, or **mode.** These are defined as follows.

Median and Mode

The **sample median** m is the middle score (in the case of an odd-sized sample) or average of the two middle scores (in the case of an even-sized sample) when the scores in a sample are arranged in ascending order.

A **sample mode** is a score that appears most often in the collection. (There may be more than one mode in a sample.)

As before, we refer to the **population median** and **population mode** if the sample consists of the data from the entire population.

Quick Examples

1. The sample median of $2, -3, -1, 4, 2$ is found by first arranging the scores in ascending order: $-3, -1, 2, 2, 4$ and then selecting the middle (third) score: $m = 2$. Its mode is also 2 because this is the score that appears most often.
2. The sample $2, 5, 6, -1, 0, 6$ has median $m = (2 + 5)/2 = 3.5$ and mode 6.

The mean tends to give more weight to scores that are farther away from the center than does the median. For example, if you take the largest score in a collection of more than one number and make it larger, the mean will increase, but the median will remain the same. For this reason the median is often preferred for collections that contain a wide range of scores. The mode can sometimes lie far from the center and is thus used less often as an indication of where the center of a sample lies.

Example 1 • Teenage Spending

A 10-year survey of spending patterns of U.S. teenagers yielded the following figures (in billions of dollars spent in a year): 90, 90, 85, 80, 80, 80, 80, 85, 90, 100.[5] Compute and interpret the mean, median, and mode and illustrate the data on a graph.

Solution The *mean* is given by

$$\bar{x} = \frac{\sum_i x_i}{n}$$

$$= \frac{90 + 90 + 85 + 80 + 80 + 80 + 80 + 85 + 90 + 100}{10} = \frac{860}{10} = 86$$

Thus, spending by teenagers averaged $86 billion per year.

For the *median,* we arrange the sample data in ascending order:

80 80 80 80 85 85 90 90 90 100

We then take the average of the two middle scores:

$$m = \frac{85 + 85}{2} = 85$$

This means that in half the years in question teenagers spent $85 billion or less and in half they spent $85 billion or more.

For the *mode* we choose the score (or scores) that occurs most frequently: $80 billion. Thus, teenagers spent $80 billion per year more often than any other amount.

The frequency histogram in Figure 4 illustrates these three measures.

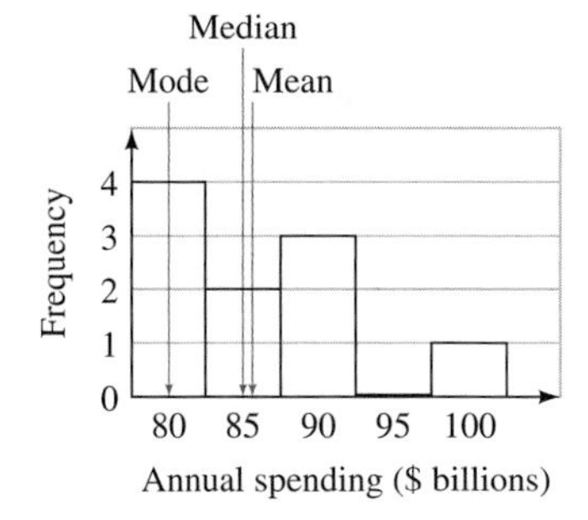

Figure 4

Before we go on . . . There is a nice geometric interpretation of the difference between the median and mean. The median line shown in the figure divides the total area of the histogram into two equal pieces, whereas the mean line passes through its "center of gravity"; if you placed the histogram on a knife edge along the mean line, it would balance.

Expected Value of a Random Variable

Now, instead of looking at a sample of values of a given random variable, let's look at the probability distribution of the random variable itself and see if we can predict the sample mean without actually taking a sample. This prediction is what we call the *expected value* of the random variable.

Example 2 • Expected Value of a Random Variable

Suppose you roll a fair die a large number of times. What do you expect to be the average of the numbers that face up?

---

[5]Spending figures are rounded and cover the years 1988–1997. SOURCES: Rand Youth Poll/Teenage Research Unlimited/*New York Times,* March 14, 1998, p. D1.

Solution Suppose we take a sample of n rolls of the die (where n is large). Because the probability of rolling a 1 is 1/6, we would expect that we would roll a 1 one-sixth of the time, or $n/6$ times. Similarly, each other number should also appear $n/6$ times. The frequency table should then look like this:

x	1	2	3	4	5	6
Number of Times x Is Rolled (frequency)	$\frac{n}{6}$	$\frac{n}{6}$	$\frac{n}{6}$	$\frac{n}{6}$	$\frac{n}{6}$	$\frac{n}{6}$

Note that we would not really expect the scores to be evenly distributed in practice, although for very large values of n we would expect the frequencies to vary only by a small percentage. To calculate the sample mean, we would add all the scores and divide by the sample size. Now, the table tells us that there are $n/6$ 1s, $n/6$ 2s, $n/6$ 3s, and so on, up to $n/6$ 6s. Adding all these gives

$$\sum_i x_i = \frac{n}{6} \cdot 1 + \frac{n}{6} \cdot 2 + \frac{n}{6} \cdot 3 + \frac{n}{6} \cdot 4 + \frac{n}{6} \cdot 5 + \frac{n}{6} \cdot 6$$

(Notice that we can obtain this number by multiplying the frequencies by the values of X and then adding.) So, the mean is

$$\bar{x} = \frac{\sum_i x_i}{n}$$

$$= \frac{\frac{n}{6} \cdot 1 + \frac{n}{6} \cdot 2 + \frac{n}{6} \cdot 3 + \frac{n}{6} \cdot 4 + \frac{n}{6} \cdot 5 + \frac{n}{6} \cdot 6}{n}$$

$$= \frac{1}{6} \cdot 1 + \frac{1}{6} \cdot 2 + \frac{1}{6} \cdot 3 + \frac{1}{6} \cdot 4 + \frac{1}{6} \cdot 5 + \frac{1}{6} \cdot 6 \qquad \text{Divide top and bottom by } n.$$

$$= 3.5$$

This is the average value we expect to get after a large number of rolls or, in short, the **expected value** of a roll of the die. More precisely, we say that this is the expected value of the random variable X whose value is the number we get by rolling a die. Notice that n, the number of rolls, does not appear in the expected value. In fact, we could redo the calculation more simply by dividing the frequencies in the table by n *before* adding. Doing this replaces the frequencies with the *probabilities,* 1/6; that is, it *replaces the frequency distribution with the probability distribution.*

x	1	2	3	4	5	6
$P(X = x)$	$\frac{1}{6}$	$\frac{1}{6}$	$\frac{1}{6}$	$\frac{1}{6}$	$\frac{1}{6}$	$\frac{1}{6}$

The expected value of X is then the sum of the products $x \cdot P(X = x)$. This is how we will compute it from now on.

Expected Value of a Random Variable

If X is a finite random variable that takes on the values $x_1, x_2, \ldots, x_n$, then the **expected value** of X, written $E(X)$ or μ, is

$$\mu = E(X) = x_1 \cdot P(X = x_1) + x_2 \cdot P(X = x_2) + \cdots + x_n \cdot P(X = x_n)$$
$$= \textstyle\sum_i x_i \cdot P(X = x_i)$$

We interpret the expected value of X as a prediction of the mean of a large random sample of measurements of X; in other words, it is what we "expect" the mean of a large number of scores to be. To compute the expected value from the probability distribution of X, we multiply the values of X by their probabilities and add the results.

Quick Example

If X has the distribution shown,

x	-1	0	4	5
$P(X = x)$	.3	.5	.1	.1

then $\mu = E(X) = -1(.3) + 0(.5) + 4(.1) + 5(.1) = -.3 + 0 + .4 + .5 = .6.$

Question Does this mean that the expected value of X is nothing more than the mean of a large random sample of measurements of X?

Answer Not exactly. Sample means—even the means of very large samples—may vary from sample to sample, so which one would we select for the expected value? What does happen is that the means of larger and larger random samples will get closer and closer to a particular number, just as the numbers 1.5, 1.05, 1.005, 1.0005, . . . get closer and closer to 1. It is that *limiting* value (1 in the case illustrated) that is the expected value.

Question Is this connected with population mean? If I take X to be the salary of a randomly selected worker, shouldn't the expected value of X just be the average salary—that is, the population mean of all salaries?

Answer Correct. Every population mean is the expected value of some random variable, as you have just illustrated: Start with any finite population and measure some quantity—salary, say—associated with each member. Now consider the experiment of choosing a random member of the population and take X to be the associated salary. The outcomes of the experiment are equally likely: You are equally likely to choose any one member as any other. Thus, the probability of any particular salary (value of X) is proportional to the number of members who earn that salary. The result is that the expected value of X is the same as the population mean of the salaries. In this way, we think of expected value as generalizing the concept of a population mean. That is why we use the same notation μ for both the expected value and the population mean.

This generalization is a strict one: Not every random variable X results from measuring some quantity associated with members of a finite population. For instance, if we roll a (possibly loaded) die, we cannot take the population to consist of the six outcomes because they will not be equally likely if the die is loaded. We might consider rolling it in all possible directions, on all possible surfaces, in all possible weather conditions, and so on, to cover the entire "population of rollings of the die." Such a population would be infinite, however, and the best you can do is calculate the mean of larger and larger samples, giving the expected value.

Question So, if an experiment consists of choosing a member of a population at random and if X is a certain numerical quantity associated with each member, then the expected value of X is just the population mean of those quantities?

Answer Correct, and we can think of all population means as expected values in this way.

Question What about the median and mode of a random variable?

Answer Although we do not use them in this text, these can be defined as follows. The **median** of a random variable X is the average of all values m of X such that $P(X \leq m) \geq 1/2$ and $P(X \geq m) \geq 1/2$. A **mode** of X is a value m such that $P(X = m)$ is largest.

Example 3 • Sports Injuries

According to historical data, the number of injuries that a member of the Enormous State University women's soccer team will sustain during a typical season is given by the following probability distribution table:

Injuries	0	1	2	3	4	5	6
Probability	.2	.2	.22	.2	.15	.01	.02

If X denotes the number of injuries sustained by a player during one season, compute $E(X)$ and interpret the result.

Solution We can compute the expected value using the following "spreadsheet" approach: Take the probability distribution table, add another row in which we compute the product $xP(X = x)$, and then add these products together:

x	0	1	2	3	4	5	6	
$P(X = x)$	.2	.2	.22	.2	.15	.01	.02	**Total**
$xP(X = x)$	0	.2	.44	.6	.6	.05	.12	2.01

The total of the entries in the bottom row is the expected value. Thus,

$$E(X) = 2.01$$

We interpret the result as follows: If many soccer players are observed for a season, the average number of injuries each will sustain is about two.

Graphing Calculator: Finding the Expected Value of a Probability Distribution

To obtain the expected value of a probability distribution on the TI-83, press STAT, select EDIT, and then press ENTER to obtain the LIST screen, which shows three columns labeled L_1, L_2, L_3. (Three more columns are off the screen.) Clear columns L_1 and L_2 if they are not already clear. (Select the heading of a column and press CLEAR ENTER to clear it.) Enter the values of X in the column under L_1 (pressing ENTER after each entry) and enter the probabilities in the column under L_2. Then, on the home screen, you can obtain the expected value as

`sum(L₁*L₂)` L_1 is 2nd 1; L_2 is 2nd 2; sum is found in 2nd STAT under MATH.

Excel: Finding the Expected Value of a Probability Distribution

As the method we used suggests, the calculation of the expected value from the probability distribution is particularly easy to do using a spreadsheet program. The following worksheet shows one way to do it. [The first two columns contain the probability distribution of X; the quantities $xP(X = x)$ are summed in cell C9.]

	A	B	C
1	x	P(X=x)	x*P(X=x)
2	0	0.2	=A2*B2
3	1	0.2	
4	2	0.22	
5	3	0.2	
6	4	0.15	
7	5	0.01	
8	6	0.02	
9			=SUM(C2:C8)

An alternative is to use the SUMPRODUCT function in Excel: Once we enter the first two columns above, the formula

```
=SUMPRODUCT(A2:A8,B2:B8)
```

computes the sum of the products of corresponding entries in the columns, giving us the expected value.

Example 4 • Roulette

A roulette wheel (of the kind used in the United States) has the numbers 1 through 36, 0 and 00. A bet on a single number pays 35 to 1. This means that if you place a \$1 bet on a single number and win (your number comes up), you get your \$1 back plus \$35 (that is, you gain \$35). If you lose, you lose the \$1 you bet. What is the expected gain from a \$1 bet on a single number?

Solution The probability of winning is 1/38, so the probability of losing is 37/38. Let X be the gain from a \$1 bet. X has two possible values: $X = -1$ if you lose, and $X = 35$ if you win. $P(X = -1) = 37/38$ and $P(X = 35) = 1/38$. This probability distribution and the calculation of the expected value are given in the following table:

x	-1	35	
$P(X = x)$	$\frac{37}{38}$	$\frac{1}{38}$	**Total**
$xP(X = x)$	$-\frac{37}{38}$	$\frac{35}{38}$	$-\frac{2}{38}$

So, we expect to average a small loss of $2/38 \approx \$0.0526$ on each spin of the wheel.

Before we go on . . . Of course, you cannot actually lose the expected \$0.0526 on one spin of the wheel. However, if you play many times, this is what you expect your *average* loss per bet to be. For example, if you played 100 times, you could expect to lose about $100 \times 0.0526 = \$5.26$.

A betting game in which the expected value is zero is called a **fair game.** For example, if you and I flip a coin and I give you \$1 each time it comes up heads but you give me \$1 each time it comes up tails, then the game is fair. Over the long run, we expect to come out even. On the other hand, a game like roulette in which the expected value is not zero is **biased** in favor of the house. Most casino games are slightly biased in favor of the house.[6] Thus, most gamblers will lose only a small amount, and many gamblers will actually win something (and return to play some more). However, when the earnings are averaged over the huge numbers of people playing, the house is guaranteed to come out ahead. This is how casinos make (lots of) money.

Expected Value of a Binomial Random Variable

Suppose you guess all the answers to the questions on a multiple-choice test. What score can you expect to get? This scenario is an example of a sequence of Bernoulli trials (see Section 8.2), and the number of correct guesses is therefore a binomial random variable whose expected value we wish to know. There is a simple formula for the expected value of a binomial random variable.

Mean of Binomial Random Variable

If X is the binomial random variable associated with n independent Bernoulli trials, each with probability p of success, then the expected value of X is

$$\mu = E(X) = np$$

Quick Example

If X is the number of successes in 20 Bernoulli trials with $p = .7$, then the expected number of successes is $\mu = E(X) = (20)(.7) = 14$.

Question Where does this formula come from?

Answer We *could* use the formula for expected value and compute the sum

$$E(X) = 0C(n, 0)p^0q^n + 1C(n, 1)p^1q^{n-1} + 2C(n, 2)p^2q^{n-2} + \cdots + nC(n, n)p^nq^0$$

directly (using the binomial theorem), but this is one of the many places in mathematics where a less direct approach is much easier. X is the number of successes in a sequence of n Bernoulli trials, each with probability p of success. Thus, p is the fraction of times we expect a success, so out of n trials we expect np successes. Since X counts successes, we expect the value of X to be np. (With a little more effort, this can be made into a formal proof that the sum above equals np.)

Example 5 • Guessing on an Exam

An exam has 50 multiple-choice questions, each having four choices. If a student randomly guesses on each question, how many correct answers can he expect to get?

Solution Each guess is a Bernoulli trial with probability of success 1 in 4, so $p = .25$. For a sequence of $n = 50$ trials,

$$\mu = E(X) = np = (50)(.25) = 12.5$$

The student can expect to get about 12.5 correct answers.

[6]Only rarely are games not biased in favor of the house. However, blackjack played without continuous shuffle machines can be beaten by card counting.

Question Wait a minute. How can a student get a fraction of a correct answer?

Answer Remember that the expected value is the *average* number of correct answers a student will get if he guesses on a large number of such tests. Or, we can say that if many students use this strategy of guessing, they will average about 12.5 correct answers each.

Using a Sample Mean to Estimate the Expected Value

It is not always possible to know the probability distribution of a random variable. For instance, if we take X to be the age of a randomly selected person on this planet, one could not be expected to know the probability distribution of X. However, we can still obtain a good *estimate* of the expected value of X (the population mean) by using the mean of a large random sample.

Example 6 • Estimating an Expected Value

The following table shows the (fictitious) incomes of a sample of 1000 U.S. lawyers in their first year out of law school:

Income Bracket ($)	20,000–29,999	30,000–39,999	40,000–49,999	50,000–59,999	60,000–69,999	70,000–79,999	80,000–89,999
Lawyers	20	80	230	400	170	70	30

Estimate the average of the incomes of all lawyers in their first year out of law school.

Solution We first interpret the question in terms of a random variable. Let X be the income of a lawyer selected at random from among all currently practicing first-year lawyers in the United States. We are given a sample of 1000 values of X and are asked to find the mean of X. It can be shown that, in a very precise sense, the best estimate of the mean of X is the sample mean if the sample is all that we know about X. We can calculate the sample mean just as we did in Example 2: by multiplying the incomes (using the midpoints of the income brackets) by the frequencies, then adding, and finally, dividing the sum by the sample size.

$$\bar{x} = \frac{\sum_i x_i}{n}$$

$$= \frac{(20 \cdot 25 + 80 \cdot 35 + 230 \cdot 45 + 400 \cdot 55 + 170 \cdot 65 + 70 \cdot 75 + 30 \cdot 85)1000}{1000}$$

$$= 54{,}500$$

(Alternatively, we could convert the table into a probability distribution table and then compute the associated expected value as we did in Example 3.) Thus, $E(X)$ is approximately \$54,500. That is, the average income of all currently practicing first-year lawyers in the United States is approximately \$54,500.

Guideline: Recognizing When to Compute the Mean and When to Compute the Expected Value

Question When am I supposed to compute the mean (add the values of X and divide by n), and when am I supposed to use the expected value formula?

Answer The formula for the mean (adding and dividing by the number of observations) is used to compute the mean of a sequence of random scores, or sampled values of X. If, on the other hand, you are given the probability distribution for X (even if it is only an estimated probability distribution), then you need to use the expected value formula.

8.3 EXERCISES

In Exercises 1–8, compute the mean, median, and mode of the data samples.

1. $-1, 5, 5, 7, 14$

2. $2, 6, 6, 7, -1$

3. $2, 5, 6, 7, -1, -1$

4. $3, 1, 6, -3, 0, 5$

5. $\frac{1}{2}, \frac{3}{2}, -4, \frac{5}{4}$

6. $-\frac{3}{2}, \frac{3}{8}, -1, \frac{5}{2}$

7. $2.5, -5.4, 4.1, -0.1, -0.1$

8. $4.2, -3.2, 0, 1.7, 0$

9. Give a sample of six scores with mean 1 and with median $\neq$ mean. (Arrange the scores in ascending order.)

10. Give a sample of five scores with mean 100 and median 1. (Arrange the scores in ascending order.)

In Exercises 11–16, calculate the expected value of X for the given probability distribution.

11.

x	0	1	2	3
$P(X = x)$	.5	.2	.2	.1

12.

x	1	2	3	4
$P(X = x)$	.1	.2	.5	.2

13.

x	10	20	30	40
$P(X = x)$	$\frac{15}{50}$	$\frac{20}{50}$	$\frac{10}{50}$	$\frac{5}{50}$

14.

x	2	4	6	8
$P(X = x)$	$\frac{1}{20}$	$\frac{15}{20}$	$\frac{2}{20}$	$\frac{2}{20}$

15.

x	-5	-1	0	2	5	10
$P(X = x)$	.2	.3	.2	.1	.2	0

16.

x	-20	-10	0	10	20	30
$P(X = x)$	.2	.4	.2	.1	0	.1

In Exercises 17–26, calculate the expected value of the given random variable X. [Exercises 21, 25, and 26 assume familiarity with counting arguments and probability (Section 7.4).]

17. X is the number of tails that come up when a coin is tossed twice.

18. X is the number of tails that come up when a coin is tossed three times.

19. X is the higher number when two dice are rolled.

20. X is the lower number when two dice are rolled.

21. X is the number of red marbles that Suzy has in her hand after she selects four marbles from a bag containing four red and two green marbles.

22. X is the number of green marbles that Suzy has in her hand after she selects four marbles from a bag containing three red and two green marbles.

23. Twenty darts are thrown at a dartboard. The probability of hitting a bull's-eye is .1. Let X be the number of bull's-eyes hit.

24. Thirty darts are thrown at a dartboard. The probability of hitting a bull's-eye is $\frac{1}{5}$. Let X be the number of bull's-eyes hit.

25. Select five cards without replacement from a standard deck of 52 and let X be the number of Queens you draw.

26. Select five cards without replacement from a standard deck of 52 and let X be the number of red cards you draw.

APPLICATIONS

27. SUVs—Tow Ratings Following is a sample of tow ratings (in pounds) for some popular 2000 model sport utility vehicles.

2000 2000 3000 4000 5000 6000 6000 6000 8000 8000

Compute the mean and median of the given sample. Fill in the blank: There were as many sampled SUVs with a tow rating of more than _____ pounds as there were with tow ratings below that.

Tow ratings vary considerably within each model. Figures cited are rounded. For more detailed information, consult http://www.rvsafety.com/towrate2k.htm.

28. Housing Prices Following is a sample of the percent increases in the price of a house from 1980 to 2001 in eight regions of the United States:

75 130 145 150 150 225 225 300

Compute the mean and median of the given sample. Fill in the blank: As many sampled regions in the United States had housing increases averaging more than ______% as below that.

Percentages are rounded to the nearest 25%. SOURCE: Third Quarter 2001 House Price Index, released November 30, 2001, by the Office of Federal Housing Enterprise Oversight. Online: http://www.ofheo.gov/house/3q01hpi.pdf.

29. Computer Sales Following are the percent growth rates in the number of computers sold worldwide for 1994–2000. (Worldwide sales were expected to reach 135 million computers in 2000).

20 25 20 15 15 25 20

Find the sample mean, median, and mode(s). What do your answers tell you about computer sales?

Figures were rounded to the nearest 5%, and the 2000 figures are projected. SOURCE: IDC/*New York Times*, October 9, 2000, p. C8.

30. Worker Productivity Following are worker productivity figures for the United States from the third quarter of 1998 through the second quarter of 2000:

110 111 112 112 113 116 116 118

Find the sample mean, median, and mode(s). What do your answers tell you about worker productivity?

Worker productivity per hour in nonfarm businesses, as a percentage of the 1992 productivity. SOURCE: Bureau of Labor Statistics/*New York Times*, September 7, 2000, p. C2.

31. Supermarkets A survey of U.S. supermarkets yielded the following relative frequency table, where X is the number of checkout lanes at a randomly chosen supermarket:

x	1	2	3	4	5	6	7	8	9	10
$P(X = x)$	.01	.04	.04	.08	.10	.15	.25	.20	.08	.05

SOURCES: J. T. McCalve, P. G. Benson, and T. Sincich, *Statistics for Business and Economics* (7th ed.) (Upper Saddle River, N.J.: Prentice Hall, 1998), 177; W. Chow et al., "A Model for Predicting a Supermarket's Annual Sales per Square Foot," Graduate School of Management, Rutgers University.

a. Compute $\mu = E(X)$ and interpret the result.
b. Which is larger, $P(X < \mu)$ or $P(X > \mu)$? Interpret the result.

32. Video Arcades Your company, Sonic Video, has conducted research that shows the following probability distribution, where X is the number of video arcades in a randomly chosen city with more than 500,000 inhabitants:

x	0	1	2	3	4	5	6	7	8	9
$P(X = x)$	.07	.09	.35	.25	.15	.03	.02	.02	.01	.01

a. Compute $\mu = E(X)$ and interpret the result.
b. Which is larger, $P(X < \mu)$ or $P(X > \mu)$? Interpret the result.

33. Pollen Count Allergy sufferers usually react to pollen counts above 30 grains/cubic meter. The following histogram shows the pollen counts in the New York Metropolitan region on 23 different days in 1999:

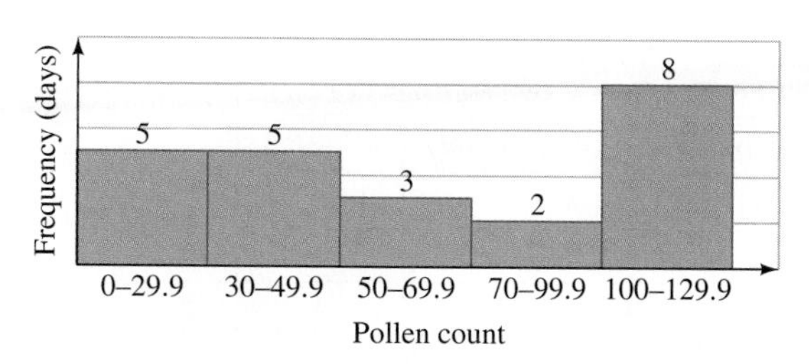

Upper limit of 129.9 was not set in original data. SOURCE: Department of Allergy and Immunology at Long Island College Hospital/*New York Times*, June 5, 1999, p. B1.

Use the estimated probability distribution based on the (rounded) midpoints of the given measurement classes to obtain an estimate of the expected pollen count (rounded to the nearest whole number) in the New York metropolitan region.

34. Pollen Count Repeat Exercise 33, using the following histogram of pollen counts in the New York metropolitan region on 20 different days in 1998:

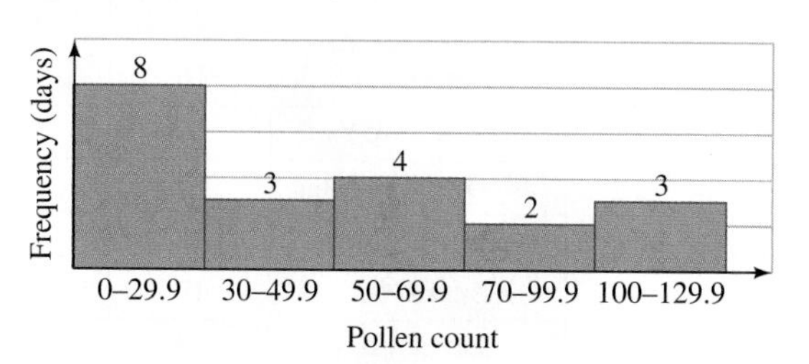

Upper limit of 129.9 was not set in original data. SOURCE: Department of Allergy and Immunology at Long Island College Hospital/*New York Times*, June 5, 1999, p. B1.

35. School Enrollment The following table shows the approximate numbers of schoolgoers in the United States (residents who attended some educational institution) in 1998, broken down by age group:

Age	3–6.9	7–12.9	13–16.9	17–22.9	23–26.9	27–42.5
Population (millions)	12	24	15	14	2	5

Data are approximate. SOURCE: *Statistical Abstract of the United States: 2000.*

Use the rounded midpoints of the given measurement classes to compute the probability distribution of the age X of a schoolgoer. (Round probabilities to two decimal places.) Hence, compute the expected value of X. What information does the expected value give about residents enrolled in schools?

36. School Enrollment Repeat Exercise 35, using the following data from 1980:

Age	3–6.9	7–12.9	13–16.9	17–22.9	23–26.9	27–42.5
Population (millions)	8	20	11	13	1	3

Data are approximate. SOURCE: *Statistical Abstract of the United States: 2000.*

37. Income Distribution The following table shows the distribution of income in the United States in 2000, based on a sample of 1000 U.S. households:

Income Bracket ($)	0–4999	5000–9999	10,000–14,999	15,000–24,999	25,000–34,999	35,000–49,999	50,000–74,999	75,000–99,999	100,000–219,999
Households	29	61	70	134	125	155	188	104	134

Based on the actual income distribution. Upper income limit is an estimate. SOURCE: U.S. Census Bureau, "Money Income in the United States: 2000," Current Population Reports, P60-213 (Washington, D.C.: Government Printing Office, 2001.)

Use this information to estimate, to the nearest $1000, the average household income in the United States.

38. Income Distribution Repeat Exercise 37, using the following data for Hispanic U.S. households:

Income Bracket ($)	0–4999	5000–9999	10,000–14,999	15,000–24,999	25,000–34,999	35,000–49,999	50,000–74,999	75,000–99,999	100,000–179,999
Households	33	73	83	182	147	177	174	74	57

Based on the actual income distribution. Upper income limit is an estimate. SOURCE: U.S. Census Bureau, "Money Income in the United States: 2000," Current Population Reports, P60-213 (Washington, D.C.: Government Printing Office, 2001.)

Highway Safety Exercises 39–42 are based on the following table, which shows crashworthiness ratings for several categories of motor vehicles. In all exercises let X be the crash-test rating of a small car, Y be the crash-test rating for a small SUV, and so on, as shown in the table:

	Number Tested	**Overall Frontal Crash-Test Rating**			
		3 (good)	**2 (acceptable)**	**1 (marginal)**	**0 (poor)**
Small Cars, X	16	1	11	2	2
Small SUVs, Y	10	1	4	4	1
Medium SUVs, Z	15	3	5	3	4
Passenger Vans, U	13	3	0	3	7
Midsize Cars, V	15	3	5	0	7
Large Cars, W	19	9	5	3	2

Ratings are by the Insurance Institute for Highway Safety. SOURCES: Stacy C. Davis and Lorena F. Truett, Oak Ridge National Laboratory, "An Analysis of the Impact of Sport Utility Vehicles in the United States," August 2000/Insurance Institute for Highway Safety. Online: http://www-cta.ornl.gov/Publications/Final SUV report.pdf http://www.highwaysafety.org/vehicle_ratings/.

39. Compute the probability distributions and expected values of X and Y. Based on the results, which of the two types of vehicle performed better in frontal crashes?

40. Compute the probability distributions and expected values of Z and V. Based on the results, which of the two types of vehicle performed better in frontal crashes?

41. Based on expected values, which of the following categories performed best in crash tests: small cars, midsize cars, or large cars?

42. Based on expected values, which of the following categories performed best in crash tests: small SUVs, medium SUVs, or passenger vans?

43. Roulette A roulette wheel has the numbers 1 through 36, 0, and 00. Half of the numbers from 1 through 36 are red, and a bet on red pays even money (that is, if you win you will get back your \$1 plus another \$1). How much do you expect to win with a \$1 bet on red?

44. Roulette A roulette wheel has the numbers 1 through 36, 0, and 00. A bet on two numbers pays 17 to 1 (that is, if one of the two numbers you bet comes up, you get back your \$1 plus another \$17). How much do you expect to win with a \$1 bet on two numbers?

45. Teenage Pastimes According to a study, the probability that a randomly selected teenager shopped at a mall at least once during a week was .63. How many teenagers in a randomly selected group of 40 would you expect to shop at a mall during the next week?

SOURCE: Rand Youth Poll/Teenage Research Unlimited/*New York Times*, March 14, 1998, p. D1.

46. Other Teenage Pastimes According to the study referred to in Exercise 45, the probability that a randomly selected teenager played a computer game at least once during a week was .48. How many teenageers in a randomly selected group of 30 would you expect to play a computer game during the next 7 days?

47. Manufacturing Your manufacturing plant produces air bags, and it is known that 10% of them are defective. A random collection of 20 air bags are tested.

a. How many of them would you expect to be defective?

b. In how large a sample would you expect to find 12 defective airbags?

48. Spiders Your pet tarantula, Spider, has a .12 probability of biting an acquaintance who comes into contact with him. Next week, you will be entertaining 20 friends (all of whom will come into contact with Spider).

a. How many guests should you expect Spider to bite?

b. At your last party, Spider bit 6 of your guests. Assuming that Spider bit the expected number of guests, how many guests did you have?

Exercises 49 and 50 assume familiarity with counting arguments and probability (Section 7.4).

49. Camping Kent's Tents has four red tents and three green tents in stock. Karin selects four of them at random. Let X be the number of red tents she selects. Give the probability distribution of X and find the expected number of red tents selected.

50. Camping Kent's Tents has five green knapsacks and four yellow ones in stock. Curt selects four of them at random. Let X be the number of green knapsacks he selects. Give the probability distribution of X, and find the expected number of green knapsacks selected.

51. Stock Portfolios You are required to choose between two stock portfolios, FastForward Funds and SolidState Securities. Stock analysts have constructed the following probability distributions for next year's rate of return for the two funds:

FastForward Funds

Rate of Return	−.4	−.3	−.2	−.1	0	.1	.2	.3	.4
Probability	.015	.025	.043	.132	.289	.323	.111	.043	.019

SolidState Securities

Rate of Return	−.4	−.3	−.2	−.1	0	.1	.2	.3	.4
Probability	.012	.023	.050	.131	.207	.330	.188	.043	.016

Which of the two funds gives the higher expected rate of return?

52. Risk Management Before making your final decision whether to invest in FastForward Funds or SolidState Securities (see Exercise 51), you consult your colleague in the risk-management department of your company. She informs you that, in the event of a stock market crash, the following probability distributions for next year's rate of return would apply:

FastForward Funds

Rate of Return	−.8	−.7	−.6	−.5	−.4	−.2	−.1	0	.1
Probability	.028	.033	.043	.233	.176	.230	.111	.044	.102

SolidState Securities

Rate of Return	−.8	−.7	−.6	−.5	−.4	−.2	−.1	0	.1
Probability	.033	.036	.038	.167	.176	.230	.211	.074	.035

Which of the two funds offers the lowest risk in case of a market crash?

53. Insurance Schemes Acme Insurance Company is launching a drive to generate greater profits, and it decides to insure racetrack drivers against wrecking their cars. The company's research shows that on average a racetrack driver races four times a year and has a 1 in 10 chance of wrecking a vehicle, worth an average of $100,000, in every race. The annual premium is $5000, and Acme automatically drops any driver who is involved in an accident (after paying for a new car) but does not refund the premium. How much profit (or loss) can the company expect to earn from a typical driver in 1 year? (*Hint:* Use a tree diagram to compute the probabilities of the various outcomes.)

54. Insurance Blue Sky Flight Insurance Company insures passengers against air disasters, charging a prospective passenger $20 for coverage on a single plane ride. In the event of a fatal air disaster, it pays out $100,000 to the named beneficiary. In the event of a nonfatal disaster, it pays out an average of $25,000 for hospital expenses. Given that the probability of a plane's crashing on a single trip is 0.000 000 87 and that a passenger involved in a plane crash has a .9 chance of being killed, determine the profit (or loss) per passenger that the insurance company expects to make on each trip. (*Hint:* Use a tree to compute the probabilities of the various outcomes.)

This was the probability of a passenger plane crashing per departure in 1990. SOURCE: National Transportation Safety Board.

COMMUNICATION AND REASONING EXERCISES

55. In a certain set of scores, as many values are above the mean as below it. It follows that
(A) The median and mean are equal.
(B) The mean and mode are equal.
(C) The mode and median are equal.
(D) The mean, mode, and median are all equal.

56. In a certain set of scores, the median occurs more often than any other score. It follows that
(A) The median and mean are equal.
(B) The mean and mode are equal.
(C) The mode and median are equal.
(D) The mean, mode, and median are all equal.

57. Your friend Charlesworth claims that the median of a collection of data is always close to the mean. Is he correct? If so, say why; if not, give an example to prove him wrong.

58. Your other friend Imogen asserts that Charlesworth is wrong and that it is the mode and the median that are always close to each other. Is she correct? If so, say why; if not, give an example to prove her wrong.

59. Must the expected number of times you hit a bull's-eye after 50 attempts always be a whole number? Explain.

60. Your statistics instructor tells you that the expected score of the upcoming midterm test is 75%. That means that 75% is the most likely score to occur, right? Explain.

61. Your grade in a recent midterm test was 80%, but the class average was 83%. Most people in the class scored better than you, right? Explain.

62. Your grade in a recent midterm test was 80%, but the class median was 100%. Your score was lower than the average score, right? Explain.

63. Slim tells you that the population mean is just the mean of a suitably large sample. Is he correct? Explain.

64. Explain how you can use a sample to estimate an expected value.

65. Sonia has just told you that the expected household income in the United States is the same as the population mean of all U.S. household incomes. Clarify her statement by describing an experiment and an associated random variable X so that the expected household income is the expected value of X.

66. If X is a random variable, what is the difference between a sample mean of measurements of X and the expected value of X? Illustrate by means of an example.

8.4 Measures of Dispersion

Variance and Standard Deviation of a Set of Scores

Your grade on a recent midterm test was 68%; the class average was 72%. How do you stand in comparison with the rest of the class? If the grades were widely scattered, then your grade may be close to the mean, and a fair number of people may have done a lot worse than you (Figure 5a). If, on the other hand, almost all the grades were within a few points of the average, then your grade may not be much higher than the lowest grade in the class (Figure 5b).

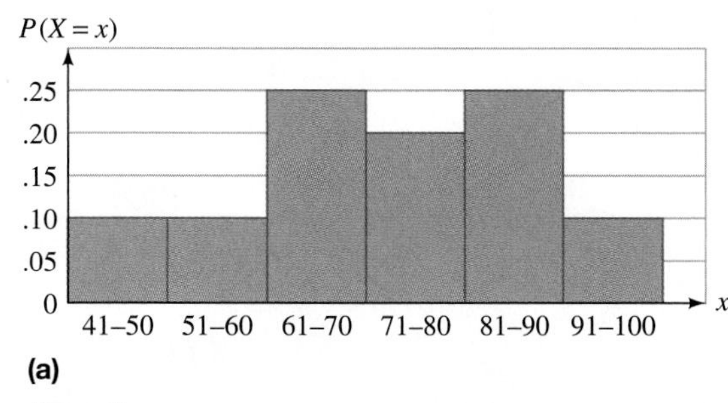

(a)

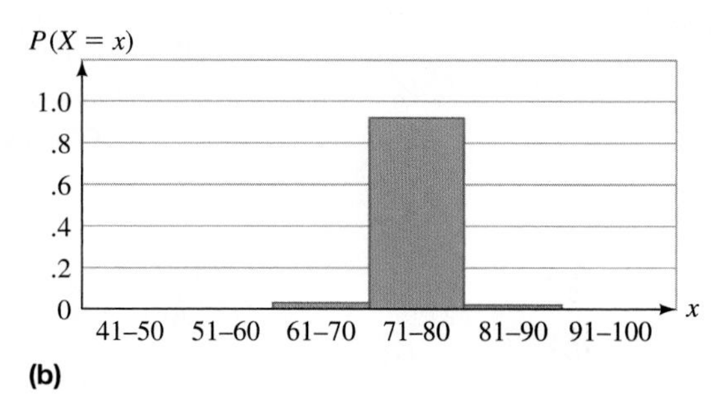

(b)

Figure 5

This scenario suggests that it would be useful to have a way of measuring not only the central tendency of a set of scores (mean, median, or mode) but also the amount of *scatter,* or *dispersion,* of the data.

Question Is there a way of measuring the dispersion of a set of scores such as the grades on an exam?

Answer If the scores are $x_1, x_2, \ldots, x_n$ and their mean is $\bar{x}$ (or μ in the case of a population mean), we are really interested in the distribution of the differences $x_i - \bar{x}$. We could compute the *average* of these differences, but this average will always be zero. (Why?) It is really the *sizes* of these differences that interests us, so we might try computing the average of the absolute values of the differences. This idea is reasonable, but it leads to technical difficulties that are avoided by a slightly different approach: The statistic we use is based on the average of the *squares* of the differences, as explained in the following definitions.

Sample Variance and Standard Deviation

The **sample variance** of the sample $x_1, x_2, \ldots, x_n$ of n values of X is given by

$$s^2 = \frac{(x_1 - \bar{x})^2 + (x_2 - \bar{x})^2 + \cdots + (x_n - \bar{x})^2}{n - 1} = \frac{\sum_{i=1}^{n} (x_i - \bar{x})^2}{n - 1}$$

The **sample standard deviation** is the square root of the sample variance:

$$s = \sqrt{s^2}$$

Population Variance and Standard Deviation

If the values $x_1, x_2, \ldots, x_n$ are all the measurements of X in the entire population, then the **population variance** is given by

$$\sigma^2 = \frac{(x_1 - \mu)^2 + (x_2 - \mu)^2 + \cdots + (x_n - \mu)^2}{n} = \frac{\sum_{i=1}^{n} (x_i - \mu)^2}{n}$$

(Remember that μ is the symbol we use for the *population* mean.) The **population standard deviation** is the square root of the population variance:

$$\sigma = \sqrt{\sigma^2}$$

Quick Examples

1. The sample variance of the scores 1, 2, 3, 4, 5 is the sum of the squares of the differences between the scores and the mean $\bar{x} = 3$, divided by $n - 1 = 4$:

$$s^2 = \frac{(1-3)^2 + (2-3)^2 + (3-3)^2 + (4-3)^2 + (5-3)^2}{4} = \frac{10}{4} = 2.5$$

$$s = \sqrt{2.5} \approx 1.58$$

2. The population variance of the scores 1, 2, 3, 4, 5 is the sum of the squares of the differences between the scores and the mean $\mu = 3$, divided by $n = 5$:

$$\sigma^2 = \frac{(1-3)^2 + (2-3)^2 + (3-3)^2 + (4-3)^2 + (5-3)^2}{5} = \frac{10}{5} = 2$$

$$\sigma = \sqrt{2} \approx 1.41$$

Question The *population* variance is the average of the squares of the differences between the values and the mean. But why do we divide by $n - 1$ instead of n when calculating the *sample* variance?

Answer In real-life applications, we would like the sample variance to approximate the population variance. We can interpret this to mean that we would like the expected value of the sample variance s^2 to be the same as the population variance σ^2. It turns out that the formula for s^2 given above is the formula that accomplishes this task. The sample variance s^2 as we have defined it is referred to by statisticians as the *unbiased estimator* of the population variance σ^2; if, instead, we divided by n in the formula for s^2, we would on average tend to underestimate the population variance. (See the online text on Sampling Distributions at the Web site for further discussion of unbiased estimators.) Note that, as the sample size gets larger and larger, the discrepancy between the formulas for s^2 and σ^2 becomes negligible; dividing by n gives almost the same answer as dividing by $n - 1$.

Example 1 • Wages

Following are the hourly worker compensation costs, in U.S. dollars, for five European countries:[7]

17 21 24 17 21

[7]1999 compensation costs in Italy, the Netherlands, Norway, Britain, and Sweden for production workers in manufacturing. SOURCE: *Report on the American Workforce,* U.S. Department of Labor, 2001, p. 200. Online: http://www.bls.gov/opub/rtaw/pdf/rtaw2001.pdf.

Compute the sample mean and standard deviation. What percentage of the scores fall within 1 standard deviation of the mean? What percentage fall within 2 standard deviations of the mean?

Solution The sample mean is

$$\bar{x} = \frac{\sum_i x_i}{n} = \frac{17 + 21 + 24 + 17 + 21}{5} = \frac{100}{5} = 20$$

The sample variance is

$$\begin{aligned} s^2 &= \frac{\sum_{i=1}^{n} (x_i - \bar{x})^2}{n - 1} \\ &= \frac{(17 - 20)^2 + (21 - 20)^2 + (24 - 20)^2 + (17 - 20)^2 + (21 - 20)^2}{4} \\ &= \frac{9 + 1 + 16 + 9 + 1}{4} = \frac{36}{4} = 9 \end{aligned}$$

Thus, the sample standard deviation is

$$s = \sqrt{9} = 3$$

To ask which scores fall "within 1 standard deviation of the mean" is to ask which scores fall in the interval $[\bar{x} - s, \bar{x} + s]$, or $[20 - 3, 20 + 3] = [17, 23]$. Four of the five scores fall in this interval, so the percentage of scores that fall within 1 standard deviation of the mean is $4/5 = 80\%$.

For 2 standard deviations of the mean, the interval in question is $[\bar{x} - 2s, \bar{x} + 2s] = [20 - 6, 20 + 6] = [14, 26]$, which includes all the scores. In other words, 100% of the scores fall within 2 standard deviations of the mean.

Question Tell me a quick way of calculating standard deviation on my graphing calculator or spreadsheet.

Answer

Graphing Calculator: Calculating a Standard Deviation

On the TI-83 enter the sample scores in list L_1 on the STAT `EDIT` screen, then go to STAT `CALC`, select `1-Var Stats`, and press ENTER. The resulting display shows, among other statistics, the sample standard deviation s as `Sx` as well as the population standard deviation σ as σ`x`.

Excel: Calculating a Standard Deviation

To compute the standard deviation of a collection of scores, set up your spreadsheet as follows:

	A
1	17
2	21
3	24
4	17
5	21
6	=STDEVA(A1:A5)

For the population standard deviation, use

`=STDEVP(A1:A5)` Population standard deviation

Question In Example 1, 80% of the scores fell within 1 standard deviation of the mean and all of them fell within 2 standard deviations of the mean. Is this typical?

Answer Actually, this is atypical due to the small number of scores involved. There are two useful methods for *estimating* the percentage of scores that fall within any number of standard deviations of the mean. The first method applies to any set of data and is due to P. L. Chebyshev (1821–1894), whereas the second applies to "nice" sets of data and is based on the normal distribution, which we discuss in Section 8.5.

Chebyshev's Rule

For any set of data, the following statements are true:

- At least 3/4 of the scores fall within 2 standard deviations of the mean (within the interval $[\bar{x} - 2s, \bar{x} + 2s]$ for samples or $[\mu - 2\sigma, \mu + 2\sigma]$ for populations).
- At least 8/9 of the scores fall within 3 standard deviations of the mean (within the interval $[\bar{x} - 3s, \bar{x} + 3s]$ for samples or $[\mu - 3\sigma, \mu + 3\sigma]$ for populations).
- At least 15/16 of the scores fall within 4 standard deviations of the mean (within the interval $[\bar{x} - 4s, \bar{x} + 4s]$ for samples or $[\mu - 4\sigma, \mu + 4\sigma]$ for populations).

⋮

- At least $1 - 1/k^2$ of the scores fall within k standard deviations of the mean (within the interval $[\bar{x} - ks, \bar{x} + ks]$ for samples or $[\mu - k\sigma, \mu + k\sigma]$ for populations).

*Empirical Rule**

For a set of data whose frequency distribution is "bell-shaped" and symmetric (as in Figure 6), the following is true:

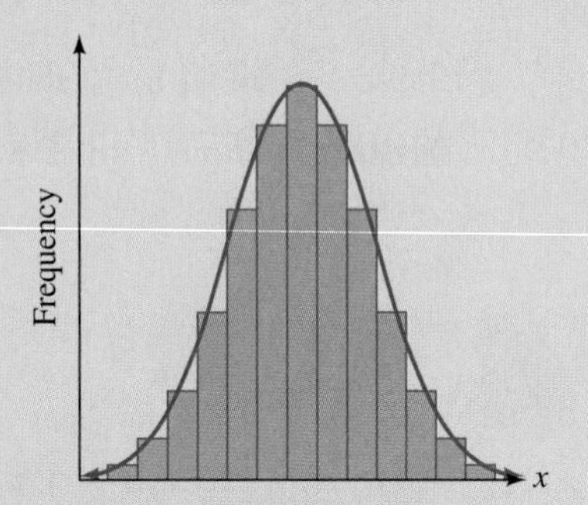

Figure 6 • Bell-shaped symmetric distribution

- Approximately 68% of the scores fall within 1 standard deviation of the mean (within the interval $[\bar{x} - s, \bar{x} + s]$ for samples or $[\mu - \sigma, \mu + \sigma]$ for populations).
- Approximately 95% of the scores fall within 2 standard deviations of the mean (within the interval $[\bar{x} - 2s, \bar{x} + 2s]$ for samples or $[\mu - 2\sigma, \mu + 2\sigma]$ for populations).
- Approximately 99.7% of the scores fall within 3 standard deviations of the mean (within the interval $[\bar{x} - 3s, \bar{x} + 3s]$ for samples or $[\mu - 3\sigma, \mu + 3\sigma]$ for populations). Thus,

Almost all the scores lie within 3 standard deviations of the mean.

---

*Unlike Chebyshev's rule, which is a precise theorem, the empirical rule is a rule of thumb that is intentionally vague about what exactly is meant by a "bell-shaped distribution" and "approximately such-and-such %." (As a result, the rule is often stated differently in different textbooks.) We will see in Section 8.5 that if the distribution is a *normal* one, the empirical rule translates to a precise statement.

Quick Examples

1. If the mean of a sample is 20 with standard deviation $s = 2$, then at least $15/16 = 93.75\%$ of the scores lie within 4 standard deviations of the mean—that is, in the interval $[12, 28]$.
2. If the mean of a sample with a bell-shaped symmetric distribution is 20 with standard deviation $s = 2$, then approximately 95% of the scores lie in the interval $[16, 24]$.

The empirical rule was not extremely accurate in Example 1. The distribution was not symmetric (sketch it to see for yourself), and the fact that there were only five scores limits the accuracy further. In practice, however, the empirical rule often gives good estimates even when the distribution is not bell-shaped or symmetric. Chebyshev's rule, on the other hand, is always valid but, as a consequence, tends to be "overcautious" and in practice underestimates the percentages.

Example 2 • Automobile Life

The average life span of a Batmobile® is 9 years, with a standard deviation of 2 years. My own Batmobile lasted less than 3 years before being condemned to the bat-junkyard.

a. Without any further knowledge about the distribution of Batmobile life spans, what can one say about the percentage of Batmobiles that last less than 3 years?

b. Refine the answer in part (a), assuming that the distribution of Batmobile life spans is bell-shaped and symmetric.

Solution

a. If we are given no further information about the distribution of Batmobile life spans, we need to use Chebyshev's rule. Since the life span of my Batmobile was more than 6 years, or 3 standard deviations, shorter than the mean, it lies outside the range $[\mu - 3\sigma, \mu + 3\sigma] = [3, 15]$. Since *at least* 8/9 of the life spans of all Batmobiles lie in this range, *at most* 1/9, or 11%, of the life spans lie outside this range (see Figure 7). Some of these, like the life span of my own Batmobile, are less than 3 years, whereas the rest are more than $\mu + 3\sigma = 15$ years.

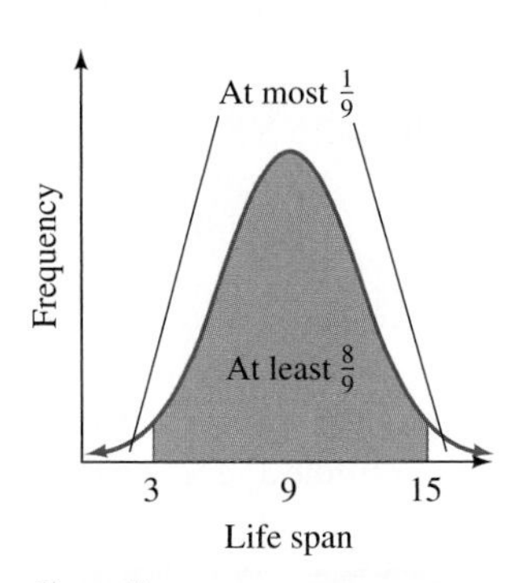

Figure 7

b. Since we know more about the distribution now than we did in part (a), we can use the empirical rule and obtain sharper results. The empirical rule predicts that approximately 99.7% of the life spans of Batmobiles lie in the range $[\mu - 3\sigma, \mu + 3\sigma] = [3, 15]$. Thus, approximately $1 - 99.7\% = 0.3\%$ of them lie outside that range. Since the distribution is symmetric, however, more can be said: Half of that 0.3%, or 0.15% of Batmobiles, will last longer than 15 years, whereas the other 0.15% are, like my own ill-fated Batmobile, doomed to a life span of less than 3 years (see Figure 8).

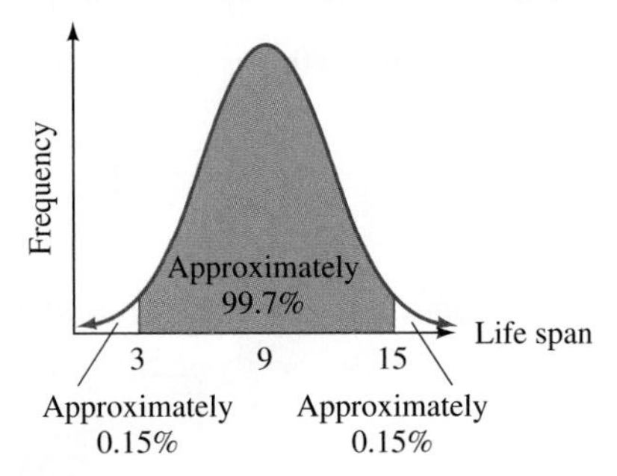

Figure 8

Variance and Standard Deviation of a Random Variable

We calculated the population variance by taking the mean of the quantities $(x_i - \mu)^2$. The x_i are the values of a random variable X. Thus, the population variance of a set of data can be written as the mean of all the values of $(X - \mu)^2$. But, for a population, the mean of all the values of $(X - \mu)^2$ is the expected value of $(X - \mu)^2$, which we can write as $E[(X - \mu)^2]$. In general, we make the following definition.

Variance and Standard Deviation of a Random Variable

If X is a finite random variable taking on values $x_1, x_2, \ldots, x_n$, then the **variance** of X is

$$\begin{aligned}\sigma^2 &= E[(X - \mu)^2] \\ &= (x_1 - \mu)^2 P(X = x_1) + (x_2 - \mu)^2 P(X = x_2) + \cdots + (x_n - \mu)^2 P(X = x_n) \\ &= \textstyle\sum_i (x_i - \mu)^2 P(X = x_i)\end{aligned}$$

The **standard deviation** of X is then the square root of the variance:

$$\sigma = \sqrt{\sigma^2}$$

To compute the variance from the probability distribution of X, first compute the expected value μ and then compute the expected value of $(X - \mu)^2$.

Quick Example

The following distribution has expected value $\mu = E(X) = 2$:

x	−1	2	3	10
P(X = x)	.3	.5	.1	.1

The variance of X is

$$\begin{aligned}\sigma^2 &= (x_1 - \mu)^2 P(X = x_1) + (x_2 - \mu)^2 P(X = x_2) + \cdots + (x_n - \mu)^2 P(X = x_n) \\ &= (-1 - 2)^2(.3) + (2 - 2)^2(.5) + (3 - 2)^2(.1) + (10 - 2)^2(.1) = 9.2\end{aligned}$$

The standard deviation of X is

$$\sigma = \sqrt{9.2} \approx 3.03$$

We can calculate the variance and standard deviation of a random variable using a "spreadsheet approach," just as we calculated the expected value in Example 3 (Section 8.3).

Example 3 • Variance of a Random Variable

Compute the variance and standard deviation for the following probability distribution:

x	10	20	30	40	50	60
P(X = x)	.2	.2	.3	.1	.1	.1

Solution We first compute the expected value, μ, in the usual way:

x	10	20	30	40	50	60	
P(X = x)	.2	.2	.3	.1	.1	.1	
xP(X = x)	2	4	9	4	5	6	$\mu = 30$

Next, we add an extra three rows:

- A row for the differences $(x - \mu)$, which we get by subtracting μ from the values of X
- A row for the squares $(x - \mu)^2$, which we obtain by squaring the values immediately above
- A row for the products $(x - \mu)^2 P(X = x)$, which we obtain by multiplying the values in the second and the fifth rows

x	10	20	30	40	50	60	
$P(X = x)$	.2	.2	.3	.1	.1	.1	
$xP(X = x)$	2	4	9	4	5	6	$\mu = 30$
$x - \mu$	−20	−10	0	10	20	30	
$(x - \mu)^2$	400	100	0	100	400	900	
$(x - \mu)^2 P(X = x)$	80	20	0	10	40	90	$\sigma^2 = 240$

The sum of the values in the last row is the variance. The standard deviation is then the square root of the variance:

$$\sigma = \sqrt{240} \approx 15.49$$

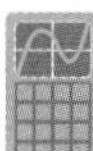

Graphing Calculator: Computing Variance

As in Example 3 in Section 8.3, begin by entering the probability distribution of X into columns `L1` and `L2` in the `LIST` screen (press STAT and select `EDIT`). Then, on the home screen, enter

`sum(L1*L2) →M` Stores the value of μ as M.

To obtain the variance, enter the following:

`sum((L1-M)^2*L2)` Computation of $\sum(x - \mu)^2 P(X = x)$

Excel: Computing Variance

As in Example 3 in Section 8.3, begin by entering the probability distribution into columns A and B and then proceed as shown:

	A	B	C	D
1	x	P(X=x)	x*P(X=x)	(x-Mu)^2 * P(X=x)
2	10	0.2	=A2*B2	=(A2-$C$8)^2*B2
3	20	0.2		
4	30	0.3		
5	40	0.1		
6	50	0.1		
7	60	0.1		
8			=SUM(C2:C7)	=SUM(D2:D7)
			Expected value	Variance

The variance then appears in cell D8.

Question Can we apply Chebyshev's rule and the empirical rule to interpret the standard deviation of a random variable?

Answer Yes. Example 4 illustrates the application of these rules to a random variable.

Example 4 • Internet Commerce

Your newly launched company, CyberPromo, sells computer games on the Internet. Statistical research indicates that the life span of an Internet marketing company such as yours is symmetrically distributed with an expected value of 30 months and standard deviation of 4 months. Complete the following sentence:

There is (at least/at most/approximately) _____ a _____% chance that CyberPromo will still be around for more than 3 years.

Solution Do we use Chebyshev's rule or the empirical rule? Since the empirical rule requires that the distribution be both symmetric and bell-shaped—not just symmetric—we cannot conclude that it applies here, so we must use Chebyshev's rule instead.

Let X be the life span of an Internet commerce site. The expected value of X is 30 months, and the hoped-for life span of CyberPromo is 36 months, which is 6 months, or $6/4 = 1.5$ standard deviations above the mean. Chebyshev's rule tells us that X is within $k = 1.5$ standard deviations of the mean at least $1 - 1/k^2$ of the time; that is,

$$P(24 \le X \le 36) \ge 1 - \frac{1}{k^2} = 1 - \frac{1}{1.5^2} \approx .56$$

In other words, at least 56% of all Internet marketing companies have life spans in the range 24–36 months. Thus, *at most* 44% have life spans outside this range. Since the distribution is symmetric, at most 22% have life spans longer than 36 months. We can complete the sentence as follows:

There is at most a 22% chance that CyberPromo will still be around for more than 3 years.

Variance and Standard Deviation of a Binomial Random Variable

We saw that there is an easy formula for the expected value of a binomial random variable: $\mu = np$, where n is the number of trials and p is the probability of success. Similarly, there is a simple formula for the variance and standard deviation.

Variance and Standard Deviation of a Binomial Random Variable

If X is a binomial random variable associated with n independent Bernoulli trials each with probability p of success, then the variance and standard deviation of X are given by

$$\sigma^2 = npq \quad \text{and} \quad \sigma = \sqrt{npq}$$

where $q = 1 - p$ is the probability of failure.

Quick Example

If X is the number of successes in 20 Bernoulli trials with $p = .7$, then the standard deviation is $\sigma = \sqrt{npq} = \sqrt{(20)(.7)(.3)} \approx 2.05$.

Note A binomial distribution is bell-shaped and (nearly) symmetric. Hence, the empirical rule applies.*

*Remember that the empirical rule only gives an *estimate* of probabilities. The larger the number of trials n, the more accurate the estimate will be. In Section 8.5 we give a more accurate approximation that takes into account the fact that the binomial distribution is not continuous.

Example 5 • Internet Commerce

You have calculated that there is a 20% chance that a hit on your Web page results in a fee paid to your company CyberPromo. Your Web page receives 25 hits per day. Let X be the number of hits that result in payment of the fee ("successful hits").

a. What are the expected value and standard deviation of X?

b. Complete the following:

On approximately 95 out of 100 days, I will get between ____ and ____ successful hits.

Solution

a. The random variable X is binomial with $n = 25$ and $p = .2$. To compute μ and σ, we use the formulas

$$\mu = np = (25)(.2) = 5 \text{ successful hits}$$

$$\sigma = \sqrt{npq} = \sqrt{(25)(.2)(.8)} = 2 \text{ hits}$$

b. Since the distribution is symmetric and bell-shaped, we can use the empirical rule, which tells us that there is approximately a 95% probability that the number of successful hits is within 2 standard deviations of the mean—that is, in the interval

$$[\mu - 2\sigma, \mu + 2\sigma] = [5 - 2(2), 5 + 2(2)] = [1, 9]$$

Thus, on approximately 95 out of 100 days, I will get between 1 and 9 successful hits.

Guideline: Recognizing When to Use the Empirical Rule or Chebyshev's Rule

Question How do I decide whether to use Chebyshev's rule or the empirical rule?

Answer Check to see whether the probability distribution you are considering is both symmetric and bell-shaped. If so, you can use the empirical rule. If not, then you must use Chebyshev's rule. For instance, if the distribution is symmetric but not known to be bell-shaped, you must use Chebyshev's rule.

8.4 EXERCISES

In Exercises 1–8, compute the (sample) variance and standard deviation of the data samples. (You calculated the means in the last exercise set. Round all answers to two decimal places.)

1. $-1, 5, 5, 7, 14$

2. $2, 6, 6, 7, -1$

3. $2, 5, 6, 7, -1, -1$

4. $3, 1, 6, -3, 0, 5$

5. $\frac{1}{2}, \frac{3}{2}, -4, \frac{5}{4}$

6. $-\frac{3}{2}, \frac{3}{8}, -1, \frac{5}{2}$

7. $2.5, -5.4, 4.1, -0.1, -0.1$

8. $4.2, -3.2, 0, 1.7, 0$

In Exercises 9–14, calculate the standard deviation of X for each probability distribution. (You calculated the expected values in the last exercise set. Round all answers to two decimal places.)

9.

x	0	1	2	3
$P(X = x)$	.5	.2	.2	.1

10.

x	1	2	3	4
$P(X = x)$	.1	.2	.5	.2

11.

x	10	20	30	40
$P(X = x)$	$\frac{3}{10}$	$\frac{2}{5}$	$\frac{1}{5}$	$\frac{1}{10}$

12.

x	2	4	6	8
$P(X = x)$	$\frac{1}{20}$	$\frac{15}{20}$	$\frac{2}{20}$	$\frac{2}{20}$

13.

x	−5	−1	0	2	5	10
$P(X = x)$	.2	.3	.2	.1	.2	0

14.

x	−20	−10	0	10	20	30
$P(X = x)$	.2	.4	.2	.1	0	.1

In Exercises 15–22, calculate the expected value, the variance, and the standard deviation of the given random variable X. (You calculated the expected values in the last exercise set. Round all answers to two decimal places.)

15. X is the number of tails that come up when a coin is tossed twice.

16. X is the number of tails that come up when a coin is tossed three times.

17. X is the highest number when two dice are rolled.

18. X is the lowest number when two dice are rolled.

19. X is the number of red marbles that Suzy has in her hand after she selects four marbles from a bag containing four red and two green marbles.

20. X is the number of green marbles that Suzy has in her hand after she selects four marbles from a bag containing three red and two green marbles.

21. Twenty darts are thrown at a dartboard. The probability of hitting a bull's-eye is .1. Let X be the number of bull's-eyes hit.

22. Thirty darts are thrown at a dartboard. The probability of hitting a bull's-eye is $\frac{1}{5}$. Let X be the number of bull's-eyes hit.

APPLICATIONS

23. Popularity Ratings In your bid to be elected class representative, you have your election committee survey five randomly chosen students in your class and ask them to rank you on a scale of 0–10. Your rankings are 3, 2, 0, 9, 1.

a. Find the sample mean and standard deviation. (Round your answers to two decimal places.)

b. Assuming that the sample mean and standard deviation are indicative of the class as a whole, in what range does the empirical rule predict that approximately 68% of the class will rank you? What other assumptions must we make to use the rule?

24. Popularity Ratings Your candidacy for elected class representative is being opposed by Slick Sally. Your election committee has surveyed six of the students in your class and had them rank Sally on a scale of 0–10. The rankings were 2, 8, 7, 10, 5, 8.

a. Find the sample mean and standard deviation. (Round your answers to two decimal places.)

b. Assuming that the sample mean and standard deviation are indicative of the class as a whole, in what range does the empirical rule predict that approximately 95% of the class will rank Sally? What other assumptions must we make to use the rule?

25. SUVs—Tow Ratings Following is a sample of tow ratings (in pounds) for some popular 2000 sport utility vehicles (SUVs):

2000 2000 3000 4000 5000 6000 6000 6000 8000 8000

a. Compute the mean and standard deviation of the given sample. (Round your answers to the nearest whole number.)

b. Assuming that the distribution of tow ratings for all popular SUVs is symmetric and bell-shaped, 68% of all SUVs have tow ratings between ____ and ____. What is the percentage of scores in the sample that fall in this range?

Tow ratings vary considerably within each model. Figures cited are rounded. For more detailed information, consult http://www.rvsafety.com/towrate2k.htm.

26. Housing Prices Following is a sample of the percent increases in the price of a house from 1980 to 2001 in eight regions of the United States:

75 125 150 150 150 225 225 300

a. Compute the mean and standard deviation of the given sample. (Round answers to the nearest whole number.)

b. Assuming that the distribution of percent housing price increases for all regions is symmetric and bell-shaped, 68% of all regions in the United States reported housing increases between ____ and ____. What is the percentage of scores in the sample that fall in this range?

Percentages are rounded to the nearest 25%. SOURCE: Third Quarter 2001 House Price Index, released November 30, 2001 by the Office of Federal Housing Enterprise Oversight. Online: http://www.ofheo.gov/house/3q01hpi.pdf.

27. SUVs Following are highway-driving gas mileages of a selection of medium-sized sport utility vehicles (SUVs):

17 18 17 18 21 16 21 18 16 14 15 22 17 19 17 18

a. Find the sample standard deviation (rounded to two decimal places).

b. In what gas-mileage range does Chebyshev's inequality predict that at least 8/9 (approximately 89%) of the selection will fall?

c. What is the actual percentage of SUV models of the sample that fall in the range predicted in part (b)? Which gives the more accurate prediction of this percentage: Chebyshev's rule or the empirical rule?

Figures are the low end of ranges for 1999 models tested. SOURCES: Stacy C. Davis and Lorena F. Truett, Oak Ridge National Laboratory, "An Analysis of the Impact of Sport Utility Vehicles in the United States," August 2000/Insurance Institute for Highway Safety. Online: http://www-cta.ornl.gov/Publications/Final SUV report.pdf.

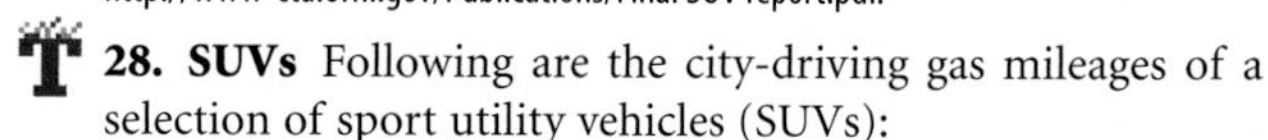

28. SUVs Following are the city-driving gas mileages of a selection of sport utility vehicles (SUVs):

14 15 14 15 13 16 12 14 19 18 16 16 12 15 15 13

a. Find the sample standard deviation (rounded to two decimal places).

b. In what gas-mileage range does Chebyshev's inequality predict that at least 75% of the selection will fall?

c. What is the actual percentage of SUV models of the sample that fall in the range predicted in part (b)? Which gives the more accurate prediction of this percentage: Chebyshev's rule or the empirical rule?

Figures are the low end of ranges for 1999 models tested. SOURCES: Stacy C. Davis and Lorena F. Truett, Oak Ridge National Laboratory, "An Analysis of the Impact of Sport Utility Vehicles in the United States," August 2000/Insurance Institute for Highway Safety. Online: http://www-cta.ornl.gov/Publications/Final SUV report.pdf.

29. Shopping Malls A survey of all the shopping malls in your region yields the following probability distribution, where X is the number of movie theater screens in a selected mall:

Movie Screens	0	1	2	3	4
Probability	.4	.1	.2	.2	.1

Compute the expected value μ and the standard deviation σ of X. (Round answers to two decimal places.) What percentage of malls have a number of movie theater screens within 2 standard deviations of μ?

30. Pastimes A survey of all the students in your school yields the following probability distribution, where X is the number of movies that a selected student has seen in the past week:

Movies	0	1	2	3	4
Probability	.5	.1	.2	.1	.1

Compute the expected value μ and the standard deviation σ of X. (Round answers to two decimal places.) For what percentage of students is X within 2 standard deviations of μ?

31. Income Distribution The following table shows the number of U.S. households at various income levels in 2000, based on a (fictitious) population of 100,000 households:

2000 Income ($ thousands)	10	25	42	66	142
Households (thousands)	20	20	20	20	20

Incomes are rounded mean income per quintile. SOURCE: U.S. Census Bureau, Current Population Survey, selected March supplements as collected in "Money Income in the United States: 2000" (Washington, D.C.: Government Printing Office, 2001), 60–213.

Compute the expected value μ and the standard deviation σ of the associated random variable X. If we define a low-income family as one whose income is more than 1 standard deviation below the mean and a high-income family as one whose income is at least 1 standard deviation above the mean, what is the income gap between high- and low-income families in the United States? (Round your answers to the nearest $1000.)

32. Income Distribution Repeat Exercise 31 using the following data for 1970:

1970 Income ($ thousands)*	7.8	21	34	47	84
Households (thousands)	20	20	20	20	20

*Incomes given in constant 2000 dollars. SOURCE: See source for Exercise 31.

33. School Enrollment The following table shows the approximate numbers of schoolgoers in the United States (residents who attended some educational institution) in 1998, broken down by age group:

Age	3–6.9	7–12.9	13–16.9	17–22.9	23–26.9	27–42.5
Population (millions)	12	24	15	14	2	5

Data are approximate. SOURCE: *Statistical Abstract of the United States: 2000.*

a. Use the rounded midpoints of the given measurement classes to compute the expected value and the standard deviation of the age X of a schoolgoer. (Round all probabilities to two decimal places.)

b. In what age interval does the empirical rule predict that 68% of all schoolgoers will fall? Estimate the actual percentage in that interval using your probability distribution.

34. School Enrollment Repeat Exercise 33, using the following data from 1980:

Age	3–6.9	7–12.9	13–16.9	17–22.9	23–26.9	27–42.5
Population (millions)	8	20	11	13	1	3

Data are approximate. SOURCE: *Statistical Abstract of the United States: 2000.*

35. Commerce You have been told that the average life span of an Internet-based company is 2 years, with a standard deviation of 0.15 year. Further, the associated distribution is highly skewed (not symmetric). Your Internet company is now 2.6 years old. What percentage of all Internet-based companies have enjoyed a life span at least as long as yours? Your answer should contain one of the following phrases: at least; at most; approximately.

36. Commerce You have been told that the average life span of a car-compounding service is 3 years, with a standard deviation of 0.2 year. Further, the associated distribution is symmetric but not bell-shaped. Your car-compounding service is exactly 2.6 years old. What fraction of car-compounding services last at most as long as yours? Your answer should contain one of the following phrases: at least; at most; approximately.

37. Batmobiles The average life span of a Batmobile is 9 years, with a standard deviation of 2 years. (See Example 2.) Further, the probability distribution of the life spans of Batmobiles is symmetric but not known to be bell-shaped.

Since my old Batmobile has been sold as bat-scrap, I have decided to purchase a new one. According to the above information, there is (A) at least, (B) at most, or (C) approximately a ____ % chance that my new Batmobile will last 13 years or more.

38. Spiderman Coupés The average life span of a Spiderman® Coupé is 8 years, with a standard deviation of 2 years. Further, the probability distribution of the life spans of Spiderman Coupés is not known to be bell-shaped or symmetric. I have just purchased a brand-new Spiderman Coupé. According to the above information, there is (A) at least, (B) at most, or (C) approximately a ____% chance that my new Spiderman Coupé will last for less than 4 years.

39. Teenage Marketing In 2000, 22% of all teenagers in the United States had checking accounts. Your bank, TeenChex Inc., is interested in targeting teenagers who do not already have a checking account.

a. If TeenChex selects a random sample of 1000 teenagers, what number of teenagers *without* checking accounts can it expect to find? What is the standard deviation of this number? (Round the standard deviation to one decimal place.)

b. Fill in the missing quantities: There is approximately a 95% chance that between ____ and ____ teenagers in the sample will not have checking accounts. (Round answers to the nearest whole number.)

SOURCE: Teenage Research Unlimited, http://www.teenresearch.com, January 25, 2001.

40. Teenage Marketing In 2000, 18% of all teenagers in the United States owned stocks or bonds. Your brokerage company, TeenStox Inc., is interested in targeting teenagers who do not already own stocks or bonds.

a. If TeenStox selects a random sample of 2000 teenagers, what number of teenagers who do *not* own stocks or bonds can it expect to find? What is the standard deviation of this number? (Round the standard deviation to one decimal place.)

b. Fill in the missing quantities: There is approximately a 99.7% chance that between ____ and ____ teenagers in the sample will not own stocks or bonds. (Round answers to the nearest whole number.)

SOURCE: Teenage Research Unlimited, http://www.teenresearch.com, January 25, 2001.

41. Teenage Pastimes According to a study, the probability that a randomly selected teenager shopped at a mall at least once during a week was .63. Let X be the number of teenagers in a randomly selected group of 40 that will shop at a mall during the next week.

a. Compute the expected value and standard deviation of X. (Round answers to two decimal places.)

b. Fill in the missing quantity: There is approximately a 2.5% chance that ____ or more students in the group will shop at a mall during the next week.

SOURCES: Rand Youth Poll/Teenage Research Unlimited/*New York Times*, March 14, 1998, p. D1.

42. Other Teenage Pastimes According to the study referred to in Exercise 41, the probability that a randomly selected teenager played a computer game at least once during a week was .48. Let X be the number of teenagers in a randomly selected group of 30 who will play a computer game during the next 7 days.

a. Compute the expected value and standard deviation of X. (Round answers to two decimal places.)

b. Fill in the missing quantity: There is approximately a 16% chance that ____ or more teenagers in the group will play a computer game during the next 7 days.

T **43. Supermarkets** A survey of U.S. supermarkets yielded the following relative frequency table, where X is the number of checkout lanes at a randomly chosen supermarket:

x	1	2	3	4	5	6	7	8	9	10
$P(X = x)$	.01	.04	.04	.08	.10	.15	.25	.20	.08	.05

SOURCES: J. T. McCalve, P. G. Benson, and T. Sincich, *Statistics for Business and Economics,* 7th ed. (Upper Saddle River, N.J.: Prentice Hall, 1998), 177; W. Chow et al. "A Model for Predicting a Supermarket's Annual Sales per Square Foot," Graduate School of Management, Rutgers University.

a. Compute the mean, variance, and standard deviation (accurate to one decimal place).

b. As financial planning manager at Express Lane Mart, you wish to install a number of checkout lanes that is in the range of at least 75% of all supermarkets. What is this range according to Chebyshev's inequality? What is the *least* number of checkout lanes you should install so as to fall within this range?

T **44. Video Arcades** Your company, Sonic Video, has conducted research that shows the following probability distribution, where X is the number of video arcades in a randomly chosen city with more than 500,000 inhabitants:

x	0	1	2	3	4	5	6	7	8	9
$P(X = x)$	.07	.09	.35	.25	.15	.03	.02	.02	.01	.01

a. Compute the mean, variance, and standard deviation (accurate to one decimal place).

b. As CEO of Startrooper Video Unlimited, you wish to install a chain of video arcades in Sleepy City. City council regulations require that the number of arcades be within the range shared by at least 75% of all cities. What is this range? What is the *largest* number of video arcades you should install so as to comply with this regulation?

Distribution of Wealth If we model after-tax household income by a normal distribution, then the figures of a 1995 study imply the information in the following table, which should be used for Exercises 45–56. Assume that the distribution of incomes in each country is bell-shaped and symmetric.

	U.S.	Canada	Switzerland	Germany	Sweden
Mean Household Income ($)	38,000	35,000	39,000	34,000	32,000
Standard Deviation ($)	21,000	17,000	16,000	14,000	11,000

The data are rounded to the nearest $1000 and based on a report published by the Luxembourg Income Study. The report shows after-tax income, including government benefits (such as food stamps) of households with children. Our figures were obtained from the published data by assuming a normal distribution of incomes. All data were based on constant 1991 U.S. dollars and converted foreign currencies (adjusted for differences in buying power). SOURCE: Luxembourg Income Study/*New York Times,* August 14, 1995, p. A9.

45. If we define a poor household as one whose after-tax income is at least 1.3 standard deviations below the mean, what is the household income of a poor family in the United States?

46. If we define a poor household as one whose after-tax income is at least 1.3 standard deviations below the mean, what is the household income of a poor family in Switzerland?

47. If we define a rich household as one whose after-tax income is at least 1.3 standard deviations above the mean, what is the household income of a rich family in the United States?

48. If we define a rich household as one whose after-tax income is at least 1.3 standard deviations above the mean, what is the household income of a rich family in Sweden?

49. Refer to Exercise 45. Which of the five countries listed has the poorest households (that is, the lowest cutoff for considering a household poor)?

50. Refer to Exercise 48. Which of the five countries listed has the wealthiest households (that is, the highest cutoff for considering a household rich)?

51. Which of the five countries listed has the largest gap between rich and poor?

52. Which of the five countries listed has the smallest gap between rich and poor?

53. What percentage of U.S. families earned an after-tax income of $17,000 or less?

54. What percentage of U.S. families earned an after-tax income of $80,000 or more?

55. What is the after-tax income range of approximately 99.7% of all Germans?

56. What is the after-tax income range of approximately 99.7% of all Swedes?

T **Aging** Exercises 57–62 are based on the following list, which shows the percentage of the aging population (residents of age 65 years and older) in each of the 50 states in 1990 and 2000:

2000

6, 9, 10, 10, 10, 11, 11, 11, 11, 11,
11, 11, 12, 12, 12, 12, 12, 12, 12, 12,
12, 12, 12, 13, 13, 13, 13, 13, 13, 13,
13, 13, 13, 13, 13, 13, 13, 14, 14, 14,
14, 14, 14, 14, 15, 15, 15, 15, 16, 18

1990

4, 9, 10, 10, 10, 10, 10, 11, 11, 11,
11, 11, 11, 11, 11, 12, 12, 12, 12, 12,
12, 13, 13, 13, 13, 13, 13, 13, 13, 13,
13, 13, 13, 13, 13, 14, 14, 14, 14, 14,
14, 14, 14, 15, 15, 15, 15, 15, 15, 18

Percentages are rounded and listed in ascending order. SOURCE: U.S. Census Bureau, Census 2000 Summary File 1. Online: http://www.census.gov/prod/2001pubs/c2kbr01-10.pdf.

57. Compute the population mean and standard deviation for the 2000 data.

58. Compute the population mean and standard deviation for the 1990 data.

59. Compare the actual percentage of states whose aging population in 2000 was within 1 standard deviation of the mean to the percentage predicted by the empirical rule. Comment on your answer.

60. Compare the actual percentage of states whose aging population in 1990 was within 1 standard deviation of the mean to the percentage predicted by the empirical rule. Comment on your answer.

61. What was the actual percentage of states whose aging population in 1990 was within 2 standard deviations of the mean? Is Chebyshev's rule valid? Explain.

62. What was the actual percentage of states whose aging population in 2000 was within 2 standard deviations of the mean? Is Chebyshev's rule valid? Explain.

COMMUNICATION AND REASONING EXERCISES

63. Which is greater: the sample standard deviation or the population standard deviation? Explain.

64. Suppose you take larger and larger samples of a given population. Would you expect the sample and population standard deviations to get closer or farther apart? Explain.

65. In one Finite Math class, the average grade was 75, and the standard deviation of the grades was 5. In another Finite Math class, the average grade was 65, and the standard deviation of the grades was 20. What conclusions can you draw about the distributions of the grades in each class?

66. You are a manager in a precision-manufacturing firm, and you must evaluate the performance of two employees. You do so by examining the quality of the parts they produce. One particular item should be 50.0 ± 0.3 millimeters (mm) long to be usable. The first employee produces parts that are an average of 50.1 mm long with a standard deviation of 0.15 mm. The second employee produces parts that are an average of 50.0 mm long with a standard deviation of 0.4 mm. Which employee do you rate higher? Why?

67. If a finite random variable has an expected value of 10 and a standard deviation of 0, what is its probability distribution?

68. If the values of X in a population consist of an equal number of 1s and -1s, what is its standard deviation?

69. Find an algebraic formula for the population standard deviation of a sample $\{x, y\}$ of two scores $(x \le y)$.

70. Find an algebraic formula for the sample standard deviation of a sample $\{x, y\}$ of two scores $(x \le y)$.

8.5 *Normal Distributions*

Continuous Random Variables

Figure 9 shows the probability distributions for the number of successes in sequences of 10 and 15 independent Bernoulli trials, each with probability of success $p = .5$.

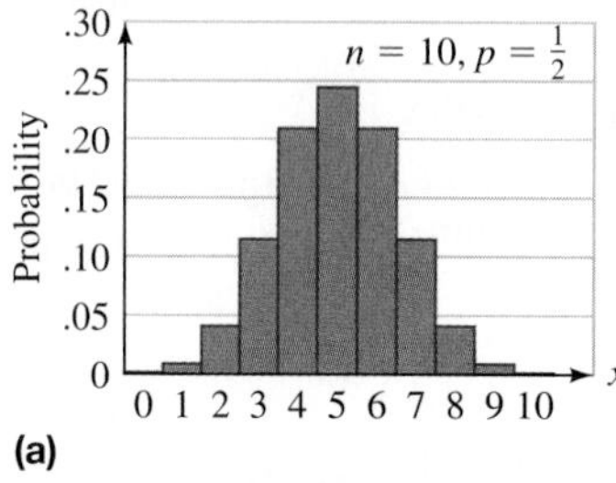

(a)

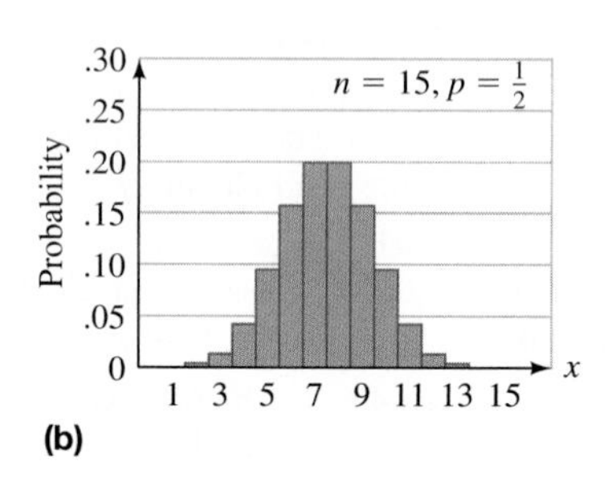

(b)

Figure 9

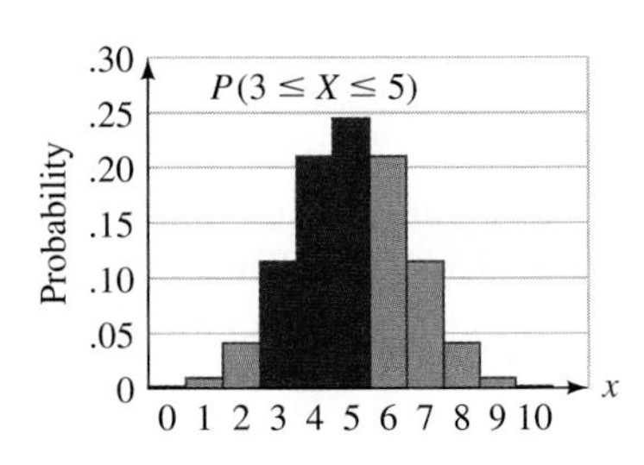

Figure 10

Since each column is 1 unit wide, its area is numerically equal to its height. Thus, the area of each rectangle can be interpreted as a probability. For example, in Figure 9a the area of the rectangle over $x = 3$ represents $P(X = 3)$. If we want to find $P(3 \leq X \leq 5)$, we can add up the areas of the three rectangles over 3, 4 and 5, shown shaded in Figure 10. Notice that, if we add up the areas of *all* the rectangles, the total is 1 because $P(0 \leq X \leq 10) = 1$. We can summarize these observations.

Properties of the Probability Distribution Histogram

In a probability distribution histogram where each column is 1 unit wide,

- The total area enclosed by the histogram is 1 square unit.
- $P(a \leq X \leq b)$ is the area enclosed by the rectangles lying between $x = a$ and $x = b$.

This discussion is motivation for considering another kind of random variable, one whose probability distribution is specified not by a bar graph as above but by the graph of a function.

Continuous Random Variable and Probability Density Function

A **continuous random variable** X may take on any real value whatsoever. The probabilities $P(a \leq X \leq b)$ are specified by means of a **probability density function,** a function whose graph lies above the x axis with the total area between the graph and the x axis being 1. The probability $P(a \leq X \leq b)$ is given by the area enclosed by the curve, the x axis, and the lines $x = a$ and $x = b$ (see Figure 11).

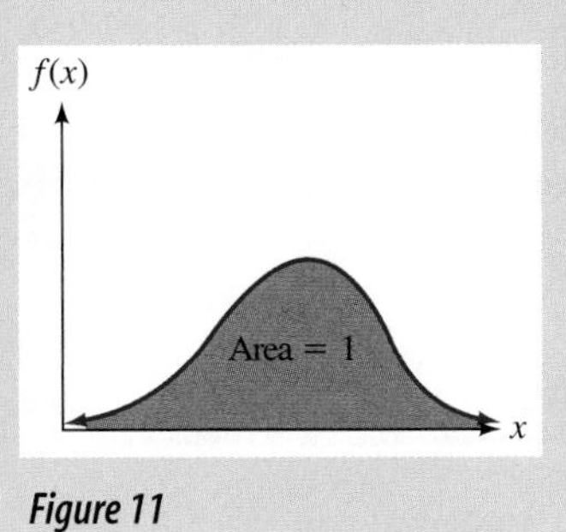

Figure 11

Question What about probabilities that are specified with strict inequalities, such as $P(a \leq X < b)$?

Answer Let's first look at a simpler case and compare $P(X \leq b)$ with $P(X < b)$. Their difference, $P(X \leq b) - P(X < b)$, is equal to $P(X = b)$. But $P(X = b) = P(b \leq X \leq b)$ is the area under the curve between the lines $x = b$ and $x = b$—no area at all! Thus, $P(X = b) = 0$, and so $P(X \leq b) = P(X < b)$. When we are measuring probabilities associated with a continuous random variable, we can be as sloppy as we like with inequalities. For example,

$$P(a \leq X \leq b) = P(a < X \leq b) = P(a \leq X < b) = P(a < X < b)$$

Normal Density Functions

Among all the possible probability density functions, there is an important class of functions called **normal density functions,** or **normal distributions.** The graph of a normal density function is bell-shaped and symmetric, as the figure below shows. The formula for a normal density function is rather complicated looking:

$$f(x) = \frac{1}{\sigma\sqrt{2\pi}} e^{-(x-\mu)^2/(2\sigma^2)}$$

The quantity μ is called the **mean** and can be any real number. The quantity σ is called the **standard deviation** and can be any positive real number. The number $e = 2.7182\ldots$ is a useful constant that shows up many places in mathematics, much as the constant π does. Finally, the constant $1/(\sigma\sqrt{2\pi})$ makes the total area come out to be 1. Statisticians rarely use the actual formula in computations; instead, they use tables or technology.

Normal Density Function and Normal Distribution

A **normal density function,** or **normal distribution,** is a function of the form

$$f(x) = \frac{1}{\sigma\sqrt{2\pi}} e^{-(x-\mu)^2/(2\sigma^2)}$$

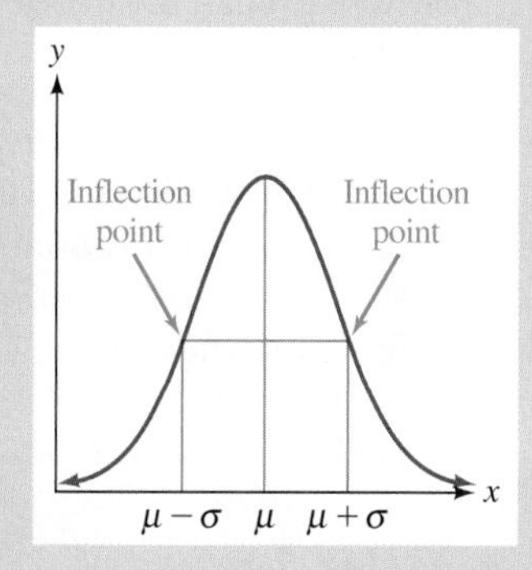

where μ is the mean and σ is the standard deviation. The *inflection points* are the points where the curve changes from bending in one direction to bending in another.*

*Pretend you are driving along the curve in a car. Then the points of inflection are the points where you change the direction in which you are steering (from left to right or right to left).

Figure 12 shows the graph of several normal density functions. The third of these has mean 0 and standard deviation 1 and is called the **standard normal distribution.** We use Z rather than X to refer to the standard normal variable.

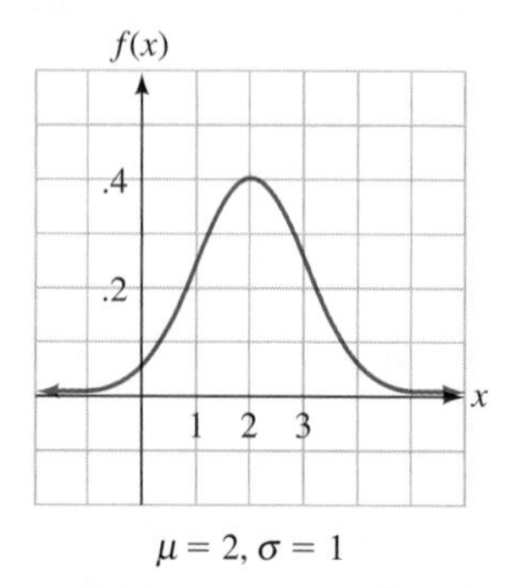

$\mu = 2, \sigma = 1$

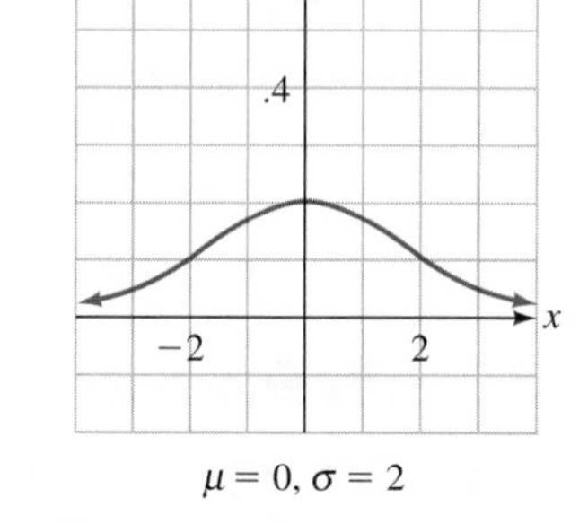

$\mu = 0, \sigma = 2$

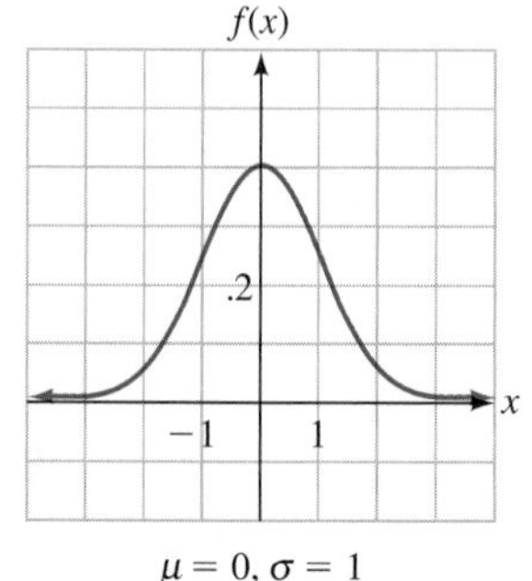

$\mu = 0, \sigma = 1$
Standard normal distribution

Figure 12 • Normal distributions

Using the Standard Normal Distribution to Calculate Probabilities

The standard normal distribution has $\mu = 0$ and $\sigma = 1$. The corresponding variable is called the **standard normal variable,** which we always denote by Z. Recall that, to calculate the probability $P(a \leq Z \leq b)$, we need to find the area under the distribution curve between the vertical lines $Z = a$ and $Z = b$. We can use the table in Appendix B to look up these areas or use technology. Here is an example.

Example 1 • Standard Normal Distribution

Let Z be the standard normal variable. Calculate the following probabilities.

a. $P(0 \leq Z \leq 2.4)$

b. $P(0 \leq Z \leq 2.43)$

c. $P(-1.37 \leq Z \leq 2.43)$

d. $P(1.37 \leq Z \leq 2.43)$

Solution

Figure 13

a. We are asking for the shaded area under the standard normal curve shown in Figure 13. We can find this area correct to four decimal places by looking at the table in the appendix, which lists the area under the standard normal curve from $z = 0$ to $z = b$ for any value of b between 0 and 3.09. To use the table, write 2.4 as 2.40 and read the entry in the row labeled 2.4 and the column labeled .00. $(2.4 + .00 = 2.40)$ Here is the relevant portion of the table:

Z	.00	.01	.02	.03
2.3	.4893	.4896	.4898	.4901
2.4	.4918	.4920	.4922	.4925
2.5	.4938	.4940	.4941	.4943

Thus, $P(0 \leq Z \leq 2.4) = .4918$.

b. The area we require can be read from the same portion of the table shown above. Write 2.43 as $2.4 + .03$ and read the entry in the row labeled 2.4 and the column labeled .03:

Z	.00	.01	.02	.03
2.3	.4893	.4896	.4898	.4901
2.4	.4918	.4920	.4922	.4925
2.5	.4938	.4940	.4941	.4943

Thus, $P(0 \leq Z \leq 2.43) = .4925$.

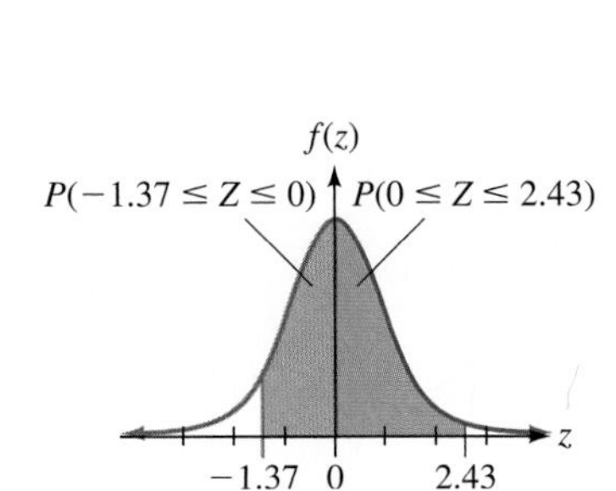

Figure 14

c. Here we cannot use the table directly because the range $-1.37 \leq Z \leq 2.43$ does not start at 0. But we can break the area up into two smaller areas that start or end at 0:

$$P(-1.37 \leq Z \leq 2.43) = P(-1.37 \leq Z \leq 0) + P(0 \leq Z \leq 2.43)$$

In terms of the graph, we are splitting the desired area into two smaller areas (Figure 14).

We already calculated the area of the right-hand piece in part (b):

$$P(0 \le Z \le 2.43) = .4925$$

For the left-hand piece, the symmetry of the normal curve tells us that

$$P(-1.37 \le Z \le 0) = P(0 \le Z \le 1.37)$$

This we can find on the table. Look at the row labeled 1.3 and the column labeled .07, and read

$$P(-1.37 \le Z \le 0) = P(0 \le Z \le 1.37) = .4147$$

Thus,

$$\begin{aligned} P(-1.37 \le Z \le 2.43) &= P(-1.37 \le Z \le 0) + P(0 \le Z \le 2.43) \\ &= .4147 + .4925 \\ &= .9072 \end{aligned}$$

d. The range $1.37 \le Z \le 2.43$ does not contain 0, so we cannot use the technique of part (c). Instead, the corresponding area can be computed as the *difference* of two areas:

$$\begin{aligned} P(1.37 \le Z \le 2.43) &= P(0 \le Z \le 2.43) - P(0 \le Z \le 1.37) \\ &= .4925 - .4147 \\ &= .0778 \end{aligned}$$

Graphing Calculator: Calculating Standard Normal Distribution

Many calculators permit you to calculate the area under the standard normal curve without using a table. On the TI-83 press [2nd] [VARS] to obtain the selection of distribution functions. The first function, `normalpdf`, gives the values of the normal density function (whose graph is the normal curve). The second, `normalcdf`, gives $P(a \le Z \le b)$. For example, to compute $P(0 \le Z \le 2.43)$, enter

```
normalcdf(0, 2.43)
```

To compute $P(-1.37 \le Z \le 2.43)$, enter

```
normalcdf(-1.37, 2.43)
```

Excel: Calculating Standard Normal Distribution

The function NORMSDIST (Normal Standard Distribution) gives the area shown on the left in Figure 15. (Tables give the area shown on the right.)

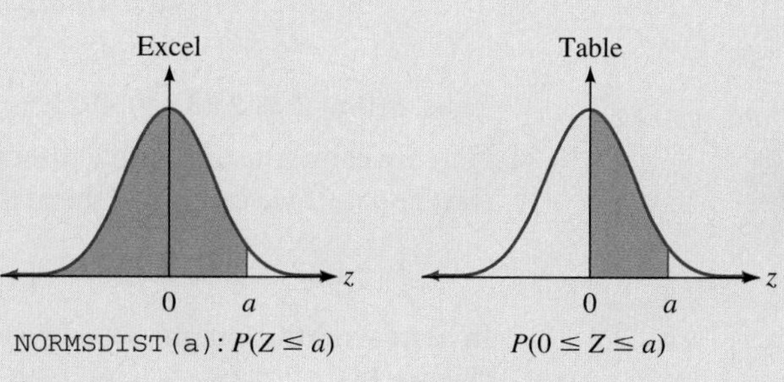

Figure 15

To compute a general area, $P(a \leq Z \leq b)$, subtract the cumulative area to a from that to b:

```
=NORMSDIST(b)-NORMSDIST(a)
```
$P(a \leq Z \leq b)$

In particular, to compute $P(0 \leq Z \leq 2.43)$, use

```
=NORMSDIST(2.43)-NORMSDIST(0)
```

and to compute $P(-1.37 \leq Z \leq 2.43)$, use

```
=NORMSDIST(2.43)-NORMSDIST(-1.37)
```

Question What about the area under *nonstandard* normal distributions? For example, if $\mu = 2$ and $\sigma = 3$, then how would we calculate $P(.5 \leq X \leq 3.2)$?

Answer We can use the following conversion formula.

Standardizing a Normal Distribution

If X has a normal distribution with mean μ and standard deviation σ, and if Z is the standard normal variable, then

$$P(a \leq X \leq b) = P\left(\frac{a-\mu}{\sigma} \leq Z \leq \frac{b-\mu}{\sigma}\right)$$

Quick Example

If $\mu = 2$ and $\sigma = 3$, then

$$P(0.5 \leq X \leq 3.2) = P\left(\frac{.5-2}{3} \leq Z \leq \frac{3.2-2}{3}\right)$$

$$= P(-0.5 \leq Z \leq 0.4) = .1915 + .1554 = .3469$$

Question Where does the standardizing formula come from?

Answer To justify it completely requires more mathematics than we will discuss here. Here is the main idea: If X is normal with mean μ and standard deviation σ, then $X - \mu$ is normal with mean 0 and standard deviation still σ, and $(X - \mu)/\sigma$ is normal with mean 0 and standard deviation 1. In other words, $(X - \mu)/\sigma = Z$. Therefore,

$$P(a \leq X \leq b) = P\left(\frac{a-\mu}{\sigma} \leq \frac{X-\mu}{\sigma} \leq \frac{b-\mu}{\sigma}\right) = P\left(\frac{a-\mu}{\sigma} \leq Z \leq \frac{b-\mu}{\sigma}\right)$$

Example 2 • Quality Control

Pressure gauges manufactured by Precision Corp. must be checked for accuracy before being placed on the market. To test a pressure gauge, a worker uses it to measure the pressure of a sample of compressed air known to be at a pressure of exactly 50 pounds per square inch. If the gauge reading is off by more than 1% (0.5 pound), it is rejected. Assuming that the reading of a pressure gauge under these circumstances is a normal random variable with mean 50 and standard deviation 0.4, find the percentage of gauges rejected.

Solution If X is the reading of the gauge, then X has a normal distribution with $\mu = 50$ and $\sigma = 0.4$. We are asking for $P(X < 49.5 \text{ or } X > 50.5) = 1 - P(49.5 \leq X \leq 50.5)$. We calculate

$$
\begin{aligned}
P(49.5 \leq X \leq 50.5) &= P\left(\frac{49.5 - 50}{0.4} \leq Z \leq \frac{50.5 - 50}{0.4}\right) && \text{Standardize.} \\
&= P(-1.25 \leq Z \leq 1.25) \\
&= 2 \cdot P(0 \leq Z \leq 1.25) \\
&= 2(.3944) = .7888
\end{aligned}
$$

So,

$$
\begin{aligned}
P(X < 49.5 \text{ or } X > 50.5) &= 1 - P(49.5 \leq X \leq 50.5) \\
&= 1 - .7888 = .2112
\end{aligned}
$$

In other words, about 21% of the gauges will be rejected.

Graphing Calculator: Standardizing a Normal Distribution

On the TI-83, the built-in normalcdf function permits us to compute $P(a \leq X \leq b)$ for nonstandard normal distributions as well. The format is

```
normalcdf(a,b,μ,σ)
```
$P(a \leq X \leq b)$

For example, we can compute $P(49.5 \leq X \leq 50.5)$ by entering

```
normalcdf(49.5,50.5,50,0.4)
```

Excel: Standardizing a Normal Distribution

We use the function NORMDIST instead of NORMSDIST. Its format is similar to NORMSDIST but includes extra arguments as shown:

```
=NORMDIST(a,μ,σ,1)
```
$P(X \leq a)$

(The last argument, set to 1, tells Excel that you want the cumulative distribution.) To compute $P(a \leq X \leq b)$, we enter the following in any vacant cell:

```
=NORMDIST(b,μ,σ,1)-NORMDIST(a,μ,σ,1)
```
$P(a \leq X \leq b)$

For example, we can compute $P(49.5 \leq X \leq 50.5)$ by entering

```
=NORMDIST(50.5,50,0.4,1)-NORMDIST(49.5,50,0.4,1)
```

Web Site: Standardizing a Normal Distribution

At the Web site, follow the path

Web site → Online Utilities → Normal Distribution Utility

This utility is pretty intuitive to use: Just enter the values of the mean, standard deviation, a and/or b and select `Compute`. You will also find a link to the normal distribution table if you wish to do computations by hand.

In many applications we need to know the probability that a value of a normal random variable will lie within 1 standard deviation of the mean, within 2 standard deviations, or within some number of standard deviations. To compute these probabilities, we first notice that if X has a normal distribution with mean μ and standard deviation σ, then

$$P(\mu - k\sigma \leq X \leq \mu + k\sigma) = P(-k \leq Z \leq k)$$

by the standardizing formula. We can compute these probabilities for various values of k using the table in Appendix B, and we obtain the following results.

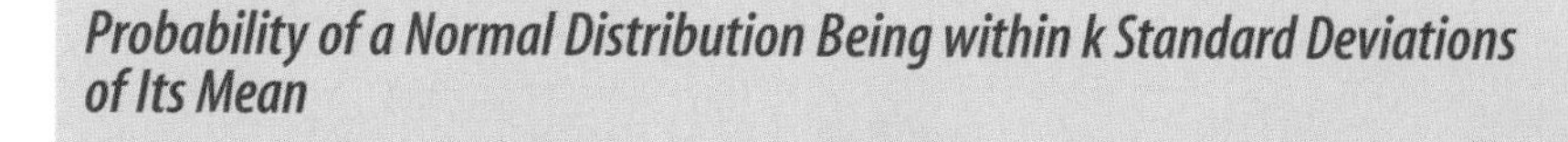

Probability of a Normal Distribution Being within k Standard Deviations of Its Mean

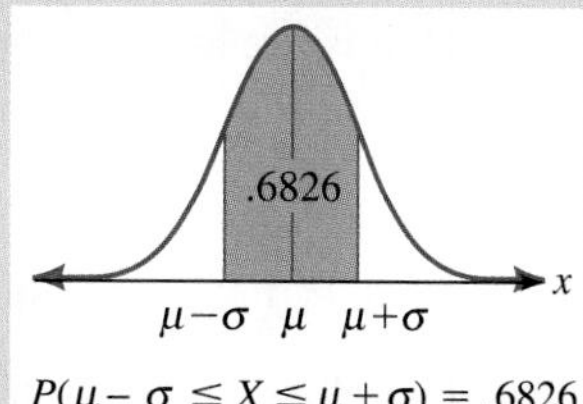

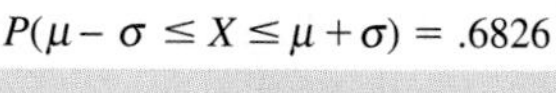

$P(\mu - \sigma \leq X \leq \mu + \sigma) = .6826$

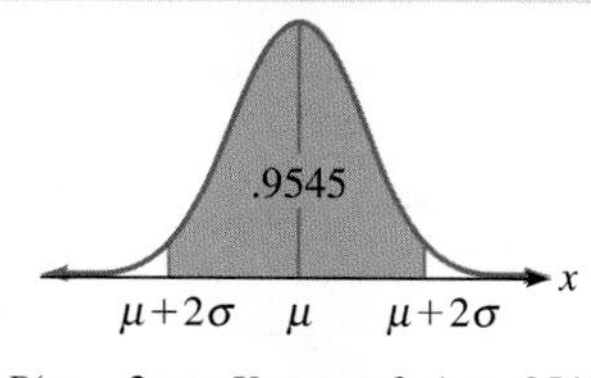

$P(\mu - 2\sigma \leq X \leq \mu + 2\sigma) = .9545$

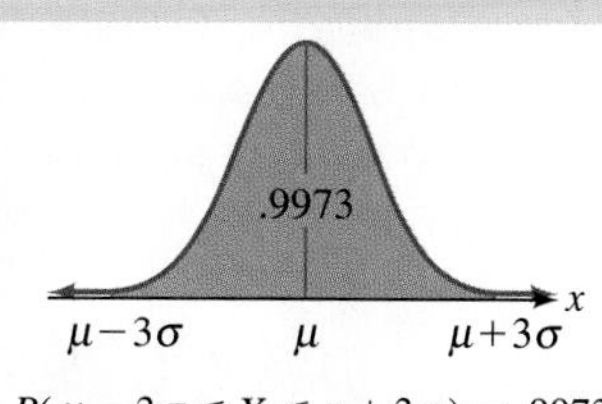

$P(\mu - 3\sigma \leq X \leq \mu + 3\sigma) = .9973$

Now you can see where the empirical rule comes from! The probabilities above are a good deal larger than the lower bounds given by Chebyshev's rule. Chebyshev's rule must work for distributions that are skewed or any shape whatsoever.

Example 3 • Loans

The values of mortgage loans made by a certain bank one year were normally distributed with a mean of \$120,000 and a standard deviation of \$40,000.

a. What is the probability that a randomly selected mortgage loan was in the range \$40,000–200,000?

b. You would like to state in your annual report that 50% of all mortgage loans were in a certain range. What is that range?

Solution

a. We are asking for the probability that a loan was within 2 standard deviations (\$80,000) of the mean. By the calculation done above, this probability is .9545.

b. We look for the k such that

$$P(120{,}000 - k \cdot 40{,}000 \leq X \leq 120{,}000 + k \cdot 40{,}000) = .5$$

Since

$$P(120{,}000 - k \cdot 40{,}000 \leq X \leq 120{,}000 + k \cdot 40{,}000) = P(-k \leq Z \leq k)$$

we look in the appendix to see for which k we have

$$P(0 \leq Z \leq k) = .25$$

That is, we look *inside* the table to see where .25 is, and find the corresponding k. We find

$$P(0 \leq Z \leq 0.67) = .2486$$

$$P(0 \leq Z \leq 0.68) = .2517$$

Therefore, the k we want is about halfway between 0.67 and 0.68, call it 0.675. This tells us that 50% of all mortgage loans were in the range

$$120{,}000 - 0.675 \cdot 40{,}000 = \$93{,}000 \quad \text{to} \quad 120{,}000 + 0.675 \cdot 40{,}000 = \$147{,}000$$

You might have noticed that the histograms of binomial distributions we have drawn (for example, those in Figure 9) have a very rough bell shape. In fact, in many cases it is possible to draw a normal curve that closely approximates a given binomial distribution.

Normal Approximation to a Binomial Distribution

If X is the number of successes in a sequence of n independent Bernoulli trials, with probability p of success in each trial, and if the range of values of X within 3 standard deviations of the mean lies entirely within the range 0 to n (the possible values of X), then

$$P(a \leq X \leq b) \approx P(a - 0.5 \leq Y \leq b + 0.5)$$

where Y has a normal distribution with the same mean and standard deviation as X; that is, $\mu = np$ and $\sigma = \sqrt{np(1-p)}$.

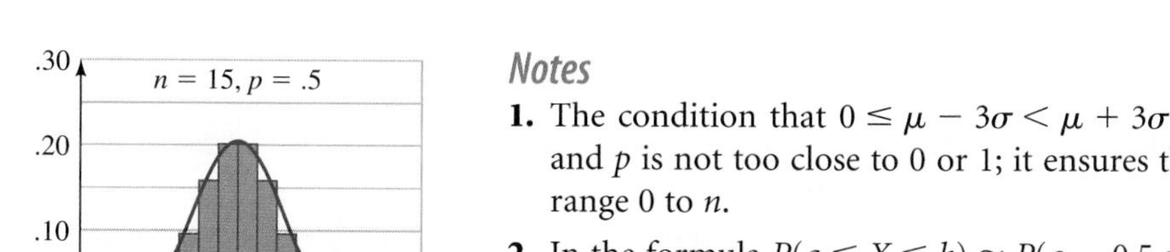

Figure 16

Notes

1. The condition that $0 \le \mu - 3\sigma < \mu + 3\sigma \le n$ is satisfied if n is sufficiently large and p is not too close to 0 or 1; it ensures that most of the normal curve lies in the range 0 to n.
2. In the formula $P(a \le X \le b) \approx P(a - 0.5 \le Y \le b + 0.5)$, we assume that a and b are integers. The use of $a - 0.5$ and $b + 0.5$ is called the **continuity correction.** To see that it is necessary, consider what would happen if you want to approximate, say, $P(X = 2) = P(2 \le X \le 2)$.

Figures 16 and 17 show two binomial distributions with their normal approximations superimposed and illustrate how closely the normal approximation fits the binomial distribution.

Example 4 • Coin Tosses

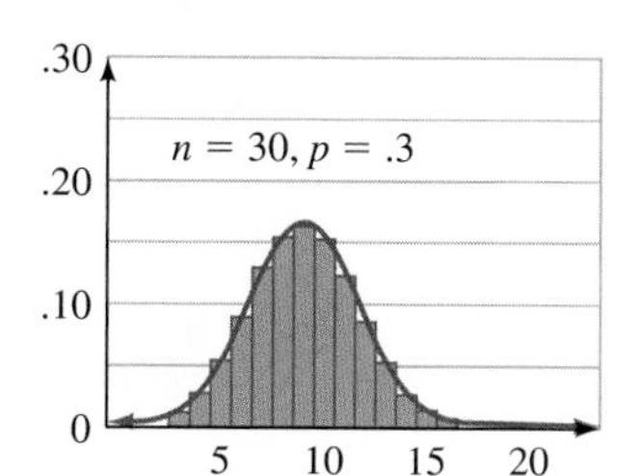

Figure 17

a. If you flip a fair coin 100 times, what is the probability of getting more than 55 heads or fewer than 45 heads?

b. What number of heads (out of 100) would make you suspect that the coin is not fair?

Solution

a. We are asking for

$$P(X < 45 \text{ or } X > 55) = 1 - P(45 \le X \le 55)$$

We *could* compute this by calculating

$$C(100, 45)(.5)^{45}(.5)^{55} + C(100, 46)(.5)^{46}(.5)^{54} + \cdots$$

but we can much more easily *approximate* it by looking at a normal distribution with mean $\mu = 50$ and standard deviation $\sigma = \sqrt{(100)(.5)(.5)} = 5$. (Notice that 3 standard deviations above and below the mean is the range 35–65, which is well within the range of possible values for X, which is 0 to 100, so the approximation should be a good one.) Let Y have this normal distribution. Then

$$\begin{aligned} P(45 \le X \le 55) &\approx P(44.5 \le Y \le 55.5) \\ &= P(-1.1 \le Z \le 1.1) \\ &= .7286 \end{aligned}$$

Therefore,

$$P(X < 45 \text{ or } X > 55) \approx 1 - .7286 = .2714$$

b. This is a deep question that touches on the concept of **statistical significance**: What evidence is strong enough to overturn a reasonable assumption (the assumption that the coin is fair)? Statisticians have developed sophisticated ways of answering this question, but we can look at one simple test now. Suppose we tossed a coin 100 times and got 66 heads. If the coin were fair, then $P(X > 65) \approx P(Y > 65.5) = P(Z > 3.1) \approx .001$. This is small enough to raise a reasonable doubt that the coin is fair. However, we should not be too surprised if we threw 56 heads because we can calculate $P(X > 55) \approx .1357$, which is not such a small probability. As we said, the actual tests of statistical significance are more sophisticated than this, but we will not go into them.

Guideline: When and When Not to Subtract from .5

Question When computing probabilities like, say, $P(Z \leq 1.2)$, $P(Z \geq 1.2)$, or $P(1.2 \leq Z \leq 2.1)$ using a table, just looking up the given values (1.2, 2.1, or whatever) is not enough. Sometimes you have to subtract from .5, sometimes not. Is there a simple rule telling me what to do when?

Answer The simplest—and also most instructive—way of knowing what to do is to draw a picture of the standard normal curve and shade in the area you are looking for. Drawing pictures also helps you come up with the following mechanical rules:

1. To compute $P(a \leq Z \leq b)$, look up the areas corresponding to $|a|$ and $|b|$ in the table. If a and b have opposite sign, add these areas. Otherwise, subtract the smaller area from the larger.
2. To compute $P(Z \leq a)$, look up the area corresponding to $|a|$. If a is positive, add .5; otherwise, subtract from .5.
3. To compute $P(Z \geq a)$, look up the area corresponding to $|a|$. If a is positive, subtract from .5; otherwise, add .5.

8.5 EXERCISES

Note: Answers for Section 8.5 were computed using the four-digit table in Appendix B and may differ slightly from the more accurate answers generated using technology.

In Exercises 1–8, Z is the standard normal distribution. Find the indicated probabilities.

1. $P(0 \leq Z \leq 0.5)$

2. $P(0 \leq Z \leq 1.5)$

3. $P(-0.71 \leq Z \leq 0.71)$

4. $P(-1.71 \leq Z \leq 1.71)$

5. $P(-0.71 \leq Z \leq 1.34)$

6. $P(-1.71 \leq Z \leq 0.23)$

7. $P(0.5 \leq Z \leq 1.5)$

8. $P(0.71 \leq Z \leq 1.82)$

In Exercises 9–14, X has a normal distribution with the given mean and standard deviation. Find the indicated probabilities.

9. $\mu = 50, \sigma = 10$; find $P(35 \leq X \leq 65)$.

10. $\mu = 40, \sigma = 20$; find $P(35 \leq X \leq 45)$.

11. $\mu = 50, \sigma = 10$; find $P(30 \leq X \leq 62)$.

12. $\mu = 40, \sigma = 20$; find $P(30 \leq X \leq 53)$.

13. $\mu = 100, \sigma = 15$; find $P(110 \leq X \leq 130)$.

14. $\mu = 100, \sigma = 15$; find $P(70 \leq X \leq 80)$.

15. Find the probability that a normal variable takes on values within 0.5 standard deviation of its mean.

16. Find the probability that a normal variable takes on values within 1.5 standard deviations of its mean.

17. Find the probability that a normal variable takes on values more than $\frac{2}{3}$ standard deviation away from its mean.

18. Find the probability that a normal variable takes on values more than $\frac{5}{3}$ standard deviations away from its mean.

19. If you roll a die 100 times, what is the probability that you will roll between 10 and 15 1s, inclusive? (Round your answer to two decimal places.)

20. If you roll a die 100 times, what is the probability that you will roll between 15 and 20 1s, inclusive? (Round your answer to two decimal places.)

21. If you roll a die 200 times, what is the probability that you will roll fewer than 25 1s? (Round your answer to two decimal places.)

22. If you roll a die 200 times, what is the probability that you will roll more than 40 1s? (Round your answer to two decimal places.)

APPLICATIONS

23. SAT Scores SAT test scores are normally distributed with a mean of 500 and a standard deviation of 100. Find the probability that a randomly chosen test taker will score 650 or higher.

24. LSAT Scores LSAT test scores are normally distributed with a mean of 500 and a standard deviation of 100. Find the probability that a randomly chosen test taker will score 250 or lower.

25. IQ Scores IQ scores (as measured by the Stanford–Binet intelligence test) are normally distributed with a mean of 100 and a standard deviation of 16. Find the approximate number of people in the United States (assuming a total population of 280,000,000) with an IQ higher than 120.

26. IQ Scores IQ scores (as measured by the Stanford–Binet intelligence test) are normally distributed with a mean of 100 and a standard deviation of 16. Find the approximate number of people in the United States (assuming a total population of 280,000,000) with an IQ higher than 140.

27. Baseball The mean batting average in major league baseball is about .250. Suppose batting averages are normally distributed, the standard deviation in the averages is 0.03, and there are 250 batters. What is the expected number of batters with an average of at least .400?

28. Baseball The mean batting average in major league baseball is about .250. Suppose batting averages are normally distributed, the standard deviation in the averages is 0.05, and there are 250 batters. What is the expected number of batters with an average of at least .400?[8]

29. Marketing Your pickle company rates its pickles on a scale of spiciness from 1 to 10. Market research shows that customer preferences for spiciness are normally distributed, with a mean of 7.5 and a standard deviation of 1. Assuming that you sell 100,000 jars of pickles, how many jars with a spiciness of 9 or above do you expect to sell?

30. Marketing Your hot sauce company rates its sauce on a scale of spiciness of 1 to 20. Market research shows that customer preferences for spiciness are normally distributed, with a mean of 12 and a standard deviation of 2.5. Assuming that you sell 300,000 bottles of sauce, how many bottles with a spiciness below 9 do you expect to sell?

Distribution of Wealth If we model after-tax household income with a normal distribution, then the figures of a 1995 study imply the information in the following table, which should be used for Exercises 31–36.[9] Assume that the distribution of incomes in each country is normal and round all percentages to the nearest whole number.

	U.S.	Canada	Switzerland	Germany	Sweden
Mean Household Income ($)	38,000	35,000	39,000	34,000	32,000
Standard Deviation ($)	21,000	17,000	16,000	14,000	11,000

SOURCE: Luxembourg Income Study/*New York Times,* August 14, 1995, p. A9.

31. What percentage of U.S. households had an income of $50,000 or more?

32. What percentage of German households had an income of $50,000 or more?

33. What percentage of Swiss households are either very wealthy (income at least $100,000) or very poor (income at most $12,000)?

34. What percentage of Swedish households are either very wealthy (income at least $100,000) or very poor (income at most $12,000)?

35. Which country has a higher proportion of very poor families (income $12,000 or less): the United States or Canada?

36. Which country has a higher proportion of very poor families (income $12,000 or less): Canada or Switzerland?

37. Comparing IQ Tests IQ scores as measured by both the Stanford–Binet intelligence test and the Wechsler intelligence test have a mean of 100. The standard deviation for the Stanford–Binet test is 16, and that for the Wechsler test is 15. For which test does a smaller percentage of test takers score less than 80? Why?

38. Comparing IQ Tests Referring to Exercise 37, for which test does a larger percentage of test takers score more than 120?

39. Product Repairs The new copier your business bought lists a mean time between failures of 6 months, with a standard deviation of 1 month. One month after a repair, it breaks down again. Is this surprising? (Assume that the times between failures are normally distributed.)

40. Product Repairs The new computer your business bought lists a mean time between failures of 1 year, with a standard deviation of 2 months. Ten months after a repair, it breaks down again. Is this surprising? (Assume that the times between failures are normally distributed.)

Software Testing Exercises 41–46 are based on the following information, gathered from student testing of a statistical software package called MODSTAT. Students were asked to complete certain tasks using the software, without any instructions. The results were as follows. (Assume that the time for each task is normally distributed.)

	Mean Time (min)	Standard Deviation
Task 1: Descriptive Analysis of Data	11.4	5.0
Task 2: Standardizing Scores	11.9	9.0
Task 3: Poisson Probability Table	7.3	3.9
Task 4: Areas under Normal Curve	9.1	5.5

Data are rounded to one decimal place. SOURCE: Joseph M. Nowakowski, *Student Evaluations of MODSTAT* (New Concord, Ohio: Muskingum College, 1997). Online: http://members.aol.com/rcknodt//pubpage.htm.

[8]The last time that a batter ended the year with an average above .400 was in 1941. The batter was Ted Williams of the Boston Red Sox, and his average was .406. Over the years, as pitching and batting have improved, the standard deviation in batting averages has declined from around 0.05 when professional baseball began to around 0.03 by the end of the twentieth century. For a very interesting discussion of statistics in baseball and in evolution, see Stephen Jay Gould, *Full House: The Spread of Excellence from Plato to Darwin,* (New York: Random House, 1997).

[9]The data are rounded to the nearest $1000 and based on a report published by the Luxembourg Income Study. The report shows after-tax income, including government benefits (such as food stamps) of households with children. Our figures were obtained from the published data by assuming a normal distribution of incomes. All data were based on constant 1991 U.S. dollars and converted foreign currencies (adjusted for differences in buying power).

41. Find the probability that a student will take at least 10 minutes to complete task 1.

42. Find the probability that a student will take at least 10 minutes to complete task 3.

43. Assuming that the time it takes a student to complete each task is independent of the others, find the probability that a student will take at least 10 minutes to complete each of tasks 1 and 2.

44. Assuming that the time it takes a student to complete each task is independent of the others, find the probability that a student will take at least 10 minutes to complete each of tasks 3 and 4.

45. It can be shown that, if X and Y are independent normal random variables with means μ_X and μ_Y and standard deviations σ_X and σ_Y, respectively, then their sum $X + Y$ is also normally distributed and has mean $\mu = \mu_X + \mu_Y$ and standard deviation $\sigma = \sqrt{\sigma_X^2 + \sigma_Y^2}$. Assuming that the time it takes a student to complete each task is independent of the others, find the probability that a student will take at least 20 minutes to complete both tasks 1 and 2.

46. Referring to Exercise 45, compute the probability that a student will take at least 20 minutes to complete both tasks 3 and 4.

47. Computers In 2001, 51% of all U.S. households had a computer. Find the probability that in a small town with 800 households at least 400 had a computer in 2001.

SOURCE: NTIA and ESA, U.S. Department of Commerce, using U.S. Bureau of the Census Current Population Survey supplements.

48. Television Ratings Based on data from Nielsen Research, there is a 15% chance that any television that is turned on during the time of the evening newscasts will be tuned to ABC's evening news show. Your company wishes to advertise on a small local station carrying ABC that serves a community with 2500 households that regularly tune in during this time slot. Find the probability that at least 400 households will be tuned in to the show.

SOURCE: Nielsen Media Research/ABC Network/*New York Times,* March 18, 2002, p. C1.

49. Aviation The probability of a plane crashing on a single trip in 1989 was .000 001 65. Find the probability that, in 100,000,000 flights, there will be fewer than 180 crashes.

SOURCE: National Transportation Safety Board.

50. Aviation The probability of a plane crashing on a single trip in 1990 was .000 000 87. Find the probability that in 100,000,000 flights there will be more than 110 crashes.

SOURCE: National Transportation Safety Board.

51. Insurance Your company issues flight insurance. You charge \$2 and, in the event of a plane crash, you will pay \$1 million to the victim or his or her family. In 1989 the probability of a plane crashing on a single trip was .000 001 65. If ten people per flight buy insurance from you, what was your probability of losing money over the course of 100 million flights in 1989? (*Hint:* First determine how many crashes there must be for you to lose money.)

SOURCE: National Transportation Safety Board.

52. Insurance Refer to Exercise 51. What is your probability of losing money over the course of 10 million flights?

SOURCE: National Transportation Safety Board.

53. Polls In a certain political poll, each person polled has a 90% probability of telling his or her real preference. Suppose 55% of the population really prefer candidate Goode and 45% prefer candidate Slick. First, find the probability that a person polled will say that he or she prefers Goode. Then, find the probability that if 1000 people are polled more than 52% will say they prefer Goode.

54. Polls In a certain political poll, each person polled has a 90% probability of telling his or her real preference. Suppose 1000 people are polled and 51% say that they prefer candidate Goode and 49% say that they prefer candidate Slick. Find the probability that Goode could do at least this well if in fact only 49% prefer Goode.

55. IQ Scores Mensa is a club for people with high IQs. To qualify you must be in the top 2% of the population. One way of qualifying is by having an IQ of at least 148, as measured by the Cattell intelligence test. Assuming that scores on this test are normally distributed with a mean of 100, what is the standard deviation? (*Hint:* Use the table in the appendix backward.)

56. SAT Scores Another way to qualify for Mensa (see Exercise 55) is to score at least 1250 on the SAT, which puts you in the top 2%. Assuming that SAT scores are normally distributed with a mean of 1000, what is the standard deviation? (See the hint for Exercise 55.)

COMMUNICATION AND REASONING EXERCISES

57. Under what assumptions are the estimates in the empirical rule exact?

58. If X is a continuous random variable, what values can the quantity $P(X = a)$ have?

59. Which is larger for a continuous random variable, $P(X \le a)$ or $P(X < a)$?

60. Which of the following is greater in general: $P(X \le b)$ or $P(a \le X \le b)$?

61. A uniform continuous distribution is one with a probability density curve that is a horizontal line. If X takes on values between the numbers a and b with a uniform distribution, find the height of its probability density curve.

62. Which would you expect to have the greater variance: the standard normal distribution or the uniform distribution taking values between -1 and 1? Explain.

63. Which would you expect to have a density curve that is higher at the mean: the standard normal distribution or a normal distribution with standard deviation 0.5? Explain.

64. Suppose students must perform two tasks: task 1 and task 2. Which of the following would you expect to have a smaller standard deviation?

(A) The time it takes a student to perform both tasks if the time it takes to complete task 2 is independent of the time it takes to complete task 1.

(B) The time it takes a student to perform both tasks if students will perform similarly in both tasks.

Explain.

CASE STUDY

Spotting Tax Fraud with Benford's Law[10]

Reproduced from the U.S. Department of the Treasury Web site

You are a tax-fraud specialist working for the IRS, and you have just been handed a portion of the tax return from Colossal Conglomerate (CC). The IRS suspects that the portion you were handed may be fraudulent and would like your opinion. Is there any mathematical test, you wonder, that can point to a suspicious tax return based on nothing more than the numbers entered?

You decide, on an impulse, to make a list of the first digits of all the numbers entered in the portion of the CC tax return (there are 625 of them). You reason that, if the tax return is an honest one, the first digits of the numbers should be uniformly distributed. More precisely, if the experiment consists of selecting a number at random from the tax return and the random variable X is defined to be the first digit of the selected number, then X should have the following probability distribution:

x	1	2	3	4	5	6	7	8	9
$P(X = x)$	$\frac{1}{9}$	$\frac{1}{9}$	$\frac{1}{9}$	$\frac{1}{9}$	$\frac{1}{9}$	$\frac{1}{9}$	$\frac{1}{9}$	$\frac{1}{9}$	$\frac{1}{9}$

You then do a quick calculation based on this probability distribution and find an expected value of $E(X) = 5$. Next, you turn to the CC tax return data and calculate the relative frequency (estimated probability) of the actual numbers in the tax return. You find the following results:

Colossal Conglomerate Return

y	1	2	3	4	5	6	7	8	9
$P(Y = y)$	.29	.10	.04	.15	.31	.08	.01	.01	.01

It certainly does look suspicious! For one thing, the digits 1 and 5 seem to occur a lot more often than any of the other digits and roughly three times what you predicted. Moreover, when you compute the expected value, you obtain $E(Y) = 3.48$, considerably lower than the value of 5 you predicted. "Gotcha!" you exclaim.

You are about to file a report recommending a detailed audit of Colossal Conglomerate when you recall an article you once read about first digits in lists of numbers. The article dealt with a remarkable discovery in 1938 by Frank Benford, a physicist at General Electric. What Benford noticed was that the pages of logarithm tables that listed numbers starting with the digits 1 and 2 tended to be more soiled and dog-eared than the pages that listed numbers starting with higher digits—say, 8. For some reason, numbers that start with low digits seemed more prevalent than numbers that start with high digits. He subsequently analyzed more than 20,000 sets of numbers, such as tables of baseball statistics, listings of widths of rivers, half-lives of radioactive elements, street addresses, and numbers in magazine articles. The result was always the same: Inexplicably, numbers that start with low digits tended to appear more frequently than those that start with high ones, with numbers beginning with the digit 1 most prevalent of all.[11] Moreover, the expected value of the first digit was not the expected 5, but 3.44.

[10]The discussion is based on the article "Following Benford's Law, or Looking Out for No. 1," Malcolm W. Browne, *New York Times*, August 4, 1998, p. F4. The use of Benford's law in detecting tax evasion is discussed in a Ph.D. dissertation by Dr. Mark J. Nigrini (Southern Methodist University, Dallas).

[11]This does not apply to all lists of numbers. For instance, a list of randomly chosen numbers between 100 and 999 will have first digits uniformly distributed between 1 and 9.

Since the first digits in the CC return have an expected value of 3.48, very close to Benford's value, it might appear that your suspicion was groundless after all. (Back to the drawing board . . .)

Out of curiosity, you decide to investigate Benford's discovery more carefully. What you find is that Benford did more than simply observe a strange phenomenon in lists of numbers. He went further and derived the following formula for the probability distribution of first digits in lists of numbers:

$$P(X = x) = \log\left(1 + \frac{1}{x}\right) \qquad (x = 1, 2, \ldots, 9)$$

You compute these probabilities and find the following distribution (the probabilities are all rounded and thus do not add to exactly 1):

x	1	2	3	4	5	6	7	8	9
P(X = x)	.30	.18	.12	.10	.08	.07	.06	.05	.05

You then enter these data along with the CC tax return data in your spreadsheet program and obtain the graph shown in Figure 18.

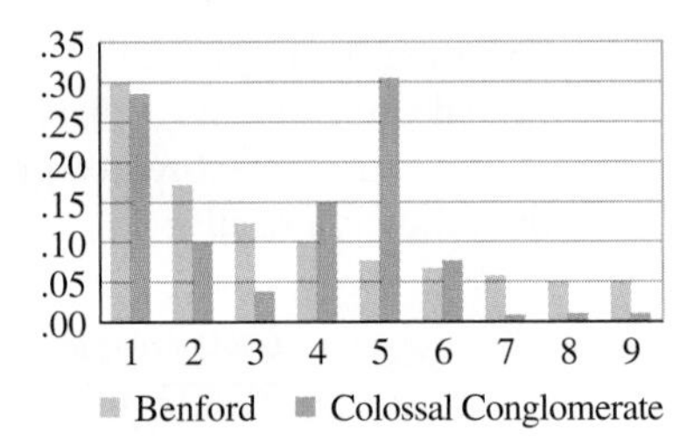

Figure 18

The graph shows something awfully suspicious happening with the digit 5. The percentage of numbers in the CC return that begin with 5 far exceeds Benford's prediction that approximately 8% of all numbers should begin with 5.

Now it seems fairly clear that you are justified in recommending Colossal Conglomerate for an audit, after all.

Question Since no given set of data can reasonably be expected to satisfy Benford's law *exactly,* how can I be certain that the CC data are not simply due to chance?

Answer You can never be 100% certain. It is certainly conceivable that the tax figures just happen to result in the "abnormal" distribution in the CC tax return. However—and this is the subject of *inferential statistics*—there is a method for deciding whether you can be, say, "95% certain" that the anomaly reflected in the data is not due to chance. To check, you must first compute a statistic that determines how far a given set of data deviates from satisfying a theoretical prediction (Benford's law, in this case). This statistic is called a **sum-of-squares error** and is given by the following formula (reminiscent of the variance):

$$\text{SSE} = n\left\{\frac{[P(y_1) - P(x_1)]^2}{P(x_i)} + \frac{[P(y_2) - P(x_2)]^2}{P(x_2)} + \cdots + \frac{[P(y_9) - P(x_9)]^2}{P(x_9)}\right\}$$

Here, n is the sample size: 625 in the case of Colossal Conglomerate. The quantities $P(x_i)$ are the theoretically predicted probabilities according to Benford's law, and the $P(y_i)$ are the probabilities in the CC return. Notice that if the CC return probabilities had exactly matched the theoretically predicted probabilities, then SSE would have been zero.

Notice also the effect of multiplying by the sample size n: The larger the sample, the more likely that the discrepancy between the $P(x_i)$ and the $P(y_i)$ is *not* due to chance. Substituting the numbers gives[12]

$$\text{SSE} \approx 625\left[\frac{(.29 - .30)^2}{.30} + \frac{(.1 - .18)^2}{.18} + \cdots + \frac{(.01 - .05)^2}{.05}\right] \approx 552$$

Question The value of SSE does seem quite large. But how can I use this figure in my report? I would like to say something impressive, such as "Based on the portion of the Colossal Conglomerate tax return analyzed, one can be 95% certain that the figures are anomalous."

Answer Statisticians use the error SSE to answer exactly such a question. What they would do is compare this figure to the largest SSE we would have expected to get by chance in 95 out of 100 selections of data that *do* satisfy Benford's law. This "biggest error" is computed using a *chi-squared* distribution and can be found in Excel by entering

```
=CHIINV(0.05,8)
```

Here, the 0.05 is $1 - 0.95$, encoding the 95% certainty, and the 8 is called the *number of degrees of freedom* = number of outcomes (9) minus 1.

You now find, using Excel, that the chi-squared figure is 15.5, meaning that the largest SSE that you could have expected purely by chance is 15.5. Since Colossal Conglomerate's error is much larger at 552, you can now justifiably say in your report that there is a 95% certainty that the figures are anomalous.[13]

EXERCISES

State whether you would expect each of the following lists of data to follow Benford's law. If the answer is no, give a reason.

1. Distances between cities in France, measured in kilometers
2. Distances between cities in France, measured in miles
3. The grades (0–100) in your math instructor's grade book
4. The Dow Jones averages for the past 100 years
5. Verbal SAT scores of college-bound high school seniors
6. Life spans of companies

Use a spreadsheet to determine whether the given distribution of first digits fail, with 95% certainty, to follow Benford's law.

7. Good Neighbor Inc.'s tax return ($n = 1000$)

y	1	2	3	4	5	6	7	8	9
$P(Y = y)$	.31	.16	.13	.11	.07	.07	.05	.06	.04

8. Honest Growth Funds Stockholder Report ($n = 400$)

y	1	2	3	4	5	6	7	8	9
$P(Y = y)$	.28	.16	.1	.11	.07	.09	.05	.07	.07

[12]If you use more accurate values for the probabilities in Benford's distribution, the value is approximately 560.

[13]What this actually means is that, if you were to do a similar analysis on a large number of *valid* tax returns, 95% of them would have values of SSE less than 15.5.

CHAPTER 8 REVIEW TEST

1. In each case find the probability distribution for the given random variable and draw a histogram.

a. A couple has two children; X = the number of boys. (Assume an equal likelihood of a child being a boy or a girl.)

b. A four-sided die (with sides numbered 1–4) is rolled twice in succession; X = the sum of the two numbers.

c. 48.2% of XBox® players are in their teens, 38.6% are in their twenties, 11.6% are in their thirties, and the rest are in their forties; X = age of an XBox player. (Use the midpoints of the measurement classes.)

d. From a bin that contains 20 defective joysticks and 30 good ones, 3 are chosen at random; X = the number of defective joysticks chosen. (Round all probabilities to four decimal places.)

e. Two dice are weighted so that each of 2, 3, 4, and 5 is half as likely to face up as each of 1 and 6; X = the number of 1s that face up when both are thrown.

2. a. Use any method to calculate the sample mean, median, and standard deviation of the following sample of scores: $-1, 2, 0, 3, 6$.

b. Give an example of a sample of four scores with mean 1 and median 0. (Arrange them in ascending order.)

c. Give an example of a sample of six scores with sample standard deviation 0 and mean 2.

d. Give an example of a population of six scores with mean 0 and population standard deviation 1.

e. Give an example of a sample of five scores with mean 0 and sample standard deviation 1.

3. A die is constructed in such a way that rolling a 6 is twice as likely as rolling each other number. That die is rolled four times. Let X be the number of times a 6 is rolled. Evaluate each of the following. (Round all answers to four decimal places.)

a. $P(X = 1)$

b. The probability that 6 comes up at most twice

c. The probability that X is at least 2

d. $P(X > 3)$

e. $P(1 \leq X \leq 3)$

4. a. A couple has three children; X = the number of girls. (Assume an equal likelihood of a child being a boy or a girl.) Find the probability distribution, expected value, and standard deviation of X and complete the following sentence: All values of X lie within ____ (whole number) standard deviations of the expected value.

b. A random variable X has the following frequency distribution:

x	−3	−2	−1	0	1	2	3
fr(X = x)	1	2	3	4	3	2	1

Find the probability distribution, expected value, and standard deviation of X and complete the following sentence: 87.5% (or 14/16) of the time, X is within ____ (round to one decimal place) standard deviations of the expected value.

c. A random variable X has expected value $\mu = 100$ and standard deviation $\sigma = 16$. Use Chebyshev's rule to find an interval in which X is guaranteed to lie with a probability of at least 90%.

d. A random variable X has a bell-shaped, symmetric distribution, with expected value $\mu = 100$ and standard deviation $\sigma = 30$. Use the empirical rule to give an interval in which X lies approximately 95% of the time.

5. In each case the mean and standard deviation of a normal variable X are given. Find the indicated probabilities.

a. X is the standard normal variable Z; $P(0 \leq Z \leq 1.5)$.

b. X is the standard normal variable Z; $P(Z \leq -1.5)$.

c. X is the standard normal variable Z; $P(|Z| \geq 2.1)$.

d. $\mu = 100, \sigma = 16$; $P(80 \leq X \leq 120)$

e. $\mu = 0, \sigma = 2$; $P(X \leq -1)$

f. $\mu = -1, \sigma = 0.5$; $P(X \geq 1)$

OHaganBooks.com—COPING WITH UNCERTAINLY

6. As a promotional gimmick, OHaganBooks.com has been selling copies of the *Encyclopedia Galactica* at an extremely low price that is changed each week at random in a nationally televised drawing. The following table summarizes the anticipated sales:

Price ($)	5.50	10	12	15
Frequency (weeks)	1	2	3	4
Weekly Sales	6200	3500	3000	1000

a. What is the expected value of the price of *Encyclopedia Galactica*?

b. What are the expected weekly sales of *Encyclopedia Galactica*?

c. What is the expected weekly revenue? (Revenue = price per copy sold × number of copies sold.)

d. True or false? If X and Y are two random variables, then $E(XY) = E(X)E(Y)$ (the expected value of the product of two random variables is the product of the expected values). Support your claim by referring to the answers for parts (a), (b), and (c).

7. Recent research shows the following number of online orders at OHaganBooks.com per million residents in 100 U.S. cities during 1 month:

Orders (per million residents)	1–2.9	3–4.9	5–6.9	7–8.9	9–10.9
Cities	25	35	15	15	10

a. Let X be the number of orders per million residents in a randomly chosen U.S. city (use rounded midpoints of the given measurement classes). Construct the probability distribution for X and then compute the expected value μ of X and standard deviation σ. (Round answers to four decimal places.)

b. What range of orders per million residents does the empirical rule predict from approximately 68% of all cities? Would you judge that the empirical rule applies? Why?

c. Refer to the frequency distribution given above. The actual percentage of cities from which you obtain between three and eight orders per million residents is (choose the correct answer that gives the most specific information)

(A) Between 50% and 65% **(B)** At least 65%
(C) At least 50% **(D)** 57.5%

8. On average, 5% of all hits by Mac OS® users and 10% of all hits by Windows® users result in orders for books at OHaganBooks.com. Due to online promotional efforts, the site traffic is 10 hits/hour by Mac OS users, and 20 hits/hour by Windows users. (Round all answers to three decimal places.)

a. What is the probability that exactly three Windows users will order books in the next hour?

b. What is the probability that at most three Windows users will order books in the next hour?

c. What is the probability that exactly one Mac OS user and three Windows users will order books in the next hour?

d. What assumption must you make to justify your calculation in part (c)?

e. How many orders for books can OHaganBooks.com expect in the next hour?

9. OHaganBooks.com has launched a subsidiary, GnuYou.com, which sells beauty products online. Most products sold by GnuYou.com are skin creams and hair products. The following table shows monthly revenues earned through sales of these products. (Assume a normal distribution. Round answers to three decimal places.)

	Skin Creams	Hair Products
Mean Monthly Revenue	\$38,000	\$34,000
Standard Deviation	\$21,000	\$14,000

a. What is the probability that GnuYou.com will sell *at least* \$50,000 worth of skin cream next month?

b. What is the probability that GnuYou.com will sell *at most* \$50,000 worth of hair products next month?

c. Which type of product is most likely to yield sales of less than \$12,000 next month? (Support your conclusion numerically.)

10. Billy Sean O'Hagan, now a senior at Suburban State University, has done exceptionally well and has just joined Mensa, a club for people with high IQs. (To qualify, you must be in the top 2% of the population.)

a. Within Mensa is a group called the 3 Sigma Club because their IQ scores are at least 3 standard deviations higher than the U.S. mean. Assuming a U.S. population of 280 million, how many people in the United States are qualified for the 3 Sigma Club? (Round your answer to the nearest 1000 people.)

b. To join Mensa (not necessarily the 3 Sigma Club), one needs an IQ of at least 132. Assuming that scores on this test are normally distributed with a mean of 100, what is the standard deviation? (Round your answer to the nearest whole number.)

c. Based on the data given in part (b), what score must Billy Sean have to get into the 3 Sigma Club? (Assume that IQ scores are normally distributed with a mean of 100 and use the rounded standard deviation.)

ADDITIONAL ONLINE REVIEW

If you follow the path

Web site → Everything for Finite Math → Chapter 8

you will find the following additional resources to help you review:

A comprehensive chapter summary (including examples and interactive features)

Additional review exercises (including interactive exercises and many with help)

A true/false chapter quiz

9

MARKOV SYSTEMS

CASE STUDY

Predicting the Price of Gold

The price of gold is currently \$320 per ounce and has been moving up and down by an average of \$2 per trading day, more or less randomly over the past several months. Precious metals analysts are predicting that the market will heat up if one of two things happens: (1) the price reaches the psychological barrier of \$400 per ounce, or (2) it drops to the current support level of \$280 per ounce. How long will it take for one or the other to happen?

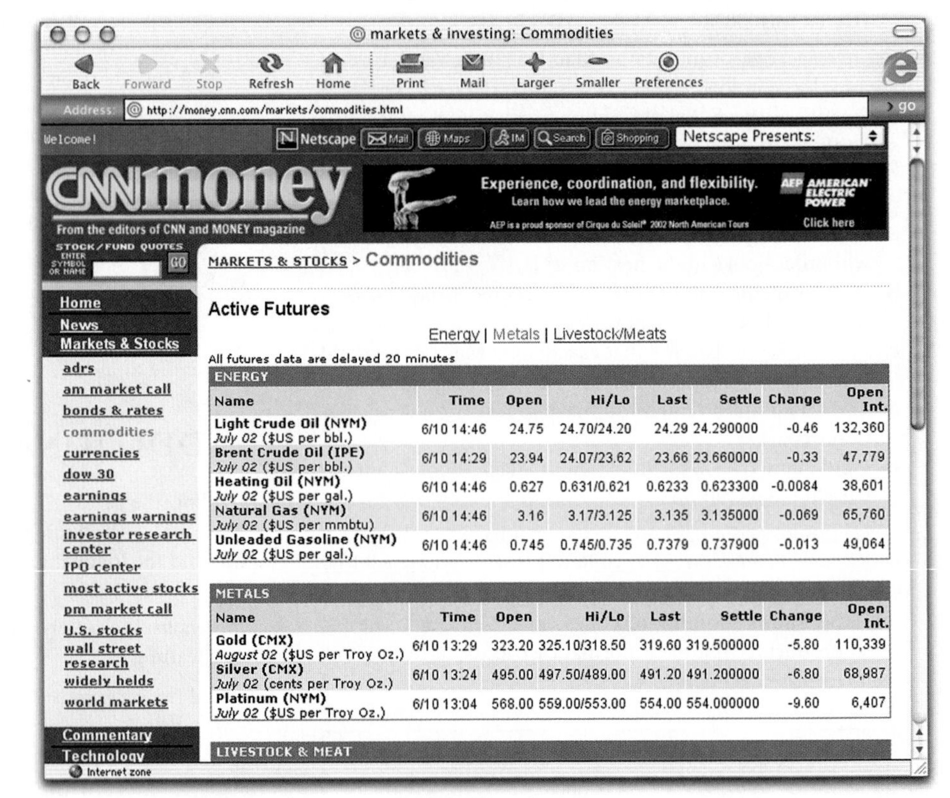

ENERGY

Name	Time	Open	Hi/Lo	Last	Settle	Change	Open Int.
Light Crude Oil (NYM) *July 02* (\$US per bbl.)	6/10 14:46	24.75	24.70/24.20	24.29	24.290000	-0.46	132,360
Brent Crude Oil (IPE) *July 02* (\$US per bbl.)	6/10 14:29	23.94	24.07/23.62	23.66	23.660000	-0.33	47,779
Heating Oil (NYM) *July 02* (\$US per gal.)	6/10 14:46	0.627	0.631/0.621	0.6233	0.623300	-0.0084	38,601
Natural Gas (NYM) *July 02* (\$US per mmbtu)	6/10 14:46	3.16	3.17/3.125	3.135	3.135000	-0.069	65,760
Unleaded Gasoline (NYM) *July 02* (\$US per gal.)	6/10 14:46	0.745	0.745/0.735	0.7379	0.737900	-0.013	49,064

METALS

Name	Time	Open	Hi/Lo	Last	Settle	Change	Open Int.
Gold (CMX) *August 02* (\$US per Troy Oz.)	6/10 13:29	323.20	325.10/318.50	319.60	319.500000	-5.80	110,339
Silver (CMX) *July 02* (cents per Troy Oz.)	6/10 13:24	495.00	497.50/489.00	491.20	491.200000	-6.80	68,987
Platinum (NYM) *July 02* (\$US per Troy Oz.)	6/10 13:04	568.00	559.00/553.00	554.00	554.000000	-9.60	6,407

INTERNET RESOURCES FOR THIS CHAPTER

At the Web site, follow the path

Web site → Everything for Finite Math → Chapter 9

where you will find links to step-by-step tutorials for the main topics in this chapter, a detailed chapter summary you can print out, a true/false quiz, and a collection of sample test questions. You will also find downloadable Excel tutorials for each section and other resources, such as an online simulation and computational tools for Markov systems.

Introduction

In this chapter we put the algebra of matrices to work in interesting ways. Many real-life situations can be modeled by processes that pass from state to state with given probabilities. A simple example of such a *Markov system* is the fluctuation of a gambler's fortune as he or she continues to bet. Other examples come from the study of trends in the commercial world, such as the situation discussed in the chapter opening. In addition to these examples, we consider the implications of brand switching, viability studies of banking institutions, dissipation, drift, and even the theory behind computers. In fact, there has recently been a great deal of interest in Markov systems in the fields of neural networks and artificial intelligence. The importance of Markov systems can hardly be overstated.

We will be using probabilities, but this chapter may be read independently of the chapter on probability theory. For the purposes of this chapter, all you need is a basic commonsense idea of what probabilities mean. For example, to say that if a fair die is cast the probability of rolling a 2 is 1/6 means that, if we rolled the die many times, we would expect that about one-sixth of the time we would roll a 2. Thus, the probability of something happening is simply the fraction of times we expect it to happen in a large number of trials.

9.1 Markov Systems

Here is a basic example that we use many times in this chapter.

Example 1 • Laundry Detergent Switching

A market analyst for Gamble Detergents is interested in whether consumers prefer powdered laundry detergents or liquid detergents. Two market surveys taken 1 year apart revealed that 20% of all powdered-detergent users had switched to liquid 1 year later and the rest were still using powder. Only 10% of liquid-detergent users had switched to powder 1 year later, with the rest still using liquid.

We analyze this example as follows. Every year a consumer may be in one of two possible **states**: He may be a powdered-detergent user or a liquid-detergent user. Let's number these states: A consumer is in state 1 if he uses powdered detergent and in state 2 if he uses liquid. There is a basic **time step** of 1 year. If a consumer happens to be in state 1 during a given year, then there is a probability of 20% = .2 (the chance that a randomly chosen powder user will switch to liquid) that he will be in state 2 the next year. We write

$$p_{12} = .2$$

to indicate that the probability of going *from* state 1 *to* state 2 in one time step is .2. The other 80% of the powder users are using powder the next year. We write

$$p_{11} = .8$$

to indicate that the probability of *staying* in state 1 from one year to the next is .8.[1] What if a consumer is in state 2? Then the probability of going to state 1 is given as 10% = .1, so the probability of remaining in state 2 is .9. Thus,

$$p_{21} = .1 \quad \text{and} \quad p_{22} = .9$$

Figure 1 shows the **state transition diagram** for this example. The numbers p_{ij}, which appear as labels on the arrows, are the **transition probabilities.**

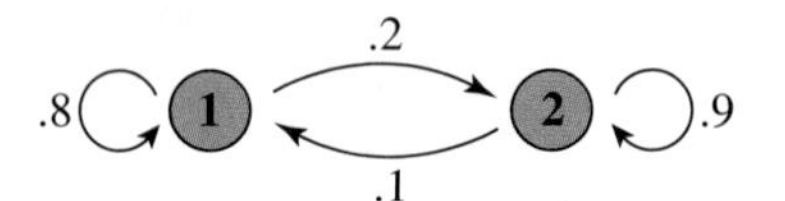

Figure 1

Markov System, States, and Transition Probabilities

A **Markov system*** (or **Markov process** or **Markov chain**) is a system that can be in one of several specified **states.** There is specified a certain **time step,** and at each step the system will randomly change states or remain where it is. The probability of going from state i to state j is a fixed number p_{ij}, called the **transition probability.**

Quick Example

The Markov system depicted in Figure 1 has two states: state 1 and state 2. The transition probabilities are as follows:

p_{11} = probability of going from state 1 to state 1 = .8

p_{12} = probability of going from state 1 to state 2 = .2

p_{21} = probability of going from state 2 to state 1 = .1

p_{22} = probability of going from state 2 to state 2 = .9

*Named after the Russian mathematician A. A. Markov (1856–1922), who first studied these nondeterministic processes.

Question In Example 1 the transition probabilities that originate in each state add up to 1. Is this always the case?

Answer Yes. If the system is in any state, then there is a 100% chance (probability of 1) that it will be in *some* state at the next time step. Look at state 1 in Example 1: If 20% of the consumers who are presently using powdered detergent will be using liquid in

[1]For those who have studied probability, these are actually *conditional* probabilities. For instance, p_{12} is the probability that the system (the consumer in this case) will go into state 2, *given that the system (the consumer) is in state 1.*

1 year, then the other 80% of them will still be using powder. This $.2 + .8 = 1$ accounts for all of them. If you have studied probability, this is just the fact that the sum of the probabilities of all possible outcomes must be 1.

The numbers p_{ij} may be conveniently arranged in a matrix, called the **transition matrix,** as follows.

Transition Matrix

The **transition matrix** associated with a given Markov system is the matrix P whose ijth entry is the transition probability p_{ij}, the transition probability of going *from* state i *to* state j. In other words, the entry in position ij is the *label on the arrow going from state i to state j* in a state transition diagram.

Thus, the transition matrix for a system with two states would be set up as follows:

$$\begin{array}{cc} & \textbf{To} \\ & \begin{array}{cc} \textbf{1} & \textbf{2} \end{array} \\ \textbf{From} \begin{array}{c} \textbf{1} \\ \textbf{2} \end{array} & \begin{bmatrix} p_{11} & p_{12} \\ p_{21} & p_{22} \end{bmatrix} \end{array} \quad \begin{array}{l} \text{Arrows originating in state 1} \\ \text{Arrows originating in state 2} \end{array}$$

Quick Example

In the system described in Example 1, the transition matrix is

$$P = \begin{bmatrix} .8 & .2 \\ .1 & .9 \end{bmatrix}$$

Note The sum of the transition probabilities that originate at any state is 1. As a consequence, *the sum of the entries in every row of a transition matrix is 1.*

Example 2 • Profitability

Two surveys of the savings and loan (S&L) industry were taken 1 year apart. The results were as follows: 10% of those S&Ls that were profitable at the time of the first survey were found to be operating at a loss by the time of the second survey, and the rest of them remained profitable. Of the S&Ls that were operating at a loss at the time of the first survey, 20% were bankrupt by the time of the second (the chief executives were bailed out by the government at taxpayers' expense), and 10% of them became profitable. Assume that once an S&L goes under, its assets are disposed so that it can never operate again. Write the transition matrix for this system.

Solution It is helpful to draw the state transition diagram before writing the matrix. We must first decide what the states are and number them. (We can number them any way we want, but different numberings will lead to different matrices.) Looking at the problem, we see three states in which an S&L can be: (1) *profitable,* (2) *losing,* and (3) *bankrupt.*[2]

The first statement, "10% of those S&Ls that were profitable at the time of the first survey were found to be operating at a loss by the time of the second survey, and the rest

[2]Although it may seem more convenient to label the states P, L, and B for profitable, losing, and bankrupt, we need to use numbers instead in order to construct the transition matrix. It is more in keeping with convention to speak of "the 1, 2 entry" of a matrix rather than the "P, L entry."

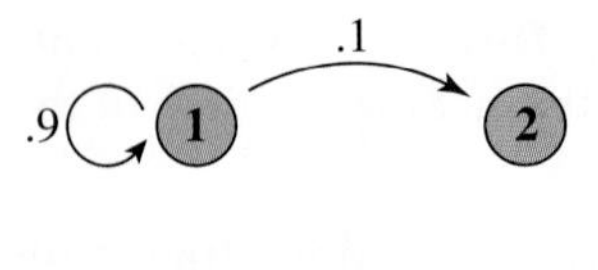

Figure 2

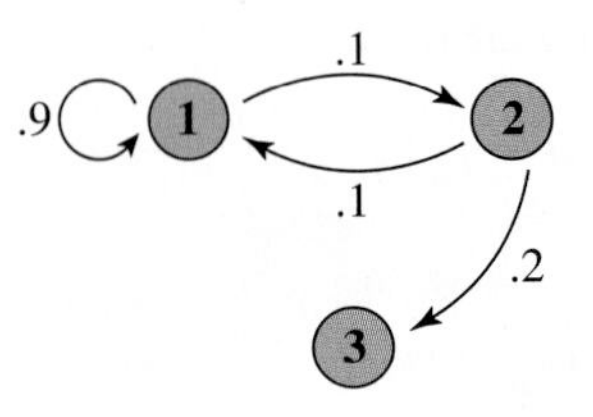

Figure 3

of them remained profitable," allows us to draw and label the arrows originating at state 1, as shown in Figure 2.

Note that, because none of the profitable S&Ls became bankrupt in one time step (1 year), the arrow from state 1 to state 3 should be labeled with a 0, but we left it out. To avoid cluttering the diagrams, we always leave out zero arrows.

The next statement, "of the S&Ls that were operating at a loss at the time of the first survey, 20% were bankrupt by the time of the second, and 10% of them became profitable," describes arrows leaving state 2, which we add to the diagram (Figure 3).

Now there seems to be a bit of a problem: The probabilities on the arrows originating at state 2 add up to .3, not 1. However, nothing is said about the arrow from state 2 to itself. Because the probabilities on the three arrows originating at state 2 must add to 1, the arrow from state 2 to itself must be labeled .7, so we get Figure 4.

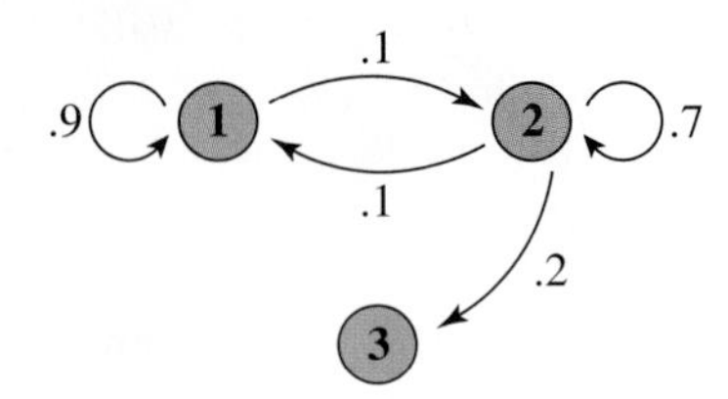

Figure 4

The last statement tells us that once an S&L enters state 3 it never leaves.[3] We call such a state an **absorbing state.** The completed diagram is shown in Figure 5.

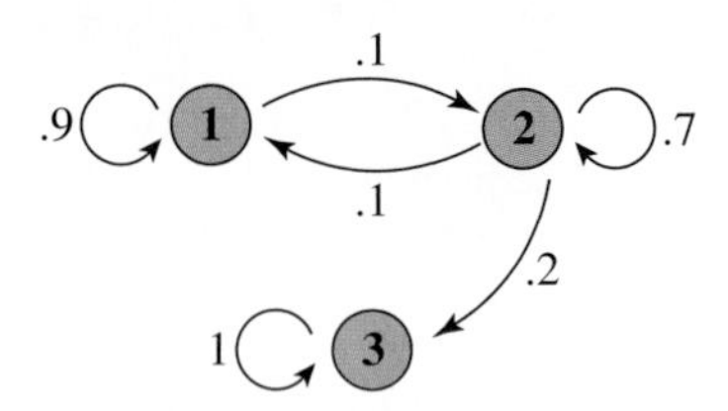

Figure 5

We can now write the transition matrix P:

$$
\begin{array}{cc}
 & \textbf{To} \\
 & \begin{array}{ccc} \textbf{1} & \textbf{2} & \textbf{3} \end{array} \\
\textbf{From} \begin{array}{c} \textbf{1} \\ \textbf{2} \\ \textbf{3} \end{array} & \begin{bmatrix} .9 & .1 & 0 \\ .1 & .7 & .2 \\ 0 & 0 & 1 \end{bmatrix}
\end{array}
$$

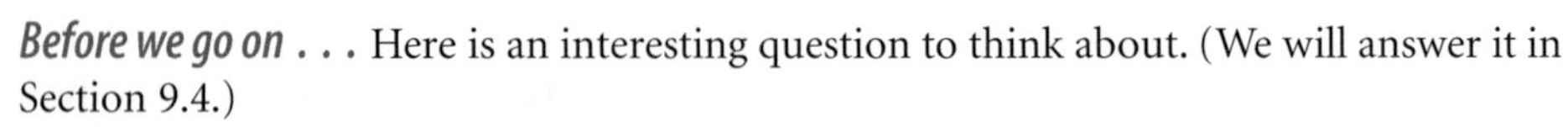

Before we go on . . . Here is an interesting question to think about. (We will answer it in Section 9.4.)

Question I wish to invest my life savings in the Lifelong Trust Savings and Loan, whose last balance sheet showed a healthy profit. How long can I expect my investment to be safe?

Web Site

If you would like to answer the question experimentally, follow the path

Web site → Online Utilities → Markov Simulation

[3]This state is much like a Roach Motel® ("Roaches check in, but they don't check out")!

where you can enter the transition matrix and watch a visual simulation showing the system hopping from state to state. Run the simulation a number of times to get an experimental estimate of the time to absorption. (Take the average time to absorption over many runs.) Also, try varying the transition probabilities to study the effect on the time to absorption.

Example 3 • Gambler's Ruin

A gambler, armed with her annual bonus of \$20, decides to play roulette, using the following scheme. At each spin of the wheel, she places \$10 on red. If red comes up, she wins an additional \$10; if black comes up, she loses \$10. For the sake of simplicity, assume that she has a probability of 1/2 of winning. (In the real game, the probability is slightly lower—a fact that many gamblers forget.) She keeps playing until she has either doubled her money or lost it all. In either case she then packs up and leaves. Model this situation as a Markov system and find the associated transition matrix.

Solution We must first decide on the states of the system. A good choice is the gambler's financial state: the amount of money she has at any stage of the game. According to her rules, she can either be broke or have any number of dollars (in multiples of 10) up to \$40. Thus, there are five states: 1 = \$0, 2 = \$10, 3 = \$20, 4 = \$30, and 5 = \$40. Since she bets \$10 each time, she moves down \$10 if she loses and up \$10 if she wins, each transition having a probability of 1/2. The state transition diagram is shown in Figure 6.

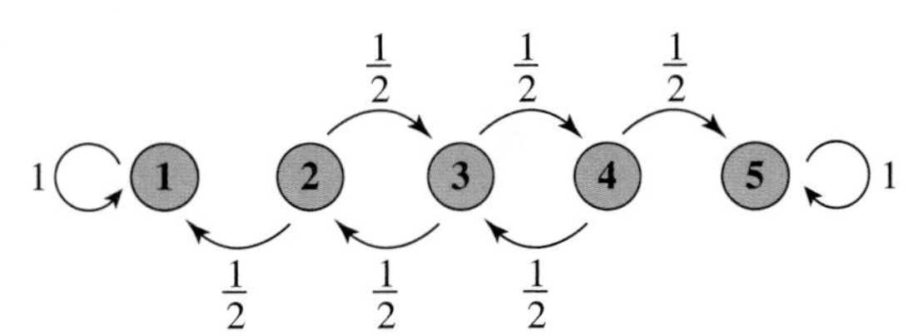

Figure 6

Note that the two end states are absorbing states: If she goes into state 1, she is broke and leaves the game; if she goes into state 5, she has doubled her money and also leaves the game. We now get the following 5 × 5 transition matrix:

$$P = \begin{bmatrix} 1 & 0 & 0 & 0 & 0 \\ \frac{1}{2} & 0 & \frac{1}{2} & 0 & 0 \\ 0 & \frac{1}{2} & 0 & \frac{1}{2} & 0 \\ 0 & 0 & \frac{1}{2} & 0 & \frac{1}{2} \\ 0 & 0 & 0 & 0 & 1 \end{bmatrix}$$

Before we go on . . . This process is also referred to as a **one-dimensional random walk with absorbing barriers.** Notice the symmetry in the transition matrix!

In Section 9.2 we will see that the transition matrix is more than just a convenient way of recording the transition probabilities and that matrix arithmetic can help us understand the behavior of Markov systems.

9.1 EXERCISES

In Exercises 1–10, write down the transition matrix associated with each state transition diagram.

1.

2.

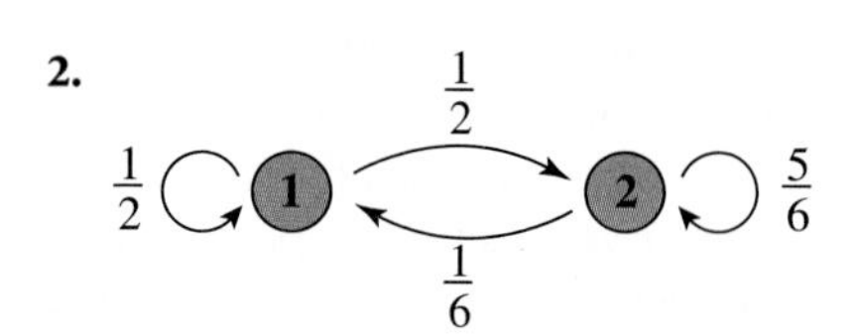

3.

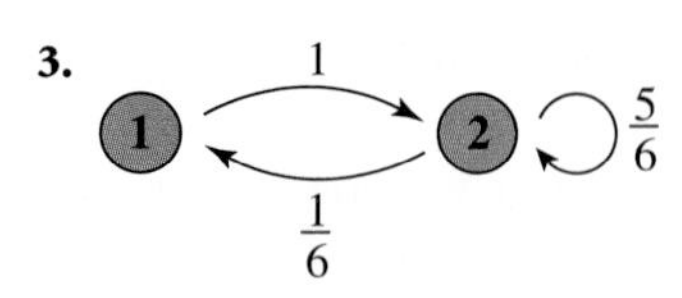

4. **5.**

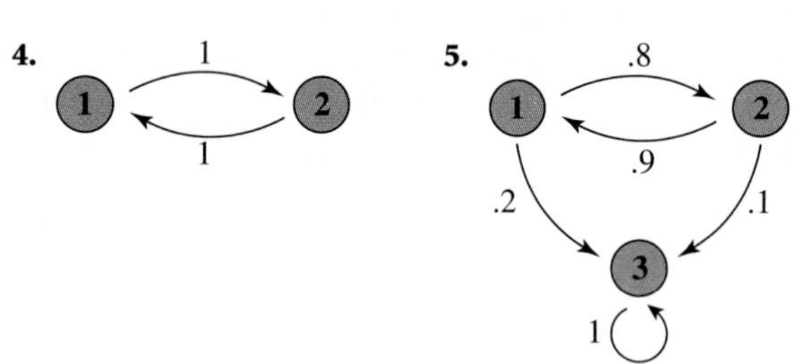

6.

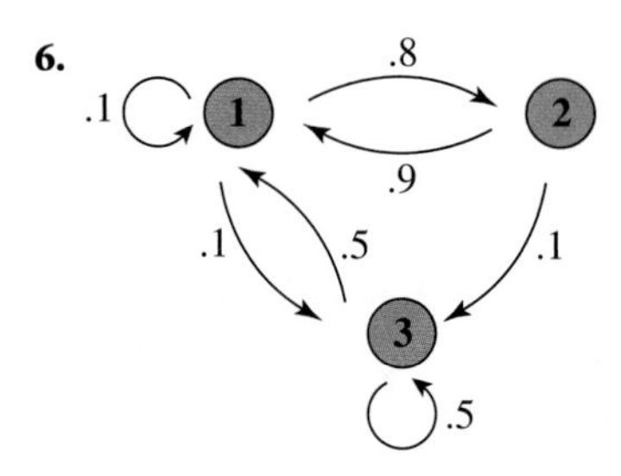

7.

8.

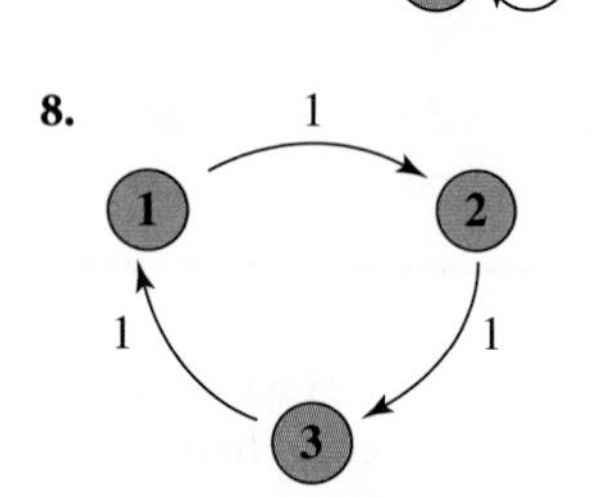

9.

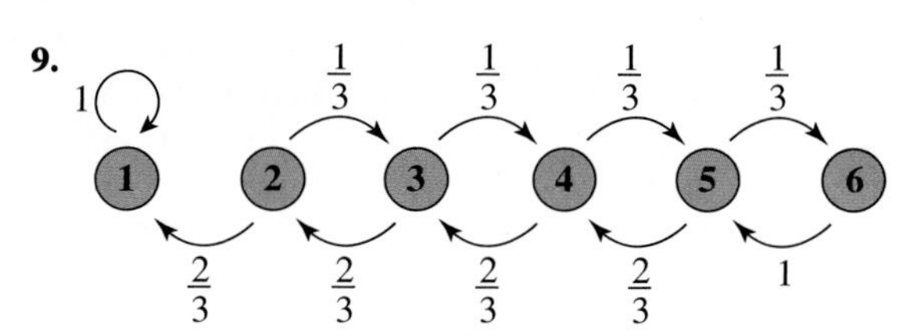

10.

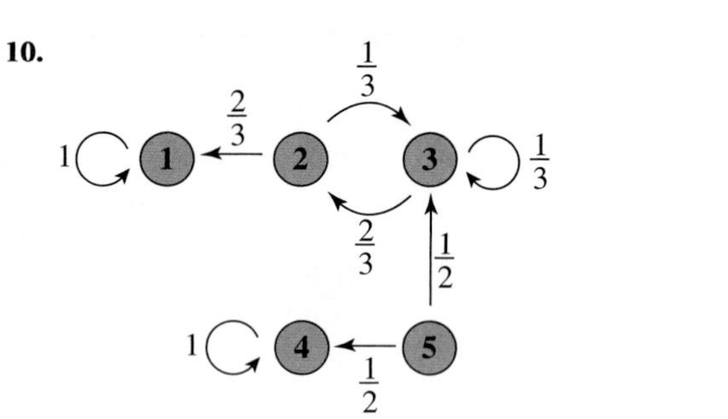

In Exercises 11–20, decide whether the given matrix is the transition matrix of a Markov system. If it is, give the associated state transition diagram.

11. $P = \begin{bmatrix} 0 & 1 \\ .2 & .8 \end{bmatrix}$ **12.** $P = \begin{bmatrix} .1 & .9 \\ .8 & .2 \end{bmatrix}$

13. $P = \begin{bmatrix} \frac{1}{2} & \frac{1}{2} \\ \frac{1}{3} & \frac{2}{3} \end{bmatrix}$ **14.** $P = \begin{bmatrix} \frac{2}{7} & \frac{5}{7} \\ \frac{2}{3} & \frac{1}{3} \end{bmatrix}$

15. $P = \begin{bmatrix} 1 & 0 & 1 \\ 0 & 1 & 1 \\ 0 & 1 & 2 \end{bmatrix}$ **16.** $P = \begin{bmatrix} .5 & .5 & 0 \\ 0 & .8 & .2 \\ 0 & 0 & 0 \end{bmatrix}$

17. $P = \begin{bmatrix} .5 & .5 & 0 \\ .1 & .1 & .8 \\ .5 & .2 & .3 \end{bmatrix}$ **18.** $P = \begin{bmatrix} \frac{2}{7} & \frac{1}{7} & \frac{4}{7} \\ 0 & \frac{5}{7} & \frac{2}{7} \\ \frac{2}{3} & \frac{1}{3} & 0 \end{bmatrix}$

19. $P = \begin{bmatrix} 0 & 1 & 0 & 0 \\ 0 & 0 & 1 & 0 \\ 0 & 0 & 0 & 1 \\ 1 & 0 & 0 & 0 \end{bmatrix}$ **20.** $P = \begin{bmatrix} 1 & 0 & 0 & 0 \\ 0 & 1 & 1 & 0 \\ 0 & 0 & 0 & 1 \\ 1 & 0 & 0 & 1 \end{bmatrix}$

APPLICATIONS

In Exercises 21–32, draw a state transition diagram and give the associated transition matrix.

21. Marketing A market survey shows that half the owners of Sorey State Boogie Boards became disenchanted with the product and switched to C&T Super Professional Boards the next surf season, and the other half remained loyal to Sorey State. On the other hand, three-quarters of the C&T Boogie Board users remained loyal to C&T, and the rest switched to Sorey State.

22. **Major Switching** At Suburban Community College, 10% of all business majors switched to another major the next semester, and the remaining 90% continued as business majors. Of all nonbusiness majors, 20% switched to a business major the following semester, and the rest did not.

23. **Pest Control** In an experiment to test the effectiveness of the latest roach trap, the Roach Resort, 50 roaches were placed in the vicinity of the trap and left there for 1 hour. At the end of the hour, it was observed that 30 of them had "checked in," and the rest were still scurrying around. (Remember that "once a roach checks in, it never checks out.")

24. **Employment** You have worked for the Department of Administrative Affairs (DAA) for 27 years, and you still have little or no idea exactly what your job entails. To make your life a little more interesting, you have decided on the following course of action. Every Friday afternoon, you will roll a die once. If you roll a 6, you will immediately quit your job, never to return. Otherwise, you will return to work the following Monday. [Use the following states: (1) employed by the DAA and (2) not employed by the DAA.]

25. **Textbook Adoptions** College instructors who adopt this book are (we hope!) twice as likely to continue to use the book the following year as they are to drop it, whereas nonusers are nine times as likely to remain nonusers the following year as they are to adopt this book.

26. **Confidence Level** Tommy the dunker's performance on the basketball court is influenced by his state of mind: If he scores he is twice as likely to score on the next shot as he is to miss, whereas if he misses a shot he is three times as likely to miss the next shot as he is to score.

27. **Genetics** For a couple who has one or more children, there is a 2/3 probability that the next child will be of the same gender as the preceding one.

    SOURCE: H. A. Hastings, Department of Mathematics, Hofstra University.

28. **Academics** The world is divided into two groups of people: those who graduated from the University of Chicago (including both authors of this book) and those who did not. According to a study, 5% of children of University of Chicago graduates attend the University of Chicago. Of these, 83% graduate. Assume that 0.0001% of the children of non-Chicago graduates go on to graduate from the University of Chicago.

    SOURCE: "Winds of Academic Change Rustle University of Chicago," *New York Times*, December 28, 1998, p. A1.

29. **Traffic Flow** The recently completed Enormous State University Monorail serves three terminals: the Main Caf Junction, the Math Department Interchange, and Field House Terminus. Prof. Hanis's diligent Finite Math students record the traffic flow and observe that half the students who board at Main Caf go to the Math Department, and the other half go to the Field House; one-third of the students who board at the Math Department go to the Main Caf, and the rest of them go to the Field House; one quarter of the students who board at the Field House go to the Main Caf, and the rest go to the Math Department.

30. **Air Traffic** A survey of air traffic at Lilliput's three major airports at Little Pebble, Little Venice, and Mini Metropolis yields the following data: One-half of the flights that depart from Little Pebble return there (due to confusion on the part of the air traffic controllers), and one-fourth of them wind up at Little Venice. Of those flights that leave Little Venice, half return there and half go to Mini Metropolis. Finally, one-fourth of the flights that leave Mini Metropolis (packed with eager Mini Metropolitans) wind up at Little Pebble, and the rest go to Little Venice.

31. **Tooth Decay** According to a University of Oregon Dental School survey of teenagers, 90% of badly decayed teeth remained in that state 6 months later, 5% of them were restored, and 5% of them were extracted. All restored teeth remained in that state (as did extracted teeth!). [Use the following states: (1) badly decayed, (2) restored, and (3) extracted.]

    SOURCE: K. H. Lu, "A Markov Chain Analysis of Caries Process with Consideration for the Effect of Restoration," *Archives of Oral Biology* 13 (1968): 1119–1132.

32. **Tooth Decay** The study cited in Exercise 31 showed that approximately 85% of healthy teeth were healthy 6 months later, 14% of them had decayed, and the rest had been restored. Approximately 2% of all decayed teeth were restored (and the rest remained decayed), and all restored teeth remained in that state. [Use the following states: (1) healthy, (2) decayed, and (3) restored.]

In Exercises 33–36, use the given information to set up a transition matrix.

33. **Random Walks** A random walk with "partially reflecting barriers" may be modeled by the following Markov chain:

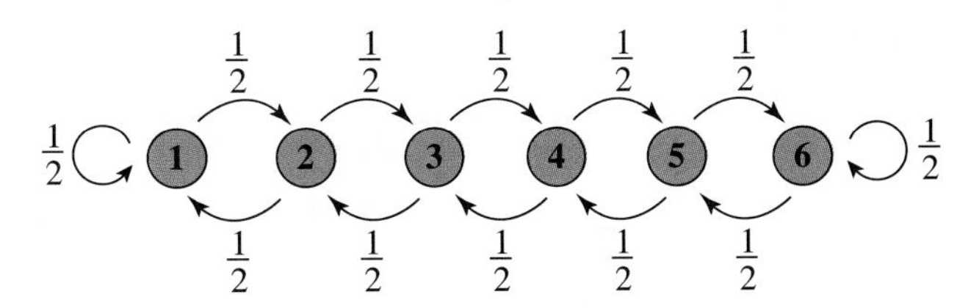

    This is a crude model for the motion of a particle along a straight line where the particle moves at random either to the left or the right at each time step. When it hits either end (state 1 or 6), it either remains there the next time step or is reflected back. This process is also known as a **discrete model of one-dimensional dissipation with partially reflecting barriers.**

34. **Random Walks** A discrete model of one-dimensional dissipation "with leftward drift" is shown in the following state transition diagram:

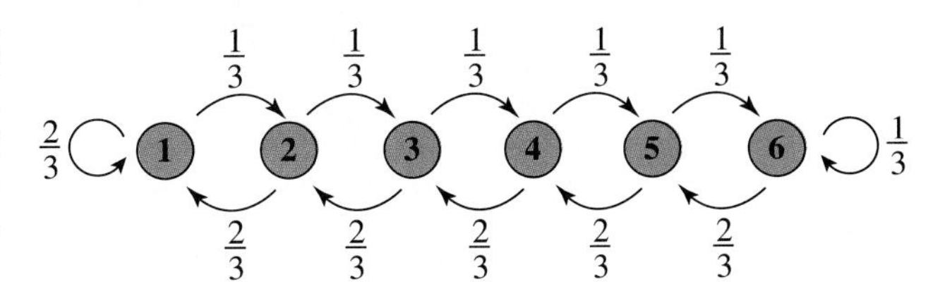

35. Population Movement In 1999, 0.15% of the population in the Northeast moved to the Midwest, 0.75% moved to the South, and 0.24% moved to the West. (The remainder stayed in the Northeast.) In the same year, 0.11% of the population in the Midwest moved to the Northeast, 0.57% moved to the South, and 0.32% moved to the West. Also, 0.18% of the population of the South moved to the Northeast, 0.42% moved to the Midwest, and 0.43% moved to the West. Finally, 0.17% of the population in the West moved to the Northeast, 0.35% moved to the Midwest, and 0.77% moved to the South.

Note that this exercise ignores migration into or out of the country. The internal migration figures and 2000 population figures are accurate. SOURCE: U.S. Census Bureau, Current Population Survey, http://www.census.gov/, March 2000.

36. Voting Shifts According to a *New York Times*/CBS News Poll taken in October 1990, 75% of people who had last voted for a Republican would continue to vote Republican, 6% would vote for a Democrat, and the rest would vote for an Independent. Of those that had last voted Democrat, 78% would continue to vote for a Democrat, 6% would vote for a Republican, and the rest would vote for an Independent. Also, 30% of Independent voters would vote for a Democrat, 29% would vote for a Republican, and the rest would vote for an Independent.

SOURCE: *New York Times*, October 12, 1990. We have taken a slight liberty in interpreting the poll; actually, the voters were classified according to party affiliation rather than according to the way they had last voted.

37. Brand Switching The following table contains data on purchases by 370 coffee drinkers:

Next Brand / *Current Brand*	**Maxwell Equation**	**Full O Choc**	**Gauss Jordan**	**Total**
Maxwell Equation	30	10	60	100
Full O Choc	30	60	30	120
Gauss Jordan	30	30	90	150
Total	90	100	180	370

Use the data in the table to set up a transition matrix for a Markov system whose states are the three brands of coffee.

38. Section Switching The following table contains data on 450 students who switched from one section of Finite Math to another last semester:

To / *From*	**Prof. Lively**	**Prof. Pomp**	**Prof. Snooze**	**Total**
Prof. Lively	30	30	90	150
Prof. Pomp	40	60	100	200
Prof. Snooze	30	30	40	100
Total	100	120	230	450

Use the data in the table to set up a transition matrix for a Markov system whose states are the three professors' sections.

COMMUNICATION AND REASONING EXERCISES

39. Describe an interesting situation that can be modeled by the transition matrix

$$P = \begin{bmatrix} .2 & .8 & 0 \\ 0 & 1 & 0 \\ .4 & .6 & 0 \end{bmatrix}$$

40. Describe an interesting situation that can be modeled by the transition matrix

$$P = \begin{bmatrix} .8 & .1 & .1 \\ 1 & 0 & 0 \\ .3 & .3 & .4 \end{bmatrix}$$

41. Describe some drawbacks to using Markov processes to model the behavior of the stock market with states (1) bull market and (2) bear market.

42. Can the repeated toss of a fair coin be modeled by a Markov process? If so, describe a model; if not, explain the reason.

43. If the transition matrix P of a Markov system has a column of zeros, what does that tell you about the corresponding state?

44. If the transition matrix P of a Markov system has a row with all but one entry zeros, what does that tell you about the corresponding state?

45. You would like to model consumer switching between short-sleeved shirts and long-sleeved shirts using a Markov system that also takes the season of purchase into account. What states could you use?

46. You would like to model a football kicker's success at field goals, using a Markov system that takes into account his success or failure in the last three kicks. How many states would you need, and what are they?

9.2 Distribution Vectors and Powers of the Transition Matrix

Let's return to Example 1 from Section 9.1.

Example 1 • Laundry Detergent Switching

A market analyst for Gamble Detergents is interested in whether consumers prefer powdered laundry detergents or liquid detergents. Two market surveys taken 1 year apart revealed that 20% of all powdered-detergent users had switched to liquid 1 year later and the rest were still using powder. Only 10% of liquid-detergent users had switched to powder 1 year later, with the rest still using liquid. At the time of the first market survey, out of 1400 consumers, 800 said they were using powder, and 600 said they were using liquid.

a. How many were using each type of detergent by the time of the second survey 1 year later?

b. Assuming the trend continues, predict how many consumers will be using each type of detergent 2 years after the initial survey.

Solution

a. In Section 9.1 we said that a consumer who uses powdered detergent is in state 1 and a consumer who uses liquid is in state 2. We saw that the transition diagram and matrix for the system are as shown in Figure 7.

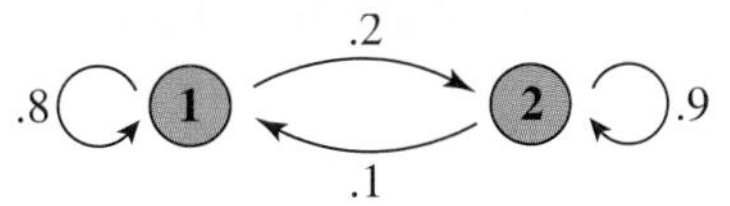

$$P = \begin{bmatrix} .8 & .2 \\ .1 & .9 \end{bmatrix}$$

Figure 7

Let's calculate how many consumers were using powder 1 year after the first survey. We know that 80% of the original 800 powder users were still using powder 1 year later and 10% of the 600 liquid users had switched to powder. Thus,

$$\text{total number of powder users after 1 year} = \underset{\substack{\uparrow \\ \text{From state 1}}}{.8 \times 800} + \underset{\substack{\uparrow \\ \text{From state 2}}}{.1 \times 600} = 700$$

This calculation is identical to the following matrix product:

$$[800 \quad 600]\underset{\substack{\uparrow \\ \text{First column of } P}}{\begin{bmatrix} .8 \\ .1 \end{bmatrix}} = [700]$$

We calculate the total number of liquid-detergent users after 1 year similarly:

$$\text{total number of liquid users after 1 year} = \underset{\substack{\uparrow \\ \text{From state 1}}}{.2 \times 800} + \underset{\substack{\uparrow \\ \text{From state 2}}}{.9 \times 600} = 700$$

Again, we have really calculated a matrix product:

$$[800 \quad 600]\underset{\substack{\uparrow \\ \text{Second column of } P}}{\begin{bmatrix} .2 \\ .9 \end{bmatrix}} = [700]$$

Thus, to get the distribution of detergent users after 1 year, all we have to do is *multiply the initial distribution* [800 600] *by the columns of P*. We can do both calculations at once by multiplying the row matrix [800 600] by the whole matrix *P*:

$$\underset{\text{Initial distribution}}{[800 \quad 600]}\underset{\text{Transition matrix}}{\begin{bmatrix} .8 & .2 \\ .1 & .9 \end{bmatrix}} = \underset{\text{Distribution after 1 step}}{[700 \quad 700]}$$

We refer to [800 600] as the **initial distribution vector** and we refer to the result [700 700] as the **distribution vector after one step.** The answer is the distribution after one step: We can expect 700 consumers to be using powdered detergent and 700 to be using liquid detergent after 1 year.

b. Now what about the distribution after *two* years? If we assume that the same fraction of consumers switch or stay put in the second year as in the first, we can simply repeat the calculation we did above, using the new distribution vector:

$$\underset{\text{Distribution after 1 step}}{[700 \quad 700]}\underset{\text{Transition matrix}}{\begin{bmatrix} .8 & .2 \\ .1 & .9 \end{bmatrix}} = \underset{\text{Distribution after 2 steps}}{[630 \quad 770]}$$

Thus, after 2 years we can expect 630 consumers to be using powdered detergent and 770 to be using liquid detergent.

Before we go on . . . Note that the sum of the entries in each of the distribution vectors is 1400, the total number of consumers surveyed.

If we wish to predict the usage after 3 or more years, we keep multiplying by *P*:

$$\underset{\text{Distribution after 2 steps}}{[630 \quad 770]}\underset{\text{Transition matrix}}{\begin{bmatrix} .8 & .2 \\ .1 & .9 \end{bmatrix}} = \underset{\text{Distribution after 3 steps}}{[581 \quad 819]}$$

$$\underset{\text{Distribution after 3 steps}}{[581 \quad 819]}\underset{\text{Transition matrix}}{\begin{bmatrix} .8 & .2 \\ .1 & .9 \end{bmatrix}} = \underset{\text{Distribution after 4 steps}}{[546.7 \quad 853.3]}$$

We do not mean that 546.7 consumers will be using powdered detergent 4 years later. If you have studied probability and statistics, you may recognize that what we are saying is that the *expected number* of consumers using powdered detergent will be 546.7. In other words, we are saying that, *on average,* 546.7 out of every 1400 consumers (or 39.05%) will be using detergent powder.

Using Technology

You are probably beginning to get the (correct) impression that there will be a great deal of matrix multiplication in this chapter, and this is where technology comes in very handy. In this example, try using technology to obtain the distribution after four steps, five steps, and so on. You may notice something interesting happening after a while. (We'll return to this in Section 9.3.) Here are some details on how to use specific technologies.

Graphing Calculator: Finding Distribution Vectors

In Chapter 3 we saw how to set up and multiply matrices. For this example we can use the matrix feature on the TI-83 to define

$$\texttt{[A]} = [800 \quad 600] \qquad \texttt{[B]} = \begin{bmatrix} 0.8 & 0.2 \\ 0.1 & 0.9 \end{bmatrix}$$

Entering [A] (obtained by pressing MATRX 1 ENTER) will show you the initial distribution. To obtain the distribution after one step, press × MATRX 2 ENTER, which has the effect of multiplying the previous answer by the transition matrix [B]. This will give you the distribution after two steps.

Now, just press ENTER repeatedly to continue multiplying by the transition matrix and obtain the distribution after any number of steps. Here is a table showing the results:

Action	Result
MATRX 1 ENTER	[800 600] Initial distribution
× MATRX 2 ENTER	[700 700] Distribution after one step
ENTER	[630 770] Distribution after two steps
ENTER	[581 819] Distribution after three steps
⋮	⋮

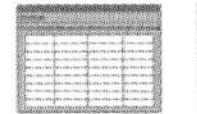

Excel: Finding Distribution Vectors

Enter the initial distribution vector in cells A1 and B1 and the transition matrix to the right of that, as shown.

	A	B	C	D	E
1	800	600		0.8	0.2
2				0.1	0.9
3					

To calculate the distribution after one step, use the array formula

```
=MMULT(A1:B1,$D$1:$E$2)
```

The absolute cell references (dollar signs) ensure that the formula always refers to the same transition matrix, even if we copy it into other cells. To use the array formula, select cells A2 and B2, where the distribution vector will go, enter this formula, and then press Control + Shift + Enter.*

	A	B	C	D	E
1	800	600		0.8	0.2
2	=MMULT(A1:B1,$D$1:$E$2)			0.1	0.9
3					

The result is the following:

	A	B	C	D	E
1	800	600		0.8	0.2
2	700	700		0.1	0.9
3					

*On a Macintosh, you can also use Command + Enter.

To calculate the distribution after two steps, select cells A2 and B2 and drag the fill handle down to copy the formula to cells A3 and B3. Note that the formula now takes the vector in A2:B2 and multiplies it by the transition matrix to get the vector in A3:B3.

	A	B	C	D	E
1	800	600		0.8	0.2
2	700	700		0.1	0.9
3	630	770			
4					

Calculations like this are easily done in Excel, using the worksheet above, by selecting cells A2 and B2 and dragging the fill handle down as far as desired.

	A	B	C	D	E
1	800	600		0.8	0.2
2	700	700		0.1	0.9
3					
4					
5					
6					

The result, in this case, is as shown:

	A	B	C	D	E
1	800	600		0.8	0.2
2	700	700		0.1	0.9
3	630	770			
4	581	819			
5	546.7	853.3			
6					

Web Site: Finding Distribution Vectors

Go to the matrix algebra tool by following the path

Web site → Online Utilities → Matrix Algebra Tool

You can then enter the transition matrix P and the initial distribution vector v in the input area and compute the various distribution vectors using the following formulas:

`v*P`	Distribution after one step
`v*P*P`	Distribution after two steps; alternative formula: `v*P^(2)`
`v*P*P*P`	Distribution after three steps; alternative formula: `v*P^(3)`
⋮	

Let's now restate symbolically what we did in Example 1. We saw that if

initial distribution $= v$

then

distribution after one step $= vP$

To get the distribution after two steps, we multiplied the answer, vP, by P, so

distribution after two steps $= (vP)P = vP^2$

distribution after three steps $= (vP^2)P = vP^3$

⋮

distribution after m steps $= vP^m$

Thus, to get the distribution after two steps, we multiply the initial distribution v by the matrix P^2. To get the distribution after three steps, we multiply v by the matrix P^3, and so on.

Question What do the entries of P^2 mean?

Answer P is the one-step transition matrix, and P^2 is the *two-step* transition matrix. In other words: The ijth entry in P^2 is the probability of going from state i to state j in *two steps.*

Generalizing, we get the following result, which summarizes what we know so far.

Powers of the Transition Matrix and the Distribution Vector after m Steps

P^m $(m = 1, 2, 3, \ldots)$ is the **m-step transition matrix.** The ijth entry in P^m is the probability of a transition from state i to state j in m steps.

If v is any distribution vector, the distribution vector after m steps is given by

$$\text{distribution after } m \text{ steps} = vP^m$$

Alternatively, multiply v on the right by P m times to get

$$\text{distribution after } m \text{ steps} = v \cdot P \cdot P \cdot \cdots \cdot P \qquad (m \text{ times})$$

Quick Examples

1. $P = \begin{bmatrix} 0 & 1 \\ .5 & .5 \end{bmatrix}$ One-step transition matrix

$P^2 = P \cdot P = \begin{bmatrix} .5 & .5 \\ .25 & .75 \end{bmatrix}$ Two-step transition matrix

$P^3 = P \cdot P^2 = \begin{bmatrix} .25 & .75 \\ .375 & .625 \end{bmatrix}$ Three-step transition matrix

The probability of going from state 1 to state 2 in two steps = (1, 2) entry of $P^2 = .5$.

The probability of going from state 1 to state 2 in three steps = (1, 2) entry of $P^3 = .75$.

2. If $v = [100 \quad 100]$ is the initial distribution, then

distribution after one step $= vP = [100 \quad 100]\begin{bmatrix} 0 & 1 \\ .5 & .5 \end{bmatrix} = [50 \quad 150]$

distribution after two steps $= vP^2 = [100 \quad 100]\begin{bmatrix} .5 & .5 \\ .25 & .75 \end{bmatrix} = [75 \quad 125]$

Alternatively, multiply vP by P: $[50 \quad 150]\begin{bmatrix} 0 & 1 \\ .5 & .5 \end{bmatrix} = [75 \quad 125]$

distribution after three steps $= vP^3 = [100 \quad 100]\begin{bmatrix} .25 & .75 \\ .375 & .625 \end{bmatrix} = [62.5 \quad 137.5]$

Alternatively, multiply vP^2 by P: $[75 \quad 125]\begin{bmatrix} 0 & 1 \\ .5 & .5 \end{bmatrix} = [62.5 \quad 137.5]$

Example 2 • Investments

In Section 9.1 we looked at an example of a Markov system in which the states represented the financial condition of a savings and loan institution, and we found the 1-year transition diagram and transition matrix shown in Figure 8.

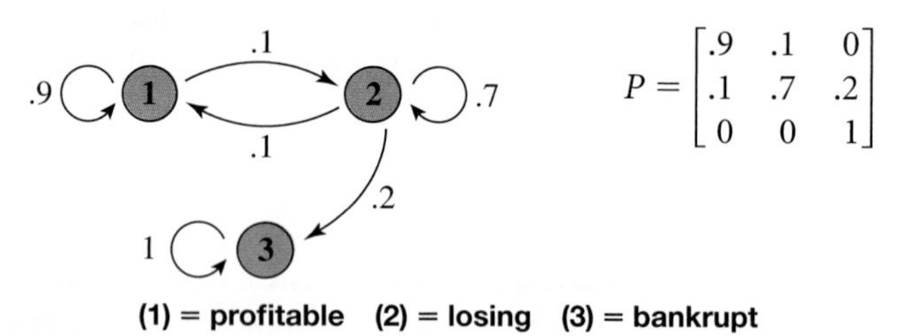

$$P = \begin{bmatrix} .9 & .1 & 0 \\ .1 & .7 & .2 \\ 0 & 0 & 1 \end{bmatrix}$$

Figure 8

a. Calculate the two- and three-step transition matrices.

b. The Solid Trust Savings and Loan is presently operating at a loss. Find the probability that it will be bankrupt in each of the next 3 years.

c. Suppose that half of the S&Ls in the country are profitable and half are losing money. Determine how many of them we can expect to be bankrupt after 3 years.

Solution

a. We are given the one-step transition matrix:

$$P = \begin{bmatrix} .9 & .1 & 0 \\ .1 & .7 & .2 \\ 0 & 0 & 1 \end{bmatrix}$$

The two-step transition matrix is

$$P^2 = P \cdot P = \begin{bmatrix} .9 & .1 & 0 \\ .1 & .7 & .2 \\ 0 & 0 & 1 \end{bmatrix}\begin{bmatrix} .9 & .1 & 0 \\ .1 & .7 & .2 \\ 0 & 0 & 1 \end{bmatrix} = \begin{bmatrix} .82 & .16 & .02 \\ .16 & .5 & .34 \\ 0 & 0 & 1 \end{bmatrix}$$

Notice that the entries in each row of P^2 add to 1, just as they do in P. Incidentally, this makes it slightly easier to calculate P^2: All we need to do is calculate two of the three entries in each row. The third entry in each row is obtained by subtracting the sum of the other two from 1. Here is the three-step transition matrix, P^3:

$$P^3 = P \cdot P^2 = \begin{bmatrix} .9 & .1 & 0 \\ .1 & .7 & .2 \\ 0 & 0 & 1 \end{bmatrix}\begin{bmatrix} .82 & .16 & .02 \\ .16 & .5 & .34 \\ 0 & 0 & 1 \end{bmatrix} = \begin{bmatrix} .754 & .194 & .052 \\ .194 & .366 & .44 \\ 0 & 0 & 1 \end{bmatrix}$$

b. We are asked for the probabilities of going from the losing state (state 2) to the bankrupt state (state 3) in one, two, and three steps. The probability of the transition from state 2 to state 3 is given by the (2, 3) entry of the appropriate transition matrix. Let's start with the one-step transition matrix P:

$$P = \begin{bmatrix} .9 & .1 & 0 \\ .1 & .7 & \boxed{.2} \\ 0 & 0 & 1 \end{bmatrix}$$

There is a .2 (or 20%) probability that a losing S&L will be bankrupt 1 year later. We have

$$P^2 = \begin{bmatrix} .82 & .16 & .02 \\ .16 & .5 & \boxed{.34} \\ 0 & 0 & 1 \end{bmatrix}$$

There is a .34 (or 34%) probability that a losing S&L will be bankrupt 2 years later. Also,

$$P^3 = \begin{bmatrix} .754 & .194 & .052 \\ .194 & .366 & \boxed{.44} \\ 0 & 0 & 1 \end{bmatrix}$$

There is a .44 (or 44%) probability that a losing S&L will be bankrupt 3 years later. Notice how the probability of going bankrupt increases from year to year!

c. For the initial distribution vector we can take

$$v = [.5 \quad .5 \quad 0]$$

representing the fact that half of the S&Ls are profitable, half are losing money, and none are (yet) bankrupt. We could now calculate vP, $(vP)P$, and finally $((vP)P)P$ by repeatedly multiplying by P as we did in Example 1. Alternatively, we could use the fact that the expected distribution after 3 years can be computed using the three-step transition matrix P^3:

$$vP^3 = [.5 \quad .5 \quad 0]\begin{bmatrix} .754 & .194 & .052 \\ .194 & .366 & .44 \\ 0 & 0 & 1 \end{bmatrix} = [.474 \quad .280 \quad .246]$$

Thus, we can expect 24.6% of the S&Ls to be bankrupt after 3 years.

Before we go on . . . We can think of the entries of the initial distribution vector v in part (c) as being the probabilities that a randomly chosen S&L is profitable, losing money, or bankrupt. We can interpret the entries of vP^3 as probabilities similarly. A vector that has nonnegative entries adding up to 1 (like v) is often called a **probability vector.**

Using Technology

It is an exceedingly tedious task to find high powers of a matrix by hand or with an ordinary calculator. However, it becomes a simple task using technology, as shown below. Use technology to calculate higher and higher powers of P in this example. Do you notice anything interesting that happens?

Graphing Calculators: Calculating High Powers

On the TI-83 you can, for instance, obtain the fourth power of the matrix `[A]` by multiplying `[A]*[A]*[A]*[A]` or by entering `[A]^4`.

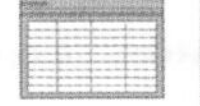

Excel: Calculating High Powers

Begin by entering P in the top left of the worksheet. To help tell where one matrix ends and the next begins, select the matrix and draw a border around its outside, as shown:

	A	B	C
1	0.9	0.1	0
2	0.1	0.7	0.2
3	0	0	1
4			

To calculate P^2, select cells A4:C6, where the result will go, and enter the following formula:

```
=MMULT(A1:C3,$A$1:$C$3)
```

Once again, the dollar signs ensure that the second argument always refers to the original transition matrix P. Remember to press Control+Shift+Enter (or Command+Enter) when entering this formula. Once again, draw a border around the result:

	A	B	C
1	0.9	0.1	0
2	0.1	0.7	0.2
3	0	0	1
4	0.82	0.16	0.02
5	0.16	0.5	0.34
6	0	0	1
7			

To calculate P^3, select P^2 and drag the fill handle down three rows:

	A	B	C	D
1	0.9	0.1	0	
2	0.1	0.7	0.2	
3	0	0	1	
4	0.82	0.16	0.02	
5	0.16	0.5	0.34	
6	0	0	1	
7				
8				
9				

The result is as follows:

	A	B	C
1	0.9	0.1	0
2	0.1	0.7	0.2
3	0	0	1
4	0.82	0.16	0.02
5	0.16	0.5	0.34
6	0	0	1
7	0.754	0.194	0.052
8	0.194	0.366	0.44
9	0	0	1
10			

Web Site: Computing High Powers

If you follow the path

Web site → Online Utilities → Matrix Algebra Tool

you can enter the transition matrix P and the distribution vector v and compute both powers of P and products of the form vP^n, using the format `v*P^(n)`. There is a more specialized utility at

Web site → Online Utilities → Markov Computation Utility

which you can use instead.

9.2 EXERCISES

In Exercises 1–22, use the given transition matrix P and initial distribution vector v to obtain (a) the two- and three-step transition matrices P^2 and P^3 and (b) the distribution vectors after two and three steps.

1. $P = \begin{bmatrix} 0 & 1 \\ 1 & 0 \end{bmatrix}$, $v = [100 \quad 0]$

2. $P = \begin{bmatrix} 0 & 1 \\ 0 & 1 \end{bmatrix}$, $v = [100 \quad 0]$

3. $P = \begin{bmatrix} .5 & .5 \\ 0 & 1 \end{bmatrix}$, $v = [100 \quad 0]$

4. $P = \begin{bmatrix} 1 & 0 \\ .5 & .5 \end{bmatrix}$, $v = [0 \quad 100]$

5. $P = \begin{bmatrix} \frac{1}{2} & \frac{1}{2} \\ 1 & 0 \end{bmatrix}$, $v = [10 \quad 5]$

6. $P = \begin{bmatrix} 0 & 1 \\ \frac{1}{4} & \frac{3}{4} \end{bmatrix}$, $v = [20 \quad 80]$

7. $P = \begin{bmatrix} \frac{3}{4} & \frac{1}{4} \\ \frac{3}{4} & \frac{1}{4} \end{bmatrix}$, $v = [50 \quad 50]$

8. $P = \begin{bmatrix} \frac{2}{3} & \frac{1}{3} \\ \frac{2}{3} & \frac{1}{3} \end{bmatrix}$, $v = [30 \quad 60]$

9. $P = \begin{bmatrix} 0 & 1 & 0 \\ 0 & 0 & 1 \\ 1 & 0 & 0 \end{bmatrix}$, $v = [0 \quad 20 \quad 0]$

10. $P = \begin{bmatrix} 0 & 1 & 0 \\ 0 & 0 & 1 \\ 0 & 0 & 1 \end{bmatrix}$, $v = [20 \quad 0 \quad 0]$

11. $P = \begin{bmatrix} .5 & .5 & 0 \\ 0 & 1 & 0 \\ 0 & .5 & .5 \end{bmatrix}$, $v = [1000 \quad 0 \quad 0]$

12. $P = \begin{bmatrix} .5 & 0 & .5 \\ 1 & 0 & 0 \\ 0 & .5 & .5 \end{bmatrix}$, $v = [0 \quad 1000 \quad 0]$

13. $P = \begin{bmatrix} 0 & 1 & 0 \\ \frac{1}{3} & \frac{1}{3} & \frac{1}{3} \\ 1 & 0 & 0 \end{bmatrix}$, $v = [900 \quad 0 \quad 900]$

14. $P = \begin{bmatrix} \frac{1}{2} & \frac{1}{2} & 0 \\ \frac{1}{2} & \frac{1}{2} & 0 \\ \frac{1}{2} & 0 & \frac{1}{2} \end{bmatrix}$, $v = [20 \quad 0 \quad 20]$

15. $P = \begin{bmatrix} .1 & .9 & 0 \\ 0 & 1 & 0 \\ 0 & .2 & .8 \end{bmatrix}$, $v = [1 \quad 0 \quad 0]$

16. $P = \begin{bmatrix} .1 & .1 & .8 \\ .5 & 0 & .5 \\ .5 & 0 & .5 \end{bmatrix}$, $v = [1 \quad 0 \quad 0]$

T For Exercises 17–22, use of technology is recommended. Round all answers to two decimal places.

17. $P = \begin{bmatrix} .2 & .8 \\ .1 & .9 \end{bmatrix}$, $v = [100 \quad 200]$

18. $P = \begin{bmatrix} .3 & .7 \\ .8 & .2 \end{bmatrix}$, $v = [200 \quad 100]$

19. $P = \begin{bmatrix} .25 & .25 & .50 \\ 0 & .35 & .65 \\ .50 & .15 & .35 \end{bmatrix}$, $v = [10 \quad 10 \quad 10]$

20. $P = \begin{bmatrix} .15 & .35 & .50 \\ 0 & .55 & .45 \\ .50 & .25 & .25 \end{bmatrix}$, $v = [1 \quad 1 \quad 0]$

21. $P = \begin{bmatrix} .5 & .25 & .25 & 0 \\ .3 & .3 & .3 & .1 \\ .1 & .2 & .1 & .6 \\ 1 & 0 & 0 & 0 \end{bmatrix}$, $v = [.5 \quad 0 \quad .5 \quad 0]$

22. $P = \begin{bmatrix} .3 & .3 & .25 & .15 \\ 0 & 1 & 0 & 0 \\ 0 & .25 & .25 & .5 \\ .1 & .1 & .1 & .7 \end{bmatrix}$, $v = [0 \quad .5 \quad 0 \quad .5]$

APPLICATIONS

Many of the following exercises involve scenarios similar to those introduced in the preceding section.

23. Marketing A market survey shows that half the owners of Sorey State Boogie Boards became disenchanted with the product and switched to C&T Super Professional Boards the next surf season, and the other half remained loyal to Sorey State. On the other hand, three-quarters of the C&T Boogie Board users remained loyal to C&T, and the rest switched to Sorey State.

a. Set these data up as a Markov transition matrix and calculate the probability that a Sorey State Board user will be using the same brand two seasons later.

b. A sample of 800 surf riders is equally split between Sorey and C&T. What do you expect the distribution to be in two seasons' time?

24. Major Switching At Suburban Community College, 10% of all business majors switched to another major the next semester, and the remaining 90% continued as business majors. Of all nonbusiness majors, 20% switched to a business major the following semester, and the rest did not.

a. Set up these data as a Markov transition matrix and calculate the probability that a business major will no longer be a business major in two semesters' time.

b. One hundred freshmen register as business majors and 500 register as nonbusiness majors. What do you expect the distribution to look like in two semesters?

25. Pest Control In an experiment to test the effectiveness of the latest roach trap, the Roach Resort, 50 roaches were placed in the vicinity of the trap and left there for 1 hour. At the end of the hour, it was observed that 30 of them had "checked in," and the rest were still scurrying around. (Remember that "once a roach checks in, it never checks out.")

a. Set up the transition matrix P for the system with decimal entries and calculate P^2 and P^3.

b. Given the initial distribution of 50 roaches outside the Resort, track what becomes of them at the end of hours 1, 2, and 3. (Round answers to the nearest roach.)

c. What do you expect to be the long-term impact on the number of roaches?

26. Employment You have worked for the Department of Administrative Affairs (DAA) for 27 years, and you still have little or no idea exactly what your job entails. To make your life a little more interesting, you have decided on the following course of action. Every Friday afternoon, you will use your desktop computer to generate a random digit from 0 to 9 (inclusive). If the digit is a 0, you will immediately quit your job, never to return. Otherwise, you will return to work the following Monday.

a. Use the states (1) employed by the DAA and (2) not employed by the DAA to set up a transition probability matrix P with decimal entries and calculate P^2 and P^3.

b. What is the probability that you will still be employed by the DAA after each of the next 3 weeks?

c. What are your long-term prospects for employment at the DAA?

27. Textbook Adoptions College instructors who adopt this book are (we hope!) twice as likely to continue to use the book the following year as they are to drop it, whereas nonusers are nine times as likely to remain nonusers the following year as they are to adopt this book. Determine the probability that a nonuser will be a user in 2 years.

28. Confidence Level Tommy the dunker's performance on the basketball court is influenced by his state of mind: If he scores, he is twice as likely to score on the next shot as he is to miss, whereas if he misses a shot, he is three times as likely to miss the next shot as he is to score. If Tommy misses a shot, find the probability that he will score two shots later.

29. Genetics For a couple who has one or more children, there is a 2/3 probability that the next child will be of the same gender as the preceding one. Hilde and Wolfgang have just had their first child, a boy. What is the probability that their third child will be a girl?

SOURCE: Conversations with H. A. Hastings, Department of Mathematics, Hofstra University.

30. Academics The world is divided into two groups of people: those who graduated from the University of Chicago (including both authors of this book) and those who did not. According to a study, 5% of children of University of Chicago graduates attend the University of Chicago. Of these, 83% graduate. Assume that 0.0001% of the children of non-Chicago graduates go on to graduate from the University of Chicago. April May is a graduate of the University of Chicago. What is the probability that her granddaughter will also graduate from the University of Chicago?

SOURCE: "Winds of Academic Change Rustle University of Chicago," *New York Times*, December 28, 1998, p. A1.

31. Investment Risk Two surveys of the S&L industry were taken 1 year apart. The results were as follows: 10% of those S&Ls that were profitable at the time of the first survey were found to be operating at a loss by the time of the second survey, and the rest of them remained profitable. Of the S&Ls that were operating at a loss at the time of the first survey, 20% were bankrupt by the time of the second survey, and 10% of them became profitable. All the bankrupt S&Ls were bailed out by the government (at taxpayers' expense) and set up once again as profitable institutions.

a. Write down the transition matrix for this system.

b. I wish to invest my life savings in Solid Trust S&L, whose recent balance sheet showed a healthy profit. Find the probability that it will be profitable 1 year, 2 years, and 3 years from now.

32. More Investment Risk Refer to Exercise 31. How would your answers be affected if the government policy was *not* to bail out failed S&Ls but simply to let them close down (as many people believe should be the case)?

33. Traffic Flow The recently completed Enormous State University Monorail serves three terminals: the Main Caf Junction, the Math Department Interchange, and the Field House Terminus. Prof. Hanis's diligent Finite Math students record the traffic flow and observe that half the students who board at the Main Caf go to the Math Department, and the other half go to the Field House; one-third of the students who board at the Math Department go to the Main Caf, and the rest of them go to the Field House; one-quarter of the students who board at the Field House go to the Main Caf, and the rest of them go to the Math Department.

a. What is the probability that a traveling student who boards at the Main Caf will be at the Math Department two trips later?

b. If 360 students board at the Math Department and spend the day riding the monorail, how are they distributed after one, two, and three trips?

34. Air Traffic A survey of air traffic at Lilliput's three major airports at Little Pebble, Little Venice, and Mini Metropolis yields the following data: One-half of the flights that depart from Little Pebble return there (due to confusion on the part of the air traffic controllers), and one-fourth of them wind up at Little Venice. Of those flights that leave Little Venice, half return there and half go to Mini Metropolis. Finally, one-fourth of the flights leaving Mini Metropolis (packed with eager Mini Metropolitans) wind up at Little Pebble, and the rest go to Little Venice.

a. Compute the probability that a passenger who boards a random flight at Little Pebble will wind up at Mini Metropolis after two trips.

b. A sample of 1600 Mini Metropolitans who take random flights is watched over a three-trip period. Where can you expect to find these travelers at the end of each trip?

35. Income Brackets The following diagram shows the movement of U.S. households among three income groups, affluent, middle class, and poor, over the 11-year period 1980–1991:

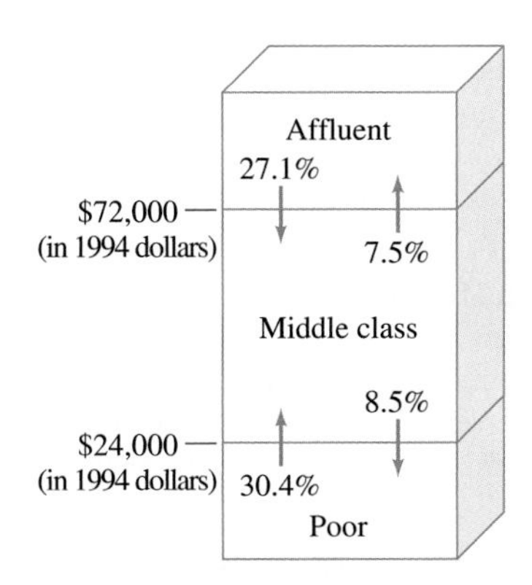

The figures are based on household after-tax income. The study, conducted by G. J. Duncan of Northwestern University and T. Smeeding of Syracuse University, is based on annual surveys of the personal finances of 5000 households since the late 1960s. (The surveys were conducted by the University of Michigan.) SOURCE: *New York Times*, June 4, 1995, p. E4.

a. Use the transitions shown in the diagram to construct a transition matrix (assuming zero probabilities for the transitions between affluent and poor).

b. Assuming that the trend shown were to continue, what percent of households classified as affluent in 1980 were predicted to become poor in 2002? (Give your answer to the nearest 0.1%.)

36. Income Brackets The following diagram shows the movement of U.S. households among three income groups, affluent, middle class, and poor, over the 12-year period 1967-1979:

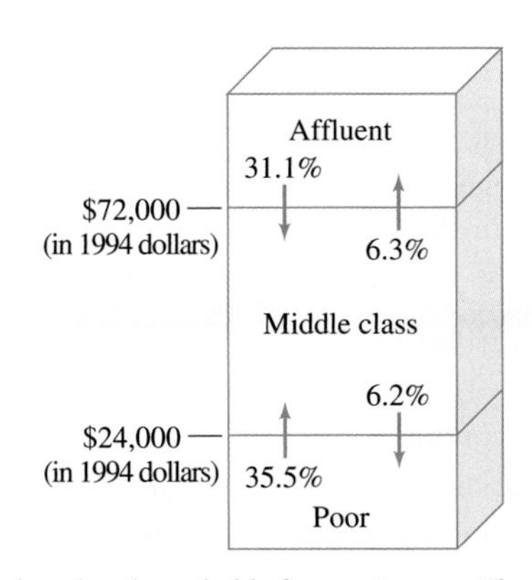

The figures are based on household after-tax income. The study, conducted by G. J. Duncan of Northwestern University and T. Smeeding of Syracuse University, is based on annual surveys of the personal finances of 5000 households since the late 1960s. (The surveys were conducted by the University of Michigan.) SOURCE: *New York Times*, June 4, 1995, p. E4.

a. Use the transitions shown in the diagram to construct a transition matrix (assuming zero probabilities for the transitions between affluent and poor).

b. Assuming that the trend shown had continued, what percent of households classified as affluent in 1967 would have been poor in 1991? (Give your answer to the nearest 0.1%.)

37. Debt Analysis You manage the credit department for a large chain of stores, and you have access to the debt history of all your customers. Of the total debt owed to your company, $4 million is in the bad-debt category (representing accounts on which nothing has been paid for more than 90 days), $5 million is in the 31–90-day category (representing accounts on which the last payment was in this period), $8 million is in the 0–30-day category, and $100 million is in the paid category. The following table gives the probabilities that a single dollar will move from one category to the next in the period of 1 month:

	To			
From	**Paid**	**0–30 Days**	**31–90 Days**	**Bad Debts**
Paid	1	0	0	0
0–30 Days	.5	.3	.2	0
31–90 Days	.2	.5	.2	.1
Bad Debts	.1	.2	0	.7

SOURCE: Based on a discussion in D. R. Anderson, D. J. Sweeny, and T. A. Williams, *An Introduction to Management Science*, 6th ed. (St. Paul: West, 1991).

a. How do you expect the company's present debts to be distributed 1 month from now?

b. What will be the distribution 2 months from now?

c. Assuming that the company sells $1 million on credit during each month, predict the distribution over the coming 2 months. (Count amounts sold on credit during a month as 0–30-day debt at the end of that month.)

38. Debt Analysis Repeat Exercise 37, using the following table:

	To			
From	**Paid**	**0–30 Days**	**31–90 Days**	**Bad Debts**
Paid	1	0	0	0
0–30 Days	.3	.5	.2	0
31–90 Days	.2	.2	.5	.1
Bad Debts	.1	.1	0	.8

39. Random Walks A random walk with partially reflecting barriers may be modeled by the following Markov chain:

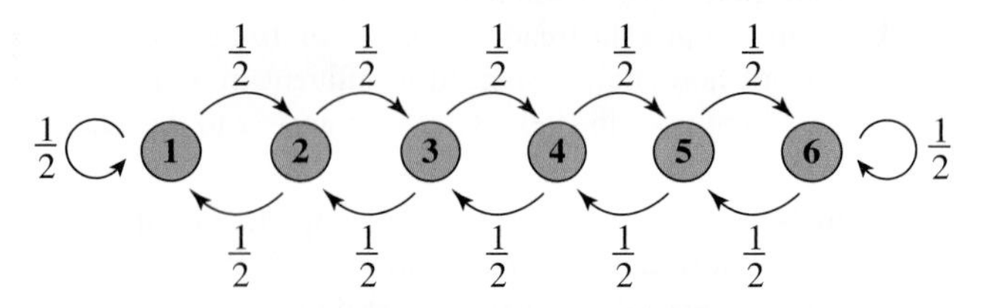

a. Write down the transition matrix for the process.

b. Compute the two-step transition matrix and use it to calculate the distribution vector after two steps if the initial distribution vector is [1 0 0 0 0 0].

c. Find the probability that a particle starting in state 5 will be in state 2 after four steps.

40. Dissipation with Drift A discrete model of one-dimensional dissipation with drift is given by the following state transition diagram:

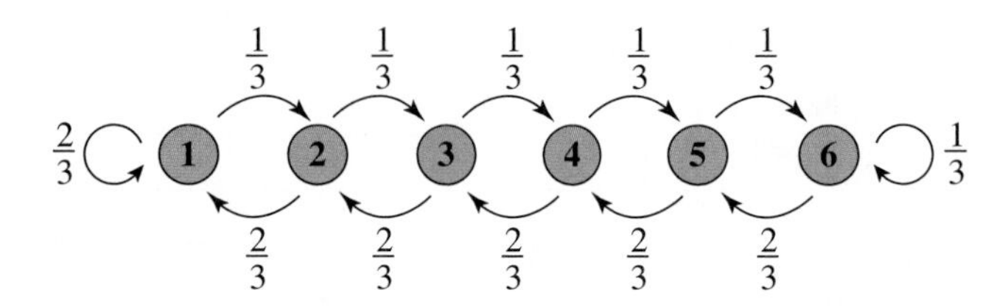

a. Write down the transition matrix for the process.

b. Compute the two-step transition matrix and use it to calculate the distribution vector after two steps if the initial distribution vector is [1 0 0 0 0 0].

c. Find the probability that a particle starting in state 1 will be in state 6 after four steps.

 In Exercises 41–48, use technology to solve.

41. Income Distribution A University of Michigan study shows the following one-generation transition probabilities among four major income groups:

	Son's Income			
Father's Income	**Bottom 10%**	**10–50%**	**50–90%**	**Top 10%**
Bottom 10%	.30	.52	.17	.01
10–50%	.10	.48	.38	.04
50–90%	.04	.38	.48	.10
Top 10%	.01	.17	.52	.30

SOURCES: Gary Solon, University of Michigan/*New York Times,* May 18, 1992, p. D5. We have adjusted some of the figures so that the probabilities add to 1; they did not do so in the original table due to rounding.

If a sample of 100 male earners is evenly distributed among the four income groups, what will the distribution look like after 1, 2, and 3 generations of eldest sons?[4] What trend does this suggest?

42. Income Distribution Refer to Exercise 41. Suppose you are a man in the top 10% income group. Find the probability that your grandson and great-grandson will be in one of the top two income groups. Contrast this with the case in which you are in the bottom 10% income group and comment on the old adage "riches to riches."

43. Population Movement In 1999 the U.S. population, broken down by regions, was 53.9 million in the Northeast, 64.3 million in the Midwest, 100.0 million in the South, and 63.3 million in the West. The matrix P below shows the population movement during the period 1999–2000:

$$P = \begin{array}{r} \\ \textbf{From NE} \\ \textbf{From MW} \\ \textbf{From S} \\ \textbf{From W} \end{array} \begin{array}{c} \begin{array}{cccc} \textbf{To NE} & \textbf{To MW} & \textbf{To S} & \textbf{To W} \end{array} \\ \begin{bmatrix} .9886 & .0015 & .0075 & .0024 \\ .0011 & .9900 & .0057 & .0032 \\ .0018 & .0042 & .9897 & .0043 \\ .0017 & .0035 & .0077 & .9871 \end{bmatrix} \end{array}$$

Use these data to predict the distribution of the population in 2000, 2001, and 2002, assuming that the 1999–2000 trend continues. (Round your final answers to two decimal places.)

Note that this exercise ignores migration into or out of the country. The internal migration figures and 2000 population figures are accurate. SOURCE: U.S. Census Bureau, Current Population Survey, http://www.census.gov/, March 2000.

44. Voting Shifts According to a *New York Times*/CBS News Poll taken in October 1990, 75% of people who had last voted for a Republican would continue to vote Republican, 6% would vote for a Democrat, and the rest would vote for an Independent. Of those who had last voted Democrat, 78% would continue to vote for a Democrat, 6% would vote for a Republican, and the rest would vote for an Independent. Also, 30% of Independent voters would vote for a Democrat, 29% would vote for a Republican, and the rest would vote for an Independent. Given a pool of 100 Republicans, 150 Democrats, and 50 Independents, use this information to predict the vote in the next election and the following two. (Round your answers to the nearest voter.)

SOURCE: *New York Times,* October 12, 1990. We have taken a slight liberty in interpreting the poll; actually, the voters were classified according to party affiliation rather than according to the way they had last voted.

45. Debt Analysis Refer to Exercise 37 and predict the company's distribution of debts 6 months from now.

46. Debt Analysis Refer to Exercise 38 and predict the company's distribution of debts 6 months from now.

[4] In this way we are simplifying the problem by tracking only one offspring per generation. Of course, it need not be the case that everyone has a son, so we can't really use this model to predict the total number of offspring in each income group unless we know how the birthrate varies with income groups.

47. Random Walks Refer to the state transition diagram in Exercise 39. Evaluate the eighth power of the associated transition matrix and use it to determine the most likely state for the system to be in after eight steps if it starts in state 3. (Round your answers to four decimal places.)

48. Dissipation with Drift Repeat Exercise 47 but refer instead to the state transition diagram in Exercise 40.

COMMUNICATION AND REASONING EXERCISES

49. Let $P = \begin{bmatrix} 0 & 0 & 1 \\ 1 & 0 & 0 \\ 0 & 1 & 0 \end{bmatrix}$. Describe *all* powers of P.

50. Let $P = \begin{bmatrix} 0 & 1 & 0 & 0 \\ 0 & 0 & 1 & 0 \\ 0 & 0 & 0 & 1 \\ 0 & 0 & 0 & 1 \end{bmatrix}$. Describe *all* powers of P.

51. You are using a Markov system to model the state of your customers' accounts with the three states (1) account is fully paid up, (2) last payment on account received 1–6 months ago, and (3) no payment received for over 6 months. You find, with P the 1-month transition matrix, that

$$P^{12} = \begin{bmatrix} .35 & .40 & .25 \\ .35 & .40 & .25 \\ .35 & .40 & .25 \end{bmatrix}$$

How do you interpret the result?

52. You are using a Markov system to model brand switching among brands 1, 2, and 3 of toothpaste and you find, with P the 1-year transition matrix, that $[100 \quad 200 \quad 300]P^5 = [300 \quad 200 \quad 100]$. How do you interpret this?

53. You wish to model car-buyers' preferences for domestic versus imported automobiles with a Markov system that has a 1-year time step and the states (1) foreign car owner and (2) domestic car owner.

a. Describe how you would use two surveys taken 1 year apart to construct the model, indicating what specific questions you would ask.

b. If you use this model to predict buying habits in 5 years' time, what underlying assumption are you making?

54. A **deterministic** Markov process is one whose transition matrix has only 0s and 1s as entries. If P is the transition matrix of such a system, what can you say about the powers of P?

T In Exercises 55–58, use technology to solve.

55. Let $P = \begin{bmatrix} .1 & .9 \\ .8 & .2 \end{bmatrix}$. Calculate P^{20}. What do you notice about the answer?

56. Repeat Exercise 55, with $P = \begin{bmatrix} .3 & .5 & .2 \\ .1 & .2 & .7 \\ .2 & .2 & .6 \end{bmatrix}$.

57. Find the sixth power of $P = \begin{bmatrix} \frac{1}{3} & \frac{2}{3} \\ \frac{1}{9} & \frac{8}{9} \end{bmatrix}$ *exactly* as follows:

a. Observe that $9P$ has only integer entries.

b. Calculate the sixth power of $9P$.

c. Use the equation $P^6 = \frac{1}{9^6}(9P)^6$ to express P^6 in terms of fractions with denominator 9^6.

58. a. Use the method of Exercise 57 to obtain the seventh power of $P = \begin{bmatrix} \frac{1}{3} & \frac{2}{3} \\ \frac{1}{2} & \frac{1}{2} \end{bmatrix}$.

b. Approximate $\frac{1}{3}$ by 0.33 and $\frac{2}{3}$ by 0.67 and calculate the seventh power of $P = \begin{bmatrix} .33 & .67 \\ .5 & .5 \end{bmatrix}$ directly. How accurate is this approximation?

9.3 Long-Range Behavior of Regular Markov Systems

We return to our laundry detergent example.

Example 1 • Laundry Detergent Switching

The state transition diagram and transition matrix for the laundry detergent example (Example 1 in both Sections 9.1 and 9.2) are shown in Figure 9.

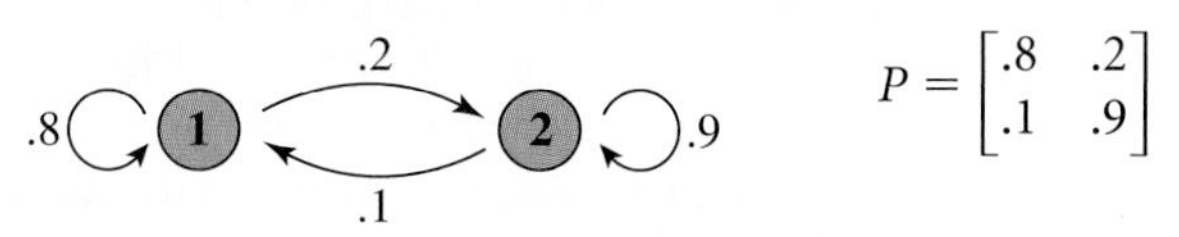

(1) = uses powder (2) = uses liquid

Figure 9

Suppose that one year 70% of consumers use powder and 30% use liquid. Assuming that the transition matrix remains valid the whole time, what will be the distribution 1 2, 3, . . . , and 50 years later?

Solution We need technology to do these calculations (see the technology discussion in Example 1 of Section 9.2), but we can start off working by hand:

$$\text{Distribution after 1 year: } [.7 \quad .3]\begin{bmatrix} .8 & .2 \\ .1 & .9 \end{bmatrix} = [.59 \quad .41]$$

$$\text{Distribution after 2 years: } [.59 \quad .41]\begin{bmatrix} .8 & .2 \\ .1 & .9 \end{bmatrix} = [.513 \quad .487]$$

$$\text{Distribution after 3 years: } [.513 \quad .487]\begin{bmatrix} .8 & .2 \\ .1 & .9 \end{bmatrix} = [.4591 \quad .5409]$$

$\vdots$

Distribution after 48 years: [.333 333 35 .666 666 65]

Distribution after 49 years: [.333 333 34 .666 666 66]

Distribution after 50 years: [.333 333 34 .666 666 66]

Thus, the distribution after 50 years is approximately [.333 333 34 .666 666 66].

Before we go on . . . Something interesting seems to be happening. The distribution seems to be getting closer and closer to

$$[.333\,333\ldots \quad .666\,666\ldots] = \begin{bmatrix} \frac{1}{3} & \frac{2}{3} \end{bmatrix}$$

Let's call this distribution vector v_∞. Notice two things about v_∞:

- The entries of v_∞ are positive and add up to 1. As we mentioned in Example 2 in Section 9.2, a vector with nonnegative entries adding up to 1 is called a **probability vector.** In this case we can interpret the entries of v_∞ as the probabilities that a randomly chosen consumer uses powdered detergent or liquid detergent.
- If we calculate $v_\infty P$, we find

$$v_\infty P = \begin{bmatrix} \frac{1}{3} & \frac{2}{3} \end{bmatrix}\begin{bmatrix} .8 & .2 \\ .1 & .9 \end{bmatrix} = \begin{bmatrix} \frac{1}{3} & \frac{2}{3} \end{bmatrix} = v_\infty$$

In other words,

$$v_\infty P = v_\infty$$

We call a vector v with the property that $vP = v$ a **steady-state vector.** If v is also a probability vector (like v_∞ above) we call it a **steady-state probability vector.**

Question Where does the name *steady-state vector* come from?

Answer If $vP = v$, then v is a distribution that will not change from time step to time step. In Example 1, since $v = [1/3 \quad 2/3]$ is a steady-state vector, if 1/3 of consumers use powdered detergent and 2/3 use liquid detergent one year, then the proportions will be the same the next year. Individual consumers may still switch from year to year, but as many will switch from powder to liquid as switch from liquid to powder, so the number using each will remain constant.

Example 2 • Laundry Detergent Switching (continued)

Using the transition matrix P from Example 1, calculate P^2 through P^{50}.

Solution As in Example 1, we compute the first few powers of P by hand and then turn to technology (see Example 2 in Section 9.2):

$$P = \begin{bmatrix} .8 & .2 \\ .1 & .9 \end{bmatrix}$$

$$P^2 = P \cdot P = \begin{bmatrix} .66 & .34 \\ .17 & .83 \end{bmatrix}$$

$$P^3 = P \cdot P^2 = \begin{bmatrix} .562 & .438 \\ .219 & .781 \end{bmatrix}$$

$$P^4 = P \cdot P^3 = \begin{bmatrix} .4934 & .5066 \\ .2533 & .7467 \end{bmatrix}$$

$$\vdots$$

$$P^{49} \approx \begin{bmatrix} .333\,333\,35 & .666\,666\,65 \\ .333\,333\,32 & .666\,666\,68 \end{bmatrix}$$

$$P^{50} \approx \begin{bmatrix} .333\,333\,35 & .666\,666\,65 \\ .333\,333\,33 & .666\,666\,67 \end{bmatrix}$$

Before we go on . . . Again, something interesting seems to be happening. The higher and higher powers of P seem to be getting closer and closer to the matrix

$$\begin{bmatrix} .333\,333\ldots & .666\,666\ldots \\ .333\,333\ldots & .666\,666\ldots \end{bmatrix} = \begin{bmatrix} \frac{1}{3} & \frac{2}{3} \\ \frac{1}{3} & \frac{2}{3} \end{bmatrix}$$

Let's call this matrix P^∞. Notice two things about P^∞:

- $P^\infty \cdot P = P^\infty$. (Check this directly.) In other words, multiplying P^∞ by P doesn't change it.
- The rows of P^∞ are the same and are copies of v_∞.

We call P^∞ the **long-term** or **steady-state transition matrix.**

Many Markov systems, including the **regular** systems we define toward the end of this section, have the interesting property exhibited in Example 1 that, no matter what the original distribution vector, after many time steps the distribution will be close to a steady-state distribution. They also have the property that the powers of the transition matrix P approach the matrix P^∞ whose rows are copies of the steady-state probability vector.

Question Can there be more than one steady-state probability vector for a Markov system?

Answer In general, yes. A simple example is the system with

$$P = \begin{bmatrix} 1 & 0 \\ 0 & 1 \end{bmatrix}$$

This system has many steady-state probability vectors, including $[1 \quad 0]$ and $[.5 \quad .5]$. In fact, any probability vector is a steady-state probability vector for this system. (Why?) We focus for the most part on systems that have only one possible steady-state vector (again, the *regular* Markov systems we talk about shortly have this property). Thus, we usually speak of *the* steady-state vector of a system.

Question Is there a way to calculate the steady-state probability vector and long-term transition matrix without having to take larger and larger powers until the matrix stops changing much?

Answer The next example shows how we can obtain a steady-state vector analytically.

Example 3 • Calculating the Steady-State Vector

Calculate the steady-state probability vector and long-term transition matrix for the transition matrix in the preceding examples:

$$P = \begin{bmatrix} .8 & .2 \\ .1 & .9 \end{bmatrix}$$

Solution We are asked to find

$$v_\infty = [x \quad y]$$

This vector must satisfy the equation

$$v_\infty P = v_\infty \quad \text{or} \quad [x \quad y]\begin{bmatrix} .8 & .2 \\ .1 & .9 \end{bmatrix} = [x \quad y]$$

Doing the matrix multiplication gives

$$[.8x + .1y \quad .2x + .9y] = [x \quad y]$$

Equating corresponding entries gives

$$.8x + .1y = x$$

$$.2x + .9y = y$$

or

$$-.2x + .1y = 0$$

$$.2x - .1y = 0$$

Now these equations tell us something about x and y, but not enough because they are really the same equation. (Do you see that?) There is one more thing we know, though: Since $[x \quad y]$ is a probability vector, its entries must add up to 1. This gives one more equation:

$$x + y = 1$$

Taking this equation together with one of the two equations above gives us the following system:

$$x + \quad y = 1$$

$$-.2x + .1y = 0$$

We now solve this system using any of the techniques we learned for solving systems of linear equations. One approach well suited to using technology is to use matrix arithmetic. Write the system as $AX = B$ where

$$A = \begin{bmatrix} 1 & 1 \\ -.2 & .1 \end{bmatrix} \qquad X = \begin{bmatrix} x \\ y \end{bmatrix} \qquad B = \begin{bmatrix} 1 \\ 0 \end{bmatrix}$$

Then

$$X = A^{-1}B = \begin{bmatrix} \frac{1}{3} & -\frac{10}{3} \\ \frac{2}{3} & \frac{10}{3} \end{bmatrix} \begin{bmatrix} 1 \\ 0 \end{bmatrix} = \begin{bmatrix} \frac{1}{3} \\ \frac{2}{3} \end{bmatrix}$$

Thus, the solution is $x = \frac{1}{3}$, and $y = \frac{2}{3}$, so the steady-state vector is

$$v_\infty = [x \quad y] = \begin{bmatrix} \frac{1}{3} & \frac{2}{3} \end{bmatrix}$$

Since the rows of the long-term transition matrix are copies of v_∞, we have

$$P^\infty = \begin{bmatrix} \frac{1}{3} & \frac{2}{3} \\ \frac{1}{3} & \frac{2}{3} \end{bmatrix}$$

as suggested in Example 1.

Graphing Calculator: Calculating the Steady-State Vector

In the TI-83 we can use the methods of Chapter 3 to compute the matrix product $A^{-1}B$: We set up the matrices

$$[A] = \begin{bmatrix} 1 & 1 \\ -.2 & .1 \end{bmatrix} \qquad [B] = \begin{bmatrix} 1 \\ 0 \end{bmatrix}$$

and, on the home screen, enter the following:

`[A]`$^{-1}$`[B]`

To convert the entries to fractions, we can follow this by the command

`►Frac` MATH ENTER ENTER

Excel: Calculating the Steady-State Vector

We enter A in cells A1:B2, B in cells D1:D2, and the formula for $X = A^{-1}B$ in a convenient location, say, B4:B5.

	A	B	C	D
1	1	1		1
2	-0.2	0.1		0
3				
4		=MMULT(MINVERSE(A1:B2),D1:D2)		
5				
6				

When we press Control + Shift + Enter, we see the result:

	A	B	C	D
1	1	1		1
2	-0.2	0.1		0
3				
4		0.33333333		
5		0.66666667		

If we want to see the answer in fraction rather than decimal form, we format the cells as fractions:

	A	B	C	D
1	1	1		1
2	-0.2	0.1		0
3				
4		1/3		
5		2/3		

Web Site: Calculating the Steady-State Vector

The Matrix Algebra Tool available at

Web site → Online Utilities → Matrix Algebra Tool

permits you to do the computation online. Once you have entered the matrices A and B, the formula

```
A^(-1)*B
```

will give you the steady-state distribution. To see the entries as fractions, check the `Fraction Mode` box before selecting `Compute`.

The method we just used works for any size transition matrix and can be summarized as follows.

Calculating the Steady-State Distribution Vector and Long-Term Transition Matrix

To calculate the steady-state probability vector for a Markov system with transition matrix P, we solve the system of equations given by

$$x + y + z + \cdots = 1$$

$$[x \quad y \quad z \quad \ldots]P = [x \quad y \quad z \quad \ldots]$$

where we use as many unknowns as there are states in the Markov system. The steady-state probability vector is then

$$v_\infty = [x \quad y \quad z \quad \ldots]$$

and the long-term transition matrix is

$$P^\infty = \begin{bmatrix} x & y & z & \ldots \\ x & y & z & \ldots \\ \vdots & \vdots & \vdots & \vdots \\ x & y & z & \ldots \end{bmatrix}$$

Interpreting the Steady-State Vector

The entries $x, y, z, \ldots$ of the steady-state vector give the long-term probabilities that the system will be in the corresponding states, or the fractions of time one can expect to find the Markov system in the corresponding states.

Note The system of equations we get using this method always has one more equation than unknowns. However, we can drop any one of the equations arising from the requirement that $v_\infty P = v_\infty$. This follows from the fact that the entries in any row of P add to 1 so that any one of the resulting equations is the negative of the sum of the others. We see this in the following example.

Example 4 • Brand Switching

Sigma Chemical's SigmaBlock sunblock cream has two major competitors on the market: Triple Ban lotion and Sun Erase lotion. SigmaBlock has enjoyed a 30% share of the market but has been unable to increase this share for years. Sigma Chemical recently hired a new advertising vice president who claimed—to justify his large salary—that he would add another 5–10% to SigmaBlock's market share. He launched an ad campaign to discourage consumers from buying Triple Ban lotion, using the ad slogan

While Triple Ban may block some rays
SigmaBlock cream works for days.
Since Sigma simply is the best,
Triple Ban . . . Go take a rest!

The campaign seems to have had mixed results, according to recent market surveys. The following table shows the percentages of customers who switched brands during the first month of the ad campaign:

	To		
From	**SigmaBlock**	**Triple Ban**	**Sun Erase**
SigmaBlock	60%	0%	40%
Triple Ban	60%	0%	40%
Sun Erase	10%	10%	80%

a. Assuming this trend continues, what are the long-term market shares of the three products? Is the new advertising vice president's large salary justified?

b. The market surveys also found that 4500 customers were using SigmaBlock, 4500 were using Triple Ban, and 6000 were using Sun Erase at the start of the ad campaign. Assuming the above trend continues, how many customers will be using each brand in the long term?

Solution

a. To predict the long-term market shares, we need to find the steady-state probability vector for the transition matrix

$$P = \begin{bmatrix} .6 & 0 & .4 \\ .6 & 0 & .4 \\ .1 & .1 & .8 \end{bmatrix}$$

Since the Markov system has three states, we let

$$v_\infty = [x \quad y \quad z]$$

and solve the system given by the following equations:

$$x + y + z = 1$$

$$[x \quad y \quad z]\begin{bmatrix} .6 & 0 & .4 \\ .6 & 0 & .4 \\ .1 & .1 & .8 \end{bmatrix} = [x \quad y \quad z]$$

That is,

$$\begin{aligned} x + \quad y + \quad z &= 1 \\ .6x + .6y + .1z &= x \\ .1z &= y \\ .4x + .4y + .8z &= z \end{aligned}$$

or

$$\begin{aligned} x + \quad y + \quad z &= 1 \\ -.4x + .6y + .1z &= 0 \\ -y + .1z &= 0 \\ .4x + .4y - .2z &= 0 \end{aligned}$$

Any one equation after the first can be dropped, so let's drop the last one. (Notice that it is the negative of the sum of the two above it—check it.) We are left with the following three equations in three unknowns:

$$\begin{aligned} x + \quad y + \quad z &= 1 \\ -.4x + .6y + .1z &= 0 \\ -y + .1z &= 0 \end{aligned}$$

We can solve this system using matrix arithmetic (or your favorite alternative method). It has the unique solution

$$[x \quad y \quad z] = \begin{bmatrix} \frac{4}{15} & \frac{1}{15} & \frac{2}{3} \end{bmatrix}$$

so that the steady-state distribution vector is

$$v_\infty = \begin{bmatrix} \frac{4}{15} & \frac{1}{15} & \frac{2}{3} \end{bmatrix}$$

and the long-term transition matrix is

$$P^\infty = \begin{bmatrix} \frac{4}{15} & \frac{1}{15} & \frac{2}{3} \\ \frac{4}{15} & \frac{1}{15} & \frac{2}{3} \\ \frac{4}{15} & \frac{1}{15} & \frac{2}{3} \end{bmatrix}$$

We can now predict that, in the long term, the market shares will be

SigmaBlock: $\frac{4}{15}$ (approximately 26.67%)

Triple Ban: $\frac{1}{15}$ (approximately 6.67%)

Sun Erase: $\frac{2}{3}$ (approximately 66.67%)

Since SigmaBlock's predicted market share is *lower* than the 30% it enjoyed before the new vice president was hired, it looks as though a certain vice president may soon be looking for employment elsewhere!

b. We are given the initial distribution vector

$$v = [4500 \quad 4500 \quad 6000]$$

To calculate the distribution after a large number of transitions, we use the formula

$$v_{\text{long-term}} = v_{\text{initial}}P^\infty = [4500 \quad 4500 \quad 6000]\begin{bmatrix} \frac{4}{15} & \frac{1}{15} & \frac{2}{3} \\ \frac{4}{15} & \frac{1}{15} & \frac{2}{3} \\ \frac{4}{15} & \frac{1}{15} & \frac{2}{3} \end{bmatrix} = [4000 \quad 1000 \quad 10{,}000]$$

so that 4000 customers will be using SigmaBlock, 1000 will be using Triple Ban, and 10,000 will be using Sun Erase.

Before we go on . . . If we used a different initial distribution vector with entries that also add to 15,000, such as $[10{,}000 \quad 2000 \quad 3000]$, we would find that we obtain the same answer for the distribution after a large number of transitions:

$$[10{,}000 \quad 2000 \quad 3000]P^{\infty} = [4000 \quad 1000 \quad 10{,}000]$$

In fact, this is $15{,}000v_{\infty}$. Why?

Question Is there always a steady-state vector and long-term transition matrix?

Answer Steady-state vectors always exist (a fact we will not prove here), but if we keep multiplying P by itself, the answers need not approach a matrix whose rows are a steady-state vector. For instance,

$$P = \begin{bmatrix} 0 & 1 \\ 1 & 0 \end{bmatrix}$$

has $[.5 \quad .5]$ as a steady-state vector, but if we take powers of P, we find that all odd powers of P are equal to P and all even powers of P are equal to the identity matrix. Thus, the powers of P do not stabilize but hop back and forth between P and the identity.

Question If the powers of P *do* stabilize, can't we simply find P^{∞} by using a calculator or Excel to keep multiplying P by itself until it stops changing? Why do we need the analytical method described above?

Answer Calculating powers is okay for *small* matrices (such as 2×2 or 3×3 matrices), provided you don't need to go to a large power. But, if the matrix P is large, it is an extremely inefficient way of going about things. The powers of P might not "settle down" (stop changing significantly) until a very large power. As a consequence, you might be forced to calculate a huge power like

$$P^{10{,}000}$$

(this is P raised to the googol power)[5] before you know what P^{∞} is to the desired degree of accuracy. Another possible problem is that, depending on exactly how you go about calculating large powers of P, rounding errors can make the calculations go awry.

Question Under what circumstances can I be sure that the higher powers of P will stabilize or "settle down"? In other words, when can I find P^{∞}?

Answer When the Markov system is *regular.* A system with transition matrix P is said to be **regular** if there is some power of P that contains *no zero entries.* For example, if P itself contains no zero entries or if P^{200} contains no zero entries, then the system represented by P is regular.

Example 5 • Regular Markov System

Determine whether the system with the given transition matrix is regular.

a. $P = \begin{bmatrix} .5 & 0 & .5 \\ .6 & 0 & .4 \\ 0 & 1 & 0 \end{bmatrix}$ **b.** $P = \begin{bmatrix} .6 & .4 \\ 0 & 1 \end{bmatrix}$

---

[5] A googol is a 1 followed by 100 zeros. Incidentally, a *googolplex* is a 1 followed by a googol zeros!

Solution

a. Here is a useful technique for determining whether a system is regular. Start by replacing every positive entry in the transition matrix with the symbol $+$:

$$P = \begin{bmatrix} + & 0 & + \\ + & 0 & + \\ 0 & + & 0 \end{bmatrix}$$

Now we can compute products symbolically:

$$P^2 = \begin{bmatrix} + & 0 & + \\ + & 0 & + \\ 0 & + & 0 \end{bmatrix}\begin{bmatrix} + & 0 & + \\ + & 0 & + \\ 0 & + & 0 \end{bmatrix} = \begin{bmatrix} + & + & + \\ + & + & + \\ + & 0 & + \end{bmatrix}$$

For example, the (1, 2) entry in the product is

$$(+ \cdot 0) + (0 \cdot 0) + (+ \cdot +) = 0 + 0 + (+) = +$$

using the fact that the product of two positive numbers is positive (not zero). Since P^2 still has a 0 in position (3, 2), we try a higher power:

$$P^4 = \begin{bmatrix} + & + & + \\ + & + & + \\ + & 0 & + \end{bmatrix}\begin{bmatrix} + & + & + \\ + & + & + \\ + & 0 & + \end{bmatrix} = \begin{bmatrix} + & + & + \\ + & + & + \\ + & + & + \end{bmatrix}$$

Thus, we know that P^4 will have all positive entries, and we conclude that the system is regular.

b. The transition matrix P has the symbolic form

$$P = \begin{bmatrix} + & + \\ 0 & + \end{bmatrix}$$

We now see that

$$P^2 = \begin{bmatrix} + & + \\ 0 & + \end{bmatrix}\begin{bmatrix} + & + \\ 0 & + \end{bmatrix} = \begin{bmatrix} + & + \\ 0 & + \end{bmatrix}$$

so that P^2 has the same symbolic form as P. A little thought will convince you that you can continue multiplying powers of P forever, but you'll keep getting back the same form, with a zero in position (2, 1). Thus, the given system is *not* a regular system.

Question What good are regular Markov systems?

Answer First, a regular Markov system has only one steady-state probability vector. Second, for a regular Markov system we are guaranteed that there is a P^∞ and that our procedure will find it. (For proofs of these facts, consult an advanced book on Markov systems.) You must always check that your system is regular before trying to compute P^∞.[6]

[6]Actually, there are systems that are *not* regular but for which we can still find a P^∞. We see an example of this in the exercises.

9.3 EXERCISES

In Exercises 1–22, use the given transition matrix P and the given initial distribution vector v to find (a) the long-term transition matrix P^∞ and (b) the long-term distribution vector associated with the given initial distribution vector v.

1. $P = \begin{bmatrix} \frac{1}{2} & \frac{1}{2} \\ 1 & 0 \end{bmatrix}, v = [10 \quad 5]$ **2.** $P = \begin{bmatrix} \frac{1}{3} & \frac{2}{3} \\ 1 & 0 \end{bmatrix}, v = [5 \quad 10]$

3. $P = \begin{bmatrix} \frac{1}{3} & \frac{2}{3} \\ \frac{1}{2} & \frac{1}{2} \end{bmatrix}, v = [0 \quad 10]$ **4.** $P = \begin{bmatrix} \frac{1}{4} & \frac{3}{4} \\ \frac{1}{2} & \frac{1}{2} \end{bmatrix}, v = [10 \quad 10]$

5. $P = \begin{bmatrix} .2 & .8 \\ .1 & .9 \end{bmatrix}, v = [100 \quad 200]$

6. $P = \begin{bmatrix} .3 & .7 \\ .8 & .2 \end{bmatrix}, v = [200 \quad 100]$

7. $P = \begin{bmatrix} .1 & .9 \\ .6 & .4 \end{bmatrix}, v = [1 \quad 0]$ **8.** $P = \begin{bmatrix} .2 & .8 \\ .7 & .3 \end{bmatrix}, v = [0 \quad 1]$

9. $P = \begin{bmatrix} \frac{1}{2} & 0 & \frac{1}{2} \\ 1 & 0 & 0 \\ 0 & \frac{1}{2} & \frac{1}{2} \end{bmatrix}, v = [10 \quad 0 \quad 0]$

10. $P = \begin{bmatrix} 0 & \frac{1}{2} & \frac{1}{2} \\ \frac{1}{2} & \frac{1}{2} & 0 \\ 1 & 0 & 0 \end{bmatrix}, v = [100 \quad 0 \quad 0]$

11. $P = \begin{bmatrix} 0 & \frac{1}{2} & \frac{1}{2} \\ \frac{1}{2} & \frac{1}{2} & 0 \\ 0 & \frac{1}{4} & \frac{3}{4} \end{bmatrix}, v = [10 \quad 10 \quad 10]$

12. $P = \begin{bmatrix} \frac{1}{4} & 0 & \frac{3}{4} \\ \frac{1}{2} & \frac{1}{4} & \frac{1}{4} \\ 0 & \frac{1}{4} & \frac{3}{4} \end{bmatrix}, v = [10 \quad 10 \quad 10]$

13. $P = \begin{bmatrix} 1 & 0 & 0 \\ .5 & .5 & 0 \\ 0 & .5 & .5 \end{bmatrix}, v = [0 \quad 20 \quad 0]$

14. $P = \begin{bmatrix} 0 & 1 & 0 \\ 0 & 0 & 1 \\ 0 & 0 & 1 \end{bmatrix}, v = [20 \quad 0 \quad 0]$

15. $P = \begin{bmatrix} .1 & .9 & 0 \\ 0 & 1 & 0 \\ 0 & .2 & .8 \end{bmatrix}, v = [1 \quad 0 \quad 0]$

16. $P = \begin{bmatrix} .1 & .1 & .8 \\ .5 & 0 & .5 \\ .5 & 0 & .5 \end{bmatrix}, v = [1 \quad 0 \quad 0]$

T Use of technology is suggested for Exercises 17–22. Round all answers to two decimal places.

17. $P = \begin{bmatrix} .55 & .45 \\ .85 & .15 \end{bmatrix}, v = [0 \quad 10]$

18. $P = \begin{bmatrix} .65 & .35 \\ .15 & .85 \end{bmatrix}, v = [10 \quad 10]$

19. $P = \begin{bmatrix} .25 & .25 & .50 \\ 0 & .35 & .65 \\ .50 & .15 & .35 \end{bmatrix}, v = [10 \quad 10 \quad 10]$

20. $P = \begin{bmatrix} .15 & .35 & .50 \\ 0 & .55 & .45 \\ .50 & .25 & .25 \end{bmatrix}, v = [1 \quad 1 \quad 0]$

21. $P = \begin{bmatrix} .5 & .25 & .25 & 0 \\ .3 & .3 & .3 & .1 \\ .1 & .2 & .1 & .6 \\ 1 & 0 & 0 & 0 \end{bmatrix}, v = [.5 \quad 0 \quad .5 \quad 0]$

22. $P = \begin{bmatrix} .3 & .3 & .25 & .15 \\ 0 & 1 & 0 & 0 \\ 0 & .25 & .25 & .5 \\ .1 & .1 & .1 & .7 \end{bmatrix}, v = [0 \quad .5 \quad 0 \quad .5]$

APPLICATIONS

Many of the following applications are new, but some of them go more deeply into exercises from preceding sections.

23. Risk Analysis An auto insurance company classifies each motorist as high risk if the motorist has had at least one moving violation during the past calendar year and low risk if the motorist has had no violations during the past calendar year. According to the company's data, a high-risk motorist has a 50% chance of remaining in the high-risk category the next year and a 50% chance of moving to the low-risk category. A low-risk motorist has a 10% chance of moving to the high-risk category the next year and a 90% chance of remaining in the low-risk category.

a. In the long term, what percentage of motorists fall into each category?

b. If the company insures 100 low-risk motorists, how many of them should it expect to fall into each category in the long term?

24. Debt Analysis A credit card company classifies its card holders as falling into one of two credit ratings: good and poor. Based on its rating criteria, the company finds that a card holder with a good credit rating has an 80% chance of remaining in that category the following year and a 20% chance of dropping into the poor category. A card holder with a poor credit rating has a 40% chance of moving into the good rating the following year and a 60% chance of remaining in the poor category.

a. In the long term, what percentage of card holders fall into each category?

b. If the company issues 100 credit cards to new customers with good credit ratings, how many of them should it expect to fall into each category in the long term?

25. **Textbook Adoptions** College instructors who adopt this book are twice as likely to continue to use the book the following year as they are to drop it, whereas nonusers are nine times as likely to remain nonusers the following year as they are to adopt this book. In the long term, what proportion of college instructors will be users of this book?

26. **Confidence Level** Tommy the dunker's performance on the basketball court is influenced by his state of mind: If he scores he is twice as likely to score on the next shot as he is to miss, whereas if he misses a shot he is three times as likely to miss the next shot as he is to score. In the long term, what percentage of shots are successful?

27. **Marital Relations** A survey of matrimonial relations taken over a 1-year period yields the following data: One-half of the happily married people stay happily married, one-fourth of the marriages wind up on the rocks, and the rest get divorced. Of those people whose marriages are on the rocks, half remain in that state, and the other half get divorced. Finally, one-fourth of the people who are divorced remarry (happily), and the rest remain single for that year.
    a. Record these data in the form of a transition matrix.
    b. A sample of 100 originally happily married couples is watched over a long time. In the long run, what fraction of the sample can be expected to be happily married?
    c. Over the long run, is a person more likely to be happily married, be in a marriage that is on the rocks, or be divorced? What fraction of people is in this likeliest state?

28. **Helicopter Traffic** The Metropolis Helicopter Service is deciding where to build its new maintenance hangar. Its flight schedules reveal the following pattern:

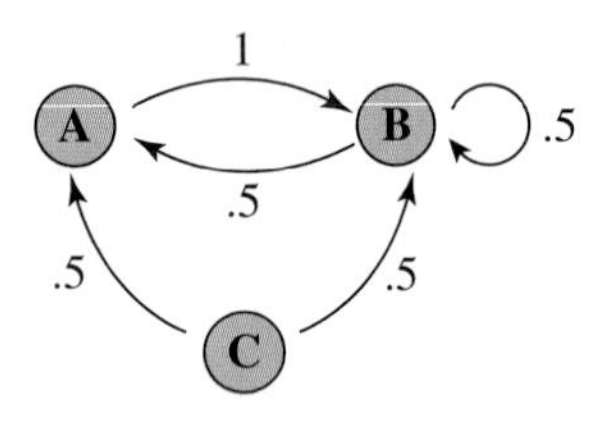

Here, A = Airport, B = Badminton Hollows, and C = Creepy Valley. The arrows represent fractions of daily nonstop flights from one city to another. (Half the flights starting at Badminton Hollows are sightseeing trips, so they depart and arrive at the same spot. Also, there appears to be a scare of some sort at Creepy Valley.)
    a. Find P^∞, the long-term transition matrix. (Ignore the fact that P may not be regular.)
    b. If Metropolis wishes to build its hangar at the busiest heliport, where should it build?

29. **Debt Analysis** As the manager of a large retailing outlet, you have classified all credit customers as falling into one of the following categories: paid up, outstanding 0–90 days, and bad debts. Based on an audit of your company's records, you have come up with the following table that gives the probabilities that a single credit customer will move from one category to the next in the period of 1 month:

	To		
From	**Paid Up**	**0–90 Days**	**Bad Debts**
Paid Up	.5	.5	0
0–90 Days	.5	.3	.2
Bad Debts	0	.5	.5

    a. How do you expect the company's credit customers to be distributed in the long term?
    b. The company's records show that 5000 customers are presently in the paid-up category, 1000 are in the 0–90 days category, and 2000 are in the bad-debts category. How do you expect them to be distributed 10 years from now?

30. **Debt Analysis** Repeat Exercise 29, using the following table:

	To		
From	**Paid Up**	**0–90 Days**	**Bad Debts**
Paid Up	.8	.2	0
0–90 Days	.5	.3	.2
Bad Debts	0	.5	.5

31. **Residence Hall Management** Matilda Waltzing is a bureaucrat in charge of three student residences: Hall A, Hall B, and Hall C. Since her job calls for little or no work, she justifies her large salary by having students moved randomly from residence hall to residence hall, according to the following scheme: Every semester she gets on the phone and orders that half of the students in Hall A be moved to Hall B with the rest to stay in Hall A, a third of the students in Hall B be moved to Hall A with the rest going to Hall C, and the population of Hall C be moved so that a quarter goes to Hall A, half goes to Hall B, and the rest stays in Hall C. After a large number of semesters, how are the (bewildered) students distributed among the three residence halls (that is, what fraction is in each)?

32. **Genetics** Whether your eyes are blue or brown appears to be controlled by a single gene. If your *genotype* is *BB* or *Bb*, you will have brown eyes; if your genotype is *bb*, you will have blue eyes (brown is *dominant;* blue is *recessive*). Now, if one parent has genotype *Bb*, the genotype of a child could be any of the three, with probabilities given in the following table:

	Offspring		
Other Parent	***BB***	***Bb***	***bb***
BB	$\frac{1}{2}$	$\frac{1}{2}$	0
Bb	$\frac{1}{4}$	$\frac{1}{2}$	$\frac{1}{4}$
bb	0	$\frac{1}{2}$	$\frac{1}{2}$

We refer to this situation as a *Bb*-crossing because one of the parents is assumed to have the genotype *Bb*. Suppose now that the child of such a pairing marries someone with genotype *Bb*, and they have a child, who marries someone with genotype *Bb*, and so on.

a. Interpret this as a Markov process by drawing the associated one-step transition diagram with states *BB*, *Bb*, and *bb*.

b. If an offspring of a *BB* parent is involved in a *Bb*-crossing, use the two-step transition matrix to find the probability that he will have a child with blue eyes.

c. After many generations of such *Bb*-crossings, what is the most likely genotype to emerge?

33. Income Brackets The following diagram shows the movement of U.S. households among three income groups, affluent, middle class, and poor, over the 11-year period 1980–1991:

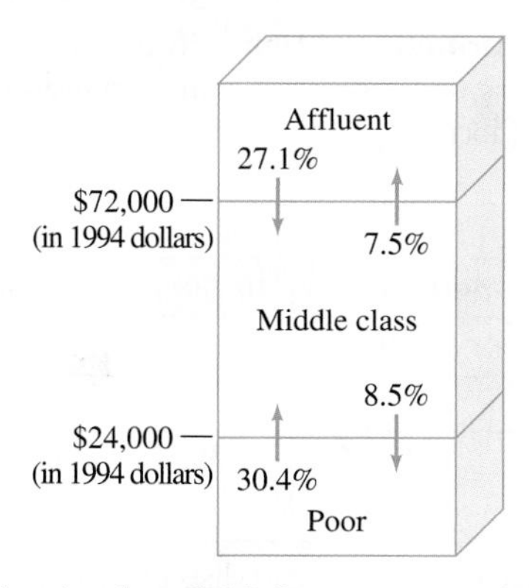

The figures are based on household after-tax income. The study, conducted by G. J. Duncan of Northwestern University and T. Smeeding of Syracuse University, is based on annual surveys of the personal finances of 5000 households since the late 1960s. (The surveys were conducted by the University of Michigan.) SOURCE: *New York Times*, June 4, 1995, p. E4.

(Assume zero probabilities for the transitions between affluent and poor.) According to the model, what percentage of all U.S. households will be in each income bracket in the long term? (Give your answer to the nearest 0.1%.)

34. Income Brackets The following diagram shows the movement of U.S. households among three income groups, affluent, middle class, and poor, over the 12-year period 1967–1979:

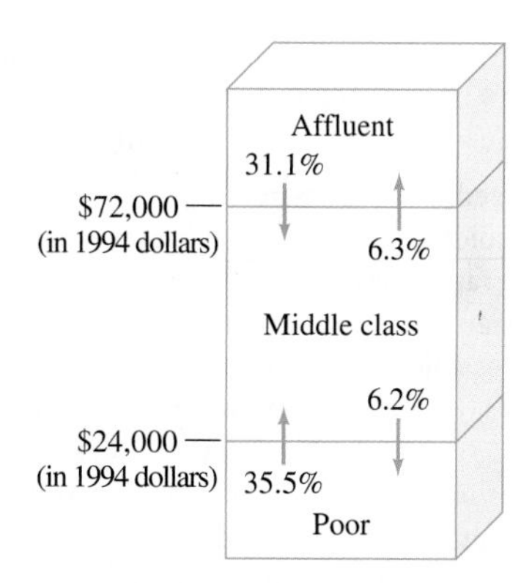

The figures are based on household after-tax income. The study, conducted by G. J. Duncan of Northwestern University and T. Smeeding of Syracuse University, is based on annual surveys of the personal finances of 5000 households since the late 1960s. (The surveys were conducted by the University of Michigan.) SOURCE: *New York Times*, June 4, 1995, p. E4.

(Assume zero probabilities for the transitions between affluent and poor.) According to the model, what percentage of all U.S. households will be in each income bracket in the long term? (Give your answer to the nearest 0.1%.)

35. Income Brackets Do Exercises 33 and 34 justify the claim that there is a tendency toward widening income inequality in the United States? Explain.

36. Income Brackets Do Exercises 33 and 34 justify the claim that there is a tendency toward increasing poverty? Explain.

37. Traffic Delays As an experienced negotiator of rush-hour traffic conditions, you have decided to use a Markov system to model your fortunes on the highway. You classify traffic on the Sheridan Expressway as being light, moderate, or heavy, and you have been tabulating your chances of passing from one state to another over time intervals of 10 minutes. You have calculated that if you are in light traffic at the start of some time period, you have a 40% chance of being in light traffic at the start of the next time period, a 30% chance of encountering moderate traffic, and a 30% chance of hitting heavy traffic. If you start in moderate traffic, there is a 60% chance that you will again be in moderate traffic 10 minutes later and a 20% chance of encountering either light or heavy traffic. If you are in heavy traffic, you will again be in heavy traffic 80% of the time, you will have progressed to moderate traffic 20% of the time, but you will never be in light traffic in the next time interval.

a. On average, what percentage of the time do you spend in heavy traffic?
b. In a 100-minute journey, how many minutes do you expect to spend in each kind of traffic situation?
c. You average 50 miles per hour (0.833 mile/minute) in light traffic, 30 mph (0.5 mile/minute) in moderate traffic, and 10 mph (0.167 mile/minute) in heavy traffic. How far do you expect to travel in 100 minutes?

38. Weather Prediction Frustrated with the poor record of weather predicting by your local TV channel, you have decided to operate your own weather prediction service using a Markov system to model the weather. To make matters as simple as possible, you decide to classify weather on a day-by-day basis as either (predominantly) sunny, (predominantly) cloudy, or (predominantly) rainy. You then go to work, gathering tons of statistics on day-to-day weather patterns, and come up with the following figures: If it is sunny on a given day, there is a probability of .6 of it being sunny the next day, a probability of .3 of it being cloudy, and a probability of .1 of rain the next day. If it is cloudy on one day, there is a probability of .5 that it will become sunny the next day, a probability of .2 that it will remain cloudy the next day, and a probability of .3 that it will rain. Starting from a rainy day, there is a probability of .2 that it will become sunny the next day, a probability of .4 that it will remain cloudy without rain, and a probability of .4 that it will continue to rain.
a. Overall, what percentage of days are rainy?
b. In a 100-day period, how many days would fall in each category?
c. On a sunny day, what would be your 3-day weather forecast? (Give the probability of each type of weather each day.)

39. Dissipation Consider the following five-state model of one-dimensional dissipation without drift:

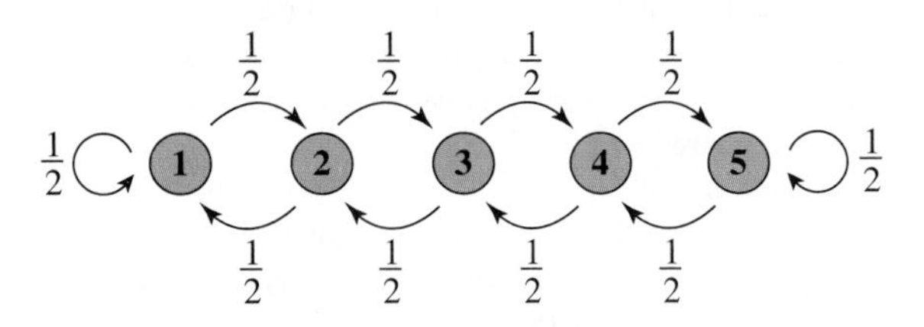

a. Write down the transition matrix for the process.
b. Compute the long-term transition matrix and use it to calculate the percentage of time the system will spend in each state.
c. Find the distribution vector after a large number of steps if the initial distribution vector is $[1 \quad 0 \quad 0 \quad 0 \quad 0]$.

40. Dissipation with Drift Consider the following five-state model of one-dimensional dissipation with drift:

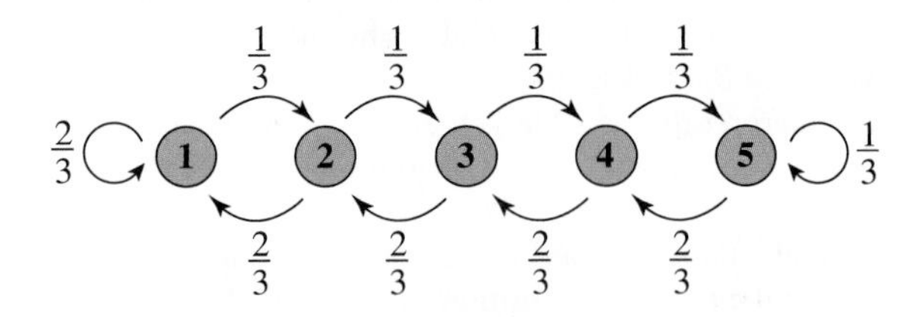

a. Write down the transition matrix for the process.
b. Compute the long-term transition matrix and use it to calculate the percentage of time the system will spend in each state.
c. Find the distribution vector after a large number of steps if the initial distribution vector is $[1 \quad 0 \quad 0 \quad 0 \quad 0]$.

T In Exercises 41–48, use technology to solve.

41. Income Distribution A University of Michigan study shows the following one-generation transition probabilities among four major income groups:

Father's Income	*Eldest Son's Income*			
	Bottom 10%	**10–50%**	**50–90%**	**Top 10%**
Bottom 10%	.30	.52	.17	.01
10–50%	.10	.48	.38	.04
50–90%	.04	.38	.48	.10
Top 10%	.01	.17	.52	.30

SOURCES: Gary Solon, University of Michigan/*New York Times*, May 18, 1992, p. D5. We have adjusted some of the figures so that the probabilities add to 1; they did not do so in the original table due to rounding.

a. If a sample of 100 male earners is evenly distributed among the four income groups, what will the distribution look like after many generations of eldest sons?[7]
b. In the long term, what percentage of male earners would you expect to find in each category? Why are the long-range figures not necessarily 10% in the lowest 10% income bracket, 40% in the 10–50% range, 40% in the 50–90% range, and 10% in the top 10% range?

[7]In this way we are simplifying the problem by tracking only one offspring per generation. Of course, it need not be the case that everyone has a son, so we can't really use this model to predict the total number of offspring in each income group unless we know how the birthrate varies with income groups.

42. Income Distribution Repeat Exercise 41, using the following data:

Father's Income	*Eldest Son's Income*			
	Bottom 10%	**10–50%**	**50–90%**	**Top 10%**
Bottom 10%	.50	.32	.17	.01
10–50%	.10	.48	.38	.04
50–90%	.04	.38	.48	.10
Top 10%	.01	.17	.32	.50

43. Population Movement In 1999 the U.S. population, broken down by regions, was 53.9 million in the Northeast, 64.3 million in the Midwest, 100.0 million in the South, and 63.3 million in the West. The matrix P below shows the population movement during the period 1999–2000:

$$P = \begin{array}{r} \\ \\ \textbf{From NE} \\ \textbf{From MW} \\ \textbf{From S} \\ \textbf{From W} \end{array} \begin{array}{c} \begin{array}{cccc} \textbf{To} & \textbf{To} & \textbf{To} & \textbf{To} \\ \textbf{NE} & \textbf{MW} & \textbf{S} & \textbf{W} \end{array} \\ \begin{bmatrix} .9886 & .0015 & .0075 & .0024 \\ .0011 & .9900 & .0057 & .0032 \\ .0018 & .0042 & .9897 & .0043 \\ .0017 & .0035 & .0077 & .9871 \end{bmatrix} \end{array}$$

Use these data to predict the long-term distribution of the U.S. population (assuming no overall population growth and that the 1999–2000 trend continues indefinitely).

Note that this exercise ignores migration into or out of the country. The internal migration figures and 2000 population figures are accurate. SOURCE: U.S. Census Bureau, Current Population Survey, http://www.census.gov/, March 2000.

44. Voting Shifts According to a *New York Times*/CBS News Poll taken in October 1990, 75% of people who had last voted for a Republican would continue to vote Republican, 6% would vote for a Democrat, and the rest would vote for an Independent. Of those who had last voted Democrat, 78% would continue to vote for a Democrat, 6% would vote for a Republican, and the rest would vote for an Independent. Also, 30% of Independent voters would vote for a Democrat, 29% would vote for a Republican, and the rest would vote for an Independent. Given a pool of 100 Republicans, 150 Democrats, and 50 Independents, use this information to predict the vote in the distant future. (Round your answers to the nearest voter.)

SOURCE: *New York Times*, October 12, 1990. We have taken a slight liberty in interpreting the poll; actually, the voters were classified according to party affiliation rather than according to the way they had last voted.

45. Debt Analysis You manage the credit department for a large chain of stores, and you have access to the debt history of all your customers. Of the total debt owed to your company, $4 million is in the bad-debt category (representing accounts on which nothing has been paid for more than 90 days), $5 million is in the 31–90-day category (representing accounts on which the last payment was in this period), $8 million is in the 0–30-day category, and $100 million is in the paid category. The following table gives the probabilities that a single dollar will move from one category to the next in the period of 1 month:

	To			
From	**Paid**	**0–30 Days**	**31–90 Days**	**Bad Debts**
Paid	1	0	0	0
0–30 Days	.5	.3	.2	0
31–90 Days	.2	.5	.2	.1
Bad Debts	0	0	0	1

SOURCE: Based on a discussion in D. R. Anderson, D. J. Sweeny, and T. A. Williams, *An Introduction to Management Science*, 6th ed. (St. Paul: West, 1991).

How do you expect the company's present debts to be distributed in the long term?

46. Debt Analysis Repeat Exercise 45, using the following table:

	To			
From	**Paid**	**0–30 Days**	**31–90 Days**	**Bad Debts**
Paid	1	0	0	0
0–30 Days	.3	.5	.2	0
31–90 Days	.2	.2	.5	.1
Bad Debts	0	0	0	1

SOURCE: Based on a discussion in D. R. Anderson, D. J. Sweeny, and T. A. Williams, *An Introduction to Management Science*, 6th ed. (St. Paul: West, 1991).

47. Random Walks Refer to the state transition diagram in Exercise 39. Construct the transition matrix for a similar ten-state dissipation model and examine its long-term behavior. Can you generalize the result to an n-state dissipation model?

48. Random Walk with Drift Repeat Exercise 47 but refer instead to the state transition diagram in Exercise 40.

COMMUNICATION AND REASONING EXERCISES

49. By referring to the interpretation of the steady-state probability vector, explain why its entries add to 1.

50. If the entries in an initial distribution vector add to 1000, use the interpretation of the corresponding long-term distribution vector to explain why its entries add to 1000.

51. Explain: If Q is a matrix whose rows are steady-state distribution vectors, then $QP = Q$.

52. Construct a four-state Markov system in such a way that both $[.5 \quad .5 \quad 0 \quad 0]$ and $[0 \quad 0 \quad .5 \quad .5]$ are steady-state vectors. (*Hint:* Try one in which no arrows link the first two states to the last two.)

53. Refer to the following state transition diagram and explain in words (without doing any calculation) why the steady-state vector has a zero in position 1:

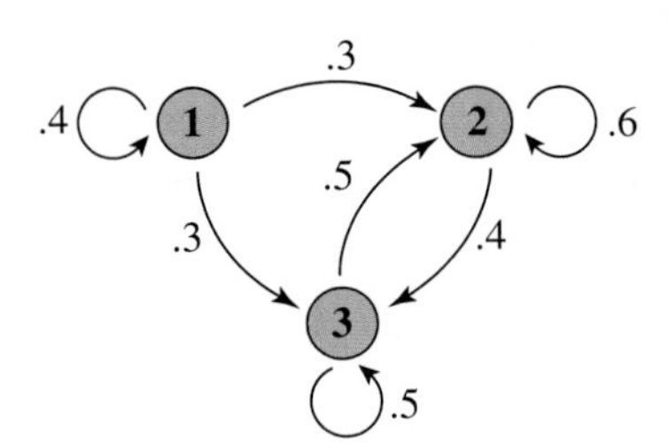

54. Without doing any calculation, find the steady-state distribution of the following system and explain the reasoning behind your claim:

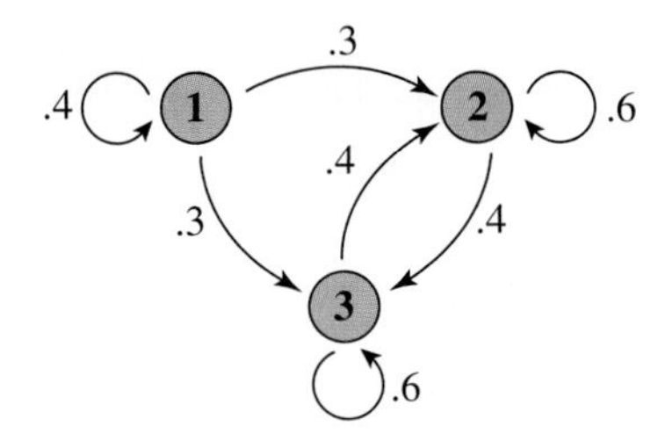

55. Construct a regular state transition diagram that possesses the steady-state vector [.3 .3 .4].

56. Construct a regular state transition diagram possessing a steady-state vector [.6 .3 0 .1].

57. Show that if a Markov system has two distinct steady-state distributions v and w, then $(v + w)/2$ is another steady-state distribution.

58. If higher and higher powers of P approach a fixed matrix Q, explain why the rows of Q must be steady-state distribution vectors.

9.4 Absorbing Markov Systems

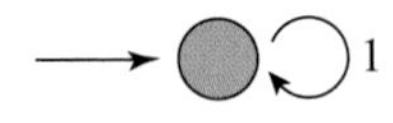

Figure 10

We now take a look at Markov systems that include absorbing states, such as Example 2 of Section 9.1. Recall that an **absorbing state** is a state from which there is a zero probability of exiting (much like a Roach Motel). Thus, an absorbing state may have one or more arrows entering it, but it has no arrows leaving it, as shown in Figure 10.

Absorbing Markov System

An **absorbing Markov system** is a Markov system that contains *at least one* absorbing state and is such that it is possible to get from each nonabsorbing state to some absorbing state in one or more time steps.

Quick Examples

1. "Roach Resort"
(Exercise 23 in Section 9.1)

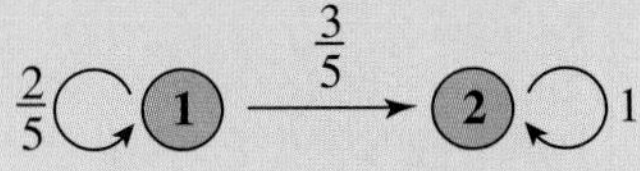

2. S&L Profitability
(Example 2 of Section 9.1)

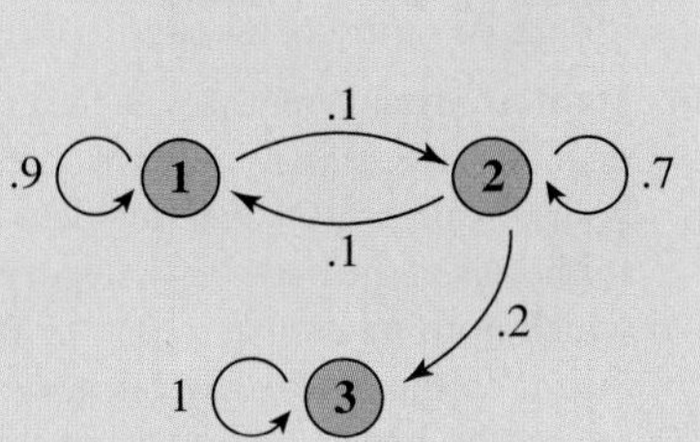

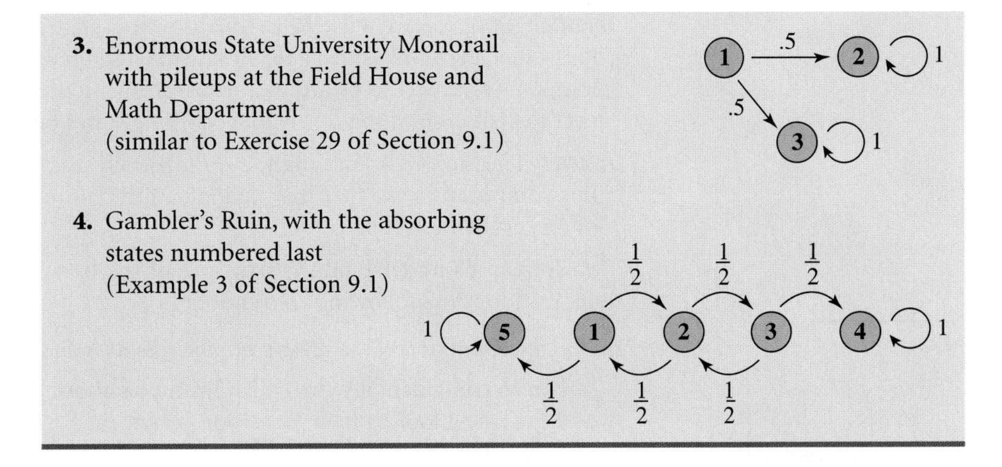

3. Enormous State University Monorail with pileups at the Field House and Math Department (similar to Exercise 29 of Section 9.1)
.5
.5
1
1
4. Gambler's Ruin, with the absorbing states numbered last (Example 3 of Section 9.1)
1/2
1/2
1/2
1
1
1/2
1/2
1/2

Notes

1. Simply having absorbing states does not make a system an absorbing system. The system shown in Figure 11 is *not* an absorbing system, even though it has two absorbing states.

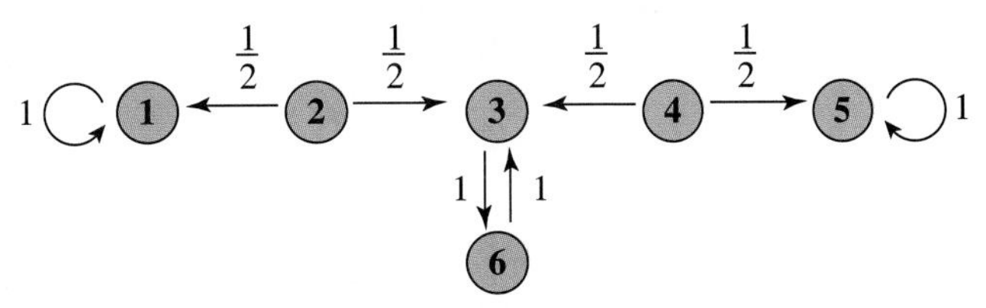

Figure 11

   The reason is that it is not possible to get from either state 3 or state 6 to any absorbing state.[8]

2. An absorbing system with more than one state is *never* regular. In the transition matrix and any of its powers, the row corresponding to an absorbing state will always have zeros, representing the zero probability of ever leaving that state.

Now just what are we to do with these absorbing systems? We might compute P^∞, but this turns out not to be very interesting. For example, if we consider the Roach Resort example, with

$$P = \begin{bmatrix} \frac{2}{5} & \frac{3}{5} \\ 0 & 1 \end{bmatrix}$$

we find[9] that

$$P^\infty = \begin{bmatrix} 0 & 1 \\ 0 & 1 \end{bmatrix}$$

All P^∞ tells us is that eventually the system will wind up in the absorbing state. This is what we expect in any absorbing system. So, we know what the long-term picture is for such systems, and it is not very interesting. There is, however, the following question.

---

[8]In a sense, states 3 and 6 together constitute an absorbing "state" because once you get into either of these states, you can never leave them. Mathematically, we could therefore treat them as a single absorbing state. The interested reader should consult a book on Markov chains.

[9]You could verify this experimentally, using a calculator or Excel, or theoretically, by computing the steady-state vector.

Question Suppose we start in a nonabsorbing state within an absorbing system. We know that, *eventually,* it will be our fate to wind up in one of the absorbing states. How long can we expect to be around before absorption? Also, how many times can we expect to visit each of the other nonabsorbing states before that happens?

Answer The answer is amazingly, even magically, easy to compute. To show how to do this computation, we first have to look a little more carefully at the transition matrix of an absorbing system.

To help us understand the structure of the transition matrix in an absorbing system, we adopt the following convention:

In an absorbing system, we always number the absorbing states last.

Not so coincidentally, the Quick Examples above have their states numbered in this way. Let's take a look at their transition matrices.

1. $P = \begin{bmatrix} \frac{2}{5} & \frac{3}{5} \\ 0 & 1 \end{bmatrix}$

2. $P = \begin{bmatrix} .9 & .1 & 0 \\ .1 & .7 & .2 \\ 0 & 0 & 1 \end{bmatrix}$

3. $P = \begin{bmatrix} 0 & .5 & .5 \\ 0 & 1 & 0 \\ 0 & 0 & 1 \end{bmatrix}$

4. $P = \begin{bmatrix} 0 & \frac{1}{2} & 0 & 0 & \frac{1}{2} \\ \frac{1}{2} & 0 & \frac{1}{2} & 0 & 0 \\ 0 & \frac{1}{2} & 0 & \frac{1}{2} & 0 \\ 0 & 0 & 0 & 1 & 0 \\ 0 & 0 & 0 & 0 & 1 \end{bmatrix}$

Notice that in every case the transition matrix P has a square identity matrix in the bottom right corner because of the absorbing states. The size of this identity matrix is the number of absorbing states. Thus, for example, when there is a single absorbing state, there is a 1×1 identity at the bottom right as in (1) and (2), whereas in (3) and (4), which each have two absorbing states, there are 2×2 identity matrices at the bottom right. Also notice that there is always a block of zeros in the bottom left, next to the identity matrix (representing the fact that you can never leave an absorbing state). Symbolically, we can break P down into the following blocks:

$$P = \begin{bmatrix} S & T \\ \mathbf{0} & I \end{bmatrix}$$

Here I is an $m \times m$ identity matrix ($m =$ the number of absorbing states), S is a square $(n - m) \times (n - m)$ matrix ($n =$ total number of states, so $n - m =$ the number of nonabsorbing states), $\mathbf{0}$ is a zero matrix, and T is an $(n - m) \times m$ matrix. The matrix S gives the transition probabilities for movement *among the nonabsorbing states* and is pivotal in answering the question asked above.

We can now answer the second part of that question.

The Fundamental Matrix and Number of Times a State Is Visited Prior to Absorption

For an absorbing Markov system, the **fundamental matrix** is the matrix

$$Q = (I - S)^{-1}$$

where I is the identity matrix with the same dimensions as S. If we are in a nonabsorbing state i at time 0, the number of time steps we can expect, on average, to land at nonabsorbing state j before absorption is given by the ijth entry in the fundamental matrix. If $i = j$, then this number includes the start time.

Quick Example

Quick Example 2 on page 538 has $Q = \begin{bmatrix} 15 & 5 \\ 5 & 5 \end{bmatrix}$. (We will compute Q for this system in Example 1.) We interpret the entries of Q as follows.

$Q_{11} = 15$: Starting in state 1, we will land in state 1 an average of 15 time steps (including the start time) before absorption.

$Q_{12} = 5$: Starting in state 1, we will land in state 2 an average of 5 time steps before absorption.

$Q_{21} = 5$: Starting in state 2, we will land in state 1 an average of 5 time steps before absorption.

$Q_{22} = 5$: Starting in state 2, we will land in state 2 an average of 5 time steps (including the start time) before absorption.

Note This does not mean that, starting in state 1, we can expect to spend 15 time steps just sitting in state 1; in fact we expect to pay 5 visits to state 2, as well, before absorption.

We justify this interpretation of Q after Example 1, in which we return to the savings and loan problem discussed in Example 2 of Section 9.1, and now find that we can answer the question posed at the end of that example.

Example 1 • Profitability

Two surveys of the S&L industry were taken 1 year apart. The results were as follows: Of those S&Ls that were profitable at the time of the first survey, 10% were found to be operating at a loss by the time of the second survey, and the rest of them remained profitable. Of the S&Ls that were operating at a loss at the time of the first survey, 20% were bankrupt by the time of the second survey (the chief executives were bailed out by the government at taxpayers' expense), and 10% of them became profitable. Assume that once an S&L goes under, its assets are disposed so that it can never operate again. I wish to invest my life savings in the Lifelong Trust Savings and Loan, whose last balance sheet showed a healthy profit. How long can I expect my investment to be safe?

Solution Recall that the three states are (1) profitable, (2) losing, and (3) bankrupt. This is exactly the Markov system described in Quick Example 2:

$$P = \begin{bmatrix} .9 & .1 & 0 \\ .1 & .7 & .2 \\ 0 & 0 & \boxed{1} \end{bmatrix}$$

so

$$S = \begin{bmatrix} .9 & .1 \\ .1 & .7 \end{bmatrix}$$

To compute the fundamental matrix, we first calculate

$$I - S = \begin{bmatrix} 1 & 0 \\ 0 & 1 \end{bmatrix} - \begin{bmatrix} .9 & .1 \\ .1 & .7 \end{bmatrix} = \begin{bmatrix} .1 & -.1 \\ -.1 & .3 \end{bmatrix}$$

Thus,

$$Q = (I - S)^{-1} = \begin{bmatrix} 15 & 5 \\ 5 & 5 \end{bmatrix}.$$

(We calculate this using Excel or the methods of Chapter 2.) Since profitability is represented by state 1, we look at the first row of Q. The $(1, 1)$ entry, 15, corresponds to the transition profitable → profitable, so it tells us that such a bank can be expected to be in the profitable state a total of 15 1-year time steps (including the first time step, in which I first invest my money) prior to going bankrupt. The $(1, 2)$ entry, 5, corresponds to the transition profitable → losing, so it tells us that a profitable bank can be expected to be in the losing state a total of 5 1-year periods prior to going bankrupt.[10] That gives a total of 20 years one can expect it to be either losing money or being profitable before going bankrupt.

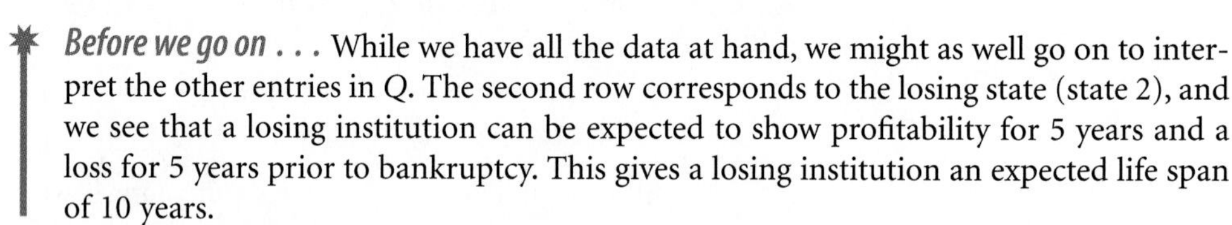

Before we go on . . . While we have all the data at hand, we might as well go on to interpret the other entries in Q. The second row corresponds to the losing state (state 2), and we see that a losing institution can be expected to show profitability for 5 years and a loss for 5 years prior to bankruptcy. This gives a losing institution an expected life span of 10 years.

Summary: Calculating the Expected Number of Steps to Absorption

1. To get the number of time steps, starting in state i, you expect to land in state j before absorption, calculate the fundamental matrix Q and take its ijth entry.
2. The *total* number of steps expected before absorption, starting in state i, equals the total number of visits you expect to make to all the nonabsorbing states. This is *the sum of all the entries in the ith row of Q.*

Question Why do the entries in the fundamental matrix give us the expected numbers of visits before absorption?

Answer Here is a sketch of the argument. Recalling how we formed the matrix S, notice that S gives the transition probabilities for movement among the nonabsorbing states. We also know that S^2 gives the two-step transition probabilities for movement among the nonabsorbing states; S^3, the three-step probabilities; and so on.

For example, the probability of getting from i to j in 20 steps (without getting absorbed) is the ijth entry of S^{20}. Other ways of interpreting that number are as the fraction of the times we will visit state j on the 20th step, if we start out in state i many

[10] To put it another way, the figures are telling us that prior to bankruptcy we can expect 15 of the company's annual reports to show a profit and 5 to show a loss.

times, or as the expected number of times we will visit state j on the 20th step if we start once in state i. It is this last interpretation that we will use.

To get the total number of times we expect to land at state j before absorption, on average, we add the number of times we expect to visit it on the first step, the number of times on the second step, and so on (we add 1 if we start out there):

total number of times we expect to land at state j if we start at state i

$= 1$ if $i = j$ — This is the ijth entry of the identity matrix. — I_{ij}

$+$ the number of times we expect to land in state j after one step — S_{ij}

$+$ the number of times we expect to land in state j after two steps — $(S^2)_{ij}$

$+$ the number of times we expect to land in state j after three steps — $(S^3)_{ij}$

$+ \cdots$

$$= I_{ij} + S_{ij} + (S^2)_{ij} + (S^3)_{ij} + \cdots$$

$= $ the ijth entry of $(I + S^1 + S^2 + S^3 + \cdots)$ — See footnote 11

Question OK, we have just found a new way to calculate the expected number of times we land at state j if we start at state i: Take the ijth entry of the matrix $(I + S^1 + S^2 + S^3 + \cdots)$. How does this justify the original claim that the expected number is also the ijth entry of $Q = (I - S)^{-1}$?

Answer Because $(I + S^1 + S^2 + S^3 + \cdots)$ is the same as $(I - S)^{-1}$ (!). To verify that, let's multiply the sum by $(I - S)$, using the distributive law. We get

$$(I - S)(I + S^1 + S^2 + S^3 + \cdots) = I - S + S - S^2 + S^2 - S^3 + S^3 - S^4 + \cdots$$

Notice that all the terms but the first cancel out, so we are left with the identity I! Thus,

$$(I - S)(I + S^1 + S^2 + S^3 + \cdots) = I$$

so $(I + S^1 + S^2 + S^3 + \cdots)$ must be the inverse of $(I - S)$. In other words,

$$(I + S^1 + S^2 + S^3 + \cdots) = (I - S)^{-1} = Q$$

and we have justified our use of the matrix Q.

Example 2 • Gambler's Ruin

Consider the Gambler's Ruin scenario from Example 3 of Section 9.1, with the absorbing states numbered last (Quick Example 4 above):

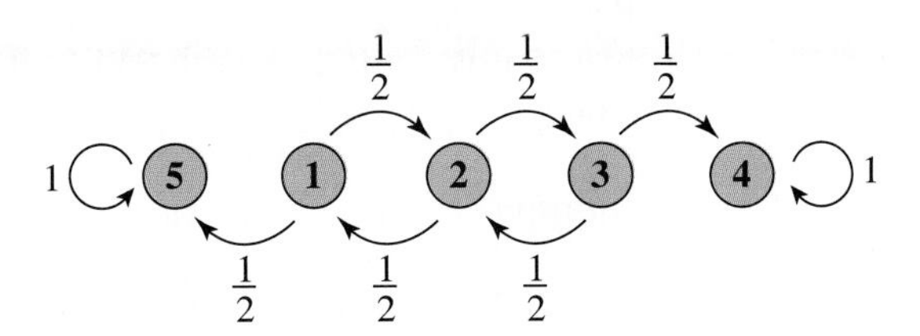

[11]Although what we are saying about these *infinite* sums of matrices is true, we will not present it as rigorously as we would in an upper-level book intended for mathematics majors. For example, we will assume that the infinite sum makes sense and that the algebra we perform with it is valid. Both of these facts require justification that we will not give.

Suppose you begin with \$20 (state 2 with this numbering). How many steps later do you expect to reach an absorbing state (that is, how long will it take you to either lose all your money or walk away with \$40)?

Solution The transition matrix and the matrix S are

$$P = \begin{bmatrix} 0 & \frac{1}{2} & 0 & 0 & \frac{1}{2} \\ \frac{1}{2} & 0 & \frac{1}{2} & 0 & 0 \\ 0 & \frac{1}{2} & 0 & \frac{1}{2} & 0 \\ 0 & 0 & 0 & 1 & 0 \\ 0 & 0 & 0 & 0 & 1 \end{bmatrix} \qquad S = \begin{bmatrix} 0 & \frac{1}{2} & 0 \\ \frac{1}{2} & 0 & \frac{1}{2} \\ 0 & \frac{1}{2} & 0 \end{bmatrix}$$

Then

$$Q = (I - S)^{-1} = \begin{bmatrix} \frac{3}{2} & 1 & \frac{1}{2} \\ 1 & 2 & 1 \\ \frac{1}{2} & 1 & \frac{3}{2} \end{bmatrix}$$

We conclude that the expected time to absorption is $1 + 2 + 1 = 4$ time steps. In other words, you can expect to play roulette a total of four times before you walk away from the table either broke or extremely happy, assuming you began with \$20.

This example raises another question: If you gamble as above, what are your chances of leaving the table broke versus leaving the table happy? We can state this question more formally.

Question Suppose we start in nonabsorbing state i of an absorbing system and there are at least two competing absorbing states (as there were in Example 2). What is the probability of winding up in a *specified* absorbing state?

Answer This is where we use the hitherto neglected portion T of the matrix P. It turns out that all we need to do is take the product QT, which is an $(n - m) \times m$ matrix, and look at row i. The entries in that row give the probabilities of winding up in each absorbing state. We'll explain why after the next example.

Example 3 • Interpreting T

Refer to Example 2. Now assume that you start with \$10 (state 1). Find the probabilities that you wind up with \$40 (state 4) and that you wind up losing all your money (state 5).

Solution Following the instructions, we calculate

$$QT = \begin{bmatrix} \frac{3}{2} & 1 & \frac{1}{2} \\ 1 & 2 & 1 \\ \frac{1}{2} & 1 & \frac{3}{2} \end{bmatrix} \begin{bmatrix} 0 & \frac{1}{2} \\ 0 & 0 \\ \frac{1}{2} & 0 \end{bmatrix} = \begin{bmatrix} \frac{1}{4} & \frac{3}{4} \\ \frac{1}{2} & \frac{1}{2} \\ \frac{3}{4} & \frac{1}{4} \end{bmatrix}$$

Since you started in state 1, look at row 1: $\begin{bmatrix} \frac{1}{4} & \frac{3}{4} \end{bmatrix}$. We interpret the entries as follows: There is a $\frac{1}{4}$ (25%) chance of winding up in the first-numbered absorbing state (state 4: walking away with a total of \$40) and a greater chance, $\frac{3}{4}$ (75%), of winding up in the second-numbered absorbing state (state 5: leaving the table broke). This seems about right because you start out with only \$10. Notice that the entries in each row of the matrix QT add up to 1. (Why?)

Question Why does $(I - S)^{-1}T$ give the probabilities of winding up in each absorbing state?

Answer We can make an argument very similar to the argument for the interpretation of $(I - S)^{-1}$. The ijth entry in T gives the probability of going from the ith nonabsorbing state to the jth absorbing state in just one step.[12] The ijth entry in ST gives the probability of going from state i to some nonabsorbing state and then to the jth absorbing state. The ijth entry of S^2T gives the probability of going from state i to two nonabsorbing states and then to the jth absorbing state, and so on. Thus, the probability of eventually winding up in the jth absorbing state, assuming that we start in state i, is the ijth entry of the sum

$$T + ST + S^2T + S^3T + \cdots = (I + S + S^2 + S^3 + \cdots)T = (I - S)^{-1}T$$

[12]Notice that, although the ith nonabsorbing state is just the state numbered i, the jth absorbing state is not the state numbered j. For example, the first absorbing state in the preceding two examples is state 4, and the second absorbing state is state 5.

9.4 EXERCISES

In Exercises 1–10, determine whether each transition diagram represents an absorbing Markov system. If not, say why not.

1.

2.

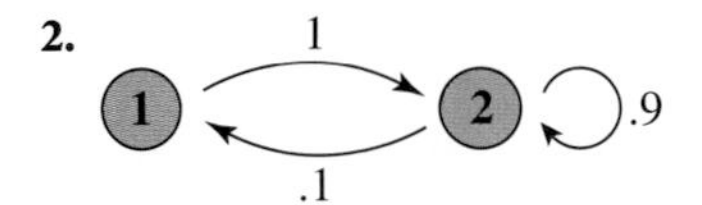

3. **4.**

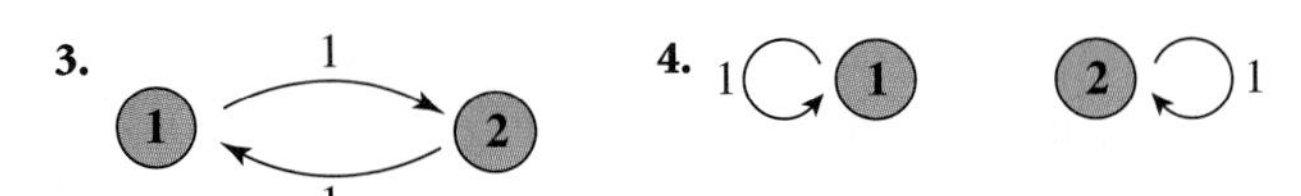

5. **6.**

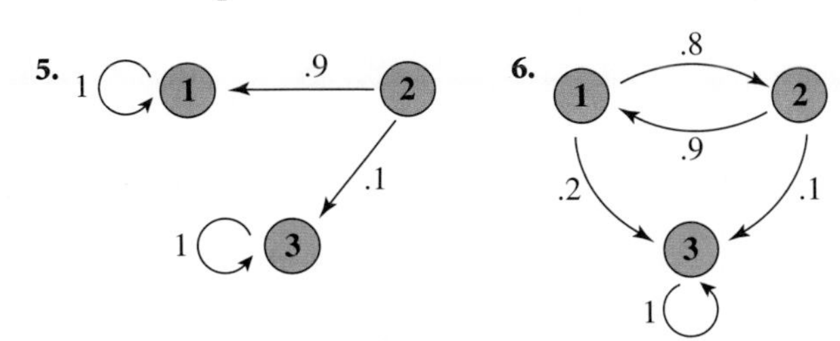

7.

8.

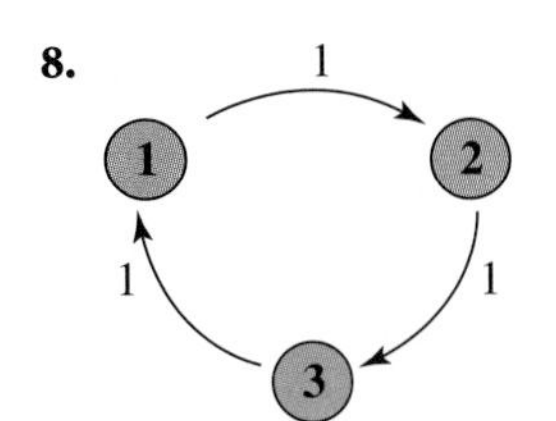

9.

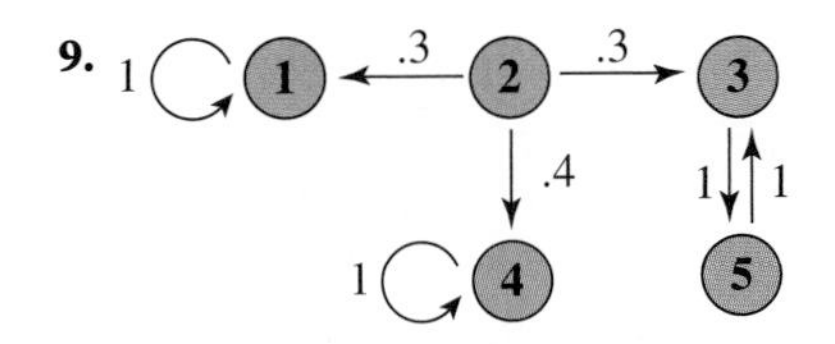

10.

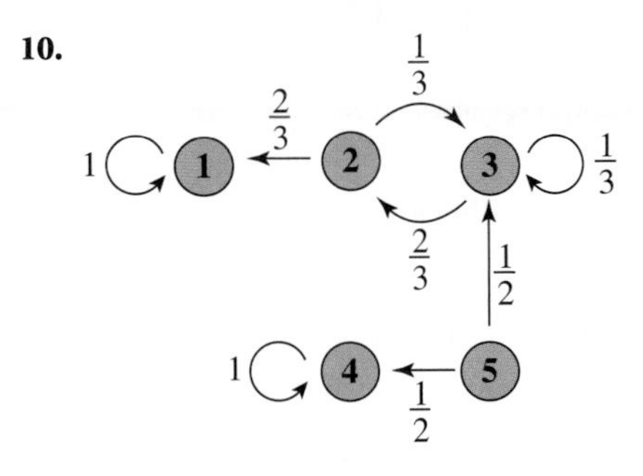

In Exercises 11–14, the transition matrix of an absorbing system is given, but the absorbing states are not numbered last. Remedy this by first filling in the entries in the blank table and then renumbering the states using the order suggested by the table. Write the new transition matrix and also the matrix S.

11. $P = \begin{bmatrix} 1 & 0 & 0 & 0 \\ 0 & 1 & 0 & 0 \\ .2 & .3 & .5 & 0 \\ .5 & .2 & .3 & 0 \end{bmatrix}$

	To			
From	**3**	**4**	**1**	**2**
3				
4				
1				
2				

12. $P = \begin{bmatrix} 1 & 0 & 0 & 0 \\ 0 & 1 & 0 & 0 \\ .1 & .1 & .2 & .6 \\ .6 & .1 & .2 & .1 \end{bmatrix}$

	To			
From	**3**	**4**	**1**	**2**
3				
4				
1				
2				

13. $P = \begin{bmatrix} 1 & 0 & 0 & 0 \\ .5 & .3 & .2 & 0 \\ .2 & .5 & .2 & .1 \\ 0 & 0 & 0 & 1 \end{bmatrix}$

	To			
From	**2**	**3**	**4**	**1**
2				
3				
4				
1				

14. $P = \begin{bmatrix} .3 & .4 & .1 & .2 \\ 0 & 1 & 0 & 0 \\ 0 & 0 & 1 & 0 \\ .2 & 0 & .4 & .4 \end{bmatrix}$

	To			
From	**4**	**1**	**2**	**3**
4				
1				
2				
3				

In Exercises 15–30, determine which of the given matrices represent absorbing Markov systems. In those cases where the system is absorbing, (a) compute the fundamental matrix Q and (b) determine the number of times, starting in each nonabsorbing state, you can expect the system to visit that same state, including the initial visit, prior to absorption.

15. $P = \begin{bmatrix} \frac{1}{2} & 0 & \frac{1}{2} \\ \frac{1}{4} & \frac{3}{4} & 0 \\ 0 & 0 & 1 \end{bmatrix}$

16. $P = \begin{bmatrix} \frac{1}{2} & \frac{1}{2} & 0 \\ 0 & \frac{3}{4} & \frac{1}{4} \\ 0 & 0 & 1 \end{bmatrix}$

17. $P = \begin{bmatrix} .3 & .7 & 0 \\ .7 & .3 & 0 \\ 0 & 0 & 1 \end{bmatrix}$

18. $P = \begin{bmatrix} 0 & .5 & .5 \\ 0 & 1 & 0 \\ 0 & 0 & 1 \end{bmatrix}$

19. $P = \begin{bmatrix} 0 & 0 & 1 \\ 0 & 1 & 0 \\ 0 & 0 & 1 \end{bmatrix}$

20. $P = \begin{bmatrix} 1 & 0 & 0 \\ 0 & 1 & 0 \\ 0 & 0 & 1 \end{bmatrix}$

21. $P = \begin{bmatrix} 0 & .5 & 0 & .5 \\ .5 & 0 & .5 & 0 \\ 0 & 0 & 1 & 0 \\ 0 & 0 & 0 & 1 \end{bmatrix}$

22. $P = \begin{bmatrix} .5 & .5 & 0 & 0 \\ .25 & 0 & .75 & 0 \\ 0 & 0 & 1 & 0 \\ 0 & 0 & 0 & 1 \end{bmatrix}$

23. $P = \begin{bmatrix} \frac{1}{3} & \frac{2}{3} & 0 & 0 \\ \frac{1}{2} & \frac{1}{2} & 0 & 0 \\ 0 & 0 & 1 & 0 \\ 0 & 0 & 0 & 1 \end{bmatrix}$

24. $P = \begin{bmatrix} \frac{1}{3} & \frac{2}{3} & 0 & 0 \\ \frac{1}{2} & \frac{1}{2} & 0 & 0 \\ \frac{1}{3} & 0 & \frac{1}{3} & \frac{1}{3} \\ 0 & 0 & 0 & 1 \end{bmatrix}$

25. $P = \begin{bmatrix} \frac{1}{2} & 0 & 0 & \frac{1}{2} & 0 \\ 0 & \frac{1}{2} & 0 & \frac{1}{2} & 0 \\ \frac{1}{2} & 0 & \frac{1}{2} & 0 & 0 \\ 0 & 0 & 0 & 1 & 0 \\ 0 & 0 & 0 & 0 & 1 \end{bmatrix}$

26. $P = \begin{bmatrix} 0 & \frac{1}{2} & 0 & 0 & \frac{1}{2} \\ \frac{1}{2} & 0 & \frac{1}{2} & 0 & 0 \\ \frac{1}{2} & 0 & 0 & \frac{1}{2} & 0 \\ 0 & 0 & 0 & 1 & 0 \\ 0 & 0 & 0 & 0 & 1 \end{bmatrix}$

T In Exercises 27–30, use technology to solve.

27. $P = \begin{bmatrix} .25 & .35 & 0 & .25 & .15 \\ .15 & .35 & .13 & .27 & .1 \\ .2 & .2 & .2 & .2 & .2 \\ 0 & 0 & 0 & 1 & 0 \\ 0 & 0 & 0 & 0 & 1 \end{bmatrix}$

28. $P = \begin{bmatrix} .25 & .35 & 0 & .25 & .15 \\ .15 & .35 & .13 & .27 & .1 \\ 1 & 0 & 0 & 0 & 0 \\ 0 & 0 & 0 & 1 & 0 \\ 0 & 0 & 0 & 0 & 1 \end{bmatrix}$

29. $P = \begin{bmatrix} 0 & .5 & 0 & 0 & 0 & 0 & 0 & 0 & 0 & 0 & .5 \\ .5 & 0 & .5 & 0 & 0 & 0 & 0 & 0 & 0 & 0 & 0 \\ 0 & .5 & 0 & .5 & 0 & 0 & 0 & 0 & 0 & 0 & 0 \\ 0 & 0 & .5 & 0 & .5 & 0 & 0 & 0 & 0 & 0 & 0 \\ 0 & 0 & 0 & .5 & 0 & .5 & 0 & 0 & 0 & 0 & 0 \\ 0 & 0 & 0 & 0 & .5 & 0 & .5 & 0 & 0 & 0 & 0 \\ 0 & 0 & 0 & 0 & 0 & .5 & 0 & .5 & 0 & 0 & 0 \\ 0 & 0 & 0 & 0 & 0 & 0 & .5 & 0 & .5 & 0 & 0 \\ 0 & 0 & 0 & 0 & 0 & 0 & 0 & .5 & 0 & .5 & 0 \\ 0 & 0 & 0 & 0 & 0 & 0 & 0 & 0 & .5 & .5 & 0 \\ 0 & 0 & 0 & 0 & 0 & 0 & 0 & 0 & 0 & 0 & 1 \end{bmatrix}$

30. $P = \begin{bmatrix} 0 & .5 & 0 & 0 & 0 & 0 & 0 & 0 & 0 & 0 & .5 \\ .5 & 0 & .5 & 0 & 0 & 0 & 0 & 0 & 0 & 0 & 0 \\ 0 & .5 & 0 & .5 & 0 & 0 & 0 & 0 & 0 & 0 & 0 \\ 0 & 0 & .5 & 0 & .5 & 0 & 0 & 0 & 0 & 0 & 0 \\ 0 & 0 & 0 & .5 & 0 & .5 & 0 & 0 & 0 & 0 & 0 \\ 0 & 0 & 0 & 0 & .5 & 0 & .5 & 0 & 0 & 0 & 0 \\ 0 & 0 & 0 & 0 & 0 & .5 & 0 & .5 & 0 & 0 & 0 \\ 0 & 0 & 0 & 0 & 0 & 0 & .5 & 0 & .5 & 0 & 0 \\ 0 & 0 & 0 & 0 & 0 & 0 & 0 & .5 & 0 & .5 & 0 \\ 0 & 0 & 0 & 0 & 0 & 0 & 0 & 0 & 0 & 1 & 0 \\ 0 & 0 & 0 & 0 & 0 & 0 & 0 & 0 & 0 & 0 & 1 \end{bmatrix}$

31. In the Markov system represented by the following transition diagram, determine how long, on average, it will take to reach the absorbing state, assuming that the system starts in state 2.

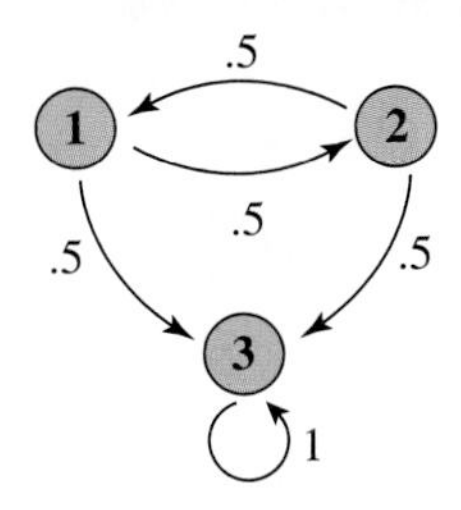

32. The following diagram represents a three-state Markov process. Assuming that the system is in state 1, how many times do you expect the system to be in that state prior to absorption?

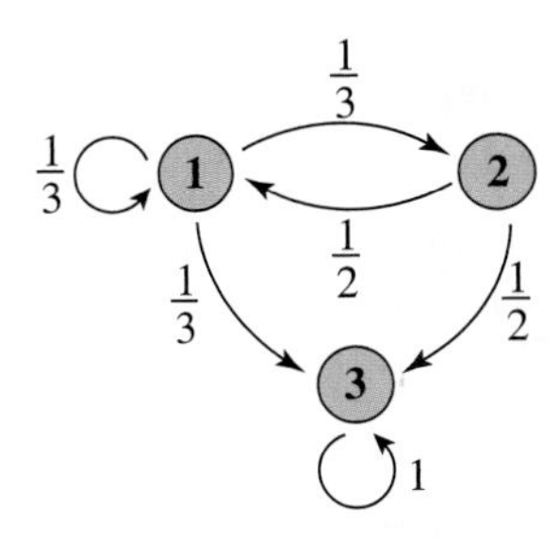

In Exercises 33–36, determine (a) the expected time until absorption from each initial nonabsorbing state and (b) the number of times, starting in each nonabsorbing state, the system can be expected to return there before absorption.

33. $P = \begin{bmatrix} \frac{1}{4} & \frac{1}{4} & \frac{1}{2} & 0 \\ \frac{1}{4} & 0 & \frac{1}{2} & \frac{1}{4} \\ 0 & 0 & 1 & 0 \\ 0 & 0 & 0 & 1 \end{bmatrix}$

34. $P = \begin{bmatrix} \frac{1}{3} & \frac{1}{3} & \frac{1}{3} & 0 \\ \frac{1}{3} & 0 & \frac{1}{3} & \frac{1}{3} \\ 0 & 0 & 1 & 0 \\ 0 & 0 & 0 & 1 \end{bmatrix}$

35. $P = \begin{bmatrix} \frac{1}{2} & 0 & 0 & \frac{1}{2} & 0 \\ 0 & \frac{1}{2} & 0 & 0 & \frac{1}{2} \\ \frac{1}{2} & 0 & \frac{1}{2} & 0 & 0 \\ 0 & 0 & 0 & 1 & 0 \\ 0 & 0 & 0 & 0 & 1 \end{bmatrix}$

36. $P = \begin{bmatrix} 0 & \frac{1}{2} & 0 & 0 & \frac{1}{2} \\ \frac{1}{2} & 0 & \frac{1}{2} & 0 & 0 \\ \frac{1}{2} & 0 & 0 & \frac{1}{2} & 0 \\ 0 & 0 & 0 & 1 & 0 \\ 0 & 0 & 0 & 0 & 1 \end{bmatrix}$

APPLICATIONS

37. Debt Analysis You manage the credit department for a large chain of stores and have access to the debt history of all your customers. Of the total debt owed to your company, $4 million is in the bad-debt category (representing accounts on which nothing has been paid for more than 90 days), $5 million is in the 31–90-day category (representing accounts on which the last payment was in this period), $8 million is in the 0–30-day category, and $100 million is in the paid category. The following table gives the probabilities that a single dollar will move from one category to the next in the period of 1 month:

	To			
From	**0–30 Days**	**31–90 Days**	**Bad Debts**	**Paid**
0–30 Days	.3	.2	0	.5
31–90 Days	.5	.2	.1	.2
Bad Debts	0	0	1	0
Paid	0	0	0	1

SOURCE: Based on a discussion in D. R. Anderson, D. J. Sweeny, and T. A. Williams, *An Introduction to Management Science*, 6th Ed. (St. Paul: West, 1991).

a. How many months do you expect it to take for a dollar in the 0–30-day category to wind up in either the paid category or the bad-debt category?
b. What percentage of the 0–30-day debt is ultimately paid?
c. What percentage of the 0–30-day debt is ultimately written off?

38. Debt Analysis Repeat Exercise 37, using the following table:

	To			
From	**0–30 Days**	**31–90 Days**	**Bad Debts**	**Paid**
0–30 Days	.5	.2	0	.3
31–90 Days	.2	.5	.1	.2
Bad Debts	0	0	1	0
Paid	0	0	0	1

39. Airplane Safety Assuming that there is a 1 in 10,000 chance of a plane crashing, use a Markov system to show that one can expect a plane crash on the 10,000th trip (on average).

40. Airplane Safety Repeat Exercise 39, given that the probability of a plane crash is 1 in 1 million.

41. Evaluating Health Care Programs The General Accounting Office, in assessing the impact of health care programs, obtained data from two groups of persons aged 65 years and older and classified the ability of each to perform 13 specific daily-living activities without assistance. This was repeated a year later, giving a Markov transition matrix. The classification was Best: able to perform all 13 activities without assistance; Next Best: able to perform the 13 activities but requiring assistance for at least one activity; Worst: Unable to perform the activities, even with assistance; and Death: died in the intervening year. The transition probabilities for the two groups were as follows:

Group I: Not Covered by Health Care Program

		Following Year Condition			
		Best	**Next Best**	**Worst**	**Death**
Current Year Condition	**Best**	.80	.10	.05	.05
	Next Best	.05	.80	.10	.05
	Worst	0	.10	.80	.10
	Death	0	0	0	1

Group II: Covered by Health Care Program

		Following Year Condition			
		Best	**Next Best**	**Worst**	**Death**
Current Year Condition	**Best**	.80	.10	.05	.05
	Next Best	.10	.80	.05	.05
	Worst	0	.15	.80	.05
	Death	0	0	0	1

For each of the two groups, calculate the expected survival time for persons in each category. Use these data to assess the effectiveness of health care programs in terms of increased life expectancy for each of the three categories.

SOURCE: Based on an extended application in D. R. Anderson, D. J. Sweeny, and T. A. Williams, *An Introduction to Management Science,* 6th ed. (St. Paul: West, 1991). The application was provided by Bill Anderson, U.S. General Accounting Office, Washington, D.C.

42. Evaluating Health Care Programs Repeat Exercise 41, using the following data:

Group I: Not Covered by Health Care Program

		Following Year Condition			
		Best	**Next Best**	**Worst**	**Death**
Current Year Condition	**Best**	.80	.10	.05	.05
	Next Best	0	.80	.10	.10
	Worst	0	.10	.80	.10
	Death	0	0	0	1

Group II: Covered by Health Care Program

		Following Year Condition			
		Best	**Next Best**	**Worst**	**Death**
Current Year Condition	**Best**	.80	.10	.05	.05
	Next Best	.10	.80	.05	.05
	Worst	0	.10	.80	.10
	Death	0	0	0	1

43. One-Dimensional Diffusion A particle undergoes a one-dimensional diffusion process modeled by the following transition diagram:

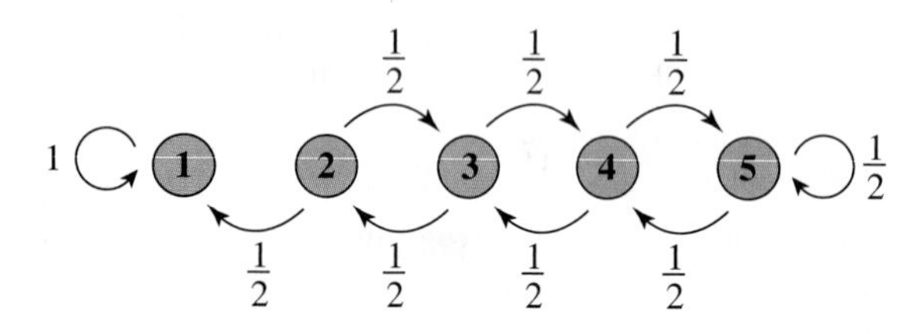

a. Determine the expected number of steps a particle starting in state 5 will take until it reaches state 1.

b. Indicate, *justifying your answer,* which of the four nonabsorbing states is the most frequently visited on average.

c. Look for a pattern in your computations and, without further calculation, write down the fundamental matrix Q for the process shown in the following diagram:

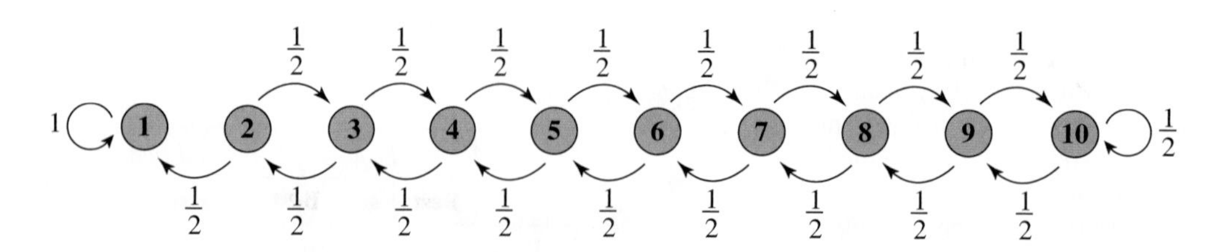

44. Random Walk with Traps The following diagram models a random walk with two "traps" (one at each end):

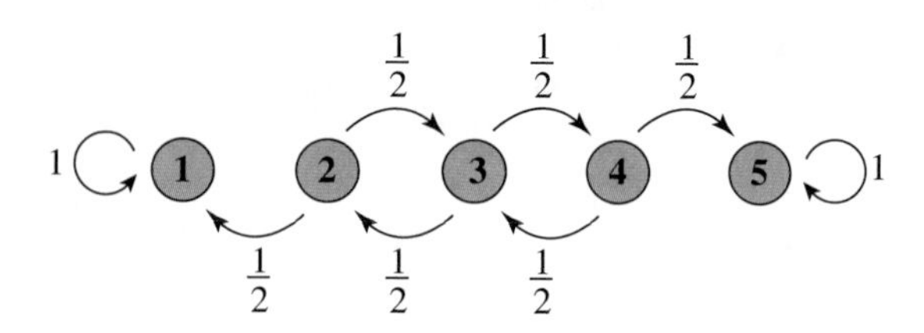

a. How long on average will it take for a particle starting in state 3 to reach one of the traps?

b. If a particle starts in state 3, how many subsequent times should one expect it to return there before entrapment?

c. Enumerate three or four possible routes from state 3 to state 1, showing the probability of each.

Exercises 45–48 require technology.

45. Rats in Mazes If a rat is placed in the maze shown in the diagram, it moves randomly from room to room. All possible directions, including staying put, are equally likely, except that once the rat gets to the cheese, it stays there and has a grand feast. How many moves do you expect it to take to reach the cheese if it starts in the top left corner?

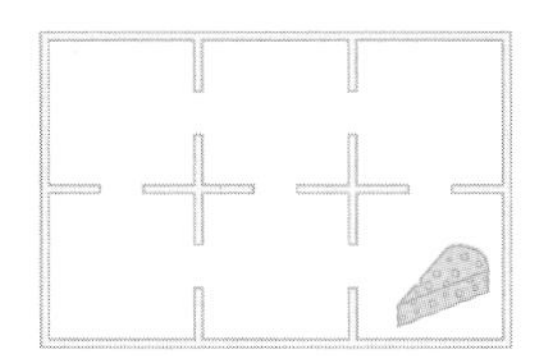

46. Rats in Mazes Repeat Exercise 45 in the event that the rat never stays in the same room until it reaches the cheese.

47. Fantasy One day Cunning the Swift summoned his followers before him. "You see before you a plan of the abode of the Mud Beast, who dwells at the center node," he declared. "All who enter his node are consumed alive and never emerge. As you well know, citizens of the empire who are sentenced to death must move among the outer nodes according to the probabilities shown until it is their sad fortune to enter the Mud Beast's lair at the center, never to return. Each day at dusk, every doomed citizen must cast the Die of Death and move as shown in the plan.

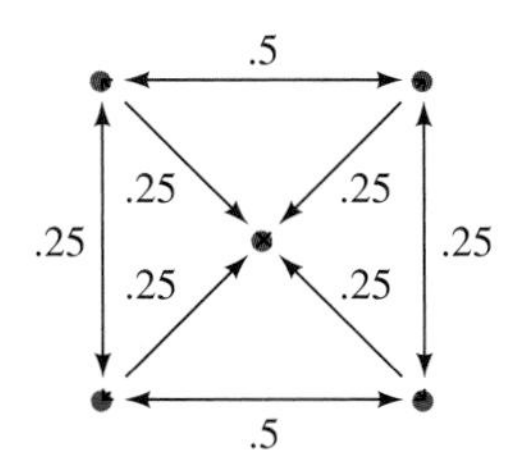

Lest you desire to join in their terrible fate, you must successfully answer the following questions. If you succeed, you will earn 10 gold pieces. If you fail, my questions will nevertheless be answered as I observe your fates!"

a. For how many days, oh my followers, does an average condemned prisoner survive?

b. How many times can one of those unfortunate souls expect to return to his starting point before meeting doom in the lair of the Mud Beast?

c. On average, how many times will a prisoner visit an adjacent node before his or her death?

d. On average, how many times will a prisoner visit the farthest node before his or her death?

e. What fraction of prisoners meet the Mud Beast after 2 days?

48. Fantasy Refer to Exercise 47.

a. How many times on average does a condemned prisoner visit an adjacent node to the north or south before meeting the Mud Beast?

b. Assuming that a prisoner begins at the northeast node, how many times on average will he or she visit the northwest node prior to meeting with doom?

c. What fraction of prisoners who begin at the northeast node return there after 2 days?

Dental Restoration According to a study at the University of Oregon Dental School, the following transition matrix models the progress of (untreated) decay on an upper second premolar over a 6-month time step. The diagrams show the status of the tooth (top view) with decayed surface(s) shaded. Exercises 49–58 are based on this system.

					Restoration	Loss of Tooth
	.90	0	0	.04	.06	0
	0	.75	0	.25	0	0
	0	0	0	.50	.50	0
	0	0	0	.90	.05	.05
Restoration	0	0	0	0	1	0
Loss of Tooth	0	0	0	0	0	1

SOURCE: K. H. Lu, "A Markov Chain Analysis of Caries Process with Consideration for the Effect of Restoration," *Archives of Oral Biology* 13 (1968) 1119–1132. The rounded data are based on a study of 184 grade and high school students. The matrix shown does not represent the entire system but only the portion of the system corresponding to two or more decayed surfaces.

49. Explain why all entries below the diagonal are zeros.

50. Experimentally, the Restoration state was found to be absorbing. Is this realistic?

51. According to the data shown, was it possible for an upper second premolar with decay on the front and rear faces (first state shown) to be lost after 6 months? 12 months? 18 months? 24 months?

52. Was a tooth with decay on three faces more likely to be lost or restored after 12 months?

Compute the fundamental matrix Q for the dental restoration system above and use it to answer the following questions.

53. How long on average does it take before a tooth with three decayed surfaces is either lost or restored?

54. How long on average does it take for a tooth with decay on the front and rear faces (first state shown) to deteriorate further?

55. Among the three states corresponding to decay on two surfaces, which takes the longest to deteriorate further?

56. Among the three states corresponding to decay on two surfaces, which deteriorates further most rapidly?

57. What percentage of teeth with decay on the front and rear faces (state 1) will ultimately be restored?

58. What percentage of teeth with decay on three faces will ultimately be restored?

COMMUNICATION AND REASONING EXERCISES

59. Let

$$P = \begin{bmatrix} 1 & 0 & 0 & 0 \\ 0 & 1 & 0 & 0 \\ .2 & .3 & .5 & 0 \\ .1 & .2 & .3 & .4 \end{bmatrix}$$

Calculate P^2, P^4, P^8, . . . by repeatedly squaring the answer until the powers of P appear to stop changing. Interpret the result.

60. Repeat Exercise 59, with

$$P = \begin{bmatrix} 1 & 0 & 0 & 0 \\ 0 & 1 & 0 & 0 \\ .1 & .2 & .3 & .4 \\ .2 & .1 & .4 & .3 \end{bmatrix}$$

61. Explain why calculation of the steady-state transition matrix is pointless for an absorbing Markov system with one absorbing state.

62. Can an absorbing Markov system also be regular? Give a reason to support your answer.

63. Find two different steady-state vectors for a Markov system with two absorbing states.

64. If the higher and higher powers of a transition matrix P approach the matrix

$$R = \begin{bmatrix} 1 & 0 & 0 & 0 \\ 0 & 1 & 0 & 0 \\ .4 & .6 & 0 & 0 \\ .4 & .6 & 0 & 0 \end{bmatrix}$$

how should you interpret the result?

65. Use an absorbing Markov system to construct a 4×4 matrix P with

$$P^3 = \begin{bmatrix} 0 & 0 & 0 & 1 \\ 0 & 0 & 0 & 1 \\ 0 & 0 & 0 & 1 \\ 0 & 0 & 0 & 1 \end{bmatrix}$$

but with $P^2 \neq P^3$.

66. Comment on the following statement in the light of absorbing Markov systems: If all forms of death other than accidental death were eliminated, some of us would be immortal.

CASE STUDY

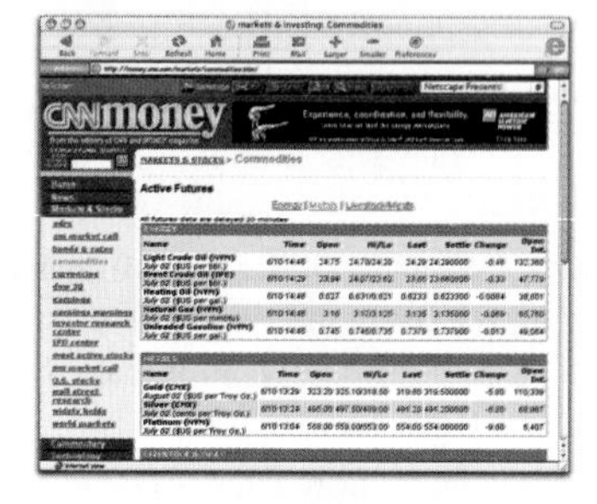

CNN MONEY IS A TRADEMARK OF TURNER BROADCASTING SYSTEM, INC. AN AOL TIME WARNER COMPANY

Predicting the Price of Gold

You are a consultant to a brokerage firm specializing in precious metals futures, and you have been supplied figures that show a rather lackluster bullion market. The price of gold is currently \$320 per ounce and has been moving up and down by an average of \$2 per trading day, more or less randomly over the past several months. Precious metals analysts are predicting that the market will heat up if one of two things happens: (1) The price reaches the psychological barrier of \$400 per ounce or (2) it drops to the current support level of \$280 per ounce. You have been asked to estimate how long it will take for one or the other to happen.

Your first impulse is to answer that, at an increase of \$2 per day, it will take $(400 - 320)/2 = 40$ trading days to reach \$400 and $(320 - 280)/2 = 20$ days to drop to \$280 per ounce, so you prepare a preliminary report stating that it will take between 20 and 40 trading days before the market heats up. However, after a little thought you realize that these answers are based on the assumption that the price of gold will continue moving in one direction, which is not what seems to be happening. In fact, what is happening is that the price of gold is moving on a random walk, with a .5 probability of moving to either a higher or lower level. The states of the system correspond to the possible prices per ounce, as shown in Figure 12.

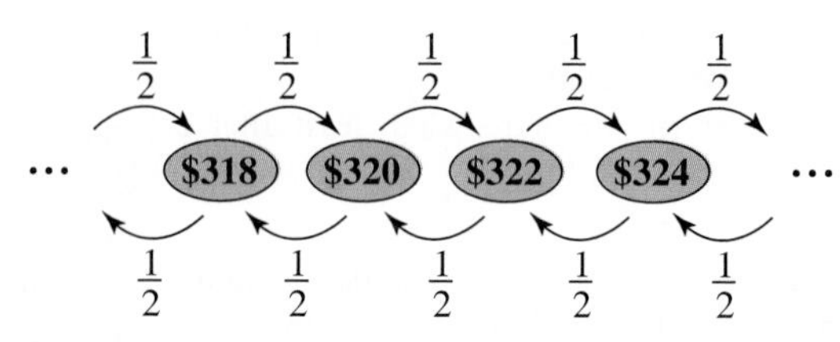

Figure 12

Since you are interested in how long, on average, the system will take to reach one of two possible states, \$400 or \$280, you decide that it will be easiest to think of them as absorbing states so that you can then compute the expected time to absorption. You number the states in the traditional manner so that you have the random walk with absorbing barriers and 61 states shown in Figure 13.

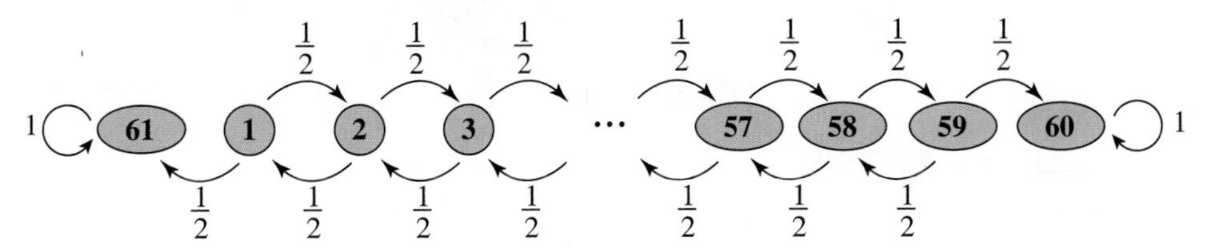

Figure 13

The task of setting up and manipulating a 61×61 matrix is daunting, even using technology, so it occurs to you to try to understand the behavior of this large Markov system by first looking at similar but smaller ones, as shown in Figure 14.

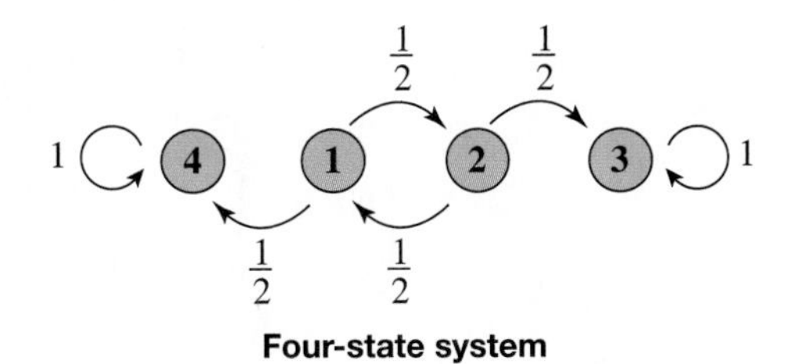

Four-state system

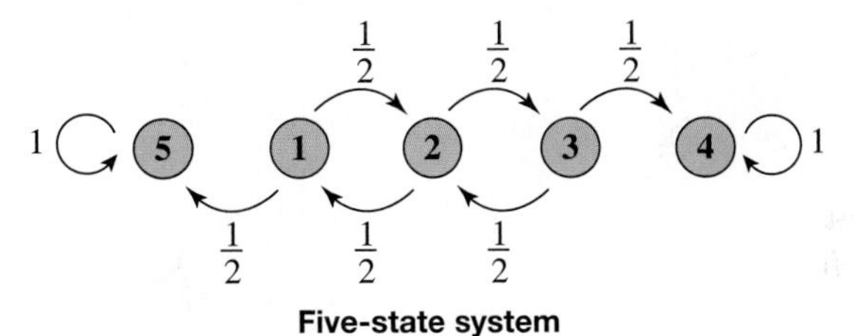

Five-state system

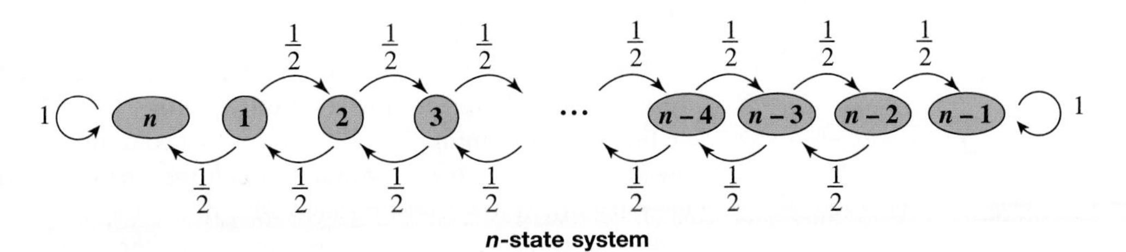

n-state system

Figure 14

Thus, you get to work on smaller systems, hoping to see a pattern that you can use to generalize to the n-state system. We have the four-state system:

$$P = \begin{bmatrix} 0 & \frac{1}{2} & 0 & \frac{1}{2} \\ \frac{1}{2} & 0 & \frac{1}{2} & 0 \\ 0 & 0 & 1 & 0 \\ 0 & 0 & 0 & 1 \end{bmatrix} \qquad S = \begin{bmatrix} 0 & \frac{1}{2} \\ \frac{1}{2} & 0 \end{bmatrix} \qquad Q = \begin{bmatrix} \frac{4}{3} & \frac{2}{3} \\ \frac{2}{3} & \frac{4}{3} \end{bmatrix} = \frac{1}{3}\begin{bmatrix} 4 & 2 \\ 2 & 4 \end{bmatrix}$$

The five-state system:

$$P = \begin{bmatrix} 0 & \frac{1}{2} & 0 & 0 & \frac{1}{2} \\ \frac{1}{2} & 0 & \frac{1}{2} & 0 & 0 \\ 0 & \frac{1}{2} & 0 & \frac{1}{2} & 0 \\ 0 & 0 & 0 & 1 & 0 \\ 0 & 0 & 0 & 0 & 1 \end{bmatrix} \qquad S = \begin{bmatrix} 0 & \frac{1}{2} & 0 \\ \frac{1}{2} & 0 & \frac{1}{2} \\ 0 & \frac{1}{2} & 0 \end{bmatrix} \qquad Q = \frac{1}{4}\begin{bmatrix} 6 & 4 & 2 \\ 4 & 8 & 4 \\ 2 & 4 & 6 \end{bmatrix}$$

The six-state system:

$$P = \begin{bmatrix} 0 & \frac{1}{2} & 0 & 0 & 0 & \frac{1}{2} \\ \frac{1}{2} & 0 & \frac{1}{2} & 0 & 0 & 0 \\ 0 & \frac{1}{2} & 0 & \frac{1}{2} & 0 & 0 \\ 0 & 0 & \frac{1}{2} & 0 & \frac{1}{2} & 0 \\ 0 & 0 & 0 & 0 & 1 & 0 \end{bmatrix} \qquad S = \begin{bmatrix} 0 & \frac{1}{2} & 0 & 0 \\ \frac{1}{2} & 0 & \frac{1}{2} & 0 \\ 0 & \frac{1}{2} & 0 & \frac{1}{2} \\ 0 & 0 & \frac{1}{2} & 0 \end{bmatrix} \qquad Q = \frac{1}{5}\begin{bmatrix} 8 & 6 & 4 & 2 \\ 6 & 12 & 8 & 4 \\ 4 & 8 & 12 & 6 \\ 2 & 4 & 6 & 8 \end{bmatrix}$$

The 11-state system:

$$Q = \frac{1}{10}\begin{bmatrix} 18 & 16 & 14 & 12 & 10 & 8 & 6 & 4 & 2 \\ 16 & 32 & 28 & 24 & 20 & 16 & 12 & 8 & 4 \\ 14 & 28 & 42 & 36 & 30 & 24 & 18 & 12 & 6 \\ 12 & 24 & 36 & 48 & 40 & 32 & 24 & 16 & 8 \\ 10 & 20 & 30 & 40 & 50 & 40 & 30 & 20 & 10 \\ 8 & 16 & 24 & 32 & 40 & 48 & 36 & 24 & 12 \\ 6 & 12 & 18 & 24 & 30 & 36 & 42 & 28 & 14 \\ 4 & 8 & 12 & 16 & 20 & 24 & 28 & 32 & 16 \\ 2 & 4 & 6 & 8 & 10 & 12 & 14 & 16 & 18 \end{bmatrix}$$

Now you begin to see a pattern: For an n-state system, the matrix Q is an $(n-2) \times (n-2)$ integer matrix multiplied by $1/(n-1)$. The top row, read right to left, lists successive multiples of 2, and the first column is the same as the first row. If you now delete both the first row and first column, the matrix that is left has the top row (again read right to left) listing successive multiples of 4, and the first column the same as the first row. Deleting these now reveals that the next row and column list multiples of 6, and so on, until you reach the lower right corner element. This pattern allows you to construct the matrix Q for any number of states.

Since you are interested in how long it takes the system to go into an absorbing state from a given initial state, you need to sum across each row. From the pattern you discovered above, you deduce the following sums:

$$\text{Row 1:} \quad \frac{1}{n-1}\,2(1+2+\cdots+n-2)$$

$$\text{Row 2:} \quad \frac{1}{n-1}\{2(2)(1+2+\cdots+n-3)+2(n-3)\}$$

$$\text{Row 3:} \quad \frac{1}{n-1}\{2(3)[1+2+\cdots+n-4]+2(n-4)[1+2]\}$$

$\vdots$

Row r:

$$\frac{1}{n-1}\{2r[1+2+\cdots+(n-r-1)]+2(n-r-1)[1+2+\cdots+(r-1)]\}$$

$$=\frac{1}{n-1}\left[2r\frac{(n-r-1)(n-r)}{2}+2(n-r-1)\frac{r(r-1)}{2}\right] \qquad \text{(ftn 13)}$$

$$=\frac{1}{n-1}r(n-r-1)[n-r+r-1]$$

$$=\frac{1}{n-1}r(n-r-1)(n-1)$$

$$=r(n-1-r)$$

This is the expected number of steps from the rth state to absorption in a system with n states.

The expected time to absorption starting in state r is $r(n-1-r)$ time steps.

Now you return to your own situation. Here, $n = 61$ (the number of states) and the gold price is currently at \$320, which is \$40 above the low barrier of \$280. You take state 1 to be \$282, state 2 to be \$284, and so on. This means that our initial state, \$320, is state 20, so you expect $20(61 - 1 - 20) = 800$ trading days until the gold market will pick up! This is a far cry from your original estimates of between 20 and 40 days.

EXERCISES

1. Give an intuitive reason why the original estimate of 20–40 days for the gold price to reach a barrier was unrealistic.
2. If the price of gold is currently \$2/ounce above the psychological barrier of \$280, how long do you expect it to take until absorption?
3. Using a ten-state system, graph the equation $e = r(10 - 1 - r)$ and determine the value of r for which e is a maximum. Interpret your answer.
4. Referring to Exercise 3 or otherwise, determine the price of gold that results in the longest expected time to absorption.
5. Referring to the original discussion, find the probability that the price of gold will ultimately fall to \$280/ounce.

---

[13]By the formula $1 + 2 + \cdots + k = \dfrac{k(k+1)}{2}$.

6. Referring to the original discussion, find the probability that the price of gold will eventually reach \$400/ounce.

7. How would your answer in the original discussion change if the price of gold moved in daily steps of \$4/ounce instead of \$2?

8. If the price of gold is now \$282/ounce, what is the probability that it will eventually fall to \$280/ounce? Why is the answer not .5?

9. Suppose \$280 was an absolute lower limit to the price of gold and, if the price of 1 ounce reached \$280, it would either stay there with a probability of .5 the next trading day or move up to \$282 with a probability of .5. How long would it then take for the price of 1 ounce of gold to reach \$400?

10. Suppose the price of gold could go up, down, or stay the same each trading day and all three were equally likely. How would this change your answer in the original discussion? (We suggest you use a calculator or computer software to experiment with this.)

CHAPTER 9 REVIEW TEST

1. A poll shows that half the consumers who use brand A switched to brand B the following year, and the other half stayed with brand A. Three-quarters of the brand B users stayed with brand B the following year, and the rest switched to brand A.

a. Give the associated Markov state distribution matrix with state 1 representing using brand A and state 2 representing using brand B.

b. Compute the associated two- and three-step transition matrices. What is the probability that a brand A user will be using brand B 3 years later?

c. If 640 consumers are presently using brand A and 320 are using brand B, how are these consumers distributed in 3 years' time?

d. In the long term, what fraction of the time will a user spend using each of the two brands?

e. Given the population distribution in part (c), how do you expect a group of 960 consumers to be divided after many years of brand switching?

2. You are sitting on your favorite couch in front of the television, and every hour on the hour, you toss a fair die once. You will get up and go to bed when you throw a 6 (and not a moment before).

a. Model the situation by a Markov process and hence calculate the probability that you will be in bed within 3 hours.

b. It is now just after midnight, and you are still on your couch. Calculate the probability that the die will instruct you to go to bed at 3 A.M. (but not before).

c. Is the answer to part (b) greater than, equal to, or smaller than the probability that the die will instruct you to go to bed at exactly 8 A.M.? Explain.

d. It is now 12:30 A.M., and you are still on your couch. When can you expect to go to bed?

3. A Markov system has the state transition diagram shown:

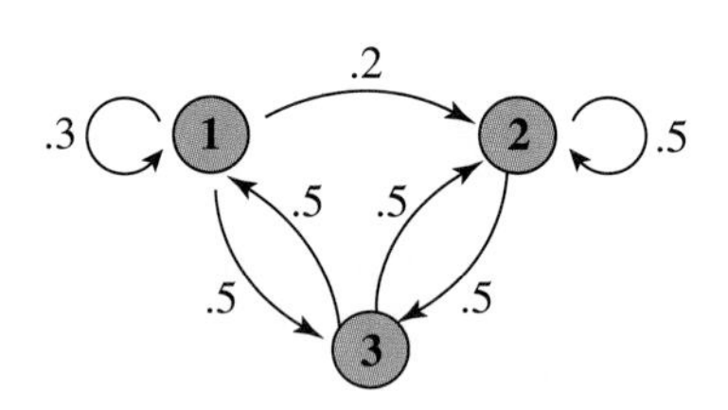

a. Write the transition matrix for the system.

b. Compute the associated two-step transition matrix.

c. What is the probability of the system being in state 2 after three steps if it starts in state 3?

d. Given the initial distribution vector $v = [100 \quad 60 \quad 80]$, calculate the distribution after two steps.

e. Given the initial distribution vector $v = [5 \quad 9 \quad 7]$, calculate the distribution after 31 steps.

4. Let $P = \begin{bmatrix} 0 & .5 & .5 & 0 \\ 0 & 0 & .5 & .5 \\ .5 & 0 & 0 & .5 \\ .5 & .5 & 0 & 0 \end{bmatrix}$.

a. Draw the associated state transition diagram.

b. Compute P^2 and P^4.

c. Without doing any computation, say what fraction of the time the system will spend in each of the four states on average. Motivate your answer by referring to the transition diagram.

d. Now write down the long-term state transition matrix without doing any computation.

5. Consider the Markov system represented by the following diagram:

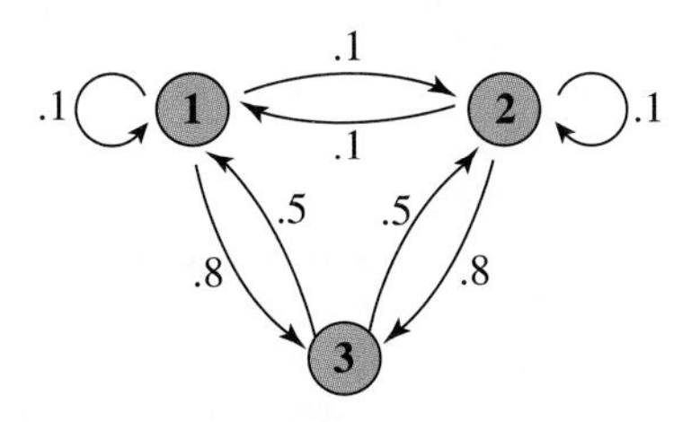

a. Compute the long-term transition matrix for this system.
b. What fraction of the time does the system spend in state 2?
c. For a system starting in state 1, which is more likely: that the system is in state 2 after 2 steps or that the system is in state 2 after 256 steps? How likely is that?
d. Given the initial distribution vector $v = [5 \quad 4 \quad 9]$, calculate its distribution after a large number of steps.
e. Consider the two-state Markov system obtained by thinking of states 1 and 2 combined as a single state A and state 3 as the single state B. What is the transition matrix for this system? What is its long-term transition matrix? (It is not necessary to do any further computation.)

6. A Markov process has the following state transition diagram:

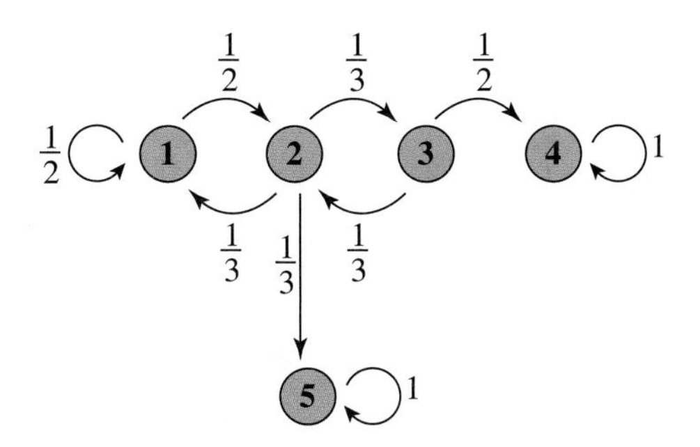

a. How many more steps on average will a particle starting in state 1 take until it becomes absorbed?
b. Which of the three nonabsorbing states is visited most frequently on average?
c. What fraction of the particles starting in state 1 are eventually absorbed in state 4?

OHaganBooks.com—LONG-TERM PLANNING

7. OHaganBooks.com has two main competitors, JungleBooks.com and FarmerBooks.com, and no other competitors of any significance. The following table shows the movement of customers during July.[14] (For instance, the first row tells us that 80% of OHaganBooks.com's customers remained loyal, 10% of them went to JungleBooks.com, and the remaining 10% went to FarmerBooks.com.)

	To		
From	**OHaganBooks**	**JungleBooks**	**FarmerBooks**
OHaganBooks	80%	10%	10%
JungleBooks	40%	60%	0%
FarmerBooks	20%	0%	80%

At the beginning of July, OHaganBooks.com had an estimated 2000 customers, and its two competitors had 4000 each.

a. Estimate the number of customers each company has at the end of July.
b. Assuming the July trends continue in August, predict the number of customers each company will have at the end of August.
c. Name one or more important factors that the Markov model does not take into account.
d. Assuming the July trend were to continue indefinitely, predict the market share enjoyed by each of the three e-commerce sites.

8. The office space layout at OHaganBooks.com is unorganized and crowded and results in frequent collisions between computer technicians rushing from place to place. The company has recently commissioned Office Synergy Consultants to propose a reorganization of the office space. The consultants tracked the comings and goings of a typical staff member at OHaganBooks.com and came up with the following (simplified) diagram:

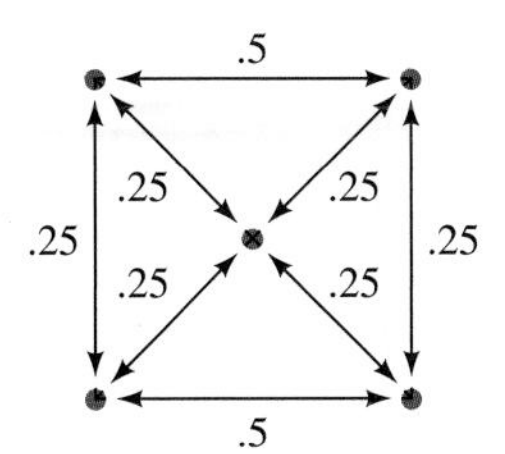

[14]By a "customer" of one of the three e-commerce sites, we mean someone who purchases more at that site than at either of the two competitors' sites.

The values on the arrows show the probabilities of the destinations when a staff member goes from one location to another. The four outside nodes represent the four work locations, and the center node represents the coffee room.

a. Assuming that a staff member is not at the coffee room, find the probability that he or she will be there after three changes of location.

b. Assuming that a staff member is at the coffee room, find the probability that he or she is not back there after three changes of location.

c. What is the probability that a worker who starts in one of the work locations is back there after three moves?

T **d.** The Human Resources Department at OHaganBooks.com suspects that, on average, the coffee room is visited more often than any of the four work locations. Is this suspicion justified? (Support your answer by calculating the fraction of times each location is visited in the long term.)

e. The local union chapter would like to ensure that coffee breaks are lengthened to twice the current length, and they claim that staff members currently spend less than 10% of the time in the coffee area. Is this claim contradicted by the data? Explain.

9. Although OHaganBooks.com has a large warehouse in Texas, many orders are for books not in stock, and so the company must attempt to locate books at another vendor and place some orders on hold. Some of these orders are eventually canceled, and others are shipped. The following table shows the transition probabilities for status of a book order at OHaganBooks.com over a 1-day period:

	To			
From	**On Hold 1–5 Days**	**On Hold >5 Days**	**Canceled**	**Shipped**
On Hold 1–5 Days	.60	.10	.05	.25
On Hold >5 Days	0	.90	.05	.05
Canceled	0	0	1	0
Shipped	0	0	0	1

a. OHaganBooks.com currently has 1600 orders on hold, of which 1000 have been on hold for more than 5 days. How many of those orders can the company expect to be canceled within 2 days?

b. Compute the fundamental matrix for the associated Markov system (with the states numbered in the order shown on the table).

c. How many more days do you expect it to take for an order currently on hold to be either shipped or canceled, assuming it has been on hold for up to 5 days?

d. How many more days do you expect it to take for an order currently on hold to be either shipped or canceled, assuming it has been on hold for more than 5 days?

e. What percentage of all orders placed on hold are ultimately shipped? (*Hint:* An order is in the first category when it is placed on hold.)

ADDITIONAL ONLINE REVIEW

If you follow the path

Web site → Everything for Finite Math → Chapter 9

you will find the following additional resources to help you review:

A comprehensive chapter summary (including examples and interactive features)

Additional review exercises (including interactive exercises and many with help)

A true/false chapter quiz

A matrix algebra tool to help you check your work

A Markov systems simulation utility

Appendix A

REAL NUMBERS

The **real numbers** are the numbers that can be written in decimal notation, including those that require an infinite decimal expansion. The set of real numbers includes all integers, positive and negative; all fractions; and the irrational numbers, those with decimal expansions that never repeat. Examples of irrational numbers are

$$\sqrt{2} = 1.414\,213\,562\,373\ldots \quad \text{and} \quad \pi = 3.141\,592\,653\,589\ldots$$

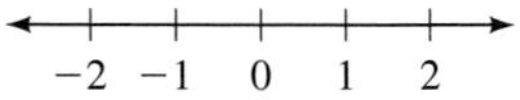

Figure 1

It is very useful to picture the real numbers as points on a line. As shown in Figure 1, larger numbers appear to the right—in the sense that if $a < b$, then the point corresponding to b is to the right of the one corresponding to a.

Intervals

Some subsets of the set of real numbers, called **intervals,** show up quite often and so we have a compact notation for them.

Interval Notation

Here is a list of types of intervals along with examples.

	Interval	Description	Picture	Example
Closed	$[a, b]$	Set of numbers x with $a \le x \le b$	 (includes endpoints)	$[0, 10]$
Open	(a, b)	Set of numbers x with $a < x < b$	 (excludes endpoints)	$(-1, 5)$
Half-Open	$(a, b]$	Set of numbers x with $a < x \le b$		$(-3, 1]$
	$[a, b)$	Set of numbers x with $a \le x < b$		$[0, 5)$
Infinite	$[a, +\infty)$	Set of numbers x with $a \le x$		$[10, +\infty)$
	$(a, +\infty)$	Set of numbers x with $a < x$		$(-3, +\infty)$

	Interval	Description	Picture	Example
Infinite (continued)	$(-\infty, b]$	Set of numbers x with $x \le b$	b	$(-\infty, -3]$
	$(-\infty, b)$	Set of numbers x with $x < b$	b	$(-\infty, 10)$
	$(-\infty, +\infty)$	Set of all real numbers		$(-\infty, +\infty)$

Operations

There are five important operations on real numbers: addition, subtraction, multiplication, division, and exponentiation. *Exponentiation* means raising a real number to a power; for instance, $3^2 = 3 \cdot 3 = 9$; $2^3 = 2 \cdot 2 \cdot 2 = 8$.

A note on technology: Most graphing calculators and spreadsheets use an asterisk * for multiplication and a caret sign ^ for exponentiation. Thus, for instance, 3×5 is entered as 3*5, $3x$ as 3*x, and 3^2 as 3^2.

When we write an expression involving two or more operations, like

$$2 \cdot 3 + 4 \quad \text{or} \quad \frac{2 \cdot 3^2 - 5}{4 - (-1)}$$

we need to agree on the order in which to do the operations. Does $2 \cdot 3 + 4$ mean $(2 \cdot 3) + 4 = 10$ or $2 \cdot (3 + 4) = 14$? We all agree to use the following rules for the order in which we do the operations.

Standard Order of Operations

Parentheses and Fraction Bars

First, calculate the values of all expressions inside parentheses or brackets, working from the innermost parentheses out, before using them in other operations. In a fraction, calculate the numerator and denominator separately before doing the division.

Quick Examples

1. $6[2 + (3 - 5) - 4] = 6[2 + (-2) - 4] = 6(-4) = -24$
2. $\dfrac{(4 - 2)}{3(-2 + 1)} = \dfrac{2}{3(-1)} = \dfrac{2}{-3} = -\dfrac{2}{3}$
3. $3/(2 + 4) = \dfrac{3}{2 + 4} = \dfrac{3}{6} = \dfrac{1}{2}$
4. $(x + 4x)/(y + 3y) = 5x/(4y)$

Exponents

Next, perform exponentiation.

Quick Examples

1. $2 + 4^2 = 2 + 16 = 18$
2. $(2 + 4)^2 = 6^2 = 36$

Note the difference.

3. $2\left(\dfrac{3}{4 - 5}\right)^2 = 2\left(\dfrac{3}{-1}\right)^2 = 2(-3)^2 = 2 \times 9 = 18$
4. $2(1 + 1/10)^2 = 2(1.1)^2 = 2 \times 1.21 = 2.42$

Appendix A

Multiplication and Division

Next, do all multiplications and divisions, from left to right.

Quick Examples

1. $2(3 - 5)/4 \cdot 2 = 2(-2)/4 \cdot 2$ — Parentheses first
   $= -4/4 \cdot 2$ — Leftmost product
   $= -1 \cdot 2 = -2$ — Multiplications and divisions, left to right
2. $2(1 + 1/10)^2 \times 2/10 = 2(1.1)^2 \times 2/10$ — Parentheses first
   $= 2 \times 1.21 \times 2/10$ — Exponent
   $= 4.84/10 = 0.484$ — Multiplications and divisions, left to right
3. $4\dfrac{2(4-2)}{3(-2 \cdot 5)} = 4\dfrac{2(2)}{3(-10)} = 4\dfrac{4}{-30} = \dfrac{16}{-30} = -\dfrac{8}{15}$

Addition and Subtraction

Last, do all additions and subtractions, from left to right.

Quick Examples

1. $2(3 - 5)^2 + 6 - 1 = 2(-2)^2 + 6 - 1 = 2(4) + 6 - 1 = 8 + 6 - 1 = 13$
2. $\left(\dfrac{1}{2}\right)^2 - (-1)^2 + 4 = \dfrac{1}{4} - 1 + 4 = -\dfrac{3}{4} + 4 = \dfrac{13}{4}$
3. $3/2 + 4 = 1.5 + 4 = 5.5$ } Note the difference.
4. $3/(2 + 4) = 3/6 = 1/2 = 0.5$ }
5. $4/2^2 + (4/2)^2 = 4/2^2 + 2^2 = 4/4 + 4 = 1 + 4 = 5$

Entering Formulas

Any good calculator or spreadsheet will respect the standard order of operations. However, we must be careful with division and exponentiation and use parentheses as necessary. The following table gives some examples of simple mathematical expressions and their equivalents in the functional format used in most graphing calculators, spreadsheets, and computer programs.

Mathematical Expression	**Formula**	**Comments**
$\dfrac{2}{3-x}$	`2/(3-x)`	Note the use of parentheses instead of the fraction bar. If we omit the parentheses, we get the expression shown next.
$\dfrac{2}{3} - x$	`2/3-x`	The calculator follows the usual order of operations.
$\dfrac{2}{3 \times 5}$	`2/(3*5)`	Putting the denominator in parentheses ensures that the multiplication is carried out first. The asterisk is usually used for multiplication in graphing calculators and computers.
$\dfrac{2}{x} \times 5$	`(2/x)*5`	Putting the fraction in parentheses ensures that it is calculated first. Some calculators will interpret `2/3*5` as $\dfrac{2}{3 \times 5}$, but `2/3(5)` as $\dfrac{2}{3} \times 5$.
$\dfrac{2-3}{4+5}$	`(2-3)/(4+5)`	Note once again the use of parentheses in place of the fraction bar.

Mathematical Expression	Formula	Comments
2^3	`2^3`	The caret ^ is commonly used to denote exponentiation.
2^{3-x}	`2^(3-x)`	Be careful to use parentheses to tell the calculator where the exponent ends. Enclose the *entire exponent* in parentheses.
$2^3 - x$	`2^3-x`	Without parentheses, the calculator will follow the usual order of operations: exponentiation and then subtraction
3×2^{-4}	`3*2^(-4)`	On some calculators, the negation key is separate from the minus key.
$2^{-4\times 3} \times 5$	`2^(-4*3)*5`	Note once again how parentheses enclose the entire exponent.
$100\left(1 + \frac{0.05}{12}\right)^{60}$	`100*(1+0.05/12)^60`	This is a typical calculation for compound interest.
$PV\left(1 + \frac{r}{m}\right)^{mt}$	`PV*(1+r/m)^(m*t)`	This is the compound interest formula. *PV* is understood to be a single number (present value) and not the product of *P* and *V* (or else we would have used `P*V`).
$\frac{2^{3-2} \times 5}{y - x}$	`2^(3-2)*5/(y-x)` or `(2^(3-2)*5)/(y-x)`	Notice again the use of parentheses to hold the denominator together. We could also have enclosed the numerator in parentheses, although this is optional. (Why?)
$\frac{2^y + 1}{2 - 4^{3x}}$	`(2^y+1)/(2-4^(3*x))`	Here, it is necessary to enclose both the numerator and the denominator in parentheses.
$2^y + \frac{1}{2} - 4^{3x}$	`2^y+1/2-4^(3*x)`	This is the effect of leaving out the parentheses around the numerator and denominator in the previous expression.

Accuracy and Rounding

When we use a calculator or computer, the results of our calculations are often given to far more decimal places than are useful. For example, suppose we are told that a square has an area of 2.0 square feet and we are asked how long its sides are. Each side is the square root of the area, which the calculator tells us is

$$\sqrt{2} \approx 1.414\ 213\ 562$$

However, the measurement of 2.0 square feet is likely accurate to only two digits, so our estimate of the lengths of the sides can be no more accurate than that. Therefore, we round the answer to two digits:

$$\text{length of one side} \approx 1.4 \text{ feet}$$

The digits that follow 1.4 are meaningless. The following guide makes these ideas more precise.

Significant Digits, Decimal Places, and Rounding

The number of **significant digits** in a decimal representation of a number is the number of digits that are not leading zeros after the decimal point (as in .0005) or trailing zeros before the decimal point (as in 5,400,000). We say that a value is **accurate to *n* significant digits** if only the first n significant digits are meaningful.

When to Round

After doing a computation in which all the quantities are accurate to no more than n significant digits, round the final result to n significant digits.

Quick Examples

1. 0.000 67 has two significant digits.` — The 000 before 67 are leading zeros.
2. 0.000 670 has three significant digits. — The 0 after 67 is significant.

Appendix A

3. 5,400,000 has two or more significant digits. We can't say how many of the zeros are trailing.*

4. 5,400,001 has 7 significant digits. The string of zeros is not trailing.

5. Rounding 63,918 to three significant digits gives 63,900.

6. Rounding 63,958 to three significant digits gives 64,000.

7. $\pi = 3.141\ 592\ 653\ldots$ and $\frac{22}{7} = 3.142\ 857\ 142\ldots$. Therefore, $\frac{22}{7}$ is an approximation of π that is accurate to only three significant digits (3.14).

8. $4.02(1 + 0.02)^{1.4} \approx 4.13$ We rounded to three significant digits.

*If we obtained 5,400,000 by rounding 5,401,011, then it has three significant digits because the zero after the 4 is significant. On the other hand, if we obtained it by rounding 5,411,234, then it has only two significant digits. The use of scientific notation avoids this ambiguity: 5.40×10^6 (or 5.40 E6 on a calculator or computer) is accurate to three digits and 5.4×10^6 is accurate to two.

One more point, though: If, in a long calculation, you round the intermediate results, your final answer may be even less accurate than you think. As a general rule,

When calculating, don't round intermediate results. Rather, use the most accurate results obtainable or have your calculator or computer store them for you.

When you are done with the calculation, *then* round your answer to the appropriate number of digits of accuracy.

EXERCISES

In Exercises 1–24, calculate each expression, giving the answer as a whole number or a fraction in lowest terms.

1. $2[4 + (-1)](2 \cdot -4)$

2. $3 + [(4 - 2) \cdot 9]$

3. `20/(3*4)-1`

4. `2-(3*4)/10`

5. $\dfrac{3 + \{[3 + (-5)]\}}{3 - 2 \times 2}$

6. $\dfrac{12 - (1 - 4)}{2(5 - 1) \cdot 2 - 1}$

7. `(2-5*(-1))/1-2*(-1)`

8. `2-5*(-1)/(1-2*(-1))`

9. $2 \cdot (-1)^2/2$

10. $2 + 4 \cdot 3^2$

11. $2 \cdot 4^2 + 1$

12. $1 - 3 \cdot (-2)^2 \times 2$

13. `3^2+2^2+1`

14. `2^(2^2-2)`

15. $\dfrac{3 - 2(-3)^2}{-6(4 - 1)^2}$

16. $\dfrac{1 - 2(1 - 4)^2}{2(5 - 1)^2 \cdot 2}$

17. `10*(1+1/10)^3`

18. `121/(1+1/10)^2`

19. $3\left[\dfrac{-2 \cdot 3^2}{-(4 - 1)^2}\right]$

20. $-\left[\dfrac{8(1 - 4)^2}{-9(5 - 1)^2}\right]$

21. $3\left[1 - \left(-\dfrac{1}{2}\right)^2\right]^2 + 1$

22. $3\left[\dfrac{1}{9} - \left(\dfrac{2}{3}\right)^2\right]^2 + 1$

23. `(1/2)^2-1/2^2`

24. `2/(1^2)-(2/1)^2`

In Exercises 25–50, convert each expression into its technology formula equivalent as in the table in the text.

25. $3 \times (2 - 5)$

26. $4 + \frac{5}{9}$

27. $\dfrac{3}{2 - 5}$

28. $\dfrac{4 - 1}{3}$

29. $\dfrac{3 - 1}{8 + 6}$

30. $3 + \dfrac{3}{2 - 9}$

31. $3 - \dfrac{4 + 7}{8}$

32. $\dfrac{4 \times 2}{\left(\frac{2}{3}\right)}$

33. $\dfrac{2}{3 + x} - xy^2$

34. $3 + \dfrac{3 + x}{xy}$

35. $3.1x^3 - 4x^{-2} - \dfrac{60}{x^2 - 1}$

36. $2.1x^{-3} - x^{-1} + \dfrac{x^2 - 3}{2}$

37. $\dfrac{\left(\frac{2}{3}\right)}{5}$

38. $\dfrac{2}{\left(\frac{3}{5}\right)}$

39. $3^{4-5} \times 6$

40. $\dfrac{2}{3 + 5^{7-9}}$

41. $3\left(1 + \dfrac{4}{100}\right)^{-3}$

42. $3\left(\dfrac{1 + 4}{100}\right)^{-3}$

43. $3^{2x-1} + 4^x - 1$

44. $2^{x^2} - (2^{2x})^2$

45. 2^{2x^2-x+1}

46. $2^{2x^2-x} + 1$

47. $\dfrac{4e^{-2x}}{2 - 3e^{-2x}}$

48. $\dfrac{e^{2x} + e^{-2x}}{e^{2x} - e^{-2x}}$

49. $3\left[1 - \left(-\dfrac{1}{2}\right)^2\right]^2 + 1$

50. $3\left[\dfrac{1}{9} - \left(\dfrac{2}{3}\right)^2\right]^2 + 1$

Appendix A

Appendix B

AREA UNDER A NORMAL CURVE

Z	0.00	0.01	0.02	0.03
2.3	.4893	.4896	.4898	.4901
2.4	.4918	.4920	.4922	.4925
2.5	.4938	.4940	.4941	.4943

The table below gives the probabilities $P(0 \leq Z \leq b)$ where Z is a standard normal variable. For example, to find $P(0 \leq Z \leq 2.43)$, write 2.43 as $2.4 + 0.03$, and read the entry in the row labeled 2.4 and the column labeled 0.03. From the portion of the table shown at left, you will see that $P(0 \leq Z \leq 2.43) = .4925$.

Z	0.00	0.01	0.02	0.03	0.04	0.05	0.06	0.07	0.08	0.09
0.0	.0000	.0040	.0080	.0120	.0160	.0199	.0239	.0279	.0319	.0359
0.1	.0398	.0438	.0478	.0517	.0557	.0596	.0636	.0675	.0714	.0753
0.2	.0793	.0832	.0871	.0910	.0948	.0987	.1026	.1064	.1103	.1141
0.3	.1179	.1217	.1255	.1293	.1331	.1368	.1406	.1443	.1480	.1517
0.4	.1554	.1591	.1628	.1664	.1700	.1736	.1772	.1808	.1844	.1879
0.5	.1915	.1950	.1985	.2019	.2054	.2088	.2123	.2157	.2190	.2224
0.6	.2257	.2291	.2324	.2357	.2389	.2422	.2454	.2486	.2517	.2549
0.7	.2580	.2611	.2642	.2673	.2704	.2734	.2764	.2794	.2823	.2852
0.8	.2881	.2910	.2939	.2967	.2995	.3023	.3051	.3078	.3106	.3133
0.9	.3159	.3186	.3212	.3238	.3264	.3289	.3315	.3340	.3365	.3389
1.0	.3413	.3438	.3461	.3485	.3508	.3531	.3554	.3577	.3599	.3621
1.1	.3643	.3665	.3686	.3708	.3729	.3749	.3770	.3790	.3810	.3830
1.2	.3849	.3869	.3888	.3907	.3925	.3944	.3962	.3980	.3997	.4015
1.3	.4032	.4049	.4066	.4082	.4099	.4115	.4131	.4147	.4162	.4177
1.4	.4192	.4207	.4222	.4236	.4251	.4265	.4279	.4292	.4306	.4319
1.5	.4332	.4345	.4357	.4370	.4382	.4394	.4406	.4418	.4429	.4441
1.6	.4452	.4463	.4474	.4484	.4495	.4505	.4515	.4525	.4535	.4545
1.7	.4554	.4564	.4573	.4582	.4591	.4599	.4608	.4616	.4625	.4633
1.8	.4641	.4649	.4656	.4664	.4671	.4678	.4686	.4693	.4699	.4706
1.9	.4713	.4719	.4726	.4732	.4738	.4744	.4750	.4756	.4761	.4767
2.0	.4772	.4778	.4783	.4788	.4793	.4798	.4803	.4808	.4812	.4817
2.1	.4821	.4826	.4830	.4834	.4838	.4842	.4846	.4850	.4854	.4857
2.2	.4861	.4864	.4868	.4871	.4875	.4878	.4881	.4884	.4887	.4890
2.3	.4893	.4896	.4898	.4901	.4904	.4906	.4909	.4911	.4913	.4916
2.4	.4918	.4920	.4922	.4925	.4927	.4929	.4931	.4932	.4934	.4936
2.5	.4938	.4940	.4941	.4943	.4945	.4946	.4948	.4949	.4951	.4952
2.6	.4953	.4955	.4956	.4957	.4959	.4960	.4961	.4962	.4963	.4964
2.7	.4965	.4966	.4967	.4968	.4969	.4970	.4971	.4972	.4973	.4974
2.8	.4974	.4975	.4976	.4977	.4977	.4978	.4979	.4979	.4980	.4981
2.9	.4981	.4982	.4982	.4983	.4984	.4984	.4985	.4985	.4986	.4986
3.0	.4987	.4987	.4987	.4988	.4988	.4989	.4989	.4989	.4990	.4990

ANSWERS TO SELECTED EXERCISES

Chapter 1

Exercises 1.1

1. a. 2 **b.** 0.5 **3. a.** -1.5 **b.** 8 **c.** -8 **5. a.** -7 **b.** -3 **c.** 1 **d.** $4y - 3$ **e.** $4(a + b) - 3$ **7. a.** 3 **b.** 6 **c.** 2 **d.** 6 **e.** $a^2 + 2a + 3$ **f.** $(x + h)^2 + 2(x + h) + 3$ **9. a.** 2 **b.** 0 **c.** 65/4 **d.** $x^2 + 1/x$ **e.** $(s + h)^2 + 1/(s + h)$ **f.** $(s + h)^2 + 1/(s + h) - (s^2 + 1/s)$ **11. a.** 1 **b.** 1 **c.** 0 **d.** 27 **13. a.** Yes; $f(4) = 63/16$ **b.** Not defined **c.** Not defined **15. a.** Not defined **b.** Not defined **c.** Yes; $f(-10) = 0$ **17. a.** $h(2x + h)$ **b.** $2x + h$ **19. a.** $-h(2x + h)$ **b.** $-(2x + h)$

21. `0.1*x^2-4*x+5`

x	0	1	2	3	4	5	6	7	8	9	10
f(*x*)	5	1.1	-2.6	-6.1	-9.4	-12.5	-15.4	-18.1	-20.6	-22.9	-25

23. `(x^2-1)/(x^2+1)`

x	0.5	1.5	2.5	3.5	4.5	5.5	6.5	7.5	8.5	9.5	10.5
h(*x*)	-0.6000	0.3846	0.7241	0.8491	0.9059	0.9360	0.9538	0.9651	0.9727	0.9781	0.9820

25. a. $P(5) = 117$, $P(10) = 132$, and $P(9.5) \approx 131$. Approximately 117 million people were employed in the United States on July 1, 1995; 132 million people on July 1, 2000; and 131 million people on January 1, 2000. **b.** [5, 11] **27. a.** $C(0) = 800$, $C(4) = 2800$, $C(5) = 4100$. Thus, there were 800 coffee shops in 1990, 2800 in 1994, and 4100 in 1995. **b.** 1996 **c.**

t	0	1	2	3	4	5	6	7	8	9	10
C(*t*)	800	1300	1800	2300	2800	4100	5400	6700	8000	9300	10,600

29. a. `(0.08*t+0.6)*(t<8)+(0.355*t-1.6)*(t>=8)`

t	0	1	2	3	4	5	6	7	8	9	10	11
C(*t*)	0.6	0.68	0.76	0.84	0.92	1	1.08	1.16	1.24	1.595	1.95	2.305

b. 0.355 **31.** $T(26{,}000) = \$600 + 0.15(26{,}000 - 6000) = \$3600.00$; $T(65{,}000) = \$3757.50 + 0.27(65{,}000 - 27{,}050) = \$14{,}004$ **33. a.** 358,600 **b.** 361,200 **c.** \$6 **35. a.** $0 \le t \le 8$ **b.** $t \ge 0$ is not an appropriate domain because it would predict investments in South Africa into the indefinite future with no basis. (It would also lead to preposterous results for large values of *t*.) **37. a.** (2) **b.** \$36.8 billion **39. a.** \$12,000 **b.** $N(q) = 2000 + 100q^2 - 500q$; $N(20) = \$32{,}000$ **41. a.** `100*(1-12200/t^4.48)` **b.**

t	9	10	11	12	13	14	15	16	17	18	19	20
p(*t*)	35.2	59.6	73.6	82.2	87.5	91.1	93.4	95.1	96.3	97.1	97.7	98.2

c. 82.2% **d.** 14 months **43.** *t*; *m* **45.** $y(x) = 4x^2 - 2$, or $f(x) = 4x^2 - 2$ **47.** $N = 200 + 10t$ (N = number of sound files, t = time in days) **49.** As the text reminds us: To evaluate *f* of a quantity (such as $x + h$) replace *x* everywhere by the *whole quantity* $x + h$, getting $f(x + h) = (x + h)^2 - 1$. **51.** False: Functions with infinitely many points in their domain [such as $f(x) = x^2$] cannot be specified numerically.

Exercises 1.2

1. a. 20 **b.** 30 **c.** 30 **d.** 20 **e.** 0 **3. a.** -1 **b.** 1.25 **c.** 0 **d.** 1 **e.** 0 **5. a.** (I) **b.** (IV) **c.** (V) **d.** (VI) **e.** (III) **f.** (II)

7.

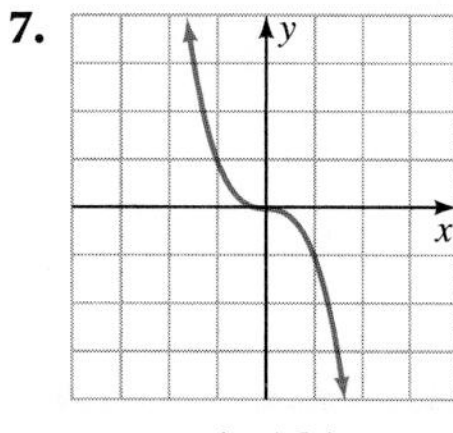

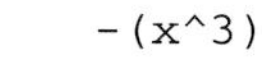

`-(x^3)`

9.

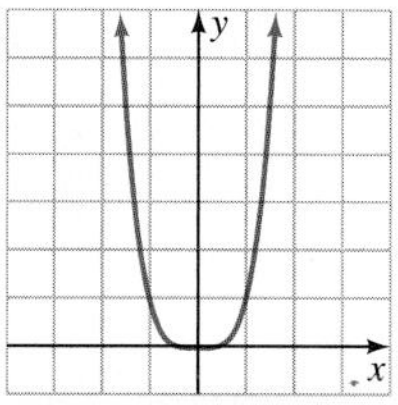

`x^4`

11.

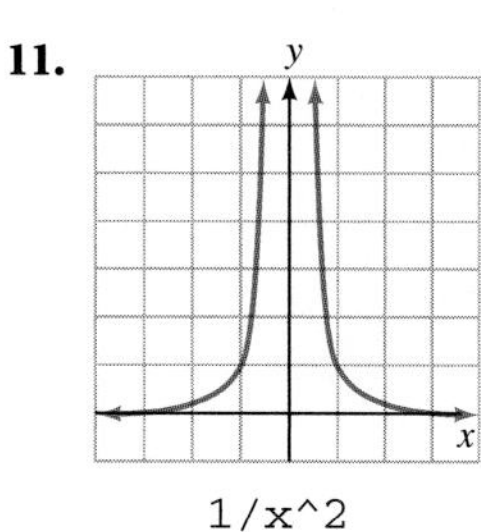

`1/x^2`

13. a. -1 **b.** 2 **c.** 2

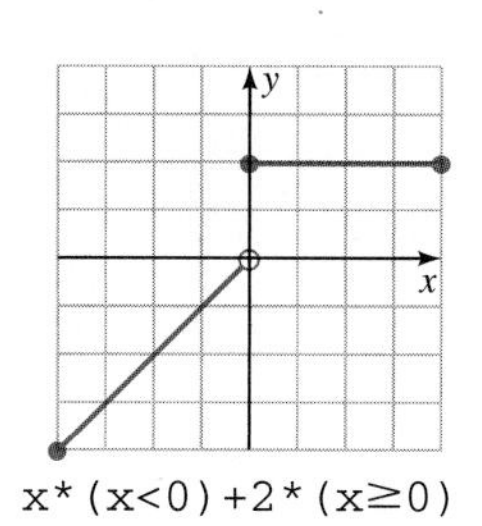

`x*(x<0)+2*(x≥0)`

15. a. 1 **b.** 0 **c.** 1

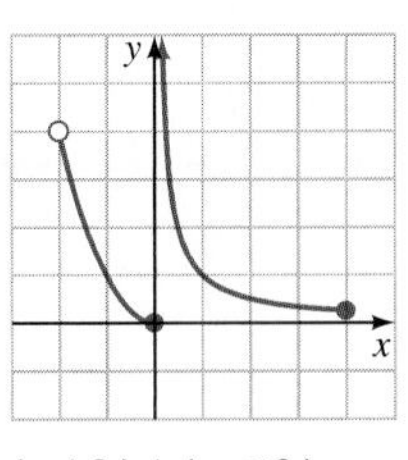

```
(x^2)*(x≤0)+
(1/x)*(0<x)
```

17. a. 0 **b.** 2 **c.** 3 **d.** 3

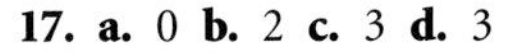

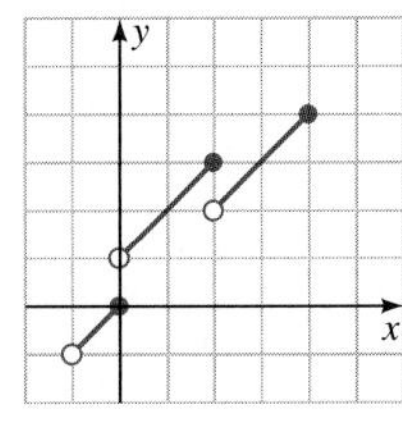

```
x*(x≤0)
+(x+1)*(0<x)*(x≤2)
+x*(2<x)
```

19. $f(6) \approx 2000$, $f(9) \approx 2800$, $f(7.5) \approx 2500$. In 1996, 2,000,000 SUVs were sold. In 1999, 2,800,000 were sold, and in the year beginning July 1997, 2,500,000 were sold. **21.** $f(6) - f(5)$; SUV sales increased more from 1995 to 1996 than from 1999 to 2000. **23. a.** $[-1.5, 1.5]$ **b.** $N(-0.5) \approx 131$, $N(0) \approx 132$, $N(1) \approx 132$. In July 1999 approximately 131 million people were employed. In January 2000 and January 2001, approximately 132 million people were employed. **c.** $[0.5, 1.5]$; employment was falling during the period July 2000–July 2001. **25. a.** (C) **b.** $20.80/shirt if the team buys 70 shirts **b.** Graph:

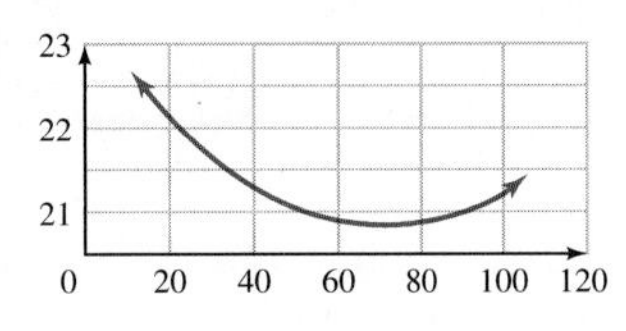

27. A quadratic model (B) is the best choice; the other models either predict perpetually increasing tourism or perpetually decreasing tourism. In fact, a parabola passes exactly through the three data points. **29. a.** `100*(1-12200/t^4.48)` **b.** Graph:

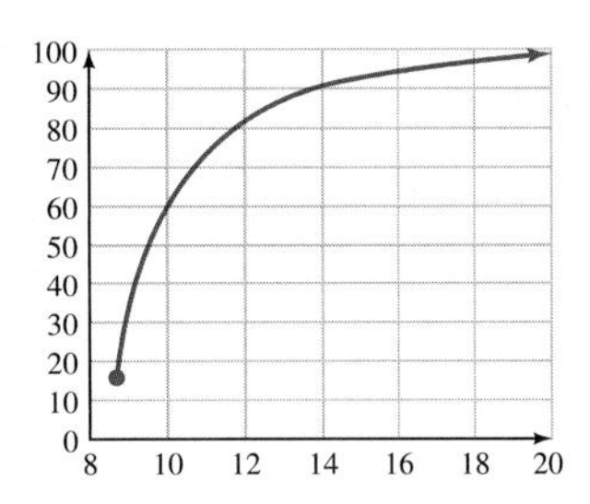

c. 82% **d.** 14 months **31.** 1999 Graph:

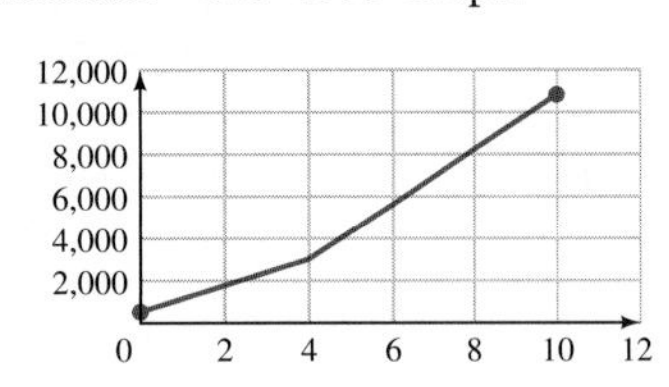

33. a. `(0.08*t+0.6)*(t<8)+(0.355*t-1.6)*(t>=8)` **b.** Graph:

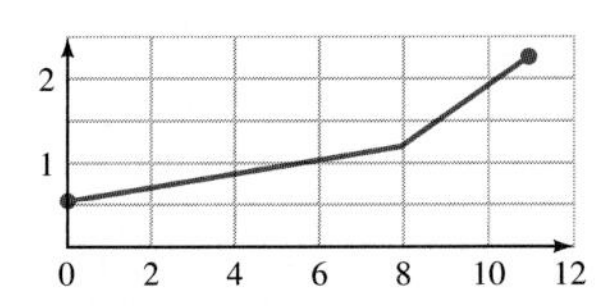

35. True. We can construct a table of values from any graph by reading off a set of values. **37.** False. In a numerically specified function, only certain values of the function are specified, giving only certain points on the graph. **39.** They are different portions of the graph of the associated equation $y = f(x)$. **41.** The graph of $g(x)$ is the same as the graph of $f(x)$ but shifted 5 units to the right.

Exercises 1.3

1.

x	-1	0	1
y	5	8	11

$m = 3$

3.

x	2	3	5
$f(x)$	-1	-2	-4

$m = -1$

5.

x	-2	0	2
$f(x)$	4	7	10

$m = 3/2$

7. $f(x) = -x/2 - 2$ **9.** $f(0) = -5$, $f(x) = -x - 5$ **11.** f is linear: $f(x) = 4x + 6$ **13.** g is linear: $g(x) = 2x - 1$ **15.** $-3/2$ **17.** 1/6 **19.** Undefined **21.** 0 **23.** $-4/3$

25.

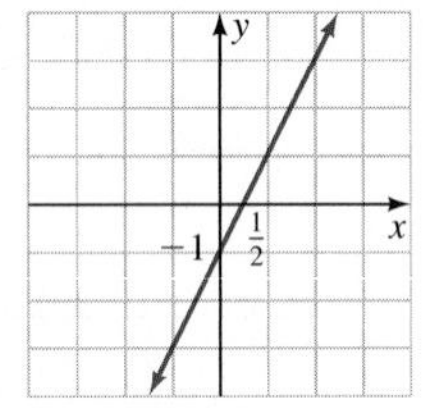

27.

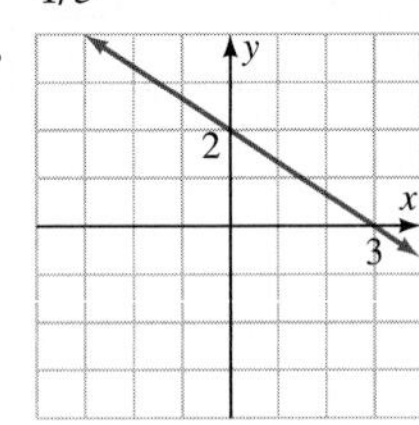

29.

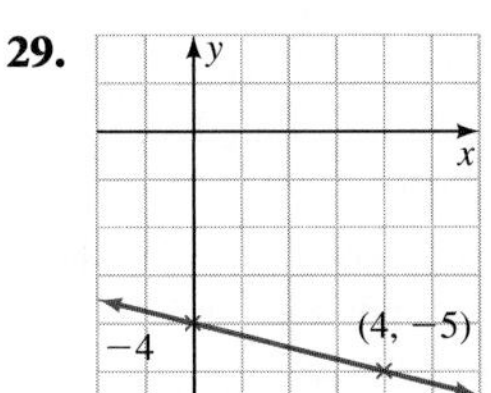

31.

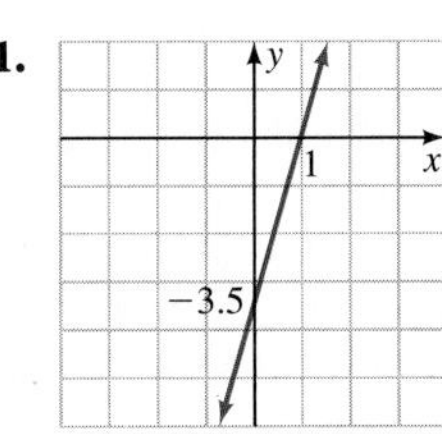

33.

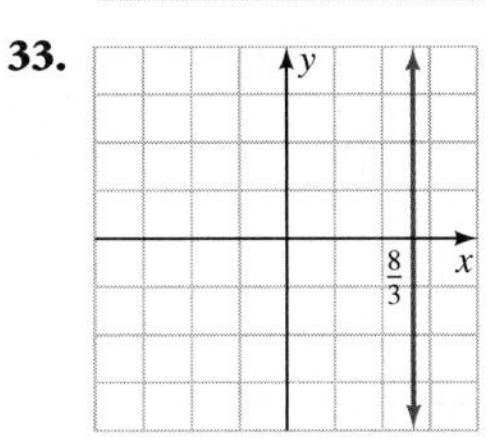

35.

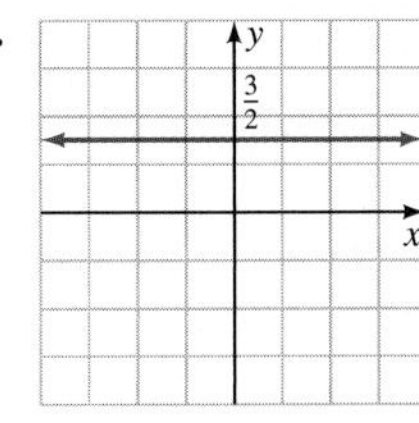

37.

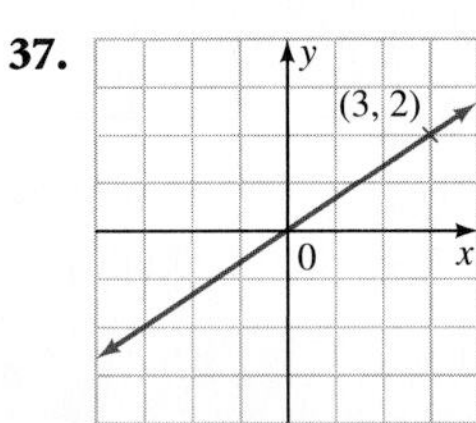

39. 2 **41.** 2 **43.** -2 **45.** Undefined **47** 1.5 **49.** -0.09 **51.** 1/2 **53.** $(d-b)/(c-a)$ **55. a.** 1 **b.** 1/2 **c.** 0 **d.** 3 **e.** $-1/3$ **f.** -1 **g.** Undefined **h.** $-1/4$ **i.** -2 **57.** $y = 3x$

59. $y = \frac{1}{4}x - 1$ **61.** $y = 10x - 203.5$ **63.** $y = -5x + 6$

65. $y = -3x + 2.25$ **67.** $y = -x + 12$ **69.** $y = 2x + 4$ **71.** Compute the corresponding successive changes Δx in x and Δy in y and compute the ratios $\Delta y/\Delta x$. If the answer is always the same number, then the values in the table come from a linear function. **73.** $f(x) = -\frac{a}{b}x + \frac{c}{b}$. If $b = 0$, then $\frac{a}{b}$ is undefined, and y cannot be specified as a function of x. (The graph of the resulting equation would be a vertical line.) **75.** Slope, 3 **77.** If m is positive, then y will increase as x increases; if m is negative, then y will decrease as x increases; if m is zero, then y will not change as x changes. **79.** The slope increases because an increase in the y coordinate of the second point increases Δy while leaving Δx fixed.

Exercises 1.4

1. $C(x) = 1500x + 1200$ per day **a.** \$5700 **b.** \$1500 **c.** \$1500 **3.** Fixed cost = \$8000, marginal cost = \$25 per bicycle **5. a.** $C(x) = 0.4x + 70$, $R(x) = 0.5x$, $P(x) = 0.1x - 70$ **b.** $P(500) = -20$; a loss of \$20 **c.** 700 copies **7.** $q = -40p + 2000$ **9.** $q = -0.15p + 56.25$. The demand equation predicts that demand increases as the price goes down. However, the first-quarter 1998 price was lower than the third-quarter 1997 price, and yet the demand was also lower. **11. a.** Demand: $q = -60p + 150$; supply: $q = 80p - 60$ **b.** \$1.50 each **13. a.** (1996, 125) and (1997, 135) or (1998, 140) and (1999, 150) **b.** The number of new in-ground pools increased most rapidly during the periods 1996–1997 and 1998–1999, when it rose by 10,000 new pools in a year. **15.** $N = 400 + 50t$ million transactions. The slope gives the additional number of online shopping transactions per year and is measured in (millions of) transactions per year. **17. a.** $s = 14.4t + 240$; Medicare spending is predicted to rise at a rate of \$14.4 billion/year **b.** \$816 billion **19. a.** 2.5 feet/second **b.** 20 feet along the track **c.** After 6 seconds **21. a.** 130 miles/hour **b.** $s = 130t - 1300$ **23.** $F = 1.8C + 32$; 86°F; 72°F; 14°F; 7°F **25.** $I(N) = 0.05N + 50{,}000$; $N = \$1{,}000{,}000$; marginal income is m = 5¢/dollar of net profit. **27. a.** $c = 0.05n + 1.7$ **b.** The slope represents the increase in daily circulation per additional newspaper that a company publishes. **c.** The model predicts that circulation would increase to 6.35 million. **29.** $w = 2n - 58$; 42 billion pounds **31.** $c = 0.075m - 1.5$; 0.75 pound **33.** $T(r) = (1/4)r + 45$; $T(100) = 70$°F **35.** $P(x) = 100x - 5132$, with domain [0, 405]. For profit, $x \geq 52$. **37.** 5000 units **39.** $FC/(SP - VC)$ **41.** $P(x) = 579.7x - 20{,}000$, with domain $x \geq 0$; $x = 34.50$ g/day for break even **43.** Increasing by \$355,000/year **45. a.** $p = 10t + 15$ **b.** $p = 15t + 10$ **c.** $p = \begin{cases} 10t + 15 & \text{if } 0 \leq t < 1 \\ 15t + 10 & \text{if } 1 \leq t \leq 3 \end{cases}$

d. 40% **47.** $C(t) = \begin{cases} -1400t + 30{,}000 & \text{if } 0 \leq t \leq 5 \\ 7400t - 14{,}000 & \text{if } 5 < t \leq 10 \end{cases}$; $C(3) =$ 25,800 students

49. $d(r) = \begin{cases} -40r + 74 & \text{if } 1.1 \leq r \leq 1.3 \\ \frac{130r}{3} - \frac{103}{3} & \text{if } 1.3 < r \leq 1.6 \end{cases}$; $d(1) = 34\%$

51. Bootlags per Martian yen; bootlags **53.** (B) **55.** It must increase by 10 units/day, including the third. **57.** Increasing the number of items from the break even results in a profit: Since the slope of the revenue graph is larger than the slope of the cost graph, it is higher than the cost graph to the right of the point of intersection, and hence corresponds to a profit.

Exercises 1.5

1. 6 **3.** 86 **5. a.** 0.5 (better fit) **b.** 0.75 **7. a.** 27.42 **b.** 27.16 (better fit)

9.

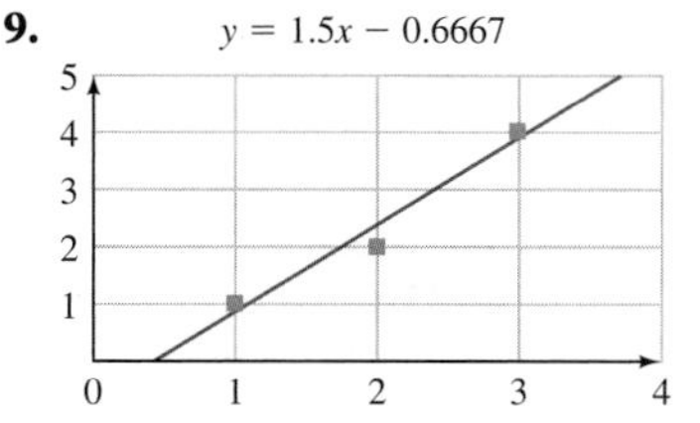

11.

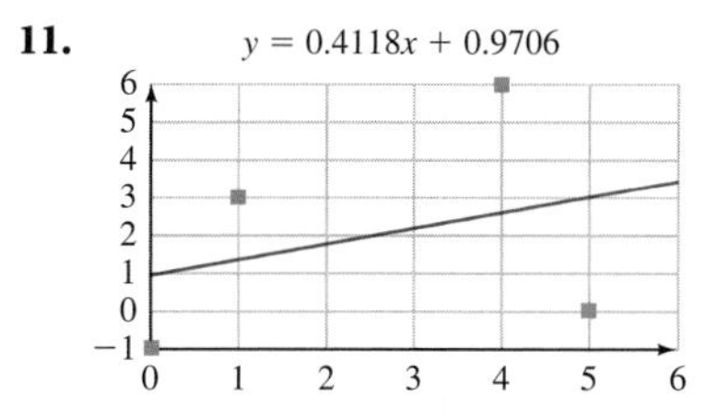

13. a. $r = 0.9959$ (best, not perfect) **b.** $r = 0.9538$ **c.** $r = 0.3273$ (worst) **15.** $y = -0.141t + 6.84$, $y(25) \approx \$3.3$ billion **17.** $y = 1.46x + 6.97$, $y(10) \approx 22$

19. a.

$y = -0.34t + 5.92$

b. Camera sales were declining at a rate of 0.34 million/year. **c.** Use of such a linear model to make predictions is not reasonable because the possible factors that may influence sales are not taken into account; 4.2 million cameras **d.** 5.8 million cameras. This is large compared with the other residues, confirming the point raised in part (c). **21.** 7500 phone lines **23.** Increasing at the rate of 0.27 year/year **25. a.** The graph suggests that the percentage rate decreases as the size of the 1991 homicide rate increases.

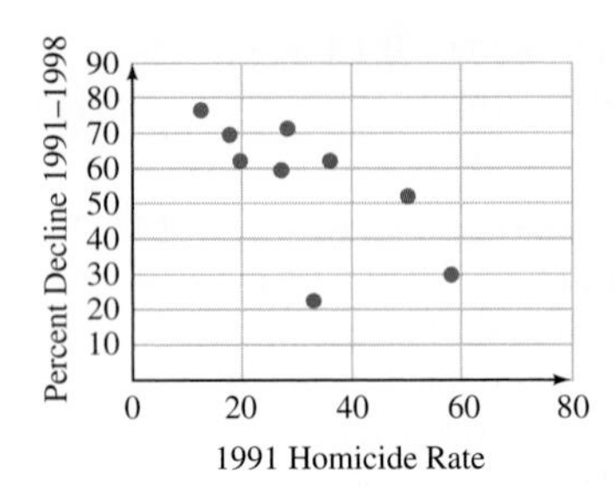

b. Regression equation: $y \approx -0.9148x + 85.441$; $r \approx -0.689$. The slope is negative, supporting the observation that the percentage decreases with increasing 1991 homicide rate. The value of r suggests a reasonable fit. **c.** The percent decrease in the homicide rate declines by 0.91% for every 1-point increase in the 1991 homicide rate. **d.** 48.8% **27.** The line that passes through (a, b) and (c, d) gives a sum-of-squares error SSE = 0, which is the smallest value possible. **29.** 0 **31.** The regression line is the line passing through the given points. **33.** No. The regression line through $(-1, 1)$, $(0, 0)$, and $(1, 1)$ passes through none of these points.

Chapter 1 Review Test

1. a.

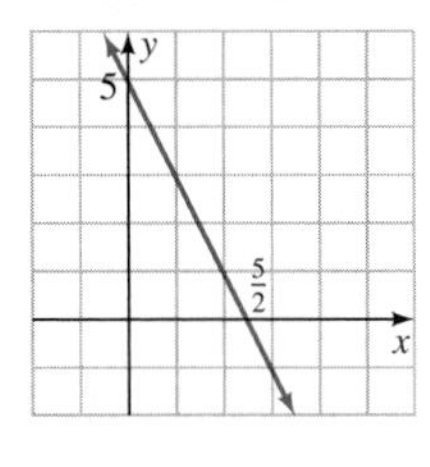

b. y 6 x −4

c.

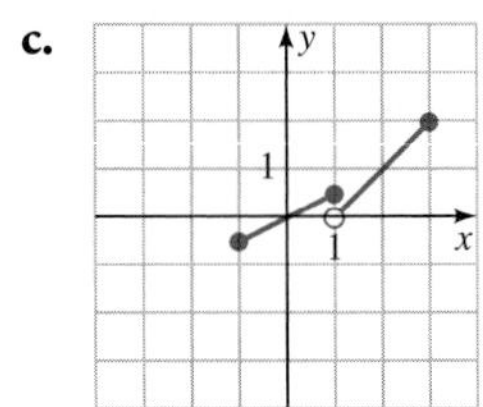

d.

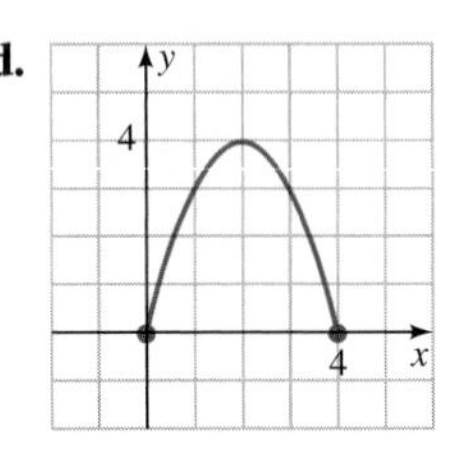

2. Absolute value, linear, linear, exponential, quadratic **3. a.** $y = -3x + 11$ **b.** $y = -x + 1$ **c.** $y = (1/2)x + 3/2$ **d.** $y = (1/2)x - 1$ **4. a.** 500 hits: 10 books/day; 1000 hits: 20 books/day; 1500 hits: 32.5 books/day **b.** Book sales are increasing at a rate of 0.025 book/hit when the number of hits is between 1000 and 2000 each day **c.** 1400 hits/day **5. a.** (A) **b.** (A) Leveling off, (B) Rising, (C) Rising (they begin to fall after 7 months), (D) Rising **6. a.** $h = 0.05c + 1800$ **b.** 2100 hits/day **c.** \$14,000/month **7. a.** 2080 hits/day **b.** Probably not. This model predicts that Web-site traffic will start to decrease as advertising increases beyond \$8500/month and then drop toward zero. **8. a.** Cost: $C = 900 + 4x$; revenue: $R = 5.5x$; profit: $P = 1.5x - 900$ **b.** 600 novels/month **c.** 900 novels/month **9. a.** $q = -60p + 950$ **b.** 50 novels/month **c.** \$10, for a profit of \$1200 **10. a.** $q = -53.3945p + 909.8165$ **b.** 483 novels/month

Chapter 2

Exercises 2.1

1. (2, 2) **3.** (3, 1) **5.** (6, 6) **7.** (5/3, −4/3) **9.** (0, −2) **11.** $(x, (1 - 2x)/3)$ or $\left(\frac{1}{2}(1 - 3y), y\right)$ **13.** No solution **15.** (0.3, −1.1) **17.** (116.6, −69.7) **19.** (3.3, 1.8) **21.** (3.4, 1.9) **23.** 200 quarts of vanilla and 100 quarts of mocha **25.** Two servings of Mixed Cereal and one serving of Mango Tropical Fruit **27. a.** Four servings of beans and five slices of bread **b.** No. One of the variables in the solution of the system has a negative value. **29.** Let x = no. of servings of Cell-Tech and y = no. servings of RiboForce HP. $10x + 5y = 80$; $200x = 1000$; mix five servings of Cell-Tech and six servings of RiboForce HP for a cost of \$20.60. **31.** 100 CSCO, 200 AOL **33.** 242 for and 193 against **35.** Five soccer games and seven football games **37.** Seven brand X **39.** \$1.50 each **41.** Demand: $q = -4p + 47$; supply: $q = 4p - 29$; equilibrium price: \$9.50 **43.** 33 pairs of dirty socks and 11 T-shirts **45.** \$1200 **47.** \$40 **49.** 55 widgets **51.** A system of three equations in two unknowns will have a unique solution if either (1) the three corresponding lines intersect in a single point or (2) two of the equations correspond to the same line and the third line intersects it in a single point. **53.** Yes. Even if two lines have negative slope, they will still intersect if the slopes differ. **55.** You cannot round up both of them because there will not be sufficient eggs and cream. Rounding down both answers will ensure that you will not run out of ingredients. It may be possible to round one answer down and the other up, and this should be tried. **57.** (B) **59.** (B) **61.** It is very likely. Two randomly chosen straight lines are unlikely to be parallel.

Exercises 2.2

1. (3, 1) **3.** (6, 6) **5.** $\left(\frac{1}{2}(1 - 3y), y\right)$; y arbitrary **7.** No solution **9.** (1/4, 3/4) **11.** No solution **13.** (10/3, 1/3) **15.** (4, 4, 4) **17.** $\left(-1, -3, \frac{1}{2}\right)$ **19.** (z, z, z); z arbitrary **21.** No solution **23.** (−1, 1, 1) **25.** $(1, z - 2, z)$; z arbitrary **27.** $(4 + y, y, -1)$; y arbitrary **29.** $(4 - y/3 + z/3, y, z)$; y, z arbitrary **31.** (−17, 20, −2) **33.** $\left(-\frac{3}{2}, 0, \frac{1}{2}, 0\right)$ **35.** $(-3z, 1 - 2z, z, 0)$; z arbitrary **37.** $\left(\frac{1}{5}(7 - 17z + 8w), \frac{1}{5}(1 - 6z - 6w), z, w\right)$; z, w arbitrary **39.** (1, 2, 3, 4, 5) **41.** $(-2, -2 + z - u, z, u, 0)$; z, u arbitrary **43.** (16, 12/7, −162/7, −88/7) **45.** (−8/15, 7/15, 7/15, 7/15, 7/15) **47.** (1.0, 1.4, 0.2) **49.** (−5.5, −0.9, −7.4, −6.6) **51.** A *pivot* is an entry in a matrix that is selected to "clear a column"; that is, use the row operations of a certain type to obtain zeros everywhere above and below it. *Pivoting* is the procedure of clearing a column, using a designated pivot. **53.** $2R_1 + 5R_4$, or $6R_1 + 15R_4$ (which is less desirable). **55.** It will include a row of zeros. **57.** The claim is wrong. If there are more equations than unknowns, there can be a unique solution as well as row(s) of zeros in the reduced matrix, as in Example 6. **59.** Two **61.** The number of pivots must equal the number of variables because no variable will be used as a parameter. **63.** A simple example is $x = 1$; $y - z = 1$; $x + y - z = 2$.

Exercises 2.3

1. 100 batches of vanilla, 50 batches of mocha, and 100 batches of strawberry **3.** Three sections of Finite Math, two sections of Applied Calculus, and one section of Computer Methods **5.** Five of each **7.** Seven of each **9.** 22 tons from Cheesy Cream, 56 tons from Super Smooth & Sons, and 22 tons from Bagel's Best Friend **11.** 10 evil sorcerers, 50 warriors, and 500 orcs **13.** \$3.6 billion for rock music, \$1.8 billion for rap music, and \$0.4 billion for classical music **15.** The third equation is $x_1 + x_2 + x_3 + x_4 = 1$. Solution: $x_1 = 0.39 - 0.36x_4$; $x_2 = 0.53 - 1.14x_4$; $x_3 = 0.08 + 0.5x_4$. General Mills is most impacted by "Other." **17.** \$5000 in PNF, \$2000 in CMBFX, \$2000 in SFCOX **19.** 100 IBM, 20 HWP, 80 DELL **21.** 200 Democrats, 20 Republicans, 13 of other parties **23. a.** Brooklyn to Manhattan, 500 books; Brooklyn to Long Island, 500 books; Queens to Manhattan, 1000 books; Queens to Long Island, 1000 books **b.** Brooklyn to Manhattan, 1000 books; Brooklyn to Long Island, none; Queens to Manhattan, 500 books; Queens to Long Island, 1500 books. Total cost = \$8000. **25. a.** The associated system of equations has infinitely many solutions. **b.** No; the associated system of equations still has infinitely many solutions. **c.** Yes; North America to Australia, 440,000; North America to South Africa, 190,000; Europe to Australia, 950,000; Europe to South Africa, 950,000 **27. a.** $x + y = 14{,}000$; $z + w = 95{,}000$, $x + z = 63{,}550$, $y + w = 45{,}450$. The system does not have a unique solution, indicating that the given data are insufficient to obtain the missing data. **b.** $(x, y, z, w) = (5600, 8400, 57{,}950, 37{,}050)$ **29.** Yes; \$20 million in company X; \$5 million in company Y, \$10 million in company Z, and \$30 million in company W **31. a.** No; the general solution is Eastward Blvd., $S + 200$; Northwest La., $S + 50$; Southwest La., S, where S is arbitrary. Thus, it would suffice to know the traffic along Southwest La. **b.** Yes, because it leads to the solution Eastward Blvd., 260; Northwest La., 110; Southwest La., 60 **33. a.** No; the corresponding system of equations is underdetermined. The net flow of traffic along any of the three stretches of Broadway would suffice. **b.** West **35.** \$10 billion **37.** It donated \$600 to each of the MPBF and the SCN, and \$1200 to the NYJ. **39.** x = water, y = gray matter, z = tumor **41.** x = water, y = bone, z = tumor, u = air **43.** Tumor **45.** United, 120; American, 40; Southwest, 50 **47.** It is not realistic to expect to use exactly all the ingredients. Solutions of the associated system may involve negative numbers or not exist. Only solutions with nonnegative values for all the unknowns correspond to being able to use all the ingredients. **49.** Yes; $x = 100$ **51.** Yes; $0.3x - 0.7y + 0.3z = 0$ is one form of the equation. **53.** No; represented by an inequality rather than an equation. **55.** Answers will vary.

Chapter 2 Review Test

1. a. (6/5, 7/5) **b.** (5/3, 1/3) **c.** $(3y/2, y)$; y arbitrary **d.** No solution **e.** $(-0.7, 1.7)$ **f.** (1/30, 0) **2. a.** $(-1, -1, -1)$ **b.** (1, 2, 3) **c.** $(z - 2, 4(z - 1), z)$; z arbitrary **d.** $(-1 - y + 2z, y, z)$; y, z arbitrary **e.** No solution **f.** $(2 - w, w, -1 + 2w, w)$; w arbitrary **3. a.** $-40°$ **b.** 320°F (160°C) **c.** It is impossible; setting $F = 1.8C$ leads to an inconsistent system of equations. **4. a.** $x + y + z + w = 10$; linear **b.** $x - 3y - 3z = 0$; linear **c.** $w = 0$; linear **d.** $x - y^2 - z^2 - w^2 = 0$; nonlinear **e.** $-1.3y + z = 0$ or $1.3y - z = 0$; linear **f.** $-0.7y + z = 0$ or $0.7y - z = 0$; linear **5. a.** 550 packages from Duffin House, 350 from Higgins Press **b.** 400 packages from Duffin House, 200 from Higgins Press **c.** 600 packages from Duffin House, 200 from Higgins Press **6.** 5000 hits/day at OHaganBooks.com, 1250 at JungleBooks.com, 3750 at FarmerBooks.com **7.** DHS, 1000 shares; HPR, 600 shares; SPUB, 400 shares **8.** Billy Sean is forced to take exactly the following combination: Liberal Arts, 52 credits; Sciences, 12 credits; Fine Arts, 12 credits; Mathematics, 48 credits. **9.** New York to OHaganBooks.com, 450 packages; New York to FantasyBooks.com, 50 packages; Illinois to OHaganBooks.com, 150 packages; Illinois to FantasyBooks.com, 150 packages **10. a.** $x = 100$, $y = 100 + w$, $z = 300 - w$; w arbitrary **b.** 100 book orders/day **c.** 300 book orders/day **d.** $x = 100$, $y = 400$, $z = 0$, $w = 300$ **e.** 100 book orders/day

Chapter 3

Exercises 3.1

1. 1×4; 0 **3.** 4×1; 5/2 **5.** $p \times q$; e_{22} **7.** 2×2; 3 **9.** $1 \times n$; d_r **11.** $x = 1, y = 2, z = 3, w = 4$

13. $\begin{bmatrix} 0.25 & -2 \\ 1 & 0.5 \\ -2 & 5 \end{bmatrix}$ **15.** $\begin{bmatrix} -0.75 & -1 \\ 0 & -0.5 \\ -1 & 6 \end{bmatrix}$ **17.** $\begin{bmatrix} -1 & -1 \\ 1 & -1 \\ -1 & 5 \end{bmatrix}$

19. $\begin{bmatrix} 0 & 2 & -2 \\ -2 & 0 & 4 \end{bmatrix}$ **21.** $\begin{bmatrix} 4 & -1 & -1 \\ 5 & 1 & 0 \end{bmatrix}$

23. $\begin{bmatrix} -2 + x & 0 & 1 + w \\ -5 + z & 3 + r & 2 \end{bmatrix}$ **25.** $\begin{bmatrix} -1 & -2 & 1 \\ -5 & 5 & -3 \end{bmatrix}$

27. $\begin{bmatrix} 9 & 15 \\ 0 & -3 \\ -3 & 3 \end{bmatrix}$ **29.** $\begin{bmatrix} -8.5 & -22.35 & -24.4 \\ 54.2 & 20 & 42.2 \end{bmatrix}$

31. $\begin{bmatrix} 1.54 & 8.58 \\ 5.94 & 0 \\ 6.16 & 7.26 \end{bmatrix}$ **33.** $\begin{bmatrix} 7.38 & 76.96 \\ 20.33 & 0 \\ 29.12 & 39.92 \end{bmatrix}$

35. $\begin{bmatrix} -19.85 & 115.82 \\ -50.935 & 46 \\ -57.24 & 94.62 \end{bmatrix}$

37. a. Sales in 2000 = sales in 1999 + increase in 2000 = [330,000 100,000 20,000] + [10,000 0 0] = [340,000 100,000 20,000] **b.** Sales in 2001 = sales in 2000 + increase in 2001 = [340,000 100,000 20,000] + [−30,000 −10,000 10,000] = [310,000 90,000 30,000]

39. Sales = $\begin{bmatrix} 700 & 1300 & 2000 \\ 400 & 300 & 500 \end{bmatrix}$;

inventory − sales = $\begin{bmatrix} 300 & 700 & 3000 \\ 600 & 4700 & 1500 \end{bmatrix}$

41. a. Use =

	Proc	Mem	Tubes
Pom II	2	16	20
Pom Classic	1	4	40

;

inventory = $\begin{bmatrix} 500 & 5000 & 10{,}000 \\ 200 & 2000 & 20{,}000 \end{bmatrix}$;

inventory − 100 · use = $\begin{bmatrix} 300 & 3400 & 8{,}000 \\ 100 & 1600 & 16{,}000 \end{bmatrix}$

b. After 4 months

43. Total bankruptcy filings = filings in Manhattan + filings in Brooklyn + filings in Newark = [150 250 150 100 150] + [300 400 300 200 250] + [250 400 250 200 200] = [700 1050 700 500 600]
45. Filings in Brooklyn − filings in Newark = [300 400 300 200 250] − [250 400 250 200 200] = [50 0 50 0 50]. The difference was greatest in January 2001, July 2001, and January 2002.
47. Profit = revenue − cost;

	2001	2002	2003
Full Boots	\$8000	\$7200	\$8800
Half Boots	\$5600	\$5760	\$7040
Sandals	\$2800	\$3500	\$4000

49. a. $A = \begin{bmatrix} 440 & 190 \\ 950 & 950 \\ 1790 & 200 \end{bmatrix}$, $D = \begin{bmatrix} -20 & 40 \\ 50 & 50 \\ 0 & 100 \end{bmatrix}$, 2008 tourism = $A + D = \begin{bmatrix} 420 & 230 \\ 1000 & 1000 \\ 1790 & 300 \end{bmatrix}$ **b.** $\begin{bmatrix} -20 & 40 \\ 50 & 50 \\ 0 & 100 \end{bmatrix}$
51. 1980 distribution = A = [49.1 58.9 75.4 43.2]; 1990 distribution = B = [50.8 59.7 85.4 52.8]; net change 1980 to 1990 = $B - A$ = [1.7 0.8 10 9.6] (all net increases) **53.** The ijth entry of the sum $A + B$ is obtained by adding the ijth entries of A and B. **55.** It would have zeros down the main diagonal: $A = \begin{bmatrix} 0 & \# & \# & \# & \# \\ \# & 0 & \# & \# & \# \\ \# & \# & 0 & \# & \# \\ \# & \# & \# & 0 & \# \\ \# & \# & \# & \# & 0 \end{bmatrix}$. The symbol # indicates an arbitrary number. **57.** $(A^T)_{ij} = A_{ji}$ **59.** Answers will vary. **a.** $\begin{bmatrix} 0 & -4 \\ 4 & 0 \end{bmatrix}$ **b.** $\begin{bmatrix} 0 & -4 & 5 \\ 4 & 0 & 1 \\ -5 & -1 & 0 \end{bmatrix}$ **61.** The associativity of matrix addition is a consequence of the associativity of addition of numbers because we add matrices by adding the corresponding entries (which are real numbers).

Exercises 3.2

1. [13] **3.** [5/6] **5.** $[-2y + z]$ **7.** Undefined
9. [3 0 −6 −2] **11.** [−6 37 7]
13. $\begin{bmatrix} -4 & -7 & -1 \\ 9 & 17 & 0 \end{bmatrix}$ **15.** $\begin{bmatrix} 0 & 1 \\ 0 & 0 \end{bmatrix}$ **17.** $\begin{bmatrix} 1 & -1 \\ 1 & -1 \end{bmatrix}$ **19.** $\begin{bmatrix} 0 & 0 \\ 0 & 0 \end{bmatrix}$

21. Undefined **23.** $\begin{bmatrix} 1 & -5 & 3 \\ 0 & 0 & 9 \\ 0 & 4 & 1 \end{bmatrix}$ **25.** $\begin{bmatrix} 3 \\ -4 \\ 0 \\ 3 \end{bmatrix}$

27. $\begin{bmatrix} 0.23 & 5.36 & -21.65 \\ -13.18 & -5.82 & -16.62 \\ -11.21 & -9.9 & 0.99 \\ -2.1 & 2.34 & 2.46 \end{bmatrix}$

29. $A^2 = \begin{bmatrix} 0 & 0 & 1 & 2 \\ 0 & 0 & 0 & 1 \\ 0 & 0 & 0 & 0 \\ 0 & 0 & 0 & 0 \end{bmatrix}$, $A^3 = \begin{bmatrix} 0 & 0 & 0 & 1 \\ 0 & 0 & 0 & 0 \\ 0 & 0 & 0 & 0 \\ 0 & 0 & 0 & 0 \end{bmatrix}$, $A^4 = \begin{bmatrix} 0 & 0 & 0 & 0 \\ 0 & 0 & 0 & 0 \\ 0 & 0 & 0 & 0 \\ 0 & 0 & 0 & 0 \end{bmatrix}$, $A^{100} = \begin{bmatrix} 0 & 0 & 0 & 0 \\ 0 & 0 & 0 & 0 \\ 0 & 0 & 0 & 0 \\ 0 & 0 & 0 & 0 \end{bmatrix}$ **31.** $\begin{bmatrix} 4 & -1 \\ -1 & -7 \end{bmatrix}$

33. $\begin{bmatrix} 4 & -1 \\ -12 & 2 \end{bmatrix}$ **35.** $\begin{bmatrix} -2 & 1 & -2 \\ 10 & -2 & 2 \\ -10 & 2 & -2 \end{bmatrix}$

37. $\begin{bmatrix} -2 + x - z & 2 - r & -6 + w \\ 10 + 2z & -2 + 2r & 10 \\ -10 - 2z & 2 - 2r & -10 \end{bmatrix}$

39. a.–d. $P^2 = P^4 = P^8 = P^{1000} = \begin{bmatrix} 0.2 & 0.8 \\ 0.2 & 0.8 \end{bmatrix}$

41. a. $P^2 = \begin{bmatrix} 0.01 & 0.99 \\ 0 & 1 \end{bmatrix}$ **b.** $P^4 = \begin{bmatrix} 0.0001 & 0.9999 \\ 0 & 1 \end{bmatrix}$
c. and **d.** $P^8 \approx P^{1000} \approx \begin{bmatrix} 0 & 1 \\ 0 & 1 \end{bmatrix}$

43. a.–d. $P^2 = P^4 = P^8 = P^{1000} = \begin{bmatrix} 0.25 & 0.25 & 0.50 \\ 0.25 & 0.25 & 0.50 \\ 0.25 & 0.25 & 0.50 \end{bmatrix}$

45. $2x - y + 4z = 3;\ -4x + 3y/4 + z/3 = -1;\ -3x = 0$
47. $x - y + w = -1;\ x + y + 2z + 4w = 2$
49. $\begin{bmatrix} 1 & -1 \\ 2 & -1 \end{bmatrix}\begin{bmatrix} x \\ y \end{bmatrix} = \begin{bmatrix} 4 \\ 0 \end{bmatrix}$ **51.** $\begin{bmatrix} 1 & 1 & -1 \\ 2 & 1 & 1 \\ \frac{3}{4} & 0 & \frac{1}{2} \end{bmatrix}\begin{bmatrix} x \\ y \\ z \end{bmatrix} = \begin{bmatrix} 8 \\ 4 \\ 1 \end{bmatrix}$

53. Revenue = price × quantity = $[15\ \ 10\ \ 12]\begin{bmatrix} 50 \\ 40 \\ 30 \end{bmatrix} = [1510]$

55. Price: $\begin{matrix} \text{Hard} \\ \text{Soft} \\ \text{Plastic} \end{matrix}\begin{bmatrix} 30 \\ 10 \\ 15 \end{bmatrix}$; $\begin{bmatrix} 700 & 1300 & 2000 \\ 400 & 300 & 500 \end{bmatrix}\begin{bmatrix} 30 \\ 10 \\ 15 \end{bmatrix} = \begin{bmatrix} \$64{,}000 \\ \$22{,}500 \end{bmatrix}$

57. Number of books = number of books/editor × number of editors = $[3\ \ 3.5\ \ 5\ \ 5.2]\begin{bmatrix} 16{,}000 \\ 15{,}000 \\ 12{,}500 \\ 13{,}000 \end{bmatrix}$ = 230,600 new books

59. $\begin{bmatrix} 2 & 16 & 20 \\ 1 & 4 & 40 \end{bmatrix}\begin{bmatrix} 100 & 150 \\ 50 & 40 \\ 10 & 15 \end{bmatrix} = \begin{bmatrix} \$1200 & \$1240 \\ \$700 & \$910 \end{bmatrix}$

61. [1.2 1.0], which represents the amount, in billions of pounds, by which cheese production in north central states exceeded that in western states. **63.** Number of bankruptcy filings handled by firm = percentage handled by firm × total number = $[0.10\ \ 0.05\ \ 0.20]\begin{bmatrix} 150 & 150 & 150 \\ 300 & 300 & 250 \\ 250 & 250 & 200 \end{bmatrix} = [80\ \ 80\ \ 67.5]$
65. The number of filings in Manhattan and Brooklyn combined in each of the months shown

67. $[1 \quad -1 \quad 1]\begin{bmatrix}150 & 150 & 150\\300 & 300 & 250\\250 & 250 & 200\end{bmatrix}\begin{bmatrix}1\\1\\1\end{bmatrix} = [300]$

69. $AB = \begin{bmatrix}29.6\\85.5\\97.5\end{bmatrix}$, $AC = \begin{bmatrix}22 & 7.6\\47.5 & 38\\89.5 & 8\end{bmatrix}$. The entries of AB give the number of people from each of the three regions who settle in Australia or South Africa, and the entries in AC break those figures down further into settlers in South Africa and settlers in Australia. **71.** Distribution in 1999 = $A = [53.9 \quad 64.3 \quad 100.0 \quad 63.3]$; distribution in 2000 = $A \cdot P \approx [53.6 \quad 64.4 \quad 100.2 \quad 63.2]$ **73.** Answers will vary. One example: $A = [1 \quad 2]$; $B = \begin{bmatrix}1 & 2 & 3\\4 & 5 & 6\end{bmatrix}$. Another example: $A = [1]$; $B = [1 \quad 2]$. **75.** The claim is correct. Every matrix equation represents the equality of two matrices. Equating the corresponding entries gives a system of equations. **77.** Here is a possible scenario: costs of items A, B, and C in 1995 = $[10 \quad 20 \quad 30]$, percent increases in these costs in 1996 = $[0.5 \quad 0.1 \quad 0.20]$, actual increases in costs = $[10 \times 0.5 \quad 20 \times 0.1 \quad 30 \times 0.20]$. **79.** It produces a matrix whose ij entry is the product of the ij entries of the two matrices.

Exercises 3.3

1. Yes **3.** Yes **5.** No **7.** $\begin{bmatrix}-1 & 1\\2 & -1\end{bmatrix}$ **9.** $\begin{bmatrix}0 & 1\\1 & 0\end{bmatrix}$

11. $\begin{bmatrix}1 & -1\\-1 & 2\end{bmatrix}$ **13.** Singular **15.** $\begin{bmatrix}1 & -1 & 0\\0 & 1 & -1\\0 & 0 & 1\end{bmatrix}$

17. $\begin{bmatrix}1 & -1 & 1\\\frac{1}{2} & 0 & -\frac{1}{2}\\-\frac{1}{2} & 1 & -\frac{1}{2}\end{bmatrix}$ **19.** $\begin{bmatrix}1 & \frac{1}{3} & -\frac{1}{3}\\1 & -\frac{2}{3} & -\frac{1}{3}\\-1 & \frac{1}{3} & \frac{2}{3}\end{bmatrix}$ **21.** Singular

23. $\begin{bmatrix}0 & 1 & -2 & 1\\0 & 1 & -1 & 0\\1 & -1 & 2 & -1\\0 & 1 & -1 & 1\end{bmatrix}$ **25.** $\begin{bmatrix}1 & -2 & 1 & 0\\0 & 1 & -2 & 1\\0 & 0 & 1 & -2\\0 & 0 & 0 & 1\end{bmatrix}$

27. -2; $\begin{bmatrix}\frac{1}{2} & \frac{1}{2}\\\frac{1}{2} & -\frac{1}{2}\end{bmatrix}$ **29.** -2; $\begin{bmatrix}-2 & 1\\\frac{3}{2} & -\frac{1}{2}\end{bmatrix}$ **31.** $\frac{1}{36}$; $\begin{bmatrix}6 & 6\\0 & 6\end{bmatrix}$

33. 0; singular **35.** $\begin{bmatrix}0.38 & 0.45\\0.49 & -0.41\end{bmatrix}$ **37.** $\begin{bmatrix}0.00 & -0.99\\0.81 & 2.87\end{bmatrix}$

39. Singular **41.** $\begin{bmatrix}91.35 & -8.65 & 0 & -71.30\\-0.07 & -0.07 & 0 & 2.49\\2.60 & 2.60 & -4.35 & 1.37\\2.69 & 2.69 & 0 & -2.10\end{bmatrix}$

43. $(5/2, 3/2)$ **45.** $(0, -2)$ **47.** $(6, 6, 6)$ **49. a.** $(10, -5, -3)$ **b.** $(6, 1, 5)$ **c.** $(0, 0, 0)$ **51. a.** 10/3 servings of beans, and 5/6 slices of bread **b.** $\begin{bmatrix}-\frac{1}{2} & \frac{1}{6}\\\frac{7}{8} & -\frac{5}{24}\end{bmatrix}\begin{bmatrix}A\\B\end{bmatrix} = \begin{bmatrix}-\frac{A}{2} + \frac{B}{6}\\\frac{7A}{8} - \frac{5B}{24}\end{bmatrix}$; that is, $-\frac{A}{2} + \frac{B}{6}$ servings of beans and $\frac{7A}{8} - \frac{5B}{24}$ slices of bread **53. a.** 100 batches of vanilla, 50 batches of mocha, 100 batches of strawberry **b.** 100 batches of vanilla, no mocha, 200 batches of strawberry **c.** $\begin{bmatrix}1 & -\frac{1}{3} & -\frac{1}{3}\\-1 & 0 & 1\\0 & \frac{2}{3} & -\frac{1}{3}\end{bmatrix}\begin{bmatrix}A\\B\\C\end{bmatrix}$, or $A - \frac{B}{3} - \frac{C}{3}$ batches of vanilla, $-A + C$ batches of mocha, and $\frac{2B}{3} - \frac{C}{3}$ batches of strawberry **55.** \$5000 in PNF, \$2000 in CMBFX, \$2000 in SFCOX **57.** 100 IBM, 20 HWP, 80 DELL **59.** Distribution in 1999 = $A = [53.9 \quad 64.3 \quad 100.0 \quad 63.3]$; distribution in 1998 = $A \cdot P^{-1} \approx [54.2 \quad 64.2 \quad 99.8 \quad 63.4]$ **61. a.** $(-0.7071, 3.5355)$ **b.** R^2, R^3 **c.** R^{-1} **63.** $[37 \quad 81 \quad 40 \quad 80 \quad 15 \quad 45 \quad 40 \quad 96 \quad 29 \quad 59 \quad 4 \quad 8]$ **65.** CORRECT ANSWER **67.** (A) **69.** The inverse does not exist—the matrix is singular. (If two rows of a matrix are the same, then row-reducing it will lead to a row of zeros, and so it cannot be reduced to the identity.) **73.** If a square matrix A reduces to one with a row of zeros, then it cannot have an inverse. The reason is that, if A has an inverse, then every system of equations $AX = B$ has a unique solution, namely, $X = A^{-1}B$. But if A reduces to a matrix with a row of zeros, then such a system has either infinitely many solutions or no solution at all. **75.** If A *were* invertible, then multiplying both sides of the equation $AB = O$ on the left by A^{-1} would yield $B = O$, contradicting the fact that B is not the zero matrix. Thus, A cannot possibly be invertible. A similar argument shows that B cannot be invertible. **77.** $(AB)(B^{-1}A^{-1}) = A(BB^{-1})A^{-1} = AIA^{-1} = AA^{-1} = I$

Exercises 3.4

1. a. 0.8 **b.** 0.2 **c.** 0.05 **3.** $\begin{bmatrix}0.2 & 0.1\\0.5 & 0\end{bmatrix}$ **5.** $[52{,}000 \quad 40{,}000]^T$ **7.** $[50{,}000 \quad 50{,}000]^T$ **9.** $[2560 \quad 2800 \quad 4000]^T$ **11.** $[27{,}000 \quad 28{,}000 \quad 17{,}000]^T$ **13.** Increase of 100 units in each sector **15.** Increase of $[1.5 \quad 0.2 \quad 0.1]^T$; the ith column of $(I - A)^{-1}$ gives the change in production necessary to meet an increase in external demand of 1 unit for the product of sector i.

17. $A = \begin{bmatrix}0.2 & 0.4 & 0.5\\0 & 0.8 & 0\\0 & 0.2 & 0.5\end{bmatrix}$ **19.** Main DR, \$80,000; Bits & Bytes, \$38,000 **21.** Equipment sector production, approximately \$86,000 million; components sector production, approximately \$140,000 million **23. a.** 0.006 **b.** textiles; clothing and footwear

25. Columns of $\begin{bmatrix}1140.99 & 2.05 & 13.17 & 20.87\\332.10 & 1047.34 & 26.05 & 111.18\\0.12 & 0.13 & 1031.19 & 1.35\\93.88 & 95.69 & 215.50 & 1016.15\end{bmatrix}$ (in millions of dollars) **27. a.** \$0.78 **b.** Other food products **29.** It would mean that all the sectors require neither their own product nor the product of any other sector. **31.** It would mean that all the output of that sector was used internally in the economy; none of the output was available for export, and no importing was necessary. **33.** It means that an increase in demand for one sector (the column sector) has no effect on the production of another sector (the row sector). **35.** Usually, to produce 1 unit of one sector requires less than 1 unit of input from another. We would expect then that an increase in demand of 1 unit for one sector would require a smaller increase in production in another sector.

Chapter 3 Review Test

1. a. Undefined **b.** $\begin{bmatrix} 4 & 4 & 4 \\ 3 & 3 & 3 \end{bmatrix}$ **c.** $\begin{bmatrix} 1 & 8 \\ 5 & 11 \\ 6 & 13 \end{bmatrix}$ **d.** Undefined

e. $\begin{bmatrix} 1 & 3 \\ 2 & 3 \\ 3 & 3 \end{bmatrix}$ **f.** Undefined **g.** $\begin{bmatrix} 1 & -2 \\ 0 & 1 \end{bmatrix}$ **h.** $\begin{bmatrix} 1 & -3 \\ 0 & 1 \end{bmatrix}$

2. a. $\begin{bmatrix} 1 & 1 \\ 0 & 1 \end{bmatrix}$ **b.** Singular **c.** $\begin{bmatrix} 1 & -\frac{1}{2} & -\frac{5}{2} \\ 0 & \frac{1}{4} & -\frac{1}{4} \\ 0 & 0 & 1 \end{bmatrix}$

d. $\begin{bmatrix} 1 & 0 & -2 & 1 \\ \frac{1}{3} & -\frac{1}{3} & \frac{4}{3} & -\frac{7}{3} \\ -\frac{2}{3} & \frac{2}{3} & -\frac{2}{3} & \frac{5}{3} \\ \frac{1}{3} & -\frac{1}{3} & \frac{1}{3} & -\frac{1}{3} \end{bmatrix}$ **e.** Singular **f.** $\begin{bmatrix} 0 & 1 & 0 & 0 \\ 1 & 0 & 0 & 0 \\ 0 & 0 & 0 & 1 \\ 0 & 0 & 1 & 0 \end{bmatrix}$

3. a. $\begin{bmatrix} 1 & 2 \\ 3 & 4 \end{bmatrix}\begin{bmatrix} x \\ y \end{bmatrix} = \begin{bmatrix} 0 \\ 2 \end{bmatrix}; \begin{bmatrix} x \\ y \end{bmatrix} = \begin{bmatrix} 2 \\ -1 \end{bmatrix}$

b. $\begin{bmatrix} 1 & 1 & 1 \\ 0 & 1 & 2 \\ 0 & 1 & -1 \end{bmatrix}\begin{bmatrix} x \\ y \\ z \end{bmatrix} = \begin{bmatrix} 3 \\ 4 \\ 1 \end{bmatrix}; \begin{bmatrix} x \\ y \\ z \end{bmatrix} = \begin{bmatrix} 0 \\ 2 \\ 1 \end{bmatrix}$

c. $\begin{bmatrix} 1 & 1 & 1 \\ 1 & 2 & 1 \\ 1 & 1 & 2 \end{bmatrix}\begin{bmatrix} x \\ y \\ z \end{bmatrix} = \begin{bmatrix} 2 \\ 3 \\ 1 \end{bmatrix}; \begin{bmatrix} x \\ y \\ z \end{bmatrix} = \begin{bmatrix} 2 \\ 1 \\ -1 \end{bmatrix}$

d. $\begin{bmatrix} 1 & 1 & 0 & 0 \\ 0 & 1 & 1 & 0 \\ 0 & 0 & 1 & 1 \\ 1 & 0 & 0 & -1 \end{bmatrix}\begin{bmatrix} x \\ y \\ z \\ w \end{bmatrix} = \begin{bmatrix} 0 \\ 1 \\ 0 \\ 3 \end{bmatrix}; \begin{bmatrix} x \\ y \\ z \\ w \end{bmatrix} = \begin{bmatrix} 1 \\ -1 \\ 2 \\ -2 \end{bmatrix}$

4. a. Inventory − sales = $\begin{bmatrix} 2500 & 4000 & 3000 \\ 1500 & 3000 & 1000 \end{bmatrix} - \begin{bmatrix} 300 & 500 & 100 \\ 100 & 450 & 200 \end{bmatrix} = \begin{bmatrix} 2200 & 3500 & 2900 \\ 1400 & 2550 & 800 \end{bmatrix}$

b. Inventory − sales + restock = $\begin{bmatrix} 2200 & 3500 & 2900 \\ 1400 & 2550 & 800 \end{bmatrix} - 1.2\begin{bmatrix} 300 & 500 & 100 \\ 100 & 450 & 200 \end{bmatrix} + \begin{bmatrix} 3000 & 4000 & 2250 \\ 3000 & 4000 & 2250 \end{bmatrix} = \begin{bmatrix} 4840 & 6900 & 5030 \\ 4280 & 6010 & 2810 \end{bmatrix}$

c. $N = \begin{bmatrix} 2200 & 3500 & 2900 \\ 1400 & 2550 & 800 \end{bmatrix} - x\begin{bmatrix} 280 & 550 & 100 \\ 50 & 500 & 120 \end{bmatrix}$

5. a. Revenue = quantity × price = $\begin{bmatrix} 280 & 550 & 100 \\ 50 & 500 & 120 \end{bmatrix}\begin{bmatrix} 5 \\ 6 \\ 5.5 \end{bmatrix} = \begin{bmatrix} 5250 \\ 3910 \end{bmatrix}$ Texas, Nevada

b. $\begin{bmatrix} 280 & 550 & 100 \\ 50 & 500 & 120 \end{bmatrix}\left(\begin{bmatrix} 5 \\ 6 \\ 5.5 \end{bmatrix} - \begin{bmatrix} 2 \\ 3.5 \\ 1.5 \end{bmatrix}\right) = \begin{bmatrix} 2615 \\ 1880 \end{bmatrix}$ Texas, Nevada

6. a. $[2000 \quad 4000 \quad 4000]\begin{bmatrix} 0.8 & 0.1 & 0.1 \\ 0.4 & 0.6 & 0 \\ 0.2 & 0 & 0.8 \end{bmatrix} = [4000 \quad 2600 \quad 3400]$

b. $[4000 \quad 2600 \quad 3400]\begin{bmatrix} 0.8 & 0.1 & 0.1 \\ 0.4 & 0.6 & 0 \\ 0.2 & 0 & 0.8 \end{bmatrix} = [4920 \quad 1960 \quad 3120]$

c. Here are three. (1) It is possible for someone to be a customer at two different enterprises. (2) Some customers may stop using all three of the companies. (3) New customers can enter the field. **7. a.** July 1, 1000 shares; August 1, 2000 shares; September 1, 2000 shares **b.** Loss = number of shares × (purchase price − dividends − selling price) = $[1000 \quad 2000 \quad 2000] \times \left(\begin{bmatrix} 20 \\ 10 \\ 5 \end{bmatrix} - \begin{bmatrix} 0.10 \\ 0.10 \\ 0 \end{bmatrix} - \begin{bmatrix} 3 \\ 1 \\ 1 \end{bmatrix}\right) = [42{,}700]$ **8. a.** $\begin{bmatrix} \frac{19}{17} & \frac{10}{17} \\ \frac{1}{85} & \frac{18}{17} \end{bmatrix}$ $10/17 \approx$ \$0.588 worth of paper must be produced in order to meet a \$1 increase in the demand for books. **b.** \$1190 worth of paper, \$1802 worth of books **c.** paper, \$350,000; books, \$185,000

Chapter 4

Exercises 4.1

1.

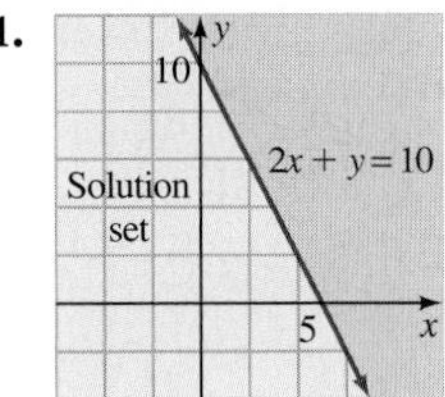

Unbounded

3.

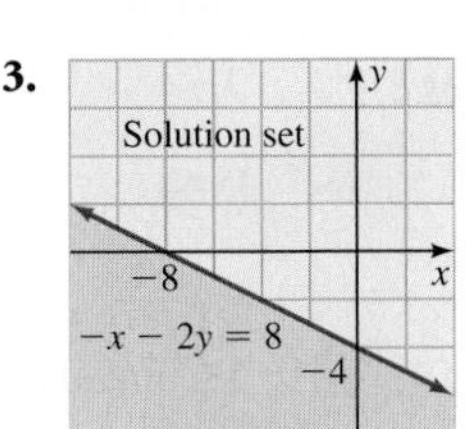

Unbounded

5.

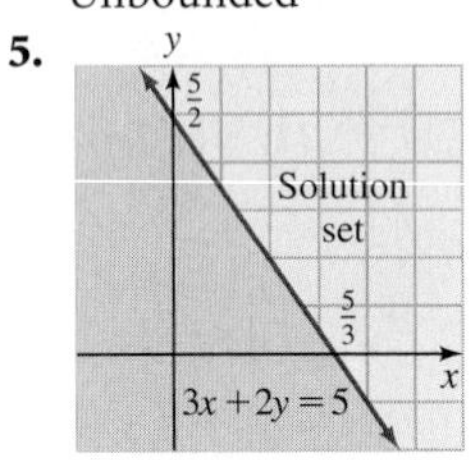

Unbounded

7.

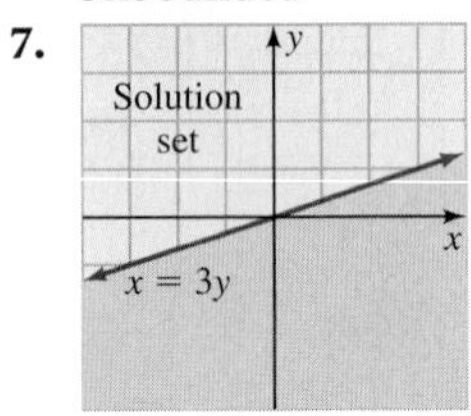

Unbounded

9.

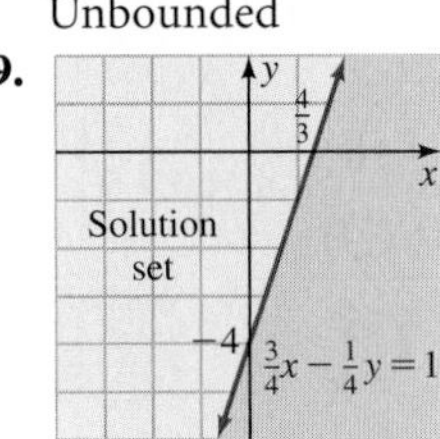

Unbounded

11.

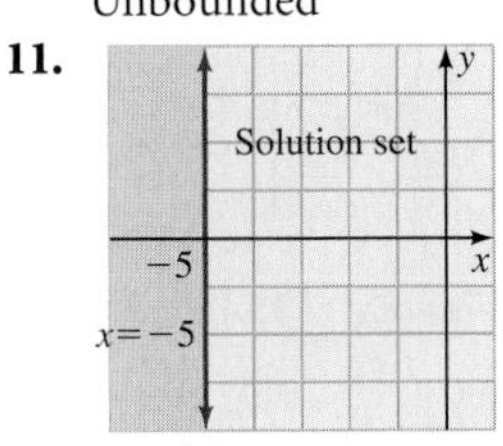

Unbounded

13.

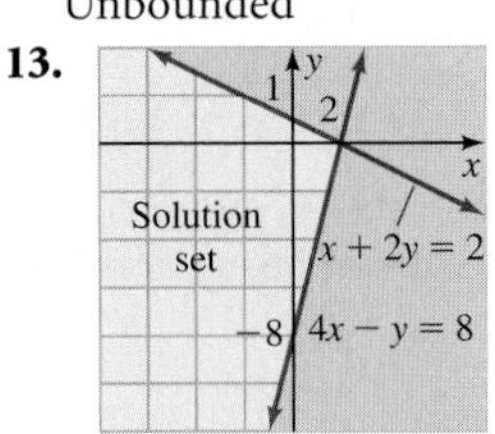

Unbounded; corner point: (2, 0)

15.

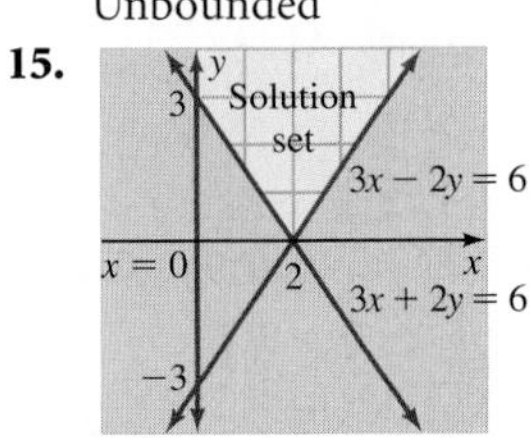

Unbounded; corner points: (2, 0), (0, 3)

17. 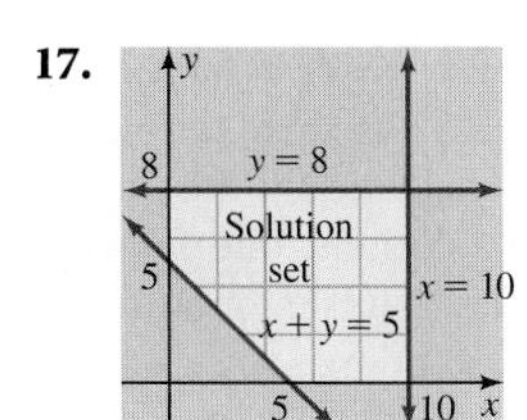

Bounded; corner points: (5, 0), (10, 0), (10, 8), (0, 8), (0, 5)

19. 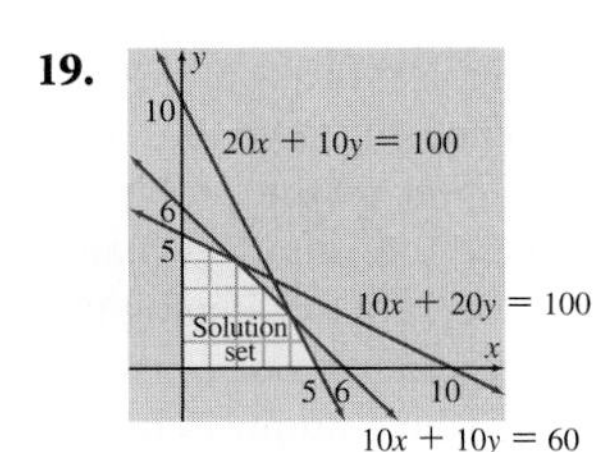

Bounded; corner points: (0, 0), (5, 0), (0, 5), (2, 4), (4, 2)

21. 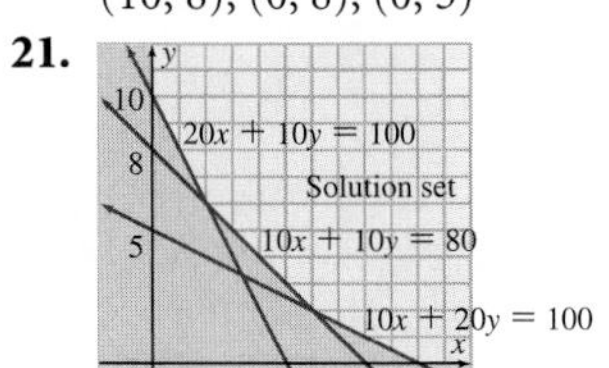

Unbounded; corner points: (0, 10), (10, 0), (2, 6), (6, 2)

23. 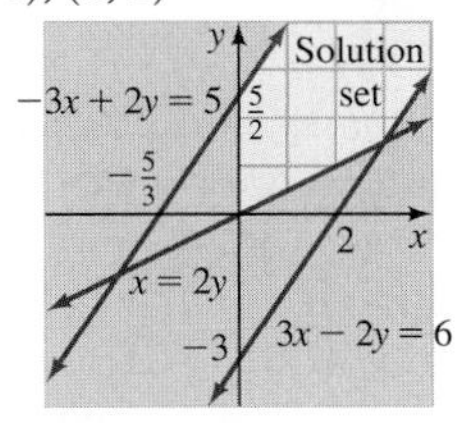

Unbounded; corner points: (0, 0), (0, 5/2), (3, 3/2)

25. 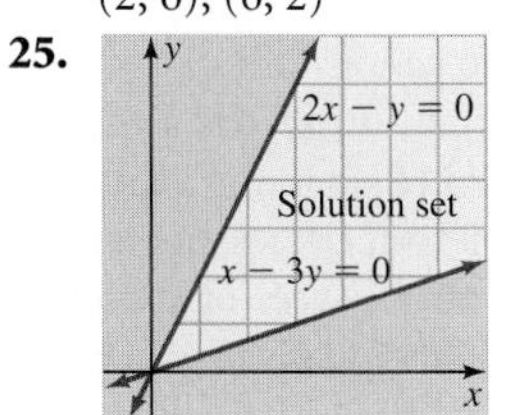

Unbounded; Corner point: (0, 0)

27. 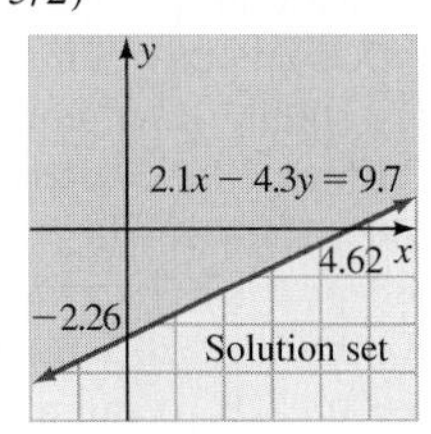

29. 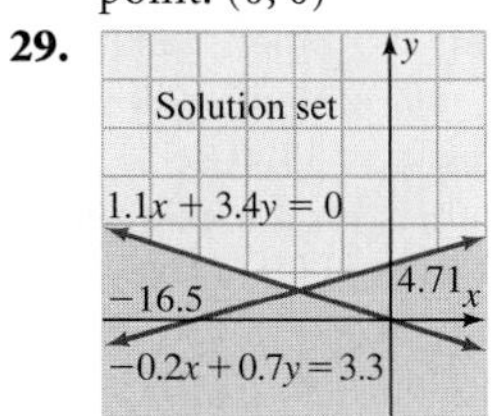

Corner point: (−7.74, 2.50)

31.

7.5x − 4.4y = 5.7
y
1.18
Solution set
4.1x − 4.3y = 4.4
0.76
x
1.07 2.33
−1.02
4.3x + 8.5y = 10
−1.30

Corner points: (0.36, −0.68), (1.12, 0.61)

33. Let $x =$ no. of quarts of Creamy Vanilla, $y =$ no. of quarts of Continental Mocha.

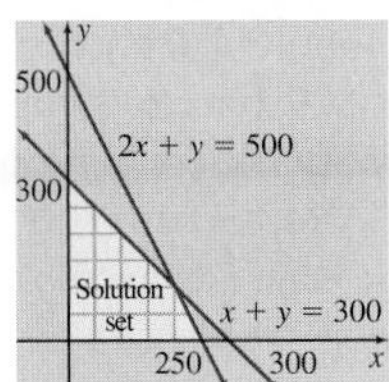

Corner points: (0, 0), (250, 0), (0, 300), (200, 100)

35. Let $x =$ no. of ounces of chicken, $y =$ no. of ounces of grain.

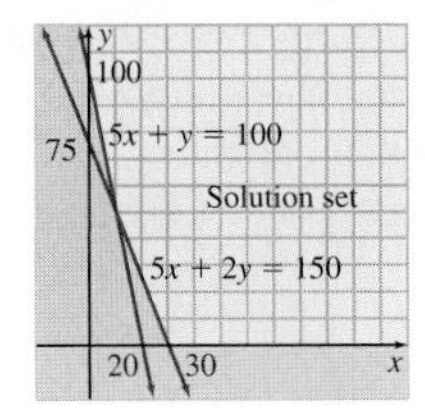

Corner points: (30, 0), (10, 50), (0, 100)

37. Let $x =$ no. of servings of Mixed Cereal for Baby, $y =$ no. of servings of Mango Tropical Fruit Dessert.

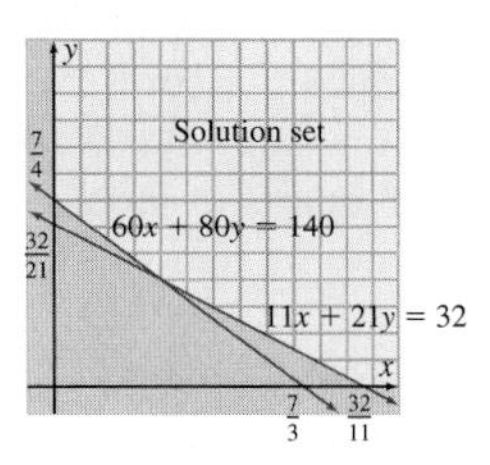

Corner points: (0, 7/4), (1, 1), (32/11, 0)

39. Let $x =$ no. of dollars in PNF, $y =$ no. of dollars in PYSDX.

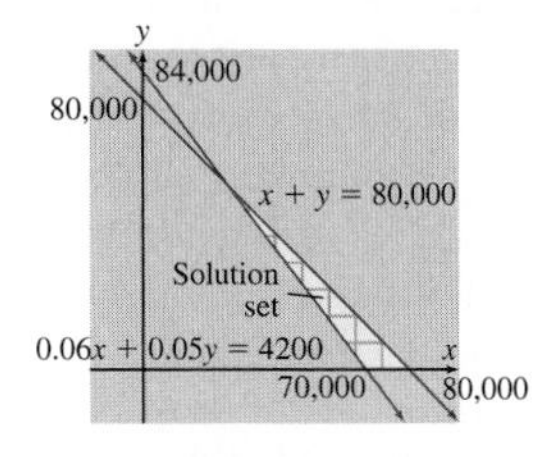

Corner points: (70,000, 0) (80,000, 0) (20,000, 60,000)

41. Let $x =$ no. of shares of MO, $y =$ no. of shares of RJR.

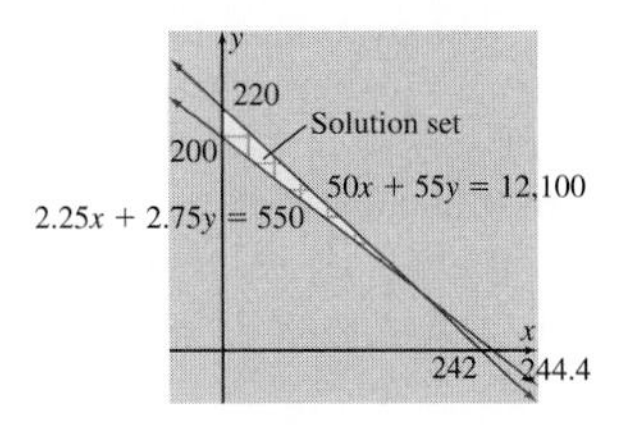

Corner points: (0, 200), (0, 220), (220, 20)

43. Let $x =$ no. of full-page ads in *Sports Illustrated*, $y =$ no. of full-page ads in *GQ*.

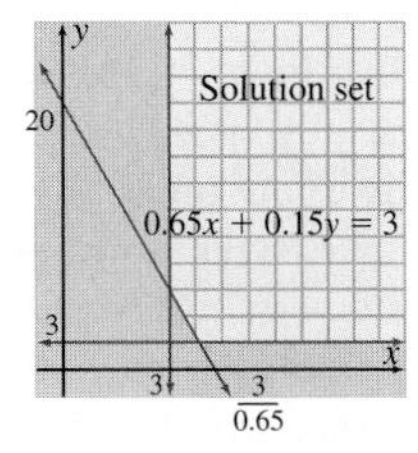

Corner points: (3, 7), (4, 3) (rounded)

45. An example is $x \geq 0, y \geq 0, x + y \geq 1$. **47.** The given triangle can be described as the solution set of the system $x \geq 0, y \geq 0, x + 2y \leq 2$. **49.** (C) **51.** (B) **53.** There are no feasible solutions; that is, it is impossible to satisfy all the constraints. **57.** (Answers may vary.) One limitation is that the method is only suitable for situations with two unknown quantities. Accuracy is also limited when graphing.

Exercises 4.2

1. $p = 6,\ x = 3,\ y = 3$ **3.** $c = 4,\ x = 2,\ y = 2$ **5.** $p = 24,\ x = 7, y = 3$ **7.** $p = 16, x = 4, y = 2$ **9.** $c = 1.8, x = 6, y = 2$ **11.** Max: $p = 16,\ x = 4,\ y = 6$; min: $p = 2,\ x = 2,\ y = 0$ **13.** No optimal solution; objective function unbounded

Answers to Selected Exercises

15. $c = 28$; $(x, y) = (14, 0)$ and $(6, 4)$ and the line connecting them. **17.** $c = 3, x = 3, y = 2$ **19.** No solution; feasible region empty **21.** Let x = no. of quarts of vanilla and y = no. of quarts of mocha. Maximize $p = 3x + 2y$ subject to $2x + y \le 500$, $3x + 3y \le 900$, $x \ge 0$, and $y \ge 0$. You should make 200 quarts of vanilla and 100 quarts of mocha. **23.** Let x = no. of ounces of chicken, y = no. of ounces of grain. Minimize $c = 10x + y$ subject to $10x + 2y \ge 200$, $5x + 2y \ge 150$, $x \ge 0$, and $y \ge 0$. Ruff Inc. should use 100 ounces of grain and no chicken. **25.** Let x = no. of servings cereal, y = no. of servings dessert. Minimize $C = 30x + 50y$ subject to $11x + 21y \ge 32$, $3x + 4y \ge 7$, $x \ge 0$, $y \ge 0$. Feed your child one serving of cereal and one serving of dessert. **27.** Let x = no. of servings of Cell-Tech, y = no. of servings of RiboForce HP. Minimize $c = 2.20x + 1.60y$ subject to $10x + 5y \ge 80$, $2x + y \ge 10$, $75x + 15y \le 750$, $200x \le 1000$. Mix five servings of Cell-Tech and six servings of RiboForce HP for a cost of \$20.60. **29.** Let x = no. of IBM shares, y = no. of HWP shares. Maximize $p = 4.00x + 0.40y$ subject to $100x + 20y \le 10{,}000$, $(0.005)(100x) + (0.015)(20y) \ge 100$. Buy 50 shares of IBM and 250 shares of HWP for maximum company earnings of \$300. **31.** Let x = no. of shares of MO, y = no. of shares of RJR. Minimize $c = 2.0x + 3.0y$ subject to $50x + 55y \le 12{,}100$, $2.25x + 2.75y \ge 550$. Buy 220 shares of MO and 20 shares of RJR. The minimum total risk index is $c = 500$. **33. a.** New York: Gas sales are 800,000 gallons per gas station per year, with revenues of \$960,000; Connecticut: Gas sales per gas station per year are 1,000,000 gallons/year with revenues of \$1,200,000. **b.** With x = no. of gas stations in New York and y = no. in Connecticut, minimize $160{,}000x + 300{,}000y$ subject to $x + y \le 20$, $960{,}000x + 1{,}200{,}000y \ge 20{,}400{,}000$. Solution: 15 in New York and 5 in Connecticut. **35.** Let x = no. of Dracula Salamis and y = no. of Frankenstein. Maximize $P = x + 3y$ subject to $x + 2y \le 1000$, $3x + 2y \le 2400$, $y \le 2x$, $x \ge 0$, $y \ge 0$. You should make 200 Dracula Salamis and 400 Frankenstein Sausages, for a profit of \$1400. **37.** Let x = no. of spots on *Becker* and y = no. of spots on *The Simpsons.* Maximize $V = 8.3x + 7.5y$ subject to $x + y \ge 30$, $2x + 1.5y \le 70$, $-x + y \le 0$, $x \ge 0$, $y \ge 0$. You should purchase 20 spots on *Becker* and 20 spots on *The Simpsons.* **39.** 100 hours/week for new customers and 60 hours/week for old customers. **41.** Let x = no. of hours spent in battle instruction per week, y = no. of hours spent per week in diplomacy instruction. Maximize $P = 50x + 40y$ subject to $x + y \le 50$, $x \ge 2y$, $y \ge 10$, $10x + 5y \ge 400$, $x \ge 0$, $y \ge 0$. He should instruct in diplomacy for 10 hours/week and in battle for 40 hours/week, giving a weekly profit of 2400 ducats. **43.** Let x = no. of sleep spells and y = no. of shock spells. Minimize $c = 500x + 750y$ subject to $2x + 3y \ge 1440$, $3x + 2y \ge 1200$, $x - 3y \le 0$, $-x + 2y \le 0$, $x \ge 0$, $y \ge 0$. Gillian could expend a minimum of 360,000 pico-shirleys of energy by using 480 sleep spells and 160 shock spells. (There is actually a whole line of solutions joining the one above with $x = 2880/7$, $y = 1440/7$.) **45.** (A) **47.** Every point along the line connecting them is also an optimal solution. **49.** Answers will vary. **51.** Answers will vary. **53.** A simple example is the following: Maximize profit $p = 2x + y$ subject to $x \ge 0$, $y \ge 0$. Then p can be made as large as we like by choosing large values of x and/or y. Thus, there is no optimal solution to the problem. **55.** Mathematically this means that there are infinitely many possible solutions: one for each point along the line joining the two corner points in question. In practice, select those points with integer solutions (since x and y must be whole numbers in this problem) that are in the feasible region and close to this line and choose the one that gives the largest profit.

Exercises 4.3

1. $p = 8; x = 4, y = 0$ **3.** $p = 4; x = 4, y = 0$ **5.** $p = 80; x = 10, y = 0, z = 10$ **7.** $p = 53;\ x = 5,\ y = 0,\ z = 3$ **9.** $z = 14{,}500;\ x_1 = 0,\ x_2 = 500/3,\ x_3 = 5000/3$ **11.** $p = 6; x = 2, y = 1, z = 0, w = 3$ **13.** $p = 7; x = 1, y = 0, z = 2, w = 0, v = 4$ (or $x = 1, y = 0, z = 2, w = 1, v = 3$) **15.** $p = 21; x = 0, y = 2.27, z = 5.73$ **17.** $p = 4.52;\ x = 1,\ y = 0,\ z = 0.67,\ w = 1.52$ **19.** $p = 7.7;\ x = 1.1,\ y = 0,\ z = 2.2,\ w = 0,\ v = 4.4$ **21.** You should purchase 500 calculus texts, no history texts, and no marketing texts. The maximum profit is \$5000/semester. **23.** They make a maximum profit of \$650 by making 100 gallons of PineOrange, 200 gallons of PineKiwi, and 150 gallons of OrangeKiwi. **25.** They should offer no Ancient History, 30 sections of Medieval History, and 15 sections of Modern History, for a profit of \$1,050,000. There will be 500 students without classes, but all sections and professors are used. **27.** Plant 80 acres of tomatoes and leave the other 20 acres unplanted. This will give you a profit of \$160,000. **29.** It can make a profit of \$10,000 by selling 1000 servings of Granola, 500 servings of Nutty Granola, and no Nuttiest Granola. It is left with 2000 ounces of almonds. **31.** Allocate 5 million gallons to process A and 45 million gallons to process C. Another solution: Allocate 10 million gallons to process B and 40 million gallons to process C. **33.** Use 15 servings of RiboForce HP and none of the others for a maximum of 75 grams creatine. **35.** She is wrong; you should buy 100 shares of IBM and no others. **37.** Allocate \$2,250,000 to automobile loans, \$500,000 to signature loans, and \$2,250,000 to any combination of furniture loans and other secured loans. **39.** Invest \$75,000 in Universal and none in the rest. Another optimal solution is to invest \$18,750 in Universal and \$75,000 in EMI. **41.** Tucson to Honolulu, 290 boards; Tucson to Venice Beach, 330 boards; Toronto to Honolulu, 0 boards; Toronto to Venice Beach, 200 boards—giving 820 boards shipped. **43.** Fly ten people from Chicago to Los Angeles, five people from Chicago to New York, and ten people from Denver to New York. **45.** Yes; the given problem can be stated as maximize $p = 3x - 2y$ subject to $-x + y - z \le 0$, $x - y - z \le 6$ **47.** The graphical method applies only to LP problems in two unknowns, whereas the simplex method can be used to solve LP problems with any number of unknowns. **49.** She is correct. Since there are only two constraints, there can be only two active variables, giving two or fewer nonzero values for the unknowns at each stage. **51.** A basic solution to a system of linear equations is a solution in which all the nonpivotal variables are taken to be zero; that is, all variables whose value is arbitrary are assigned the value zero. **53.** No. Let's assume for the sake of simplicity that all the pivots are 1s. (They may certainly be changed to 1s without affecting the value of any of the variables.) Since the entry at the bottom of the pivot column is negative, the bottom row gets replaced by itself plus a positive multiple of the pivot row. The value of the objective function (bottom-right entry) is thus replaced by itself plus a positive multiple of the nonnegative rightmost entry of the pivot row. Therefore, it cannot decrease.

Exercises 4.4

1. $p = 20/3$; $x = 4/3$, $y = 16/3$ **3.** $p = 850/3$; $x = 50/3$, $y = 25/3$ **5.** $p = 750$; $x = 0$, $y = 150$, $z = 0$ **7.** $p = 135$; $x = 0$, $y = 25$, $z = 0$, $w = 15$ **9.** $c = 80$; $x = 20/3$, $y = 20/3$ **11.** $c = 100$; $x = 0$, $y = 100$, $z = 0$ **13.** $c = 111$; $x = 1$, $y = 1$, $z = 1$ **15.** $c = 200$; $x = 200$, $y = 0$, $z = 0$, $w = 0$ **17.** $p = 136.75$; $x = 0$, $y = 25.25$, $z = 0$, $w = 15.25$ **19.** $c = 66.67$; $x = 0$, $y = 66.67$, $z = 0$ **21.** $c = -250$; $x = 0$, $y = 500$, $z = 500$, $w = 1500$ **23.** 10,000 quarts of orange juice and 2000 quarts of orange concentrate **25.** One serving of cereal, one serving of juice, and no dessert **27.** 10 mailings to the East Coast, none to the Midwest, 10 to the West Coast. Cost: \$900. Another solution resulting in the same cost is no mailings to the East Coast, 15 to the Midwest, and none to the West Coast. **29.** 15 bundles from Nadir, 5 from Sonny, and none from Blunt. Cost: \$70,000. Another solution resulting in the same cost is 10 bundles from Nadir, none from Sony, and 10 from Blunt. **31.** Stock 10,000 rock CDs, 5000 rap CDs, and 5000 classical CDs for a maximum anticipated revenue of \$255,000 **33. a.** Build one convention-style hotel, four vacation-style hotels, and two small motels. The total cost will amount to \$188 million. **b.** Since 20% of this is \$37.6 million, you will still be covered by the subsidy. **35.** Mix six servings of Ribo-Force HP and ten servings of Creatine Transport for a cost of \$15.60. **37.** Hire no more cardiologists, 12 rehabilitation specialists, and 5 infectious disease specialists. **39.** No hamburger, 3 cups powdered milk, 4 eggs, to give $c = 65¢$/serving. *Comment:* This is hardly a hamburger! **41.** Tucson to Honolulu, 500 boards/week; Tucson to Venice Beach, 120 boards/week; Toronto to Honolulu, 0 board/week; Toronto to Venice Beach, 410 boards/week. Minimum weekly cost is \$9700. **43.** Fly ten people from Chicago to Los Angeles, five from Chicago to New York, none from Denver to Los Angeles, ten from Denver to New York at a total cost of \$4520. **45.** \$2500 from Congressional Integrity Bank, \$0 from Citizens' Trust, \$7500 from Checks R Us. **47.** The solution $x = 0$, $y = 0$, ... represented by the initial tableau may not be feasible. In phase I we use pivoting to arrive at a basic solution that is feasible. **49.** The basic solution corresponding to the initial tableau has all the unknowns equal to zero, and this is not a feasible solution since it does not satisfy the given inequality. **51.** (C) **53.** Answers may vary. Examples are Exercises 1 and 2. **55.** Answers may vary. A simple example is maximize $p = x + y$ subject to $x + y \le 10$, $x + y \ge 20$, $x \ge 0$, $y \ge 0$.

Exercises 4.5

1. Minimize $c = 6s + 2t$ subject to $s - t \ge 2$, $2s + t \ge 1$, $s \ge 0$, $t \ge 0$ **3.** Maximize $p = 100x + 50y$ subject to $x + 2y \le 2$, $x + y \le 1$, $x \le 3$, $x \ge 0$, $y \ge 0$. **5.** Minimize $c = 3s + 4t + 5u + 6v$ subject to $s + u + v \ge 1$, $s + t + v \ge 1$, $s + t + u \ge 1$, $t + u + v \ge 1$, $s \ge 0$, $t \ge 0$, $u \ge 0$, $v \ge 0$. **7.** Maximize $p = 1000x + 2000y + 500z$ subject to $5x + z \le 1$, $-x + z \le 3$, $y \le 1$, $x - y \le 0$, $x \ge 0$, $y \ge 0$, $z \ge 0$. **9.** $c = 4$; $s = 2$, $t = 2$ **11.** $c = 80$; $s = 20/3$, $t = 20/3$ **13.** $c = 1.8$; $s = 6$, $t = 2$ **15.** $c = 25$; $s = 5$, $t = 15$ **17.** $c = 30$; $s = 30$, $t = 0$, $u = 0$ **19.** $c = 100$; $s = 0$, $t = 100$, $u = 0$ **21.** $c = 30$; $s = 10$, $t = 10$, $u = 10$ **23.** Four ounces each of fish and cornmeal, for a total cost of 40¢/can; 5/12¢ per gram of protein, 5/12¢ per gram of fat. **25.** 100 ounces of grain and no chicken, for a total cost of \$1; 1/2¢ per gram of protein, 0¢ per gram of fat. **27.** One serving of cereal, one serving of juice, and no dessert for a total cost of 37¢; 1/6¢ per calorie and 17/120¢ per percent U.S. RDA of vitamin C **29.** Ten mailings to the East Coast, none to the Midwest, 10 to the West Coast. Cost: \$900; 20¢/Democrat and 40¢/Republican. *Or* 15 mailings to the Midwest and no mailing to the coasts. Cost: \$900; 20¢/Democrat and 40¢/Republican. **31.** Gillian should use 480 sleep spells and 160 shock spells, costing 360,000 pico-shirleys of energy *or* 2880/7 sleep spells and 1440/7 shock spells. **33.** The dual of a standard minimization problem satisfying the nonnegative objective condition is a standard maximization problem, which can be solved using the standard simplex algorithm, thus avoiding the need to do phase I. **35.** Answers will vary. An example is minimize $c = x - y$ subject to $x - y \ge 100$, $x + y \ge 200$, $x \ge 0$, $y \ge 0$. This problem can be solved using the techniques in Section 4.4. **37.** Build one convention-style hotel, four vacation-style hotels, and two small motels. **39.** Answers will vary.

Chapter 4 Review Test

1. a. Unbounded

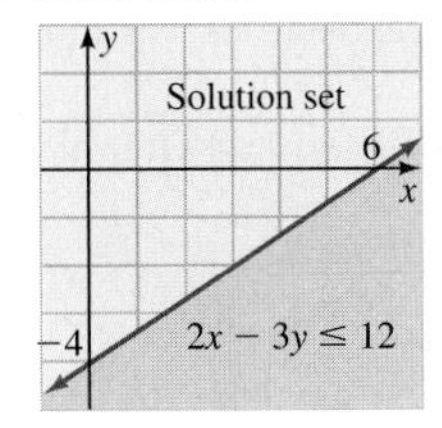

b. Unbounded

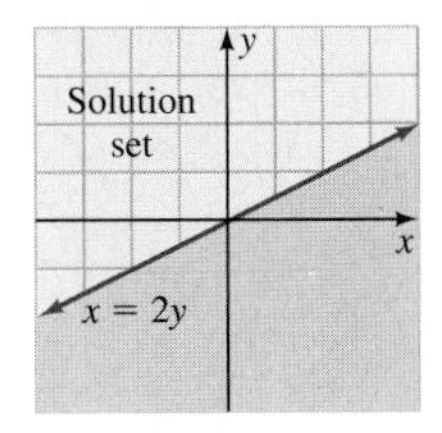

c. Bounded

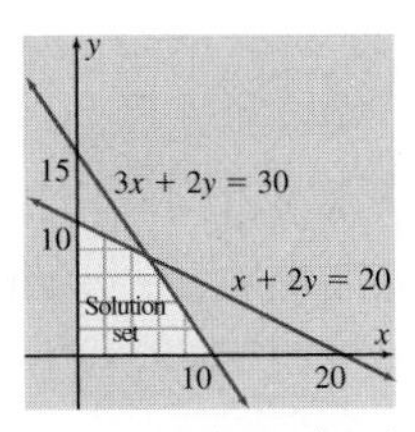

Corner points: (0, 0), (0, 10), (5, 15/2), (10, 0)

d. Unbounded

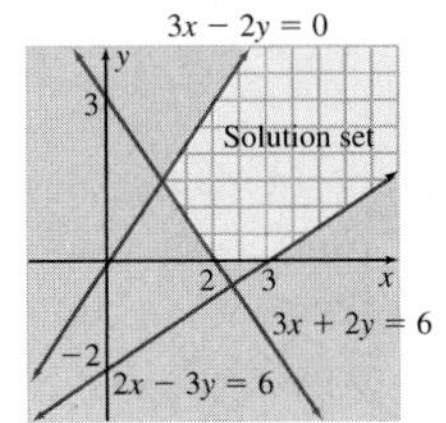

Corner points: (2, 0), (3, 0) (1, 3/2)

2. a. $p = 21; x = 9, y = 3$ **b.** $p = 60; x = 0, y = 20$ **c.** $c = 22; x = 8, y = 6$ **d.** $c = 9/2; x = 1, y = 3/2$ **3. a.** $p = 45; x = 0, y = 15, z = 15$ **b.** $p = 36; x = 24, y = 12, z = 0$ **c.** $p = 220; x = 20, y = 20, z = 60$ **d.** $c = 30; x = 30, y = 0, z = 0$ **e.** No solution; feasible region empty **f.** No solution; feasible region unbounded **4. a.** $c = 60; x = 24, y = 12$ **b.** $c = 140; x = 0, y = 20, z = 60$ **c.** $c = 20; x = 0, y = 20$ **d.** $c = 200/3; x = 100/3, y = 0, z = 0$ **5. a.** (A) **b.** (B) **c.** 35 **6. a.** 400 packages from each for a minimum cost of \$52,000 **b.** (B), (D) **c.** 450 packages from Duffin House and 375 from Higgins Press for a minimum cost of \$52,500 **d.** Same solution as (c) **7.** Duffin should print no paperbacks, 300 quality paperbacks, and 1900 hardcovers for a total daily profit of \$6300. **8.** $c = 90{,}000; x = 0, y = 600, z = 0$ **9. a.** Let x = no. of Science credits, y = no. of Fine Arts credits, z = no. of Liberal Arts credits, and w = no. of Math credits. Minimize $C = 300x + 300y + 200z + 200w$ subject to $x + y + z + w \geq 120, x - y \geq 0, -2x + w \leq 0, -y + 3z - 3w \leq 0, x \geq 0, y \geq 0, z \geq 0, w \geq 0$. **b.** Billy Sean should take the following combination: Sciences, 24 credits; Fine Arts, no credits; Liberal Arts, 48 credits; Mathematics, 48 credits—for a total cost of \$26,400. **10.** Smallest cost is \$20,000. New York to OHaganBooks.com, 600 packages; New York to FantasyBooks.com, no packages; Illinois to OHaganBooks.com, no packages; Illinois to FantasyBooks.com, 200 packages.

Chapter 5

Exercises 5.1

1. $INT = \$120$, $FV = \$2120$ **3.** $INT = \$505$, $FV = \$20{,}705$ **5.** $INT = \$250$, $FV = \$10{,}250$ **7.** \$787.40 **9.** 5% **11.** In 2 years **13.** 4.531% **15.** 86.07% **17.** –21.94% **19.** 1999–2000, 86.07% **21.** No. Simple interest contraction is linear, decreasing by the same amount each year. However, Sony's net income decreased by 43 billion yen from 1997 to 1998 and by 57 billion yen from 1998 to 1999, so the decrease was not linear. **23.** 9.2% **25.** 3,260,000 **27.** $P = 500 + 46t$ (t = time in years since 1950) Graph:

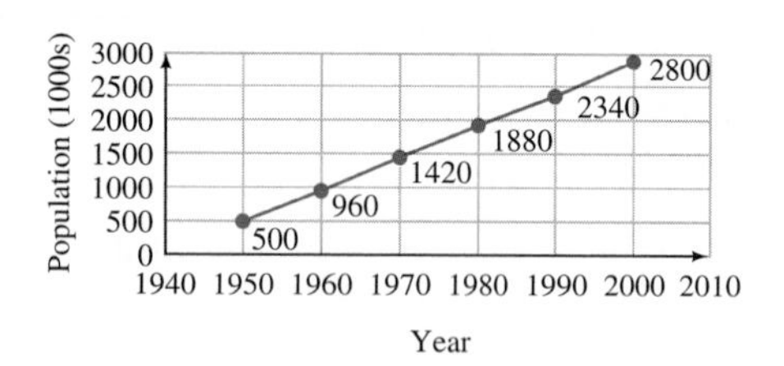

29. Graph (A) is the only possible choice because the equation $FV = PV(1 + rt) = PV + PVrt$ gives the future value as a linear function of time. **31.** Wrong. In simple interest growth, the change each year is a fixed percentage of the *starting* value, not the preceding year's value. (Also see the next exercise.) **33.** Simple interest is always calculated on a constant amount, PV. If interest is paid into your account, then the amount on which interest is calculated does not remain constant.

Exercises 5.2

1. \$13,439.16 **3.** \$11,327.08 **5.** \$19,154.30 **7.** \$12,709.44 **9.** \$613.91 **11.** \$810.65 **13.** \$1227.74 **15.** 5.09% **17.** 10.47% **19.** 10.52% **21.** \$268.99 **23.** \$2491.75 **25.** \$2927.15 **27.** \$21,161.79 **29.** \$163,414.56 **31.** \$55,526.45 per year **33.** \$174,110 **35.** \$750.00 **37.** \$27,171.92 **39.** \$111,678.96 **41.** \$1039.21 **43.** The one earning 11.9% compounded monthly **45.** Yes. The investment will have grown to about \$150,281 million. **47.** 147 reals **49.** 744 pesos **51.** 1224 pesos **53.** The Ecuadorian investment is better; it is worth 1.01614 units of currency (in constant units) per unit invested as opposed to 1.01262 units for Chile. **55. a.** \$1510.31 **b.** \$54,701.29 **c.** 23.51%

57. **59.**

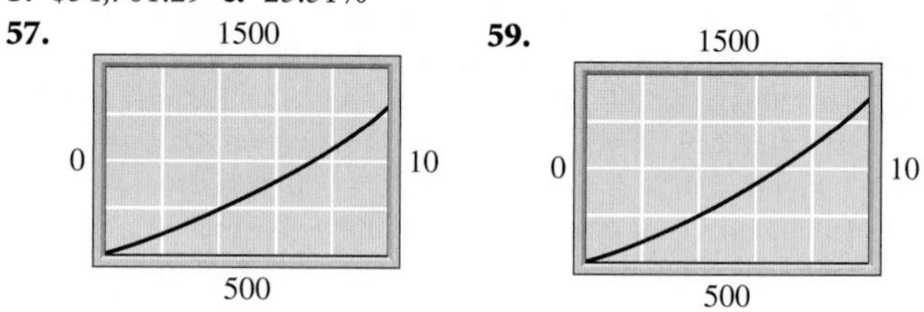

61.

Years	1	2	3	4	5	6	7
Value (\$)	1050	1103	1158	1216	1276	1340	1407

63. 31 years; about \$26,100 **65.** 2.3 years **67.** The function $y = P(1 + \frac{r}{m})^{mx}$ is not a linear function of x, but an exponential function. Thus, its graph is not a straight line. **69.** Wrong. Its growth is exponential and can be modeled by $0.01(1.10)^t$. **71.** The graphs are the same because the formulas give the same function of x; a compound interest investment behaves as though it were being compounded once a year at the effective rate. **73.** The effective rate exceeds the nominal rate when the interest is compounded more than once a year, because then interest is being paid on interest accumulated during each year, resulting in a larger effective rate. Conversely, if the interest is compounded less often than once a year, the effective rate is less than the nominal rate. **75.** Compare their future values in constant dollars. The investment with the larger future value is the better investment. **77.** The graphs are approaching a particular curve as m gets larger, approximately the curve given by the largest two values of m.

Exercises 5.3

1. \$15,528.23 **3.** \$171,793.82 **5.** \$23,763.28 **7.** \$147.05 **9.** \$491.12 **11.** \$105.38 **13.** \$90,155.46 **15.** \$69,610.99 **17.** \$95,647.68 **19.** \$554.60 **21.** \$1366.41 **23.** \$524.14 **25.** \$248.85 **27.** \$1984.65 **29.** \$999.61 **31.** \$998.47 **33.** 3.617% **35.** 3.059% **37.** \$973.54 **39.** \$7451.49 **41.** You should take the loan from Solid Savings & Loan; it will have payments of \$248.85/month. The payments on the other loan would be more than \$300/month. **43.** Answers using correctly rounded intermediate results:

Year	Interest ($)	Payment on Principal ($)
1	3934.98	1798.98
2	3785.69	1948.27
3	3623.97	2109.99
4	3448.84	2285.12
5	3259.19	2474.77
6	3053.77	2680.19
7	2831.32	2902.64
8	2590.39	3143.57
9	2329.48	3404.48
10	2046.91	3687.05
11	1740.88	3993.08
12	1409.47	4324.49
13	1050.54	4683.42
14	661.81	5072.15
15	240.84	5491.80

45. First 5 years: \$402.62/month; last 25 years: \$601.73 **47.** Original monthly payments were \$824.79. The new monthly payments will be \$613.46. You will save \$36,481.77 in interest. **49.** 13 years **51.** 4.5 years **53.** 24 years **55.** He is wrong because his estimate ignores the interest that will be earned by your annuity—both while it is increasing and while it is decreasing. Your payments will be considerably smaller (depending on the interest earned). **57.** He is not correct. For instance, the payments on a \$100,000 10-year mortgage at 12% are \$1434.71, whereas for a 20-year mortgage at the same rate, they are \$1101.09, which is a lot more than half the 10-year mortgage payment.

Chapter 5 Review Test

1. a. \$1187.50 **b.** No **c.** 17 years **d.** \$6305.80 **e.** \$25.71 **2. a.** \$5555.56 **b.** 4.5% **c.** In 4 years **d.** \$9895.00; 6.37% **3. a.** \$8027.12 **b.** \$13,601.87 **c.** \$12,319.63 **d.** At the end of the third quarter in 2009 ($t = 6.75$) **4. a.** \$5900.42; \$500.42 of that is interest. **b.** \$199.34 **c.** \$191.57 **d.** \$880.52; \$9564.14; \$196,987.20 **5. a.** 2003

Year	2000	2001	2002	2003	2004
Revenue	\$180,000	\$216,000	\$259,200	\$311,040	\$373,248

b. \$15,305 **c.** At least 52,515 shares **6. a.** Industrial: \$3234.94; expansion: \$3346.56 **b.** \$224,111 **c.** 3.59% **7. a.** \$420,275 **b.** \$140,778 **c.** \$1453.06 **d.** \$2239.90/month **8. a.** \$53,055.66 **b.** \$53,949.84 **c.** 5.99%

Chapter 6

Exercises 6.1

1. $F =$ {spring, summer, fall, winter} **3.** $I = \{1, 2, 3, 4, 5, 6\}$ **5.** $A = \{1, 2, 3\}$ **7.** $B = \{2, 4, 6, 8\}$ **9. a.** $S =$ {(H, H), (H, T), (T, H), (T, T)} **b.** $S =$ {(H, H), (H, T), (T, T)} **11.** $S = \{(1, 5), (2, 4), (3, 3), (4, 2), (5, 1)\}$ **13.** $S = \{(1, 5), (2, 4), (3, 3)\}$ **15.** $S = \varnothing$

17.

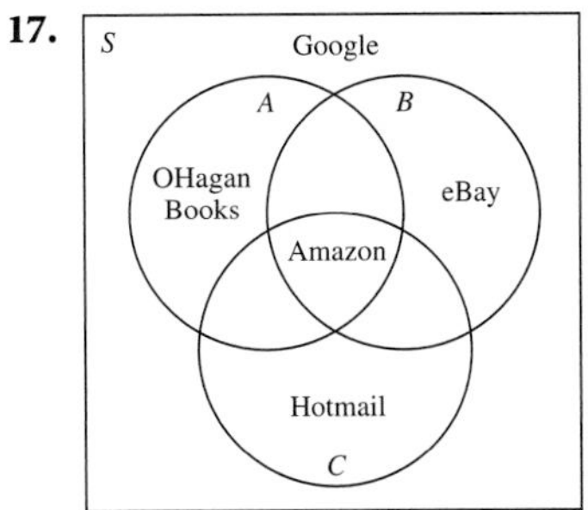

19.

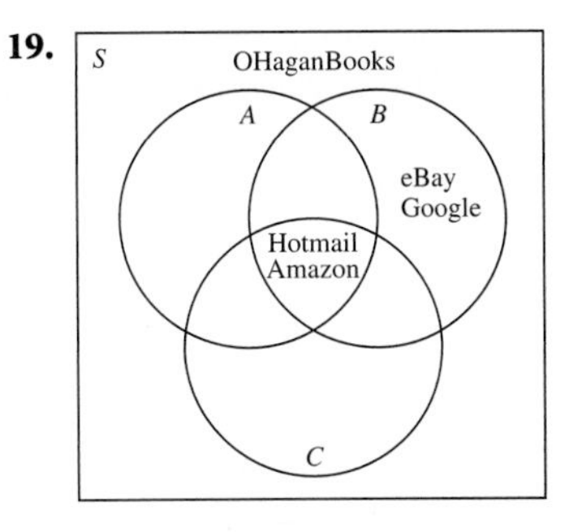

21. A **23.** A **25.** {June, Janet, Jill, Justin, Jeffrey, Jello, Sally, Solly, Molly, Jolly} **27.** {Jello} **29.** $\varnothing$ **31.** {Jello} **33.** {Janet, Justin, Jello, Sally, Solly, Molly, Jolly} **35.** {(small, triangle), (small, square), (medium, triangle), (medium, square), (large, triangle), (large, square)} **37.** {(small, blue), (small, green), (medium, blue), (medium, green), (large, blue), (large, green)}

39.

	A	B	C
1		Triangle	Square
2	Blue	Blue Triangle	Blue Square
3	Green	Green Triangle	Green Square

41.

	A	B	C
1		Blue	Green
2	Small	Small Blue	Small Green
3	Medium	Medium Blue	Medium Green
4	Large	Large Blue	Large Green

43. $B \times A =$ {1H, 1T, 2H, 2T, 3H, 3T, 4H, 4T, 5H, 5T, 6H, 6T} **45.** $A \times A \times A =$ {HHH, HHT, HTH, HTT, THH, THT, TTH, TTT} **47.** {(1, 1), (1, 3), (1, 5), (3, 1), (3, 3), (3, 5), (5, 1), (5, 3), (5, 5)} **49.** $\varnothing$ **51.** {(1, 1), (1, 3), (1, 5), (3, 1), (3, 3), (3, 5), (5, 1), (5, 3), (5, 5), (2, 2), (2, 4), (2, 6), (4, 2), (4, 4), (4, 6), (6, 2), (6, 4), (6, 6)} **61.** $A \cap B =$ {Acme, Crafts} **63.** $B \cup C =$ {Acme, Brothers, Crafts, Dion, Effigy, Global, Hilbert} **65.** $A' \cap C =$ {Dion, Hilbert} **67.** $A \cap B' \cap C' = \varnothing$

69.

	A	B	C	D
1		Used Boats	New Boats	Accessories
2	1998	1998 Used	1998 New	1998 Acc.
3	1999	1999 Used	1999 New	1999 Acc.
4	2000	2000 Used	2000 New	2000 Acc.
5	2001	2001 Used	2001 New	2001 Acc.

{1998, 1999, 2000, 2001} × {Used Boats, New Boats, Accessories}

71. Let $A = \{1\}$, $B = \{2\}$, and $C = \{1, 2\}$. Then $(A \cap B) \cup C = \{1, 2\}$ but $A \cap (B \cup C) = \{1\}$. In general, $A \cap (B \cup C)$ must be a subset of A, but $(A \cap B) \cup C$ need not be; also, $(A \cap B) \cup C$ must contain C as a subset, but $A \cap (B \cup C)$ need not. **73.** (B) **75.** A universal set is a set containing all "things" currently under consideration. When discussing sets of positive integers, the universe might be the set of all positive integers, or the set of all integers (positive, negative, and zero), or any other set containing the set of all positive integers. **77.** A is the set of suppliers who deliver components on time, B is the set of suppliers whose components are known to be of high quality, and C is the set of suppliers who do not promptly replace defective components. **79.** Let A = movies that are violent, B = movies that are shorter than 2 hours, C = movies that have a tragic ending, and D = movies that have an unexpected ending. The given sentence can be rewritten as "She prefers movies in $A' \cap B \cap (C \cup D)'$." It can also be rewritten as "She prefers movies in $A' \cap B \cap C' \cap D'$."

Exercises 6.2

1. 9 **3.** 7 **5.** 4 **7.** $n(A \cup B) = 7, n(A) + n(B) - n(A \cap B) = 4 + 5 - 2 = 7$ **9.** 4 **11.** 18 **13.** 72 **15.** 60 **17.** 20 **19.** 6 **21.** 9 **23.** 4 **25.** $n[(A \cap B)'] = 9$, $n(A') + n(B') - n[(A \cup B)'] = 6 + 7 - 4 = 9$

27.

S; A; B; C; 4; 4; 8; 2; 10; 6; 10; 6

29.

S; A; B; C; 3; 4; 5; 0; 3; 10; 15; 100

31. 7068 **33.** 2 **35.** 93 **37.** $C \cap N$ is the set of authors who are both successful and new. $C \cup N$ is the set of authors who are either successful or new (or both). $n(C) = 30$; $n(N) = 20$; $n(C \cap N) = 5$; $n(C \cup N) = 45$; $45 = 30 + 20 - 5$ **39.** $C \cap N'$ is the set of authors who are successful but not new. $n(C \cap N') = 25$ **41.** 31.25%; 83.33% **43.** $N \cap C$; $n(N \cap C) = 8$ billion **45.** $C \cap N'$; $n(C \cap N') = 13$ billion **47.** $A \cap (N \cup U)$; $n[A \cap (N \cup U)] = 14$ billion **49.** $V \cap I' n(V \cap I') = 15$ **51.** 80; the number of stocks that were either not pharmaceutical stocks or were unchanged in value after a year (or both) **53.** 3/8; the fraction of Internet stocks that increased in value **55. a.** 931 **b.** 382

57. a.

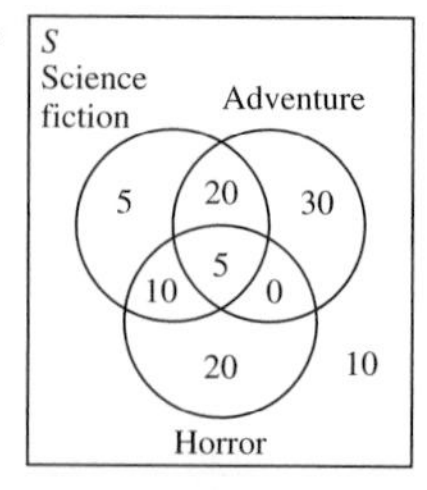

b. 37.5%

59. 17 **61.** When $A \cap B \neq \varnothing$ **63.** When $B \subseteq A$ **65.** The number of elements in the Cartesian product of two finite sets is the product of the number of elements in the two sets. **67.** Answers will vary. **69.** $n(A \cup B \cup C) = n(A) + n(B) + n(C) - n(A \cap B) - n(B \cap C) - n(A \cap C) + n(A \cap B \cap C)$

Exercises 6.3

1. $2 + 3 + 5 = 10$ **3.** $2 \times 3 \times 5 = 30$ **5.** 6 outcomes **7.** 15 outcomes **9.** 13 outcomes **11.** 25 outcomes **13.** 4 **15.** $2 \times 2 \times 2 \times 2 = 16$ **17.** $6 \times 5 = 30$ **19.** $3 \times 3 + 2 \times 2 = 13$ **21.** $(2 \times 5) + (2 \times 2 \times 2) = 18$ **23.** $2^{10} \times 5^2 = 1024 \times 25 = 25{,}600$ **25.** $2^8 + 5^5 = 3381$ **27. a.** $4 \times 3 \times 8 \times 3 = 288$ **b.** $4 \times 3 \times 8 \times 3 = 288$ **29.** $2^8 = 256$ **31.** 10 **33.** 286 **35.** 4 **37. a.** $8 \times 10^6 = 8{,}000{,}000$ **b.** $10^4 + 10^4 + 10^4 = 30{,}000$ **c.** $8 \times 9^6 = 4{,}251{,}528$ **39. a.** $4^3 = 64$ **b.** 4^n **c.** $4^{2.1 \times 10^{10}}$ **41. a.** $16^6 = 16{,}777{,}216$ **b.** $16^3 = 4096$ **c.** $16^2 = 256$ **d.** $3 \times 16^2 - 2 = 766$ **43.** $(10 \times 9 \times 8 \times 7 \times 6 \times 5 \times 4) \times (8 \times 7 \times 6 \times 5) = 1{,}016{,}064{,}000$ possible casts **45. a.** $26^3 \times 10^3 = 17{,}576{,}000$ **b.** $26^2 \times 23 \times 10^3 = 15{,}548{,}000$ **c.** $15{,}548{,}000 - 3 \times 10^3 = 15{,}545{,}000$ **47. a.** $4 \times 1 = 4$ **b.** 4 **c.** There would be an infinite number of routes. **49. a.** $6 \times 3 \times 2 \times 2 \times 1 \times 1 = 72$ **b.** $3 \times 3 \times 2 \times 2 \times 1 \times 1 = 36$ **51.** $2 \times 2 \times 2 \times 2 \times 2 \times 3 = 96$ paintings **53. a.** $4 \times 9 = 36$ **b.** $36 + 1 = 37$ **55.** 4 **57.** Step 1: Choose a day of the week on which Jan. 1 will fall: seven choices. Step 2: Decide whether or not it is a Leap Year: two choices. Total: $7 \times 2 = 14$ possible calendars **59.** Step 1: Choose a position in the left–right direction: m choices. Step 2: Choose a position in the front–back direction: n choices. Step 3: Choose a position in the up–down direction: r choices. Hence, there are $m \cdot n \cdot r$ possible outcomes. **61.** 1900 **63.** Product **65.** The decision algorithm produces every pair of shirts twice, first in one order and then in the other. **67.** Think of placing the five squares in a row of five empty slots. Step1: Choose a slot for the blue square: five choices. Step 2: Choose a slot for the green square: four choices. Step 3: Choose the remaining three slots for the yellow squares: 1 choice. Hence, there are 20 possible five-square sequences.

Exercises 6.4

1. 720 **3.** 56 **5.** 360 **7.** 15 **9.** 3 **11.** 45 **13.** 20 **15.** 4950 **17.** 360 **19.** 35 **21.** $5! = 120$ **23.** 120 **25.** 20 **27.** $6 \times C(5, 3) \times C(2, 2) = 60$ **29.** $4! = 24$ **31.** $C(10, 4) = 210$ **33.** $C(3, 3)C(7, 1) = 7$ **35.** $C(7, 4) = 35$ **37.** $3 \times 2 \times 2 \times 2 = 24$ **39.** $C(3, 2)C(7, 3) + C(3, 3)C(7, 2) = 126$ **41.** $C(2, 1)C(8, 4) + C(8, 5) = 196$ **43.** $C(1, 1)C(7, 4) + C(2, 1)C(7, 4) = 105$ **45.** $C(13, 2)C(4, 2)C(4, 2) \times 44 = 123{,}552$ **47.** $13 \times C(4, 2)C(12, 3) \times 4 \times 4 \times 4 = 1{,}098{,}240$ **49.** $10 \times 4 \times 4 \times 4 \times 4 \times 4 - 10 \times 4 = 10{,}200$ **51.** $\dfrac{C(30, 5) \times 5^{25}}{6^{30}} = 0.192$ **53.** $\dfrac{C(30, 15) \times 3^{15} \times 3^{15}}{6^{30}} = 0.144$ **55. a.** 23! **b.** 18! **c.** $19 \times 18!$ **57.** Step 1: Select Boondoggle as a do-nothing member. Step 2: Select the chief investigator from the Party Party. Step 3: Select the assistant investigators from the Study Party. Step 4: Select the rabble rousers. Step 5: Select the other two do-nothings. We thus get $C(1, 1)C(9, 1)C(9, 2)C(15, 2)C(13, 2)$. **59.** $C(11, 1)C(10, 4)C(6, 4)C(2, 2)$ **61.** $C(11, 2)C(9, 1)C(8, 1)C(7, 3)C(4, 1)C(3, 1)C(2, 1)C(1, 1)$ **63.** $C(10, 2)C(8, 4)C(4, 1)C(3, 1)C(2, 1)C(1, 1)$ **65.** $2^8 + 5^5 + 5! = 3501$ **67.** (A) **69.** (D) **71. a.** 9880 **b.** 1560 **c.** $9880 + 1560 + 40 = 11{,}480$ **73. a.** $C(20, 2) = 190$ **b.** $C(n, 2)$ **75.** The multiplication principle because the multiplication principle can be used to solve all problems that call for the formulas for permutations. **77.** Urge your friend not to focus on formulas but instead learn to formulate decision algorithms and use the principles of counting. **79.** 80; 32,768

Chapter 6 Review Test

1. a. $N = \{-3, -2, -1\}$ **b.** $S = \{$HHHHH, HHHHT, HHHTH, HHHTT, HHTHH, HHTHT, HHTTH, HHTTT, HTHHH, HTHHT, HTHTH, HTHTT, HTTHH, HTTHT, HTTTH, HTTTT, THHHH, THHHT, THHTH, THHTT, THTHH, THTHT, THTTH, THTTT, TTHHH, TTHHT, TTHTH, TTHTT, TTTHH, TTTHT, TTTTH, TTTTT$\}$ **c.** $S = \{(1, 2), (1, 3), (1, 4), (1, 5), (1, 6), (2, 1), (2, 3), (2, 4), (2, 5), (2, 6), (3, 1), (3, 2), (3, 4), (3, 5), (3, 6), (4, 1), (4, 2), (4, 3), (4, 5), (4, 6), (5, 1), (5, 2), (5, 3), (5, 4), (5, 6), (6, 1), (6, 2), (6, 3), (6, 4), (6, 5)\}$ **d.** $(A \cap B) \cup C = \{1, 2, 3, 4, 5, 6, 7\}, A \cap (B \cup C) = \{1, 2, 3, 4, 5\}$ **e.** $A \cup B' = \{a, b, d\}$, $A \times B' = \{(a, a), (a, d), (b, a), (b, d)\}$ **2. a.** $A \cap B'$ **b.** $A \times B$ **c.** $E' \cup F$ **d.** $(P \cap E' \cap Q)'$ or $P' \cup E \cup Q'$ **3. a.** $n(A \cup B) = n(A) + n(B) - n(A \cap B)$; 16 **b.** $n(A \cup B) = n(A) + n(B) - n(A \cap B)$, $n(C') = n(S) - n(C)$; 100 **c.** $n(A \times B) = n(A)n(B)$; 15 **d.** $n(A \times B) = n(A)n(B)$, $n(A \cup B) = n(A) + n(B) - n(A \cap B)$, $n(A') = n(S) - n(A)$; 21 **4. a.** $2C(4, 2)C(4, 3)$ **b.** $C(12, 1)C(4, 2)C(11, 3)C(4, 1)C(4, 1)C(4, 1)$ **c.** $C(12, 1)C(4, 3)C(11, 2)C(4, 1)C(4, 1)$ **d.** $C(4, 1)C(10, 1)$ **5. a.** $C(12, 5) = 792$ **b.** $C(4, 4)C(8, 1) = 8$ **c.** $C(12, 5) - C(4, 4)C(8, 1) = 784$ **d.** $C(3, 2)C(9, 3) + C(3, 3)C(9, 2) = 288$ **e.** $C(4, 0)C(5, 5) + C(4, 1)C(5, 4) = 21$ **6. a.** The set of books that are either science fiction or stored in Texas (or both); $n(S \cup T) = 112{,}000$ **b.** The set of horror books in California; $n(H \cap C) = 12{,}000$ **c.** The set of books that are either stored in California or not science fiction; $n(C \cup S') = 175{,}000$ **d.** The romance books stored in Texas together with the horror books; $n[(R \cap T) \cup H] = 59{,}000$ **e.** The romance books that are also horror books or stored in Texas; $n[R \cap (T \cup H)] = 20{,}000$ **f.** The science fiction books stored in Washington together with the horror books not stored in California; $n[(S \cap W) \cup (H \cap C')] = 37{,}000$ **7. a.** 1000 **b.** 3400 **c.** FarmerBooks.com; 1800 **d.** FarmerBooks.com; 5400 **e.** JungleBooks.com; 3500 **8. a.** $26 \times 26 \times 26 = 17{,}576$ **b.** $26 \times 25 \times 24 = 15{,}600$ **c.** $26 \times 25 \times 9 \times 10 = 58{,}500$ **d.** Two letters, four digits; 2,948,400 **9. a.** 60,000 **b.** 28,000 **c.** 19,600

Chapter 7

Exercises 7.1

1. $S = \{$HH, HT, TH, TT$\}$; $E = \{$HH, HT, TH$\}$ **3.** $S = \{$HHH, HHT, HTH, HTT, THH, THT, TTH, TTT$\}$; $E = \{$HTT, THT, TTH, TTT$\}$

5. $$S = \left\{\begin{matrix} (1,1) & (1,2) & (1,3) & (1,4) & (1,5) & (1,6) \\ (2,1) & (2,2) & (2,3) & (2,4) & (2,5) & (2,6) \\ (3,1) & (3,2) & (3,3) & (3,4) & (3,5) & (3,6) \\ (4,1) & (4,2) & (4,3) & (4,4) & (4,5) & (4,6) \\ (5,1) & (5,2) & (5,3) & (5,4) & (5,5) & (5,6) \\ (6,1) & (6,2) & (6,3) & (6,4) & (6,5) & (6,6) \end{matrix}\right\};$$

$E = \{(1, 4), (2, 3), (3, 2), (4, 1)\}$

7. $$S = \left\{\begin{matrix} (1,1) & (1,2) & (1,3) & (1,4) & (1,5) & (1,6) \\ & (2,2) & (2,3) & (2,4) & (2,5) & (2,6) \\ & & (3,3) & (3,4) & (3,5) & (3,6) \\ & & & (4,4) & (4,5) & (4,6) \\ & & & & (5,5) & (5,6) \\ & & & & & (6,6) \end{matrix}\right\};$$

$E = \{(1, 3), (2, 2)\}$

9. S as in Exercise 7; $E = \{(2, 2), (2, 3), (2, 5), (3, 3), (3, 5), (5, 5)\}$ **11.** $S = \{$m, o, z, a, r, t$\}$; $E = \{$o, a$\}$ **13.** $S = \{$(s, o), (s, r), (s, e), (o, s), (o, r), (o, e), (r, s), (r, o), (r, e), (e, s), (e, o), (e, r)$\}$; $E = \{$(o, s), (o, r), (o, e), (e, s), (e, o), (e, r)$\}$ **15.** $S = \{01, 02, 03, 04, 10, 12, 13, 14, 20, 21, 23, 24, 30, 31, 32, 34, 40, 41, 42, 43\}$; $E = \{10, 20, 21, 30, 31, 32, 40, 41, 42, 43\}$ **17.** $S = \{$domestic car, imported car, van, antique car, antique truck$\}$; $E = \{$van, antique truck$\}$ **19. a.** All sets of 4 gummy bears chosen from the packet of 12 **b.** All sets of 4 gummy bears in which 2 are strawberry and 2 are black currant **21. a.** All lists of 14 people chosen from 20 **b.** All lists of 14 people chosen from 20, in which Colin Powell occupies the first position. **23.** $A \cap B$; $n(A \cap B) = 1$ **25.** B'; $n(B') = 33$ **27.** $B' \cap D'$; $n(B' \cap D') = 2$ **29.** $C \cup B$; $n(C \cup B) = 12$ **31.** $W \cap I$ **33.** $E \cup I'$ **35.** $I \cup (W \cap E')$ **37.** $E = \{$New England, Pacific, Middle Atlantic$\}$ **39.** $E \cup F$ is the event that you choose a region that saw an increase in housing prices of 9% or more or is on the East Coast; $E \cup F = \{$Pacific, New England, Middle Atlantic, South Atlantic$\}$. $E \cap F$ is the event that you choose a region that saw an increase in housing prices of 9% or more and is on the East Coast; $E \cap F = \{$New England, Middle Atlantic$\}$. **41. a.** Mutually exclusive **b.** Not mutually exclusive **43.** $S \cap N$ is the event that an author is successful and new. $S \cup N$ is the event that an author is either successful or new. $n(S \cap N) = 5$; $n(S \cup N) = 45$ **45.** N and E **47.** $S \cap N'$ is the event that an author is successful but not a new author. $n(S \cap N') = 25$ **49.** 31.25%; 83.33% **51.** $V \cap I'$; $n(V \cap I') = 15$ **53.** 80; the number of stocks that were either not pharmaceutical stocks or were unchanged in value after a year (or both). **55.** P and E, P and I, E and I, N and E, V and N, V and D, N and D **57.** 3/8; the fraction of Internet stocks that increased in value **59. a.** $E' \cap H$ **b.** $E \cup H$ **c.** $(E \cup G)' = E' \cap G'$ **61. a.** $\{9\}$ **b.** $\{6\}$ **63. a.** The dog's "fight" drive is weakest. **b.** The dog's "fight" and "flight" drives are either both strongest or both weakest. **c.** Either the dog's "fight" drive is strongest or its "flight" drive is strongest. **65.** $C(6, 4) = 15$; $C(1, 1)C(5, 3) = 10$ **67. a.** $n(S) = P(7, 3) = 210$ **b.** $E \cap F$ is the event that Celera wins and Electoral College is in second or third place. In other words, it is the set of all lists of three horses in which Celera is first and Electoral College is second or third. $n(E \cap F) = 10$. **69.** $C(8, 3) = 56$ **71.** $C(4, 1)C(2, 1)C(2, 1) = 16$ **73.** Subset of the sample space **75.** E and F do not both occur. **77.** True. Consider the following experiment: Select an element of the set S at random. **79.** Answers may vary. Cast a die and record the remainder when the number facing up is divided by 2. **81.** Yes. For instance, $E = \{(2, 5), (5, 1)\}$ and $F = \{(4, 3)\}$ are two such events.

Exercises 7.2

1. .4 **3.** .8

5.

Outcome	HH	HT	TH	TT
Probability	.275	.2375	.3	.1875

7. .575 **9.** The second coin *seems* slightly biased in favor of heads, since heads comes up approximately 58% of the time. On the other hand, it is conceivable that the coin is fair and that heads came up 58% of the time purely by chance. Deciding which conclusion is more reasonable requires some knowledge of inferential statistics. **15. a.** .10 **b.** .94

17. a.

Test Rating	3	2	1	0
Probability	.1	.4	.4	.1

b. .5

19. a. College degree, Internet user: .19; college degree, nonuser: .07; no college degree, Internet user: .25; no college degree, nonuser: .49 **b.** .45 (.44 if rounded answers are used) **21.** .25 **23.** .2 **25.** .7 **27.** 5/6 **29.** 5/16

31.

Outcome	U	C	R
Probability	.2	.64	.16

33.

Conventional	No pesticide	Single pesticide	Multiple pesticide
Probability	.27	.13	.60
Organic	No pesticide	Single pesticide	Multiple pesticide
Probability	.77	.13	.10

35.

Outcome	Low	Middle	High
Probability	.4	.3	.3

37. P(false negative) = 10/400 = .025, P(false positive) = 10/200 = .05 **41.** The fraction of times E occurs **43.** For a (large) number of days, record the temperature prediction for the next day and then check the actual temperature the next day. Record whether the prediction was accurate (within, say, 2°F of the actual temperature). The fraction of times the prediction was accurate is the estimated probability. **45.** He is wrong. It is possible to have a run of losses of any length. Tony may have grounds to *suspect* that the game is rigged but no proof.

Exercises 7.3

1. $P(E) = 1/4$ **3.** $P(E) = 1$ **5.** $P(E) = 3/4$ **7.** $P(E) = 3/4$ **9.** $P(E) = 1/2$ **11.** $P(E) = 1/9$ **13.** $P(E) = 0$ **15.** $P(E) = 1/4$ **17.** 1/12; {(4, 4), (2, 3), (3, 2)}

19.

Outcome	1	2	3	4	5	6
Probability	$\frac{1}{9}$	$\frac{2}{9}$	$\frac{1}{9}$	$\frac{2}{9}$	$\frac{1}{9}$	$\frac{2}{9}$

$P(\{1, 2, 3\}) = \frac{4}{9}$

21.

Outcome	1	2	3	4
Probability	$\frac{8}{15}$	$\frac{4}{15}$	$\frac{2}{15}$	$\frac{1}{15}$

23. $P(e) = .2$ **a.** .9 **b.** .95 **c.** .1 **d.** .8 **25.** .13 **27.** .42 **29.** .55

31.

Outcome	Hispanic or Latino	White (not Hispanic)	African American	Asian	Other
Probability	.42	.37	.09	.08	.04

P(neither White nor Asian) = .55

33. 960 **35. a.** S = {stock market success, sold to other concern, fail}

b.

Outcome	Stock market success	Sold to other concern	Fail
Probability	.2	.3	.5

c. .5 **37.** .256 **39.** .515 **41.** .498 **43.** .431 **45.** .125

47. a.

Outcome	NAFTA	Asia	Europe
Probability	$\frac{5}{17}$	$\frac{7}{34}$	$\frac{1}{2}$

b. 17 million **49.** $P(1) = P(6) = 1/10$; $P(2) = P(3) = P(4) = P(5) = 1/5$, $P(\text{odd}) = 1/2$ **51.** $P(1, 1) = P(2, 2) = \ldots = P(6, 6) = 1/66$; $P(1, 2) = \ldots = P(6, 5) = 1/33$, $P(\text{odd sum}) = 6/11$ **53.** $P(2) = 15/38$; $P(4) = 3/38$, $P(1) = P(3) = P(5) = P(6) = 5/38$, $P(\text{odd}) = 15/38$ **55.** $C(6, 4) = 15$; $C(1, 1)C(5, 3) = 10$; 2/3 **57.** Estimated; theoretical; number of trials gets larger. **59.** Wrong. For a pair of fair dice, the theoretical probability of a pair of matching numbers is 1/6, as Ruth says. However, it is quite possible, although not very likely, that if you cast a pair of fair dice 20 times, you will never obtain a matching pair (in fact, there is approximately a 2.6% chance that this will happen). In general, a nontrivial claim about theoretical probability can never be absolutely validated or refuted experimentally. All we can say is that the evidence suggests that the dice are not fair. **61.** Zero. According to the assumption, no matter how many thunderstorms occur, lightning cannot strike your favorite spot more than once, and so, after n trials the estimated probability will never exceed $1/n$, and so will approach zero as the number of trials gets large. **63.** It is an estimated probability. Roughly, the weather service is saying that half the time these weather conditions were seen before, it rained. **65.** Wrong. The assertion amounts to the claim that the estimated probability of such an event must always be zero. What must be true is that, as the number of trials gets larger, the estimated probability of the event must approach zero. That does not mean that the estimated probability *is* zero.

Exercises 7.4

1. 1/42 **3.** 7/9 **5.** 1/7 **7.** 1/2 **9.** 41/42 **11.** 1/21 **13.** 2/7 **15.** 2/7 **17.** .4226 **19.** .0475 **21.** .0020 **23.** Probability of being a big winner = $1/C(50, 5)$ = 1/2,118,760 ≈ .000 000 472. Probability of being a small-fry winner = $C(5, 4)C(45, 1)/C(50, 5)$ = 225/2,118,760 ≈ .000 106 194. Probability of being either a winner or a small-fry winner = 226/2,118,760 ≈ .000 106 666. **25. a.** $C(600, 300)/C(700, 400)$ **b.** $C(699, 399)/C(700, 400)$ or 400/700 **27.** $P(10, 3)/10^3$ = 18/25 = .72 **29.** $1/27^{39}$ **31.** $8!/8^8$ **33.** 1/8 **35.** $1/(2^8 \times 5^5 \times 5!)$ **37.** 1/8 **39.** 37/10,000 **41. a.** $C(6, 1)C(6, 1)C(10, 2)C(8, 5)$ = 90,720 **b.** $C(1, 1)C(1, 1)C(10, 2)C(8, 5) + C(5, 1)C(6, 1)C(9, 2)C(7, 5)$ = 25,200 **c.** 25,200/90,720 = 25/90 ≈

.28 **43.** 1/7 **45.** The four outcomes listed are not equally likely; for example, (red, blue) can occur in four ways. The methods of this section yield a probability for (red, blue) of $C(2, 2)/C(4, 2) = 1/6$ **47.** No. If we do not pay attention to order, the probability is $C(5, 2)/C(9, 2) = 10/36 = 5/18$. If we do pay attention to order, the probability is $P(5, 2)/P(9, 2) = 20/72 = 5/18$ again. The difference between permutations and combinations cancels when we compute the probability.

Exercises 7.5

1. .65 **3.** .1 **5.** .7 **7.** .4 **9.** .25 **11.** 1.0 **13.** .3 **15.** 1.0 **17.** Yes **19.** No; $P(A \cup B)$ should be $\leq P(A) + P(B)$. **21.** Yes **23.** No; $P(A \cup B)$ should be $\geq P(B)$. **25.** .46 **27.** 5/6 **29.** 32% **31.** 84% **33.** .09 **35.** .87 **37.** All of them **39.** (C) **41.** .0151 **43. a.** .83 **b.** .67 **c.** .64 **d.** .19 **45.** .884 **47.** They are mutually exclusive. **49.** Let $S = \{1, 2, 3\}$ with $P(1) = 1/2$, $P(2) = 0$, and $P(3) = 1/2$. Let $A = \{1, 2\}$ and $B = \{2, 3\}$. Then $A \cap B \neq \varnothing$, but $P(A) + P(B) = P(A \cup B)$. **51.** When $A \cap B = \varnothing$ we have $P(A \cap B) = P(\varnothing) = 0$, so $P(A \cup B) = P(A) + P(B) - P(A \cap B) = P(A) + P(B) - 0 = P(A) + P(B)$. **53.** $P(A \cup B \cup C) = P(A) + P(B) + P(C) - P(A \cap B) - P(A \cap C) - P(B \cap C) + P(A \cap B \cap C)$

Exercises 7.6

1. 1/10 **3.** 1/5 **5.** 2/9 **7.** 1/84 **9.** 5/21 **11.** 24/175

13. **15.**

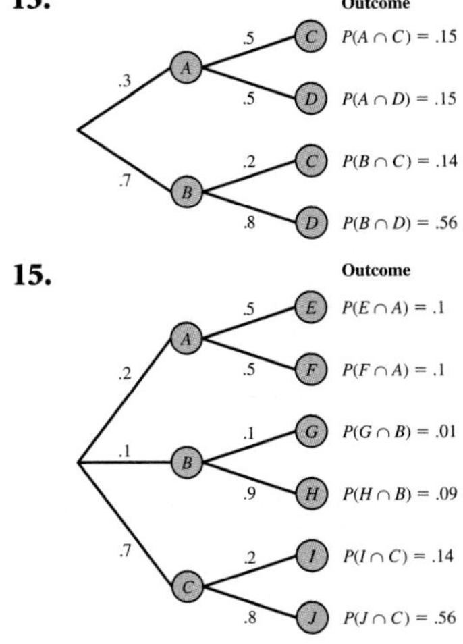

17. (B) **19.** (C) **21.** $\frac{1}{2} \cdot \frac{1}{2} = \frac{1}{4}$, independent **23.** $\frac{5}{18} \cdot \frac{1}{2} \neq \frac{1}{9}$, dependent **25.** $\frac{25}{36} \cdot \frac{5}{18} \neq \frac{2}{9}$, dependent **27.** $(1/2)^{11} = 1/2048$ **29.** $6/7 \approx .86$ **31.** .00015 **33.** .106 **35.** 5/6 **37.** 3/4 **39.** 11/16 **41.** 11/14 **43. a.** .59 **b.** \$35,000 or more: $P(\text{Internet user} \mid < \$35{,}000) \approx .27 < P(\text{Internet user} \mid \geq \$35{,}000) \approx .59$ **45.** .77 **47.** .31 **49.** .37 **51.** .97 **53.** The claim is false. The probability that an unemployed person has 1 to 3 years of college is .25, whereas the corresponding figure for an employed person is .28. **55.** $P(K \mid D) = 1.31P(K \mid D')$ **57.** $P(R \mid J)$ **59.** (D) **61. a.** .000 057 **b.** .015 043

63.

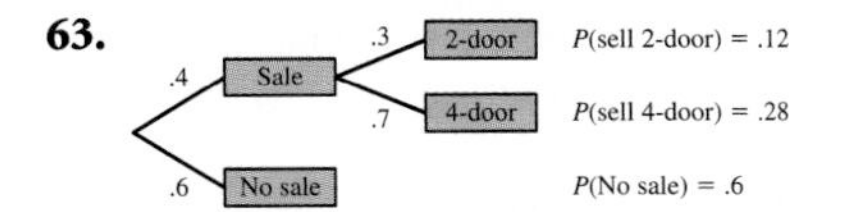

65.

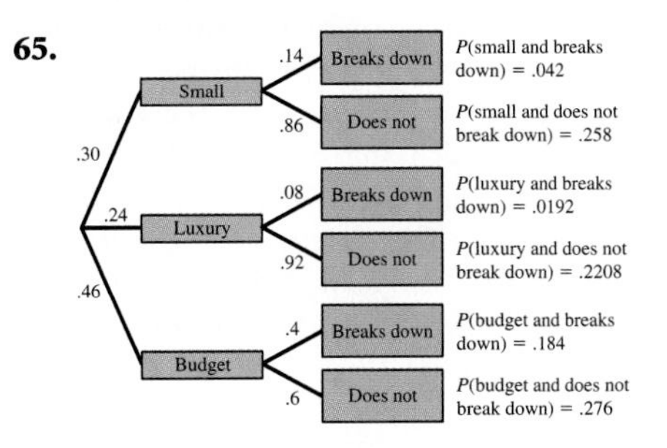

67. .25

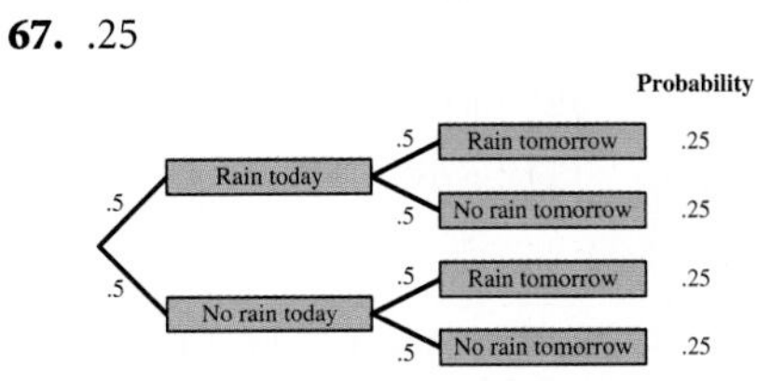

69. 11% **71.** .631 **73.** Not independent; $P(\text{giving up} \mid \text{used brand X}) = .1$ is larger than $P(\text{giving up})$ **75. a.** $P(I \mid T) > P(I)$ **b.** It was ineffective. **77.** The probability you seek is $P(E \mid F)$, or should be. If, for example, you were going to place a wager on whether E occurs or not, it is crucial to know that the sample space has been reduced to F (you know that F did occur). If you base your wager on $P(E)$ rather than $P(E \mid F)$, you will misjudge your likelihood of winning. **79.** Answers will vary. Here is a simple one. E: the first toss is a head; F: the second toss is a head; G: the third toss is a head. **81.** If $A \subseteq B$ then $A \cap B = A$, so $P(A \cap B) = P(A)$ and $P(A \mid B) = P(A \cap B)/P(B) = P(A)/P(B)$. **83.** Your friend is correct. If A and B are mutually exclusive, then $P(A \cap B) = 0$. On the other hand, if A and B are independent, then $P(A \cap B) = P(A)P(B)$. Thus, $P(A)P(B) = 0$. If a product is 0, then one of the factors must be 0, so either $P(A) = 0$ or $P(B) = 0$. Thus, it cannot be true that A and B are mutually exclusive, have nonzero probabilities, and are independent all at the same time. **85.** $P(A' \cap B') = 1 - P(A \cup B) = 1 - [P(A) + P(B) - P(A \cap B)] = 1 - [P(A) + P(B) - P(A)P(B)] = [1 - P(A)][1 - P(B)] = P(A')P(B')$

Exercises 7.7

1. .4 **3.** .7887 **5.** .7442 **7.** .1163 **9.** .1724 **11.** .7097 **13. a.** .029 **b.** .057; the probability is almost doubled. **15. a.** 14.43%; **b.** 19.81% of single homeowners have pools. Thus, they should go after the single homeowners. **17.** 26.8% **19.** 9 **21.** .9310 **23.** 1.76% **25.** .20 **27.** 90%. **29.** 12% **31.** .40% **33.** K: child killed; D: air bag deployed; $P(K \mid D) = 1.31P(K \mid D')$; $P(D \mid K) = 1.31(.25)/[1.31(.25) + .75] = .30$ **35.** Show him an example like Example 1 of this section, in which $P(T \mid A) = .95$ but $P(A \mid T) \approx .64$. **37.** Suppose the steroid test gives 10% false negatives and only 0.1% of the tested population uses steroids. Then the probability that an athlete uses steroids, given that he or she has tested positive, is $[(.9)(.001)]/[(.9)(.001) + (.01)(.999)] \approx .083$. **39.** The reasoning is flawed. Let A be the event that a Democrat agrees with Safire's column and let F and M be the events that a Democrat reader is female and male, respectively. Then A. D. makes the following argument: $P(M \mid A) = .9$, $P(F \mid A') = .9$. Therefore, $P(A \mid M) = .9$. According to Bayes' theorem, we cannot conclude anything about $P(A \mid M)$ unless we know $P(A)$, the percentage of all Democrats who agreed with Safire's column. This was not given. **41.** Draw a tree in which the first branching shows which of R_1, R_2, or R_3 occurred, and the second branching shows which of T or T' then occurred. There are three final outcomes in which T occurs: $P(R_1 \cap T) = P(T \mid R_1)P(R_1)$, $P(R_2 \cap T) = P(T \mid R_2)P(R_2)$, and $P(R_3 \cap T) = P(T \mid R_3)P(R_3)$. In only one of these, the first, does R_1 occur. Thus,

$$P(R_1 \mid T) = \frac{P(R_1 \cap T)}{P(T)}$$

$$= \frac{P(T \mid R_1)P(R_1)}{P(T \mid R_1)P(R_1) + P(T \mid R_2)P(R_2) + P(T \mid R_3)P(R_3)}$$

Chapter 7 Review Test

1. a. $n(S) = 8$; $E = \{$HHT, HTH, HTT, THH, THT, TTH, TTT$\}$; $P(E) = 7/8$ **b.** $n(S) = 16$; $E = \{$HTTT, THTT, TTHT, TTTH, TTTT$\}$; $P(E) = 5/16$ **c.** $n(S) = 36$; $E = \{(1, 6), (2, 5), (3, 4), (4, 3), (5, 2), (6, 1)\}$; $P(E) = 1/6$ **d.** $n(S) = 6$; $E = \{2\}$; $P(E) = 1/8$ **e.** $n(S) = 21$; $E = \{(1, 6), (2, 5), (3, 4)\}$; $P(E) = 1/6$ **2. a.** .76 **b.** .9 **c.** .25 **d.** .8 **3. a.** .5 **b.** 7/12 **c.** 7/15 **d.** .4 **4. a.** 8/792 **b.** 48/792 **c.** 1 **d.** 288/792 **e.** 21/792 **5. a.** $C(8, 5)/C(52, 5)$ **b.** $C(12, 5)/C(52, 5)$ **c.** $C(4, 3)C(1, 1)C(3, 1)/C(52, 5)$ **d.** $C(6, 1)C(5, 1)C(4, 3)C(4, 2)/C(52, 5)$ **e.** $C(13, 1)C(2, 1)C(12, 2)/C(52, 5)$ **6. a.** 1/5; dependent **b.** 0; dependent **c.** 1/6; independent **d.** 1/5; dependent **e.** 1; dependent **f.** 1/3; dependent **7. a.** 14/25 **b.** 3/40 **c.** 15/94 **d.** 5/11 **e.** 79/167 **f.** 6/11 **8. a.** 98% **b.** 6.9% **c.** .931 **d.** $P(H \cap C)$, since $P(H \mid C) > P(H)$ gives $P(H \cap C) > P(H)P(C)$. **e.** 0.75% **9. a.** 16 **b.** .0049 **c.** .2462

Chapter 8

Exercises 8.1

1. Finite; $\{2, 3, \ldots, 12\}$ **3.** Discrete infinite; $\{0, 1, -1, 2, -2, \ldots\}$ (negative profits indicate loss) **5.** Continuous; X can assume any value between 0 and 60. **7.** Finite; $\{0, 1, 2, \ldots, 10\}$ **9.** Discrete infinite; $\{k/1, k/4, k/9, k/16, \ldots\}$ **11. a.** $S = \{$HH, HT, TH, TT$\}$ **b.** X is the rule that assigns to each outcome the number of tails.

c.

Outcome	HH	HT	TH	TT
Value of X	0	1	1	2

13. a. $S = \{(1, 1), (1, 2), \ldots, (1, 6), (2, 1), (2, 2), \ldots, (6, 6)\}$ **b.** X is the rule that assigns to each outcome the sum of the two numbers.

c.

Outcome	(1, 1)	(1, 2)	(1, 3)	. . .	(6, 6)
Value of X	2	3	4	. . .	12

15. a. $S = \{(4, 0), (3, 1), (2, 2)\}$ [listed in order (red, green)] **b.** X is the rule that assigns to each outcome the number of red marbles.

c.

Outcome	(4, 0)	(3, 1)	(2, 2)
Value of X	4	3	2

17. a. S = the set of students in the study group **b.** X is the rule that assigns to each student his or her final exam score. **c.** The values of X, in the order given, are 89%, 85%, 95%, 63%, 92%, 80%. **19. a.** $P(X = 8) = P(X = 6) = .3$ **b.** .7

21.

x	1	2	3	4	5	6
P(X = x)	$\frac{1}{6}$	$\frac{1}{6}$	$\frac{1}{6}$	$\frac{1}{6}$	$\frac{1}{6}$	$\frac{1}{6}$

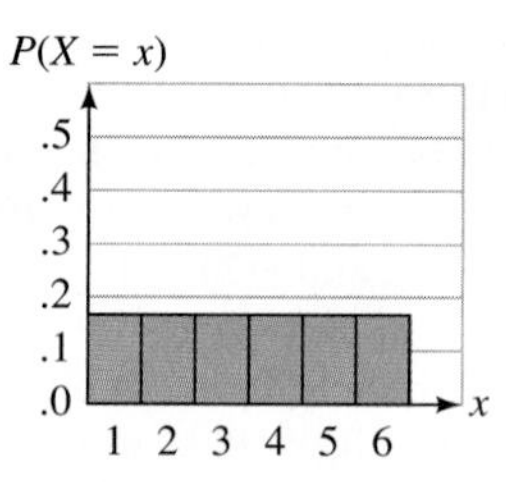

23.

x	0	1	4	9
P(X = x)	$\frac{1}{8}$	$\frac{3}{8}$	$\frac{3}{8}$	$\frac{1}{8}$

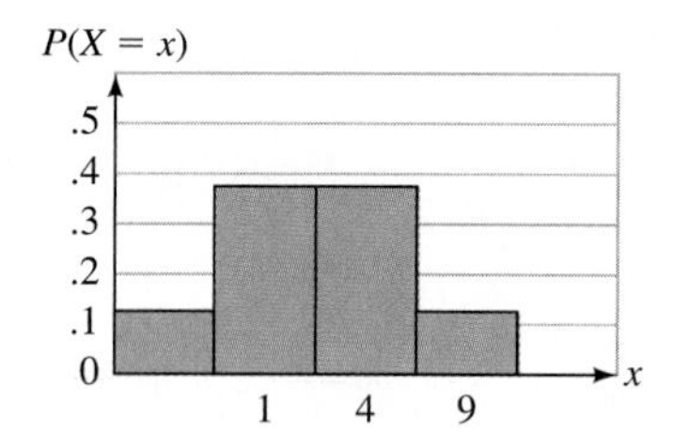

25.

x	2	3	4	5	6	7	8	9	10	11	12
P(X = x)	$\frac{1}{36}$	$\frac{2}{36}$	$\frac{3}{36}$	$\frac{4}{36}$	$\frac{5}{36}$	$\frac{6}{36}$	$\frac{5}{36}$	$\frac{4}{36}$	$\frac{3}{36}$	$\frac{2}{36}$	$\frac{1}{36}$

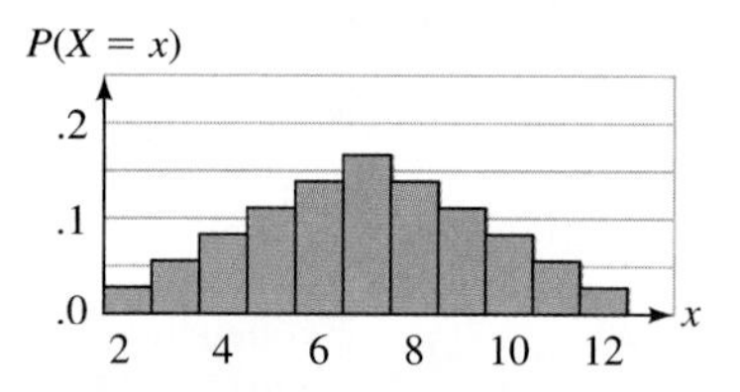

27.

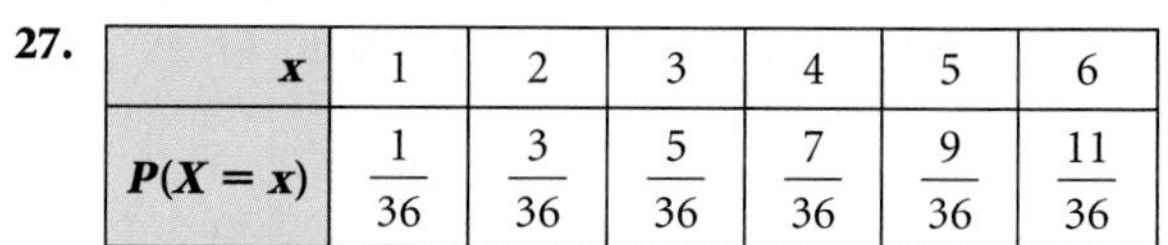

x	1	2	3	4	5	6
P(X = x)	$\frac{1}{36}$	$\frac{3}{36}$	$\frac{5}{36}$	$\frac{7}{36}$	$\frac{9}{36}$	$\frac{11}{36}$

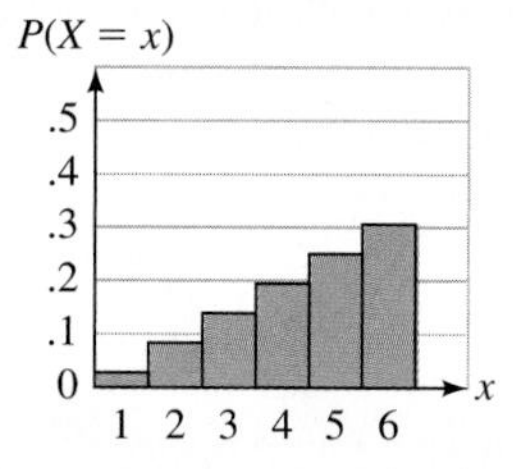

29. a. 2000, 3000, 4000, 5000, 6000, 7000, 8000 (7000 is optional)

b.

x	2000	3000	4000	5000	6000	7000	8000
Freq.	2	1	1	1	2	0	3
P(X = x)	.2	.1	.1	.1	.2	0	.3

c. $P(X \leq 5000) = .5$

31. The random variable is X = pollen count on a given day.

x	15	40	60	85	115
P(X = x)	$\frac{5}{23}$	$\frac{5}{23}$	$\frac{3}{23}$	$\frac{2}{23}$	$\frac{8}{23}$

33.

Class	1.1–2.0	2.1–3.0	3.1–4.0
Freq.	4	7	9

x	1.5	2.5	3.5
P(X = x)	.20	.35	.45

35. 1910/2000 = 95.5%

37.

x	3	2	1	0
P(X = x)	.0625	.6875	.125	.125

39. .75. The probability that a randomly selected small car is rated Good or Acceptable is .75. **41.** $P(Y \geq 2) = .50$, $P(Z \geq 2) \approx .53$, suggesting that medium SUVs are safer than small SUVs in frontal crashes. **43.** Small cars **45.** .375

47.

x	1	2	3	4
P(X = x)	$\frac{4}{35}$	$\frac{18}{35}$	$\frac{12}{35}$	$\frac{1}{35}$

; $P(X \geq 2) = 31/35 \approx .886$

49. a. Probability distribution:

x	2.5	7.5	12.5	20	30	42.5	62.5	87.5	160
P(X = x)	.029	.061	.07	.134	.125	.155	.189	.104	.134

b. .427; histogram:

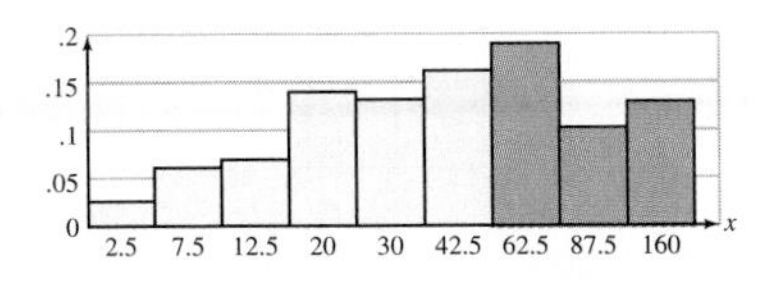

51. Answers will vary. **53.** No; for instance, if X is the number of times you must toss a coin until heads comes up, then X is infinite but not continuous. **55.** By measuring the values of X for a large number of outcomes and then using the estimated probability (relative frequency) **57.** Here is an example: let X be the number of days a diligent student waits before beginning to study for an exam scheduled in 10 days' time. **59.** Answers will vary. If we are interested in exact page counts, then the number of possible values is very large and the values are (relatively speaking) close together, so using a continuous random variable might be advantageous. In general, the finer and more numerous the measurement classes, the more likely it becomes that a continuous random variable could be advantageous.

Exercises 8.2

1. .0729 **3.** .59049 **5.** .00001 **7.** .99144 **9.** .00856 **11.** .08192 **13.** .90112 **15.** .0016 **17.** .720896

19.

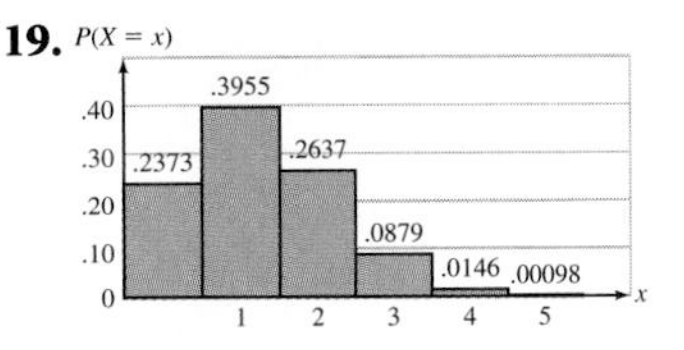

21.

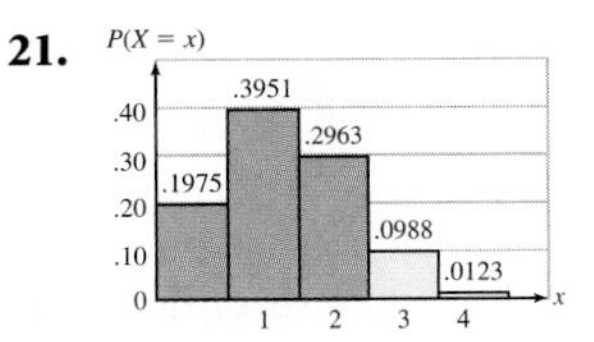

$P(X \leq 2) = .8889$

23. .2637 **25.** .8926 **27.** .875 **29. a.** .0081 **b.** .08146 **31.** .41 **33.** .000298 **35.** .8321 **37. a.** 21 **b.** 20 **c.** The graph for $n = 50$ trials is more widely distributed than the graph for $n = 20$. **39.** 69 trials **41.** $.562 \times 10^{-5}$ **43.** No; in a sequence of Bernoulli trials, the occurrence of one success does not effect the probability of success on the next attempt. **45.** No; if life is a sequence of Bernoulli trials, then the occurrence of one misfortune ("success") does not affect the probability of a misfortune on the next trial. Hence, misfortunes may very well not "occur in threes." **47.** The probability of selecting a red marble changes after each selection, as the number of marbles left in the bag decreases. This violates the requirement that, in a sequence of Bernoulli trials, the probability of "success" does not change.

Exercises 8.3

1. $\bar{x} = 6$, median = 5, mode = 5 **3.** $\bar{x} = 3$, median= 3.5, mode = −1 **5.** $\bar{x} = -0.1875$, median = 0.875, every value is a mode **7.** $\bar{x} = 0.2$, median = −0.1, mode = −0.1 **9.** Answers will vary. Two examples are 0, 0, 0, 0, 0, 6 and 0, 0, 0, 1, 2, 3. **11.** 0.9 **13.** 21 **15.** −0.1 **17.** 1 **19.** 4.472 **21.** 2.667 **23.** 2 **25.** 0.385 **27.** $\bar{x} = 5000$, $m = 5500$; 5500 **29.** All three are equal to 20. The average annual percent growth in computer sales for the years 1994–2000 was 20%. It was more than 20% in as many years as it was below 20%, and the most likely rate of growth was also 20%. **31. a.** 6.5; there were an average of 6.5 checkout lanes in a supermarket that was surveyed. **b.** $P(X < \mu) = .42$; $P(X > \mu) = .58$, and is thus larger. Most supermarkets have more than the average number of checkout lanes. **33.** 67

35.

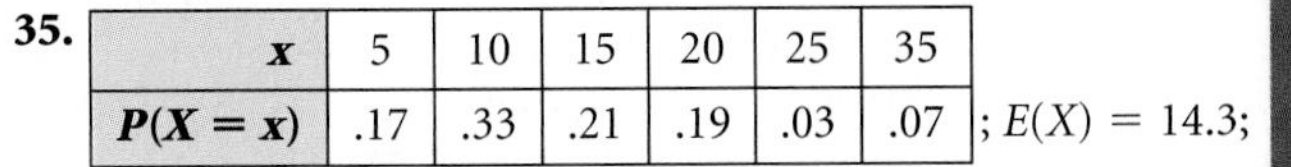

x	5	10	15	20	25	35
P(X = x)	.17	.33	.21	.19	.03	.07

; $E(X) = 14.3$; the average age of a schoolgoer in 1998 was 14.3.

37. $57,000

39.

x	3	2	1	0
$P(X = x)$	.0625	.6875	.125	.125

; $E(X) = 1.6875$

y	3	2	1	0
$P(Y = y)$	.1	.4	.4	.1

; $E(Y) = 1.5$; small cars

41. Large cars **43.** Expect to lose 5.3¢ **45.** 25.2 students **47. a.** 2 defective air bags **b.** 120 air bags
49.

x	1	2	3	4
$P(X = x)$	$\frac{4}{35}$	$\frac{18}{35}$	$\frac{12}{35}$	$\frac{1}{35}$

; $E(X) = 16/7 \approx 2.2857$ tents

51. FastForward, 3.97%; SolidState, 5.51%; SolidState gives the higher expected return. **53.** A loss of \$29,390 **55.** (A) **57.** He is wrong; for example, the collection 0, 0, 300 has mean 100 and median 0. **59.** No. The expected number of times you will hit the dartboard is the average number of times you will hit the bull's-eye per 50 shots; the average of a set of whole numbers need not be a whole number. **61.** Wrong. It might be the case that only a small fraction of people in the class scored better than you but received exceptionally high scores that raised the class average. Suppose, for instance, that ten people are in the class. Four received 100%, you received 80%, and the rest received 70%. Then the class average is 83%, five people have lower scores than you, but only four have higher scores. **63.** No; the mean of a very large sample is only an *estimate* of the population mean. The means of larger and larger samples *approach* the population mean as the sample size increases. **65.** Select a U.S. household at random and let X be the income of that household. The expected value of X is then the population mean of all U.S. household incomes.

Exercises 8.4

1. $s^2 = 29$; $s = 5.39$ **3.** $s^2 = 12.4$; $s = 3.52$ **5.** $s^2 = 6.64$; $s = 2.58$ **7.** $s^2 = 13.01$; $s = 3.61$ **9.** 1.04 **11.** 9.43 **13.** 3.27 **15.** Expected value = 1, variance = 0.5, standard deviation = 0.71 **17.** Expected value = 4.47, variance = 1.97, standard deviation = 1.40 **19.** Expected value = 2.67, variance = 0.36, standard deviation = 0.60 **21.** Expected value = 2, variance = 1.8, standard deviation = 1.34 **23. a.** $\bar{x} = 3$, $s = 3.54$ **b.** [0, 6.54]; we must assume that the population distribution is bell-shaped and symmetric. **25. a.** $\bar{x} = 5000$, $s \approx 2211$ **b.** 2789, 7211, 60% **27. a.** 2.18 **b.** [11.22, 24.28] **c.** 100%; empirical rule **29.** $\mu = 1.5$, $\sigma = 1.43$; 100% **31.** $\mu = \$57{,}000$, $\sigma = \$46{,}000$; income gap = \$93,000 **33. a.** $E(X) = 14.3$, $\sigma \approx 7.7$ **b.** [6.6, 22.0]; about 74% **35.** At most 6.25% **37.** At most 12.5% **39. a.** $\mu = 780$, $\sigma \approx 13.1$ **b.** 754, 806 **41. a.** $\mu = 25.2$ students, $\sigma = 3.05$ **b.** 31 **43. a.** $\mu = 6.5$, $\sigma^2 = 4.0$, $\sigma = 2.0$ **b.** [2.5, 10.5]; three checkout lanes **45.** \$10,700 or less **47.** \$65,300 or more **49.** U.S. **51.** U.S. **53.** 16% **55.** 0–\$76,000 **57.** $\mu = 12.56\%$, $\sigma \approx 1.8885\%$ **59.** 78%; the empirical rule predicts 68%. The associated probability distribution is roughly bell-shaped but not symmetric. **61.** 96%. Chebyshev's rule is valid because it predicts that *at least* 75% of the scores are in this range. **63.** The sample standard deviation is bigger; the formula for sample standard deviation involves division by the smaller term $n - 1$ instead of n, which makes the resulting number larger. **65.** The grades in the first class were clustered fairly close to 75. By Chebyshev's inequality, at least 88% of the class had grades in the range 60–90. On the other hand, the grades in the second class were widely dispersed. The second class had a much wider spread of ability than did the first class. **67.** The variable must take on only the value 10, with probability 1. **69.** $(y - x)/2$

Exercises 8.5

Note: Answers for Section 8.5 were computed using the four-digit table in the appendix and may differ slightly from the more accurate answers generated using technology.
1. .1915 **3.** .5222 **5.** .6710 **7.** .2417 **9.** .8664 **11.** .8621 **13.** .2286 **15.** .3830 **17.** .5028 **19.** .35 **21.** .05 **23.** .5 − .4332 = .0668 **25.** 29,600,000 **27.** 0 **29.** About 6680 **31.** 28% **33.** 5% **35.** The U.S. **37.** Wechsler. Since this test has a smaller standard deviation, a greater percentage of scores fall within 20 points of the mean. **39.** This is surprising because the time between failures was more than 5 standard deviations away from the mean, which happens with an extremely small probability. **41.** .6103 **43.** .6103 × .5832 = .3559 **45.** .6255 **47.** .7257 **49.** .8708 **51.** .0029 **53.** Probability that a person will say Goode = .54. Probability that Goode polls more than 52% ≈ .8925. **55.** 23.4 **57.** When the distribution is normal **59.** Neither; they are equal. **61.** $1/(b - a)$ **63.** A normal distribution with standard deviation 0.5, since it is narrower near the mean, but must enclose the same amount of area as the standard curve, and so it must be higher.

Chapter 8 Review Test

1. a.

x	0	1	2
$P(X = x)$	$\frac{1}{4}$	$\frac{1}{2}$	$\frac{1}{4}$

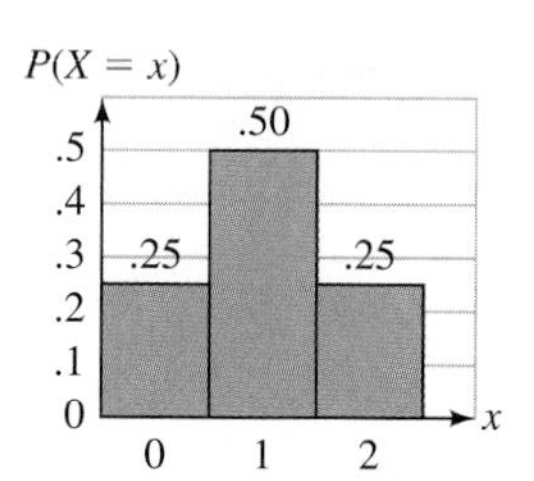

b.

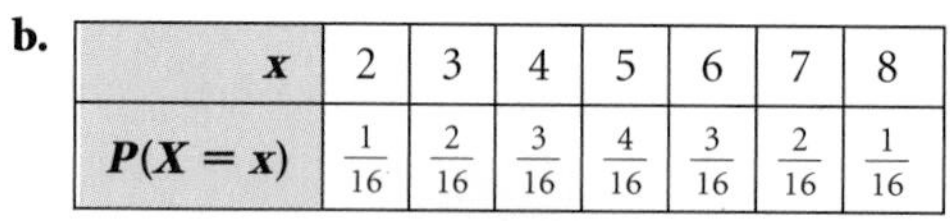

x	2	3	4	5	6	7	8
$P(X = x)$	$\frac{1}{16}$	$\frac{2}{16}$	$\frac{3}{16}$	$\frac{4}{16}$	$\frac{3}{16}$	$\frac{2}{16}$	$\frac{1}{16}$

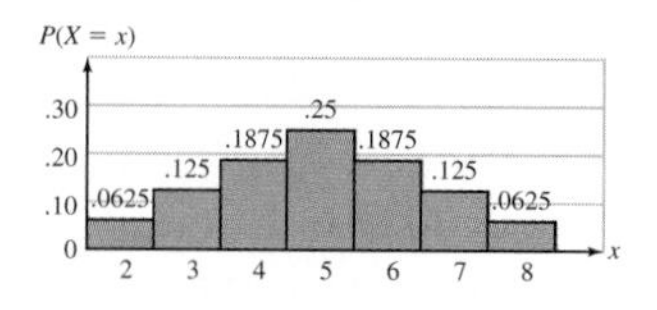

c.

x	15	25	35	45
$P(X = x)$	.482	.386	.116	.016

d.

x	0	1	2	3
$P(X = x)$	.2071	.4439	.2908	.0582

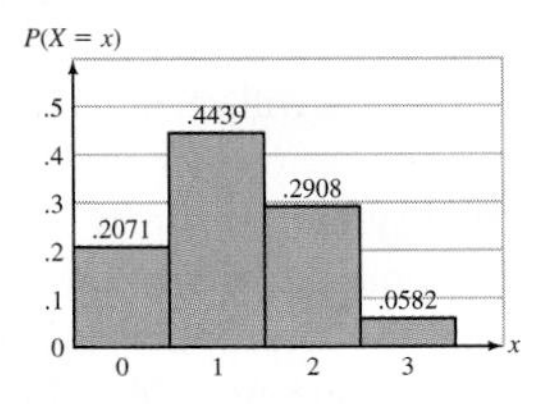

e.

x	0	1	2
$P(X = x)$	$\frac{9}{16}$	$\frac{6}{16}$	$\frac{1}{16}$

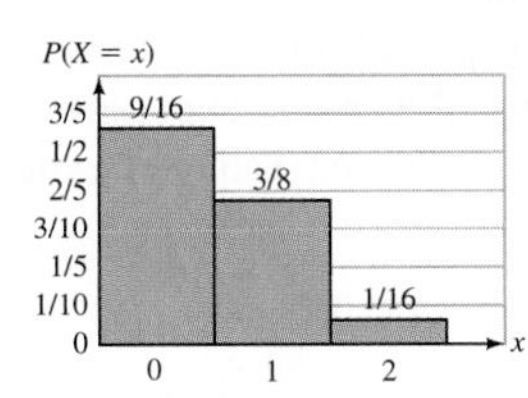

2. a. $\bar{x} = 2$, $m = 2$, $s \approx 2.7386$ **b.** Two examples are 0, 0, 0, 4 and −1, −1, 1, 5. **c.** 2, 2, 2, 2, 2, 2 **d.** An example is −1, −1, −1, 1, 1, 1. **e.** An example is −1, −1, 0, 1, 1.

3. a. .4165 **b.** .9267 **c.** .3232 **d.** .0067 **e.** .7330

4. a.

x	0	1	2	3
$P(X = x)$	$\frac{1}{8}$	$\frac{3}{8}$	$\frac{3}{8}$	$\frac{1}{8}$

; $\mu = 1.5$ $\sigma = 0.8660$; within 2 standard deviations of the mean

b.

x	−3	−2	−1	0	1	2	3
$P(X = x)$	$\frac{1}{16}$	$\frac{2}{16}$	$\frac{3}{16}$	$\frac{4}{16}$	$\frac{3}{16}$	$\frac{2}{16}$	$\frac{1}{16}$

; $\mu = 0$, $\sigma = 1.5811$; within 1.3 standard deviations of the mean **c.** [49.4, 150.6] **d.** [40, 160] **5. a.** .4332 **b.** .0668 **c.** .0358 **d.** .7888 **e.** .3085 **f.** .0000 **6. a.** \$12.15 **b.** 2620 copies **c.** \$27,210 **d.** False; let X = price and Y = weekly sales. Then weekly revenue = XY. However, $27{,}210 \neq 12.15 \times 2620$. In other words, $E(XY) \neq E(X)E(Y)$.

7. a.

x	2	4	6	8	10
$P(X = x)$	.25	.35	.15	.15	.1

; $\mu = 5$, $\sigma = 2.5690$ **b.** Between 2.431 and 7.569 orders per million residents; the empirical rule does not apply because the distribution is not symmetric. **c.** (A) **8. a.** .190 **b.** .867 **c.** .060 **d.** The event that a Mac OS visitor to the site orders books is independent of the event that a Windows visitor to the site orders books. **e.** (.05)(10) + (.10)(20) = 2.5 **9. a.** .284 **b.** .873 **c.** Skin cream, with a probability of approximately .108, compared with hair products, with a probability of approximately .058. **10. a.** Using normal distribution table: 364,000 people; more accurate answer: 378,000 people **b.** 16 **c.** 148

Chapter 9

Exercises 9.1

1. $\begin{bmatrix} \frac{1}{4} & \frac{3}{4} \\ \frac{1}{2} & \frac{1}{2} \end{bmatrix}$

3. $\begin{bmatrix} 0 & 1 \\ \frac{1}{6} & \frac{5}{6} \end{bmatrix}$

5. $\begin{bmatrix} 0 & .8 & .2 \\ .9 & 0 & .1 \\ 0 & 0 & 1 \end{bmatrix}$

7. $\begin{bmatrix} 1 & 0 & 0 \\ 0 & 1 & 0 \\ 0 & 0 & 1 \end{bmatrix}$

9. $\begin{bmatrix} 1 & 0 & 0 & 0 & 0 & 0 \\ \frac{2}{3} & 0 & \frac{1}{3} & 0 & 0 & 0 \\ 0 & \frac{2}{3} & 0 & \frac{1}{3} & 0 & 0 \\ 0 & 0 & \frac{2}{3} & 0 & \frac{1}{3} & 0 \\ 0 & 0 & 0 & \frac{2}{3} & 0 & \frac{1}{3} \\ 0 & 0 & 0 & 0 & 1 & 0 \end{bmatrix}$

11. Yes:

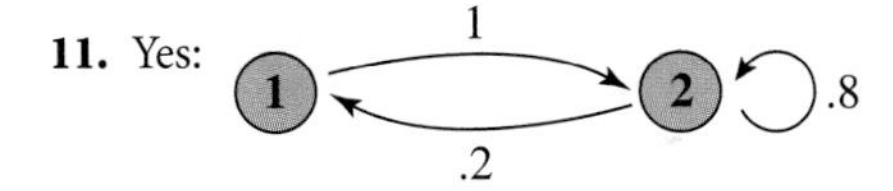

13. Yes:

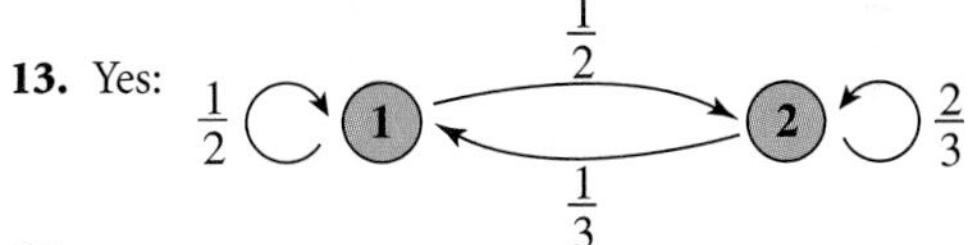

15. no

17. Yes:

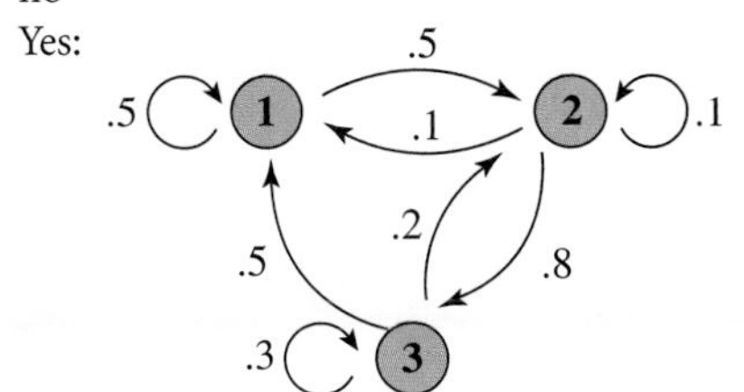

19. Yes:

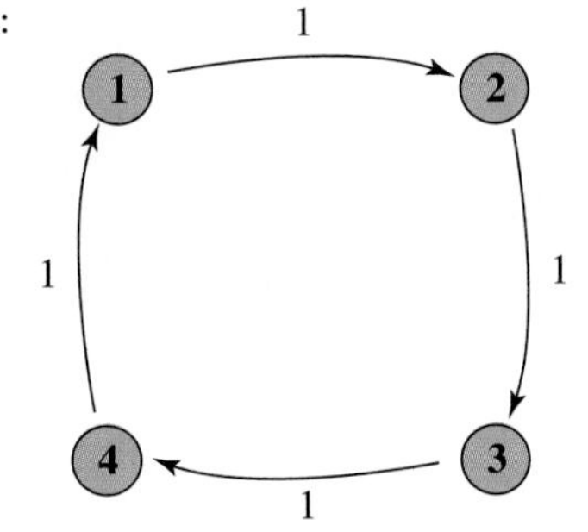

21. 1 = Sorey State, 2 = C&T

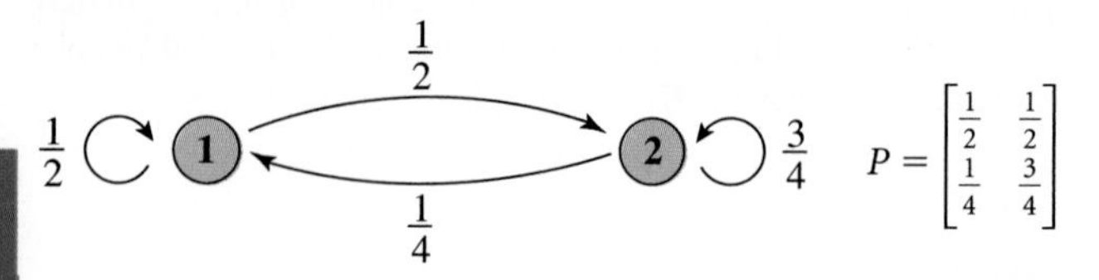

$P = \begin{bmatrix} \frac{1}{2} & \frac{1}{2} \\ \frac{1}{4} & \frac{3}{4} \end{bmatrix}$

23. 1 = not checked in, 2 = checked in

$P = \begin{bmatrix} \frac{2}{5} & \frac{3}{5} \\ 0 & 1 \end{bmatrix}$

25. 1 = user, 2 = nonuser

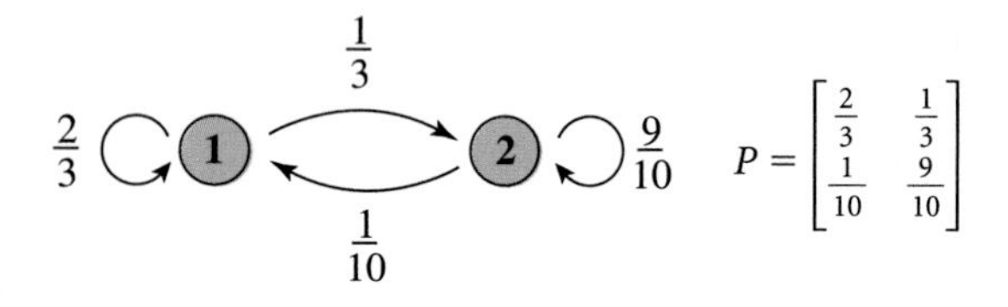

$P = \begin{bmatrix} \frac{2}{3} & \frac{1}{3} \\ \frac{1}{10} & \frac{9}{10} \end{bmatrix}$

27. 1 = male, 2 = female

$P = \begin{bmatrix} \frac{2}{3} & \frac{1}{3} \\ \frac{1}{3} & \frac{2}{3} \end{bmatrix}$

29. 1 = Main Caf, 2 = Math Dept., 3 = Field House

$P = \begin{bmatrix} 0 & \frac{1}{2} & \frac{1}{2} \\ \frac{1}{3} & 0 & \frac{2}{3} \\ \frac{1}{4} & \frac{3}{4} & 0 \end{bmatrix}$

31.

$P = \begin{bmatrix} .9 & .05 & .05 \\ 0 & 1 & 0 \\ 0 & 0 & 1 \end{bmatrix}$

33. $P = \begin{bmatrix} \frac{1}{2} & \frac{1}{2} & 0 & 0 & 0 & 0 \\ \frac{1}{2} & 0 & \frac{1}{2} & 0 & 0 & 0 \\ 0 & \frac{1}{2} & 0 & \frac{1}{2} & 0 & 0 \\ 0 & 0 & \frac{1}{2} & 0 & \frac{1}{2} & 0 \\ 0 & 0 & 0 & \frac{1}{2} & 0 & \frac{1}{2} \\ 0 & 0 & 0 & 0 & \frac{1}{2} & \frac{1}{2} \end{bmatrix}$

35. 1 = Northeast, 2 = Midwest, 3 = South, 4 = West;

$P = \begin{bmatrix} .9886 & .0015 & .0075 & .0024 \\ .0011 & .9900 & .0057 & .0032 \\ .0018 & .0042 & .9897 & .0043 \\ .0017 & .0035 & .0077 & .9871 \end{bmatrix}$

37. 1 = Maxwell Equation, 2 = Full O Choc, 3 = Gauss Jordan;

$P = \begin{bmatrix} .3 & .1 & .6 \\ .25 & .5 & .25 \\ .2 & .2 & .6 \end{bmatrix}$ **39.** Answers will vary. **41.** There are two assumptions made by Markov systems that may not be true about the stock market: the assumption that the transition probabilities do not change over time and the assumption that the transition probability depends only on the current state. **43.** That there is zero probability of entering or staying in that state **45.** Short-Sleeved Winter, Short-Sleeved Spring, Short-Sleeved Summer, Short-Sleeved Fall, Long-Sleeved Winter, Long-Sleeved Spring, Long-Sleeved Summer, Long-Sleeved Fall

Exercises 9.2

1. a. $P^2 = \begin{bmatrix} 1 & 0 \\ 0 & 1 \end{bmatrix}$, $P^3 = \begin{bmatrix} 0 & 1 \\ 1 & 0 \end{bmatrix}$

b. After two steps: [100 0]; after three steps: [0 100]

3. a. $P^2 = \begin{bmatrix} .25 & .75 \\ 0 & 1 \end{bmatrix}$, $P^3 = \begin{bmatrix} .125 & .875 \\ 0 & 1 \end{bmatrix}$

b. After two steps:[25 75]; after three steps: [12.5 87.5]

5. a. $P^2 = \begin{bmatrix} \frac{3}{4} & \frac{1}{4} \\ \frac{1}{2} & \frac{1}{2} \end{bmatrix}$, $P^3 = \begin{bmatrix} \frac{5}{8} & \frac{3}{8} \\ \frac{3}{4} & \frac{1}{4} \end{bmatrix}$

b. After two steps: [10 5]; after three steps: [10 5]

7. a. $P^2 = \begin{bmatrix} \frac{3}{4} & \frac{1}{4} \\ \frac{3}{4} & \frac{1}{4} \end{bmatrix}$, $P^3 = \begin{bmatrix} \frac{3}{4} & \frac{1}{4} \\ \frac{3}{4} & \frac{1}{4} \end{bmatrix}$

b. After two steps: [75 25]; after three steps: [75 25]

9. a. $P^2 = ; \begin{bmatrix} 0 & 0 & 1 \\ 1 & 0 & 0 \\ 0 & 1 & 0 \end{bmatrix}$, $P^3 = \begin{bmatrix} 1 & 0 & 0 \\ 0 & 1 & 0 \\ 0 & 0 & 1 \end{bmatrix}$

b. After two steps:[20 0 0]; after three steps: [0 20 0]

11. a. $P^2 = \begin{bmatrix} .25 & .75 & 0 \\ 0 & 1 & 0 \\ 0 & .75 & .25 \end{bmatrix}$, $P^3 = \begin{bmatrix} .125 & .875 & 0 \\ 0 & 1 & 0 \\ 0 & .875 & .125 \end{bmatrix}$

b. After two steps: [250 750 0]; after three steps: [125 875 0]

13. a. $P^2 = \begin{bmatrix} \frac{1}{3} & \frac{1}{3} & \frac{1}{3} \\ \frac{4}{9} & \frac{4}{9} & \frac{1}{9} \\ 0 & 1 & 0 \end{bmatrix}$, $P^3 = \begin{bmatrix} \frac{4}{9} & \frac{4}{9} & \frac{1}{9} \\ \frac{7}{27} & \frac{16}{27} & \frac{4}{27} \\ \frac{1}{3} & \frac{1}{3} & \frac{1}{3} \end{bmatrix}$ **b.** After two steps: [300 1200 300]; after three steps: [700 700 400]

15. a. $P^2 = \begin{bmatrix} .01 & .99 & 0 \\ 0 & 1 & 0 \\ 0 & .36 & .64 \end{bmatrix}$, $P^3 = \begin{bmatrix} .001 & .999 & 0 \\ 0 & 1 & 0 \\ 0 & .488 & .512 \end{bmatrix}$ **b.** After two steps: [0.01 0.99 0]; after three steps [0.001 0.999 0]

17. a. $P^2 = \begin{bmatrix} .12 & .88 \\ .11 & .89 \end{bmatrix}$, $P^3 = \begin{bmatrix} .11 & .89 \\ .11 & .89 \end{bmatrix}$ **b.** After two steps: [34 266]; after three steps [33.4 266.6]

19. a. $P^2 = \begin{bmatrix} .31 & .23 & .46 \\ .33 & .22 & .46 \\ .30 & .23 & .47 \end{bmatrix}$, $P^3 = \begin{bmatrix} .31 & .23 & .46 \\ .31 & .23 & .46 \\ .31 & .23 & .46 \end{bmatrix}$ **b.** After two steps: [9.38 6.75 13.88]; after three steps [9.28 6.79 13.93]

21. a. $P^2 = \begin{bmatrix} .35 & .25 & .23 & .18 \\ .37 & .23 & .20 & .21 \\ .72 & .11 & .10 & .08 \\ .50 & .25 & .25 & 0 \end{bmatrix}$, $P^3 = \begin{bmatrix} .45 & .21 & .19 & .16 \\ .48 & .20 & .18 & .14 \\ .48 & .23 & .22 & .07 \\ .35 & .25 & .23 & .18 \end{bmatrix}$
b. After two steps: [0.54 0.18 0.16 0.13]; after three steps: [0.46 0.22 0.20 0.11]

23. a. 1 = Sorey State, 2 = C&T; $P = \begin{bmatrix} \frac{1}{2} & \frac{1}{2} \\ \frac{1}{4} & \frac{3}{4} \end{bmatrix}$; 3/8 = 0.375
b. 275 Sorey State, 525 C&T **25. a.** 1 = not checked in; 2 = checked in; $P = \begin{bmatrix} .4 & .6 \\ 0 & 1 \end{bmatrix}$, $P^2 = \begin{bmatrix} .16 & .84 \\ 0 & 1 \end{bmatrix}$, $P^3 = \begin{bmatrix} .064 & .936 \\ 0 & 1 \end{bmatrix}$
b. Hour 1: [20 30]; hour 2: [8 42]; hour 3: [3 47] **c.** Eventually, all the roaches will have checked in. **27.** 47/300 ≈ 0.156 667 **29.** 4/9 (Note that the third child corresponds to two transition steps.) **31. a.** 1 = profitable, 2 = losing, 3 = bankrupt; $P = \begin{bmatrix} .9 & .1 & 0 \\ .1 & .7 & .2 \\ 1 & 0 & 0 \end{bmatrix}$ **b.** Profitable after 1 year: .9; 2 years: .82; 3 years: .774 **33. a.** 3/8
b.

	Main Caf	Math Dept.	Field House
1 trip	120	0	240
2 trips	60	240	60
3 trips	95	75	190

35. a. $P = \begin{bmatrix} .729 & .271 & 0 \\ .075 & .84 & .085 \\ 0 & .304 & .696 \end{bmatrix}$ **b.** 2.3% **37. a.** Paid: $105.4 million; 0–30 days: $5.7 million; 31–90 days: $2.6 million; bad debts: $3.3 million **b.** Paid: $109.1 million; 0–30 days: $3.67 million; 31–90 days: $1.66 million; bad debts: $2.57 million **c.** After 1 month: paid: $105.4 million; 0–30 days: $6.7 million; 31–90 days: $2.6 million; bad debts: $3.3 million; after 2 months: paid: $109.6 million; 0–30 days: $4.97 million; 31–90 days: $1.86 million; bad debts: $2.57 million

39. a. $P = \begin{bmatrix} \frac{1}{2} & \frac{1}{2} & 0 & 0 & 0 & 0 \\ \frac{1}{2} & 0 & \frac{1}{2} & 0 & 0 & 0 \\ 0 & \frac{1}{2} & 0 & \frac{1}{2} & 0 & 0 \\ 0 & 0 & \frac{1}{2} & 0 & \frac{1}{2} & 0 \\ 0 & 0 & 0 & \frac{1}{2} & 0 & \frac{1}{2} \\ 0 & 0 & 0 & 0 & \frac{1}{2} & \frac{1}{2} \end{bmatrix}$

b. $P^2 = \begin{bmatrix} \frac{1}{2} & \frac{1}{4} & \frac{1}{4} & 0 & 0 & 0 \\ \frac{1}{4} & \frac{1}{2} & 0 & \frac{1}{4} & 0 & 0 \\ \frac{1}{4} & 0 & \frac{1}{2} & 0 & \frac{1}{4} & 0 \\ 0 & \frac{1}{4} & 0 & \frac{1}{2} & 0 & \frac{1}{4} \\ 0 & 0 & \frac{1}{4} & 0 & \frac{1}{2} & \frac{1}{4} \\ 0 & 0 & 0 & \frac{1}{4} & \frac{1}{4} & \frac{1}{2} \end{bmatrix}$; distribution after two steps = $[\frac{1}{2} \quad \frac{1}{4} \quad \frac{1}{4} \quad 0 \quad 0 \quad 0]$ **c.** 0 **41.** [11.25 38.75 38.75 11.25], [8.9125 41.0875 41.0875 8.9125], [8.515 41.485 41.485 8.515]. There is a tendency to be drawn toward the middle-income groups. **43.** *2000:* Northeast, 53.64 million; Midwest, 64.38 million; South, 100.23 million; West, 63.25 million. *2001:* Northeast, 53.39 million; Midwest, 64.46 million; South, 100.45 million; West, 63.20 million. *2002:* Northeast, 53.14 million; Midwest, 64.54 million; South, 100.67 million; West, 63.15 million. **45.** Paid: $115.05 million; 0–30: $0.80 million; 31–90: $0.32 million; bad debts: $0.82 million

47. $P^8 = \begin{bmatrix} .2734 & .2188 & .2188 & .1133 & .1133 & .0625 \\ .2188 & .2734 & .1133 & .2188 & .0625 & .1133 \\ .2188 & .1133 & .2734 & .0625 & .2188 & .1133 \\ .1133 & .2188 & .0625 & .2734 & .1133 & .2188 \\ .1133 & .0625 & .2188 & .1133 & .2734 & .2188 \\ .0625 & .1133 & .1133 & .2188 & .2188 & .2734 \end{bmatrix}$

Starting in state 3, after eight steps the most likely state is state 3.

49. The powers of P are $P^2 = \begin{bmatrix} 0 & 1 & 0 \\ 0 & 0 & 1 \\ 1 & 0 & 0 \end{bmatrix}$, $P^3 = \begin{bmatrix} 1 & 0 & 0 \\ 0 & 1 & 0 \\ 0 & 0 & 1 \end{bmatrix}$, and then $P^4 = P$, $P^5 = P^2$, and so on.
51. After 1 year, 35% of all accounts will be paid up, 40% will have had their last payment received 1–6 months ago, and 25% will have had no payment for more than 6 months. **53. a.** If you survey the same people and correlate the results, you can ask the first year what kind of car they own then and then the second year ask the same question. The (experimental) transition probabilities would then be computed as follows: p_{11} = number of people who owned a foreign car both years ÷ the number of people who owned a foreign car the first year; p_{12} = number of people who owned a foreign car the first year and a domestic car the second ÷ the number of people who owned a foreign car the first year; and so on. **b.** You are assuming that the transition probabilities remain constant from year to year and are not affected, for example, by the changing car market.

55. $P^{20} = \begin{bmatrix} .471011 & .528989 \\ .470213 & .529787 \end{bmatrix}$. The rows are very nearly equal.

57. a. $9P = \begin{bmatrix} 3 & 6 \\ 1 & 8 \end{bmatrix}$ **b.** $(9P)^6 = \begin{bmatrix} 75{,}975 & 455{,}466 \\ 75{,}911 & 455{,}530 \end{bmatrix}$

c. $P^6 = \begin{bmatrix} \frac{75{,}975}{531{,}441} & \frac{455{,}466}{531{,}441} \\ \frac{75{,}911}{531{,}441} & \frac{455{,}530}{531{,}441} \end{bmatrix}$

Exercises 9.3

1. a. $P^\infty = \begin{bmatrix} \frac{2}{3} & \frac{1}{3} \\ \frac{2}{3} & \frac{1}{3} \end{bmatrix}$ **b.** $vP^\infty = [10 \quad 5]$ **3. a.** $P^\infty = \begin{bmatrix} \frac{3}{7} & \frac{4}{7} \\ \frac{3}{7} & \frac{4}{7} \end{bmatrix}$

b. $vP^\infty = \begin{bmatrix} \frac{30}{7} & \frac{40}{7} \end{bmatrix}$ **5. a.** $P^\infty = \begin{bmatrix} \frac{1}{9} & \frac{8}{9} \\ \frac{1}{9} & \frac{8}{9} \end{bmatrix}$ **b.** $vP^\infty = \begin{bmatrix} \frac{100}{3} & \frac{800}{3} \end{bmatrix}$

7. a. $P^\infty = \begin{bmatrix} \frac{2}{5} & \frac{3}{5} \\ \frac{2}{5} & \frac{3}{5} \end{bmatrix}$ **b.** $vP^\infty = \begin{bmatrix} \frac{2}{5} & \frac{3}{5} \end{bmatrix}$

9. a. $P^\infty = \begin{bmatrix} \frac{2}{5} & \frac{1}{5} & \frac{2}{5} \\ \frac{2}{5} & \frac{1}{5} & \frac{2}{5} \\ \frac{2}{5} & \frac{1}{5} & \frac{2}{5} \end{bmatrix}$ **b.** $vP^\infty = [4 \quad 2 \quad 4]$

11. a. $P^\infty = \begin{bmatrix} \frac{1}{5} & \frac{2}{5} & \frac{2}{5} \\ \frac{1}{5} & \frac{2}{5} & \frac{2}{5} \\ \frac{1}{5} & \frac{2}{5} & \frac{2}{5} \end{bmatrix}$ **b.** $vP^\infty = [6 \quad 12 \quad 12]$

13. a. $P^\infty = \begin{bmatrix} 1 & 0 & 0 \\ 1 & 0 & 0 \\ 1 & 0 & 0 \end{bmatrix}$ **b.** $vP^\infty = [20 \quad 0 \quad 0]$

15. a. $P^\infty = \begin{bmatrix} 0 & 1 & 0 \\ 0 & 1 & 0 \\ 0 & 1 & 0 \end{bmatrix}$ **b.** $vP^\infty = [0 \quad 1 \quad 0]$

17. a. $P^\infty = \begin{bmatrix} .65 & .35 \\ .65 & .35 \end{bmatrix}$ **b.** $vP^\infty = [6.54 \quad 3.46]$

19. a. $P^\infty = \begin{bmatrix} .31 & .23 & .46 \\ .31 & .23 & .46 \\ .31 & .23 & .46 \end{bmatrix}$ **b.** $vP^\infty = [9.29 \quad 6.79 \quad 13.93]$

21. a. $P^\infty = \begin{bmatrix} .45 & .22 & .20 & .14 \\ .45 & .22 & .20 & .14 \\ .45 & .22 & .20 & .14 \\ .45 & .22 & .20 & .14 \end{bmatrix}$

b. $vP^\infty = [.45 \quad .22 \quad .20 \quad .14]$
23. a. 16.67% fall into the high-risk category and 83.33% into the low-risk category. **b.** 17 will fall into the high-risk category and 83 into the low-risk category (rounded to the nearest person). **25.** 3/13

27. a. $P = \begin{bmatrix} \frac{1}{2} & \frac{1}{4} & \frac{1}{4} \\ 0 & \frac{1}{2} & \frac{1}{2} \\ \frac{1}{4} & 0 & \frac{3}{4} \end{bmatrix}$ **b.** 2/7 are happily married in a given year **c.** The predominant state is divorced, accounting for 4/7 of the sample. **29. a.** 41.67% of the customers will be in the paid-up category, 41.67% in the 0–90-days category, and 16.67% in the bad-debt category. **b.** 3333 customers will be in the paid-up category, 3333 in the 0–90-days category, and 1333 in the bad-debt category. (Answers rounded to the nearest customer) **31.** Hall A: 10/27, or 37.04%; Hall B: 1/3, or 33.33%; Hall C: 8/27, or 29.63% **33.** Affluent: 17.8%; middle class: 64.3%; poor: 18.0% **35.** The claim is justified by the data. According to the models, the following long-term trends are predicted: *1967–1979:* affluent, 14.7%; middle class, 72.6%; poor, 12.7%. *1981–1992:* affluent, 17.8%; middle class, 64.3%; poor, 18.0%. These figures suggest a declining middle class and growing affluent and poor-income groups and hence a tendency toward widening income inequality. **37. a.** 52.94% **b.** 11.76 minutes in light traffic, 35.29 minutes in moderate traffic, and 52.94 minutes in heavy traffic **c.** 36.28 miles

39. a. $P = \begin{bmatrix} \frac{1}{2} & \frac{1}{2} & 0 & 0 & 0 \\ \frac{1}{2} & 0 & \frac{1}{2} & 0 & 0 \\ 0 & \frac{1}{2} & 0 & \frac{1}{2} & 0 \\ 0 & 0 & \frac{1}{2} & 0 & \frac{1}{2} \\ 0 & 0 & 0 & \frac{1}{2} & \frac{1}{2} \end{bmatrix}$ **b.** $P^\infty = \begin{bmatrix} \frac{1}{5} & \frac{1}{5} & \frac{1}{5} & \frac{1}{5} & \frac{1}{5} \\ \frac{1}{5} & \frac{1}{5} & \frac{1}{5} & \frac{1}{5} & \frac{1}{5} \\ \frac{1}{5} & \frac{1}{5} & \frac{1}{5} & \frac{1}{5} & \frac{1}{5} \\ \frac{1}{5} & \frac{1}{5} & \frac{1}{5} & \frac{1}{5} & \frac{1}{5} \\ \frac{1}{5} & \frac{1}{5} & \frac{1}{5} & \frac{1}{5} & \frac{1}{5} \end{bmatrix}$

Thus, the system spends an average of 1/5 of the time in each state. **c.** $\begin{bmatrix} \frac{1}{5} & \frac{1}{5} & \frac{1}{5} & \frac{1}{5} & \frac{1}{5} \end{bmatrix}$ **41.** Long-term income distribution (top to bottom): [8.43 41.57 41.57 8.43] **b.** The percentages in each category are given by the same numbers. **43.** Northeast: 34.0; Midwest: 73.7; South: 111.8; West: 61.9 **45.** $111.9 million paid and $5.1 million in bad debts.
47. [.1 .1 ... .1] This generalizes to $\begin{bmatrix} \frac{1}{n} & \frac{1}{n} & \cdots & \frac{1}{n} \end{bmatrix}$ for n states. **49.** There are really two interpretations of the steady-state distribution vector. First, it gives the fractions of a population that will be found in the various states after a long time. Second, it gives the probabilities of finding the system in the various states after a long time. In either case, since the members of the population or the system must be in one of the states, the fractions or the probabilities have to add up to 1. **51.** If q is a row of Q, then by assumption, $qP = q$. Thus, when we multiply the rows of Q by P, nothing changes, and $QP = Q$. **53.** At each step only .4 of the population in state 1 remains there, and nothing enters from any other state. Thus, when the first entry in the steady-state distribution vector is multiplied by .4, it must remain unchanged. The only number for which this true is 0. **55.** An example is

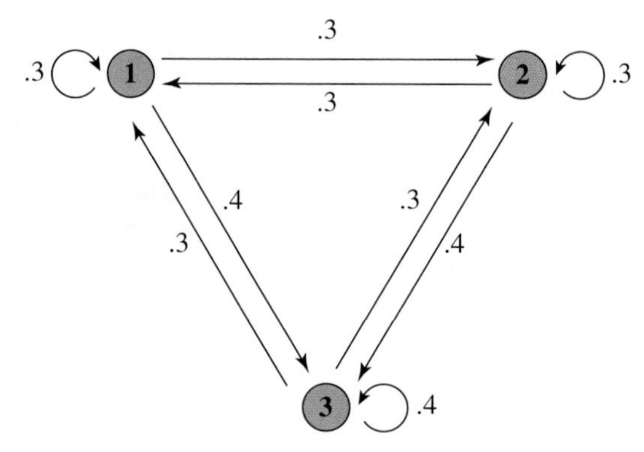

57. If $vP = v$ and $wP = w$, then $\frac{1}{2}(v + w)P = \frac{1}{2}vP + \frac{1}{2}wP = \frac{1}{2}v + \frac{1}{2}w = \frac{1}{2}(v + w)$. Further, if the entries of v and w add up to 1, then so do the entries of $(v + w)/2$.

Exercises 9.4

1. Absorbing **3.** Not absorbing; there are no absorbing states. **5.** Absorbing **7.** Absorbing **9.** Not absorbing; it is impossible to get from either state 3 or state 5 to an absorbing state.

11.

	To			
From	**3**	**4**	**1**	**2**
3	.5	0	.2	.3
4	.3	0	.5	.2
1	0	0	1	0
2	0	0	0	1

$$P = \begin{bmatrix} .5 & 0 & .2 & .3 \\ .3 & 0 & .5 & .2 \\ 0 & 0 & 1 & 0 \\ 0 & 0 & 0 & 1 \end{bmatrix}, S = \begin{bmatrix} .5 & 0 \\ .3 & 0 \end{bmatrix}$$

13.

	To			
From	**2**	**3**	**4**	**1**
2	.3	.2	0	.5
3	.5	.2	.1	.2
4	0	0	1	0
1	0	0	0	1

$$P = \begin{bmatrix} .3 & .2 & 0 & .5 \\ .5 & .2 & .1 & .2 \\ 0 & 0 & 1 & 0 \\ 0 & 0 & 0 & 1 \end{bmatrix}, S = \begin{bmatrix} .3 & .2 \\ .5 & .2 \end{bmatrix}$$

15. Absorbing **a.** $Q = \begin{bmatrix} 2 & 0 \\ 2 & 4 \end{bmatrix}$ **b.** State 1: visit twice; state 2: visit four times **17.** Not absorbing **19.** Absorbing **a.** $Q = [1]$ **b.** State 1: visit once **21.** Absorbing **a.** $Q = \begin{bmatrix} \frac{4}{3} & \frac{2}{3} \\ \frac{2}{3} & \frac{4}{3} \end{bmatrix}$ **b.** State 1: visit 4/3 times; state 2: visit 4/3 times **23.** Not absorbing **25.** Absorbing **a.** $Q = \begin{bmatrix} 2 & 0 & 0 \\ 0 & 2 & 0 \\ 2 & 0 & 2 \end{bmatrix}$ **b.** State 1: visit twice; state 2: visit twice; state 3: visit twice **27.** Absorbing **a.** $Q = \begin{bmatrix} 1.5466 & .8766 & .1425 \\ .4571 & 1.8785 & .3053 \\ .5009 & .6888 & 1.3619 \end{bmatrix}$ **b.** State 1: visit 1.5466 times; state 2: visit 1.8785 times; state 3: visit 1.3619 times

29. a. $$Q = \begin{bmatrix} 2 & 2 & 2 & 2 & 2 & 2 & 2 & 2 & 2 & 2 \\ 2 & 4 & 4 & 4 & 4 & 4 & 4 & 4 & 4 & 4 \\ 2 & 4 & 6 & 6 & 6 & 6 & 6 & 6 & 6 & 6 \\ 2 & 4 & 6 & 8 & 8 & 8 & 8 & 8 & 8 & 8 \\ 2 & 4 & 6 & 8 & 10 & 10 & 10 & 10 & 10 & 10 \\ 2 & 4 & 6 & 8 & 10 & 12 & 12 & 12 & 12 & 12 \\ 2 & 4 & 6 & 8 & 10 & 12 & 14 & 14 & 14 & 14 \\ 2 & 4 & 6 & 8 & 10 & 12 & 14 & 16 & 16 & 16 \\ 2 & 4 & 6 & 8 & 10 & 12 & 14 & 16 & 18 & 18 \\ 2 & 4 & 6 & 8 & 10 & 12 & 14 & 16 & 18 & 20 \end{bmatrix}$$

b. State 1: visit twice; state 2: visit 4 times; state 3: visit 6 times; . . . ; state 10: visit 20 times **31.** 2 time steps **33. a.** From state 1 to absorption: 20/11; from state 2 to absorption: 16/11 **b.** State 1: 5/11; state 2: 1/11 **35. a.** From state 1 to absorption: 2; from state 2 to absorption: 2; from state 3 to absorption: 4 **b.** State 1: once; state 2: once; state 3: once **37. a.** 2.17 months **b.** 95.65% **c.** 4.35% **41.** *Group I:* Best, 15.79 years; Next Best, 15.26; Worst, 12.63. *Group II:* Best, 20 years; Next Best, 20; Worst, 20. *Increased Life Expectancy Is:* Best, 4.21 years; Next Best, 4.74; Worst, 7.37 **43. a.** 20 steps. **b.** We add the columns of Q to find out how many times each state is visited; the most frequently visited state is thus state 5.

c. $$Q = \begin{bmatrix} 2 & 2 & 2 & 2 & 2 & 2 & 2 & 2 & 2 \\ 2 & 4 & 4 & 4 & 4 & 4 & 4 & 4 & 4 \\ 2 & 4 & 6 & 6 & 6 & 6 & 6 & 6 & 6 \\ 2 & 4 & 6 & 8 & 8 & 8 & 8 & 8 & 8 \\ 2 & 4 & 6 & 8 & 10 & 10 & 10 & 10 & 10 \\ 2 & 4 & 6 & 8 & 10 & 12 & 12 & 12 & 12 \\ 2 & 4 & 6 & 8 & 10 & 12 & 14 & 14 & 14 \\ 2 & 4 & 6 & 8 & 10 & 12 & 14 & 16 & 16 \\ 2 & 4 & 6 & 8 & 10 & 12 & 14 & 16 & 18 \end{bmatrix}$$

45. 14 moves **47. a.** 4 days **b.** 71/105 **c.** 12/7 **d.** 64/105 **e.** 3/16 **49.** Entries below the diagonal correspond to probabilities of a tooth becoming less decayed. **51.** According to the data, an upper second premolar with decay on the front and rear can be lost after 1 year but not before. (It can progress to state 4 in the first 6 months and then be lost 6 months later.) **53.** ten time steps, or 5 years **55.** The first state **57.** 80% **59.** For large values of n,

$$P^n \approx \begin{bmatrix} 1 & 0 & 0 & 0 \\ 0 & 1 & 0 & 0 \\ .4 & .6 & 0 & 0 \\ .366\,667 & .633\,333 & 0 & 0 \end{bmatrix}$$

We interpret this as follows: If the system starts in state 3, the long-term probability of ending up in state 1 is .4, of ending up in state 2 is .6, and of being in either state 3 or state 4 is 0. If the system starts in state 4, the long-term probability of ending up in state 1 is .366 667 (which is probably 11/30), of ending up in state 2 is .633 333 (probably 19/30), and of being in either state 3 or state 4 is 0. **61.** Such a system must eventually end up in the absorbing state. Thus, the steady-state distribution matrix is all zero except for a single column of 1s corresponding to the absorbing state. **63.** Taking the system in Exercise 59, for example, both $[1\ \ 0\ \ 0\ \ 0]$ and $[0\ \ 1\ \ 0\ \ 0]$ are steady-state vectors. **65.** An example is

$$P = \begin{bmatrix} 0 & 1 & 0 & 0 \\ 0 & 0 & 1 & 0 \\ 0 & 0 & 0 & 1 \\ 0 & 0 & 0 & 1 \end{bmatrix}$$

Chapter 9 Review Test

1. a. $P = \begin{bmatrix} \frac{1}{2} & \frac{1}{2} \\ \frac{1}{4} & \frac{3}{4} \end{bmatrix}$ **b.** $P^2 = \begin{bmatrix} \frac{3}{8} & \frac{5}{8} \\ \frac{5}{16} & \frac{11}{16} \end{bmatrix}$, $P^3 = \begin{bmatrix} \frac{11}{32} & \frac{21}{32} \\ \frac{21}{64} & \frac{43}{64} \end{bmatrix}$; 21/32
c. Brand A: 325; brand B: 635 **d.** Brand A: 1/3; brand B: 2/3
e. Brand A: 320; brand B: 640
2. a. State 1 = Watching Television; State 2 = Going to Bed. $P = \begin{bmatrix} \frac{5}{6} & \frac{1}{6} \\ 0 & 1 \end{bmatrix}$; 91/216 **b.** 25/216 **c.** Greater than. Going to bed at precisely 8 A.M. requires a sequence of seven numbers other than 6 followed by a 6 in eight throws of the die. That is less likely than going to bed at 3 A.M.: a sequence of two numbers other than a 6 followed by a 6 in three throws of the die. **d.** 6 A.M.

3. a. $P = \begin{bmatrix} .3 & .2 & .5 \\ 0 & .5 & .5 \\ .5 & .5 & 0 \end{bmatrix}$ **b.** $P^2 = \begin{bmatrix} .34 & .41 & .25 \\ .25 & .5 & .25 \\ .15 & .35 & .5 \end{bmatrix}$ **c.** .455
d. $[61\ \ 99\ \ 80]$ **e.** $[5\ \ 9\ \ 7]$
4. a. All arrows have value .5.

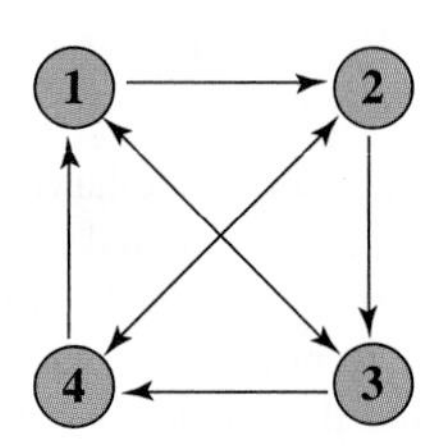

b. $P^2 = \begin{bmatrix} .25 & 0 & .25 & .5 \\ .5 & .25 & 0 & .25 \\ .25 & .5 & .25 & 0 \\ 0 & .25 & .5 & .25 \end{bmatrix}$, $P^4 = \begin{bmatrix} .125 & .25 & .375 & .25 \\ .25 & .125 & .25 & .375 \\ .375 & .25 & .125 & .25 \\ .25 & .375 & .25 & .125 \end{bmatrix}$

c. The system will spend 1/4 of the time in each state on average due to the symmetry of the state transition diagram.

d. $P^\infty = \begin{bmatrix} .25 & .25 & .25 & .25 \\ .25 & .25 & .25 & .25 \\ .25 & .25 & .25 & .25 \\ .25 & .25 & .25 & .25 \end{bmatrix}$

5. a. $\begin{bmatrix} \frac{5}{18} & \frac{5}{18} & \frac{4}{9} \\ \frac{5}{18} & \frac{5}{18} & \frac{4}{9} \\ \frac{5}{18} & \frac{5}{18} & \frac{4}{9} \end{bmatrix}$ **b.** 5/18 **c.** That the system is in state 2 after two steps, with a probability of 21/50 **d.** $[5\ \ 5\ \ 8]$

e. $P = \begin{bmatrix} .2 & .8 \\ 1 & 0 \end{bmatrix}$, $P^\infty = \begin{bmatrix} \frac{5}{9} & \frac{4}{9} \\ \frac{5}{9} & \frac{4}{9} \end{bmatrix}$

6. a. Five steps **b.** State 1 (an average of 16/3 times) **c.** 1/3
7. a. $[4000\ 2600\ 3400]$ **b.** $[4920\ \ 1960\ \ 3120]$ **c.** Here are three: (1) it is possible for someone to be a customer at two different enterprises; (2) some customers may stop using all three of the companies; (3) new customers can enter the field.
d. OHaganBooks.com: 4/7; JungleBooks.com: 1/7; FarmerBooks.com: 2/7
8. a. 13/64 **b.** 13/16 **c.** 3/32 **d.** No. Each location (including the coffee room) is visited 1/5 of the time. **e.** The model includes no information about the length of time a staff member spends at each location, and hence nothing can be said about the length of time a staff members spends at the coffee room.

9. a. 146 **b.** $Q = \begin{bmatrix} 2.5 & .2.5 \\ 0 & 10 \end{bmatrix}$ **c.** 4 days **d.** 9 days **e.** 75%

Appendix

1. −48 **3.** 2/3 **5.** −1 **7.** 9 **9.** 1 **11.** 33 **13.** 14 **15.** 5/18 **17.** 1331/1000 **19.** 6 **21.** 43/16 **23.** 0 **25.** `3*(2-5)` **27.** `3/(2-5)` **29.** `(3-1)/(8+6)` **31.** `3-(4+7)/8` **33.** `2/(3+x)-x*y^2` **35.** `3.1*x^3-4*x^(-2)-60/(x^2-1)` **37.** `(2/3)/5` **39.** `3^(4-5)*6` **41.** `3*(1+4/100)^(-3)` **43.** `3^(2*x-1)+4^x-1` **45.** `2^(2*x^2-x+1)` **47.** `4*e^(-2*x)/(2-3*e^(-2*x))` or `4*(e^(-2*x))/(2-3*e^(-2*x))` **49.** `3*(1-(-1/2)^2)^2+1`

INDEX

N

O

P

Index

Q

R

S

T

U

V

X

Y

Z

INDEX OF COMPANIES, PRODUCTS, AND AGENCIES